NUMERICAL METHODS IN ATMOSPHERIC AND OCEANIC MODELLING
THE ANDRÉ J. ROBERT MEMORIAL VOLUME

André J. Robert (1929–1993)

Numerical Methods in Atmospheric and Oceanic Modelling
The André J. Robert Memorial Volume

edited by
Charles A. Lin, McGill University
René Laprise, Université du Québec à Montréal
Harold Ritchie, Atmospheric Environment Service/
Service de l'environnement atmosphérique

Canadian Meteorological and Oceanographic Society
La société canadienne de météorologie et d'océanographie

NRC Research Press
Presses scientifiques du CNRC

ISBN 0-9698414-4-2

Published by the
Canadian Meteorological and Oceanographic Society
La société canadienne de météorologie et d'océanographie
Suite 112, McDonald Building, 150 Louis Pasteur, Ottawa, Ontario, Canada K1N 6N5

in association with the

NRC Research Press
Presses scientifiques du CNRC
1200 Montreal Road, Building M-55, Ottawa, Ontario, Canada K1A 0R6

Printed in Canada by the University of Toronto Press, Toronto, Ontario

---

Canadian Cataloguing in Publication Data

Main entry under title:
Numerical methods in atmospheric and oceanic
modelling: the André J. Robert memorial volume

A collection of papers based on presentations at the André J. Robert Memorial Symposium on Numerical Methods in Atmospheric and Oceanic Sciences, held at Université du Québec à Montréal, 5–7 October, 1994.
Co-published by: National Research Council of Canada.
Includes summaries and supplement in French.
ISBN 0-9698414-4-2

1. Meteorology – Mathematical models – Congresses. 2. Oceanography – Mathematical Models – Congresses. I. Lin, Charles A. (Charles Augustin), 1953– II. Laprise, René, 1952- III. Ritchie, Harold, 1950– IV. Robert, André J., d. 1993. V. Canadian Meteorological and Oceanographic Society. VI. National Research Council of Canada

QC851.N84 1997 551.46'001'5118 C97-900713-5

Cover artwork by Manon Pineault.

Contents

Foreword

The idea to publish this book arose from the success of the André J. Robert Memorial Symposium on Numerical Methods in Atmospheric and Oceanic Modelling, held, 5–7 October 1994 at the Université du Québec à Montréal, Canada, where the late Dr. Robert held a faculty position. The Symposium was organised by the Atmospheric Environment Service, Environment Canada. The book contains peer reviewed versions of papers presented at the Symposium, a transcript of a 1987 interview with Dr. Robert and, as a supplement, four previously unpublished papers by Dr. Robert. The book is published as a companion volume to Atmosphere-Ocean. As Dr. Robert published some of his most influential work in Atmosphere-Ocean, this memorial volume is a most fitting tribute.

Throughout his career, Dr. Robert worked on numerical methods to improve the efficiency and accuracy of numerical weather prediction and climate models. Among his major contributions are the first integration of a global spectral primitive equations model, the development of an efficient time filter, and pioneering studies in the use of the semi-implicit and semi-Lagrangian methods. The numerical schemes that he developed have now been implemented in major numerical weather prediction and climate research centres around the world. Dr. Robert also contributed to the development of the Mesoscale Compressible Community (MC2) model, the first efficient model based on the fully elastic non-hydrostatic equations that could in principle be used at all scales. This model is now used in many research centres in Canada and abroad. Through his published works, teaching activities, research seminars and participation in international forums, Dr. Robert has exercised a lasting influence on many scientists and on the science of numerical modelling.

The contributing authors are from laboratories from around the world. Their readiness to participate in the Symposium and to submit papers to the present volume is a tribute to the significance of Dr. Robert's contributions. The papers report on or review original state-of-the-art research findings in numerical modelling. This monograph should therefore be of interest to students and researchers in numerical weather prediction, climate simulation, dynamic meteorology and ocean modelling.

We would like to thank all authors for their excellent collaboration which made possible this monograph. We appreciate the encouragement and advice of Dr. Richard Asselin, Director of Publications for the Canadian Meteorological and Oceanographic Society; Gerald Neville, Head of the Monograph Publishing Program of the NRC Research Press; and Dr. Michel Béland, Director of Centre de recherche en calcul appliqué (CERCA). Financial support from the Natural Sciences and Engineering Research Council, the Atmospheric Environment Service, the Canadian Meteorological and Oceanographic Society and NRC Research Press is gratefully acknowledged.

Ms. Claire Poulin of CERCA provided word processing assistance with the previously unpublished papers of Dr. Robert. Last but not least, a special acknowledgement is due Ms. Sheila Bourque, who meticulously worked through each paper as the technical editor.

Charles A. Lin, McGill University
René Laprise, Université du Québec à Montréal
Harold Ritchie, Atmospheric Environment Service
Editors

Avant-propos

L'idée de publier ce livre est venue du succès du Colloque à la mémoire d'André J. Robert, sur les méthodes numériques en modélisation atmosphérique et océanique, qui a eu lieu les 5–7 octobre 1994 à l'Université du Québec à Montréal, là où le regretté Dr Robert était professeur. Le Colloque a été organisé par le Service de l'environnement atmosphérique, Environnement Canada. Le livre contient des versions soumises à un comité de lecture des articles présentés au Colloque, une transcription d'un entretien en 1987 avec le Dr Robert, et en supplément, quatre articles précédemment non-publiés du Dr Robert. Le livre est publié comme compagnon de la revue Atmosphère-Océan. Puisque le Dr Robert a publié certains de ses articles les plus importants dans Atmosphère-Océan, ce volume à sa mémoire est un hommage très approprié.

Le Dr Robert a consacré sa carrière aux méthodes numériques, afin d'améliorer l'efficacité et la précision des prévisions numériques météorologiques ainsi que des simulations climatiques. Parmi ses grandes contributions notons la première intégration d'un modèle global spectral aux équations primitives, la réalisation d'un filtre temporel efficace, et des études d'avant garde sur l'application des méthodes semi-implicite et semi-Lagrangienne. Les méthodes numériques qu'il a développées sont utilisées dans les grands centres de prévision numérique météorologique et de recherche climatique dans le monde entier. Le Dr Robert a aussi contribué au développement du modèle "Mesoscale Compressible Community (MC2)", le premier modèle efficace aux équations entièrement élastiques non-hydrostatiques qui peut en principe être utilisé à toutes les échelles. Ce modèle est maintenant utilisé dans plusieurs centres de recherche au Canada et à l'étranger. Par ses publications, son enseignement, ses séminaires et sa participation aux conférences internationales, le Dr Robert a eu une influence profonde sur de nombreux scientifiques et sur la science de modélisation numérique.

Les auteurs proviennent de laboratoires du monde entier. Leur enthousiasme à participer au Colloque et à soumettre des articles pour ce volume est un hommage à l'importance des contributions du Dr Robert. Les articles présentent d'importantes découvertes originales dans la recherche en modélisation numérique. Cette monographie devrait donc être d'intérêt pour les étudiants et les chercheurs en prévision numérique météorologique, en simulation du climat, en météorologie dynamique et en modélisation de l'océan.

Nous voulons remercier tous les auteurs pour leur excellente collaboration, qui a rendu possible cette monographie. Nous avons apprécié l'encouragement et les conseils du Dr Richard Asselin, Directeur des publications de la Société canadienne de météorologie et d'océanographie; de Gerald Neville, Directeur du Programme des publications de monographies du CNRC et du Dr Michel Béland, Directeur du Centre de recherche en calcul appliqué (CERCA). Nos remerciements vont au Conseil de recherches en sciences naturelles et en génie, au Service de l'environnement atmosphérique, à la Société canadienne de météorologie et d'océanographie et aux Presses scientifiques du CNRC pour leur appui financier.

Mme Claire Poulin du CERCA a fourni de l'aide avec le traitement de texte des articles précédemment non-publiés du Dr Robert. Les derniers, mais non les moindres remerciements sont adressés à Mme Sheila Bourque qui a vérifié si méticuleusement chaque article, en tant qu'éditrice technique.

Charles A. Lin, Université McGill
René Laprise, Université du Québec à Montréal
Harold Ritchie, Service de l'environnement atmosphérique
Éditeurs

Curriculum Vitae of André J. Robert

Born 28 April 1929, New York, U.S.A.
Immigrated to Canada, May 1937
Deceased 18 November 1993, Montréal, Québec, Canada

Education
1952 B.Sc., Mathematics, Université Laval
1953 M.Sc., Meteorology, University of Toronto
1965 Ph.D., Meteorology, McGill University

Professional positions
1952–1959 Atmospheric Environment Service
Forecaster
1959–1970 Atmospheric Environment Service
Scientist, Division de recherche en prévision numérique (RPN)
1970–1971 McGill University
Professor, Department of Meteorology
1971–1987 Atmospheric Environment Service
Chief, RPN, 1971–1973
Director, Canadian Meteorological Centre, 1973–1980
Senior Scientist, RPN, 1980–1987
1987–1993 Université du Québec à Montréal
Professor, Department of Physics

Professional appointments
1968–1969 Lecturer in training courses, U.S. National Meteorological Center, Washington, D.C.
1968–1972 Member, Subcommittee in Meteorology and Atmospheric Sciences, National Research Council
1970–1976 Member, Working Group on Numerical Experimentation, World Meteorological Organisation (WMO)
1971–1973 Chair, Working Group on Numerical Experimentation, WMO
1972–1973 President, Canadian Meteorological and Oceanographic Society
1972–1976 Editor, Research Activities in Atmospheric and Oceanic Modelling, WMO
1972 Chair, WMO Study Group Conference on the Parametrization of Subgrid Scale Physical Processes, Leningrad, USSR
1973 Chair, WMO Study Group Conference on Modelling Aspects of GATE, Tallahassee, Florida

1974	Chair, International Symposium on Spectral Methods in Numerical Weather Prediction, Copenhagen, Denmark
1975	Member of Canadian delegation, 7th Congress of the WMO
1980–1983	Member, Grant Selection Committee for Meteorology, Aeronomy and Astronomy, Natural Sciences and Engineering Research Council

Honours

1967, 1971	President's Prize, Canadian Meteorological and Oceanographic Society
1968	Fellow, American Meteorological Society
1981	Second Half Century Award, American Meteorological Society
1982	Fellow, Royal Society of Canada
1986	Patterson Medal, Atmospheric Environment Service
1987–1993	Emeritus Scientist, Atmospheric Environment Service

Curriculum vitae de André J. Robert

Né le 28 avril 1929, New York, États-Unis
Immigré au Canada en mai 1937
Décédé le 18 novembre 1993, Montréal, Québec, Canada

Formation académique
1952 B.Sc., Mathématiques, Université Laval
1953 M.Sc., Météorologie, Université de Toronto
1965 Ph.D., Météorologie, Université McGill

Expérience professionnelle
1952–1959 Service de l'environnement atmosphérique
Prévisionniste
1959–1970 Service de l'environnement atmosphérique
Chercheur, Division de recherche en prévision numérique (RPN)
1970–1971 Université McGill
Professeur, Département de météorologie
1971–1987 Service de l'environnement atmosphérique
Chef, RPN, 1971–1973
Directeur, Centre météorologique canadien, 1973–1980
Chercheur principal, RPN, 1980–1987
1987–1993 Université du Québec à Montréal
Professeur, Département de physique

Distinctions professionnelles
1968–1969 Chargé de cours de formation, U.S. National Meteorological Center, Washington, D.C.
1968–1972 Membre, Sous-comité en météorologie et sciences atmosphériques, Conseil national de recherches
1970–1976 Membre, Working Group on Numerical Experimentation, Organisation météorologique mondiale (OMM)
1971–1973 Président, Working Group on Numerical Experimentation, OMM
1972–1973 Président, Société canadienne de météorologie et d'océanographie
1972–1976 Éditeur, Research Activities in Atmospheric and Oceanic Modelling, OMM
1972 Président, Study Group Conference on the Parametrization of Sub-grid Scale Physical Processes (OMM), Leningrad, URSS
1973 Président, Study Group Conference on Modelling Aspects of GATE (OMM), Tallahassee, Floride

1974	Président, International Symposium on Spectral Methods in Numerical Weather Prediction, Copenhague, Danemark
1975	Membre de la délégation canadienne, 7ème Congrès de l'OMM
1980–1983	Membre, Comité de sélection de subventions en météorologie, en aéronomie et en astronomie, Conseil de recherches en sciences naturelles et en génie

Distinctions

1967, 1971	Prix-du président, Société canadienne de météorologie et d'océanographie
1968	Fellow, American Meteorological Society
1981	Second Half Century Award, American Meteorological Society
1982	Membre, Société royale du Canada
1986	Médaille Patterson, Service de l'environnement atmosphérique
1987–1993	Chercheur émérite, Service de l'environnement atmosphérique

Primary publications of André J. Robert / Publications principales de André J. Robert

Robert, A. 1997. Incompressible homogeneous fluids. (this volume).

———. 1997. Erreurs induites par le schème Lagrangien dans les ondes stationnaires. (ce volume).

———. 1997. Comportement du schème Lagrangien dans une équation différentielle relativement simple. (ce volume).

——— and M. Lepine. 1997. An anomaly in the behaviour of the time filter used with the leapfrog scheme in atmospheric models. (this volume).

Ritchie, H. and A. Robert. 1997. A historical perspective on numerical weather prediction: A 1987 interview with André Robert. (this volume).

Xu, Weimin; C. Lin and A. Robert. 1997. A C-grid ocean general circulation model: Model formulation and frictional parametrizations. (this volume).

Rivest, C.; A. Staniforth and A. Robert. 1994. Spurious resonant response of semi-Lagrangian discretizations to orographic forcing: diagnosis and solution. *Mon. Weather Rev.* **122**: 366–376.

Robert, A. 1993. Bubble convection experiments with a semi-implicit formulation of the Euler equations. *J. Atmos. Sci.* **50(13)**: 1865–1873.

Tanguay, M.; E. Yakimiv, H. Ritchie and A. Robert. 1992. Advantages of spatial averaging in semi-implicit semi-Lagrangian schemes. *Mon. Weather Rev.* **120(1)**: 115–123.

Jakimow, G.; E. Yakimiw and A. Robert. 1992. An implicit formulation for horizontal diffusion in gridpoint models. *Mon. Weather Rev.* **120(1)**: 124–130.

Tanguay, M.; A. Robert and R. Laprise. 1990. A semi-implicit semi-Lagrangian fully compressible regional forecast model. *Mon. Weather Rev.* **118(10)**: 1970–1980.

Yakimiw, E. and A. Robert. 1990. Validation experiments for a nested grid-point regional forecast model. ATMOSPHERE-OCEAN, **28(4)**: 466–472.

Tanguay, M. and A. Robert. 1990. An efficient optimum interpolation analysis. ATMOSPHERE-OCEAN, **28(3)**: 365–377.

Desharnais, F. and A. Robert. 1990. Errors near the poles generated by a semi-Lagrangian integration scheme in a global spectral model. ATMOSPHERE-OCEAN, **28(2)**: 162–176.

Robert, A. and E. Yakimiw. 1986. Identification and elimination of an inflow boundary computational solution in limited area model integrations. ATMOSPHERE-OCEAN, **24(4)**: 369–385.

Tanguay, M. and A. Robert. 1986. Elimination of the Helmholtz equation associated with the semi-implicit scheme in a grid point model of the shallow water equations. *Mon. Weather Rev.* **114(11)**: 2154–2162.

Yakimiw, E. and A. Robert. 1986. Accuracy and stability analysis of a fully implicit scheme for the shallow-water equations. *Mon. Weather Rev.* **114(1)**: 240–244.

Robert, A.; T.L. Yee and H. Ritchie. 1985. A semi-Lagrangian and semi-implicit numerical integration scheme for multilevel atmospheric models. *Mon. Weather Rev.* **113**: 388–394.

———. 1983. The design of efficient time integration schemes for the primitive equations. *In*: Proceedings of the Seminar on Numerical Methods for Weather Prediction. European Centre for Medium-Range Weather Forecasts. Vol. 2, pp. 193–200.

———; T.L. Yee and H. Ritchie. 1983. Application of the semi-implicit and semi-Lagrangian integration scheme to a limited area model. *In*: Proceedings of the Seminar on Numerical Methods for Weather Prediction. European Centre for Medium-Range Weather Forecasts. Vol. 2, pp. 201–212.

———. 1982. A semi-implicit and semi-Lagrangian numerical integration scheme for the primitive meteorological equations. *J. Meteorol. Soc. Jpn.* **60**: 319–325.

———. 1981. A stable numerical integration scheme for the primitive meteorological equations. ATMOSPHERE-OCEAN, **19**: 319–325.

———. 1979. The Semi-implicit method. Numerical Methods used in Atmospheric Models. World Meteorological Organisation, GARP Publication, Series No. 17, Chapter 8, pp. 419–437.

———. 1975. Sensitivity experiments for the development of NWP models. *In*: Proceedings of the Eleventh Stanstead Seminar. Department of Meteorology, McGill University, pp. 68–81.

———. 1974. GARP activities related to computational considerations. World Meteorological Organisation Working Group on Numerical Experimentation, Report No. 4, pp. 2–23.

———. 1973. Computational resolution requirements for accurate medium-range numerical predictions. Symposium on finite difference and spectral methods for atmosphere and ocean dynamics problems. Siberian Branch of the Academy of Sciences, Novosibirsk, U.S.S.R., pp. 82–102.

———; J. Henderson and C. Turnbull. 1972. An implicit time integration scheme for baroclinic models of the atmosphere. *Mon. Weather Rev.* **100(5)**: 329–335.

———. 1971. Truncation errors in a filtered barotropic model. *In*: Proceedings of the Ninth Stanstead Seminar. Department of Meteorology, McGill University, pp. 13–28.

Kwizak, M. and A. Robert. 1971. A semi-implicit scheme for grid point atmospheric models of the primitive equations. *Mon. Weather Rev.* **99(1)**: 32–36.

Robert, A.; F.G. Shuman and J.P. Gerrity. 1970. On partial difference equations in mathematical physics. *Mon. Weather Rev.* **98(1)**: 1–6.

———. 1969. An unstable solution to the problem of advection by the finite-difference Eulerian method. Office Note No. 30, National Meteorological Center, Washington, D.C.

———. 1969. Forecast experiments with a spectral model. *In*: Proceedings of Seminars on the Middle-Atmosphere. Stanstead, Québec. Department of Meteorology, McGill University, pp. 69–82.

———. 1968. Résultats de quelques applications récentes de la méthode spectrale. La Météorologie. Paris, France. pp. 453–469.

———. 1968. The treatment of moisture and precipitation in atmospheric models integrated by the spectral method. *J. Appl. Met.* **7**: 730–735.

———. 1968. The integration of a spectral model of the atmosphere by the implicit method. Proceedings of the International Symposium on Numerical Weather Prediction. Tokyo. pp. 19–25.

———. 1968. Integration of a spectral barotropic model from global 500-mb charts. *Mon. Weather Rev.* **96**: 83–85.

———. 1967. The incorporation of precipitation into a spectral model of the atmosphere. Proceedings of the Seminar on the Middle-Atmosphere. Stanstead, Québec. Department of Meteorology, McGill University, pp. 91–114.

———. 1966. The integration of a low order spectral form of the primitive meteorological equations. *J. Meteorol. Soc. Jpn.* **44**: 237–245.

———. 1965. The integration of the spectral form of the primitive equations. Proceedings of the Symposium on the Dynamics of Large-Scale Atmospheric Processes. Academy of Sciences of the U.S.S.R., Moscow, pp. 66–69.

———. 1965. The behaviour of planetary waves in an atmospheric model based on spherical harmonics. McGill University. Publication in Meteorology No. 77.

———. 1963. A baroclinic model for the Canadian numerical weather prediction program. Seminars on the Stratosphere and Mesosphere and Polar Meteorology. Stanstead, Québec. Department of Meteorology, McGill University, pp. 83–88.

———. 1963. Baroclinic experiments with a four-level statistical dynamical model. Meteorological Memoirs No. 15.

——— and M. Kwizak. 1963. An evaluation of simple non-geostrophic forecasts. Meteorological Memoirs No. 13.

NUMERICAL METHODS IN ATMOSPHERIC AND OCEANIC MODELLING
THE ANDRÉ J. ROBERT MEMORIAL VOLUME

A Historical Perspective on Numerical Weather Prediction: A 1987 Interview with André Robert

Harold Ritchie and André Robert*
Recherche en prévision numérique
Service de l'environnement atmosphérique
Dorval, Québec, Canada H9P 1J3

[Original manuscript received 22 February 1995; in revised form 28 August 1995]

ABSTRACT *In view of André Robert's outstanding scientific contributions, he was selected for a personal interview as part of the Oral History Project of the Canadian Meteorological Service. The interview covered a variety of topics including André's education, early career, the beginnings of numerical weather prediction (NWP) in Canada, the evolution of NWP models at the Canadian Meteorological Centre, André's general research methodology and his views on the impact and future of NWP. These phases are illustrated by many personal glimpses and anecdotes such as puzzles that his relatives challenged him with in his youth, his explanation of why "a lot of time in my life was spent trying to avoid reading literature", and his expectation that "one day we are going to see forecasts of such a high quality that these will be even more accurate than the observations". In addition to capturing André's characteristically clear and unconventional insights, the contents also provide a very informative complement to the more technical details of his contributions by setting the historical perspective in which they occurred. The interview was conducted in November 1987, when André was on the point of retiring from the Atmospheric Environment Service and becoming a professor in the Physics Department at l'Université du Québec à Montréal. Some minor editing and updating have been done, but changes have been minimized to avoid interfering with André's typically expressive style.*

RÉSUMÉ *Étant donné ses contributions scientifiques exceptionnelles, le docteur André Robert a été choisi pour une entrevue personnelle dans le cadre du Projet d'Histoire Orale du Service Météorologique Canadien. L'entrevue a inclus une variété de sujets tels son éducation, le début de sa carrière, le commencement de la prévision numérique du temps (PNT) au Canada, l'évolution des modèles de PNT au Centre Météorologique Canadien, sa méthodologie générale de recherche, et ses idées sur l'impact et l'avenir de la PNT. Ces phases sont illustrées par plusieurs visions momentanées et anecdotes tel que des casses-tête posés par les membres de sa famille pendant sa jeunesse, son explication de pourquoi "j'ai passé beaucoup de temps dans ma vie à éviter de lire les publications", et son attente que "un jour on verra des prévisions de telle qualité qu'elles seront même plus précises que les observations". En plus de capturer ces vues claires et originales qui sont caractéristiques d'André dans ces domaines, cette entrevue fournit aussi des renseignements très instructifs pour compléter les aspects plus techniques de ses contributions en établissant la perspective historique dans laquelle elles se sont passées. L'entrevue s'est tenue en novembre 1987, au*

*Most recent affiliation: Département de physique, l'Université du Québec à Montréal.

moment où André était au point de prendre sa retraite du Service de l'environnement atmosphérique pour devenir professeur au département de physique à l'Université du Québec à Montréal. On a fait des mises au point mineures, mais les changements ont été minimisés pour éviter d'interférer avec la façon très expressive dans laquelle André s'est exprimé.

---

1 Introduction and biography

This is another in the series of personal interviews with outstanding figures in Canadian meteorology as part of the Oral History Project of the Canadian Meteorological Service. Today is 27 November 1987 and I have the privilege of interviewing Dr. André Robert at Recherche en prévision numérique, Service de l'environnement atmosphérique in Dorval, Québec. The interviewer is Harold Ritchie.

André Robert, the son of Mathias Robert and Irène Grindler, was born in New York city in 1929. The Roberts emigrated to Grand'Mère, Québec in 1937, and André became a naturalized Canadian citizen in 1967. He received his primary and secondary education at l'Académie du Sacré-Coeur in Grand'Mère, and graduated from Laval University in 1952 with a Bachelor of Science degree with specialization in Mathematics. He was hired by the Atmospheric Environment Service (AES) and attended the University of Toronto where he obtained a masters degree in Meteorology in 1953. That same year he married Marguerite Mercier, and André and Marguerite have two daughters, Claire born in 1954, and Lise born in 1958.

André worked as a forecaster in Dorval, where he prepared international aviation forecasts and public forecasts for the Québec region. In 1959 he was transferred to the Division de Recherche en prévision numérique where he worked on developing atmospheric models needed for short- and medium-range forecasts. A little later he undertook studies at McGill University, which he completed in 1965 when he received a Ph.D. for his thesis on the behaviour of planetary waves in an atmospheric model based on spherical harmonics (Robert, 1965). This was the first successful integration of a spectral model of the atmosphere formulated from the complete set of meteorological equations.

He was a Professor in the Meteorology Department at McGill from 1970 to 1971, was promoted to Chief of Recherche en prévision numérique in 1971, to Director of the Canadian Meteorological Centre in 1973, and to Senior Scientist at Recherche en prévision numérique in 1980. In 1987 André was appointed Emeritus Research Scientist with the AES and became a professor in the Physics Department at l'Université du Québec à Montréal.

Dr. Robert is a pioneer of Numerical Weather Prediction in Canada and has earned an international reputation for his work in this field. He formulated, constructed and implemented the first Canadian model, a barotropic grid point one in 1963, and the second Canadian model, a baroclinic grid point one in 1968. Largely due to his pivotal work on the spectral method in the 1960's, Canada was the first country, starting in 1976 (Daley et al., 1976), to use a spectral model to produce operational forecasts. He devised and implemented the semi-implicit scheme, which was tested for the first time in 1967 (first documented in Robert, 1969). This method permits the time step in atmospheric models to be increased by a factor of six, without reducing the accuracy of the forecasts. Since the efficiency of the models also increases by a factor of six, this method has been widely adopted in both gridpoint and spectral models, and is currently used by more than half of the meteorological centres around the world.

More recently Dr. Robert combined the semi-implicit scheme with existing Lagrangian techniques to permit the time step in atmospheric models to be increased by a further factor of six, again without reducing the accuracy of the forecasts (Robert, 1981, 1982; Robert et al., 1985). This method was applied operationally in Ireland in 1982 (Bates and McDonald, 1982), in Canada in 1988 (Tanguay et al., 1989), and is spreading gradually to other meteorological centres around the world. He further extended this approach to the non-hydrostatic equations (Tanguay et al., 1990), making it applicable to a wide range of fluid dynamics problems.

Dr. Robert has received considerable recognition as a scientist. He has been active in many world organizations, and has served on numerous national and international committees. He has supervised or consulted on a variety of government and university research projects. In 1972 he served as President of the Canadian Meteorological Society, with which he has had an especially close association. He received the President's Prize of the Canadian Meteorological Society in 1967 and in 1971, the American Meteorological Society's Half Century Award in 1981 and the Atmospheric Environment Service's Patterson Medal in 1986. In recognition of his outstanding contributions as a scientist, he was elected as a Fellow of the American Meteorological Society in 1968, and as a Member of the Royal Society of Canada in 1982.

2 Education

QUESTION: Could you comment on your primary and secondary education in Grand'Mère? In particular, were there people or events in your early years that had a strong influence on you, in science or in any other fields, while you were growing up?

ANSWER: Yes, in fact my primary education really started in New York City, where I completed Grade 1 and Grade 2, and then it continued in Grand'Mère. My secondary education was also completed in Grand'Mère.

During the years of my secondary education I had an uncle, who was a Catholic priest stationed at a convent for girls, where girls were getting their secondary education. He was also a radio amateur and had fairly expensive equipment which would enable him to communicate with other people all around the world. I was very interested in that kind of activity and I used to go and see him and spend long evenings with him trying to communicate with people in other countries. But it turned out that my uncle was also a great reader and he had an extensive collection of books, some of which I was really interested in, primarily those on the subject of astronomy. But he also had some books on mathematics, and I had some kind of an interest in mathematics because I was good at school in that subject. So one day, he took one of his books, which in this case was on calculus and differential equations, and he said: you should have a look at this. I took the book and I read it and I developed a very keen interest in calculus and differential equations. This was at the age of twelve, and this interest was to the extent that when I was going back to school, studying algebra and geometry, and was given problems, quite frequently I would use techniques based on calculus or differential equations to arrive at solutions for these problems. My professors, who didn't know anything about calculus or differential equations, would look at me and say, "What is all this about? This doesn't look like geometry anymore!"

I must mention, that my uncle, who was my father's brother, was not the only one who had a scientific interest. My father himself was a great reader. He knew the dictionary by heart. You could mention any word in French and he could give you the dictionary definition just like that. He had five other brothers who had received a fairly good high school education and it is because of their father, that is, my grandfather on my father's side, who was a reasonably rich person, who more or less looked like some sort of an aristocrat, who had given his children the best education you could possibly give. So on my father's side, all his brothers and sisters were people that I always found very exciting because they would always ask me very interesting questions as a kid, for which I obviously didn't have an answer. But that forced me to think about all kinds of things, and most of the time, these had some scientific connotations. Many of them had the habit of giving me little problems: in a yard, with a certain number of rabbits and a certain number of chickens, together you have thirty animals and all together you have forty-two legs; how many rabbits and how many chickens are there? And they were giving me those problems when I was a kid, maybe ten years old. I used to spend a lot of time trying to figure out how I could get the answer; and I used to turn out an answer quite fast. So obviously, the next time I met them, they would try to give me a problem that would be even a little more complicated.

But it is really at an early age, somewhere between twelve and fourteen years old, that I developed a keen interest in mathematics. It is really around that time that I decided that later on I would go to university and probably try to take a degree in science rather than do what my father wanted me to do. He was a mechanic and owned a garage. He had a large number of employees and he was making a fair amount of money and wanted me to take his place later on. I tried. I liked to work on motors and on cars but this wasn't really my interest. My interest in mathematical problems and mathematics or astronomy really exceeded my interest in playing with motors. And that's all. So already at a very young age, my decision had been made that later on I would go to university and try to take a degree in science.

QUESTION: Were there any of your undergraduate teachers, or courses, or books that made a particularly strong impression on you at Laval?

ANSWER: I had a teacher by the name of Adrien Pouliot, who was the Dean of the Faculty of Science, who really made a very strong impression on me. I have never seen a mathematician teaching mathematics with so much enthusiasm and so much motivation. Everything that came out of him sounded so clear to me and so easy to understand that I was saying to myself that later on I must be a clear man, I will have to do a lot of vulgarization to try to make mathematics understandable by everybody. This man really did make a very significant impression on me.

Besides the fact of being a mathematician, this man was impressive in all aspects. As a human being he was out of this world. He was a very absent-minded person, and because of this he was always doing strange things that were absolutely unbelievable. For example, he was a governor of the Canadian CBC, which had meetings occasionally in Montréal. Once he travelled from Québec City to Montréal to attend a meeting. He went with his car and with his wife. When he came out of the meeting he bought his train ticket and came back to Québec City by train, and forgot both his wife and his car in Montréal. So this man was exceptional in all aspects: he was a very good mathematician, a very good teacher and he had a very strange personality.

3 Early career

QUESTION: You graduated from Laval University with a Bachelor of Science degree, in Mathematics as you mentioned. At what point did you decide you wanted to make your career in meteorology, and why?

ANSWER: In the last week that I was at Laval University, I started looking for a job. The university wanted to give me a fellowship to go and study in Europe. It was a fellowship of the order of fifteen hundred dollars, which was not very much, and I wanted to get married. I was saying: well, fifteen hundred dollars as a married couple will be tough in Europe. So, I said: maybe I should look for a job and started looking for a job. I found a poster in the hall somewhere indicating that the Federal Government needed some one hundred and twenty-five meteorologists to work at weather offices. I was not interested in that, but I read the poster more carefully. They were saying they wanted graduates in physics or in mathematics and that they were prepared to send these candidates, with full pay for one year, to take a masters degree in meteorology. So I decided to enquire about that. I thought here is a way of getting a masters degree in meteorology, in a subject that seemed to have some mathematics, because they were asking for mathematicians and physicists, with full pay. So I said to myself: I can get that degree. If I don't like the job they are going to offer me, there were no strings attached: one could leave immediately after obtaining the degree and find a job elsewhere. I decided to try that, just for the fun of it.

As soon as I went to Toronto to take my masters degree, I was amazed at the amount of mathematics that there was in meteorology, especially fluid dynamics. I developed a keen interest in fluid dynamics. I decided, during the first month that

I was at the University of Toronto to take my degree, that I was going to stay in meteorology.

QUESTION: So at that time, were there any other occupations that you considered seriously?
ANSWER: No. There were two or three positions in the field of astronomy that were being offered by the National Research Council and I think one university. The number of jobs in the field of mathematics was not very numerous. They were very few. There was one job at the National Defence Research Centre in Val Cartier that was available, but I didn't think my chances of getting it were very good. So the number of jobs was not very high, except for those offers in meteorology where the number one hundred and twenty-five struck me as a field in full expansion. I said there must be something really going on there. That excited my curiosity and it is for that reason that I decided to have a try at meteorology.

QUESTION: Did you have specific expectations of what your job would involve?
ANSWER: No, I simply could not understand why they were trying to recruit mathematicians and physicists, because, as far as I was concerned, weather forecasts and horoscopes were all the same thing. They were in a little corner on the front page of the newspaper: here was the horoscope and beside that you see a little barometer with fair, bad, stormy. I thought they were two activities in the same class. I couldn't see how there could be any science in the field of meteorology. So I was really impressed when I went to the University of Toronto and we started purchasing our first text books and we started looking, attending our first courses. I was amazed to see how at least they were trying to be very, very scientific, with dynamic meteorology, a little bit, a very little bit of Numerical Weather Prediction because it was just starting in those years, a lot of synoptic meteorology, thermodynamics and subjects like that I found to be quite fascinating.

QUESTION: Did you find that your training in Toronto prepared you for your work as a forecaster or more as a researcher?
ANSWER: Yes, the training in Toronto, the entire masters degree was oriented towards preparing people to become weather forecasters. But that does not mean that there was no science. I mean dynamic meteorology was taught to students and it was well taught. There was a lot of mathematics, the handling of the meteorological equations and geostrophic approximation, how one could compute winds from the pressure gradients, the circulation theorem and a lot of things to me had a very high mathematical content and I found them to be very, very interesting. So even though the course was preparation to become a weather forecaster, that is, there was a lot of time devoted to analyzing weather maps and so on, the course itself, I think, was a very good one.

QUESTION: Between 1953 and 1959 you were an operational forecaster in Dorval. Did you enjoy forecasting?

ANSWER: Yes and no. What I really enjoyed was what was going on in meteorology. The first thing I did, when I became a weather forecaster, was to ask to be responsible for the library that we had at the forecast office. We had a reasonable library with most American and some European journals available, and a lot of books available in this library. I wanted to be responsible for this. I wanted to make sure that everything was kept in good shape, that books would not be stolen, that they would come back to the library, that people would bring them back, and that if there were new journals coming out that we would like to make sure that we would have them. I used to spend a lot of my spare time, when it was very quiet in the weather office, in the library reading.

I was particularly interested in Numerical Weather Prediction. That was in the early 1950's when Numerical Weather Prediction was just starting. First experiments had been done by people like Charney, Smagorinsky, Shuman, Gambo and others at the Institute for Advanced Studies in Princeton. I was very interested in those experiments and I was very interested in the optimism that these people had for the future of Numerical Weather Prediction. I considered, as a forecaster, that what we were doing at the time appeared to me a waste of time, because what was coming was so fascinating compared to what we were trying to do at the time. I could see that all the future was five years beyond. I was saying to myself: in five or ten years it is going to be so different from what we are doing now that we should forget about the present situation and concentrate on the situation that will exist in the next five or ten years. I used to spend a lot of time discussing that with my colleagues in the office, and that tended to demoralize a lot of them because I was telling them that everything we were doing at the time could simply go into the wastepaper basket. Models are going to come out very soon and once these models come out, our job is not needed anymore. The efforts we spend trying to predict what tomorrow's weather is going to be, will be done by the models. That used to demoralize people. But me, the strange thing about that is, I was optimistic about the whole thing because that's the field in which I wanted to work, while my colleagues were pessimistic about the whole thing, because that meant the disappearance of their job, or that is the way they sensed the whole thing.

QUESTION: Did you have a chance, at that time, to do project work, conduct research and write papers?

ANSWER: Yes. As a matter of fact, during the years that I was a forecaster, the Canadian Meteorological Centre was located in the same building. The Canadian Meteorological Centre was opened during the years that I was a weather forecaster. The idea was that the Canadian Centre would produce a national analysis that could be used by the various weather offices, because prior to that the Ontario region could issue some weather forecasts and the Québec region could issue some weather forecasts. In some cases, you had Hull on one side of the river and Ottawa on the other side of the river, and the forecasts for Hull and Ottawa were completely inconsistent. So across the boundaries of regions you frequently had discontinuities in the weather forecasts. So they created the Canadian Meteorological Centre to produce a national analysis hoping that this would influence the forecasts produced at the local centres. So that was the original goal of the Canadian Meteorological Centre. But, at the Canadian Meteorological Centre, they realized that Numerical Weather Prediction was coming and they wanted to prepare for it. Unfortunately, they did not have anybody at that centre who knew anything about Numerical Weather Prediction. During that period I presented a number of seminars at the Canadian Meteorological Centre in the field of Numerical Weather Prediction. In fact, that is how, later on, I was transferred to the Canadian Meteorological Centre to develop work in the field of Numerical Weather Prediction, when our current division was created.

QUESTION: Looking back, do you feel that your experience as a forecaster, getting first hand synoptic experience, has had an important impact on your career?
ANSWER: Yes, it was extremely important when I started in the field of Numerical Weather Prediction. Especially when I was involved with colleagues, scientists that had followed an entirely different route, that came directly from university, graduates in the field of mathematics, and tried to start directly in the field of Numerical Weather Prediction. I found that I had a sense for a prediction that was right or that was wrong that was much better than those people who had absolutely no sense for that. They had maybe, in some cases, a better knowledge of mathematics than I may have had, but I had a sense for meteorology that these people did not. I think that even this continues to exist today: those who have taken their Ph.D. in Meteorology compared to those who have taken their Ph.D. in Physics or in Mathematics. When these people engage in Numerical Weather Prediction, I do think that a person with a Ph.D. in Meteorology has a slight advantage over the others.

QUESTION: What was your first real research project?
ANSWER: My first project consisted of constructing a barotropic model. What they wanted was to construct the model that was being used by the National Meteorological Center (NMC) as its first operational model. This barotropic model was implemented at the NMC in early 1958, and in 1959 we were trying to reconstruct that model here, in Montréal, to run on a local computer. We wanted to duplicate that model exactly as it was being run in Washington. That was my first assignment. I found it very interesting because it smoothed the way for me into the field of Numerical Weather Prediction through things that were actually being used operationally.

4 Beginning of numerical weather prediction in Canada

QUESTION: So that coincided then with your transfer to the Dynamical Prediction Research Division?
ANSWER: Yes. My transfer took place in 1959. This is the year that the division was created and I was the first employee to move into that division. I was the one who received the first assignment and the first assignment was to construct a model.

QUESTION: Was there any particular person that influenced you to go in that direction or made it easier for you to make that transfer?
ANSWER: Not really, but there was one person in the Division. The person in charge of the Division used to be Dr. Boville. At that time he was a meteorologist, he was in charge of that division, and asked for educational leave to go to McGill, where he took his Ph.D. He was replaced by the only employee of the Division, who's name was Michael Kwizak, who became Acting Division Chief during Boville's absence. Michael Kwizak was very interested in Numerical Weather Prediction. The first action he did in the first week that he was Acting Chief of the Division is he wrote a letter to Toronto to the Director of the Meteorological Service asking for some funds to try to do some Numerical Weather Prediction at Canadair, which was one of the first companies in Montréal that had a computer. He wanted to rent some time on that computer and try something. Surprisingly he got the money almost right away and he got three times as much as he asked for. Mike Kwizak was very interested in Numerical Weather Prediction and he was very keen on getting me into that division. So we were very good friends for a fairly lengthy period of time. He had some influence on me and he helped me a lot at the beginning of my career, making sure that all the help that I needed was being provided to me.

QUESTION: I noticed in an article entitled "A baroclinic model for the Canadian numerical weather prediction program", that you presented at the Stanstead Seminar in 1963 (Robert, 1963), I think you were reporting on this work when you said: "Early in 1959 a small group of meteorologists located at the Central Analysis Office in Montréal, and working under the direction of M. Kwizak, started testing models of the atmosphere in preparation for an operational Numerical Weather Prediction Program." So that actually marked the beginning of Numerical Weather Prediction in Canada?
ANSWER: Yes, that's right. The operational program itself started in 1963. The first model that we developed was the first model to be implemented. The barotropic model was run until 1968 roughly, when it was replaced by the four level baroclinic model, which I also built, with some assistance.

QUESTION: You mentioned Mike Kwizak and yourself. Who else was involved?
ANSWER: In the early years, in 1959, three positions were filled. One was offered to a person by the name of Robert Strachan and another one was Amos Eddy. Strachan

was asked to write programs to decode data, the data processing programs to decode meteorological observations, and Amos Eddy was asked to build a program to do objective analysis. Cressman's technique was used at that time. That was essentially the group that was created almost immediately after we got the money to use a computer. As a matter of fact, the computer we did use was a computer that had just been purchased or rented by McGill University. That is, in fact, the computer that we did start using.

QUESTION: It was essentially then Mike Kwizak who initiated the Numerical Weather Prediction?
ANSWER: Yes, he is really the person that was at the bottom of everything. He has to be given credit for that. He had to fight the Director of the Canadian Meteorological Centre who really did not believe in Numerical Weather Prediction.

QUESTION: Did you have a sense, at that time, that you were participating in history in the making?
ANSWER: As far as Canada is concerned, yes. As far as activities in this field around the world: no, not at that time. We were just trying to catch up with other nations. We felt that we were about five years behind the Americans. My main goal was to catch up, try to get ahead of them if that were at all possible.

QUESTION: What was the mood in the group? Was there a lot of enthusiasm?
ANSWER: Oh yes, it was very enthusiastic. Playing with computers was a lot of fun in those days and for that reason everybody was highly motivated. Also, we thought that Numerical Weather Prediction would really eventually cause a revolution in meteorology. So, for that reason, we were also very highly motivated. That feeling existed everywhere around the world at that time.

QUESTION: What were the working conditions like?
ANSWER: They were absolutely horrible in one sense. We used to work from 10 o'clock in the evening until 10 o'clock in the morning. I worked like that for over a year and a half: twelve hours a day at night. The reason that we chose to work that way was because the first computer we had, as I said was at McGill University, and during the daytime there were just too many students around the computer to make it workable at all. So we opted to work late at night and during the night, and do all our work then. The advantage of that is, for instance the baroclinic models that we built, we had two of them: one used to take sixteen hours of computer time to produce a twenty-four hour forecast and the other one took twenty-eight hours of computer time to produce a twenty-four hour forecast. As much as possible, we always tryed to produce these forecasts with as few interruptions as possible. With a twelve-hour period and two sessions; we could run one case for the model that took sixteen hours. Even with the barotropic model, it was taking us three hours to produce an integration and there again, we preferred to go there, spend the whole

night and do two, three or four of these integrations if we could. During the night, working on a computer, it was absolutely wonderful: because it was so quiet, we could operate the computer ourselves. We just had a technician hanging close by, just in case the computer broke down. Other than that, there was absolutely nobody in the building and it was a very good way to do it. We decided democratically to do it that way. We were not forced to do it. But those were the kinds of working conditions that we lived with for a period of time.

I would like to mention exchanges between various nations. When I was a member of the World Meteorological Organization Working Group on Numerical Experimentation, one subject that used to come up quite frequently was how freely meteorologists around the world could exchange ideas amongst themselves. They were saying that this was one of the rare fields or activities where we could talk meteorology with the Russians, with the Chinese, with almost anybody and there would be nobody around trying to prevent us from saying what we wanted to say. In other words there was absolutely no secrecy, as far as meteorology was concerned. We could get papers or books from the Russians, and they could take journals from the United States and they could move across the borders without any restriction at all. As a matter of fact, it seemed that the governments encouraged these exchanges in the field of meteorology. So, I think that exchanges, in general, in our field of activity were quite excellent.

5 Numerical weather prediction at the Canadian Meteorological Centre

QUESTION: In 1973 you were promoted to Director of the Canadian Meteorological Centre. Was there a special reason that you wanted that position?

ANSWER: Yes, because the CMC was blocking us. Everything that RPN was doing, every research and every piece of development that was going on here was being blocked off by the CMC. It took seven years for them to implement our first baroclinic model. They spent seven years evaluating and reluctantly saying: well we can't take it the way it is, it's not good enough and so on. When I was Division Chief I had a lot of discussions with the Director of the CMC at the time, and I always found a lot of resistance to the changes that we were advocating. So when the position became vacant, I decided to take it and to bring the CMC into modern days. Charts were still being drawn by hand or being plotted by hand, everything was backward compared to other centres around the world. So I decided to buy a lot of equipment, to use the techniques that can be implemented on the kind of computers that we were using, to have something where Numerical Weather Prediction could really play its role. That was the main reason: it was to change the attitude that existed within that organization.

QUESTION: Were you planning to embark on an administrative or management career?

ANSWER: Not at all. Not at all. As soon as I became Director, within a month I submitted a plan to my boss in Toronto, which consisted of automating seventy-five positions out of a hundred and forty; replacing these people by machines: curve-plotters and things like that, which was accepted just like that. My director general said this is a marvellous plan, you go ahead and carry it out and I will give you all the support you need. So I went through the first forty-five positions without any great difficulty, over a period of about three and a half, four years. But for the next thirty I started running into a lot of resistance because here I was aiming at professionals, trying to reduce their numbers and trying to have their work done by computers. I started running into difficulties with my director general: he started not accepting these changes that I wanted to make. So it is somewhere around that time that I said: okay, I have done what I could, now everything is standing in my way, so I will just go back to research.

I was not considering being at the CMC for the rest of my life, even though I might have opted to do so, because I was still capable of progressing but more slowly in the later years. But, what happened around 1979, 1980, while I was director of the CMC, I got this idea of associating a semi-Lagrangian scheme with the semi-implicit technique and this was bothering me for a little while, until I started realizing this was really something. So then I asked the Deputy Minister, Art Collin at the time, I said: would it be possible for me to go back to research? Could this be arranged? And it was done fairly quickly. So I came back to research immediately. I even started working on the semi-Lagrangian scheme while I was Director at the CMC. So when I came back I just went at this on a full-time basis.

QUESTION: So it seems that the resistance to accepting Numerical Weather Prediction products came essentially from people whose jobs would be very strongly affected by it?

ANSWER: Yes. In Canada and in other countries as well. I think in the United States the institution of the Numerical Weather Prediction program was not all that easy. A lot of places in the United States resisted that. It was just a natural resistance to change.

QUESTION: Would you say that Numerical Weather Prediction actually has been a break-through in meteorology?

ANSWER: I think so, especially when you look at verification scores. NMC in Washington started verification in 1947 and in the period from 1947 to 1958 the S1 scores were stable, somewhere around seventy-five points. In 1958, when Numerical Weather Prediction started at NMC, and in the years that followed you could see the forecasts improving, the scores dropping. They started dropping in a linear fashion and they kept on dropping ever since that time. Even today they are still dropping at almost the same rate: a rate of roughly one or one and a half points per year. If you look at the kind of scores that you obtain today for these forecasts, nobody ever achieved a forecast of that quality in the late 50's, that is by just producing a forecast by hand. If you look at the models we are using nowadays, and apply them to a situation in 1955 that is an analysis; they do produce the same kind of scores that we obtain now: scores around forty-five points or something like that. So it's not a question of improved data, it's really the numerical models. At NMC, as soon as people started using models, the subjective forecasts started improving very rapidly, because people were getting a lot of good points out of the numerical prediction that they were introducing into their subjective forecasts, that enabled them to start improving their scores. So I think that, by and large, Numerical Weather Prediction has introduced a new direction in meteorology that is going towards perfect forecasts. Slowly, much more slowly than we expected, we thought it would be a revolution and that, overnight, forecasts would improve by a considerable amount, but it's rather been an evolution. That evolution has gone on for the last thirty years and I think it's going to keep on going for the next fifty years, until we reach quasi perfection or what some people call the ultimate limit of predictability.

QUESTION: I have some questions I would like to ask you on that one just a bit later. I think we have covered this to some extent already but, could you briefly review the different numerical weather models that have been used operationally in Canada, mentioning what motivated going from one model to the next, and which ones produced the greatest improvement?
ANSWER: The first model was a non-divergent barotropic model with a stabilization term, it was an exact replica of the model used in Washington. The second model was a four-level baroclinic model of the potential vorticity equation. The initial suggestion was that we should use in Canada the model that was developed by Wiin-Nielsen for NMC, which was slightly different. But I had suggested that it would be more economical to integrate the potential vorticity equation and that the forecast would be just as good. Both models were developed, Wiin-Nielsen's model was developed and this one was developed, but the model of the potential vorticity equation turned out to be more efficient. So that was the one that was implemented and that was the second model used in Canada. The third model was the original version of the spectral model, truncated at wavenumber 20, rhomboidal 20, and the next model after that was the finite element one. So these are essentially the four models that we have used up to now for operational weather forecasting.

INTERVIEWER: There has been this continuous improvement in verification scores as we've gone from one to the next.
ANSWER: All throughout that period, almost everywhere in the world, at least in Canada and the United States. I think the same thing could be said about some other industrialized nations like England, France and the European Centre.

QUESTION: Are you aware of any one model that produced a drastic improvement or has it just been gradual?
ANSWER: Yes, it is the first integrations of primitive equations model. It was immediately after the first primitive equations model was implemented in Washington,

that very drastic improvements took place, and very, very early the decision was made that we would never go back to filtered models.

6 Improving numerical weather prediction models

QUESTION: You were mentioning that while you were Director of the Canadian Meterological Centre, you realized there was a potential of combining the semi-implicit scheme with the Lagrangian technique. What was it that led you to look into this idea, into this method?

ANSWER: Because I had been advocating for a long time that the semi-implicit technique was only a half-solution to a problem. That the semi-implicit technique enabled us to use larger time steps, but the time steps that we were using with the semi-implicit technique were still shorter than what I felt we could use. When looking at truncation errors associated with time integration in the semi-implicit scheme and the space truncation errors, it seemed obvious to me that our time steps were still too short. So one had to look for ways of trying to further enlarge the time steps and obviously fully implicit schemes appeared to be one possibility, but they were so difficult to use that it didn't seem to be a practical way of doing it. So I was looking for alternatives to fully implicit schemes when the idea of semi-Lagrangian techniques associated with semi-implicit techniques sort of passed through my head. I started thinking about it and started realizing that it might be a possible way of solving the problems. After thinking about it for a number of months, trying to analyze this on paper, I finally concluded that yes it was a very interesting possible solution.

QUESTION: In an article entitled "Sensitivity Experiments for the Development of Numerical Weather Prediction Models" that you presented at the Stanstead Seminar in 1975 (Robert, 1975), you presented a table that itemized, in order of importance, the causes of errors in state-of-the-art models at that time. Has much progress been made in reducing errors since then?

ANSWER: Yes, especially the numerical errors which at that time represented somewhere between forty-five and fifty percent of all the errors that are encountered in short-range forecasts. That fraction has been reduced considerably. Some of the best models that exist today, like the spectral model at the European Centre, some of the models that are used in other countries as well, in some of these models the level of numerical errors coming from discretization in space and time, and initialization; the level of those errors is now down below fifteen percent. So, nowadays, we are in the situation where most of the errors in short-range weather forecasts arise from the fact that physical processes are not parametrized well enough. This is a complete reversal of the situation that existed fifteen, twenty years ago, because the additional resolution has reduced truncation errors. People also had demonstrated that when you reduce the grid length by a factor of two, that there again, the motion of systems, cyclones and anti-cyclones, again would be increased by thirty percent and this was due to the coarse grid. So there were a lot of studies made by different people showing, in some way or another, that the resolution we were using in space and in the vertical, and the way we were starting our integrations in some cases, were leading to fairly large errors and as a matter of fact explained nearly half the errors that we were observing in short-range forecasts. So people had some knowledge of this and they were working very hard. It is the computers that were not ready. We did not have computers powerful enough to reduce the spacing between grid points or to increase the number of levels in the vertical. But we knew that this is what we had to do. So, in that sense, we were aware of these things.

QUESTION: In that same article, and I think you mentioned just a few moment ago, it was estimated that, at the current rate of improvement, it would take another fifty years for models to reach the limits imposed by the theory of predictability. First of all, do you think that a precise limit to predictability exists, and why?
ANSWER: I would say that the meteorological community at the present time considers that there is what one can call an ultimate limit to predictability. People will mention experiments that have been performed by certain groups some years ago, when we were trying to plan FGGE, the First Global GARP Experiment, which demonstrated that having very small differences at initial time, when you integrate numerical models these differences increase with time to the point where, after twenty days, the two integrations are just as different as two randomly selected charts or sets of analyses for the atmosphere. So there have been studies tending to demonstrate that we should be able to produce forecasts up to fifteen, twenty days without too much difficulty. But it seems to be virtually impossible to go beyond that. That seems to be the current belief in the meteorological community at the present time, even though a few people are saying: well, that maybe it might be possible to go quite a bit beyond that. For instance, people do talk about seasonal forecasts, but here generally they are talking about forecasting or predicting only the planetary scales, or averages over a very, very large region. I think everyone realizes that predictability is something that depends on scale. For instance, you cannot expect to be able to predict a severe cumulonimbus for periods of seven days. You can talk about a limit of predictability maybe of three hours, or six hours, or something like that, that one could make a fairly good forecast. We are talking here about deterministic forecasts. So predictability for very, very fine scale events is going to be very short in time, whereas predictability for very large scale events is going to be much longer in time. But that's all based on the idea that two slightly different integrations will diverge, and diverge fairly rapidly.

But this is looking at the problem as an initial value problem. If we talk about integrations which are guided over a period of ten days, where there is a lot of data available, lots of observations available, when after ten days we just let this integration go, then I think we will achieve predictability, rather than on the order of fifteen to twenty days, we will probably be able to reach periods of thirty to forty, or fifty days. But that's an entirely different approach to Numerical Weather

Prediction. You run a model for a long period of time, up to right now, and force the model to follow that situation over a period of ten or twenty days, and then you let it go from the present time to twenty, thirty, forty days into the future. That is going to be a very difficult problem to tackle. To make sure that, during that period of ten or twenty days that you integrate over the past, that the model state becomes the average of what has happened over that period of time and that at each instant it fits as well as possible. It's a form of initialization that people really haven't been looking at very much and I think it is a very difficult problem. It relates somehow to some of the data assimilation experiments that some people are trying to do today. But I think that ultimately it's going to result in the lengthening of what we consider to be the ultimate limit of predictability.

QUESTION: So the current estimate of a two or three week limit shouldn't really prevent us from issuing monthly or seasonal outlooks?
ANSWER: No. At least we still have a long way to go to achieve the goal that people think is achievable at the present time. I think there is still a possibility of reducing errors in forecasts by a factor of three. Consequently forecasts that have a scale up to about six or seven days, I think we can go to something like twenty days. We still have that goal, which is probably going to take another fifty years to reach. But I think even once we have reached that goal, we will find ways of going beyond that.

INTERVIEWER: So your updated estimate of how soon models will reach their limit is fifty years plus.
ANSWER: Yes, it is really fifty years plus. And the plus could be another hundred years or more.

QUESTION: Are you concerned about the public's general perception of our ability to predict the weather? Do you think the public may have unreasonable expectations of what we can forecast?
ANSWER: They seem to be unreasonable to us, but they are not unreasonable to me because that is exactly the kind of expectations that I have: the same as the public. I will only be satisfied once we produce exact forecasts, just like astronomers tell you when the next eclipse is going to take place three years in advance. I think that is the kind of goal we should aim for in Numerical Weather Prediction: produce exact forecasts as long in advance as we possibly can. I am sure that for forecasts on the order of three days, one day we are going to see forecasts of such a high quality that these will be even more accurate than the observations. I think then we will obtain a lot of credibility from the general public, if we can predict more accurately than the observations. I don't think that the public wants anything of that type of accuracy, but that is something we can do and I think by then we will be in a situation where the public will start being satisfied with us. In other words,

failures will occur so rarely, once every five years or something like that, that by then people will excuse us when it does happen.

7 General research methodology

QUESTION: I was wondering if we could shift to a sort of a different line of discussion and talk about general research methodology. The list of publications that you provided with your curriculum vitae seems a little short in view of the tremendous impact that your career is having. For example, I count nine first author papers in refereed journals, accounting for roughly sixty-nine pages. That indicates, I think, that you make an extremely judicious choice of research topics to work in. Do you have any comments on that? Do you have any sort of special guidelines that you use in choosing your research direction?

ANSWER: Maybe I use guidelines or maybe I don't. I know that everyone who tries to orient his work has difficulties. I remember these sorts of difficulties when I was starting work on my Ph.D. Every student would take the first year to read the literature quite a bit and try to find something in the literature which they felt they could continue to the next level; and consequently obtain their Ph.D. for that kind of work.

Generally, when we started, our goals were very high. As a Ph.D. my contribution is going to be a major contribution. So you search the literature for places where you could come out with a major contribution. But with major contributions there is always a risk that it will back-fire on you and that consequently you risk not getting your Ph.D. at all. So people after a little while start saying, "Well, maybe I should play it safe and maybe I should settle for a minor contribution." And ultimately that is what happens to most people who do their Ph.D. work; they settle for a very, very minor contribution but one which they are almost sure that they can carry all the way through. That has been the sort of consideration that I have always given to the choice of a subject. But my tendency has always been to go to the other end of the spectrum. If I feel that something might turn out to be a major contribution, but involves a lot of risk, I have never been afraid of the risk; I will take it and I would even take the attitude that if I could see an even greater contribution and something that appeared to be almost impossible to complete, as long as I could see that there is a five percent chance that it would work, I would go ahead and try. I would not worry about the risk at all. So I always evaluated my possible contributions. I would always neglect the risky side of it, and just look at the ultimate use that could be made by the community, how much could be beneficial to the community. So in that sense, I have never been afraid of doing things which appeared to be very, very risky. Because of that, I have not published very much. Because many of the things would involve two, three, four, five years of work before I could actually convince the scientific community that this was a worthwhile achievement. So I always opted for one good paper every ten years as being much better than ten papers every year. So for that reason I haven't published very much. And in my case, because of some luck I presume, it has been a profitable strategy. It has worked out reasonably well.

QUESTION: What does it take to keep abreast of what is happening and also to be able to be ahead of change? How do you generally learn about important new developments and manage to be on the forefront of research?
ANSWER: Some people used to say that I was an iconoclast, in the sense that with people like Charney and Rossby, I used to criticize very, very strongly some aspects of their work. Sometimes people were just saying, "Oh, you are criticizing these people just because they are great names." I have had this attitude all the way along. As a matter of fact when I was young, one approach I took to my own training, is that if somebody in the world has done something that is highly original, it is unwise for me to read this material. The only thing that is important for me is to know that he has done it. Then I should try to do it myself. This is how I tackled the problem of solving third degree equations for instance. I was learning at school how to find the solution of a second degree equation, but the darn book said that third and fourth degree equations had been solved, but never gave the method. So acting on the fact that I knew that somebody had solved the problem, I decided to try it. It took me over a year of work to find the technique, but the darn guy who did solve it, it took him two thousand years to do it. So, in that sense, the approach I had towards learning has always been the following: when I go to university and I know that next week the professor is going to be talking about something and I can't figure out how people arrived at that conclusion, I am going to rush and try to do it myself before the professor gives his lecture. So what I used to do is read text books in a strange way, read the introduction to each chapter and stop right there, before any development would start, and say, "Okay, the introduction tells me roughly what is going to be in that chapter. I've got to work it out myself before I start reading it." What I found out by doing this sort of thing is that I was trying to put myself on a par with people who have done development work or have done original things in the past. And also, I was finding out that, by doing that, the kind of methods I was using to arrive at the same answer quite frequently was a totally different approach to what people had taken before.

The other thing I had noticed before, is that when you read about the work of another person who had been innovative and had done something absolutely fantastic, when you read about this, I usually found out that it was so interesting and so well done that you could not conceive of any other way of doing it. Consequently you were in a position where you found yourself incapable of doing what this man had done. Whereas if you had tried to do it yourself and you had succeeded, first you found that the method you used is entirely different, so you see the thing. So a lot of time in my life was spent trying to avoid reading literature. Just reading the abstract of a paper and if that stirred my interest, I said, "Okay, I am going to work it out myself and I'll read the paper afterwards to just confirm whether he used the same technique, or whether he was wrong or right or so on." And I have done a lot of that in my lifetime; that is, working through equations hundreds and hundreds of times, sometimes repeating a lot of things. I have done a lot of that in my life; trying to understand what fluid dynamics is all about and what these equations are trying to do, and so on. So I have done a lot of personal work of that type and I think that has helped me develop an aptitude to do things that have some originality.

QUESTION: You have an important influence on the Canadian Meteorological Service, not only through your research contributions, but also by the encouragement and training that you give to others. Do you think scientists in general devote enough time and effort to training?

ANSWER: I would say that they probably do not devote enough time to training. You look at a field of activity like ours, there are hundreds of scientists around the world engaged in Numerical Weather Prediction, probably somewhere around eight or nine hundred. How many of those have written text books? How many of those have taken very very seriously the idea of training younger people? Not very many, probably not enough, way not enough. So I do... Training was one of my goals in life when I started to engage in a research career. I used to say to myself, "I am going to try to be as productive as I can possibly be in research up to about the age of forty-five or fifty. After that, I am going to try to pass my knowledge, my experience on to younger people." I thought that was the ideal research career, maybe twenty, twenty-five years of productivity and then another ten to fifteen years where you prepare the younger generation. That was my concept of research activity. I know there are some people in science who operate that way, there are some people who have been engaged in teaching from their very first years. People like Lorenz, for instance, always stayed at the university and has always been engaged in teaching, in training younger people, always mixed teaching and research together. I don't think one has to go that far, but I think there are probably too many people who do only research and refuse to do any teaching or to make any effort to pass their experience on to other people. I think this is a very important aspect: training the younger people and getting involved in some kind of teaching.

QUESTION: Do you find that research is a high pressure profession? And, if so, have you developed techniques or are you aware of techniques that scientists use in dealing with the pressure?

ANSWER: The work that I have been doing throughout my career, I used to tend to look at it as maybe high pressure work in the first years. There was something that my colleagues used to look at me and laugh sometimes, when we started using the first computer, we were all very conscious of how expensive it was. And when we ran our models on the computer we would sit in the computer room and watch the computer going, and if for some reason or another it stopped, then my two colleagues would start running around, jumping over machines, getting to the keyboard and start typing in, and I would stand there, sit in my chair, as if I was paralyzed. And people would say, "What's wrong with you, can't you join in and help us?" And I'd say, "No, no, no, no. That's not the way to go at it." So after four, five, ten minutes of keying at the keyboard, then I would move in and just press one key and the whole thing would go. I would start to explain to people, when you are under pressure or something happens, you have to decide: will a quick move solve the problem or would it not be better to think a little bit and then move. And most of the time, when you look at these kinds of situations, quite frequently, a few minutes of using your intelligence may be a wise thing to do before you actually make a move. So that has been my reaction in general to pressures of all kinds, whether they're pressures that occur at an instant or pressures that occur over a long period of time; that is, trying to develop in one year the kind of thing that would take you three years to do. Generally I would agree to do these kinds of things, but do it in such a way that it will be done in a year, but maybe at a lower level of quality or just a preliminary thing that will need to be cleaned up later on, or something like that. So in general I always felt very happy in the field of research because even though there were attempts to impose pressure on us, that I generally got along quite well.

One thing I would like to say about pressures is that there are a lot of people who complain about the existence of strong pressures on what they are doing. Well, what I would like to mention to them is that those pressures do not come from the organization they are working in. They do not come from the conditions under which they are working. Those pressures come from themselves. It is people who fix their own goals in research and try to achieve them. Quite frequently those goals are unrealistic. And then when the person sees that he cannot make his deadline, or that he sees that he is wasting time, or that he starts slowing down, he starts feeling pressure, but he is the one who laid down the goal, not the organization. So to those who complain about pressures, I normally say, "Well, all you have to do is change your goals, nobody is going to say a word. Just change your goals and that will relieve you of all the pressures that you are feeling." So I think in research, pressures are really self-created most of the time. So one way to avoid them is to be a little more lenient about objectives and be a lot happier when you have reached your objective in half of the time that you had set down to do it.

QUESTION: Did you ever think of changing your research to some other specialty, or even some other field entirely?

ANSWER: No, never. I have seen a number of colleagues who after a number of years in research could no longer stand research and did go into other fields, but I have never found myself in that kind of situation.

QUESTION: Some research scientists move frequently, but you have spent virtually your whole career in one city. Is there a special reason for that?

ANSWER: Yes, it's because I am French speaking. My wife is French speaking, my kids are French speaking, and it would be very difficult for my family to go far away from Québec. I decided at the very early stages in my career that I would try to make my wife and kids as happy as possible, and might even sacrifice my own career if necessary. Fortunately I was able to do some fairly good work in Québec and I was fortunate because of the existence of our group, but I think that by and large, people should not take this kind of attitude. People like Phil Merilees, who have worked in a large number of places, who have been very mobile, (having been in places like NCAR, Florida State University, University of Michigan, RPN, McGill University, Headquarters in Toronto) who have gone to many places, diversify their interests and suffer from inbreeding much less than people who stay in the same organization throughout their whole career. So what I have done should not be an example to be followed by other people. It's because of the language problems.

8 General discussion

QUESTION: What would you say have been some of the most exciting, stimulating, or satisfying points in your career?

ANSWER: Well, probably one of the events that occurred in my career that I still remember and that has affected me, is when I went to the annual meeting of the American Meteorological Society held in Denver, I think in 1967, in which I gave my first presentation outside of Canada. This was a presentation of the results that I had obtained with my first integration with the spectral model, that is my Ph.D. work. It happened that my presentation was just before coffee-break, in mid-morning, and it happened that Joe Smagorinsky was the chairman of that session. So my presentation went on and was completed. After my presentation was completed Joe came forward and asked people if they had any questions, and questions started coming. Joe was getting people to ask their questions, and I was trying to provide answers, and Joe all of a sudden realized we were running into coffee-break. So Smagorinsky made one suggestion, he said, "Well, people should have their coffee-break, but let's carry on with the questions after coffee-break and move a paper into the next session if necessary." So the whole thing broke on that, and we all went for coffee, and then we came back in, and Joe re-opened the session, "So okay, now we will finish the question period." But questions went on, and on, and on for almost an hour and a half, such that there were no presentations after coffee-break. The interest in the work that I had done with spectral models was so high, I mean, people like Lorenz were asking me questions; Norman Phillips was asking me questions; Fred Shuman was asking me questions, people were really interested in this work. And that was what impressed me most, the interest that this thing was generating in the community at that time. So that was one of the most impressive events that occurred during my career.

QUESTION: Whose work would you say has had the greatest influence on your own, either in a personal way, or through publications?

ANSWER: I would say Marchuk. I was very impressed when I read a paper by Marchuk (Marchuk,1965), where he tried to integrate the primitive equations by what one would call a fully-implicit scheme. He admitted right away that his fully-implicit scheme was very time consuming. As a matter of fact, when he was making his comparisons against an explicit scheme and running his fully-implicit scheme with a time step four times larger, in spite of the fact that the time step was four times larger, his integration took much more time than the explicit integration. So it was a very, very expensive technique. But Marchuk wanted to demonstrate that implicit schemes are feasible, that there is nothing wrong with increasing the time step, that it can be done and it can be done with no loss of accuracy. That is what he was really trying to demonstrate. He recognized that his technique was less efficient, but that's not what he worried about at the time. So when I read that paper, I said, "Okay, yes, it can be done, now all we need to do is find an efficient way of doing it." And that's when I decided to start to concentrate on finding ways of making implicit schemes efficient. And that's how the idea of using semi-implicit schemes came about. So really, that has been the big influence in my life. It came almost suddenly, because on one of my visits to NMC in Washington, one day Marchuk came and he gave a presentation at NMC about splitting techniques and about implicit treatment of terms and equations and it's immediately after that, that I tried to get whatever papers had been published by Marchuk and tried to look into his way. But that is probably the most significant influence during my career. It came from him.

QUESTION: What do you see as the primary challenge for the Atmospheric Environment Service from now until the year 2000?
ANSWER: I tend to think that one of the major changes that is going to take place in meteorology in the coming years, as far as I can see, relates to data, rather than the physics of the atmosphere, or the numerical aspects or the dynamical aspects of the atmosphere. I have the impression that satellites, meteorological satellites, things like Doppler radars for instance, and various techniques for sounding the atmosphere, lidars and so on, are slowly transforming meteorology. It has been very difficult up to now to look at numerical integrations over a very, very small domain, the size of a cloud-cluster or that sort of thing, because there is not that much information available, sometimes you have to go to special international programs to produce very high density data. But it seems that this is happening without the need of international programs, instruments are being constructed and generated. As I was saying, Doppler radars that can give you wind direction and wind speed, over a very, very fine grid if you want to look at that, instruments that enable you to measure the humidity in the atmosphere, its distribution in three dimensions and so on. I think these are slowly going to revolutionize Numerical Weather Prediction and enable us to integrate models with very fine grids that it would be useless to try to do at the present time, unless you have a special observational program for a short period of time to generate the data you need to

do this. But I think this is going to come naturally through the kind of developments that are taking place in meteorology at the present time. So my impression is that it's on the data side that everything is changing most rapidly at the present time.

QUESTION: During your career, have you had activities outside working hours that have been especially important to you that you would like to mention?
ANSWER: My activities outside of my working hours have been more oriented towards relaxation activities; that is, family life, reading my newspaper, watching television a little bit, resting from my day's work. That has been one of my techniques to be able to stand pressure, because sometimes you come back from the office, after a hard day's work you feel tired and if you don't do something then, the next morning, you're not going to be in good shape to get back to it. So generally speaking, when I get out of the office and go home I say, "Okay, now I've got to recuperate and be in good shape for the next morning." There have been frequent situations where that was the case, so my outside activities have always been oriented towards recuperation. I do a lot of walking: every evening I walk for about an hour and a half and that is another way of relaxing, because then, not even my family is there. I am alone on the street; I can meditate; I can think: I can sort of prepare, do some planning for my future work, or evaluate what I have been doing in the last week or two, and decide whether I am going in the wrong direction or not. I can do a lot of thinking and meditation. And I do that every evening. I walk for about an hour and a half, seven days a week, three hundred and sixty-five days a year. I never miss that, there is no snow storm or anything that can stop me from doing that and I enjoy it.

QUESTION: Are there any areas that we have left out? Is there something of interest that you think you would like to add before we conclude?
ANSWER: Not really. I think we have covered everything that I could really mention. There may have been a few incidents here and there that I don't remember at the present time that might have been of some interest, but I think we really have covered everything.

9 Conclusion

INTERVIEWER: Well André, I want to thank you very much for granting this interview. You have been both entertaining and informative. I thank you for taking the time out to re-live some of the highlights of a truly remarkable career with the Canadian Meteorological Service and to let us in on some of your reflections about meteorology.

Acknowledgements

We are grateful to the Oral History Project of the Canadian Meteorological Service for initiating this interview, with particular thanks to Dave Phillips for his advice and technical assistance in its preparation. The transcript was originally prepared by

Diane Trudeau. A draft version of this paper was thoughtfully reviewed by Michel Béland and Andrew Staniforth.

References

BATES, J.R. and A. MCDONALD. 1982. Multiply-upstream, semi-Lagrangian advective schemes: Analysis and application to a multilevel primitive equation model. *Mon. Weather Rev.* **110**: 1831–1842.

DALEY, R.; C. GIRARD, J. HENDERSON and I. SIMMONDS. 1976. Short term forecasting with a multilevel spectral primitive equations model. ATMOSPHERE, **14**: 98–134.

MARCHUK, G.I. 1965. A new approach to the numerical solution of differential equations of atmospheric processes. World Meteorological Organisation Technical Note No. 66, Geneva, Switzerland, pp. 286–294.

ROBERT, A. 1963. A baroclinic model for the Canadian numerical weather prediction program. *In*: Stanstead Seminars on the Stratosphere and Mesosphere and Polar Meteorology, Department of Meteorology, McGill University, pp. 83–88.

———. 1965. The behaviour of planetary waves in an atmospheric model based on spherical harmonics. Publications in Meteorology No. 77, Department of Meteorology, McGill University, 84 pp.

———. 1969. The integration of a spectral model of the atmosphere by the implicit method. *In*: Proc. WMO/IUGG Symposium on Numerical Weather Prediction in Tokyo, Jpn. Meteorol. Agency, Tokyo, Japan, pp. VII.19–VII.24.

———. 1975. Sensitivity experiments for the development of NWP models. *In*: Proc. eleventh Stanstead Seminar, Department of Meteorology, McGill University, pp. 68–81.

———. 1981. A stable numerical integration scheme for the primitive meteorological equations. ATMOSPHERE-OCEAN, **19**: 35–46.

———. 1982. A semi-Lagrangian and semi-implicit numerical integration scheme for the primitive meteorological equations. *J. Meteorol. Soc. Jpn.* **60**: 319–325.

———; T.L. YEE and H. RITCHIE. 1985. A semi-Lagrangian and semi-implicit numerical integration scheme for multilevel atmospheric models. *Mon. Weather Rev.* **113**: 388–394.

TANGUAY, M.; A. SIMARD and A. STANIFORTH. 1989. A three-dimensional semi-Lagrangian scheme for the Canadian regional finite-element forecast model. *Mon. Weather Rev.* **117**: 1861–1871.

———; A. ROBERT and R. LAPRISE. 1990. A semi-implicit semi-Lagrangian fully compressible regional forecast model. *Mon. Weather Rev.* **118**: 1970–1980.

André Robert (1929–1993): His Pioneering Contributions to Numerical Modelling

Andrew Staniforth
Recherche en prévision numérique
Service de l'environnement atmosphérique
2121 Route Trans-canadienne, local 500, Dorval, Québec, Canada H9P 1J3

[Original manuscript received 16 January 1995; in revised form 17 July 1995]

ABSTRACT *The pioneering contributions of André Robert to numerical modelling are reviewed. Highlights include: his early work on spectral modelling; the introduction of a weak time filter to permit extended integrations of models for climate modelling; the development of highly-efficient semi-implicit semi-Lagrangian time integration schemes; the extension of these techniques for efficiently integrating the hydrostatic primitive equations to the fully-elastic Euler equations; and the identification of and solution to deficiencies in the traditional approaches to applying lateral boundary conditions in limited area models. These contributions all have one thing in common. They are the result of having reduced a complex problem to its essence in order to properly understand its origin, and thereby develop a conceptually simple solution that not only addresses the problem in its simplified context but also in the original complex one. His inspirational contributions have profoundly influenced, and continue to influence, the development by the international geophysical fluid dynamics community of atmospheric and oceanic models for applications and research in weather, climate and air quality. André Robert was a true pioneer of numerical modelling and will be sorely missed, but his contributions will endure for many years to come.*

RÉSUMÉ *On retrace les contributions d'avant-garde d'André Robert à la modélisation numérique. Parmi les plus notables: son travail original sur la modélisation spectrale; l'introduction d'un filtre temporel faible permettant des intégrations prolongées pour la modélisation du climat; l'élaboration de schémas d'intégration temporelle semi-implicites et semi-lagrangiens de haut rendement; l'extension de ces techniques d'intégration efficaces des équations hydrostatiques primitives pour les appliquer aux équations eulériennes pleinement élastiques; et l'identification des irrégularités dans les approches traditionnelles à l'application des conditions aux limites latérales dans les modèles régionaux, et leurs solutions. Ces contributions ont toutes quelque chose en commun: elles sont le résultat de la rationalisation d'un problème complexe à sa plus simple expression afin de pouvoir comprendre son origine et, par la suite, d'élaborer une solution guidée par des concepts simples qui ne fait pas qu'examiner le problème dans sa forme simplifiée mais aussi dans sa forme originale complexe. Son inspiration a profondément influencé, et continue d'influencer, la communauté internationale de la dynamique des fluides géophysiques dans ses activités d'application et de recherche sur le développement des modèles atmosphériques et océaniques en météorologie, climatologie et qualité de l'air. André Robert fut un défricheur de la modélisation numérique et son départ se fait grandement sentir; cependant, ses travaux nous influenceront pour encore bien longtemps.*

---

1 Introduction

Some of the following introductory remarks are drawn from the fascinating 1987 interview abridged in Ritchie and Robert (this volume), and the Merilees and Laprise (1994) and Laprise (1994) necrologies. Although André Robert is well known as an outstanding Canadian scientist, he was in fact born in New York City in 1929 to French-Quebec parents Mathias Robert and Irène Grindler.

After two years of primary education in English in New York he moved, with his family, to the Province of Quebec, and was educated exclusively in French until he obtained his B.Sc. from Laval University in 1952. André was then offered a fellowship to continue his studies in Europe, but by then had fallen in love and wished to get married. So he decided to take advantage of a no-strings-attached incentive to recruitment offered by the Meteorological Branch of the Department of Transport and followed a 1-year M.Sc. course at the University of Toronto at full salary. He found the course and subject matter to be inspirational and he spent the rest of his life associated closely with the Canadian meteorological service throughout its various metamorphoses. Despite many opportunities to move elsewhere, at a higher level of salary and funding, he was a committed family man and French Canadian (Ritchie and Robert, this volume): 'I am French speaking. My wife is French speaking. My kids are French speaking and it would be very difficult for my family to go far away from Quebec. I decided at the very early stages in my career that I would try to make my wife and kids as happy as possible, and might even sacrifice my own career if necessary.'

Fortunately it wasn't necessary, and he was able to not only greatly influence the evolution of the Canadian weather service but also that of the other major national meteorological services throughout the world. Paradoxically he only became a Canadian citizen in 1967 at the age of 38. This was after being greatly surprised to learn (when applying for a Canadian passport to go to the former Soviet Union on official business as a Canadian Government employee) that the Government of Canada considered him to be a U.S. citizen by virtue of his place of birth rather than by his upbringing in Quebec and his parents' citizenship.

After obtaining his M.Sc. he became an operational weather forecaster and moved on to a career in research by becoming the very first employee, in 1959, of the Operational Development and Evaluation Group (the forerunner of RPN, Division de Recherche en prévision numérique), eventually obtaining his Ph.D. from McGill University in 1965 after part-time study. His career as a research scientist was interrupted in 1973 by a 7-year stint as Director of the Canadian Meteorological Centre (CMC), during which time he radically changed the way it did business by introducing new technology to automate many activities, particularly the labour-intensive production of weather maps. He officially retired, although in name only, from the Atmospheric Environment Service in 1987 after 35 years of loyal and highly-productive service, and became both its first Emeritus Scientist and a Professor at l'Université du Québec à Montréal until his untimely death.

The purpose of this paper is: to review André Robert's most significant scientific contributions; and to show how he was an outstanding reductionist who had the gift of reducing complex problems down to their very essence, thereby obtaining insight and solving them in their simplified context, followed by a demonstration of the chosen solution in the original complex context.

There are a number of different ways to review a scientist's life work. The approach taken here is to present, in roughly chronological order, the motivation behind his most significant contributions together with their scientific essence, mirroring as much as possible André's research philosophy. This might be characterized as understanding well and exploiting the fundamental properties of the underlying continuous problem. From time to time this review is interspersed with anecdotal essay-style comments, in order to illustrate some of the ways in which he personally interacted with and influenced his peers, particularly those at RPN, and thereby provide a more personal and informal historical context for his work. These comments are drawn, for the most part, from the author's twenty-year association with André, but some have been gleaned from other sources.

A key ingredient to André's work was his high level of creativity. His interest in science was very much stimulated in his early adolescence by members of his family, who would give him mathematical problems to solve. He remembered (Ritchie and Robert, this volume) being particularly influenced scientifically by one of his uncles, a Catholic priest and teacher. André must have been a precocious child and was particularly interested in mathematics, mastering calculus at the tender age of 12. His creativity was very probably stimulated not only by his family environment as an adolescent, but by his approach to his own education. Rather than passively reading and understanding the solutions to problems found by the great scientists of the past, he would deliberately nurture his creativity by first reading only the part where the problem was defined. He would then attempt to solve it himself, which he frequently did, and often by a different approach than the text-book one. In a world of exploding information, and educational systems that often emphasize rote learning, this is a very different and laudable approach and reminiscent of a quote from Einstein's autobiographical notes: 'It is in fact nothing short of a miracle that the modern methods of instruction have not yet entirely strangled the holy curiosity of enquiry; for this delicate little plant, aside from stimulation, stands mainly in need of freedom; without this it goes to wreck and ruin without fail.'

2 The early years

From 1953–1959 André was employed as an operational weather forecaster for international aviation and the Quebec public, and recollected that he had mixed feelings about this period. It did enable him to develop his practical insight into weather forecasting, but he found it frustrating because he was reading up on numerical weather prediction (NWP) in his spare time and immediately recognized its future importance and wished to contribute. It was fortunate that he was working in the same building at Dorval airport as CAO (the Central Analysis Office, the forerunner of the Canadian Meteorological Centre), since this gave him access to the library where he spent as much time as possible. He gave seminars at CAO on what he read, and this led to his being transferred there in 1959 when, following the early successes of the U.S. Joint Prediction Unit in Suitland, Maryland, it was decided to create a new group (Operational Development and Evaluation) to get Canada into the promising area of NWP.

Over the next three years, André was instrumental in the development, by this new group, of a series of first barotropic, then filtered-baroclinic models, and it was during this time that he developed his insight into numerical methods and computers. The first of the barotropic models developed by this fledgling group, '... for the prime purpose of gaining experience in programming, numerical analysis, mathematical modelling, and testing the feasibility of NWP on a medium-size computer', is described in Kwizak (1961). It is humbling to note the rudimentary level of computer power (far less than today's microcomputers) available to the pioneers of numerical weather prediction and what they were nevertheless able to achieve. Both the barotropic and filtered-baroclinic models were developed using a 28×32 grid (with $\Delta x = 381$ km) on McGill University's IBM 650 computer. This machine had only 60 words of high-speed core memory storage, 3 indexing registers, a 2000-word drum and 2 magnetic tape units (one for the program, the other for the data), and all code had to be written in machine language! In September 1962 a Bendix G-20 computer was installed at CAO and this led to Canada's first operational NWP model being implemented in December 1963. This model (Kwizak and Robert, 1963) was an evolved version of the non-divergent barotropic model described in Kwizak (1961) that included topographical effects and a stabilization of the planetary waves. It used a 1709-point octagonal grid (with $\Delta x = 381$ km), and the initial conditions were obtained using the objective analysis method described in Kruger (1964), based apparently on earlier work described in an unpublished manuscript by Eddy, McClellan and Robert (1961, unpublished manuscript).

Remarkably, by 1962 a 4-level filtered-baroclinic model had also been developed (Robert, 1963a,b) at CAO that also used a 28×32 grid. In Robert (1963b) it was stated that it was hoped that a version of this model would operationally replace the barotropic one in the winter of 1964, but this was unfortunately not to be. It was not until 1969, after upgrading CAO's computer in August 1967 to an IBM 360-65, that an evolved version was eventually implemented over an enlarged 51×55 grid. In fact, this 7-year delay in implementing a baroclinic model very much motivated André to become Director of the Canadian Meteorological Centre (the former CAO) in order to break down its resistance to change and introduce new technology more rapidly. In the meantime he turned his attention to spectral modelling and the primitive equations, and it was for this work he received his Ph.D. from McGill University.

3 Spectral Modelling

The principal results of his Ph.D. thesis (Robert, 1965) are summarised in Robert (1966). This work was motivated by André's recognition that the spectral method (i.e., an expansion of variables over the globe in a finite series of spherical harmonics) offered several potential advantages for global forecasting over the generally-adopted finite-difference methods. These include: no pole problem due to the convergence of the meridians; an isotropic representation when using triangular truncation which results in enhanced stability and longer time steps; much smaller phase errors; no aliasing of nonlinear terms; and better respect of the conservation properties of the underlying continuous governing equations. This had to be weighed against two very important difficulties: the prohibitive computational cost of the interaction-coefficient method for computing nonlinear terms that had emerged from several studies (Silberman, 1954; Platzman, 1960; Baer and Platzman, 1961; Baer, 1964); and the efficient extension of the method to the primitive equations where vector quantities appear in addition to scalar ones. Regarding this latter difficulty, André noted that while previous studies (e.g. Kubota, 1960) indicated that the generalization should be possible when using the differentiated (i.e., vorticity and divergence) form of the equations rather than the primitive (i.e., velocity component) one, this appeared to make the spectral method yet more cumbersome and expensive. The key issue identified was that while it is appropriate to expand scalars such as vorticity and stream function in terms of a spherical harmonic series, it is not appropriate to do so for vector quantities such as the two horizontal wind components.

The essence of his approach (Robert, 1965, 1966, 1968b) is to distinguish between true and pseudo scalars. The true scalars are expanded in a truncated spherical harmonic series. The pseudo scalars u and v (the two horizontal wind components) are replaced by the wind images $U = u\cos\theta$, $V = v\cos\theta$, where θ is latitude, which are true scalars and can therefore be similarly expanded in terms of spherical harmonics. This approach significantly reduces the complexity of the interaction coefficients, at the price of introducing an orthogonalization procedure, and is equivalent to using interaction coefficients, but less expensive. It permitted him to build a rudimentary global climate model, and thereby perform the *first ever spectral integration of the baroclinic equations* albeit at very low (T4, L5) resolution: i.e., the spherical harmonic series were triangularly truncated at wavenumber four, and five levels were used in the vertical. Nevertheless this proof-of-concept study paved the way for the future success of the spectral method.

There was however another fundamental problem that had to be overcome, and that was how to treat moisture and its prediction in spectral models, where the condensation rate is a threshold-controlled process occurring not in spectral space but in real space. This was addressed in Robert (1968b,c) by using relative humidity rather than specific humidity as the moisture variable, and by representing the condensation rate in terms of a continuous function which can be spectrally represented, rather than in terms of a Heaviside function associated with a threshold. This approach was ingenious (and is currently being rediscovered in the context of replacing thresholds in physical parametrizations by sharply-varying continuous functions to facilitate the definition of their adjoints), but was quickly superseded by the advent of the spectral transform method discussed below.

In Robert (1966), he foresaw the use of the spectral method for extended weather forecasting and climate simulation, and to investigate the energetics of the atmosphere and the behaviour of large-scale waves, since these are low-resolution global applications and computationally tractable using interaction coefficients. However, he was somewhat pessimistic about its use for operational weather forecasting, but two years later in a follow-up paper (Robert, 1968a), whose purpose was to '... present the results of a few global forecasts prepared from real data with the hope that this will generate an interest in the method and encourage its development', he was more optimistic: 'The atmospheric simulators of the spectral type will soon start competing with the classical models that use the gridpoint method.'

Nonetheless, the spectral method still remained more costly than finite-difference ones. However, a scant two years later the spectral transform was independently introduced by Eliassen et al. (1970) and Orszag (1970), and this breakthrough finally provided the impetus for the spectral method to profit from his pioneering work and find wide application. Today, as is well known, it is used by a majority of the national (Australia, Canada, China, France, Germany, India, Japan, Russia, U.S.) and multi-international (ECMWF) operational centres that produce global weather forecasts, and by groups in both these and other countries for climate modelling. A review of the emergence of the spectral method may be found in Bourke (1988).

4 Time filtering

The time filter, which is variously known today as the Robert, Asselin and Robert-Asselin filter, was first introduced in André's Ph.D. thesis (Robert, 1965). Although André considered it to be a minor contribution (Laprise, 1994), it has nevertheless found widespread application and is used today by almost all weather forecast and climate models in the world. The original motivation for this filter is probably not widely known – it was to stably treat dissipative terms in a centred manner in a leapfrog time scheme. This was accomplished via a two-step procedure, which he analysed. First a fully-centred leapfrog time step (including the dissipation terms) using available values is made, which would be unstable if continued. Then a weak time filter is applied to stabilize the scheme, and the two-step procedure is repeated for all future time steps.

Early atmospheric models would go unstable after a few weeks of integration following the onset of a time decoupling of the solution. The reason his time filter has found such wide application is that it circumvents this problem and enables leapfrog-based (including semi-implicit semi-Lagrangian) schemes to stably integrate for long periods of time. It is a highly-discriminating filter (Robert, 1965; Asselin,1972) that heavily damps the computational modes while only lightly damping the physical ones. It is instructive to sketch the arguments that led to the development and analysis of this filter as it illustrates André's penchant for maximum simplification while still retaining the scientific essence of the problem. Consider the simple model equation

$$\frac{dF}{dt} = i\omega F, \tag{1}$$

where ω is real. This is a simple harmonic oscillator and a *1st-order differential* equation having solution

$$F = F_0 e^{i\omega t}, \tag{2}$$

where $F = F_0$ at the initial time $t = 0$. A leapfrog approximation to (1) is

$$\frac{F(t+\Delta t) - F(t-\Delta t)}{2\Delta t} = i\omega F(t). \tag{3}$$

This is a recursive *2nd-order difference* equation, and as such has *two* independent solutions: a physical mode (the discrete analogue of the exact solution's only mode), plus a spurious computational one. Also, it not only needs an initial value (as for the exact solution), but also a starting procedure (e.g., a forward time step), because the difference equation (3) is of 2nd order whereas the differential equation (1) is only of 1st order.

Now the initial conditions project onto both the physical and computational modes. For this linear model, the physical and computational modes are independent, and their amplitudes are determined for all time entirely by the initial conditions and starting procedure. However, for a *nonlinear* model, energy is transferred between the physical and computational modes and this usually results in the well-known problem of time step decoupling. The simple solution proposed by André was to introduce a weak time filter via the following 2-step procedure:

(i) make a provisional ($\tilde{\ }$) time step using available information:

$$\frac{\tilde{F}(t+\Delta t) - F(t-\Delta t)}{2\Delta t} = i\omega \tilde{F}(t), \tag{4}$$

(ii) apply the weak time filter

$$F(t) = \tilde{F}(t) + \alpha[\tilde{F}(t+\Delta t) - 2\tilde{F}(t) + F(t-\Delta t)], \tag{5}$$

to obtain the definitive $F(t)$, where $\alpha(\geq 0)$ is the filter coefficient.

Typically a filter coefficient of order $\alpha \sim 0.1$ is used. The linear analyses of Robert (1965) and Asselin (1972) indicate that provided the filter coefficient α is of this order, then the filter should: discriminate well between the physical and computational solutions; minimally damp the physical solution; only marginally further restrict the time step when using a leapfrog-based time scheme; and should behave monotonically with respect to α (i.e., the closer α is to zero, the less damping there is of the physical mode).

Regarding this latter property André serendipitously discovered, while giving an NWP course in 1986, that it doesn't necessarily hold for nonlinear systems. To motivate the need for a time filter he examined the following simple set of equations for the rigid nonlinear pendulum:

$$\frac{dv}{dt} = -g \sin \theta, \tag{6}$$

$$\frac{d\theta}{dt} = \frac{v}{l}, \tag{7}$$

where v is the velocity of the pendulum bob, θ is the angle of the pendulum with respect to the vertical, l is its length, and g is the acceleration due to gravity. For small values of θ, $\sin \theta$ in (6) can be replaced by θ, the equations become linear, and the pendulum oscillates at the frequency $\sqrt{g/l}$. André discretized (6)–(7) using a leapfrog scheme and his time filter, and wrote a short program that integrates the equations and displays the results graphically. This he furnished to his students for them to conduct a series of experiments by varying the parameters.

The first test was to run an integration without a time filter (i.e., $\alpha = 0$) to show that the solution starts off by oscillating correctly for the first several periods, and that time step decoupling gradually sets in as energy is transferred from the physical to the computational mode. By integrating long enough, the students discovered that all of the energy is eventually transferred to the computational mode, and that by integrating an equal time further, all of the energy is gradually transferred back to the physical mode. The second test was to show that by running a set of experiments with different values of the filter coefficient α, the decoupling could be avoided at the expense of some damping of the solution, and to show that the smaller the filter coefficient the weaker the damping. André was greatly surprised when one of his students, Mario Lépine, showed him the result of an experiment where he had inadvertently set the value of α to be 0.001 rather than 0.01. Paradoxically, the run with the *smaller coefficient* gave *greater damping*. So André carefully analysed the problem and resolved the paradox in an unpublished document: a slightly-revised version may be found in this issue (Robert and Lépine, this volume).

The following conclusions emerge from the analysis:

(i) consistent with linear theory, for α greater than a critical value $\alpha_{crit} \approx 0.001$, the physical and computational modes are damped at different rates, with the computational solution being more heavily damped than the physical one;
(ii) for $\alpha = \alpha_{crit}$, a bifurcation of behaviour occurs;

(iii) for $\alpha < \alpha_{crit}$, both the physical and computational modes are heavily damped, and by the same amount (which explains the paradox), but this damping diminishes as α decreases away from α_{crit}, with zero damping in the limit $\alpha = 0$;
(iv) for the examined problem, $\alpha \approx 0.01$ is optimal, since it minimally damps the physical mode while heavily damping the computational one; and
(v) for nonlinear problems, caution must be applied when applying the Robert time filter for very small values of the filter coefficient since an anomalous behaviour may occur.

5 The semi-implicit scheme

During one of his visits in the 60's to the U.S. National Meteorological Center, André attended a seminar on implicit methods given by the celebrated Soviet scientist G.I. Marchuk. This started him thinking. He reasoned that although implicit methods are more costly than explicit ones because of the nonlinear terms, Marchuk's work (e.g., Marchuk, 1965) demonstrated that they can be both accurate and stable with long time steps. Therefore the challenge was to find a way of making implicit schemes cost effective, and this led (see Ritchie and Robert, this volume) to what is now known as the semi-implicit scheme.

André recognized that in a primitive equations model, the fastest-moving gravity modes carry very little energy when compared to the Rossby modes (at least for synoptic-scale motions), yet it is they that most constrain the time step in an explicit time scheme due to stability considerations. This is perhaps most simply illustrated (Robert, 1979) by considering the model equation (cf. Eq. (1))

$$\frac{dF}{dt} = i(\omega_R + \omega_G)F, \tag{8}$$

where ω_R and ω_G are respectively the Rossby and gravity wave frequencies, with $|\omega_R| < |\omega_G|$, and by contrasting the leapfrog and semi-implicit approximations to it.

A leapfrog approximation (the workhorse method before the advent of the semi-implicit scheme) to (8) is

$$\frac{F(t + \Delta t) - F(t - \Delta t)}{2\Delta t} = i(\omega_R + \omega_G)F(t). \tag{9}$$

Writing $F(t)$ as $F(t) = F_0 \exp(i\omega t)$, it can then be shown that (9) is stable provided

$$|\omega_R + \omega_G|\Delta t \leq 1. \tag{10}$$

This means that the time step is simultaneously constrained by both the fastest Rossby and gravity wave frequencies, and this is due to the time-explicit treatment of the corresponding terms embodied in the right-hand side of (9).

The basic idea behind the semi-implicit scheme is to treat in a *time-implicit* manner the *linear* part of the terms that govern gravitational oscillations, and to treat the other terms (including the *nonlinear* perturbations) in a *time-explicit* manner. Applying this principle to the model equation (8) gives the semi-implicit approximation

$$\frac{F(t+\Delta t) - F(t-\Delta t)}{2\Delta t} = i\omega_R F(t) + i\omega_G \left[\frac{F(t-\Delta t) + F(t+\Delta t)}{2}\right], \qquad (11)$$

and this scheme is stable provided

$$|\omega_R|\Delta t \leq 1. \qquad (12)$$

Comparing the stability condition (12) for the semi-implicit scheme to that of (10) for the leapfrog scheme, it is clear that the restriction on the time step due to the propagation of gravitational oscillations has been removed by adopting a time-implicit treatment of them.

While the crux of the issue is well illustrated by the above model problem, how the method is applied in practice is rendered more evident by examining the following semi-implicit discretization of the shallow-water equations (cf. Robert, 1969a):

$$u_t + \overline{\phi_x}^t = fv - (uu_x + vu_y), \qquad (13)$$

$$v_t + \overline{\phi_y}^t = -fu - (uv_x + vv_y), \qquad (14)$$

$$\phi_t + \Phi_0 \overline{(u_x + v_y)}^t = -(\phi u)_x - (\phi v)_y, \qquad (15)$$

where

$$F_t = \frac{F(t+\Delta t) - F(t-\Delta t)}{2\Delta t}, \qquad (16)$$

$$\overline{F}^t = \left[\frac{F(t-\Delta t) + F(t+\Delta t)}{2}\right], \qquad (17)$$

Φ_0 is the mean geopotential height of the fluid, ϕ is the perturbation about it, and the other symbols have their usual meaning. The terms on the left-hand sides of (13)–(15) are all *linear* and *implicitly* treated as time differences and averages using values at times $t - \Delta t$ and $t + \Delta t$, and are the terms that govern small-amplitude linear oscillations about a state at rest in a non-rotating fluid whose surface is at constant height above a flat surface. All terms on the right-hand sides of (13)–(15), the majority of which are *nonlinear*, are *explicitly* treated at the intermediate time level t. Thus the semi-implicit scheme yields the best of two worlds. It has the advantage of the longer time steps of an implicit scheme, but because the nonlinear terms are treated explicitly, it is only a little more costly than the explicit leapfrog scheme. The overhead of the semi-implicit scheme is the cost of solving the Helmholtz problem for ϕ which results from a manipulation of the coupled set of equations (13)–(15). This is negligible for a spectral model and affordable for other spatial discretization methods, and efficiency is therefore greatly enhanced by the semi-implicit scheme.

Regarding its stability, it is instructive to linearize (13)–(15) about a state in uniform motion ($u = U = constant$, $v = 0$) with constant geopotential height Φ_0 in the absence of the Earth's rotation ($f = 0$). This amounts to replacing the right-hand sides of (13)–(15) respectively by $-Uu_x$, $-Uv_x$ and $-U\phi_x$. By expanding the dependent variables as

$$F = F_0 e^{i(kx+ly+\omega t)}, \tag{18}$$

it is easily shown (Robert, 1979) that the scheme is stable in this case provided

$$|U| \leq \sqrt{\Phi_0}, \tag{19}$$

and

$$|kU\Delta t| \leq 1. \tag{20}$$

The first of these two conditions is satisfied since the Rossby modes propagate much more slowly than the gravity modes. The second is the usual Courant number stability criterion due to advection that is also encountered in filtered models, and condition (20) corresponds to condition (12) for our model problem where ω_R is identified with $-kU$. This illustrates the relevance of the model problem to the examination of numerical methods for the atmosphere (note that ω_G may be identified with $\pm k\sqrt{\Phi_0}$).

Another way of interpreting the semi-implicit discretization is that it, in effect, slows down the speed of propagation of the gravitational oscillations to be comparable to that of the Rossby modes, and thereby permits much longer time steps with negligible loss of accuracy. This slowing-down of the gravity waves is not a problem for two reasons: they carry relatively little energy at the synoptic scale, and the density of the observational network is insufficient to unambiguously determine them at initial time.

Historically, the first description of the semi-implicit method in the literature may be found (Robert, 1969a) in the proceedings of an international symposium on NWP held in Tokyo, where a generic formulation was given for both spectral and gridpoint discretizations of the shallow-water equations together with a demonstration of its success using a spectral model. André did not attend this meeting and his results were presented by his longstanding colleague and division chief Mike Kwizak. The gridpoint shallow-water demonstration followed two years later (Kwizak and Robert, 1971) and was the basis for Kwizak's Ph.D. thesis. The first application to a hydrostatic primitive equations model was given in Robert et al. (1972), again for a gridpoint model. Since that time, the semi-implicit scheme has been widely adopted and is used today in the vast majority of NWP and climate models throughout the world.

6 Semi-implicit semi-Lagrangian schemes

Soon after revolutionising operational NWP by introducing the semi-implicit scheme, André became Director of the Canadian Meteorological Centre in 1973 and was for the most part lost to research for the following seven years. However, towards the end of this period, he had the idea (Ritchie and Robert, this volume) of associating a semi-Lagrangian treatment of advection with a semi-implicit treatment of the terms that govern the propagation of the gravitational oscillations, and decided to return to research. The present author conjectures that this probably happened while he was preparing his review (Robert, 1979) of semi-implicit schemes for a WMO (World Meteorological Organisation) publication on numerical methods. His idea (Robert, 1981) was the consequence of recognising and connecting the following four facts:

(i) a semi-implicit treatment of the terms governing gravitational oscillations addresses the most serious time step restriction due to the conditional stability of an explicit leapfrog model;
(ii) there remains the Courant number restriction on the time step of an explicit Eulerian treatment of advection, be it spectral, finite difference, or finite element;
(iii) the temporal truncation errors are still much smaller than the spatial ones, and thus the time step can be increased with no loss of accuracy provided that a stable and accurate (and preferably efficient!) method to do so can be found; and
(iv) a semi-Lagrangian treatment of advection (cf. Wiin-Nielsen, 1959; Krishnamurti, 1962; Sawyer, 1963) is stable for Courant numbers ($U\Delta t/\Delta x$) much greater than the restricted value (of order unity) of an Eulerian treatment.

A first attempt to demonstrate the value of coupling a semi-Lagrangian treatment of advection with a semi-implicit scheme was made in Robert (1981). André first applied the method to the non-divergent barotropic vorticity equation and it worked very well. He then wrote the shallow-water equations in differentiated (vorticity-divergence) form and adopted a semi-Lagrangian treatment for the advection of vorticity, but approximated the divergence and continuity equations in the usual manner via a semi-implicit Eulerian discretization. Again it seemed to work fine, but in reality the attempt was only partially successful. Stability was indeed demonstrated, but accuracy turned out to be a problem, and the present author inadvertently participated in the discovery and diagnosis of this.

André had been piquing my curiosity for a while with the news that he had found a way of stably and efficiently quadrupling the time step of atmospheric models, and that I would find it most interesting. All would be revealed soon in a paper he was writing, a copy of which he gave me the same day he submitted it. The simplicity and power of the idea immediately convinced me to apply it to an existing shallow-water finite-element model, which at the time seemed to be made to measure since it was already formulated using the differentiated form of the equations. Within a week, the first results were available. The good news was that the results agreed well with the Eulerian control when run with the usual stability-limited time step of the Eulerian version, and that it was stable with long time steps. The bad news was that it developed serious errors in the large scales when run at long time step, in contradiction with André's findings. His reaction was predictable: go find your coding error young man! After a month of painstaking verification and experimentation (including an exchange of interpolation routines with him) I was convinced that there was no coding error and, being somewhat suspicious of limited-area models (more on this later!), I asked André if by any chance he had any dissipative mechanisms in his model. The reply – only three (!): the usual sponge at the open lateral boundaries plus a divergence damping and a time filter to compensate for the absence of an initialisation procedure. By the next day André had run his model without these and it was immediately obvious that it behaved as badly as mine.

Two problems were identified. The first was that the Coriolis terms were treated explicitly, and thus there was a weak instability when running with a 2-h time step because it marginally exceeded the stable limit of $6/\pi \approx 1$ h 55 mins. The second, and more serious, was that it was insufficient to apply the semi-Lagrangian technique to only the vorticity equation. At first André thought that all that was required was to apply a semi-Lagrangian treatment to the other two prognostic equations, but this was also inaccurate and can be attributed to divergence not being a naturally-advected quantity. However, momentum is, and this led him to a reformulation using the *primitive* (i.e., undifferentiated) form of the equations. This appeared in Robert (1982) where it was demonstrated that not only are semi-Lagrangian methods stable with long time steps, but that this can indeed be achieved without loss of accuracy; something claimed but erroneously demonstrated in Robert (1981). An overhead is incurred by the upstream interpolations, but efficiency is nevertheless much enhanced. It is unfortunate that it is insufficient to just handle vorticity advection with the semi-Lagrangian scheme since this would greatly reduce the number of interpolations and therefore reduce the overhead. Nevertheless the efficiency gains when introducing semi-Lagrangian advection in a semi-implicit model are still substantial.

The following forced advection problem in multi dimensions, drawn from the Staniforth and Côté (1991) review of semi-Lagrangian integration schemes, provides a fairly simple manner of presenting the basic idea behind the semi-Lagrangian method. Consider:

$$\frac{dF}{dt} + G(\mathbf{x}, t) = R(\mathbf{x}, t), \tag{21}$$

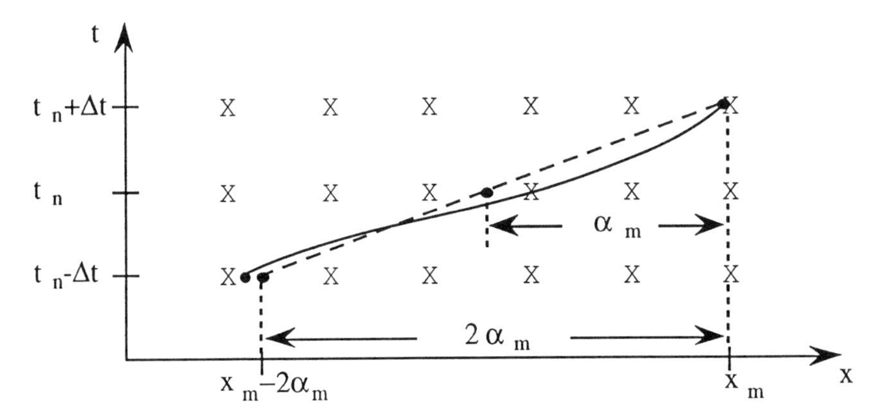

Fig. 1 Schematic for 3-time-level advection. Actual (solid curve) and approximated (dashed line) trajectories that arrive at meshpoint x_m at time $t_n + \Delta t$. Here α_m is the distance the particle is displaced in x in time Δt.

where

$$\frac{dF}{dt} = \frac{\partial F}{\partial t} + \mathbf{V}(\mathbf{x},\, t) \cdot \boldsymbol{\nabla} F, \tag{22}$$

$$\frac{d\mathbf{x}}{dt} = \mathbf{V}(\mathbf{x},\, t). \tag{23}$$

Here, $\mathbf{x}$ is the position vector (in 1-, 2- or 3-d), $\boldsymbol{\nabla}$ is the gradient operator, and G and R are forcing terms. A semi-Lagrangian approximation to (21) and (23) is then:

$$\frac{dF}{dt} + \overline{G}^t = R^0, \tag{24}$$

$$\boldsymbol{\alpha} = \Delta t \mathbf{V}(\mathbf{x} - \boldsymbol{\alpha},\, t), \tag{25}$$

where

$$\frac{dF}{dt} = \frac{F^+ - F^-}{2\Delta t}, \tag{26}$$

$$\overline{G}^t = \left(\frac{G^+ + G^-}{2}\right). \tag{27}$$

Here the superscripts "+", "0" and "−" respectively denote evaluation at: the arrival point $(\mathbf{x}, t + \Delta t)$, the trajectory midpoint $(\mathbf{x} - \boldsymbol{\alpha}, t)$ and the departure point $(\mathbf{x} - 2\boldsymbol{\alpha}, t - \Delta t)$. $\mathbf{x}$ is now an arbitrary point of the 1-, 2- or 3-d mesh. This is illustrated schematically in Fig. 1 for 1-d.

The above is a centred $O(\Delta t^2)$ approximation to (21) and (23), where G is evaluated as the time average of its values at the endpoints of the trajectory, and

R is evaluated at the midpoint of the trajectory. The trajectories are calculated (Robert, 1981) by iteratively solving (25) for the vector displacements $\boldsymbol{\alpha}$. If G is known (we assume that R is known since it involves evaluation at time t), then the algorithm proceeds as follows:

(i) Solve (25) iteratively for the vector displacements $\boldsymbol{\alpha}$ for all meshpoints $\mathbf{x}$ using some initial guess (usually its value at the previous time step), and an interpolation formula.

(ii) Evaluate $F-\Delta tG$ at upstream points $\mathbf{x}-2\boldsymbol{\alpha}$ at time $t-\Delta t$ using an interpolation formula. Evaluate $2\Delta tR$ at the midpoints $\mathbf{x}-\boldsymbol{\alpha}$ of the trajectories at time t using an interpolation formula.

(iii) Evaluate F at arrival points $\mathbf{x}$ at time $t+\Delta t$ using

$$\begin{aligned} F(\mathbf{x},\, t+\Delta t) &= (F-\Delta tG)|_{\mathbf{x}-2\boldsymbol{\alpha},\, t-\Delta t} + 2\Delta tR|_{\mathbf{x}-\boldsymbol{\alpha},\, t} - \Delta tG|_{\mathbf{x},\, t+\Delta t} \\ &= (F-\Delta tG)^{-} + 2\Delta tR^{0} - \Delta tG^{+} \end{aligned} \tag{28}$$

If G is not known at time $t+\Delta t$ (for instance if it involves another dependent variable in a set of coupled equations), then this leads to a coupling to other equations. This is the case for the shallow-water equations (13)–(15), which may be rewritten in the form of the prototypical forced advection equation (21) and then discretized as:

$$\frac{du}{dt} + \overline{\phi_x}^{t} = (fv)^0, \tag{29}$$

$$\frac{dv}{dt} + \overline{\phi_y}^{t} = -(fu)^0, \tag{30}$$

$$\frac{d\phi}{dt} + \Phi_0\overline{(u_x+v_y)}^{t} = -[\phi(u_x+v_y)]^0, \tag{31}$$

where the d/dt and $\overline{(\ )}^t$ operators are again defined by (26)–(27). The $\overline{(\ )}^t$ operator operates on the same terms as in a semi-implicit Eulerian model, the only difference being that the time average is now evaluated along the space-time trajectory that arrives at a given meshpoint rather than being simply evaluated as a temporal average at the same meshpoint. This semi-implicit semi-Lagrangian discretization of the shallow-water equations couples the equations in an analogous manner to the semi-implicit Eulerian one of the previous section and leads to a Helmholtz problem for ϕ of the same form as before. A linear stability analysis of (29)–(31) shows that the scheme should be stable provided $|f\Delta t| \leq 1$, which is fairly lenient since it permits time steps almost as long as two hours. This restriction can be eliminated (Robert, 1982) by time averaging the Coriolis terms along the trajectory, at the cost of generating a somewhat more complicated elliptic boundary-value problem.

Having demonstrated that the association of a semi-Lagrangian treatment of advection with a semi-implicit scheme works well for the shallow-water equations (Robert, 1982), the next step (Robert et al., 1985) was to demonstrate that it also works well in the context of the baroclinic primitive equations. This important step was accomplished by treating the *horizontal* advection of quantities using the semi-Lagrangian technique in a model with ten equally-spaced levels. This permits much longer time steps and circumvents the time step restriction of an Eulerian scheme due to the jets, which are particularly intense in wintertime, but does not address the stability limitation due to an explicit treatment of *vertical* advection.

By this time André had generated a great deal of interest at RPN (and elsewhere!) in semi-Lagrangian techniques, and was very influential in RPN's renaissance as a centre for developing innovative numerical methods, a reputation built on André's contributions of the 60's and early 70's. He initiated a number of projects to explore further extensions of the method and to examine and evaluate possible competing methods. He proposed and supervised several important contributions from Hal Ritchie. The first of these (Ritchie, 1986) was the non-interpolating semi-Lagrangian scheme, which decomposes the trajectory vector into the sum of two vectors, one of which goes to the nearest meshpoint, the other being the residual. Advection along the first trajectory is performed via a semi-Lagrangian technique that displaces a field from one meshpoint to another (and therefore requires no interpolation), while the advection along the second vector is performed via an undamped three-time-level Eulerian approach such that the residual Courant number is always less than one. Thus the attractive stability properties of interpolating semi-Lagrangian advection are maintained but with negligible consequent damping, even at low resolution, making it an attractive alternative for climate simulations. It is also computationally a little more efficient and is currently used in the vertical in the operational global spectral forecast model at CMC and was also used until recently at ECMWF (European Centre for Medium-range Weather Forecasts).

The second thrust of Hal's work that André initiated was to extend the semi-Lagrangian method to spherical geometry. At the time, it was generally believed that semi-Lagrangian methods were restricted to finite-difference and finite-element models using map projections of the globe. Hal first demonstrated (Ritchie, 1987) that it is possible to passively advect a scalar over the pole using semi-Lagrangian advection with time steps far exceeding the limiting time step of Eulerian advection schemes. He then introduced semi-Lagrangian advection into a semi-implicit spectral shallow-water model (Ritchie, 1988) and demonstrated its advantages here too. During the course of this work he encountered an instability, ultimately shown to be due to a metric term associated with momentum advection, and this held up development for a number of months. Now André was an outstanding troubleshooter of numerical instabilities, and so he came to the rescue. He diagnosed the source of the problem and proposed a tangent-plane algorithm as a solution (Desharnais and Robert, 1990), and this did the trick. It then inspired another colleague, Jean Côté, to generalise the approach by using the undetermined Lagrange multiplier method (Côté, 1988), which is equivalent to the independently-devised Bates et al. (1990) algorithm. It also stimulated Jean and the present author to double the efficiency of the Ritchie (1988) model by replacing its three-time-level scheme by a two-time-level one (Côté and Staniforth, 1988), and to further improve efficiency by using a reduced-size Gaussian grid. Hal then went on to implement the semi-Lagrangian scheme in the operational spectral hydrostatic primitive equations model of RPN (Ritchie, 1991), and was also instrumental in its introduction into ECMWF's operational model (Ritchie et al., 1995).

Another example within RPN of the domino effect of André's re-introduction of the semi-Lagrangian method to the meteorological modelling community, is that it led to its being proposed as a transport algorithm for air-quality applications (Pudykiewicz and Staniforth, 1984), based on its very attractive numerical dispersion properties. Some years later this eventually led to a highly-flexible environmental emergency response model (Pudykiewicz, 1990) being introduced into operations at the CMC, and to its increasing use for research within Environment Canada's Air Quality Branch.

At the same time as André was directly and indirectly stimulating further work at RPN in semi-Lagrangian methods, he was also investigating possible competing and alternative methods. The first of these was the Cohn et al. (1985) fully-implicit scheme, and he was very enthusiastic about this scheme to the point of insisting that no further semi-Lagrangian projects be initiated within RPN until it had been thoroughly examined. His reasoning was that the Cohn et al. (1985) scheme was a two-time-level highly-stable $O(\Delta t^2)$ scheme, and as such was potentially twice as efficient as three-time-level semi-Lagrangian schemes. He thereby proceeded to define a new challenge – to either verify that the Cohn et al. scheme was the one sought, or if it wasn't, to find it. (Aside: although there were already some two-time-level semi-Lagrangian schemes around at the time, e.g., Bates and McDonald (1982), he concluded that these didn't qualify since they were insufficiently accurate at large time step.) He ultimately concluded (Yakimiw and Robert, 1986) after careful analysis that the Cohn et al. (1985) scheme was fatally flawed, diplomatically worded in the paper as '... the proposed scheme may not be as attractive as originally anticipated'. In a related follow-up paper (Tanguay and Robert, 1986), another variant of the fractional-step method was examined which circumvents the problem identified in Yakimiw and Robert (1986) and additionally eliminates the need to solve a Helmholtz equation. It was shown that while the proposed scheme works well with moderate time steps, it also breaks down with the longer time steps of a semi-Lagrangian scheme, leading to the conclusion that it will be very difficult to find a variant of the fractional-step method that will work well with long time steps.

In the meantime there were two dissenters within RPN regarding André's 'no new semi-Lagrangian projects' policy, Clive Temperton and myself, who didn't want to wait before following up on my failed 1980 attempt to introduce a semi-Lagrangian scheme into RPN's shallow-water finite-element model. So we didn't, and carried the work out in secret. It was then written up and submitted in November 1985 for publication (Staniforth and Temperton, 1986), and the first André knew about it was when we also announced a seminar on the subject, which he attended two weeks later. Although he was at first highly unamused by our prank, he nevertheless very graciously commended us the next day on our initiative! This was one of André's endearing qualities: he could have cross words with you one day, and by the next it was water under the bridge. By way of atonement Clive and I decided to present the idea behind our intended follow-up work in another seminar the following month, before actually trying it. This was a smart move on our part. André not only immediately understood the idea and that it would probably work, but suggested a refinement that turned out to be a little better than what was proposed. The idea (Temperton and Staniforth, 1987; but also independently proposed in McDonald and Bates, 1987) was to extrapolate the winds to obtain $O(\Delta t^2)$-accurate trajectories and thereby address André's challenge of finding a stable, accurate and efficient two-time-level semi-implicit semi-Lagrangian scheme.

Later, André greatly surprised many of us (particularly myself), by initiating and leading a major project, ultimately successful, to improve the performance of our operational Regional Finite-Element Model and make it competitive with the Nested Grid Model at the U.S. National Meteorological Center. This included, as a substantial sub-project, the introduction of the semi-Lagrangian scheme based on the surreptitiously-conducted Staniforth and Temperton (1986) work for the shallow-water version! The project was well underway when I inherited it the following year (1987) when André retired from RPN and accepted a professorial appointment at l'Université du Québec à Montréal. It ultimately led to the Tanguay et al. (1989) paper, where it was argued that the semi-Lagrangian scheme should also be applied in the *vertical*, tying up a loose end left in Robert et al. (1985). Although this project would never have started without André, and he was instrumental in its success, he insisted on not coauthoring the paper, preferring thereby to further the career progression of younger scientists. This was typical of his generosity which has benefitted many people over the years.

At a workshop that I attended at ECMWF, Coiffier et al. (1987) presented some rather disturbing results indicating that all was not necessarily well in the wonderful world of semi-Lagrangian methods, specifically in the presence of mountains. They kindly provided a copy of their preprint and I passed on a copy to André. In Montreal we were already aware of some work of Kaas (1987), who had found a similar problem. Kaas attributed this to a lack of balance for the forced stationary response between two large terms of opposite sign in the discrete approximation, due to their being evaluated at different geographical points, and proposed a spatial averaging of terms at the intermediate time step as a solution. What was disturbing about the Coiffier et al. paper was that a careful reading of their analysis showed that Kaas' proposed solution only partially addressed the problem and that spurious numerical resonance is not eliminated. Some time later André mentioned that not only did he agree that there was a problem, but believed that he had found a solution which he would work on further in the not-too-distant future, and there the matter lay for several years.

In the meantime André proposed extending the spatial averaging technique of Kaas (1987) to include the evaluation of all terms at the intermediate time step, and it was demonstrated in Tanguay et al. (1992) that not only is it computationally more efficient to handle these terms this way, but enigmatically and in apparent contradiction with the Coiffier et al. (1987) analysis, the spurious resonance problem seems to be quite well addressed in the gridpoint model but less so in the spectral model. I found this intriguing and so, with the benefit of knowing that André claimed to have found a solution to the spurious orographic-resonance problem, I carefully reread the Tanguay et al. (1992) paper. The answer then finally emerged from its hiding place in the two appendices. The first of these is basically a rehash of the Coiffier et al. analysis using a centred scheme in the presence of orography, that reveals the source of the spurious resonance and explains why the spatial averaging should mitigate but not eliminate the problem in the spectral model. The second one shows the effect, *in the absence of forcing*, of an $O(\Delta t)$ off-centring of the time scheme, ostensibly introduced in the gridpoint model to compensate for the lack of an initialisation procedure. It is only when the two analyses are put together to examine off-centring and forcing *simultaneously*, that it becomes clear why the gridpoint model should respond much better to the orographic forcing than the spectral one. It is the temporal off-centring that is more efficient in addressing the spurious-resonance problem than the spatial averaging. The disadvantage, however, of the temporal off-centring is that it introduces an $O(\epsilon\Delta t)$ error, where ϵ is the off-centring parameter, which although apparently small does in fact degrade accuracy and overly restricts the size of the time step. From there it is a short step to the conclusion that an $O(\Delta t^2)$ off-centring should address the problem without degrading accuracy. The end result of all this was that André came into my office one day while I was discussing this with Chantal Rivest, a post-doctoral fellow at RPN at the time. We agreed to collaborate and this led to the last paper (Rivest et al., 1994), published posthumously, that André would submit for publication. In this paper the orographic resonance problem is isolated, explained by analysis, and addressed in the context of a shallow-water model. André considered this work to be of some importance, arguing that the semi-implicit semi-Lagrangian method would be of very limited value if it could not be made to work well for forced problems.

The semi-implicit semi-Lagrangian method has become increasingly popular since André introduced it over ten years ago. It is now used at a number of operational NWP centres (e.g., CMC, ECMWF, Irish Meteorological Service) and in some climate models (Max Planck Institute, NASA's Goddard Laboratory for Atmospheres, U.S. National Center for Atmospheric Research), and many other weather forecast and climate centres are currently working towards doing so.

7 Lateral boundary conditions for limited area models

André and I strongly disagreed on regional modelling issues for many years. At the end of any seminar that I gave on the development of a variable-resolution regional finite-element model, he would invariably state that a model whose domain was quasi-hemispheric could never compete with a limited-area one for regional forecasting over Canada! I would counter that no-one had ever convincingly demonstrated that a hydrostatic primitive-equations limited-area model, driven by a lower-resolution model run over a much larger domain, could successfully reproduce the results of a control model run at equivalently-high resolution over this larger domain. Not only that, but the Oliger and Sundström (1978) paper indicates that it was unlikely anyone ever would, and there matters lay for many years.

In the meantime André decided to further develop the Robert et al. (1985) limited-area model, with a view to eventually using it for routine weather forecasting. He found that the model's 24-h forecasts seemed to be adversely affected by the lateral boundaries, and he initially ascribed this to an overly-small domain size (constrained by a combination of the available computer memory and its use by the model code). To estimate the optimal domain size needed for 24-h forecasting, he therefore performed a series of integrations at fixed meshlength (190.5 km) with ever-larger domains centred on a given area of interest, using a shallow-water equations version of his model to circumvent the computer memory limitations of his baroclinic model. André was greatly surprised and disappointed to find (the present author much less so!) that the forecasts did not meet an acid test for a successful limited-area model – they did not become insensitive to the influence of the lateral boundaries, as they should, when these were progressively moved away from the area of interest, even when the resulting domain covered the entire Northern Hemisphere. This naturally called for further investigation.

André argued that, 'Since most nesting strategies contain difficulties that cannot be easily identified when they are considered in the framework of realistic models in three dimensions, it might be preferable to start out with simple problems' – Robert and Yakimiw (1986). His approach was his usual one: to reproduce the essence of the behaviour in a much-simplified context; to understand and address it in this context; and to then generalise the approach and apply it in the full context. He therefore defined two simple *linear* constant-coefficient problems having easily-determined *exact* solutions, and used these to evaluate various proposed strategies. These two problems provide *necessary* but not sufficient conditions that any proposed nesting strategy must satisfy. In other words, if any proposed strategy fails these tests, then this is sufficient reason to discard it from further consideration.

The first problem considered (Robert and Yakimiw, 1986) was to integrate the 1-d non-divergent barotropic vorticity equation on a β-plane:

$$\zeta_t + U\zeta_x + \beta\psi_x = 0, \qquad (32)$$

where $\zeta = \psi_{xx}$ is the relative vorticity, $\beta = (2\Omega/a)\cos\theta$ is the Rossby parameter, U is the mean eastward wind, a is the Earth's radius, Ω is its angular velocity and θ is latitude. This problem was used to identify some of the difficulties encountered by various nesting strategies and to motivate an alternative approach to address them.

His alternative approach consists of flattening the initial fields in a small region adjacent to the upstream and downstream lateral boundaries. It was shown that this worked very well for this simple problem and then its complexity was increased by adding eastward- and westward-propagating gravity waves to the Rossby wave of the first problem.

Thus the second problem considered (Robert and Yakimiw, 1986) was to integrate the 1-d constant-coefficient shallow-water equations:

$$u_t + Uu_x + \phi_x - fv = 0, \tag{33}$$

$$v_t + Uv_x + fu = 0, \tag{34}$$

$$\phi_t + U\phi_x - fUv + \Phi_0 u_x = 0, \tag{35}$$

where all symbols have been previously defined, and the third term of (35) reflects the geostrophic balance $d\Phi_0/dy = -fU$ of the basic state. This problem was used to demonstrate that the strategy of flattening the initial fields close to the lateral boundaries, motivated by the simpler first problem, also works very well for this somewhat more complex problem. It was also used to evaluate the popular nesting strategies of Williamson and Browning (1974), Perkey and Kreitzberg (1976) and Davies (1976), which were introduced in the seventies and variants of which are still used in almost all of today's limited-area models. It was found that *none of these methods works acceptably well* for this simple 1-d constant-coefficient problem! However, it was noted that all of these methods would work well for this simple problem if the initial fields are flattened in the recommended manner. The present author believes that this result deserves to be better known and appreciated in the mesoscale-modelling community. André also argued that it is necessary to damp the shortest scales near outflow boundaries, and that the Davies (1976) approach is probably the best one inasmuch as it does this and it has also been designed to adjust to inaccuracies in the forecast of the driving model.

It was noted in Robert and Yakimiw (1986) that while a nesting strategy may perform well in simple tests, this does not guarantee that it will work well in a complete model. In a brief research note (Yakimiw and Robert, 1990), some follow-up work was presented in the somewhat more realistic context of the 2-d nonlinear shallow-water equations using 500-mb data. The proposed nesting strategy, including the flattening of initial fields in a buffer region adjacent to the lateral boundaries, was shown to also work well in this context, inasmuch as the limited-area model can successfully reproduce the control forecast over an interior subdomain that shrinks to zero size as a function of time due to the inward propagation of errors from the lateral boundaries. However, some caution needs to be exercised in this regard.

The forecasts are only acceptably accurate after 24 h over a small percentage of the 10,000 km × 10,000 km limited-area integration domain. As noted in the paper, error is continuously advected inward from the lateral boundaries at the advective wind speed: an estimated 3,500 km after 24 h from the western boundary, and at least 2,500 km from the eastern one. After 24 h, the sub-domain where the forecast is valid (optimistically estimated to be an approximately 6,000 km × 6,000 km sub-domain) has shrunk to at best 36% of the total size of the integration domain. This doesn't seem so bad until one realises that the 500-mb windspeed of the experiment underestimates, by approximately a factor of two, the windspeed of the tropospheric jet in the more realistic context of a baroclinic model, in which case the sub-domain of forecast validity would have shrunk to zero by 24 h! Another way of looking at, and appreciating the importance of, the inward propagation of information from lateral boundaries may be found in Chouinard et al. (1994). There a methodology was introduced for more precisely estimating the minimum domain size required by a limited-area model for given initial conditions. This methodology is predicated on the argument that information propagates inwards from the lateral boundaries at a speed no less than that of the local wind, and it was applied to a case used for an international modelling intercomparison project to justify the domain size used for the experiments.

In the present author's opinion, a majority of limited-area modellers are overly optimistic about just how small they can make their integration domains while maintaining the validity of their results over their area of interest. Certainly André's opinion on this issue changed substantially as a result of his above-described work, and this in turn also modified my own. From our diametrically-opposed viewpoints, André's and my own had almost converged by the time of his death. He appreciated that the problem of nesting was more difficult than he had realised, while I had come to appreciate that it was not as difficult as I had at first thought. In the end we agreed that the inward propagation of the error from lateral boundaries is fast enough that it imposes a serious limitation on the efficiency of driven limited-area models. We also agreed that the best way to reduce its severity is to use variable resolution in some way outside the area of interest, and he supervised the introduction of this option into his baroclinic model (Ben Hadj Tahar, 1993). However, we still differed on whether this should be done in the context of a driven limited-area model or a variable-resolution global model.

8 Non-hydrostatic modelling

As computers become ever more powerful, it becomes possible to run models at higher and higher resolution. A time is approaching (Daley, 1988) when it will be possible to run current hydrostatic baroclinic primitive-equation weather forecast models at resolutions for which the hydrostatic assumption can no longer be assumed to hold. This motivates the need to *efficiently* integrate non-hydrostatic systems of equations for real-time forecasting applications over large domains, and André realised this. He recognised that:

(i) a semi-Lagrangian treatment of advection coupled to a semi-implicit treatment of the terms governing gravitational oscillations results in a highly-efficient integration scheme for integrating hydrostatic primitive-equation models;

(ii) it is highly desirable to use the *fully-elastic Euler equations* for small-scale modelling, thereby including non-hydrostatic effects without making approximations such as the anelastic and Boussinesq ones;
(iii) the Euler equations admit *acoustic* modes, which travel much faster than either Rossby or gravity modes, and consequently if care is not exercised, the limiting time step will be even more restrictive than that associated with an explicit primitive equations model;
(iv) the challenge is therefore to stably treat the elastic modes in an Euler equations model in an affordable manner; and
(v) since the elastic modes carry very little energy, it is permissible to slow them down by the use of a time-implicit treatment of the terms responsible for their existence, by analogy with the retarding of the gravity modes by the semi-implicit scheme.

This led André to extend the semi-implicit semi-Lagrangian methodology of Robert et al. (1985) for the hydrostatic primitive equations, to the Euler ones. In Tanguay et al. (1990), it is demonstrated that this can indeed be accomplished and for *negligible* additional cost, and this work represents a major breakthrough towards highly-efficient non-hydrostatic forecast models. Since this time the model has been further developed and has become the Mesoscale Compressible Community (MC2) Model of the Cooperative Centre for Research in Mesometeorology: see the Laprise et al. (this volume) paper in this issue for further information including model formulation.

While it was generally accepted that a semi-implicit semi-Lagrangian model of the Euler equations is appropriate for flows at scales as small as the mesoscale, it was less obvious that this is also true at the convective scale. The concern here is the suitability of the statically-stable basic state (required by the semi-implicit method) for simulations in which the environment is statically unstable, or at most neutrally stable. To address this important issue André performed a series of controlled bubble-convection experiments with his model in Robert (1993), and showed that it performs at least as well as, and in many respects better than, competing non-hydrostatic models, but does so in a much more economical manner. The conclusion was then drawn that a model based on the Euler equations is universal in the sense that it can be used at all meteorological scales. He followed this up by applying his model to a family of problems covering scales from as small as the laboratory scale right up to the hemispheric scale: René Laprise showed a number of these results during his presentation at the Memorial Symposium.

André's semi-implicit semi-Lagrangian formulation for integrating the fully-compressible Euler equations is likely to have a lasting impact. It permits a unification of global, regional, mesoscale, and microscale models within a single highly-efficient dynamical framework. This work has already spawned a number of major research efforts (e.g., Bubnová et al., 1995; Côté et al., this volume; Cullen et al., this volume): it is likely to be a fruitful area of research for quite some time to come, and to thereby lead to yet more efficient and complete meteorological models.

9 Miscellanea

Some of André's contributions do not fall naturally into any of the above sections and are therefore discussed briefly here. The first of these is a theory of nonlinear computational instability (Robert et al., 1970) which generalizes the work of Robert (1969b). In this latter, André showed that contrary to popular belief, the stability of the following 2nd-order finite difference approximation

$$F(x, t + \Delta t) = F(x, t - \Delta t) - \left[U(x, t) \frac{\Delta t}{\Delta x} \right] [F(x + \Delta x, t) - F(x - \Delta x, t)], \quad (36)$$

to the passive advection equation in 1-d, is not guaranteed by simply respecting the well-known Courant number condition

$$\left| U(x, t) \frac{\Delta t}{\Delta x} \right| \leq 1. \quad (37)$$

He did this by demonstrating that if $U(x, t)$ is chosen to be a constant multiple of a pure $2\Delta t$ wave (i.e., $U \propto (-1)^n$ where n is the time step number), and its amplitude is chosen sufficiently small to satisfy (37), then (36) is nevertheless *unconditionally unstable*. Note that the condition (37) is usually derived under the assumption of $U(x, t) \equiv U_0 = constant$, i.e., for uniform flow, which makes (37) a necessary condition but says nothing about its sufficiency. Thus the Robert (1969b) analysis proves (by the above-stated counter-example) that the *necessary* stability condition (37) is in fact *insufficient*.

In Robert et al. (1970) the approach is generalized to include coupled sets of equations and a more general advecting velocity of the form

$$U(x, t) = U_0 + U_1(-1)^j + U_2(-1)^n + U_3(-1)^{j+n}, \quad (38)$$

where j and n are respectively the gridpoint and time step numbers, and the coefficients U_k are constant. The form (38) for $U(x, t)$ contains contributions to the advecting velocity that are spatially and/or temporally decoupled. It includes not only the usual case of a constant contribution (for which $U_1 = U_2 = U_3 = 0$) that leads to the conditional stability condition (37), and the unconditionally unstable example (for which $U_0 = U_1 = U_3 = 0$) of Robert (1969b), but also the cases examined by Phillips (1959) for $U_0 = U_2 = 0$ and Ritchmyer (1963) for $U_2 = 0$.

The Robert et al. (1970) analysis is interesting because it illustrates how the nonlinear shallow-water equations can go unstable even when the CFL (Courant-Friedrichs-Levy) condition (37) is satisfied. It is shown that instability is caused by both spatial and temporal high frequencies that appear due to nonlinear interactions. It is also shown that it is insufficient in a nonlinear equation or set of equations to suppress only the high-wavenumber spatial modes in the undifferentiated factors in the equations, but it is also necessary to suppress the high-frequency temporal modes – one way of achieving this is by the use of the Robert time filter discussed above in Section 2.

The second miscellaneous contribution is the work, summarized in Robert (1976), on the breakdown of forecast error as a function of source. It was performed in the context of André's involvement with WMO's Working Group on Numerical Experimentation (WGNE). Here he underlined the importance of estimating the relative importance of the various sources of error with a view to ensuring that the largest sources are assigned the highest priority, and that resources not be wasted on those that contribute very little. The approach taken was to perform a set of sensitivity experiments and to use these to make the assessments as quantitative as possible. The results are by now of course very dated, but they stimulated much discussion and were widely distributed. At the time it was very clear that the horizontal truncation error was the most important error source. Coincidentally, during the preparation of this paper a British journalist was visiting Canada and wished to interview André, unaware that he had passed away. He was doing a piece on weather forecasting and wanted to know if André's result of twenty years ago, viz. the importance of the horizontal truncation error, was still valid – apparently his boss had done a similar piece twenty years ago and had interviewed André!

The third miscellaneous contribution is the work of Tanguay and Robert (1990), on a more efficient way of analysing data using the so-called optimum interpolation method. It was clear from André's comments at RPN's weekly internal seminars that he was very knowledgeable about data assimilation, yet he did relatively little work in this area since his first contribution (Eddy et al., 1961) during the early days of his research career. In the early seventies he did analyse a dataset for distribution to the community for a forecast intercomparison project conducted under the auspices of the WGNE. The resulting analyses were affectionately known as the "gravity-wave analyses" because of their dynamical imbalance due to the lack of a suitable initialization procedure in their preparation. In Tanguay and Robert (1990) the autocorrelation function for a univariate analysis scheme is approximated by a truncated Taylor series expansion of the Gaussian hill function. When applied to the analysis of precipitation data over N. America, it is shown that this leads to an almost identical analysis to that obtained from univariate optimum interpolation but at much-reduced cost. The present author suspects that this would have been an important contribution had it been made 10 or 15 years earlier, but to date it seems to have had little impact on the data assimilation community.

The last contribution considered here is that of Jakimow et al. (1992). This study at first glance seems to be nothing more than a justification for the need to treat horizontal diffusion in an implicit way in the context of a finite-difference model, and a demonstration of how to reduce its cost by replacing the discrete Laplacian operator by a spatially-split approximation. However, it is argued that forecasts lose their predictability as a function of scale, with the smallest scales being the least predictable, and that waves that have lost predictability should therefore be eliminated from the forecast. It is further argued that horizontal diffusion is a good way to accomplish this, and that it should be considered to be a numerical rather than a physical process until such time as models become sufficiently skilful (but no criterion was given for this). An analysis is then presented to show how displacement errors can be partially compensated for during the forecast in such a way as to reduce the error growth, and this is used to motivate the use of a Laplacian horizontal diffusion with a coefficient that increases linearly in time from zero. Compared to the usual formulation with a time-independent coefficient, there is thus less (more) diffusion during the early (late) part of the forecast when it has high (low) information content. Consequently the forecasts might be expected to be at least as accurate during the early part of the forecast and more accurate towards the end. The spectral slope of fields for the later part of the forecast is of course wrong, but if the objective is to obtain the best forecast (in the sense of minimizing r.m.s. errors) for a given forecast period, then this presumably isn't important: it would not of course be acceptable in the context of a climate simulation. In a moist baroclinic model, the increased horizontal diffusion would also presumably reduce the vertical motion, leading to an underestimate of the precipitation amounts.

10 Concluding remarks

In this scientific paper/essay, I have reviewed André Robert's pioneering contributions to numerical modelling, and these are summarized in the abstract. When reviewing the pioneering contributions of a distinguished scientist, there is a certain degree of arbitrariness in the selection of the material and in its manner of presentation. To make this review less dry and more personal, I have endeavoured to weave together André's scientific contributions in an interesting and coherent way, to put them into a historical context, and to do so from the personal (and admittedly biased) perspective of a well-informed witness. It is hoped that the reader will find that these objectives have been at least partially met. I will conclude with a few further personal reflections to illustrate some of André's personal qualities.

André always seemed to have a different perspective to others on issues. He loved a debate and was always ready to challenge people's positions and stimulate them intellectually. Philip Merilees gave two good examples of this when he addressed attendees at the Memorial Symposium Banquet. He recalled a debate reported in the Proceedings of the Ninth Stanstead Seminar (1972; Merilees, editor) on the relative merits of finite-difference and spectral methods. At the time André was a strong proponent of the spectral method, but was co-opted at the last minute to captain the opposing finite-difference team. This he proceeded to do so convincingly that his team won the debate! The second example recounted by Phil was that during his sabbatical year at McGill, André once complained to him that he never got any work done on campus since he spent most of his time in scientific debate with students! I personally recall many a Friday when I spent the better part of the afternoon in discussion with him. André was also always willing to give constructive scientific criticism and advice to others when asked to do so, which was frequently. I had particular cause to appreciate this during the two-and-a-half years I spent as RPN's Division Chief. Despite a previous history of us not always sharing the same viewpoint (and also of forcefully debating our differences), André was most supportive during this period and made a difficult task that much easier.

In a world where the publish-or-perish syndrome is ubiquitous, André's independence was most refreshing. He was a person who was very selective in his choice of problem; willing to make the necessary large investment in time and effort; and willing to take the risk that it would come to nought, and even in the event it was successful, that it would only add one or two entries to his list of journal papers. Since embarking on a research career in 1959, André had only published twenty journal papers in all (plus one in press) by his death in 1993: four in the sixties, three in the seventies, six in the eighties, and eight in the nineties. For a scientist of his great stature and influence, this is a remarkably and almost unbelievably small number. In 1987 André stated (Ritchie and Robert, this volume): 'I always opted for one good paper every ten years as being much better than ten papers every year', and I admire him very much for this personal commitment to excellence. He believed that there is far too much emphasis on quantity to the detriment of quality, and that this acts as a serious deterrent to undertaking the kind of high-risk/high-return research with which his name is associated.

It was a privilege to have known André as a colleague, to have witnessed history in the making (without sometimes realising it), and to have benefited from his enriching presence. He had the knack of identifying the most important problems and was the consummate problem solver. He put and kept RPN on the meteorological map, and he was the person most responsible for creating the stimulating environment in which we work today. In conclusion, I can think of no more fitting an epitaph to André Robert, Emeritus Scientist, than the following:

> 'He has passed away and will be sorely missed by us all, but his contributions to Science remain. Farewell, old friend.'

Acknowledgements

The thoughtful reviews by Michel Béland, Sylvie Gravel, René Laprise and Herschel Mitchell of a draft of this paper are gratefully acknowledged.

References

ASSELIN, R. 1972. Frequency filter for time integrations. *Mon. Weather Rev.* **100**: 487–490.

BAER, F. 1964. Integration with the spectral vorticity equation. *J. Atmos. Sci.* **21**: 260–276.

——— and G. PLATZMAN. 1961. A procedure for numerical integration of the spectral vorticity equation. *J. Meteorol.* **18**: 393–401.

BATES, J.R. and A. MCDONALD. 1982. Multiply-upstream, semi-Lagrangian advective schemes: analysis and application to a multi-level primitive equation model. *Mon. Weather Rev.* **112**: 1831–1842.

———; F.H.M. SEMAZZI, R.W. HIGGINS and S.R.M. BARROS. 1990. Integration of the shallow-water equations on the sphere using a vector semi-Lagrangian scheme with a multigrid solver. *Mon. Weather Rev.* **118**: 1615–1627.

BEN HADJ TAHAR, N. 1993. Introduction d'une maille variable dans l'horizontale d'un modèle non-hydrostatique. M.Sc. thesis, Université du Québec à Montréal, Montréal, P.Q., 100 pp.

BOURKE, W. 1988. *Physically-Based Modelling and Simulation of Climate and Climatic Change – Part 1*, M.E. Schlesinger (Ed.), Kluwer Academic Publishers, pp. 169–220.

BUBNOVA, R.; G. HELLO, P. BÉNARD and J.-F. GELEYN.

1995. Integration of the fully elastic equations cast in hydrostatic pressure terrain-following coordinate in the framework of the ARPEGE/-Aladin NWP system. *Mon. Weather Rev.* **129**: 515–535.

CHOUINARD, C.; J. MAILHOT, H.L. MITCHELL, A. STANIFORTH and R. HOGUE. 1994. The Canadian regional data assimilation system: operational and research applications. *Mon. Weather Rev.* **122**: 1306–1325.

COHN, S.E.; D. DEE, E. ISAACSON, D. MARCHESIN and G. ZWAS. 1985. A fully implicit scheme for the barotropic primitive equations. *Mon. Weather Rev.* **113**: 436–448.

COIFFIER, J.; P. CHAPELET and N. MARIE. 1987. Study of various quasi-Lagrangian techniques for numerical models. *In*: Proc. ECMWF Workshop on Techniques for Horizontal Discretization in Numerical Weather Prediction Models, 2–4 November 1987, European Centre for Medium-range Weather Forecasts, Shinfield Park, Reading, U.K., pp. 19–46.

CÔTÉ, J. 1988. A Lagrange multiplier approach for the metric terms of semi-Lagrangian models on the sphere. *Q. J. R. Meterol. Soc.* **114**: 1347–1352.

——— and A. STANIFORTH. 1988. A two-time-level semi-Lagrangian semi-implicit scheme for spectral models. *Mon. Weather Rev.* **116**: 2003–2012.

DALEY, R.W. 1988. The normal modes of the spherical non-hydrostatic equations with applications to the filtering of acoustic modes. *Tellus*, **40A**: 96–106.

DAVIES, H.C. 1976. A lateral boundary formulation for multi-level prediction models. *Q. J. R. Meterol. Soc.* **102**: 405–418.

DESHARNAIS, F. and A. ROBERT. 1990. Errors near the poles generated by a semi-Lagrangian integration scheme in a global spectral model. ATMOSPHERE-OCEAN, **28**: 162–176.

ELIASSEN, E.; B. MACHENHAUER and E. RASMUSSEN. 1970. On a numerical method for integration of the hydrodynamical equations with a spectral representation of the horizontal fields. Rep. No. 2, Institut for Teoretisk Meteorologi, Köbenhavns Iniversitet, Denmark.

JAKIMOW, G.; E. YAKIMIW and A. ROBERT. 1992. An implicit formulation for horizontal diffusion in grid point models. *Mon. Weather Rev.* **120**: 124–130.

KAAS, E. 1987. The construction of and tests with a multi-level, semi-Lagrangian and semi-implicit limited area model. Diploma thesis, Geophysics Institute, Copenhagen University, Copenhagen, Denmark, 117 pp.

KRISHNAMURTI, T.N. 1962. Numerical integration of primitive equations by a quasi-Lagrangian advective scheme. *J. Appl. Meterol.* **1**: 508–521.

KRUGER, H.B. 1964. A statistical-dynamical objective analysis scheme. Canadian Meteorol. Memoirs, No. 18, Meteorological Branch, Dept. of Transport, Toronto, Ontario, 39 pp.

KUBOTA, S. 1960. Surface spherical harmonic representation of systems of equations for analysis. Papers in Meteorol. and Geophys., Meteorol. Res. Inst., Tokyo, pp. 145–166.

KWIZAK, M. 1961. A report on the solution of the barotropic geostrophic model on a medium-sized computer. Canadian Meteorol. Memoirs, No. 9, Meteorological Branch, Dept. of Transport, Toronto, Ontario, 30 pp.

——— and A. ROBERT. 1963. An evaluation of simple non-geostrophic forecasts. Canadian Meteorol. Memoirs, No. 13, Meteorological Branch, Dept. of Transport, Toronto, Ontario, 27 pp.

——— and ———. 1971. A semi-implicit scheme for grid point atmospheric models of the primitive equations. *Mon. Weather Rev.* **99**: 32–36.

LAPRISE, R. 1994. André J. Robert, 1929–1993: Original version (in French): *Le Climat*, **12**: 81–94. English translation: Can. Meteorol. Ocean. Soc. Bull. Vol. 22 pp. 12–17.

MARCHUK, G.I. 1965. A new approach to the numerical solution of differential equations of atmospheric processes. World Meteorological Organisation Technical Note no. 66, Geneva, Switzerland, pp. 286–294.

MCDONALD, A. and J.R. BATES. 1987. Improving the estimate of the departure point position in a two time-level semi-Lagrangian and semi-implicit model. *Mon. Weather Rev.* **115**: 737–739.

MERILEES, P. (Ed.) 1972. Proceedings of the 9th Stanstead Seminar. Pub. in Meteorol. No. 103, Dept. of Meteorology, McGill U., Montreal, 136 pp.

——— and R. LAPRISE. 1994. Necrology: André J. Robert, 1929–1993. *Bull. Am. Meteorol. Soc.* **75**: 456.

OLIGER, J. and A. SUNDSTRÖM. 1978. Theoretical and practical aspects of some initial boundary value problems in fluid dynamics. *SIAM J. Appl. Math.* **35**: 419–446.

ORSZAG, S.A. 1970. Transform method for calculation of vector-coupled sums: Application to the spectral form of the vorticity equation. *J. Atmos. Sci.* **27**: 890–895.

PERKEY, D.J. and C. KREITZBERG. 1976. A time-dependent lateral boundary scheme for limited-area primitive equation models. *Mon. Weather Rev.* **104**: 744–755.

PHILLIPS, N.A. 1959. An example of non-linear computational instability. In: *The Atmosphere in Motion*, Rockefeller Institute Press, N.Y., pp. 501–504.

PLATZMAN, G.W. 1960. The spectral form of the vorticity equation. *J. Meteorol.* **17**: 635–644.

PUDYKIEWICZ, J. 1990. A predictive atmospheric tracer model. *J. Meteorol. Soc. Jpn.* **68**: 213–225.

——— and A. STANIFORTH. 1984. Some properties and comparative performance of the semi-Lagrangian method of Robert in the solution of the advection-diffusion equation. ATMOSPHERE-OCEAN, **22**: 283–308.

RITCHIE, H. 1986. Eliminating the interpolation associated with the semi-Lagrangian scheme. *Mon. Weather Rev.* **114**: 135–146.

———. 1987. Semi-Lagrangian advection on a Gaussian grid. *Mon. Weather Rev.* **115**: 608–619.

———. 1988. Application of the semi-Lagrangian method to a spectral model of the shallow-water equations. *Mon. Weather Rev.* **116**: 1587–1598.

———. 1991. Application of the semi-Lagrangian method to a multi-level spectral primitive equations model. *Q. J. R. Meteorol. Soc.* **117**: 91–106.

———; C. TEMPERTON, A SIMMONS, M. HORTAL, T. DAVIES, D. DENT and M. HAMRUD. 1995. Implementation of the semi-Lagrangian method in a high resolution version of the ECMWF forecast model. *Mon. Weather Rev.* **123**: 489–514.

RITCHMYER, R.D. 1963. A survey of difference methods for non-steady fluid dynamics. NCAR Technical Notes 63-2, National Center for Atmospheric Research, Boulder, Colorado, 25 pp.

RIVEST C.; A. STANIFORTH and A. ROBERT. 1994. Spurious resonant response of semi-Lagrangian discretizations to orographic forcing: diagnosis and solution. *Mon. Weather Rev.* **122**: 366–376.

ROBERT, A. 1963a. Baroclinic prediction experiments with a four level statistical-dynamical model. Canadian Meteorol. Memoirs, No. 15, Meteorological Branch, Dept. of Transport, Toronto, Ontario, 32 pp.

———. 1963b. A baroclinic model for the Canadian numerical weather prediction program. Stanstead Seminars on the Stratosphere and Mesosphere and Polar Meteorology, Dept. of Meteorology, McGill U., Montreal, pp. 83–88.

———. 1965. The behaviour of planetary waves in an atmospheric model based on spherical harmonics. Pub. in Meteorol. No. 77, Dept. of Meteorology, McGill U., Montreal, 84 pp.

———. 1966. The integration of a low order spectral form of the primitive meteorological equations. *J. Meteorol. Soc. Jpn.* **44**: 237–245.

———. 1968a. Integration of a spectral barotropic model from global 500-mb charts. *Mon. Weather Rev.* **96**: 83–85.

———. 1968b. Résultats de quelques applications récentes de la méthode spectrale. *La météorologie*, **2**: 453–469.

———. 1968c. The treatment of moisture and precipitation in atmospheric models integrated by the spectral method. *J. Appl. Meteorol.* **7**: 730–735.

———. 1969a. The integration of a spectral model of the atmosphere by the implicit method. *In*: Proc. of WMO/ IUGG Symposium on NWP in Tokyo, Japan Meteorol. Agency, Tokyo, Japan, pp. VII.19–VII.24.

———. 1969b. An unstable solution to the problem of advection by the finite difference Eulerian method. Office Note #30, National Meteorological Center, U.S. Dept. of Commerce, Washington, D.C. 3 pp.

———. 1976. Sensitivity experiments for the development of NWP models. *In*: Proc. of the 11th Stanstead Seminar, Pub. in Meteorology No. 114, McGill U., Montreal, pp. 68–81.

———. 1979. The semi-implicit method. Numerical Methods used in Atnospheric Models, GARP Pub. Series no. 17, Chapter 8, World Meteorological Organisation, Geneva, Switzerland, pp. 417–437.

———. 1981. A stable numerical integration scheme for the primitive meteorological equations. ATMOSPHERE-OCEAN, **19**: 35–46.

———. 1982. A semi-Lagrangian and semi-implicit numerical integration scheme for the primitive meteorological equations. *J. Meteorol. Soc. Jpn.* **60**: 319–325.

———. 1993. Bubble convection experiments with a semi-implicit formulation of the Euler equations. *J. Atmos. Sci.* **50**: 1865–1873.

———; F.G. SHUMAN and J.P. GERRITY JR. 1970. On partial difference equations in mathematical physics. *Mon. Weather Rev.* **98**: 1–6.

———; J. HENDERSON and C. TURNBULL. 1972. An implicit time integration scheme for baroclinic models of the atmosphere. *Mon. Weather Rev.* **100**: 329–335.

———; T.L. YEE and H. RITCHIE. 1985. A semi-Lagrangian and semi-implicit numerical integration scheme for multilevel atmospheric models. *Mon. Weather Rev.* **113**: 388–394.

——— and E. YAKIMIW. 1986. Identification and elimination of an inflow boundary computational solution in limited area model integrations. ATMOSPHERE-OCEAN, **24**: 369–385.

SAWYER, J.S. 1963. A semi-Lagrangian method of solving the vorticity advection equation. *Tellus*, **15**: 336–342.

SILBERMAN, I.S. 1954. Planetary waves in the atmosphere. *J. Meteorol.* **11**: 27–34.

STANIFORTH, A. and C. TEMPERTON. 1986. Semi-implicit semi-Lagrangian integration schemes for a barotropic finite-element regional model. *Mon. Weather Rev.* **114**: 2078–2090.

——— and J. CÔTÉ. 1991. Semi-Lagrangian integration schemes for atmospheric models – a review. *Mon. Weather Rev.* **119**: 2206–2223.

TANGUAY, M. and A. ROBERT. 1986. Elimination of the Helmholtz equation associated with the semi-implicit scheme in a grid point model of the shallow water equations. *Mon. Weather Rev.* **114**: 2154–2162.

———; A. SIMARD and A. STANIFORTH. 1989. A three-dimensional semi-Lagrangian scheme for the Canadian regional finite-element forecast model. *Mon. Weather Rev.* **117**: 1861–1871.

——— and A. ROBERT. 1990. An efficient optimum interpolation analysis scheme. ATMOSPHERE-OCEAN, **28**: 365–377.

———; ——— and R. LAPRISE. 1990. A semi-implicit semi-Lagrangian fully compressible regional forecast model. *Mon. Weather Rev.* **118**: 1970–1980.

———; E. YAKIMIW, H. RITCHIE and A. ROBERT. 1992. Advantages of spatial averaging in semi-implicit semi-Lagrangian schemes. *Mon. Weather Rev.* **120**: 113–123.

TEMPERTON, C. and A. STANIFORTH. 1987. An efficient two-time-level semi-Lagrangian semi-implicit integration scheme. *Q. J. R. Meteorol. Soc.* **113**: 1025–1039.

WIIN-NIELSEN, A. 1959. On the application of trajectory methods in numerical forecasting. *Tellus*, **11**: 180–196.

WILLIAMSON, D.L. and G.L. BROWNING. 1974. Formulation of the lateral boundary conditions for the NCAR limited area model. *J. Appl. Meteorol.* **13**: 8–16.

YAKIMIW, E. and A. ROBERT. 1986. Accuracy and stability analysis of a fully implicit scheme for the shallow-water equations. *Mon. Weather Rev.* **114**: 240–244.

——— and ———. 1990. Validation experiments for a nested grid-point regional forecast model. ATMOSPHERE-OCEAN, **28**: 466–472.

Simulation of Stratospheric Vortex Erosion using Three Different Global Shallow Water Numerical Models

J.R. Bates* and Yong Li
NASA Goddard Laboratory for Atmospheres
Greenbelt, Maryland, U.S.A.

[Original manuscript received 8 December 1994; in revised form 26 August 1995]

ABSTRACT *Simulation of planetary wave breaking and polar vortex erosion in the stratosphere has been carried out using three different global shallow water numerical models. The models are an Eulerian spectral model, a semi-Lagrangian finite difference model based on a vector discretization of the momentum equation and a semi-Lagrangian finite difference model based on the potential vorticity and divergence equations.*

The results of the integrations indicate advantages of the potential vorticity-based model in the vortex erosion problem.

RÉSUMÉ *On utilise trois modèles numériques mondiaux basés sur les équations de St-Venant pour simuler le déferlement d'ondes planétaires et l'érosion de la circulation polaire dans la stratosphère: le modèle spectral eulérien; le modèle à différences finies semi-lagrangien basé sur une discrétisation vectorielle de l'équation de la quantité de mouvement; et le modèle à différences finies semi-lagrangien basé sur les équations du tourbillon potentiel et de la divergence. Les résultats des intégrations indiquent les avantages du modèle basé sur le tourbillon potentiel dans le problème de l'érosion de la circulation.*

1 Introduction

The study of the stratospheric circulation has until recent years been focused largely on observing and accounting for phenomena described in terms of temperature, geopotential and wind. Beginning in the 1980s, the focus has shifted to studing the dynamics of the stratosphere in terms of potential vorticity (PV) as the basic dy-

Corresponding author address: Dr. Yong Li, Data Assimilation Office, NASA Goddard Space Flight Center, Code 910.3, Greenbelt, MD 20771, U.S.A.

*Current affliliation: University of Copenhagen, Geophysics Department, Juliane Maries Vej 30, DK-2100 Copenhagen O. Denmark.

namical variable. Many new insights have thus been gained, and phenomena have been discovered that can be missed entirely if attention is confined to the more conventional dynamical variables. By examining daily maps of PV in the middle stratosphere, McIntyre and Palmer (1983) discovered the phenomena of Rossby wave breaking, polar vortex erosion and the formation of a mid-latitude surf zone in the winter stratosphere. The satellite observations that made the PV maps possible were sufficient to reveal only the grossest features of these phenomena. Insight into the details of the breaking Rossby wave regime has been gained through high resolution numerical simulations, beginning with the pioneering work of Juckes and McIntyre (1987). Using a non-divergent barotropic model with quasi-topographic forcing of zonal wavenumber 1, this study showed the generation through nonlinear processes of a range of features from wavenumber 2 down to scales well below observational resolution: irreversible deformation of the material contours surrounding the polar vortex, the erosion of the vortex through the extrusion and peeling of PV filaments into the surf zone and the essential stability of the vortex core were found. The stability of the vortex core has important implications for the springtime Antarctic ozone depletion; without it, the temperatures within the vortex could not fall to the levels required for the formation of the polar stratospheric clouds with the consequent heterogeneous chemical reactions that liberate chlorine. Evidence for the existence of the predicted fine-scale features surrounding the vortex has subsequently been provided by aircraft observations of chemical tracers (e.g., Schoeberl et al., 1992).

Since the work of Juckes and McIntyre, a number of further studies of stratospheric dynamics using high resolution one-layer models have been carried out (Juckes, 1989; Salby et al., 1990a, 1990b, 1990c; O'Sullivan and Salby, 1990; Polvani and Plumb, 1992; Waugh, 1993; Yoden and Ishioka, 1993; Norton, 1994; Waugh and Plumb, 1994; Polvani et al., 1995). These have extended the study of vortex erosion in a number of ways: by using the shallow water equations, by ranging more generally through parameter space, and by introducing contour dynamics as a method of viewing the flow in greater detail (though with the restriction that the flow be balanced).

In the present paper, we use two recently developed semi-Lagrangian shallow water models to study the problem of stratospheric vortex erosion. We compare our results with those of a standard Eulerian spectral model. One of the semi-Lagrangian models is based on using the potential vorticity equation as a member of the governing equation set. It is reasonable to expect that such a model may have advantages in describing the evolution of potential vorticity and therefore in studying the stratospheric vortex erosion problem. In the case of the spectral model, we also pay considerable attention to the effects of the hyperdiffusion coefficient, whose value is shown to have a significant influence on the results obtained.

The three models are briefly described in Section 2, our numerical results are presented in Section 3 and the conclusions are given in Section 4.

2 Description of the models

We describe the models in the chronological order of their development. In each of the two semi-Lagrangian models the trajectory calculations are performed as in McDonald and Bates (1989), using bilinear interpolation to the trajectory midpoints. Bicubic interpolation is used in obtaining quantities at the departure points.

a *The Eulerian Spectral Model*

The spectral model, which is based on the formulation of Bourke (1972), was provided by NCAR (Hack and Jakob, 1992; Jakob et al., 1993). The prediction equations are the vorticity, divergence and continuity equations in Eulerian form:

$$\frac{\partial \eta}{\partial t} = -\frac{1}{a(1-\mu^2)}\frac{\partial}{\partial \lambda}(U\eta) - \frac{1}{a}\frac{\partial}{\partial \mu}(V\eta) + F_{\eta}^{\mathrm{Diff}} \tag{1}$$

$$\frac{\partial D}{\partial t} = \frac{1}{a(1-\mu^2)}\frac{\partial}{\partial \lambda}(V\eta) - \frac{1}{a}\frac{\partial}{\partial \mu}(U\eta) - \nabla^2\left(\Phi_s + \Phi' + \frac{U^2+V^2}{2(1-\mu^2)}\right) + F_{D}^{\mathrm{Diff}} \tag{2}$$

$$\frac{\partial \Phi'}{\partial t} = -\frac{1}{a(1-\mu^2)}\frac{\partial}{\partial \lambda}(U\Phi') - \frac{1}{a}\frac{\partial}{\partial \mu}(V\Phi') - \bar{\Phi}D + F_{\Phi}^{\mathrm{Diff}} \tag{3}$$

where

$$\eta \equiv \zeta + f = \frac{1}{a(1-\mu^2)}\frac{\partial V}{\partial \lambda} - \frac{1}{a}\frac{\partial U}{\partial \mu} + f$$

$$D = \frac{1}{a(1-\mu^2)}\frac{\partial U}{\partial \lambda} + \frac{1}{a}\frac{\partial V}{\partial \mu}$$

$$U \equiv u \cos\phi$$

$$V \equiv v \cos\phi$$

$$\mu = \sin\phi.$$

The geopotential Φ corresponding to the depth of the fluid has been expressed as $(\bar{\Phi}+\Phi')$, where $\bar{\Phi}$ is a time-invariant spatial mean and Φ' is the deviation therefrom; Φ_s is the geopotential corresponding to the height of the orography. The diffusion terms are given by

$$F_{\eta}^{\mathrm{Diff}} = -K_4\left[\nabla^4\eta - \frac{4}{a^4}\eta\right],$$

$$F_{D}^{\mathrm{Diff}} = -K_4\left[\nabla^4 D - \frac{4}{a^4}D\right],$$

$$F_{\Phi}^{\mathrm{Diff}} = -K_4\nabla^4(\Phi' + \Phi_s).$$

The linear correction term has been added to the vorticity and divergence operator to prevent damping of solid body rotation. A value of the diffusion coefficient K_4 corresponding to a 27-h damping time for the shortest resolved scales has been found to give realistic energy spectra (Jakob et al., 1993). We refer to this value of the diffusion coefficient as the recommended value.

Introducing a streamfunction ψ and velocity potential χ, the prognostic variables η and D become

$$\eta = \nabla^2\psi + f \tag{4}$$

$$D = \nabla^2\chi. \tag{5}$$

The equations of motion are transformed to spectral space using spherical harmonic functions, with triangular truncation. A three-time-level semi-implicit time differencing scheme is used in which the gravity wave and diffusion terms are treated implicitly and the remaining terms explicitly. All nonlinear terms (which are expressed at the middle time level) are evaluated in gridpoint space using the transform method with a Gaussian grid (Orszag, 1970; Eliasen et al., 1970). There is provision for the application of a linear time filter (Asselin, 1972) at the completion of each time step.

Since the advection terms are treated explicitly, the time step of the spectral model is limited by an advective stability criterion.

b *The Semi-Lagrangian* (u, v) *Model*

The semi-Lagrangian (u, v) model is a version of the two-time-level C-grid finite difference shallow water model of Bates et al. (1990), modified to include uncentred differencing in time and a treatment of the nonlinear term that accords with the method of Bates et al. (1993). (These are hereafter referred to as BSHB and BMH, respectively.)

The analytical governing equations are

$$\left(\frac{d\mathbf{V}}{dt}\right)_H = -\nabla\Phi_T - f\mathbf{k}\times\mathbf{V} \tag{6}$$

$$\frac{d\Phi}{dt} = -\Phi D \tag{7}$$

where $\mathbf{V}$ is the horizontal velocity vector and $\Phi_T(= \Phi + \Phi_s)$ is the geopotential height of the free surface. The momentum equation (6) is discretized in vector form as follows:

$$\left(\frac{\mathbf{V}^{n+1} - \mathbf{V}_*^n}{\Delta t}\right)_H = -\frac{1}{2}[(1+\varepsilon)(\nabla\Phi_T + f\mathbf{k}\times\mathbf{V})^{n+1} + (1-\varepsilon)(\nabla\Phi_T + f\mathbf{k}\times\mathbf{V})_*^n]_H \tag{8}$$

where $(\cdots)^{n+1}$ represents a value at an arrival point at the new time level, $(\cdots)_*^n$ represents a value at the corresponding departure point at the old time level, and ε is the uncentring parameter. Projecting (8) onto the horizontal at the midpoint of the trajectory and resolving into components, we solve for $(u, v)^{n+1}$ and hence taking the divergence (see BSHB and BMH for details) arrive at

$$D^{n+1} = -\tau L(\Phi')^{n+1} - \tau L(\Phi_s) + \tilde{A} \tag{9}$$

where $\tau = (1 + \varepsilon)\Delta t/2$ and

$$L = \frac{g}{a^2 \cos^2 \phi} \frac{\partial^2}{\partial \lambda^2} - \frac{1}{a^2 \cos \phi} \frac{\partial(gF)}{\partial \phi} \frac{\partial}{\partial \lambda} + \frac{1}{a^2 \cos \phi} \frac{\partial}{\partial \phi} \left(g \cos \phi \frac{\partial}{\partial \phi} \right)$$

$$\tilde{A} = \frac{1}{a \cos \phi} \frac{\partial}{\partial \lambda} [g(A_\lambda + FA_\phi)] + \frac{1}{a \cos \phi} \frac{\partial}{\partial \phi} [g \cos \phi (A_\phi - FA_\lambda)]$$

The quantities $(g, F, A_\lambda, A_\phi)$ are, mutatis mutandis, as in BMH.

The continuity equation (7) is written

$$\frac{d\Phi'}{dt} = -\bar{\Phi} D - N \tag{10}$$

where $N(= \Phi' D)$ is the nonlinear term. This is discretized as follows:

$$\frac{(\Phi')^{n+1} - (\Phi')_*^n}{\Delta t} = -\frac{\bar{\Phi}}{2}[(1 + \varepsilon)D^{n+1} + (1 - \varepsilon)D_*^n] - \frac{1}{2}[\tilde{N}^{n+1} + N_*^n] \tag{11}$$

The nonlinear term at the time level $(n + 1)$ is obtained by forward extrapolation

$$\tilde{N}^{n+1} = 2N^n - N^{n-1}$$

Elimination of D^{n+1} from (9) and (11) leads to a single elliptic equation on the sphere for $(\Phi')^{n+1}$:

$$\left[L - \frac{1}{\tau^2 \bar{\Phi}} \right] (\Phi')^{n+1} = H \tag{12}$$

where H consists of known quantities. Equation (12) is solved using the direct solver of Moorthi and Higgins (1993). The remaining details of the method of solution are as in BSHB.

This model does not have any stability criteria of the CFL kind, and the time step can be chosen on the basis of accuracy alone.

c *The Semi-Lagrangian PV-D Model*

The PV-D model is the two-time-level finite difference model developed by Bates et al. (1995) (hereafter BLBMR). It is based on a semi-Lagrangian discretization of the potential vorticity equation

$$\frac{d}{dt}\left(\frac{\zeta+f}{\Phi}\right)=0 \tag{13}$$

With a modification needed to stabilize Rossby waves, the discretized form of (13) becomes

$$\left(\frac{\zeta+f}{\Phi}\right)^{n+1}+\frac{\Delta t}{2\Phi_*^n}(\beta v)^{n+1}=r_1 \tag{14}$$

where r_1 is a known quantity (see BLBMR for details).

The companion to the PV equation is the divergence equation, which is obtained by taking the divergence of the vector form of (8) after it has been projected onto the horizontal at the trajectory midpoint (with ε omitted). This gives

$$\left(D+\frac{\Delta t}{2}\{\beta u-f\zeta+\nabla^2\Phi_T\}\right)^{n+1}=r_2 \tag{15}$$

where r_2 is known.

The continuity equation (7) is discretized without using any forward extrapolation of the nonlinear term; thus,

$$\frac{\Phi^{n+1}-\Phi_*^n}{\Delta t}=-\frac{1}{2}[(\Phi D)^{n+1}+(\Phi D)_*^n]$$

i.e.,

$$\Phi^{n+1}\left(1+\frac{\Delta t}{2}D^{n+1}\right)=r_3 \tag{16}$$

where r_3 is known.

Introducing a stream function and velocity potential, the governing equations (14), (15) and (16) can be written

$$\left(\frac{\nabla^2\psi+f}{\Phi}\right)^{n+1}+\frac{\beta\Delta t}{2\Phi_*^n}\left(\frac{1}{a\cos\phi}\frac{\partial\psi}{\partial\lambda}+\frac{1}{a}\frac{\partial\chi}{\partial\phi}\right)^{n+1}=r_1 \tag{17}$$

$$\left(\nabla^2\chi+\frac{\Delta t}{2}\left[\nabla^2\Phi-f\nabla^2\psi+\beta\left\{\frac{1}{a\cos\phi}\frac{\partial\chi}{\partial\lambda}-\frac{1}{a}\frac{\partial\psi}{\partial\phi}\right\}\right]\right)^{n+1}=r_2 \tag{18}$$

$$-\frac{\Delta t}{2}\nabla^2\Phi_s \tag{18}$$

$$\Phi^{n+1}\left(1+\frac{\Delta t}{2}\nabla^2\chi\right)^{n+1}=r_3 \tag{19}$$

These equations are spatially discretized using second-order finite differences on an unstaggered grid, and solved using the efficient nonlinear multigrid method of Ruge et al. (1996).

In contrast to the spectral model, which treats all advective nonlinearities and the nonlinear term on the right-hand side of the continuity equation explicitly at the middle time level, and to the (u, v) model, which treats the advective nonlinearities in a semi-Lagrangian manner while using forward time-extrapolation for the $\Phi' D$ term, the PV-D model treats all nonlinearities in a semi-Lagrangian semi-implicit way. It can therefore be expected to differ to some degree from the other two models in simulating nonlinear flow. The motivation for developing the PV-D model was the expectation that, by using PV as a prognostic variable and treating the nonlinearities in the above manner, the model might simulate the evolution of PV more accurately in high nonlinear situations.

The PV-D model does not have any explicit noise suppressors, such as the ∇^4 diffusion in the spectral model or the ε-uncentring in the (u, v) model. The model's only diffusivity is that inherent in the numerical formulation, primarily that associated with the bicubic interpolations to the trajectory departure points. We note that this diffusivity acts directly on the PV.

Like the (u, v) model, the PV-D model does not have any stability criterion of the CFL kind, so its time step can again be chosen on the basis of accuracy alone.

3 Numerical integrations

The three models described above are each integrated for a period of 24 days from an initial state intended to be representative of a layer of the middle stratosphere in winter. The depth of the layer is chosen to be 6 km at the south pole. The initial wind field, featuring a strong westerly jet with a maximum of 50 m s^{-1} at 60°N, is shown in Fig. 1. The wind passes through zero at 10°N and remains easterly from there to the south pole. The initial geopotential field, calculated on the basis of a gradient wind balance assumption, is shown in Fig. 2. The corresponding PV field is shown in Fig. 3.

A quasi-topographic forcing is introduced by allowing the wind to blow over a time-dependent orography of zonal wavenumber 1, having the form

$$h(\lambda, \phi, t) = H_s A(t) B(\phi) \sin \lambda \tag{20}$$

We choose $H_s = 0.72$ km and assign $A(t)$ and $B(\phi)$ the forms shown in Figs 4 and 5, respectively. (The analytical formulae for $A(t)$ and $B(\phi)$ are given in the appendix. Note that $A(t)$ repeats itself for $t > 20$ days, and that $B(\phi) = 0$ in the Southern Hemisphere.)

Each model is integrated at three different spatial resolutions: (128×65), (256×129) and (512×257) for the finite difference models (corresponding, respectively, to approximately $\Delta\lambda = \Delta\phi = 2.8°$, 1.4° and 0.7°), and T42, T85 and T170 for the spectral model. Since the PV-D model is the one of primary interest, we present the results of that model first, giving the greatest detail.

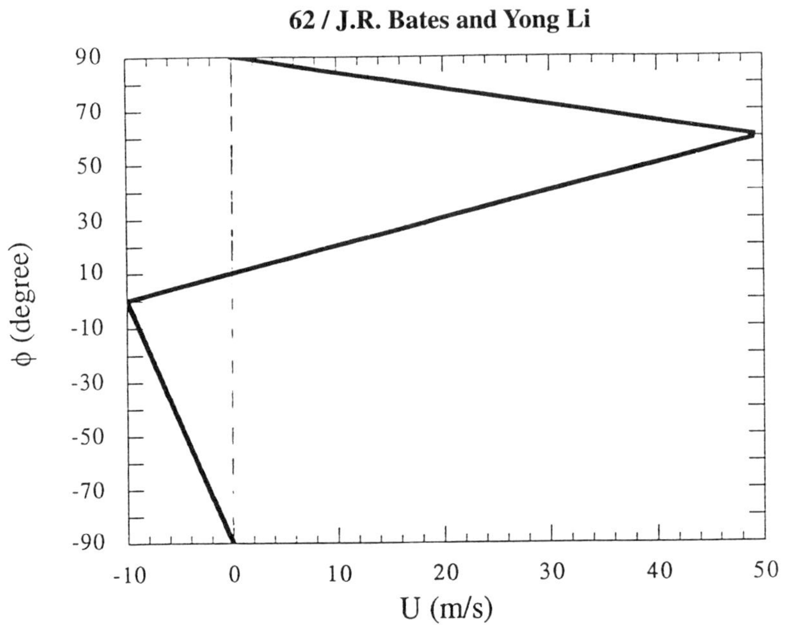

Fig. 1 The initial wind field.

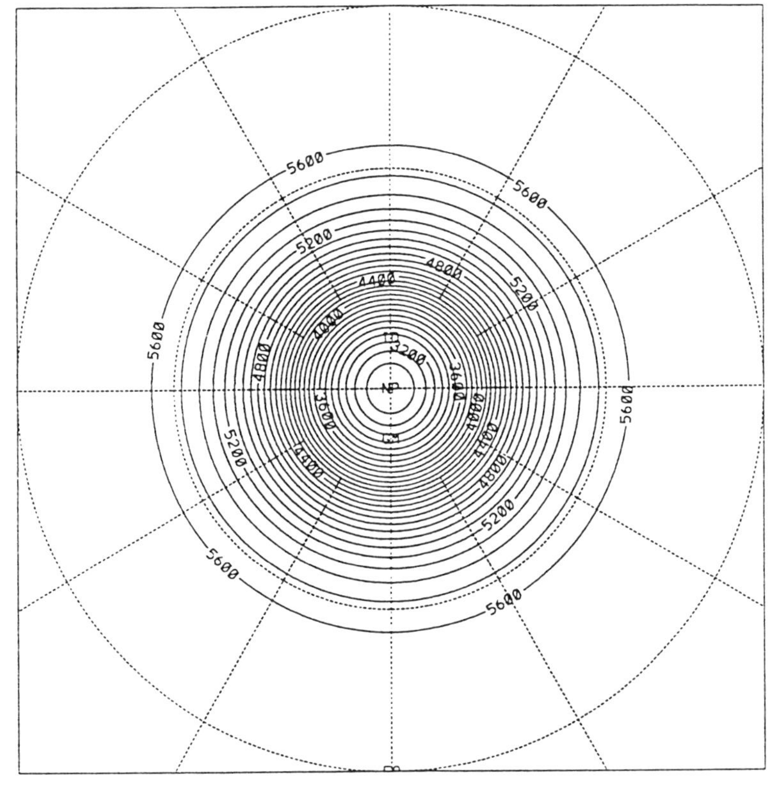

Fig. 2 The initial geopotential height field (Northern Hemisphere). Contour interval = 100 m.

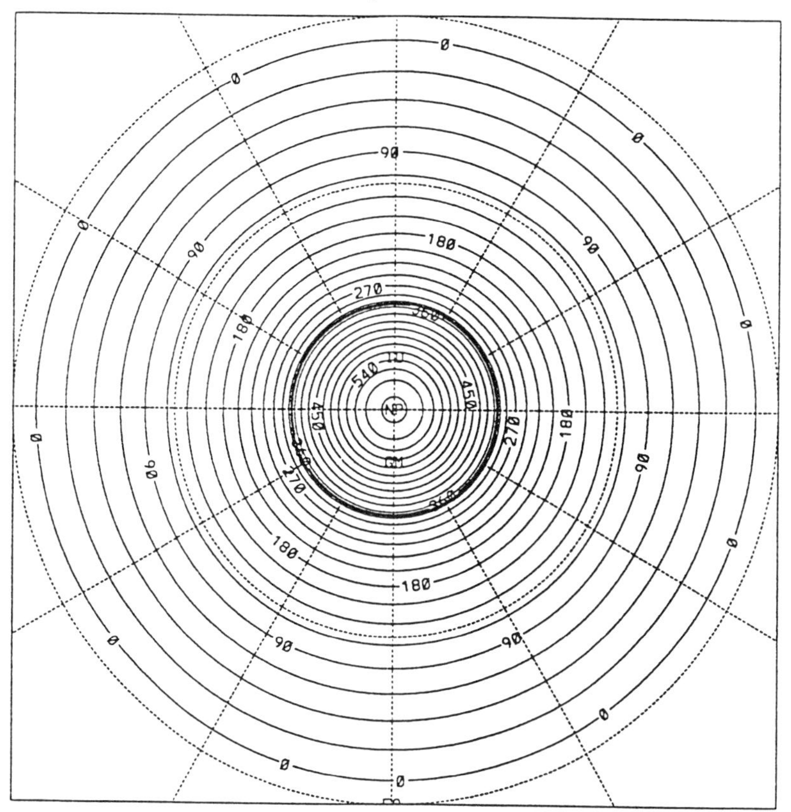

Fig. 3 The initial PV field (Northern Hemisphere). Contour interval = 22.5 in units of 10^{-11} m^{-2} s.

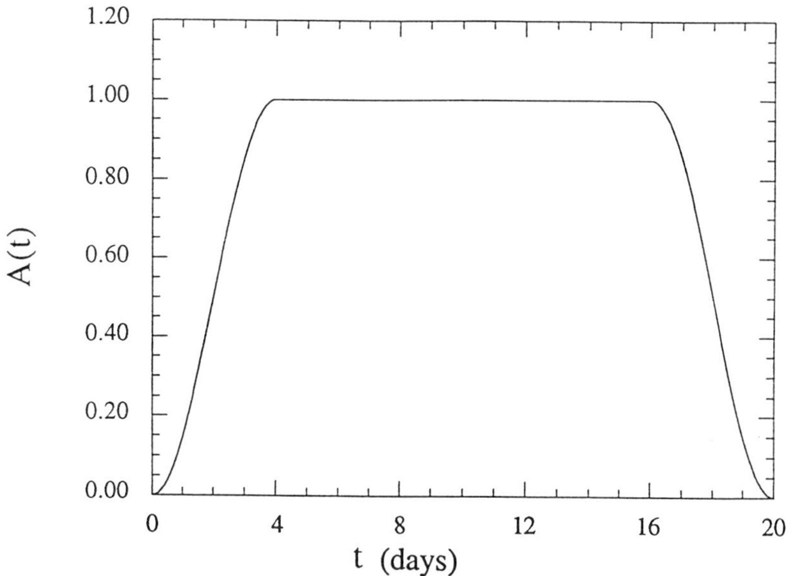

Fig. 4 The time-dependent factor $A(t)$ of the quasi-topographic forcing.

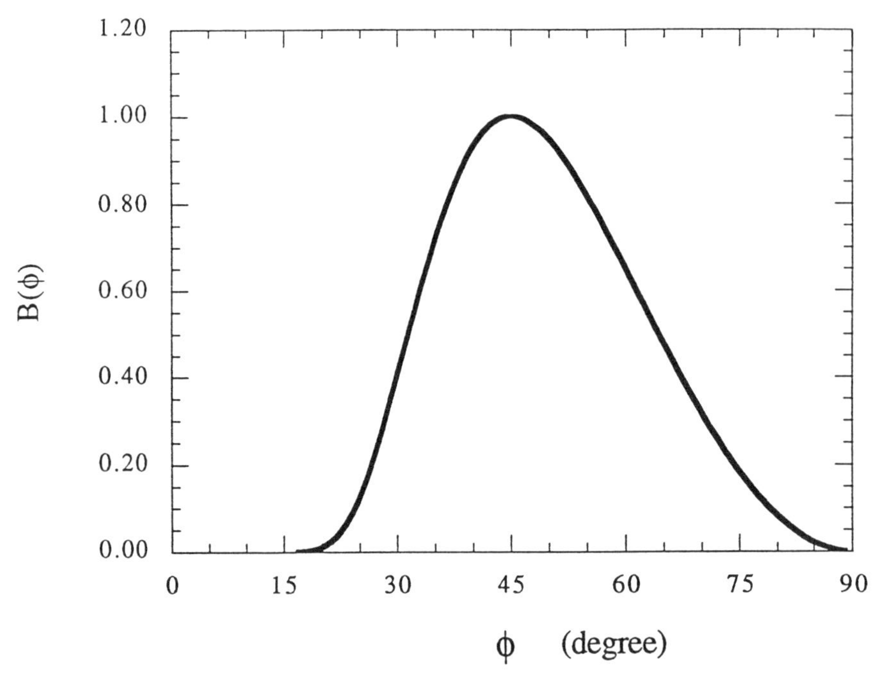

Fig. 5 The latitude-dependent factor $B(\phi)$ of the quasi-topographic forcing in the Northern Hemisphere.

a *PV-D Model Results*

For each of the three spatial resolutions given above, the PV-D model is integrated using a time step of 30 min. Having chosen the method of performing the trajectory calculations and assigned the order of the interpolations, the model has no other disposable parameters.

The evolution of the geopotential field for the high resolution case is shown in Fig. 6. The basic stability of the polar vortex is apparent, but otherwise nothing very remarkable appears to be occurring.

The corresponding evolution of the PV field is shown in Fig. 7. Here, by contrast, we see the formation of sharp PV gradients, clear evidence of Rossby wave breaking and the formation of the mid-latitude surf zone.

The PV fields at day 24 as given by the PV-D model at the three resolutions are shown in Fig. 8. We see that at the (128×65) resolution much of the interesting detail does not appear, while at the intermediate resolution of (256×129) the sharp PV gradients and some of the other fine-scale features of the flow seen in the high resolution case have become apparent.

b (u, v) *Model Results*

As with the PV-D model, the (u, v) model is integrated at the three specified spatial resolutions using a time step of 30 min. In order to achieve a noise-free integration,

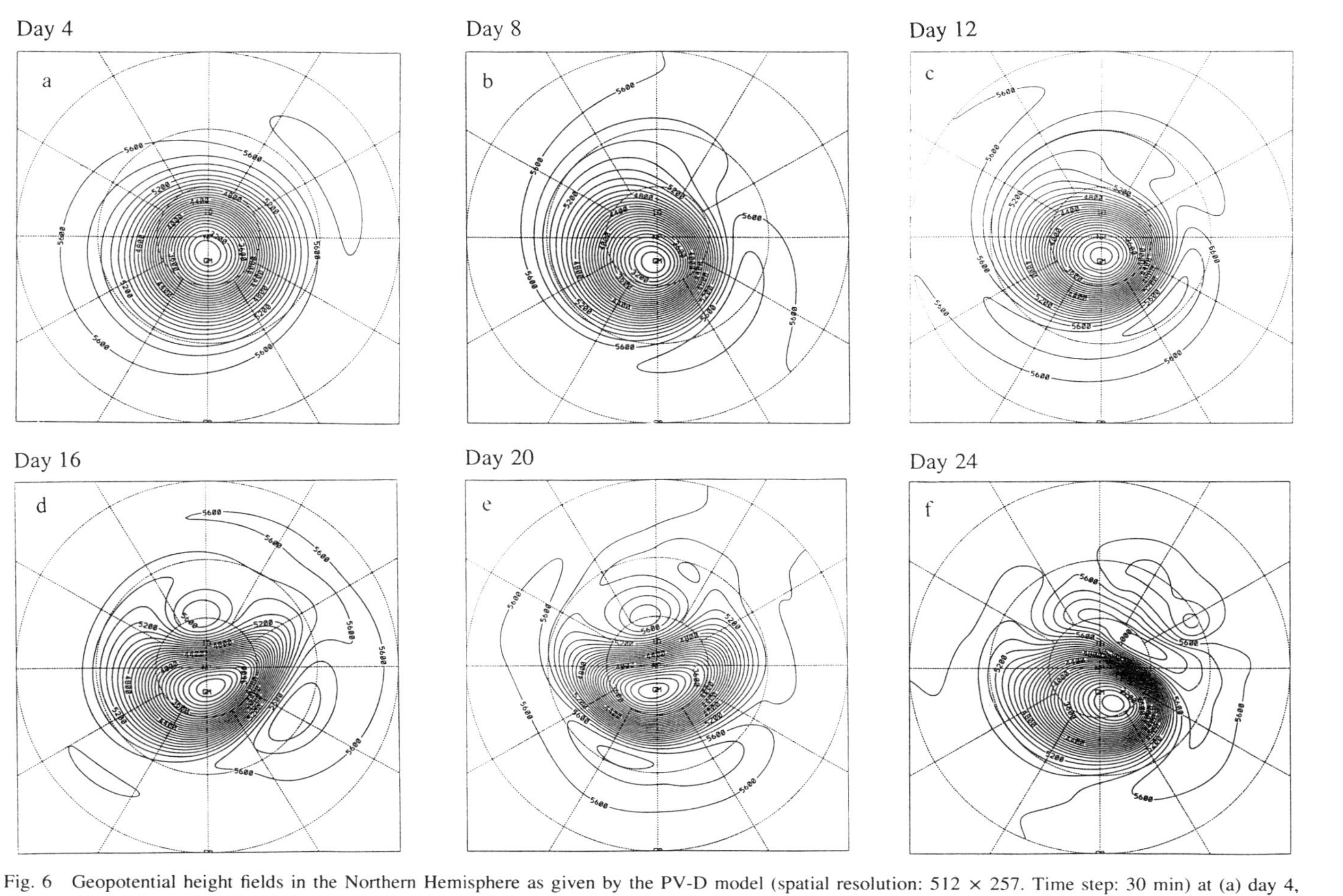

Fig. 6 Geopotential height fields in the Northern Hemisphere as given by the PV-D model (spatial resolution: 512 × 257. Time step: 30 min) at (a) day 4, (b) day 8, (c) day 12, (d) day 16, (e) day 20 and (f) day 24. Contour interval = 100 m.

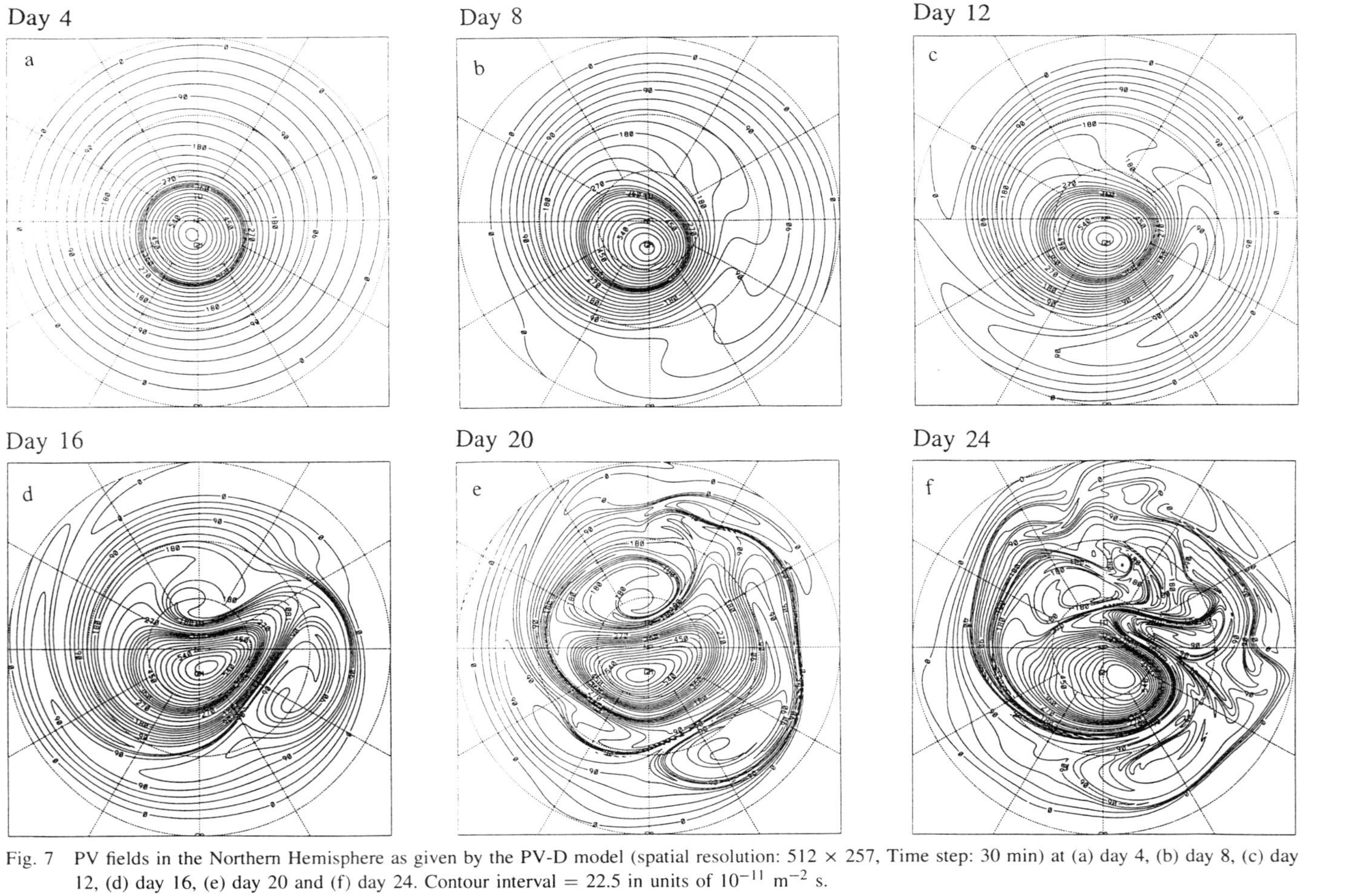

Fig. 7 PV fields in the Northern Hemisphere as given by the PV-D model (spatial resolution: 512 × 257, Time step: 30 min) at (a) day 4, (b) day 8, (c) day 12, (d) day 16, (e) day 20 and (f) day 24. Contour interval = 22.5 in units of 10^{-11} m^{-2} s.

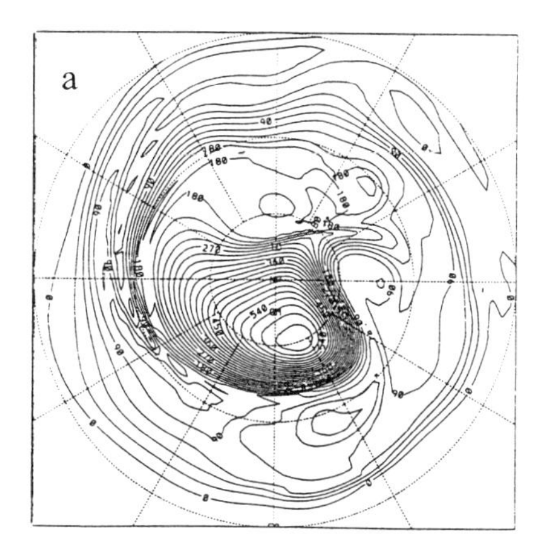

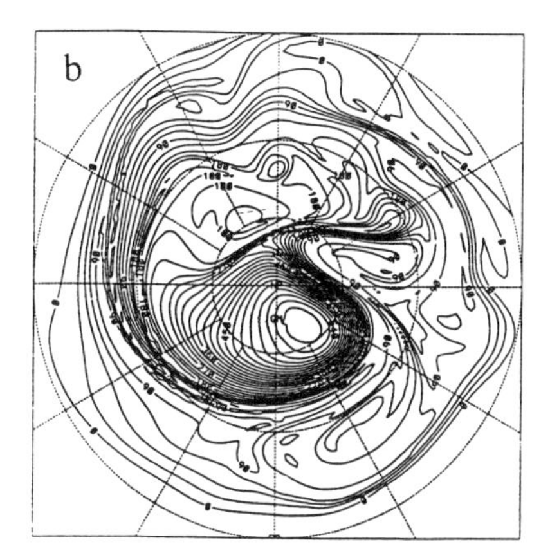

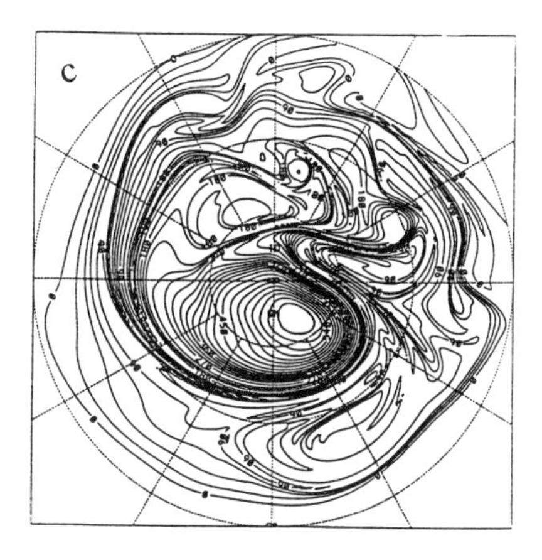

Fig. 8 PV fields in the Northern Hemisphere at day 24 as given by the PV-D model at resolutions of (a) 128 × 65, (b) 256 × 129 and (c) 512 × 257. (Δt = 30 min in each case.) Contour interval = 22.5 in units of 10^{-11} m^{-2} s.

it is necessary to choose a non-zero value of the uncentring parameter ε. It is found that the required value increases with increasing resolution (the values 0.08, 0.10 and 0.15 were used in the low, medium and high resolution cases, respectively).

The PV fields at day 24 for the three spatial resolutions of the (u, v) model are shown in Fig. 9. It can be seen that the PV gradients again become sharper, and more small-scale features occur, as the resolution increases. The simulations are generally similar to those given by the PV-D model, but the gradients of PV are less sharp and some of the fine-scale details of the flow shown by the PV-D model do not appear.

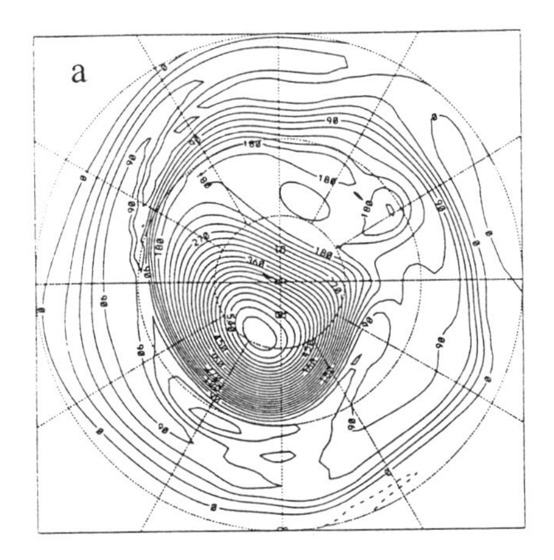

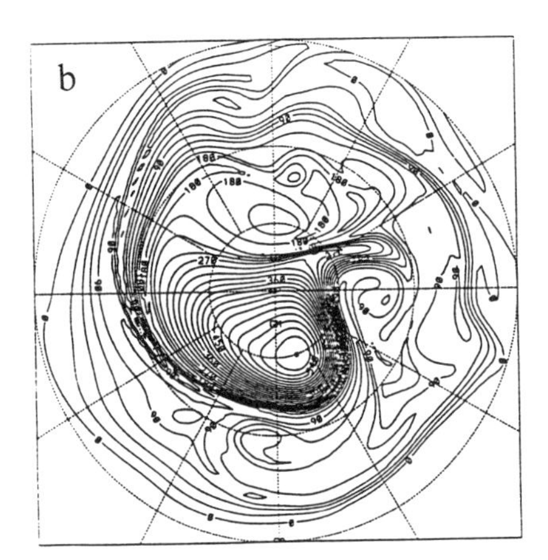

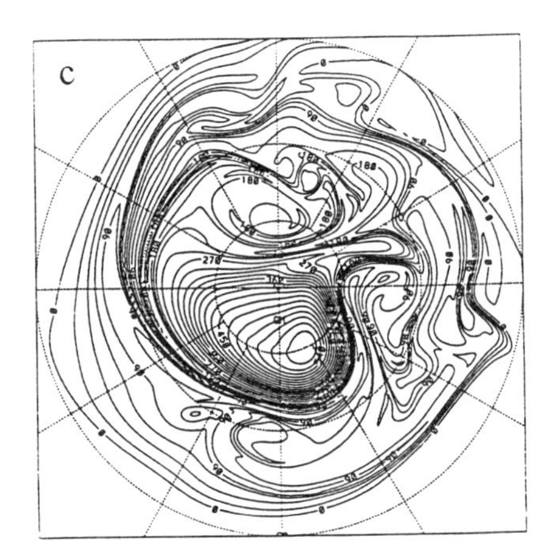

Fig. 9 PV fields in the Northern Hemisphere at day 24 as given by the (u, v) model. (a) Resolution 128×65, $\varepsilon = 0.08$, (b) Resolution 256×129, $\varepsilon = 0.10$, and (c) Resolution 512×257, $\varepsilon = 0.15$. ($\Delta t = 30$ min in all cases.) Contour interval $= 22.5$ in units of 10^{-11} m^{-2} s.

c *Spectral Model Results*

The spectral model, being subject to a CFL condition, must be integrated with shorter time steps than the semi-Lagrangian models. We use $\Delta t = 5$ min for the T42 and T85 resolutions, and $\Delta t = 3$ min for the T170 resolution. The coefficient of the Asselin time filter is set to zero in all cases for which the results are shown.

We first integrate the model with the diffusion coefficient K_4 taken at its recommended values (giving a damping time of 27 h for the shortest resolved scales in each case). The PV fields at day 24 are shown in Fig. 10. It can be seen that the fields are very noisy, though the general features are similar to those given by

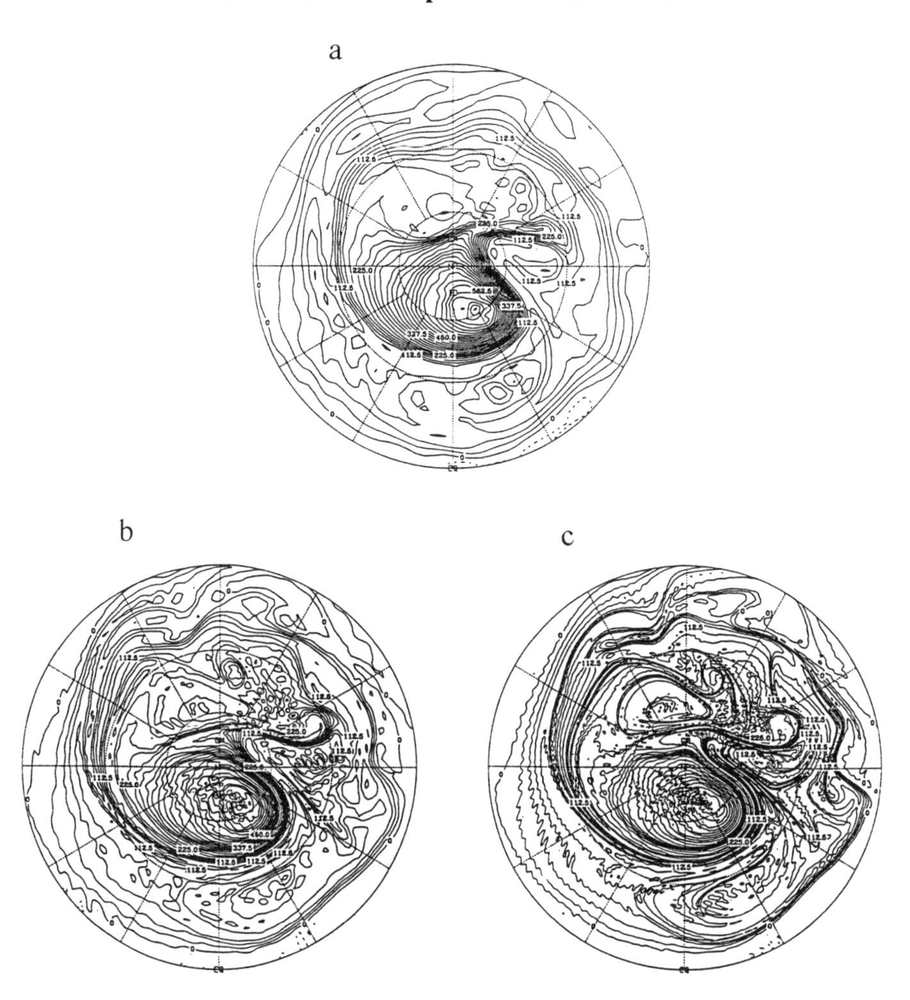

Fig. 10 PV fields in the Northern Hemisphere at day 24 as given by the spectral model. (a) Resolution T42, Δt = 5 min, $K_4 = 5.0 \times 10^{15}$ m^4 s^{-1}, (b) Resolution T85, Δt = 5 min, $K_4 = 3.1 \times 10^{14}$ m^4 s^{-1}, and (c) Resolution T170, Δt = 3 min, $K_4 = 2.0 \times 10^{13}$ m^4 s^{-1}. Contour interval = 22.5 in units of 10^{-11} m^{-2} s.

the PV-D model. (The noise cannot be suppressed by using the Asselin time filter, even with a coefficient as large as 0.1.)

The integrations are repeated with K_4 increased by a factor of 10 above its recommended values. The PV fields at day 24 in this case are shown in Fig. 11. The noisiness seen in Fig. 10 has now disappeared. It can be seen that in the two lower resolution cases the fields are smoothed by comparison with those given

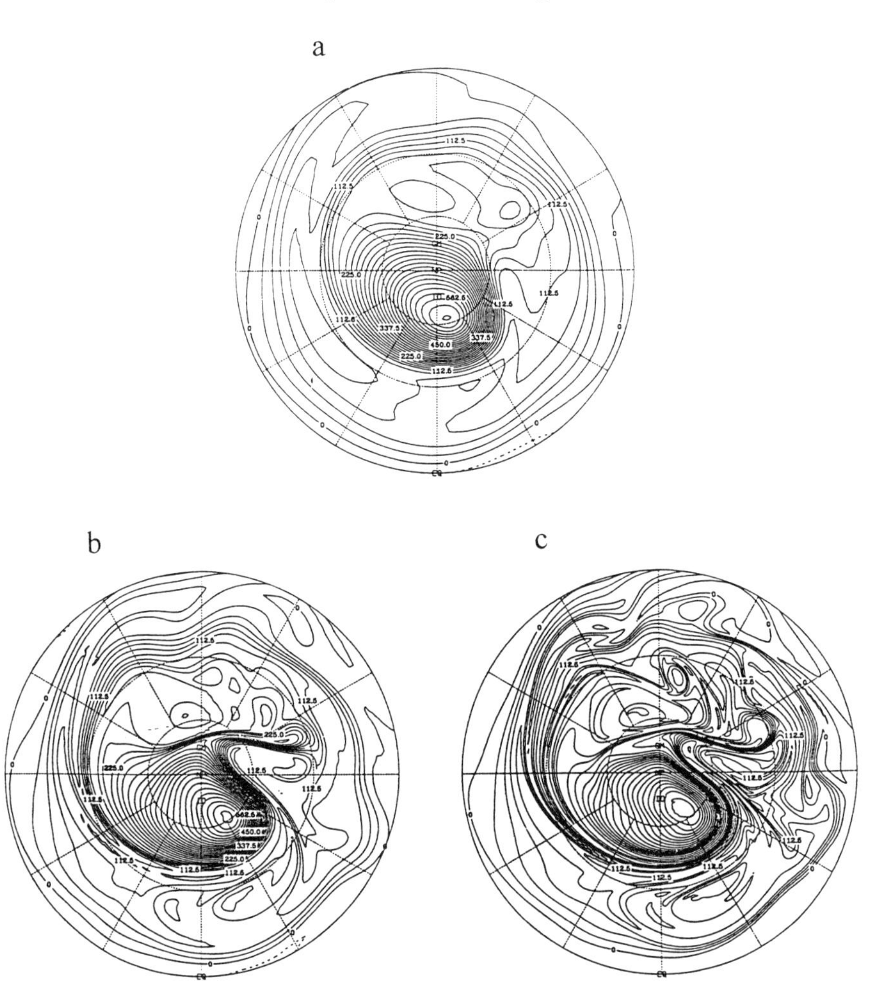

Fig. 11 As in Fig. 10, but with the values of K_4 increased by a factor of 10.

by the PV-D model. In the high resolution case the field is similar to that given by the PV-D model.

The results of increasing K_4 by a factor of 100 above its recommended values are shown in Fig. 12. The PV field has now in all cases lost many of its fine-scale features, though remarkably strong gradients of PV still remain at the edge of the vortex. These strong gradients may be a spurious product of the hyperviscosity: it has been shown by Mariotti et al. (1994), in a study of 2D vortex erosion, that hyperviscosity can induce spurious upgradient pumping of vorticity near a vortex edge.

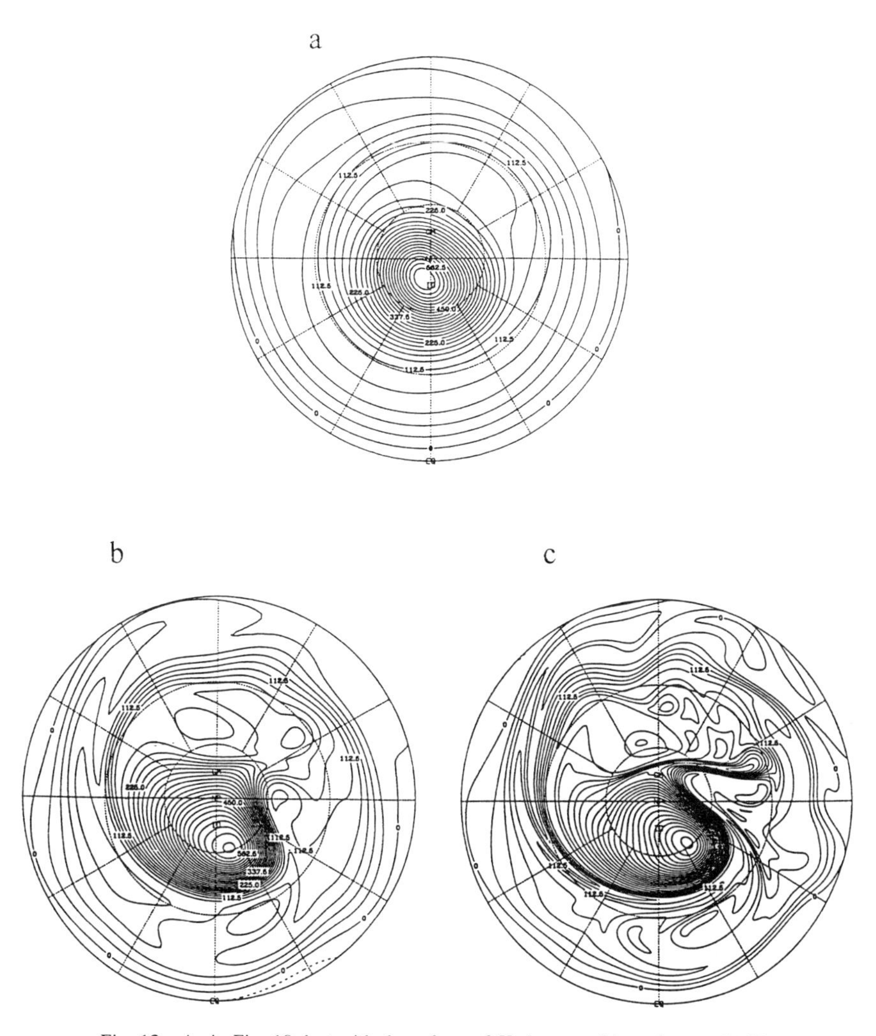

Fig. 12 As in Fig. 10, but with the values of K_4 increased by a factor of 100.

d *Computational Efficiency of the Three Models*

The spectral model is Eulerian and the time step allowed is restricted by the CFL stability criterion, while the other two models are semi-Lagrangian and computational stability is unconditional. For this reason, direct comparison may too obviously favour the two semi-Lagrangian models. It is nevertheless of interest to present some quantitative results. The CPU times needed for the 24-day integrations in a single processor of a CRAY C98 computer are presented in Table 1. These results are obtained with the three different resolutions used for the numer-

TABLE 1. CPU times for the three different models (in seconds).

	PV-D model	(u, v) model	Spectral model
Low Res.	65.7	44.5	412.1
Medium Res.	203.0	175.8	2162.4
High Res.	748.2	709.8	21475.2

TABLE 2. The low, medium and high resolution cases.

	PV-D and (u, v) models	Spectral model
Low Res.	128×65, $\Delta t = 30$ mins	T42, $\Delta t = 5$ mins
Medium Res.	256×129, $\Delta t = 30$ min	T85, $\Delta t = 5$ mins
High Res.	512×257, $\Delta t = 30$ mins	T170, $\Delta t = 3$ mins

ical experiments presented previously in this paper, which, for the convenience of comparison, are listed in Table 2.

4 Conclusions

Three different global shallow water numerical models, the spectral model, the (u, v) model and the PV-D model, are integrated to 24 days to simulate planetary wave breaking and polar vortex erosion in the stratosphere.

The spectral model simulation of the planetary wave breaking is highly dependent on the coefficient chosen for the hyperdiffusion. It was found that a value of the ∇^4 diffusion coefficient corresponding to a 2.7-h damping time for the shortest scales appears to be optimal. Integrating the model with a value ten times less results in a noisy simulation while integrating with a value ten times greater gives over-smoothing. The noise cannot be suppressed by an Asselin time filter even with a damping coefficient as large as 0.1.

The (u, v) model, with a minimum value for the uncentring parameter required to ensure a stable integration, gives a simulation of the potential vorticity field in which some of the detailed features of the wave breaking are missing.

Of the three models, the PV-D model appears to give the best simulation. Fine features of the planetary wave breaking and vortex erosion are revealed in the potential vorticity field. No computational noise appears at any of the three resolutions used despite the absence of any explicit numerical damping.

The PV-D and (u, v) models, which are approximately equivalent in computational cost, are much more efficient than the spectral model. This is particularly marked at high spatial resolution.

Acknowledgments

The first author wishes to acknowledge many stimulating discussions and fruitful correspondence with André Robert over the years.

We are grateful to Drs. Richard Rood and Alan Plumb for useful discussions of the present paper. Our thanks to Mrs. Q. Philpot for typing the manuscript and to Mrs. L. Rumburg for drafting the figures. This research has been supported by the NASA Global Atmospheric Modeling and Analysis Program under Grant 578-41-16-20.

Appendix

The formula for $A(t)$, corresponding to that of Norton (1994), is

$$A(t) = \begin{cases} 0.5[1 - \cos(\pi t/4)] & (0 \leqslant t < 4 \text{ days}) \\ 1.0 & (4 \leqslant t < 16 \text{ days}) \\ 0.5[1 + \cos(\pi(t - 16)/4)] & (16 \leqslant t < 20 \text{ days}) \end{cases}$$

with $A(t + 20) = A(t)$.

The formula for $B(\phi)$ is

$$B(\phi) = \begin{cases} 0.0 & \left(-\frac{\pi}{2} \leqslant t \leqslant 0\right) \\ Ye^{(1-Y)} & \left(0 < t \leqslant \frac{\pi}{2}\right) \end{cases}$$

with

$$Y = \left[\frac{\cot(\phi)}{\cot(\phi_c)}\right]^2$$

where $\phi_c = \frac{\pi}{4}$.

References

ASSELIN, R. 1972. Frequency filter for time integrations. *Mon. Weather Rev.* **100**: 487–490.

BATES, J.R.; Y. LI, A. BRANDT, S.F. MCCORMICK and J. RUGE. 1995. A global shallow-water numerical model based on the semi-Lagrangian advection of potential vorticity. *Q. J. R. Meteorol. Soc.* **121**: 1981–2005.

———; S. MOORTHI and R.W. HIGGINS. 1993. A global multilevel atmospheric model using a vector semi Lagrangian finite-difference scheme. Part 1: Adiabatic formulation. *Mon. Weather Rev.* **121**: 244–263.

———; F.H.M. SEMAZZI, R.W. HIGGINS and S.R.M. BARROS. 1990. Integration of the shallow water equations on the sphere using a vector semi-Lagrangian scheme with a multigrid solver. *Mon. Weather Rev.* **118**: 1615–1627.

BOURKE, W. 1972. An efficient, one-level, primitive equation spectral model. *Mon. Weather Rev.* **100**: 683–689.

ELIASEN, E.; B. MACHENHAUER and E. RASMUSSEN. 1970. On a numerical method for integration of the hydrodynamical equations with a spectral representation of the horizontal fields. Report No. 2, Institute for Theoretical Meteorology, University of Copehagen, 35 pp.

HACK, J.J. and R. JAKOB. 1992. Description of a global shallow water model based on the spectral transform method. NCAR Technical Note, NCAR/TN-343+STR, 39 pp.

JACOB, R.; J.J. HACK and D.L. WILLIAMSON. 1993. Solutions to the shallow water test set using the spectral transform method. NCAR Technical Note, NCAR/TN 388+STR, 82 pp.

JUCKES, M.N. and M.E. MCINTYRE. 1987. A high-resolution one-layer model of breaking planetary waves in the stratosphere. *Nature*, **328**: 590–596.

———. 1989. A shallow water model of the winter stratosphere. *J. Atmos. Sci.* **46**: 2934–2955.

MARIOTTI, A.; B. LEGRAS and D.G. DRITSCHEL. 1994. Vortex stripping and the erosion of coherent structures in two-dimensional flows. *Phys. Fluids*, **6**: 3954–3962.

MCDONALD, A. and J.R. BATES. 1989. Semi-Lagrangian integration of a gridpoint shallow water model on the sphere. *Mon. Weather Rev.* **117**: 130–137.

MCINTYRE, M.E. and T.N. PALMER. 1983. Breaking planetary waves in the stratosphere. *Nature*, **305**: 593–600.

MOORTHI, S. and R.W. HIGGINS. 1993. Application of fast Fourier transforms to the direct solution of a class of two-dimensional separable elliptic equations on the sphere. *Mon. Weather Rev.* **121**: 290–296.

NORTON, W. 1994. Breaking Rossby waves in a model stratosphere diagnosed by a vortex-following coordinate system and a technique for advecting material contours. *J. Atmos. Sci.* **51**: 654–673.

ORSZAG, S.A. 1970. Transform method for the calculation of vector coupled sums: Application to the spectral form of the vorticity equation. *J. Atmos. Sci.* **27**: 890–895.

O'SULLIVAN, D. and M.L. SALBY. 1990. Coupling of the quasi-biennial oscillation and the extratropical circulation in the stratosphere through planetary wave transport. *J. Atmos. Sci.* **47**: 650–673.

POLVANI, L.M. and R.A. PLUMB. 1992. Rossby wave breaking, microbreaking, filamentation and secondary vortex formation: The dynamics of a perturbed vortex. *J. Atmos. Sci.* **49**: 462–476.

———; D.W. WAUGH and R.A. PLUMB. 1995. On the subtropical edge of the stratospheric surf zone. *J. Atmos. Sci.* **52**: 1288–1309.

RUGE, J.; Y. LI, S.F. MCCORMICK, A. BRANDT and J.R. BATES. 1996. A nonlinear multigrid solver for a semi-Lagrangian potential vorticity-based shallow water model on the sphere. *SIAM J. Sci. Comput.*

SALBY, M.L.; D. O'SULLIVAN, R.R. GARCIA and P. CALLAGHAN. 1990a. Air motion accompanying the development of a planetary wave critical layer. *J. Atmos. Sci.* **47**: 1179–1204.

———; R.R. GARCIA, D. O'SULLIVAN and P. CALLAGHAN, 1990b: The interaction of horizontal eddy transport and thermal drive in the stratosphere. *J. Atmos. Sci.*, **47**, 1647–1665.

———; ———, ——— and J. TRIBBIA. 1990. Global transport calculations with an equivalent barotropic system. *J. Atmos. Sci.* **47**: 188–214.

SCHOEBERL, M.R.; L.R. LAIT, P.A. NEWMAN and J.E. ROSENFEIELD. 1992. The structure of the polar vortex. *J. Geophys. Res.* **97**: No. D8, 7859–7882.

WAUGH, D.W. 1993. Contour surgery simulations of a forced polar vortex. *J. Atmos. Sci.* **50**: 714–730.

——— and R.A. PLUMB. 1994. Contour advection with surgery: A technique for investigating finescale structure in tracer transport. *J. Atmos. Sci.* **51**: 530–540.

YODEN, S. and K. ISHIOKA. 1993. A numerical experiment on the breakdown of a polar vortex due to forced Rossby waves. *J. Meteorol. Soc. Jpn.* **71**: 59–72.

A Semi-Lagrangian NWP Model for Real-time and Research Applications: Evaluation in Single- and Multi-Processor Environments

L.M. Leslie
School of Mathematics, The University of NSW
Sydney, NSW, Australia 2052

and

R.J. Purser
UCAR Visiting Scientist,
National Meteorological Center,
Washington, D.C.

[Original manuscript received 9 December 1994; in revised form 29 August 1995]

ABSTRACT *A semi-Lagrangian numerical weather prediction (NWP) model developed for both real-time prediction and for research simulations has been evaluated. The model is second-order in time, high-order (≥3) in space, and employs a non-staggered grid in both the horizontal and vertical.*

A version of the model which is third-order in space was compared with two Eulerian models: the Australian Bureau of Meteorology's current operational regional model which is quasi-second-order in time and space, and also a new version of this model with a third-order upwind scheme that was developed for use as the Australian Bureau of Meteorology's next operational limited-area NWP model.

In a three month trial of twice-daily 48-hour forecasts it was found that both the third-order models were significantly more skillful than the current operational model, as measured by the standard performance statistics such as the S_1 *skill score (Teweles and Wobus, 1954), and root-mean-square (RMS) errors. The specific implications of this greater accuracy were examined in case studies of severe weather events.*

The semi-Lagrangian model also has been adapted to global form and run on a daily basis out to 5 days using archived operational data over a period of almost 6 months.

Finally, the semi-Lagrangian model code was parallelized on a workstation cluster and also on a scalable parallel computer, and it was found that the model was well-suited to parallelization on both computer platforms.

RÉSUMÉ *On a évalué un modèle semi-lagrangien de prévision numérique conçu pour la prévision en temps réel et pour la recherche en simulation. Le modèle est de second ordre dans le temps et d'ordre élevé (≥3) dans l'espace, employant une grille non décalée autant dans l'horizontale que dans la verticale.*

On compare une version du modèle de troisième ordre dans l'espace avec deux modèles eulériens: le modèle régional présentement en exploitation à l'Australian Bureau of Meteorology (ABM), de quasi-second ordre dans le temps et l'espace; et une nouvelle version développée comme le prochain modèle d'exploitation à régions restreintes de l'ABM, ayant un schéma de troisième ordre à pas spacial vers l'amont.

Durant un essai de trois mois de prévisions biquotidiennes de 48 heures, on a trouvé que, selon les normes de performance statistiques comme l'indice d'habileté S_1 (Teweles et Wobus, 1954) et les erreurs quadratiques moyennes, les deux modèles de troisième ordre étaient significativement plus habiles que le modèle présentement en exploitation. Les implications spécifiques de cette plus grande précision ont été examinées dans des études de cas de temps violents.

Le modèle seini-lagrangien a aussi été adapté à une forme mondiale et intégré chaque jour jusqu'à cinq jours, en utilisant des données d'archives opérationnelles couvrant une période de plus de six mois.

Enfin, on a parallélisé le code du modèle semi-lagrangien sur un réseau de stations de travail et sur un ordinateur parallèle à architecture évolutive et trouvé qu'il convenait bien à la parallélisation sur les deux plates-formes d'ordinateur.

---

1 Introduction

The recently formed Atmospheric Modelling Group in the School of Mathematics at The University of New South Wales has decided to develop a numerical weather prediction capability. The modelling effort is intended to provide both a real-time and research forecasting capacity anywhere on the globe, but particularly over southeastern Australia (see location map, Fig. 1a). This project is part of a larger effort by the authors to develop a high-order, mass-conserving, semi-Lagrangian model that can be run over the entire range from a high-resolution short-range limited-area model through the medium and extended range to general circulation mode.

The current operational short-range model employed by the Australian Bureau of Meteorology (BoM) has Eulerian semi-implicit numerics (McGregor et al., 1978; Leslie et al., 1985). The numerical formulation of the model relies heavily on the work of Robert and co-workers (Robert et al., 1972) and had remained largely unchanged for over 15 years until a new model was developed by the first-named author (LML). This new model was based on an explicit Eulerian algorithm, namely an odd-order (3rd or higher) upwind differencing scheme, and will be implemented as the next operational regional model by the Australian BoM (Puri, private communication). It also has been used extensively in research applications (see, for example, Leslie and Skinner, 1994; Speer and Leslie, 1994). At the same time that the new Eulerian upwind model was being tested in real-time mode, a very different research path has been taken in the development of the above-mentioned semi-Lagrangian model, particularly for very high-resolution and very long-range predictions, where the time step size becomes a critical factor. In a series of studies, the present authors have developed a sequence of increasingly accurate and efficient semi-Lagrangian models. Currently, the most advanced product of this work is a

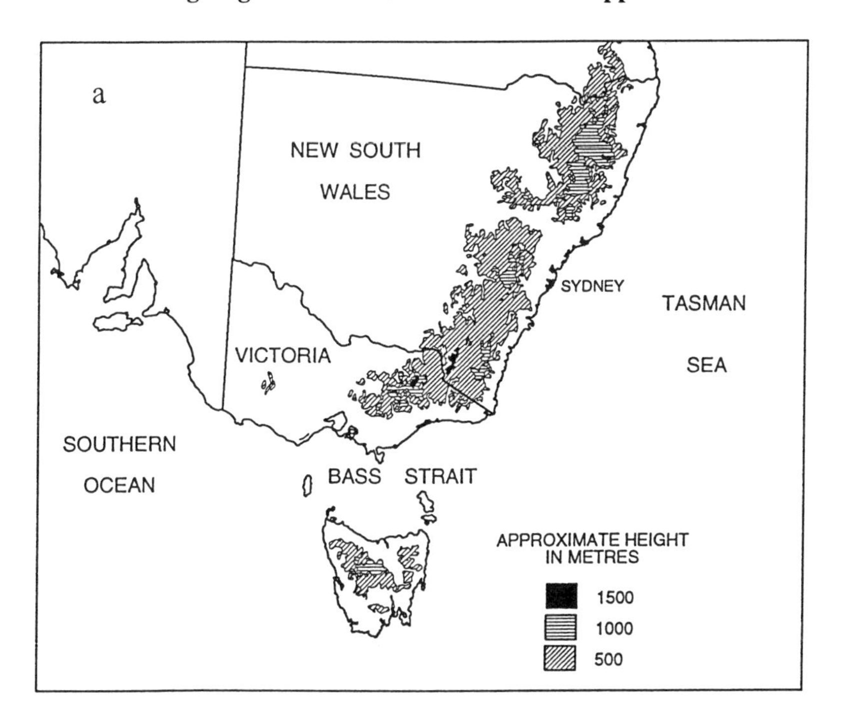

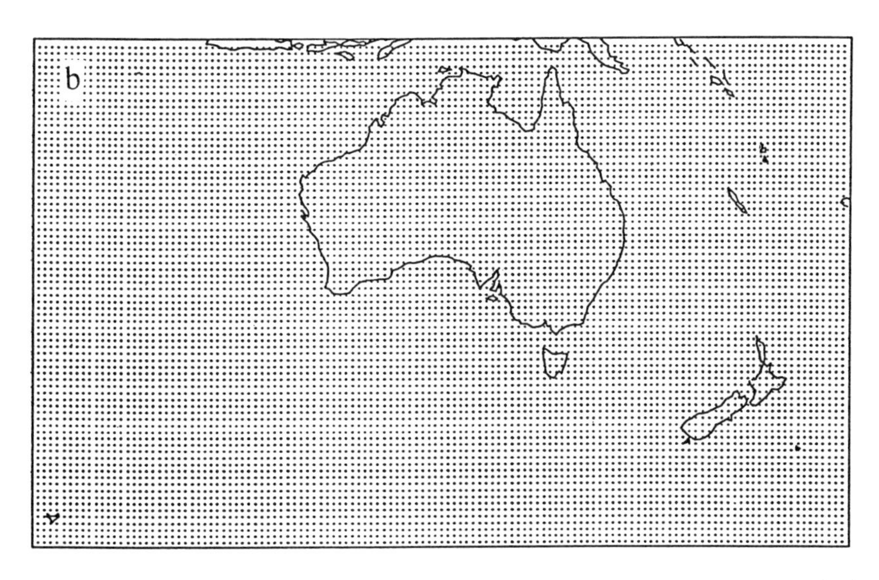

Fig. 1 a. Location map of southeastern Australia, showing area of interest for the case studies in Section 4. b. The Australian region analysis/forecast domain showing the disposition of grid points in the horizontal at 150 km resolution.

high-order (in space and time), mass-conserving, semi-implicit, semi-Lagrangian model (Leslie and Purser, 1994) that has as many equations as possible cast in conserving form. It also employs forward trajectories (Purser and Leslie, 1994) instead of the standard backward trajectories, and is cast on a grid, non-staggered in both the horizontal and the vertical. It currently has been converted to global form on a Gaussian grid and this version will be reported on in a future paper when it is running successfully as a general circulation model (GCM). However, the version that is currently used most heavily in research applications is essentially that described by Leslie and Purser (1991), which is a bi-cubic, semi-Lagrangian model using an iterative backward trajectory scheme on a non-staggered grid. It is the model which is used in this study. It has been run almost exclusively in limited-area form, but a global version (running out to 5 days) has been tested successfully on operations data over a period of almost six months.

There are three main aims in this study. Section 2 provides a brief description of each of the three models, namely, the current semi-implicit operational model, the new explicit (upwind) operational model, and the semi-Lagrangian model. In particular, some remarks are made concerning the reasons for developing the particular formation of each of the models. The first aim is to compare the three models. The comparison is presented in Section 3 where the performance of the models over the three month period December 1992 to February 1993 is summarized, using a range of standard measures of skill. A second component of Section 3 is to compare directly the accuracy of the upwind and semi-Lagrangian schemes, which are both formally second-order accurate in time and third-order accurate in space. From Section 4 onwards, attention is focused on the semi-Lagrangian model alone and the second aim of this study is to demonstrate how the accuracy of the semi-Lagrangian model is realized effectively in particular case studies. Two severe weather events of very different kinds are chosen for this purpose in Section 4. The third aim is to assess how well the semi-Lagrangian model can be adapted to a multi-processor computing environment. The adaptability of a model to emerging computer architectures is an extremely important aspect of model performance. This is achieved in Section 5 by examining the performance of the model on two quite different computing platforms, a cluster of RISC/6000 workstations and an IBM SP2 parallel computer. Finally, Section 6 discusses the results and makes some concluding comments.

2 The models

The three models are described briefly and in turn, below.

a *The current operational model*

The current operational limited-area model used by the Australian BoM is known as RASP (Regional Assimilation and Prognosis). It is described fully in the sequence of articles by McGregor et al. (1978), Leslie et al. (1985) and Mills and Seaman (1990). The model is cast in flux form, has centred (leapfrog) semi-implicit temporal differencing, and second-order energy conserving spatial differencing on an Arakawa C-grid in the horizontal. It is important to point out that the centred temporal formulation and the use of the Asselin filter to control the computational mode degrade the temporal accuracy to less than second-order. The vertical disposition to variables is on the Lorenz grid. Apart from some modifications to the representation of physical processes, the model has remained largely unchanged since the formulation of Leslie et al. (1985). The model has been run with a horizontal resolution of 10 km until May 1994 when it was reduced to 75 km. There are 17 levels in the vertical. All of the comparisons in Section 3 are at 150 km resolution.

b *The new Eulerian (upwind) model*

The new Eulerian model was selected on the basis of simplicity, accuracy and computational economy, particularly the anticipated (and subsequently realized) efficiency on emerging parallel computing platforms. The temporal differencing is second-order, two-time-level, and is optionally the Heun scheme (Mesinger and Arakawa, 1976) or the Miller-Pearce scheme (Miller and Pearce, 1974). The stability of these schemes has been analyzed extensively elsewhere and there is no need to repeat the process here. However, it should be pointed out that both schemes are weakly unstable and have slightly positive phase errors. However, the stability is very weak and the phase leads are quite small. The spatial differencing employed is a third-order upwind scheme, although it can be run at higher orders as an option. The merits of using odd-order upwind schemes have been argued for by a number of advocates (see, for example, Leonard, 1984). In essence the attractive properties of the odd-order upwind schemes are: excellent phase properties at all orders; very good amplitude properties at order 3 and higher; and a lack of oscillatory behaviour in the solution near sharp boundaries, thereby minimizing the possibility of destructive feedback from the oscillatory behaviour near sharp boundaries present in the even-order upwind schemes.

This version of the model is being implemented by the Australian BoM as its next operational regional NWP model, most likely sometime in early or mid-1995 (Puri, Australian BoM, personal communication).

c *The semi-Lagrangian model*

The formulation of the semi-Lagrangian model used in this study is also fully-documented elsewhere (Leslie and Purser, 1991). It is a two-time-level split scheme. The procedure comprises a semi-Lagrangian advection step followed by a number (usually 4 or 5) of adjustment steps. The adjustment steps use the forward-backward technique (Mesinger, 1977). The temporal differencing therefore is formally second-order. As in the upwind model a non-staggered grid is used. The interpolation scheme is third-order accurate, although it can be used optionally at any order up to eight. The model has been tested extensively using both archived

TABLE 1. S_1 skill scores for each of the three models for the period December 1992 to February 1993.

Model	December	January	February
Operational	35	39	40
Upwind	31	35	35
S-Lagrangian	31	36	34

operational data and in research mode of the simulation of a range of significant weather events.

This model also was designed to take advantage of existing vector supercomputers, but also in anticipation of merging parallel systems as parallel computers are finding widespread usage, and that trend is expected to increase from the mid-1990's onwards.

3 Results

The three models were run on a total of approximately 180 cases, at 0000 UTC and 1200 UTC from 1 December 1992 to 28 February 1993. The forecasts were all verified at 24 hours and were run at the then operational resolution of 150 km, on the grid shown in Fig. 1b. The forecast experiments run using the archived operational initial fields and nested in the archived operational lateral boundary conditions (forecasts) from the BoM's global NWP model. In that sense the comparison was a very clean one. The time steps used for the three models were the same as for the physics step (16 minutes), but were 16 minutes, 4 minutes and 32 minutes for the operational, upwind and semi-Lagrangian models respectively. The model statistics prepared were the S_1 skill scores, and the RMS values of the temperature and wind fields.

a *S_l skill scores*

The S_1 skill scores are shown in Table 1, and it is immediately clear that the third-order models have very similar errors and are far more accurate than the operational model. In each of the three months the scores for the third-order models are about 4 points lower than for the operational model. It is our experience that this difference is large enough to be obvious to the eye on a daily inspection of forecasts charts. An example will be presented below to illustrate what a four point skill score difference can look like. The other significant feature of Table 1 is that the skill scores for the third-order models are almost identical, for each of the three months.

Figures 2a and 2b are plots of the S_1 skill scores on a daily basis for December 1992. Figure 2a is a comparison of the semi-Lagrangian model and the RASP model, while Fig. 2b compares the semi-Lagrangian and upwind models. The semi-Lagrangian model is markedly more accurate than RASP on almost every day but, in contrast, is very close to the upwind model on all but a few days. The means

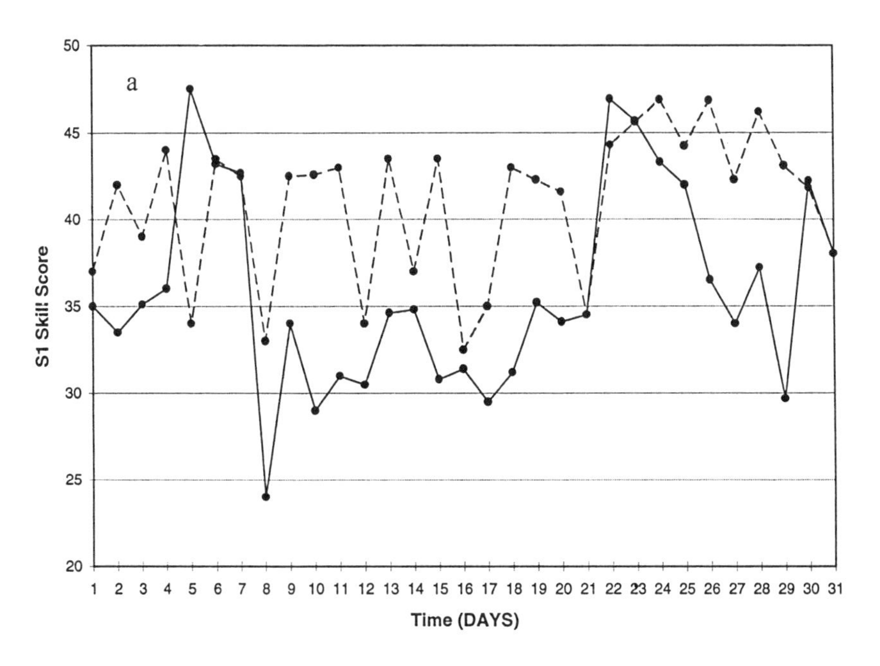

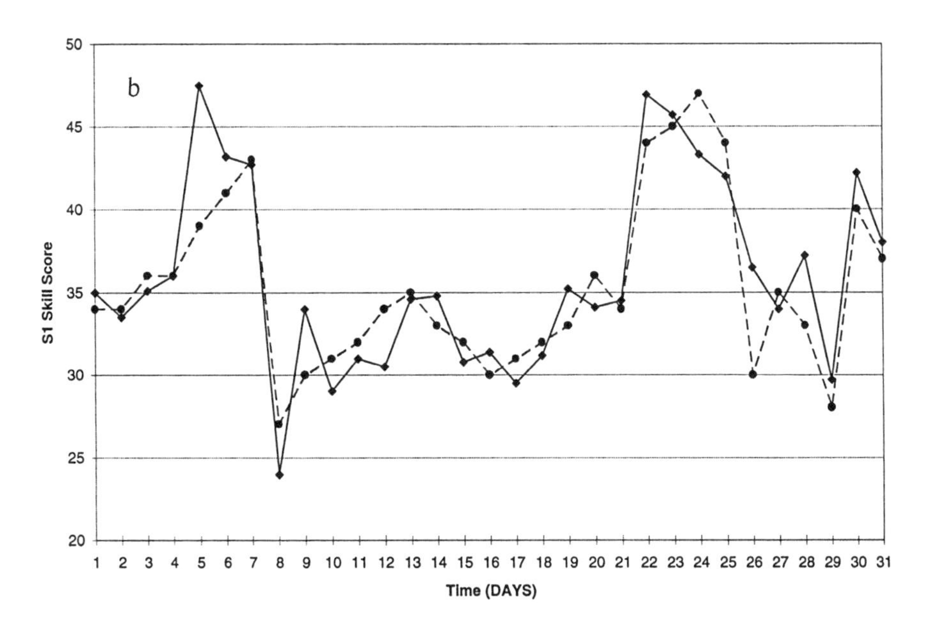

Fig. 2 a. A comparison of the daily S_1 skill scores at 0000 UTC for the operational model (dashed line) and the semi-Lagrangian model (heavy line) for December 1992. b. As in 2a except for the upwind (dashed line) and semi-Lagrangian model (heavy line).

TABLE 2. Mean RMS errors in 24-hour wind forecasts (m/s) at 850, 500 and 200 hPa levels averaged over the period December 1992 to February 1993.

Model	850 wind	500 wind	200 wind
Operational	5.7	6.5	9.1
Upwind	4.3	5.2	7.2
S-Lagrangian	4.1	5.3	7.0

TABLE 3. RMS errors (Degrees Celsius) in the 24-hour forecast temperatures at 850, 500 and 200 hPa levels for the three models, averaged over the period Dec. 1992 to Feb. 1993.

Model	850 Temp.	500 Temp.	200 Temp.
Operational	2.2	1.8	2.1
Upwind	1.8	1.5	1.9
S-Lagrangian	1.8	1.4	1.9

for the month are 35 for the operational model and 31 for each of the upwind and semi-Lagrangian models.

b *RMS wind and height errors*

As was the case for the skill scores, the RMS wind and height errors are shown in Tables 2 and 3 as a function of the month. They are also shown as functions of several selected pressure levels, the lower, middle and upper troposphere (850, 500 and 200 hPa). The reductions in model error, measured relative to radiosonde values, are once again quite large for the third-order models over the operational model. Again they are very similar in magnitude for the upwind and semi-Lagrangian models.

Turning now to the RMS temperature errors, the same pattern appears in Table 3 as for Table 2. The mean temperature errors are markedly more accurate in the 24-hour forecasts from the third-order models compared with the second-order operational model, and are very similar between the two third-order models. Again, the errors are measured relative to radiosonde values.

c *Example*

It is instructive to look at a case in detail. The particular event chosen involves a tropical cyclone which was located off the east coast of Australia at 1200 UTC 13 January 1993, and was in the process of becoming extra-tropical. During the following 24 hours the numerical analysis shows the cyclone moving SSE and the central pressure filling in from 990 hPa to 994 hPa (see Figs 3a and 3b). The 24-hour forecasts from the operational, upwind and semi-Lagrangian models are shown in Figs 3c to 3e respectively. The semi-Lagrangian and upwind models produced

a

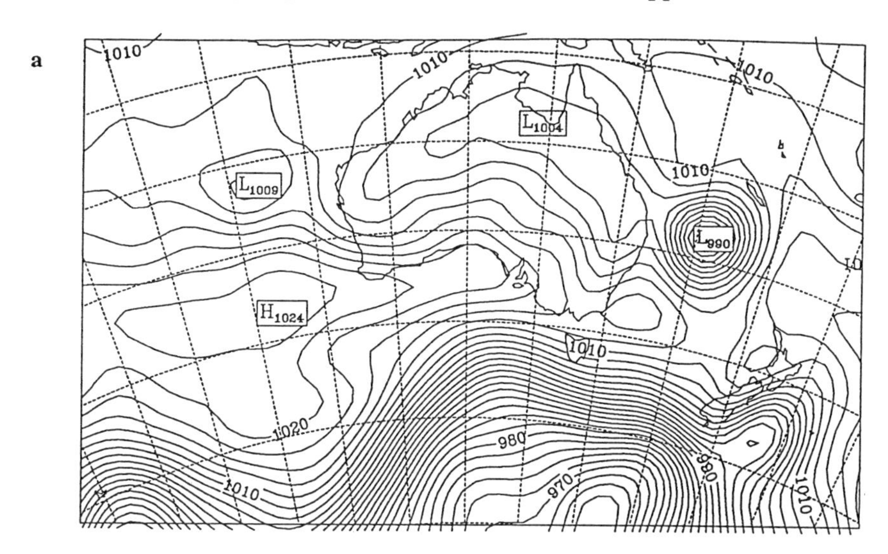

b

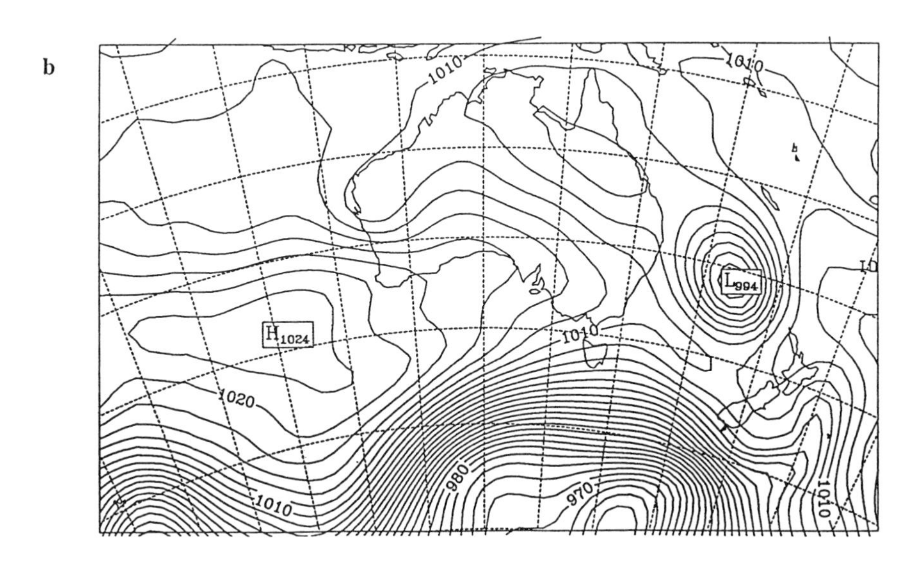

Fig. 3 a. The initial SLP analysis at 1200 UTC 13 January 1993. b. The verifying SLP analysis at 1200 UTC 14 January 1994. c. The 24-hour forecast from the operational model, valid at 1200 UTC 14 January 1994. d. As in c. except for the upwind model. e. As in c. except for the semi-Lagrangian model.

c

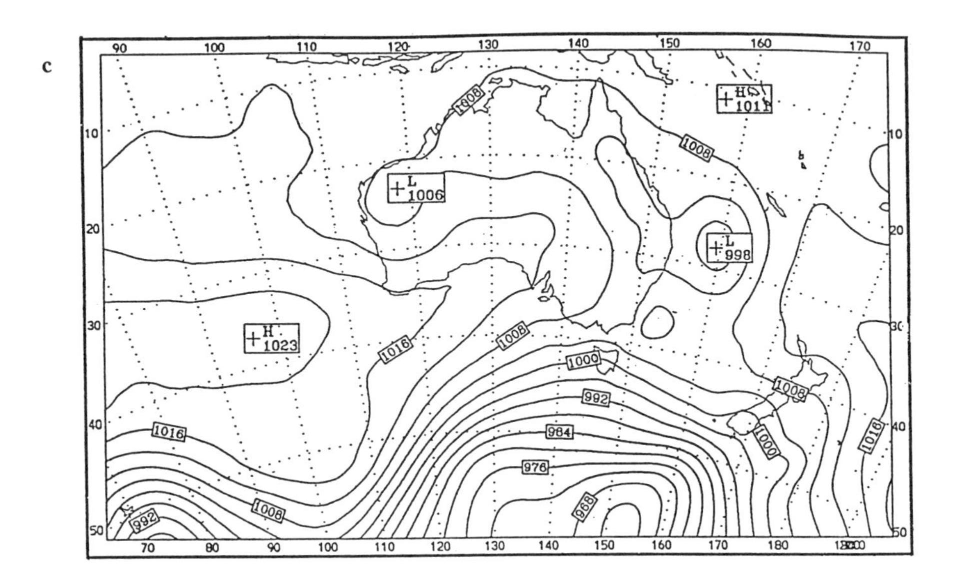

d

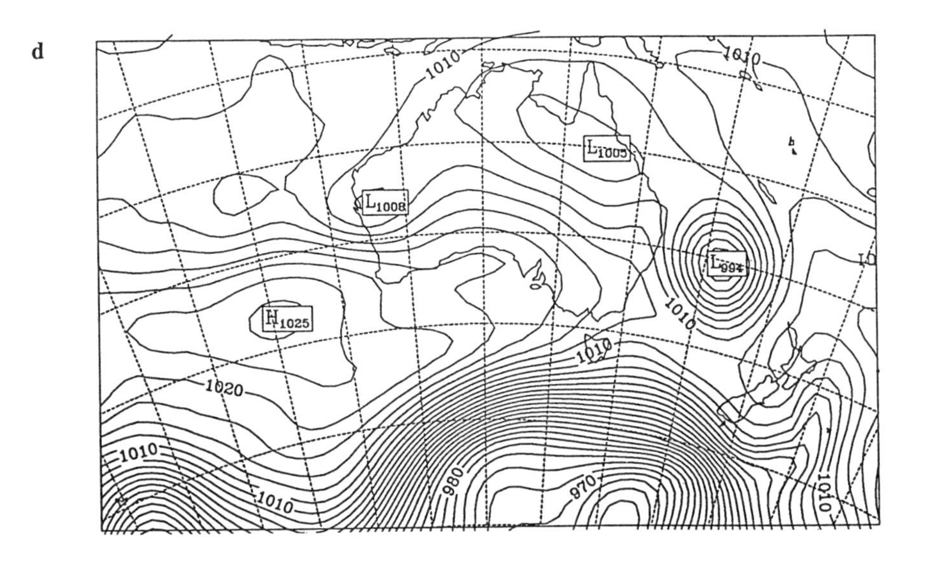

Fig. 3 *Continued.*

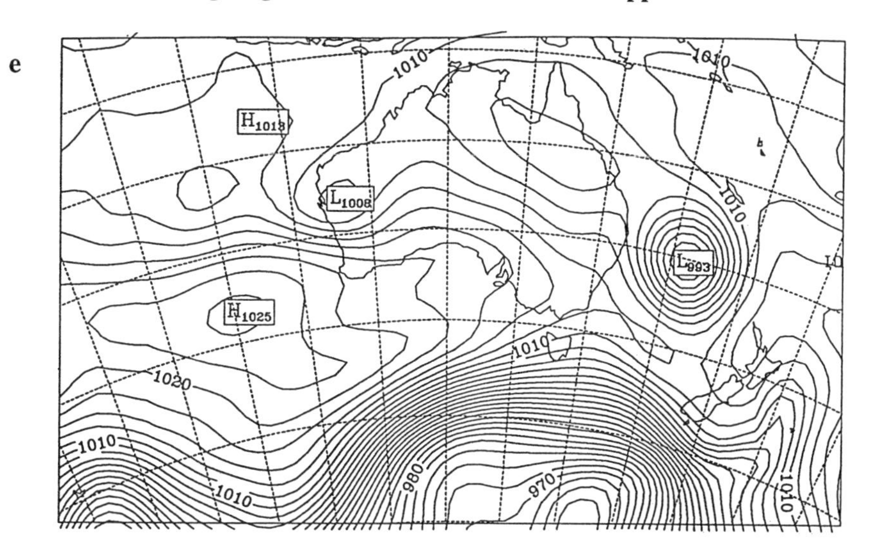

Fig. 3 *Concluded.*

forecasts that are almost identical, apart from a one hPa difference in the central pressure of the cyclone. Both models have some minor deficiencies, such as not moving the cyclone far enough to the south. However, the forecast obtained from RASP is markedly inferior, even allowing for the different contouring interval. The central pressure and position of the cyclone are quite poor, as is the prediction of the anti-cyclone to the south-west of the continent. The skill score of 38 is 5 points worse than that of the other two models, both of which had S_1 scores of 33.

d *Global integration*

The semi-Lagrangian model has been converted to a global form, in anticipation of its use for both medium and extended range forecasting. At this stage only the mass-conserving version of the model has not been adapted to the globe. In its non mass-conserving form, the global version of the semi-Lagrangian model has been run out to 5 days on a regular basis since March 1994. No concerted effort has been made to verify the model other than to record the skill scores and to monitor mean global quantities. In the first half of 1995 it is planned to begin trials of the mass-conserving form of the model in real-time at the National Meteorological Center (NMC), Washington in a comparison with existing NMC models. One example of the global model is given in Figs 4a and 4b which show, respectively, the 5-day forecast and the verifying analysis valid at 0000 UTC 18 September 1994.

4 Case studies

The coastal region of southeastern Australia (see location map, Fig. 1a) has a diverse range of synoptic and mesoscale events, and many of these have been well-

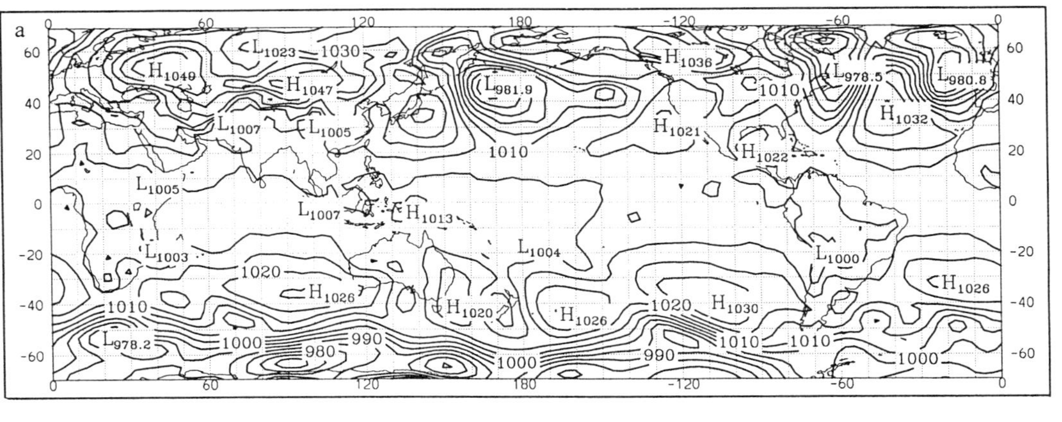

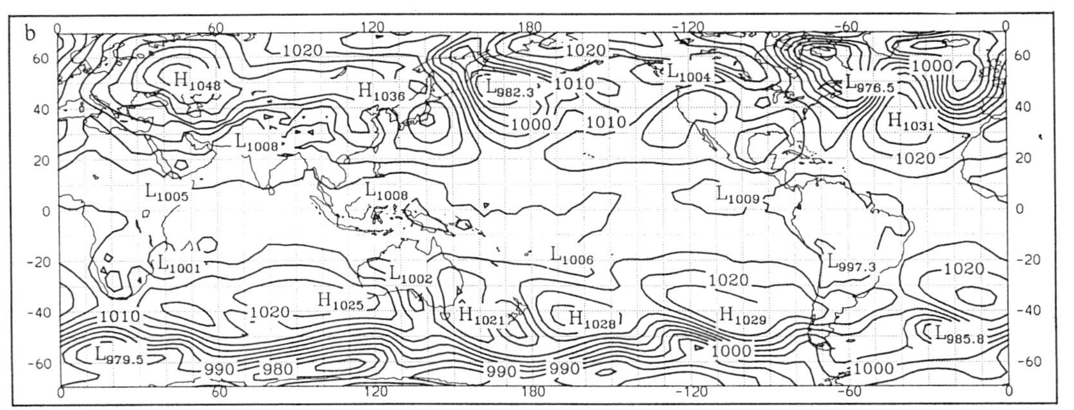

Fig. 4 a. The global analysis valid at 0000 UTC 18 September 1994. b. The 5-day semi-Lagrangian global forecast valid at 0000 UTC 18 September 1994.

documented. They include the so-called "east-coast lows" (Holland et al., 1987; Leslie et al., 1987; McInnes et al., 1993) which are synoptic scale events that in their most severe form are responsible for the major flooding events of this part of the country. They also include tropical cyclones that have become extra-tropical.

Flooding also occurs from mesoscale events and it is such an event that is the subject of the first case study, namely, a line of thunderstorms that crossed the greater Sydney metropolitan area during the early afternoon of 10 February 1990 and brought with it flash flooding. This event has been studied in considerable detail by Speer and Leslie (1994) as part of a more general investigation of mesoscale weather events over southeastern New South Wales (NSW), and the Sydney area in particular.

Another very different form of severe weather but one that is equally destructive is the strong, dry, west-to-northwest winds that produce wild-fires, or bush-fires as they are known in Australia. The most severe fire weather days are those that occur at the end of a long period of dry weather, which in turn has been preceded by a wet period. January 1994 saw the most destructive fires in the history of Sydney, with the days of January 7th and 8th the most serious. Timely indications of wind speed and direction are critical for fire authorities when these conditions are present. Forecasts for the period 1200 UTC 7 January to 1800 UTC 8 January 1994 form the basis of the second case study. It is intended also to extend investigation of this event to a major diagnostic and simulation case study.

a *The thunderstorm line of 10 February 1990*

This was a mesoscale event in which a quasi-stationary line of thunderstorms formed to the west of Sydney in a meso-low into which a strong northwards moving coastal ridge penetrated from the south. The following ingredients for convection therefore were present: high low level relative humidity, instability, and a triggering mechanism in the form of colder, drier air in the ridge. A line of convection formed almost parallel to the coastal ranges and moved very slowly eastwards across the greater Sydney area. Radar imagery (not shown) clearly indicated the line of thunderstorms as well as remnants of earlier less organized convection, at 0350 UTC. Over the Sydney area the one-hour rainfall totals for the period 1350 to 1450 local time (0450 to 0550 UTC) are presented in Fig. 5. There are two distinct maxima of over 60 mm.

For this event, the semi-Lagrangian model was run at a resolution of 15 km and with 25 levels in the vertical, and a time step reduced linearly to 16 minutes and 8 minutes respectively. From stability considerations, the reduction was not necessary, but a detailed test of efficiency was not the primary consideration here. It was decided not to have a large disparity with the physics time step size. Initial and boundary conditions were provided by telescoping down from the operational analysis resolution of 150 km to 50 km horizontal resolution and finally to 15 km. The telescoping was performed in the usual manner, namely, interpolation of fields to the higher resolutions. No additional data were included in these runs. The model

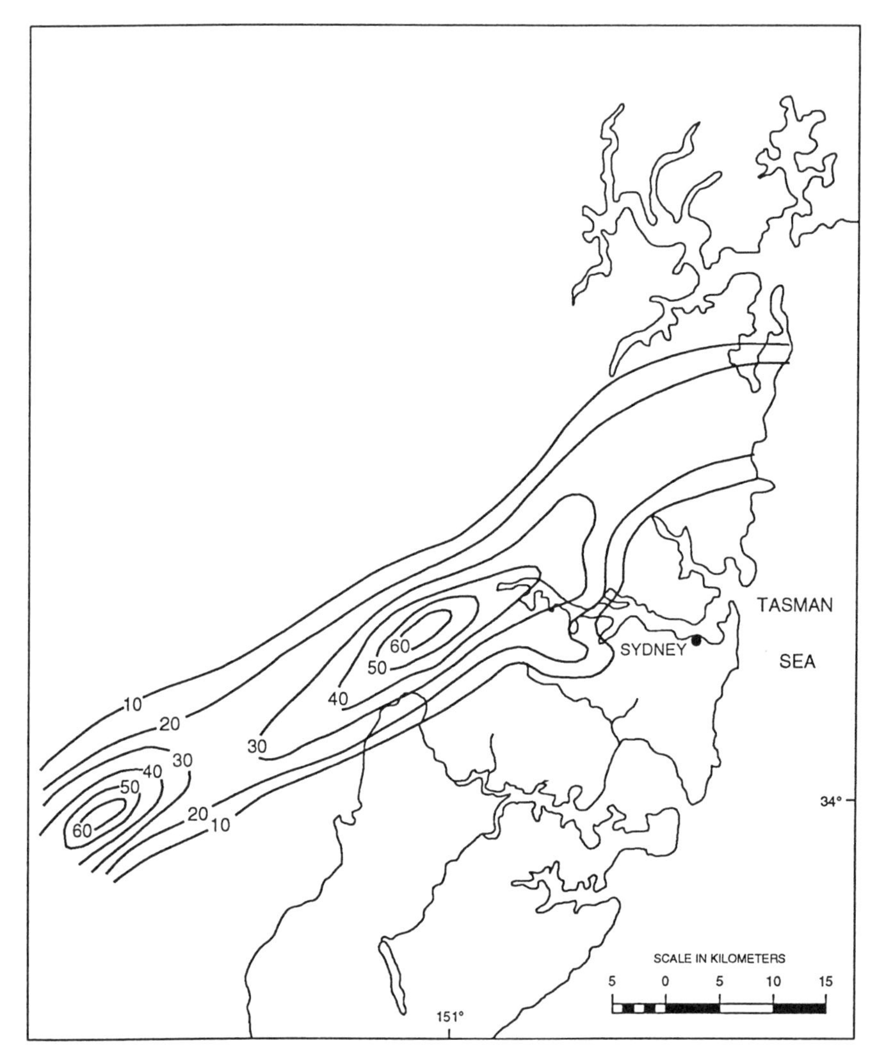

Fig. 5 The observed one-hour rainfall totals (mm) for the period 1350 to 1450 local time (approximately 0450 to 0550 UTC) on 10 February 1990. Note the double maximum.

was run for a period of 6 hours from the initial time of 0000 UTC 10 February 1990 as this period of time covered the entire life-cycle from clear skies over Sydney to the development and passage of the line of thunderstorms.

The initial fields are shown in Figs 6a to 6d. Figure 6a is the initial SLP (sea-level pressure), with little sign of the cooler air associated with the coastal ridging. The cooler air shows up in the SLP fields in that the air behind the ridge is from a cooler southern ocean air mass. However, Fig. 6b reveals a large amount of low level moisture with a large strip of the coastal region having near-surface rela-

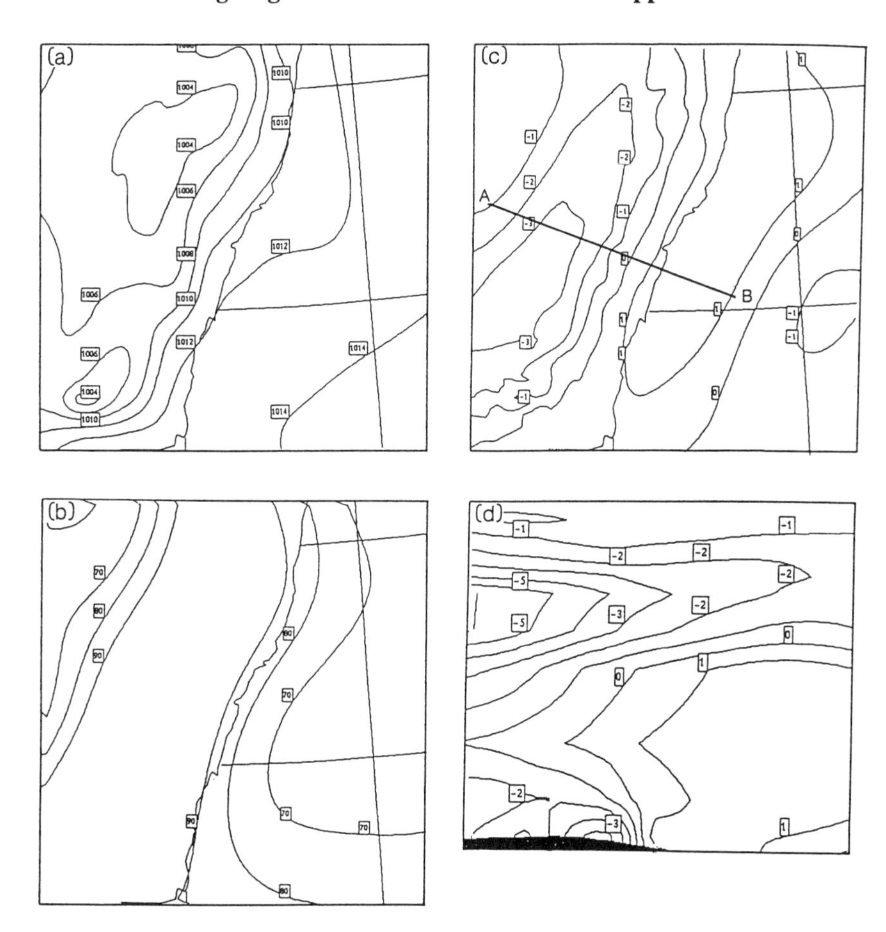

Fig. 6 The initial fields from the numerical analysis at 0000 UTC 10 February 1994. Figures a to d are, respectively: the SLP (in hPa); the relative humidity at 950 hPa (percent); the plan view of relative vorticity at 850 hPa (10^{-5}); and a cross-section of relative vorticity through the line A–B.

tive humidities above 90 per cent. Figures 6c and 6d indicate the plan view at 850 hPa and a cross-section through the line A–B, respectively, of the relative vorticity field. Note that there is an area of low-level cyclonic vorticity present. At 6 hours into the simulation the cooler, drier air from the ridging had penetrated the region and provided the lifting mechanism. The squall-line has developed dramatically in these first 6 hours. Figures 7a to 7d are valid at 0600 UTC and reveal that the line of thunderstorms has been well simulated. Figure 7a is the precipitation rate (mm/hr) with a maximum of over 70 mm/hr. Note that there is a clear double maximum, as was the case in the observed rainfall. Figures 7b and 7c are again the plan view at 850 hPa and the vertical cross-section, respectively, of the relative vorticity field, this time the cross-section being through the line C–D. Finally, Fig.

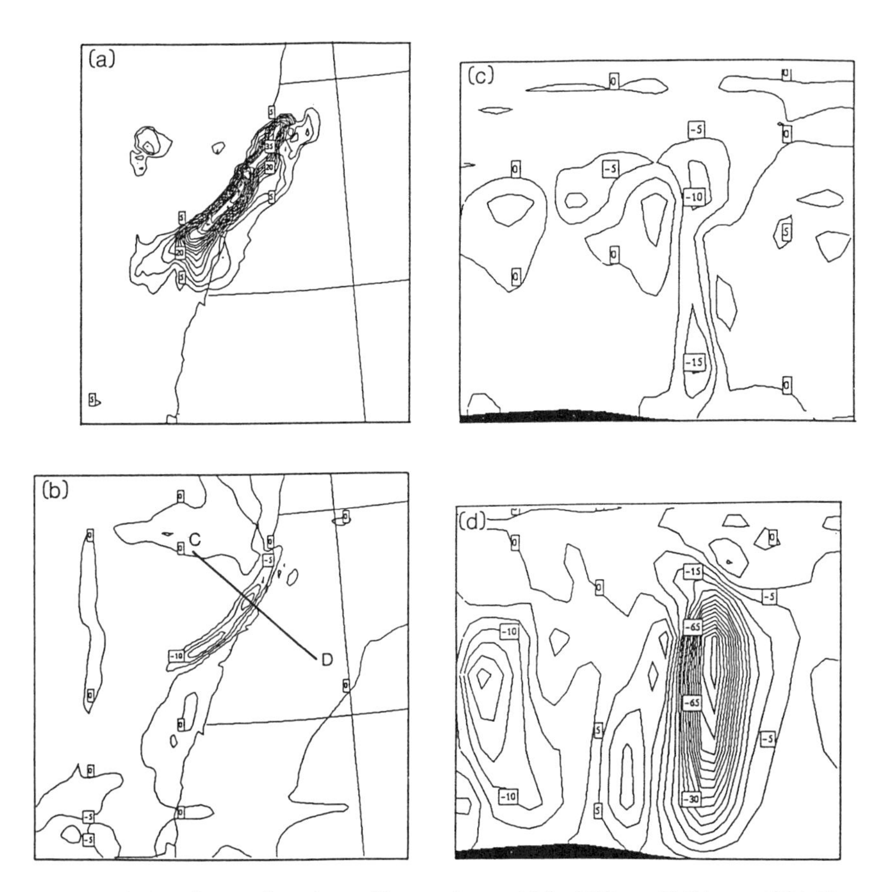

Fig. 7 The six-hour forecast from the semi-Lagrangian model for 0600 UTC 10 February 1994. Figures a to d are, respectively: the instantaneous rainfall rate (mm/hr); the plan view of relative vorticity at 850 hPa; the cross-section of relative vorticity along the line C–D; and the cross-section of vertical velocity (10^{-1} m/s) along the line C–D.

7d is the vertical cross-section through the line C–D of the vertical velocity, with a classical updraft/downdraft structure.

b *The fire weather event of 7 to 8 January 1994*

Between 27 December 1993 and 14 January 1994, eastern NSW experienced its worst bushfire period for many decades. For the greater Sydney area, with the overwhelming majority of the state's population, the worst period was 7 and 8 January. The fires presented the Australian BoM with an urgent need for timely forecasts of wind speed and direction. In particular it emphasized the important role that could be played by an operational high-resolution limited area model. The most severe individual day of the fires in Sydney was 8 January 1994. For that reason, simulations are presented for the 30-hour period starting at 1200 UTC

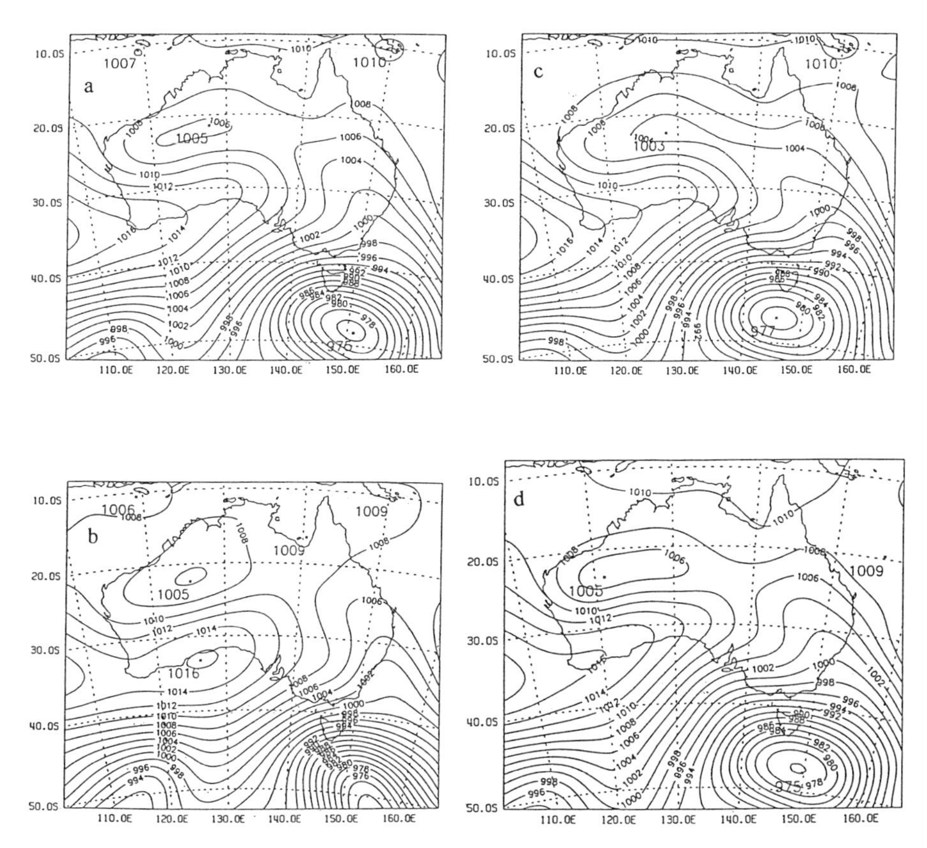

Fig. 8 Figures a and b are the 24-hour analyses from the Australian BoM operational limited area model valid at 9 am and 9 pm local time on 8 January 1994. Figures c and d are the operational model forecasts valid at the same times.

7 January to 1800 UTC 8 January 1994. Figures 8a and 8b show the SLP analyses at 0000 UTC 8 January and 1200 UTC 8 January. The 24-hour forecasts from the BoM's operational model for the corresponding times are given in Figs 8c and 8d. The operational forecasts were performed at the then operational resolution of 150 km and provide very little detail over the critical area of the NSW coast. For example, the 24-hour operational forecast for 0000 UTC (9 am local time) indicates a broad westerly airstream over NSW compared with the analysis which has lighter and more variable winds. The forecast for 1200 UTC (9 pm local time) is better but provides almost none of the detail that is critical for determining bushfire movement.

In contrast, the semi-Lagrangian model run at the much higher resolution of 50 km provides important details not captured at 150 km. Figures 9a to 9d and 10a to 10d show the sequence of six-hourly forecasts of SLP and wind vectors (using the same datasets as the operational model) for the period 1800 UTC 7 January to 1200

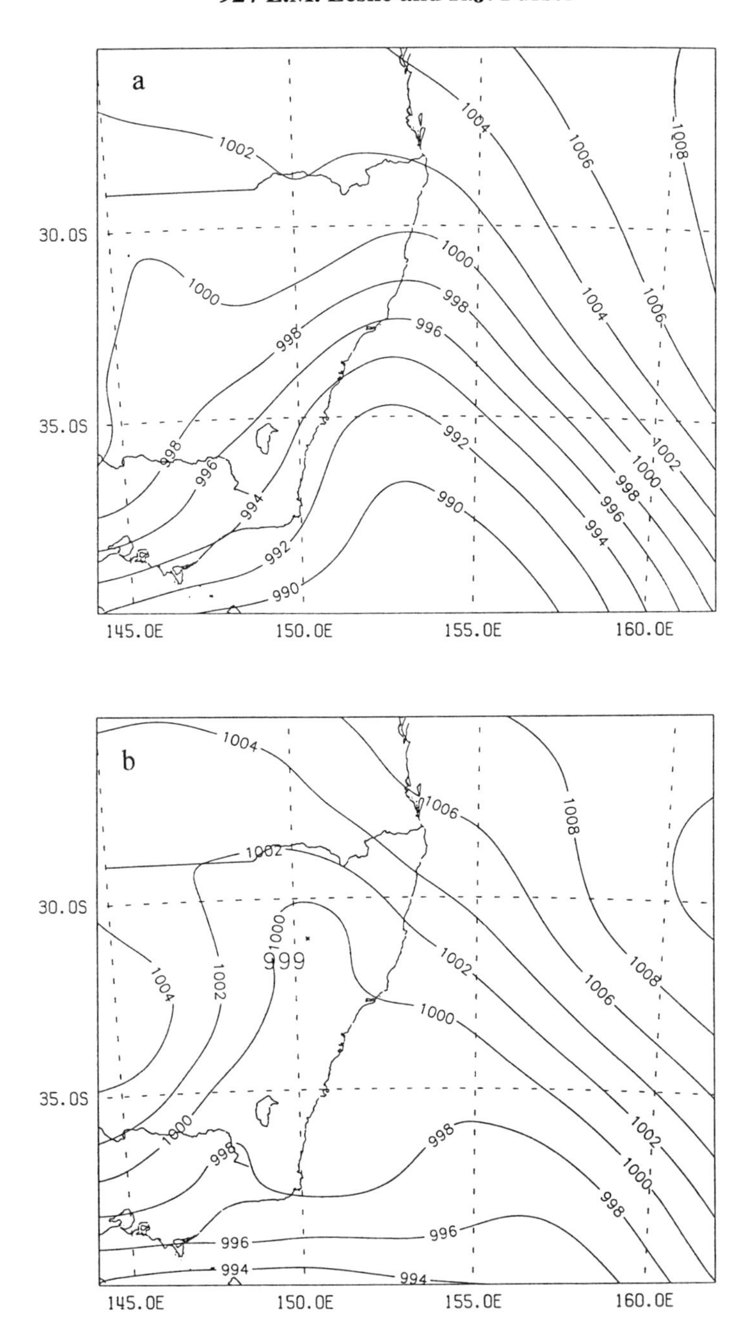

Fig. 9 A sequence of four SLP forecasts at 6-hourly intervals from 3 am local time to 9 pm local time on 8 January 1994 from the semi-Lagrangian model at 50 km resolution, and 25 levels.

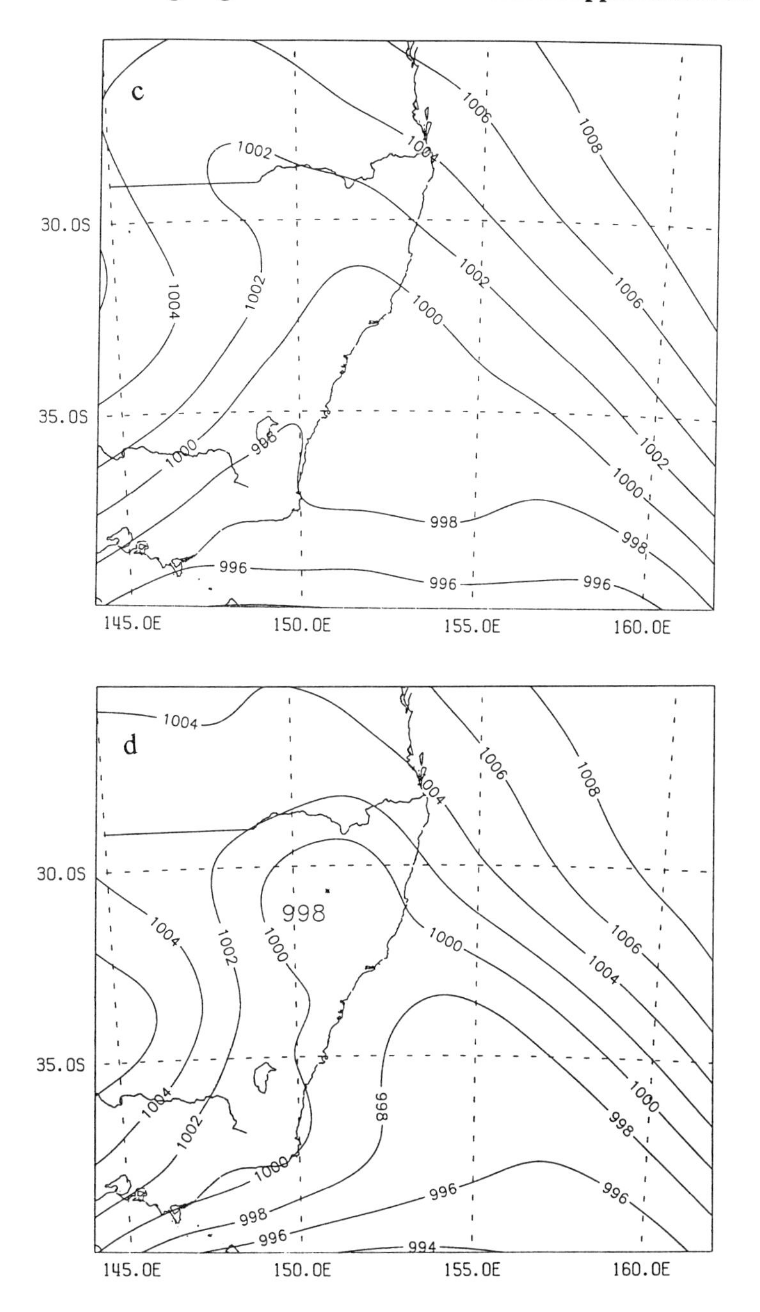

Fig. 9 *Concluded.*

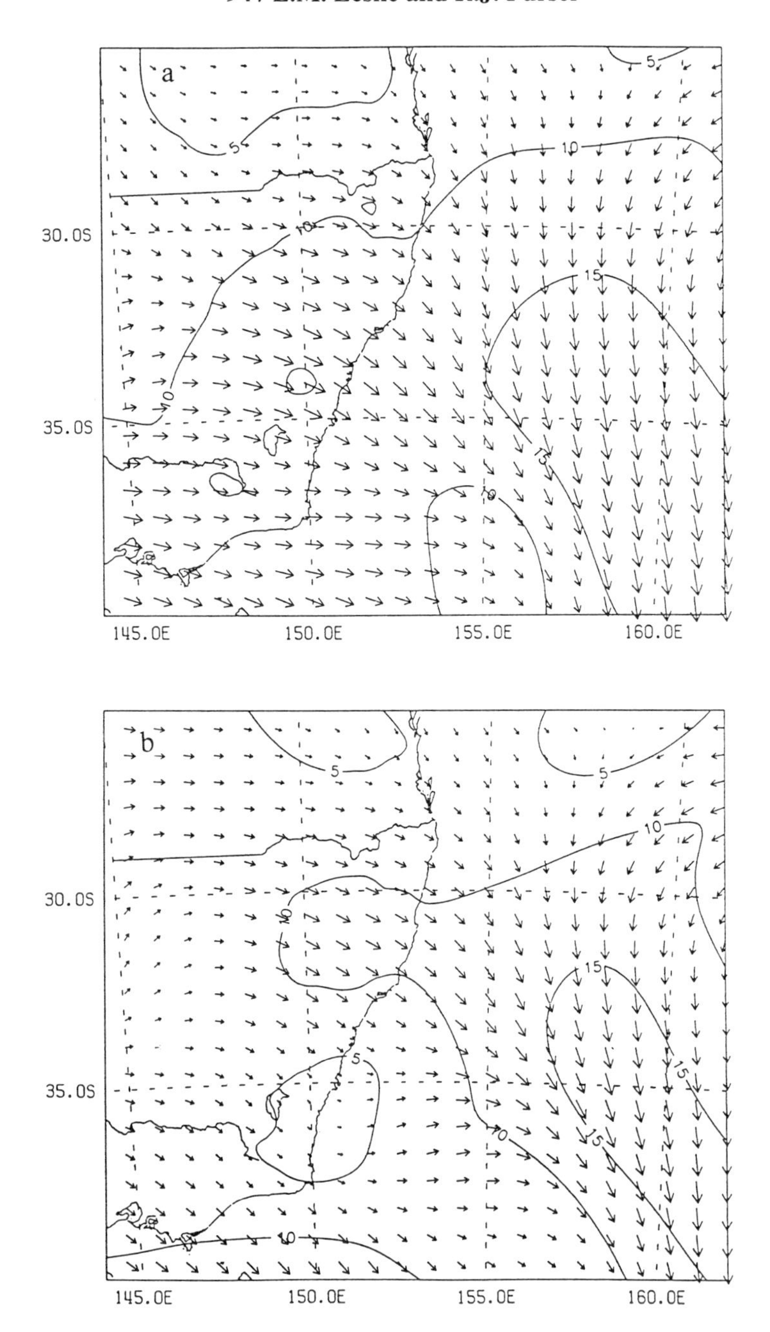

Fig. 10 As in Fig. 9, except for the wind field (vectors) in m/s.

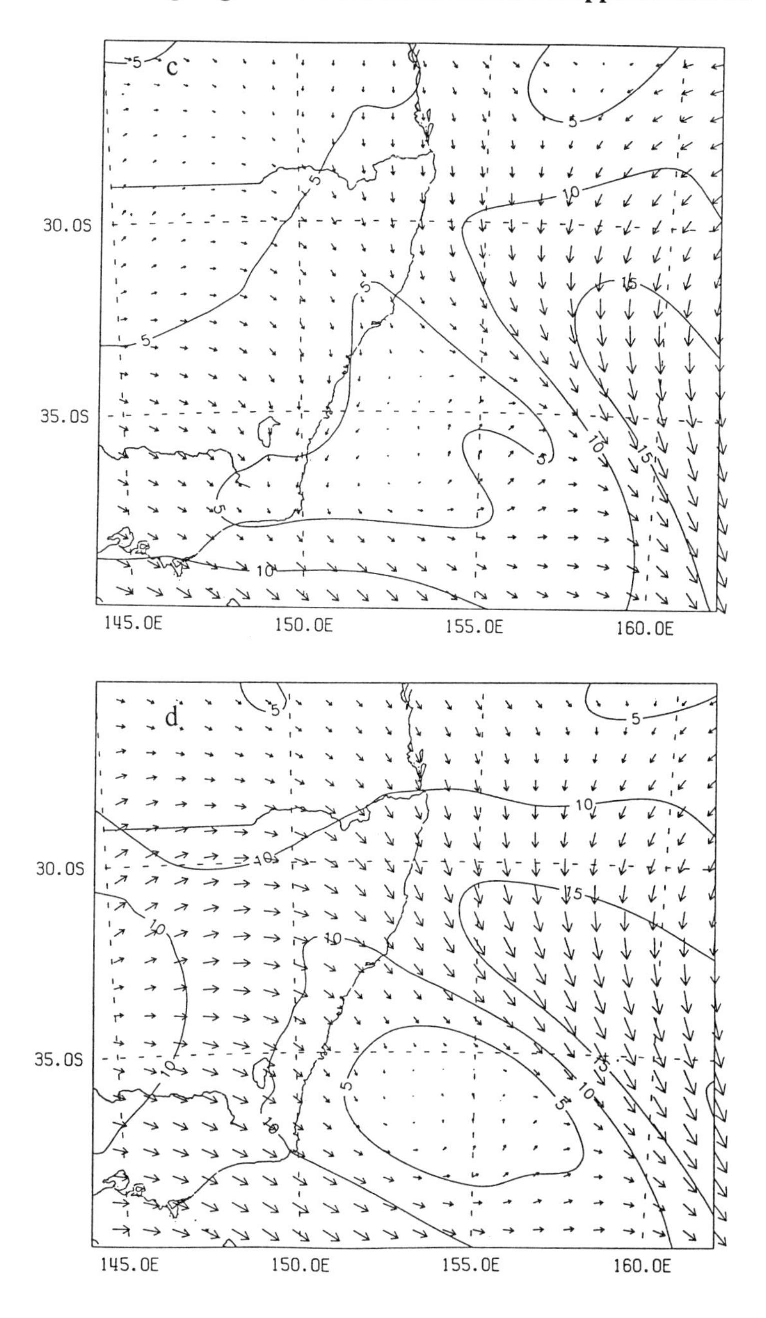

Fig. 10 *Concluded.*

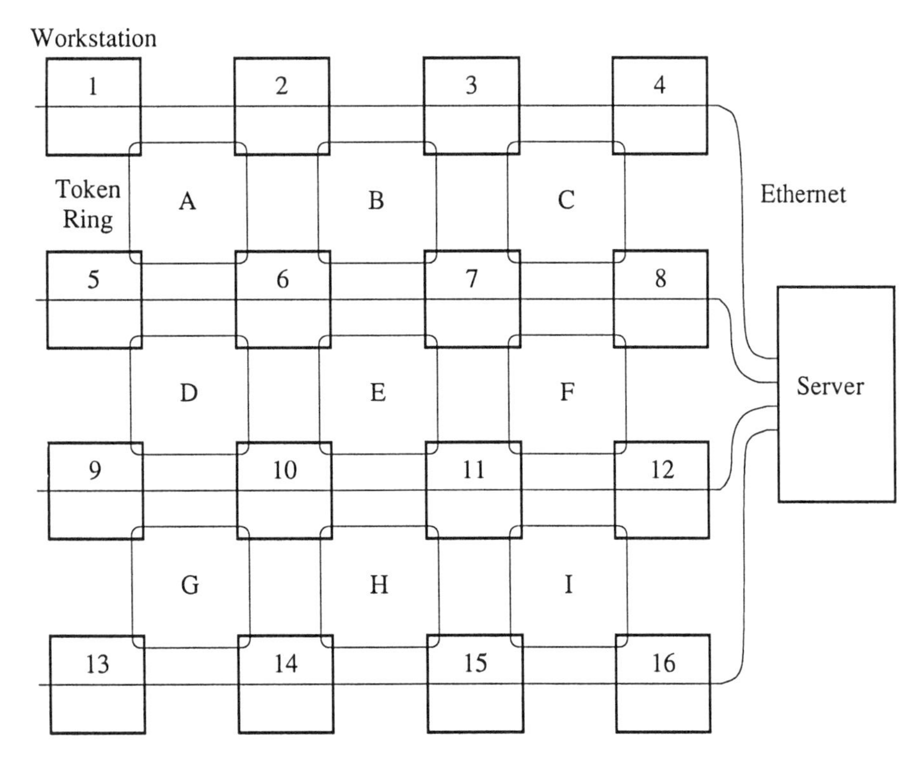

Fig. 11 Schematic illustration of the 16 node workstation cluster and server.

UTC 8 January. These early results of the fire weather simulation are sufficiently accurate to encourage further simulations down to much higher resolutions. These will be attempted with the parallelized version of the model, described below.

5 Implementation in a parallel computing environment

As mentioned above, the upwind and semi-Lagrangian models were implemented not only on scalar computers and vector supercomputers, but also on parallel computing platforms. The performance of the upwind model on a workstation cluster has been discussed elsewhere (Wightwick et al., 1995). The performance of the semi-Lagrangian model on the same workstation cluster and on a scalable parallel computer will be described now.

a *Workstation cluster*

The first parallel application of the semi-Lagrangian model was on a cluster of 16 RISC/6000 workstations topologically arranged as a 4 × 4 set connected by 4 Ethernet LANs to a server workstation. The workstation cluster is shown schematically in Fig. 11. The workstations were interconnected by Token Ring LANs in a simple mesh topology, again as shown in Fig. 11. The earliest version of the

TABLE 4. Speed-up achieved relative to one processor, for 4, 8 and 16 processors in the workstation cluster.

Resolution	4 Processor	8 Processor	16 Processor
150 km	1.99	2.29	2.74
50 km	3.01	5.98	8.66
20 km	–	7.52	11.64

cluster used the Parallel Virtual Memory (PVM) system Version 2.8 as the message passing language (later upgraded to Version 3.2). PVM is a public domain software system (Sunderam and Geist, 1990) that allows the utilization of a network of computers as a single resource. It comprises a server process, referred to as the server daemon, and a library of routines for message passing, process control and sychronization. The model code also was written in a way that attempted to keep changes from the serial version of the model to a minimum. Moreover, the "parallel code" was contained, as much as possible, in separate subroutines in order to make the code as general as possible and changing the number of processors a relatively simple task.

The model was implemented first on a single machine. The forecast was run at 150 km resolution on a domain comprising 64 × 40 points in the horizontal and 15 levels in the vertical. It was then run on 4 processors, 8 processors and 16 processors using a simple domain decomposition approach. The resolution was then increased to 50 km, and finally 20 km. The speed-up achieved on the workstation cluster is summarized in Table 4. Note that the maximum speed-up achieved was almost 12 for the 20 km, 16 processor case. Only very low speed-up values were obtained for the small grid sizes used in the 150 km resolution forecasts. Given that the communication overheads were so high with the slow message passing associated with the token rings LANs, this level of speed-up was very acceptable.

b *Parallel Supercomputer*

During the second half of 1994 the models have been implemented on an advanced parallel machine, namely the IBM PowerParallel SP2. Full details of the work, which is not fully completed, are to be published elsewhere (Wightwick et al. 1995), and only a brief description will be given of the results that have been obtained thus far. Access was restricted at this stage to only 12 processors of the SP2. However, as will be seen, about a Gflop of performance was obtained from this relatively small number of nodes.

The process of implementing the NWP model on the SP2 consisted largely of porting the PVM based version on the RISC/6000 cluster to the SP2. The changes required will not be described. In order to take advantage of the SP2 high-performance switch, it was necessary to rewrite the message passing routines in a manner that was consistent with the IBM Message Passing Library (MPL). The PVM version of the model adopted a master/slave programming approach.

The MPL uses the Single Program Multiple Data (SPMD) approach and supports a number of functions not available in PVM. Once again, speed-up was calculated relative to the performance of the model on a single SP2 node. Figure 12 shows the speed-up as a function of four model resolutions, using 1, 2, 4, 8 and 12 processors. Note that owing to memory restrictions it was not possible to run the 19 km resolution version of the model on 1 or 2 processors. The speed-up for the 19 km resolution forecasts are shown relative to the 4 processor model run.

The most important result was that near-linear and even greater than linear speed-up is achieved for a number of runs. The greater than linear speed-up is made possible by the fact that the impact of the combined caches and translation look-aside buffers of the SP2 nodes is a function of the problem size. This effect is explained in much greater detail by Wightwick et al. (1995).

6 Discussion and conclusions

A new semi-Lagrangian NWP model has been developed and tested with the intention of using it as a real-time and research model. Initial work has focused on high resolution regional modelling, but there are plans to use the model extensively in global form as a medium-range, extended term and general circulation model.

In this study the new semi-Lagrangian model was compared directly with the BoM's present operational semi-implicit model, and also with a new high-order upwind Eulerian model. The semi-Lagrangian and Eulerian models were both formally second-order in time and chosen to be third-order in space for the purposes of this comparison. It was found that the operational model was markedly inferior to the new models using traditional measures of model skill on a three-month archived operational dataset. The upwind and semi-Lagrangian models were on average 4 skill score points superior to the operational model, and had RMS errors in the 24-hour forecast height and wind fields at various levels in the atmosphere that were about 30 per cent lower. Moreover, it was also found that there was almost negligible difference between the Eulerian and semi-Lagrangian models in terms of skill. However, an accurate comparison of the elapsed times of the two approaches in terms of times required per forecast day have not been carried out in this study and no conclusions have been drawn. Examples and case studies also were presented to illustrate the performance of the semi-Lagrangian model in both limited area and global form. The authors intend to continue to use the semi-Lagrangian model for most of the future research and development work as it appears to provide the widest range of options. Nearly all of the problems associated with the semi-Lagrangian scheme have been resolved, so the advantage of being able to use a time step many times longer than that allowed by the Eulerian formulation appears to be decisive in most NWP applications. Remaining problems such as lack of formal and practical conservation properties are now being addressed by a number of researchers including the present authors.

It was anticipated that the semi-Lagrangian scheme might pose some problems for parallelization, notably in the parts of the code responsible for the interpola-

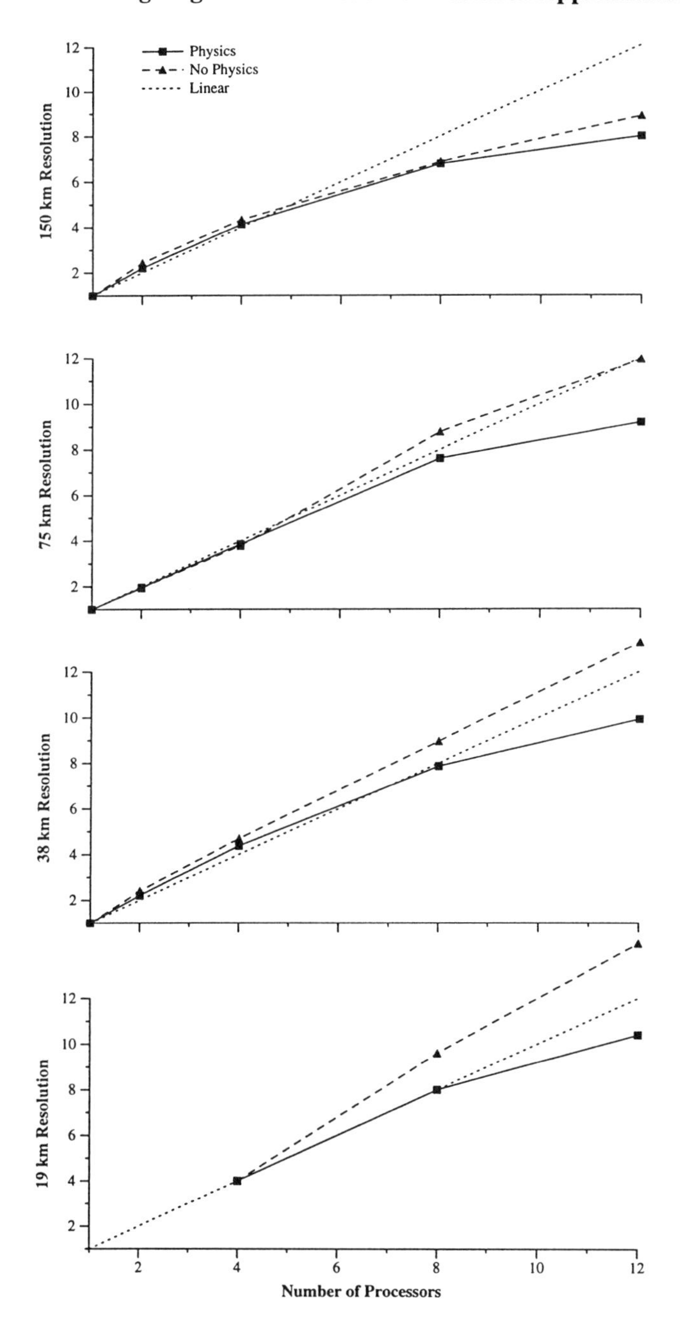

Fig. 12 Speed-up achieved on the parallel scalable computer at four different resolutions as a function of the number of nodes.

tions. However, this proved not to be the case and a quasi-linear speed-up was obtained. Additional work not reported on here (Wightwick and Leslie, personal communication) for the semi-implicit semi-Lagrangian formulation also was expected to provide additional difficulties in the solving of the Helmholtz equations associated with the semi-implicit scheme. However, the multi-grid method used to solve the Helmholtz equations also was found to parallelize easily, and yielded a quasi-linear speed-up. Future applications of the model are being made on machines with considerably greater numbers of nodes and this work will be reported on at a later time.

Acknowledgements

The authors would like to thank staff members of the Australian Bureau of Meteorology and the National Meteorological Center, Washington for helpful discussions. Glenn Wightwick of The Australian Computing and Communications Institute and of IBM Australia provided much of the expertize in the implementation of the model code on the parallel computers. RJP was supported for this project by the University Corporation for Atmospheric Research (UCAR) Visiting Scientist Program, and LML was partially supported by the Office of Naval Research under ONR Grant N00014-94-1-0556.

References

HOLLAND, G.J.; L.M. LESLIE and A.H. LYNCH. 1987. Australian east-coast cyclones. Part I: Synoptic overview and case study. *Mon. Weather Rev.* **115**: 3024–3036.

LEONARD, B.P. 1984. *Third-order upwinding as a rational basis for computational fluid dynamics. CTAC-83*, J. Noye and C.A.J. Fletcher (Eds), Elsevier North–Holland, pp. 106.

LESLIE, L.M.; G.A. MILLS, L.W. LOGAN, D.J. GAUNTLETT, G.A. KELLY, M.J. MANTON, J.L. MCGREGOR and J.M. SARDIE. 1985. A high resolution primitive equations model for operations and research. *Austral. Meteorol. Mag.* **33**: 11–35.

———; G.J. HOLLAND and A.H. LYNCH. 1987. Australian east-coast cyclones. Part II: Numerical Modelling Study. *Mon. Weather Rev.* **115**: 3037–3053.

——— and R.J. PURSER. 1991. High-order numerics in a three-dimensional time-split semi-Lagrangian forecast model. *Mon. Weather Rev.* **119**: 1612–1623.

——— and ———. 1994. A mass-conserving semi-Lagrangian scheme employing forward trajectories. Proc. 10th Conf. on Numerical Weather Prection. Portland, Oregon. 18–22 July 1994. Am. Meteorol. Soc., Boston. pp. 82–83.

——— and T.C.L. SKINNER. 1994. Numerical experiments with the West Australian summertime heat trough. *Weather Forecast.* **9**: 371–383.

MCGREGOR, J.L.; L.M. LESLIE and D.J. GAUNTLETT. 1978. The ANMRC limited area model: Consolidated formulation and operational results. *Mon. Weather Rev.* **106**: 427–438.

MCINNES, K.L.; L.M. LESLIE and J.L. MCBRIDE. 1993. Numerical simulation of cut-off lows on the Australian east coast: Sensitivity to sea-surface temperature and implications for global warming. *Int. J. Climatol.* **12**: 783–795.

MESINGER, F. 1977. Forward-backward scheme and its use in a limited-area model. *Contrib. Atmos. Phys.* **50**: 200–210.

——— and A. ARAKAWA. 1976. Numerical methods used in atmospheric models. GARP Publ. Ser. No. 17, ICSU/WMO, Geneva, Switzerland, pp. 53–56.

MILLER, M.J. and R.P. PEARCE. 1974. A three-dimensional primitive equations model of cumulus convection. *Q. J. R. Meteorol. Soc.* **100**: 133–154.

MILLS, G.A. and R.S. SEAMAN. 1990. The BMRC regional data assimilation system. *Mon. Weather Rev.* **118**: 1217–1237.

PURSER, R.J. and L.M. LESLIE. 1994. An efficient semi-Lagrangian scheme using third-order semi-implicit time integration and forward time trajectories. *Mon. Weather Rev.* **122**: 745–756.

ROBERT, A.; H. HENDERSON and C. TURNBULL. 1972. An implicit time integration scheme for baroclinic modes of the atmosphere. *Mon. Weather Rev.* **100**: 329–335.

SPEER, M.S. and L.M. LESLIE. 1994. Predictability of meso-beta convection using a high resolution NWP model. Proc. 10th Conf. on Numerical Weather Prection. Portland, Oregon. 18–22 July 1994, Am. Meteorol. Soc., Boston. pp. 472–474.

SUNDERAM, V. and G. GEIST. 1990. The PVM System. In: *Scientific Excellence in Supercomputing*, K.R. Billingsley and E. Derohanes (Eds), Baldwin Press, Athens, Georgia, pp. 779–804.

TEWELES, S. and H. WOBUS. 1954. Verification of prognostic charts. *Bull. Am. Meteorol. Soc.* **35**: 455–463.

WIGHTWICK, G.R. and L.M. LESLIE. 1992. Parallel implementation of a Numerical Weather Prediction model on a RISC/6000 cluster. Proc. Fifth Australian Supercomputer Conf., Melbourne, Australia. pp. 135–142.

———; ——— and S.F. WAIL. 1995. A numeric weather prediction model for the IBM SP2 parallel computer. First International Conference on Algorithms and Architectures for Parallel Processing. Brisbane, Australia, 19–21 April 1995, pp. 71–78.

The Implementation of the Semi-Implicit Scheme in Cell-Integrated Semi-Lagrangian Models

Bennert Machenhauer and Markus Olk
Max Planck Institute for Meteorology, Hamburg, Germany

[Original manuscript received 3 February 1995; in revised form 25 August 1995]

ABSTRACT *The cell-integrated semi-Lagrangian method, in which trajectories from the corner points of a grid cell define its extent at a previous time, may be applied to a set of equations in Lagrangian form derived from the complete set of primitive equations to construct a numerical model which conserves exactly the discrete forms of the global integral constraints of mass, momentum, entrophy and total energy. Due to the fulfilment of these integral constraints one should expect the numerical model to be absolutely stable. Experiments with a simple one-dimensional shallow water model show, however, that a time step dependent instability develops when a CFL criterion for gravity waves is exceeded. Analysis of experiments with one-dimensional shallow water models unveils the mechanism of the instability. Using again for simplicity one-dimensional models a main achievement is a successful implementation of the semi-implicit time stepping scheme in the cell-integrated models.*

RÉSUMÉ *La méthode semi-lagrangienne de cellules intégrées, dans laquelle les trajectoires des points de sommet d'une maille de grille définissent son étendue à un temps ultérieur, peut être appliquée à un ensemble d'équations sous forme lagrangienne, dérivé de l'ensemble complet des équations primitives, pour construire un modèle numérique qui conserve exactement les formes discrètes des contraintes intégrales globales de masse, de quantité de mouvement, d'entropie et d'énergie totale. En raison de l'exécution de ces contraintes intégrales, on pourrait supposer que le modèle numérique soit absolument stable; cependant, des essais avec un modèle simple à une dimension en eau peu profonde montre qu'une instabilité en fonction du pas de temps se développe lorsqu'un critère CFL pour les ondes de gravité est dépassé. Les analyses de ces essais révèlent le mécanisme de l'instabilité. Utilisant encore des modèles unidimensionnels par souci de simplicité, un résultat essentiel consiste en l'introduction d'un traitement semi-implicite du pas de temps dans les modèles de cellules integrées.*

1 Introduction

The accuracy of atmospheric numerical models has steadily improved with enhanced resolution and improvements in physical parametrization schemes. This increase in accuracy, which has taken place for the relatively high resolution weather prediction models as well as for the lower resolution climate models, has been made possible due to both steadily growing computer power and to the introduction of more efficient numerical techniques. One such numerical technique was the semi-implicit time stepping scheme, which was introduced by André Robert (Robert, 1969; Kwizak and Robert, 1971) and is used now in most atmospheric models to eliminate the Courant-Friedrichs-Levy (CFL) time step restriction due to gravity waves. In semi-implicit models typically a six-time-larger time step than in the former explicit models can be used without loss of accuracy. The time step in the usual Eulerian semi-implicit models is limited only by the advective CFL restriction. Another important new technique, also introduced in meteorological applications by Robert (1981, 1982), is a semi-Lagrangian treatment of advection offering a potential further increased efficiency by elimination of the advective CFL time step restriction. In principle, the time step in a combined semi-implicit semi-Lagrangian model can be chosen based on accuracy considerations. As developers of physical parametrizations are uncomfortable applying their schemes over time steps exceeding one hour (Williamson and Olson, 1994), this may put an upper bound to the time steps which can be used in practice. Significant computational savings can, however, be obtained in many applications, even within this limit.

The semi-Lagrangian advection offers additional advantages beyond the longer time step. It gives minimal phase error, minimizes computational dispersion, can handle sharp discontinuities and, furthermore, desirable properties such as monotonicity or, more generally, shape preservation may easily be incorporated.

A disadvantage is that the smallest scales resolved may be damped more by semi-Lagrangian methods than by some Eulerian methods. However, this does not seem to be a serious issue as it can be counteracted by a reduction or elimination of horizontal diffusion. A more serious disadvantage of semi-Lagrangian schemes today is that they do not formally conserve integral invariants as total mass or total energy. This may not be a problem in weather forecasting applications. For long simulations in climate applications, however, lack of conservation might have serious consequences. The total mass, in particular, has been found to drift significantly if no corrections are applied during longer integrations. Moorthi et al. (1995) report that the global mean surface pressure increased monotonously over a seventeen-month integration period. The rate of increase varied and at the end of this seventeen-month period it had increased 34 hPa from its initial value. In another case, reported by Machenhauer (1994), a three-month test integration with an early version of the ECMWF (European Centre for Medium-Range Forecasts) operational semi-implicit semi-Lagrangian model resulted in a systematic loss of mass corresponding to 4.5 hPa. Obviously in these cases errors in the prediction equation for surface pressure are accumulating. The mechanism(s) leading to these errors are not known and we can detect only the error in the global mean surface pressure field. Therefore, we know nothing about the three-dimensional structure of the errors in the mass field. It seems likely, however, that the accumulating large errors in the mean surface pressure are accompanied by pressure errors locally which are even larger and which may be systematically correlated with the pressure pattern. If such a correlation exists the internal dynamics of the model may be affected significantly. We do not know if this is the case, but it seems possible, and if it is the case the non-conservation is a symptom of some, perhaps more serious systematic errors.

Conservation of total mass may be obtained by a "mass-fixer" which after each time step restores the mean surface pressure to its initial value. Such a mass-fixer was tested by Moorthi et al. (1995) who repeated the above mentioned seventeen-month integration restoring, after each time step, the mean surface pressure by multiplying everywhere the preliminary calculated surface pressure by a constant factor equal to the ratio between the initial mean value and the preliminary mean value. This mass-fixer is similar to that used by (Williamson and Olson, 1994), except that they allow for variations in the total mass of water vapour. Using this form of the mass-fixer the horizontal pressure gradient is not affected by a restoration and the effect on the internal dynamics is therefore minimized. When comparing seasonally averaged fields from the above mentioned seventeen-month integrations with and without mass restoration Moorthi et al. (1995) found no significant differences. Thus, with this type of mass-fixer the restoration does not seem to significantly affect the simulated climate. This result is not so surprising as the fixer mentioned above was designed to have minimal effect on the dynamics. The restoration at each time step with the same factor everywhere is of course completely arbitrary and most likely the geographical distribution of the corrections is wrong. Recently Gravel and Staniforth (1994) have presented an alternative mass-fix procedure where the restoration of the mean pressure is made only in some specially selected points. They describe their mass-fix procedure as an extension to the shallow water equations of an algorithm introduced by Priestly (1993). The mass-fixer builds upon a mass conservation constraint imposed in the quasi-monotone semi-Lagrangian scheme of Bermejo and Staniforth (1992). This scheme in turn builds on the monotonicity constraint that an interpolated value must lie between the minimum and maximum value of the four neighbouring points. In the mass-fix procedure of Gravel and Staniforth (1994) the values are preliminarily adjusted toward linearly interpolated values at points at which a cubic interpolation violates the monotonicity constraint. These adjustments are kept as small as possible (at many points they are zero). Then the total mass is restored to its initial value by adjusting further toward the linear solution, but only in those points where such an adjustment changes the mass in the right direction and does so without violating the monotonicity constraint. Thus, the corrections applied in order to achieve conservation of mass are made only in some of those points where the monotonicity constraint is violated. Gravel and Staniforth (1993) argue that the interpolation is likely to introduce the errors that cause the lack of conservation in areas of strong gradients and that the points where they make the corrections are exactly such points. A great deal of arbitrariness is, however, still present in their procedure with regard firstly to the magnitude of correction in each point (kind of equipartition among the points chosen) and secondly by choosing not to do any mass restoring corrections in those points in which the preliminary correction to fulfil monotonicity goes in the "wrong" direction.

In the present paper we will advocate for a different approach towards incorporating conservation principles within the semi-Lagrangian framework. Namely to use special forms of the meteorological equations and special numerical schemes designed to conserve integral invariants exactly. Such a system, based upon the full set of primitive equations, was set up in Machenhauer (1994). Each of the prognostic equations in the system was written, in the form

$$\frac{d}{dt}(X\delta M) = F_X + \delta M S_X \tag{1}$$

where δM is the mass of an infinitesimal particle moving with the flow, X is unity or the mean value over the particle of a conservative variable, F_X is a flux and/or pressure term working at the surface of the parcel and S_X is a source term working inside the parcel. When X is unity ($X \equiv 1$), equation (1) is the continuity equation for which both right hand terms are zero. In the remaining equations of the system, X is specific total energy, specific angular momentum, specific entrophy, specific humidity or specific liquid water. In the semi-Lagrangian discretization of (1) grid point values of X are assumed to be mean values over the surrounding grid cell. As the moving parcel in (1) we consider the mass of air which at the end of a time step ends up in a grid cell. It is traced back in time using trajectories from the corner points of the grid cell. With a consistent evaluation of the flux terms F_X in neighbouring grid cells and a conservative remapping of X at the previous time level the integral invariants valid for (1) are also maintained for the discrete form of the system. A model based on equations of the form (1) and in the finite-difference form as indicated above will be called a cell-integrated semi-Lagrangian model.

In Machenhauer (1994) we speculated that perhaps due to the conservation properties of such a model it might be absolutely stable even with an explicit time-stepping scheme. In Section 3 we shall see, however, that experiments with simple one-dimensional versions, which are derived in Section 2, show that even such a model becomes unstable when the time step exceeds the critical value determined by the CFL criterion for gravity waves. The most unstable short waves are found to grow in amplitude even though total mass and total momentum or total mass and total energy are conserved exactly. When the amplitude has become large the trajectories begin to cross in some points which at once causes a breakdown of the conservation properties and subsequently leads to an explosion.

In order for a cell-integrated semi-Lagrangian model to be able to compete with traditional semi-Lagrangian models it is essential that either a split-explicit or a semi-implicit time-stepping scheme can be introduced in the cell-integrated system. We have tried to develop a split-explicit version of the cell-integrated model which conserves mass and momentum or mass and total energy. It turned out, however, that it was not possible to design such a model without relaxing the requirement of exact conservation of these quantities. This occurs because, when splitting the system of equations into an advective and an adjustment part, each of these subsystems does not conserve total momentum or total energy, respectively. This leaves us with the semi-implicit scheme as the only other known possibility of efficient time-stepping without abandoning the exact conservation properties. The main purpose of the present paper is to show that this can be accomplished successfully. This is done in Section 4, again using simple one-dimensional models. Finally, a summary and conclusions are given in Section 5.

After the design of the cell-integrated semi-Lagrangian system considered here it was realized that a somewhat similar system had been developed for the Navier-Stokes equations by Hirt et al. (1974) and had been applied to hydrodynamical problems. Here, a more general application to the complete system of meteorological equations is considered. Furthermore we use a semi-Lagrangian approach with a remapping every time step, whereas in Hirt et al. (1974) either an Eulerian or a fully Lagrangian approach for extended periods was used. Finally in order to be able to use larger time steps in the cell-integrated semi-Lagrangian system we introduce a semi-implicit time-stepping scheme whereas Hirt et al. (1974) introduced an iterative procedure to adjust pressure gradient forces.

More similarity, in all respects, to the cell-integrated scheme proposed here occurs in the advection schemes introduced and tested by Rancic (1992) and Laprise and Plante (1995). Although we did not know of their work when our scheme was initially developed, the cell-integrated model system we propose may be considered as an extension to the complete system of the meteorological equations of the scheme developed by Rancic (1992) and the similar one developed independently by Laprise and Plante (1995).

2 One-dimensional explicit models

We shall do experiments with simple one-dimensional versions of the cell-integrated semi-Lagrangian model. In this section we shall derive model equations based on leap-frog time differencing.

The shallow water continuity and velocity equations are

$$\frac{dh}{dt} + h\,\frac{\partial u}{\partial x} = 0 \tag{2}$$

$$\frac{du}{dt} + g\,\frac{\partial h}{\partial x} = 0 \tag{3}$$

where u is velocity, h height, x distance, t time and g gravity. We assume a periodic domain $0 \leqslant x \leqslant L$. Two different versions may be set up: A version in which mass and energy are conserved and a version in which mass and momentum are conserved. We shall call them the energy and the momentum system, respectively. We shall consider both versions since both are simplified versions of the complete three-dimensional system.

Let us consider at first the momentum system. From (2) and (3) we derive the momentum equation

$$\frac{d}{dt}(uh) + uh\frac{\partial u}{\partial x} + gh\frac{\partial h}{\partial x} = 0. \tag{4}$$

In (2) and (4) we substitute the Lagrangian expression for divergence

$$\frac{\partial u}{\partial x} = \frac{1}{\delta x}\frac{d}{dt}(\delta x) \tag{5}$$

where δx is an infinitesimal length interval. We thereby bring the momentum and continuity equation into the form of (1):

$$\frac{d}{dt}(h\delta x) = 0 \tag{6a}$$

$$\frac{d}{dt}(uh\delta x) + g\,\delta x\frac{\partial}{\partial x}\left(\frac{1}{2}h^2\right) = 0. \tag{6b}$$

The finite difference approximation of these equations is based on trajectories that end up at time $t+\Delta t$ at the cell boundaries, i.e. at the mid-points between grid points. Such trajectories are illustrated in Fig. 1. We chose a three-level leap-frog scheme with time step Δt and a uniform grid with gridlength Δx. The trajectories are then determined from velocities at time $t = n\Delta t$ using the usual iterative procedure to find the departure points at time $t - \Delta t$ which are given by

$$x_{j+1/2-2\alpha(j)} = x_{j+1/2} - 2\Delta t u^n_{j+1/2-\alpha(j)} \tag{7}$$

where $x_j = j\Delta x$ and consequently $2\alpha(j)$ is the distance in units of Δx between the departure point and the arrival point of the trajectory ending at $x_{j+1/2}$. Here the velocities are determined by cubic interpolation. The form chosen for the finite-difference analogue of (6a) is

$$h_j^{n+1}\Delta x = \bar{h}_j^-\delta x_j^- \tag{8}$$

where

$$\begin{aligned}\delta x_j^- &= x^{n-1}_{j+1/2-2\alpha(j)} - x^{n-1}_{j-1/2-2\alpha(j-1)} = \Delta x - 2\Delta t\delta u_j^0\\ \text{with } \delta u_j^0 &= u^n_{j+1/2-\alpha(j)} - u^n_{j-1/2-\alpha(j-1)}\end{aligned} \tag{9}$$

and $\bar{h}_j^-$ is the integral mean value of h^{n-1} over the interval δx_j^-:

$$\bar{h}_j^- = \frac{1}{\delta x_j^-}\int_A^B h^{n-1}(x)dx \tag{10}$$

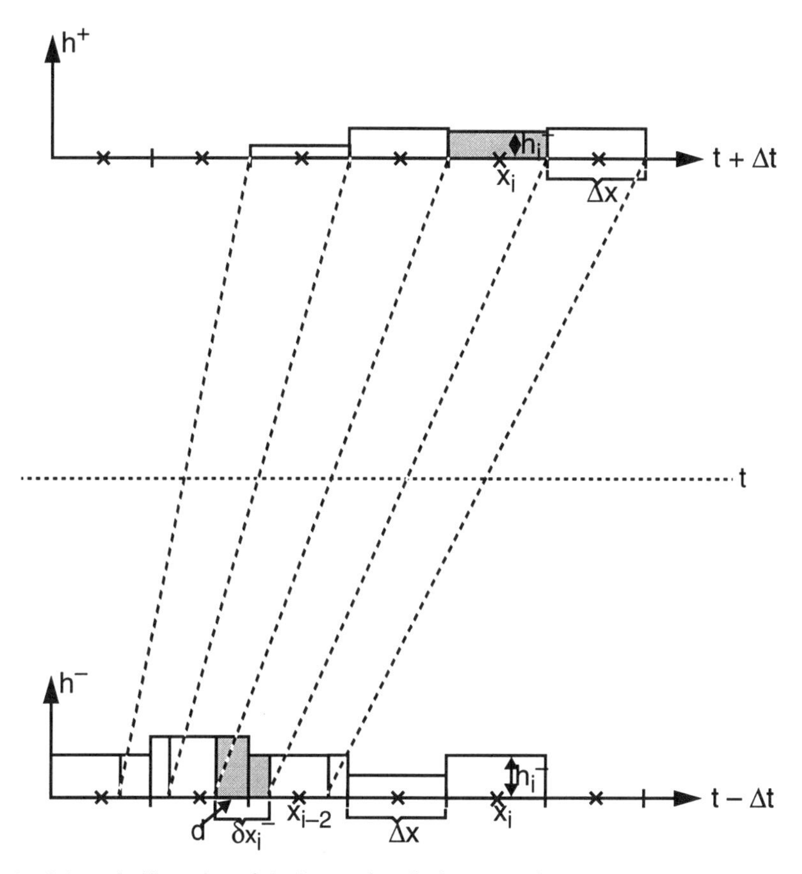

Fig. 1 Schematic illustration of the Lagrangian displacement of one-dimensional cells over a time interval of $2\Delta t$ and of quantities used in connection with the solution of the continuity equation (2). h^{+} and h^{-} are abbreviations for h^{n+1} and h^{n-1}, respectively.

with $A = x^{n-1}_{j-1/2-2\alpha(j-1)}$ and $B = x^{n-1}_{j+1/2-2\alpha(j)}$. In order to determine $\bar{h}_j^-$ the function $h^{n-1}(x)$ must be defined from the gridpoint values h_j^{n-1}. As mentioned in the introduction each of the gridpoint values are assumed to be equal to the mean value over its grid cell, i.e.

$$h_j^n = \frac{1}{\Delta x}\int_{x_{j-1/2}}^{x_{j+1/2}} h^n(x)dx. \tag{11}$$

The most simple definition of $h^{n-1}(x)$ consistent with this is to assume $h^{n-1}(x)$ in each grid cell to be piecewise constant equal to the gridpoint value, that is

$$h^n(x) = h_j^n \quad \text{for} \quad x_{j-1/2} \leqslant x \leqslant x_{j+1/2}, \tag{12}$$

as indicated in Fig. 1. The integral in (10) is then easily evaluated by a simple "length-weighted" mean, at the gridpoint x_j in the figure for instance by

$$\delta x_j^- \bar{h}_j^- = dh_{j-3}^{n-1} + (\delta x_j^- - d)h_{j-2}^{n-1} \tag{13}$$

where d is a distance defined in the figure. It is obvious that by this procedure we obtain

$$\sum_j h_j^{n+1}\Delta x = \sum_j \bar{h}_j^- \delta x_j^- = \sum_j h_j^{n-1}\Delta x \tag{14}$$

which means exact conservation of total mass. The first equality in (14) follows from (8) and the last one from (10) and (11). Calculation of mean values over a "new" set of grid cells δx_j^-, from the gridpoint values, h_j^{n-1}, in an "old" grid as we do when using (10), is called a "remapping". If the remapping satisfies the last equality in (14) it is termed "conservative". The condition for a conservative remapping, and thereby exact conservation of total mass, is that the $h^n(x)$ used in (10) must satisfy (11). The choice (12) is the most simple possibility. It is easy to show that with this choice and a constant wind (8) and (9) become equal to an upstream in space, forward in time first order advection scheme, a monotone but excessively damping scheme. Instead of the piecewise constant functions (12) a piecewise higher order polynomial, constrained to satisfy (11), may be used. The piecewise parabolic functions used by Rancic (1992) and Laprise and Plante (1995) seem to constitute a proper balance between increased accuracy and increased complexity. For simplicity we shall use the piecewise constant functions (12) and also because the advection process is of secondary importance in the experiments to be performed here.

The discrete form of the momentum equation (6b) is

$$(uh)_j^{n+1}\Delta x = \overline{(uh)}_j^- \delta x_j^- - \Delta t g\, \delta(h^2)_j^0 \tag{15}$$

where

$$\delta(h^2)_j^0 = (h_{j+1/2-\alpha(j)}^n)^2 - (h_{j-1/2-\alpha(j-1)}^n)^2 \tag{16}$$

and where $\overline{(uh)}_j^-$ is defined as an integral mean value analogous to (10) which is computed assuming that also $(uh)^{n-1}(x)$ is piecewise constant in the grid cells. For the grid point x_j in the figure for instance it becomes

$$\overline{(uh)}_j^- \delta x_j^- = (d(uh)_{j-3}^{n-1} + (\delta x_j^- - d)(uh)_{j-2}^{n-1}). \tag{17}$$

The values of h at the mid-points of the trajectories to be used in (16) are determined from the grid points by cubic interpolation.

The total momentum is conserved exactly because

$$\sum_j (uh)_j^{n+1}\Delta x = \sum_j \overline{(uh)}_j^- \delta x_j^- = \sum_j (uh)_j^{n-1}\Delta x \tag{18}$$

where the first equality follows from (15) and the last one follows from the conservative remapping of $(uh)^{n-1}$.

The prognostic equations for the momentum system, equations (8) and (15), have now been derived and we proceed with the derivation of the energy equation. We define the total energy as

$$E = \frac{1}{2} u^2 h + \frac{1}{2} g h^2. \tag{19}$$

From the original system (2) and (3) we derive the energy equation

$$\frac{dE}{dt} + E \frac{\partial u}{\partial x} + \frac{\partial}{\partial x}\left(\frac{1}{2} g h^2 u\right) = 0 \tag{20}$$

which, after substitution of (5), is brought to the form of (1):

$$\frac{d}{dt}(E \delta x) + \delta x \frac{\partial}{\partial x}\left(\frac{1}{2} g h^2 u\right) = 0. \tag{21}$$

We approximate this equation as

$$E_j^{n+1} \Delta x = \bar{E}_j^- \delta x_j^- - g \Delta t \delta (h^2 u)_j^0 \tag{22}$$

where

$$\delta(h^2 u)_j^0 = (h^n_{j+1/2-\alpha(j)})^2 u^n_{j+1/2-\alpha(j)} - (h^n_{j-1/2-\alpha(j-1)})^2 u^n_{j-1/2-\alpha(j-1)} \tag{23}$$

and, with similar assumptions as in the momentum system, $\bar{E}_j^-$ becomes a length-weighted mean value. For the case shown in the figure we obtain

$$\bar{E}_j^- \delta x_j^- = d E_{j-3}^{n-1} + (\delta x_j^- - d) E_{j-2}^{n-1}. \tag{24}$$

As for the momentum equation we use cubic interpolation in (23). From (22) and the conservative remapping of E^{n-1} we get exact conservation of energy:

$$\sum_j E_j^{n+1} \Delta x = \sum_j \bar{E}_j^- \delta x_j^- = \sum_j E_j^{n-1} \Delta x. \tag{25}$$

When using the momentum system the time-stepping procedure is straightforward. In each time step we compute at first h^{n+1} using (8) and then we use (15) to compute u_j^{n+1}, that is

$$u_j^{n+1} = \frac{(uh)_j^{n+1}}{h_j^{n+1}}.$$

For the energy system it becomes slightly more complicated. Again we compute at first h^{n+1} using (8). To obtain u_j^{n+1} we then use (22) to get E_j^{n+1} which gives

$$u_j^{n+1} = \pm \sqrt{2 E_j^{n+1} / h_j^{n+1} - g h_j^{n+1}}. \tag{26}$$

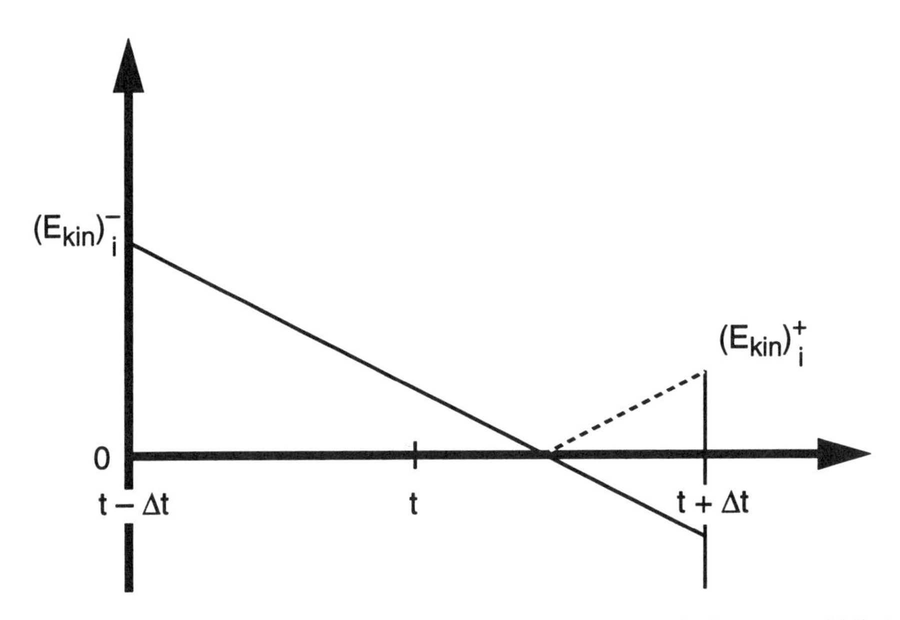

Fig. 2 Schematic diagram showing the initial erroneous variation of the kinetic energy (solid line) during a time step for a cell which should change the sign of its velocity. The dashed line indicates the corrected path to the positive value adopted.

Thus, we have to choose between the plus and minus sign in (26). Normally we choose the same sign as $\overline{(uh)}_j^-$ determined from (17). It may happen, however, that the radicant in (26) becomes negative which means that the kinetic energy at time $t + \Delta t$ has become negative. This will happen in grid cells which at the start of the time have a numerically small velocity and are being decelerated sufficiently enough during the time step to change their direction of movement. The kinetic energy of such a cell should go to zero and then increase again. Assuming that the kinetic energy $E_{kin} = 1/2\, u^2 h$ varyies linearly with time, as illustrated in Fig. 2, this may be simulated approximately simply by changing the sign of the predicted kinetic energy of every cell j for which it has become negative. As the total energy is thus increased by an amount $2|(E_{kin})_j^{n+1}|$ we reduce the predicted kinetic energy of the two neighbouring grid cells by an amount $|(E_{kin})_j^{n+1}|$ each. (If one or two of the neighbours thereby get negative energy the same procedure is used in these points.) At points where the sign of the kinetic energy has been changed we choose the opposite sign to that of $\overline{(uh)}_j^-$ in (26). By this procedure we preserve conservation of total energy and "allow" particles to change the sign of their velocities within a time step. Normally, with a realistic size of Δt and realistic atmospheric flows few points, if any, should need correction of kinetic energy. Except in unstable integrations just before the model "explodes" no corrections of kinetic energy were made in the experiments reported on later.

As for other semi-Lagrangian schemes the present scheme breaks down when any of the trajectories cross each other or in other words when any of the δx_j^- becomes negative. Using (9) it is easily shown that this condition puts a limit to the magnitude of the divergence of the velocity field which determines the trajectories, or rather to the finite difference approximation which we use for this divergence. The well known condition e.g., (Smolarkiewicz and Pudykiewicz, 1992) becomes

$$\frac{\partial u}{\partial x} \approx \frac{\delta u_j^0}{\Delta x/2} \leqslant \frac{1}{\Delta t}. \tag{27}$$

3 Stability of explicit cell-integrated models

To investigate the stability of the cell-integrated models designed above some experiments were carried out. In these experiments and those reported in the following section we used the following set up:

- A periodic domain of 64 grid points and a grid length of $\Delta x = 100$ km.
- Initial fields definded as follows:

$$u = U + u_0 \sin\left(\frac{2\pi}{L} x\right) \tag{28a}$$

$$h = H + h_0 \sin\left(\frac{2\pi}{L} x\right) \tag{28b}$$

  where $U = 10$ m/s, $H = 8000$ m, $L = 6400$ km, $u_0 = 0.5$ m/s and $h_0 = u_0\sqrt{H/g} \approx 14.28$ m. These are the initial fields of a harmonic solution with maximum wavelength to the linearized system corresponding to (2) and (3). It is one of the two gravity wave solutions with this wavelength. Its phase speed is

$$c_g = U + \sqrt{gH} \approx 290.14 \text{ m/s} \tag{29}$$

- A so-called smooth starting procedure used to compute fields at $t = \Delta t$.

In the present experiments with explicit time stepping the fields at $t = \Delta t$ were computed by an initial Euler time step over $\Delta t/2$ followed by a leap-frog step centred at $t = \Delta t/2$. In the Euler step the trajectories were computed using velocities valid at $t = 0$.

The results obtained with the cell-integrated models will be compared with those of a traditional model based on the original velocity and continuity equations (2) and (3), in the following referred to as the velocity system. The prediction equations for this model are

$$u_j^{n+1} = u_{j-2\alpha'(j)}^{n-1} - \frac{\Delta t g}{\Delta x}(\delta_{2\Delta x}(h))_{j-\alpha'(j)}^n \tag{30}$$

$$h_j^{n+1} = h_{j-2\alpha'(j)}^{n-1} - \frac{\Delta t}{\Delta x}(h\delta_{2\Delta x}(u))_{j-\alpha'(j)}^n \tag{31}$$

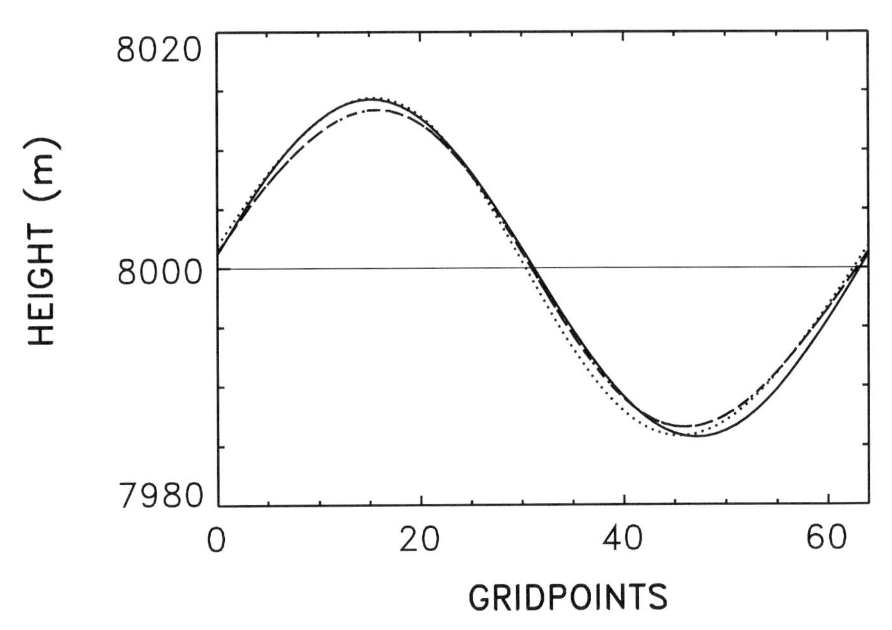

Fig. 3 The height field at time $t = 30.5$ hour for the true linear solution (solid line), and the explicit model solutions with $\Delta t = 100$ s: The traditional (dotted), the cell-integrated momentum system (dashed) and the cell-integrated energy system (dot-dashed).

where $2\alpha'(j)$ is the distance in units of Δx between departure and arrival points for the trajectory ending at gridpoint x_j. $\delta_{2\Delta x}(h)$ is a finite difference centred over $2\Delta x$. It is computed in grid points and $(\delta_{2\Delta x}(h))^n_{j-\alpha'(j)}$ is a value obtained from these by interpolation to the mid-points of the trajectories. As in the cell-integrated model, cubic interpolation is used to compute the velocities needed for the trajectory computations and to compute the terms on the right hand sides of (30) and (31).

Using at first a time step $\Delta t = 100$ s the two cell-integrated models and the traditional model were integrated for 1100 time steps or 30.5 hours. During this period the linear solution moves slightly less than 5 times through the domain. The three numerical solutions at the end of the integration period are plotted together with the linear solution in Fig. 3. Only the height field is shown. The solution of the two cell-integrated models, i.e. the dashed and dot-dashed curves in the figure, cannot be distinguished. Compared to the linear solution their amplitudes have decreased by about 5% which can be attributed almost exclusively to the smoothing due to the length weighting, i.e. expression (10) with $h^{n-1}(x)$ being piecewise constant in the grid cells, as we choose for simplicity. As mentioned above, a more sophisticated choice can reduce this damping. A similar damping is not visible in the traditional model solution. The cell-integrated model solutions deviate slightly from a sine shape being steeper between the maximum and minimum than between minimum and the maximum. We see the same deformation in the solution of the

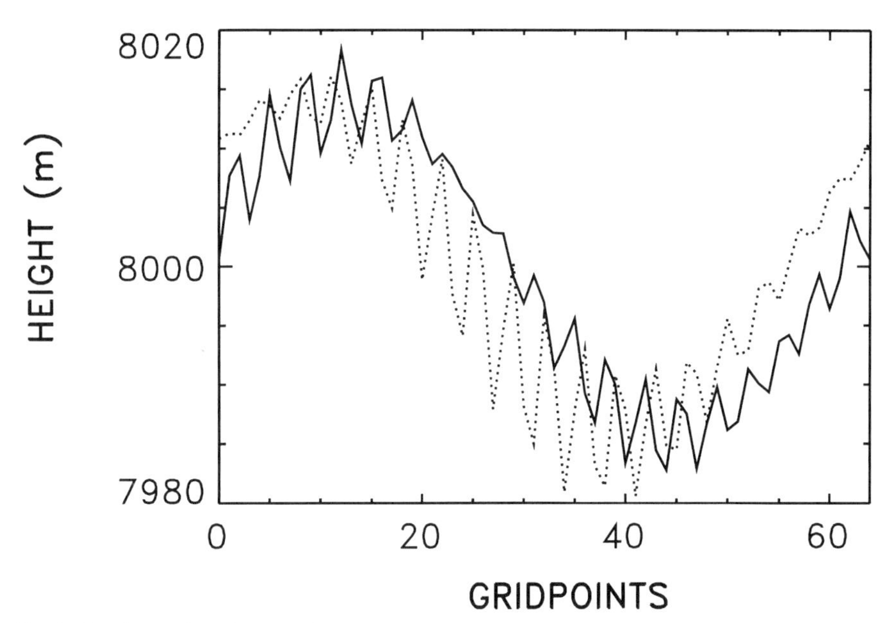

Fig. 4 The height field from explicit integrations of the cell-integrated models with a time step $\Delta t =$ 300 s: The momentum system at $t = 70\Delta t$ (solid) and the energy system at $t = 64\Delta t$ (dotted).

traditional model, which may indicate that it is a true nonlinear effect. As seen by crossing points at the 8000 m line the solutions are retarded slightly compared to the true linear solution. This retardation is largest for the traditional model solution.

One difference between the cell-integrated models solutions and the traditional model solution is, of course, that the former conserves mass and energy or mass and momentum exactly whereas the latter does not. This was verified in practice by integrating the three models for 62.5 days ($5.4 \cdot 10^6$ time steps). In these extended integrations a weak Robert-Asselin time filter with a coefficient of 0.005 was applied to the prognostic variables in order to avoid separation between values at even and odd time steps. The time filter does not affect the conservation properties of the cell-integrated models. Due to the time filter and other numerical damping mechanisms the initial perturbations were completely damped out at the end of the integration period in all three model runs. Although our models based on the simple one dimensional shallow water equations are well suited for the main objectives of the present study they are unrealistic as models of the atmosphere in most respects. The changes of mass, momentum and energy we find in the extended integrations are therefore not of particular interest and will not be listed here.

In order to test the stability of the cell-integrated models, integrations with larger Δt were performed. The resulting height fields from integrations with a time step $\Delta t = 300$ s are shown in Fig. 4, for the energy system after 64 time steps and for the momentum system after 70 time steps. Obviously in both systems an initial instability of waves with a mean wavelength around $3\Delta x$ to $4\Delta x$ is developing. A few time steps later the models "explode". This instability develops in spite of the exact conservation of total mass and total energy or total mass and total momentum in the models. What happens is the following: the short waves are growing in amplitude even though mass and momentum/energy are conserved exactly. When the amplitude has become large enough the critical value for divergence, given by (27), is exceeded locally, which causes the trajectories to cross in some cells. As a result of these crossings the conservation properties are no longer valid and immediately we see a very rapid increase in the quantities that were formerly conserved.

In order to demonstrate an equivalence between the growth of short waves in the cell-integrated systems on the one hand and a CFL-type instability of a corresponding linearized system on the other hand we have made a stability analysis of such a linearized system. The discrete continuity equation (8) may be linearized as follows. Using (9) we may write (8) as

$$h_j^{n+1}\Delta x = \bar{h}_j^- \delta x_j^- = \bar{h}_j^-(\Delta x - 2\Delta t \delta u_j^0) = \bar{h}_j^- \Delta x - 2\Delta t \bar{h}_j^- \delta u_j^0.$$

When $h_j = h'_j + H$ and $u_j = u'_j + U$ are inserted and second order terms in perturbation quantities are dropped we get the linearized continuity equation

$$h'^{n+1}_j = \bar{h}'^-_j - 2\Delta t H \frac{\delta u'^0_j}{\Delta x}, \tag{32}$$

similarly for the momentum equation (15) we obtain

$$(uh)_j^{n+1}\Delta x = \overline{(uh)}_j^- \delta x_j^- - 2\Delta t \delta (h^2)_j^0.$$

Using (9) again and the following approximations

$$(uh)_j^{n+1} = u_j^{n+1} h_j^{n+1} \cong U h'^{n+1}_j + H u'^{n+1}_j + UH$$

$$\overline{(uh)}_j^- \cong U \bar{h}'^-_j + H \bar{u}'^-_j + UH$$

$$(h^2)_j^0 \cong 2gHh'^n_j + gH^2$$

where second order terms in perturbed variables have been dropped, this equation becomes

$$\left(u'^{n+1}_j - \bar{u}'^-_j + 2\Delta t g \frac{\delta h'^0_j}{\Delta x}\right) H + \left(h'^{n+1}_j - \bar{h}'^-_j + 2\Delta t H \frac{\delta u'^0_j}{\Delta x}\right) U = 0$$

which when (32) is used gives

$$u'^{n+1}_j = \bar{u}'^-_j - 2\Delta t g \frac{\delta h'^0_j}{\Delta x}. \tag{33}$$

Dropping the perturbation marks the linearized system is

$$h_j^{n+1} - \bar{h}_j^- + 2\Delta t H \frac{\delta u_j^0}{\Delta x} = 0 \tag{34a}$$

$$u_j^{n+1} - \bar{u}_j^- + 2\Delta t g \frac{\delta h_j^0}{\Delta x} = 0. \tag{34b}$$

By a similar procedure the linearized version of the energy system may be derived. It turns out to be identical to the above system.

Using a constant basic stream velocity U in the computation of trajectories these become parallel, i.e. δx_j^- becomes constant equal to Δx and a length-weighted quantity is then simply obtained as a linear interpolation between two gridpoint values. For the height for instance

$$\bar{h}_j^- = (1-\mu)h_J^{n-1} - \mu h_{J-1}^{n-1} \tag{35}$$

where

$$\mu = \frac{2\Delta t U}{\Delta x} - \mathrm{INT}\left(\frac{2\Delta t U}{\Delta x}\right) \tag{36}$$

and

$$J = j - \mathrm{INT}\left(\frac{2\Delta t U}{\Delta x}\right). \tag{37}$$

Here INT(a) is defined as the largest integer which is $\leqslant a$. In the present case with $U = 10$ m/s, $\Delta x = 100$ km and Δt of the order 100 s, INT$(2\Delta t U/\Delta x) = 0$ and $\mu \ll 1$. The evaluation of the last term in each of the equations (34a, 34b) involves interpolations of u^n and h^n to the mid-points of the trajectories. These points lie at a distance $\mu \Delta x/2$ upstream from the mid-points between the grid points. As $\mu \ll 1$ this distance is very small so we can neglect it which simplifies the formulation of the cubic interpolation. With this approximation equations (34a, 34b) may be written

$$h_j^{n+1} - ((1-\mu)h_j^{n-1} + \mu h_{j-1}^{n-1}) + \frac{2\Delta t H}{\Delta x} \times \left(\frac{5}{8}(u_{j+1}^n - u_{j-1}^n) - \frac{1}{16}(u_{j+2}^n - u_{j-2}^n)\right) = 0 \tag{38a}$$

$$u_j^{n+1} - ((1-\mu)u_j^{n-1} + \mu u_{j-1}^{n-1}) + \frac{2\Delta t g}{\Delta x} \times \left(\frac{5}{8}(h_{j+1}^n - h_{j-1}^n) - \frac{1}{16}(h_{j+2}^n - h_{j-2}^n)\right) = 0 \tag{38b}$$

The condition for existence of harmonic solutions of the form

$$\begin{pmatrix} u_j \\ h_j \end{pmatrix}^n = \begin{pmatrix} u_k \\ h_k \end{pmatrix} e^{ik(j\Delta x - cn\Delta t)}$$

to the system (38) is found to be that

$$\sin^2(k(c-U)\Delta t) = \left(\frac{2\Delta t\sqrt{gh}}{\Delta x} F(k\Delta x) \right)^2 \tag{39}$$

where

$$F(k\Delta x) = \frac{5}{8} \sin(k\Delta x) - \frac{1}{16} \sin(2k\Delta x).$$

In the derivation of (39) we have again neglected μ as it is very small. From (39) we get the CFL condition of stability for the cell-integrated systems:

$$\frac{2\Delta t\sqrt{gh}}{\Delta x} F(k\Delta x) \leqslant 1. \tag{40}$$

And the expression for the phase speeds of the two gravity wave solutions

$$c = U \pm \frac{1}{k\Delta t} \arcsin\left(\frac{2\Delta t\sqrt{gh}}{\Delta x} F(k\Delta x) \right) \tag{41}$$

valid when (40) is satisfied.

The function $F(k\Delta x)$ and $(c-U)/(c_g-U)$ are shown in Fig. 5. The maximum value of $F(k\Delta x)$ is 0.64 which occurs for $k\Delta x = 0.55\pi$ corresponding to a wavelength of $3.6\Delta x$. Inserting the maximum value of $F(k\Delta x)$ in (40) gives the over all stability criterion

$$\Delta t \leqslant \frac{\Delta x}{1.28\sqrt{gH}} \tag{42}$$

which for the present model setup gives a maximum time step of 279 s. This agrees well with what we have found experimentally. In the results of our integrations with $\Delta t = 300$ s, which are shown in Fig. 4, we found growing waves with a wavelength around $3\Delta x$ to $4\Delta x$. This agrees well with the wavelength of $3.6\Delta x$ found above for the most unstable wave in the linearized system. Finally, the experimentally determined phase speeds agree very well with those determined from (41). These agreements show that the instability of the explicit cell-integrated semi-Lagrangian models that we found in the experiments can be identified as linear instability.

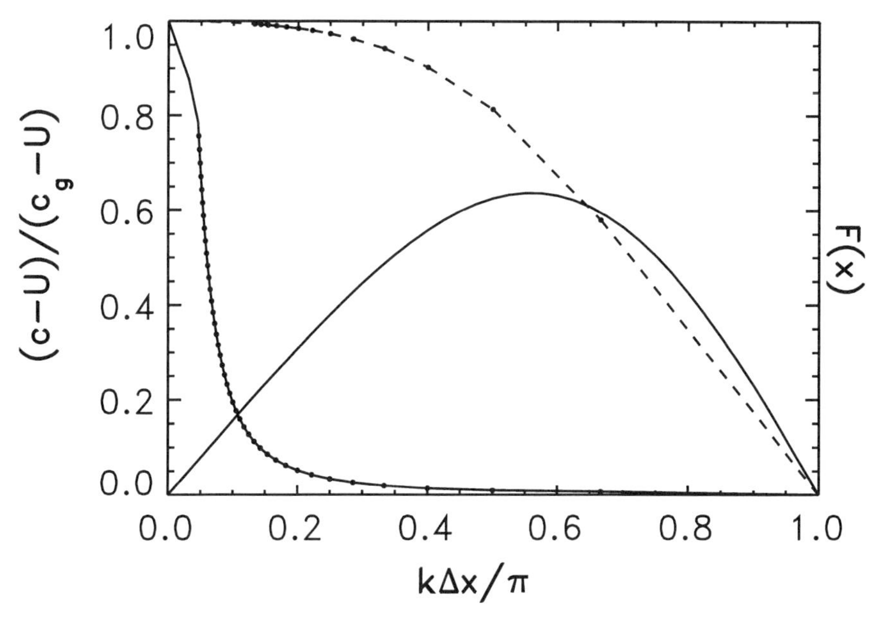

Fig. 5 The relative phase speed of harmonic solutions to the linearized cell-integrated models for the explicit model (dashed), given by (39) and for the semi-implicit model with $\Delta t = 2500$ s (solid), given by (57). At both curves are point marks at wavelengths of $3\Delta x$, $4\Delta x, \ldots$ etc. The solid curve without dots is the function $F(x)$ appearing in (39).

4 The semi-implicit formulation

If the cell-integrated semi-Lagrangian scheme is to be able to compete with traditional schemes in the context of climate modelling, formulations are needed which allow longer time steps, more reasonable in agreement with the time scale of large scale weather systems. In traditional models the most common way to eliminate the time step restriction due to linear instability of gravity waves is the use of a split-explicit or a semi-implicit time stepping scheme. As mentioned in the introduction the first alternative is not possible without losing the exact conservation properties which leaves us the second alternative, the semi-implicit scheme. We shall see how that scheme can also be used in cell-integrated semi-Lagrangian models, again for simplicity developed and tested for the one-dimensional models considered in the preceding sections.

We shall consider at first a traditional formulation corresponding to the explicit system (30) and (31). A semi-implicit system corresponding to this is obtained simply by averaging $t+t\Delta$ and $t-\Delta t$ values of the linear part of the divergence and pressure gradient terms along the trajectories instead of taking them at time t at the mid-points of the trajectories. With a slightly simplified notation the resulting equations may be written

$$u^+ = u^+_{\mathrm{exp}} - \frac{\Delta t g}{2\Delta x}(\delta_{2\Delta x} h^+ + (\delta_{2\Delta x} h)^-_D - 2(\delta_{2\Delta x} h)^0_M) \tag{43a}$$

$$h^+ = h^+_{\mathrm{exp}} - \frac{\Delta t H}{2\Delta x}(\delta_{2\Delta x} u^+ + (\delta_{2\Delta x} u)^-_D - 2(\delta_{2\Delta x} u)^0_M). \tag{43b}$$

Here "+", "−" and "0" stand for $n+1$, $n-1$ and n, respectively. "D" and "M" refer to departure and mid-points and the index j has been dropped. u^+_{exp} and h^+_{exp} are the explicitly predicted values given by (30) and (31). We may write (43a, 43b) as

$$u^+ = q_1 - \frac{\Delta t g}{2\Delta x}\,\delta_{2\Delta x} h^+ \tag{44a}$$

$$h^+ = q_2 - \frac{\Delta t H}{2\Delta x}\,\delta_{2\Delta x} u^+. \tag{44b}$$

In q_1 and q_2 we have collected terms which depend on values at $t - \Delta t$ and t. Applying the operator $\delta_{2\Delta x}$ on (44a) and substituting in (44b) gives

$$h^+ - \frac{\Delta t^2 g H}{(\Delta x)^2}\,\delta^2_{2\Delta x}(h^+) = q_2 - \frac{\Delta t H}{\Delta x}\,\delta_{2\Delta x}(q_1) \tag{45}$$

where $\delta^2_{2\Delta x} = \delta_{2\Delta x}\delta_{2\Delta x}$. The Helmholtz equation (45) can be solved to give h^+ and then (44a) can be used to determine u^+.

We shall try to set up a similar scheme for each of the cell-integrated systems. Two problems are encountered when trying to do this. The first problem that must be dealt with is the fact that the divergence, which we want to average over the time levels $t+\Delta t$ and $t-\Delta t$, has been eliminated, i.e. by the use of (4). It is hidden in the trajectory computation as may be seen from the last expression in (9):

$$\delta x_j^- = \Delta x - 2\Delta t \delta u_j^0$$

from which we may derive a finite difference expression for divergence at time level t

$$\left(\frac{\partial u}{\partial x}\right)^0_M \approx \frac{\delta u_j^0}{\frac{1}{2}(\delta x_j^- + \Delta x)} = \frac{1}{\Delta t}\,\frac{\Delta x - \delta x_j^-}{\Delta x + \delta x_j^-}.$$

Thus, in the explicit system the divergence at time level t is determined by δx_j^- and therefore by the trajectory computations. In a semi-implicit system we would like in certain terms to use trajectories which were determined by winds that were averaged over time level $t + \Delta t$ and $t - \Delta t$. That is not possible, however, as the velocities at time level $t + \Delta t$ are not available at the beginning of a time step when the trajectory computations must be made. An iterative procedure might be a possibility but we rejected that as it would be too costly. We ended up by simply neglecting the problem using just explicitly computed trajectories in all terms of the equations.

The second problem we have to deal with is caused by the fact that one of the prognostic variables is a nonlinear quantity in the basic variables u and h. We solved this problem by a linearization of the nonlinear variable at time level $t+\Delta t$.

Let us consider at first the cell-integrated momentum system. The explicit continuity equation (8) may be written

$$h^{+}_{\text{exp}}\Delta x = \bar{h}_j^{-}\Delta x - 2\Delta t(H + \bar{h}'^{-}_j)\delta u^0 \tag{46}$$

where (9) has been used. A corresponding semi-implicit equation is obtained by substituting $\frac{1}{2}(\delta u^{+} + \delta u^{-})$ for δu^0 in the linear part of the last term giving

$$h^{+} = \bar{h}_j^{-} - \frac{2\Delta t}{\Delta x}\bar{h}'^{-}_j\delta u^0 + \frac{\Delta t H}{\Delta x}(\delta u^{+} + \delta u^{-}) \tag{47}$$

where δu^{-} and δu^{+} are defined analogous to δu^0 (see equation (9)), as the difference between the velocity at the right hand trajectory point and that at the left hand trajectory point at time level $t-\Delta t$ and $t+\Delta t$, respectively. As mentioned above the trajectories referred to here are computed from velocities at time level t, as in the explicit scheme. The explicit momentum equation (15) may be written

$$(uh)^{+}_{\text{exp}}\Delta x = \overline{(uh)}_j^{-}\Delta x - 2\Delta t(UH + \overline{(uh)}'^{-}_j)\delta u^0 - \Delta t g\,\delta(h'^2 + 2h'H)^0 \tag{48}$$

and the corresponding semi-implicit one becomes

$$\begin{aligned}(uh)^{+} = \overline{(uh)}_j^{-} - \frac{2\Delta t}{\Delta x}\overline{(uh)}'^{-}_j\delta u^0 - \frac{\Delta t g}{\Delta x}\delta(h'^2)^0 \\ - \frac{\Delta t H}{\Delta x}(U(\delta u^{+} + \delta u^{-}) + g(\delta h^{+} + \delta h^{-}))\end{aligned} \tag{49}$$

The next step is to write the two semi-implicit equations (47) and (49) as

$$h^{+} = h^{+}_{\text{exp}} - \frac{\Delta t H}{\Delta x}(\delta u^{+} + \delta u^{-} - 2\delta u^0) \tag{50a}$$

$$(uh)^{+} = (uh)^{+}_{\text{exp}} - \frac{\Delta t H}{\Delta x}(U(\delta u^{+} + \delta u^{-} - 2\delta u^0) + g(\delta h^{+} + \delta h^{-} - 2\delta h^0)) \tag{50b}$$

where h^{+}_{exp} and $(uh)^{+}_{\text{exp}}$ are values determined by the explicit equations (46) and (48), respectively. We note that equations (50) are similar in their form to equations (43) of the traditional system. One difference is, however, that (50b) is nonlinear in the unknowns h^{+} and u^{+} whereas (43) were not. This prohibits us deriving a

Helmholtz equation analogous to (45), unless we carry out a linearization of (50b), we therefore do that as follows. The nonlinear term in (50b), the momentum at time level $t + \Delta t$, may be expanded as

$$(uh)^+ = ((U + u')(H + h'))^+ = (u'h')^+ + Hu'^+ + Uh^+.$$

In order to make it linear in h^+ and u'^+ we substitute $(u'h')^0$ for $(u'h')^+$ and get instead

$$(uh)^+ \approx ((u'h')^0 + Hu'^+ + Uh^+). \tag{51}$$

When (51) is inserted in (50b) it becomes after using (50a) and some algebra

$$u'^+ = \frac{1}{H}((uh)^+_{\text{exp}} - Uh^+_{\text{exp}} - (u'h')^0) - \frac{\Delta t g}{\Delta x}(\delta h^+ + \delta h^- - 2\delta h^0) \tag{52}$$

which is of a form similar to (43). We may now proceed as in the traditional system. We write (50a) and (52) as

$$h^+ = \tilde{q}_1 - \frac{\Delta t H}{\Delta x}\,\delta u^+ \tag{53a}$$

$$u'^+ = \tilde{q}_2 - \frac{\Delta t g}{\Delta x}\,\delta h^+ \tag{53b}$$

where, in $\tilde{q}_1$ and $\tilde{q}_2$, we have collected terms which depend only on values at time level $t - \Delta t$ and t. Applying the operator $\delta(\ )$ on (53b) and substituting in (53a) gives the Helmholtz equation

$$h^+ - \frac{\Delta t^2 qH}{(\Delta x)^2}\,\delta^2 h^+ = \tilde{q}_1 - \frac{\Delta t H}{\Delta x}\,\delta\tilde{q}_2 \tag{54}$$

where $\delta^2(\ ) = \delta(\delta(\ ))$. At the end of a time step (54) is solved to obtain h^+. Values of momentum and the velocity fields at time $t = t - \Delta t$ are then computed as follows. Given h^+ we determine δu^+ using (50a). We can then calculate the new values of the momentum $(uh)^+$, from (50b), and finally compute u^+ as $u^+ = (uh)^+/h^+$.

The reason for following this procedure is that by doing so we maintain conservation of mass and momentum. (50a) is satisfied because we used it to derive the Helmholtz equation from which h^+ is determined. When multiplied by Δx times a constant density ρ and then summed over all grid points (50a) gives that the total mass predicted by the semi-implicit system is equal to that predicted by the explicit system. Therefore as the explicit system conserves mass so must the semi-implicit. Similarly when the momentum is obtained from (50b) we can show by a summation over all grid points of this equation that the semi-implicit system conserves momentum.

We have derived a similar semi-implicit model for the cell-integrated energy system. The derivation is quite analogous and we end up with a Helmholtz equation which is formally equal to (54). The expression for $\tilde{q}_2$ is, however, different and instead of (50b) we have an energy equation. $\tilde{q}_2$ for the energy system becomes

$$\tilde{q}_2 = \frac{1}{UH}\left(E^+_{\text{exp}} + \frac{1}{2}gH^2 - \left(\frac{1}{2}U^2 + gH\right)h^+_{\text{exp}} - \frac{1}{2}(u'^2h + gh'^2 + 2Uu'h')^0\right) - \frac{\Delta tg}{\Delta x}(\delta u^- - 2\delta u^0) \tag{55}$$

and the energy equation substituting (50b) is

$$E^+ = E^+_{\text{exp}} - \frac{\Delta t}{\Delta x}\left(gUH(\delta h^+ + \delta h^- - 2\delta h^0) + (\frac{1}{2}U^2H + gH^2)(\delta u^+ + \delta u^- - 2\delta u^0)\right). \tag{56}$$

Results from integrations with the two semi-implicit cell-integrated models were compared with results from corresponding traditional models. The traditional models are building upon the same basic equation systems, i.e. the momentum and the energy system. The difference from the cell-integrated models is that a traditional semi-Lagrangian formulation is used instead of the cell-integrated, just as in the model derived in the beginning of this section, i.e. in (43)–(45). The derivation of the two additional traditional models proceeds completely analogously to that of the semi-implicit cell-integrated models.

Using a time step $\Delta t = 2500$ s, 25 times larger than that used in the explicit runs presented in Fig. 3, the two cell-integrated and the three traditional semi-implicit models were integrated 44 time steps or 30.5 hours. This is the same integration time as in the explicit runs during which period the linear solution to the differential equations moves slightly less than five times through the domain. The resulting height fields at the end of the integration period are shown on top of each other in Fig. 6.

The solutions of the three traditional models are indistinguishable in the figure and all appear as the solid line. It is not yet possible to see any decrease of amplitude for these solutions and they deviate only very little from the sine shape. The cell-integrated model solutions are rather similar to the traditional solutions except for a slight damping which as for the explicit solutions can be ascribed to the choice of piecewise constant functions in grid cells when computing length-weighted values. The phase is similar for all solutions. Due to the semi-implicit time stepping the phase speed c has been reduced compared to c_g, the phase speed of the true linear solution given by (29). For all five integrations we find the same relative phase speed reduction $(c - c_g)/(c_g - U) = -12.4\%$. This corresponds closely to the value computed from a solution to the linearized semi-implicit cell-

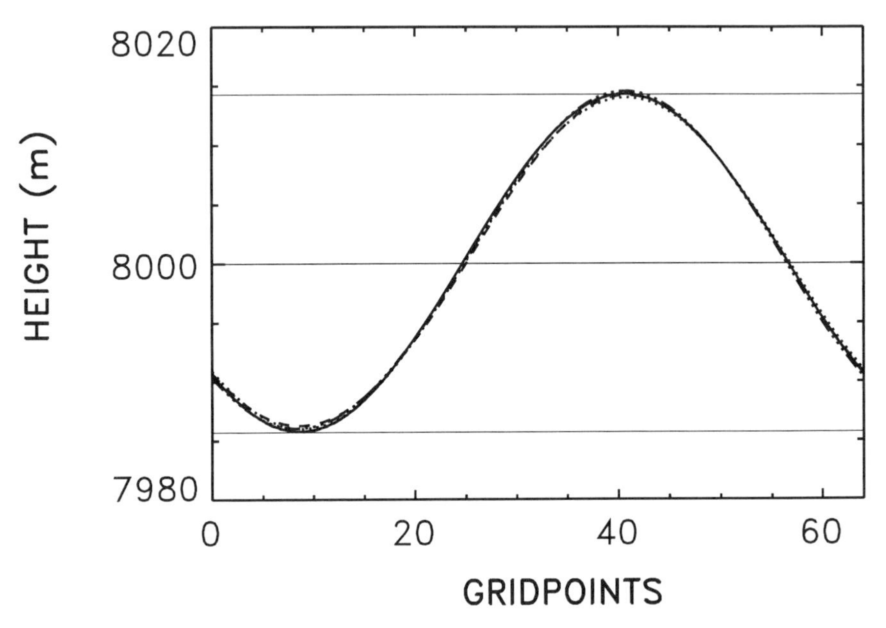

Fig. 6 The height field at time $t = 30.5$ hour for semi-implicit model solutions with $\Delta t = 2500$ s: The cell-integrated momentum system (dotted), the cell-integrated energy system (long dashed), the tradition velocity system (solid), the traditional momentum system (dot-dashed) and the traditional energy system (short dashed).

integrated momentum system corresponding to (47) and (49). The phase speed for solutions to this system is found to be

$$c = U + \frac{1}{2\Delta t k} \arctan \frac{\left(\left(\cos\left(\frac{k\Delta x}{2}\right)\right)^2 - 1\right) S}{(S^2 - 1)\cos\left(\frac{k\Delta x}{2}\right)}, \quad \text{for} \quad k\Delta x \neq \pi \tag{57}$$

where

$$S = \frac{2\Delta t\sqrt{gh}}{\Delta x} \sin \frac{k\Delta x}{2}.$$

The linearization of (47) and (49) proceeds similarly to that carried out in Section 3 for the corresponding explicit system. For the value of $\Delta t = 2500$ s as used in the present experiments we get from (36) that $\mu = 0.5$. This value was used in the derivation of (57). A curve for the relative phase speed $(c - U)/(c_g - U)$ determined from (57) is shown in Fig. 5 as a function of $k\Delta x/\pi$. It is seen that for this large time step, 2500 s, the semi-implicit scheme results in drastic reductions of the phase speed compared to that of the true linear solution, especially for short waves.

As with the explicit models the semi-implicit models were tested in integrations over 62.5 days, using again a weak Robert-Asselin time filter with a coefficient equal to 0.005. The model behaved as expected. During the long integration period the initial gravity wave was damped out completely due to the time filter, the interpolations and the length-weighted mean calculations. In the cell-integrated models the mass and momentum or mass and energy were conserved exactly and in the traditional models they were not.

5 Summary and conclusions

One of the main purposes of the present study was to investigate the stability of the cell-integrated semi-Lagrangian scheme. For that purpose two versions of the scheme were applied to the one-dimensional shallow water equations, a mass and momentum conserving version and a mass and energy conserving version. The one-dimensional shallow water models established are conveniently simple and at the same time they are well suited to the present study as they have gravity wave solutions the stability of which we wanted to study. In spite of the fact that the models conserved mass and momentum or mass and energy exactly, instabilities developed when they were integrated using the explicit leap-frog time extrapolation scheme with a time step exceeding a certain limit. The conservation properties were maintained until the time when trajectories started to cross after which the model "explodes". The CFL condition that determines the maximum permissible time step was derived from a linearized version of the models. The instabilities were found to develop in close agreement with deduction from the linearized system.

Another main purpose of our study was to find a way by which the cell-integrated semi-Lagrangian scheme could be combined with the semi-implicit time stepping. From the start this was not obvious as firstly the divergence has been eliminated and appears indirectly in the trajectory positions and secondly because some of the prognostic variables are nonlinear in the basic variables. For both the energy and momentum conserving versions, these difficulties were overcome and semi-implicit models were established which maintain the conservation properties. The results of test integrations with these models were compared with those obtained from traditional semi-Lagrangian semi-implicit models. Besides the usual traditional model with velocity and height as prognostic variables we also tested traditional models corresponding to the cell-integrated models in which momentum instead of velocity or energy instead of velocity was the second predictive variable. Results of short term integrations were very similar for the different models except for a more severe damping in the cell-integrated models, due to the assumption of piecewise constant values in the grid cells. This damping could have been reduced by assuming variations within the grid cells described by higher order polynomials.

Extended integrations of the cell-integrated models demonstrated the exact fulfilment of their conservation properties whereas the traditional models did not conserve mass, momentum or energy. We have not listed the changes of mass, momentum and energy which were found in the extended integrations because these numbers clearly are not representative of the performance of the respective schemes in realistic large scale models. One result which might be generalized is, however, that among the traditional models better conservation of energy or momentum was obtained when the energy or momentum equation respectively was chosen as a model equation.

The purpose of the test integrations performed were primarily to show that the implementation of the semi-implicit scheme in the cell-integrated systems works as satisfactorily as in the traditional semi-Lagrangian models. Generalization of the combined cell-integrated semi-implicit schemes developed here to the two- and three-dimensional model formulations which were outlined in Machenhauer (1994) now seems straight forward. As usual we shall start with the development of a two-dimensional shallow water model for which useful comparisons of performance and efficiency with traditional schemes can be made.

References

BERMEJO, R. and A. STANIFORTH. 1992. The conversion of semi-Lagrangian advection schemes to quasi-monotone schemes. *Mon. Weather Rev.* **120**: 2622–2632.

GRAVEL, S. and A. STANIFORTH. 1994. A Mass-Conserving Semi-Lagrangian Scheme for the Shallow-Water Equations. *Mon. Weather Rev.* **122**: 243–248.

HIRT, C.W.; A.A. AMSDEN and J.L. COOK. 1974. An arbitrary Lagrangian-Eulerian Computing Method for all Flow Speeds. *J. Comput. Phys.* **14**: 227–253.

KWIZAK, M. and A.J. ROBERT. 1971. A semi-implicit scheme for grid point atmospheric models of the primitive meteorological equations. *Mon. Weather Rev.* **99**: 32–36.

LAPRISE, R. and A. PLANTE. 1995. SLIC: A semi-Lagrangian integrated-by-cell mass conserving numerical transport scheme. *Mon. Weather Rev.* **123(2)**: 553–565.

MACHENHAUER, B. 1994. A note on a Mass-, Energy- and Entrophy Conserving Semi-Lagrangian and Explicit Integration Scheme for the Primitive Meteorological Equations. MPI Workshop on Semi-Lagrangian Methods, Hamburg, 8–9 Oct. 1992. MPI Report 146: pp. 73–102.

MOORTHI, S.; R.W. HIGGINS and J.R. BATES. 1995. A Global Multilevel Atmospheric Model Using a Vector Semi-Lagrangian Finite Difference Scheme. Part 2: Version with Physics. *Mon. Weather Rev.* **123**: 1523–1541.

PRIESTLY, A. 1993. A quasi-conservative version of the semi-Lagrangian advection scheme. *Mon. Weather Rev.* **121**: 621–629.

RANCIC, M. 1992. Semi-Lagrangian piecewise Biparabolic Scheme for Two-Dimensional Horizontal Advection of a Passive Scalar. *Mon. Weather Rev.* **120**: 1394–1406.

ROBERT, A. 1969. The integration of a spectral model of the atmosphere by the implicit method. *In*: Proc. of WMO/IUGG Symp. on NWP, Tokyo, 26 November – 4 December 1968, Japan Meteorol. Agency, VIL.19-VII.24.

———. 1981. A stable numerical integration scheme for the primitive meteorological equations. ATMOSPHERE-OCEAN, **19**: 35–46.

———. 1982. A semi-Lagrangian and semi-implicit numerical integration scheme for the primitive meteorological equations. *Jpn. Meteorol. Soc.* **60**: 319–325.

SMOLARKIEWICZ, P.K. and J.A. PUDYKIEWICZ. 1992. A Class of semi-Lagrangian Approximations for Fluids. *J. Atmos. Sci.* **49**: 2082–2096.

WILLIAMSON, D.L. and J.G. OLSON. 1994. Climate Simulations With a Semi-Lagrangian Version of the NCAR Community Climate Model. *Mon. Weather Rev.* **122**: 1594–1610.

On Forward-in-Time Differencing for Fluids: an Eulerian / Semi-Lagrangian Non-Hydrostatic Model for Stratified Flows

Piotr K. Smolarkiewicz
*National Center for Atmospheric Research**
Boulder, Colorado 80307

and

Len G. Margolin
Los Alamos National Laboratory†
Los Alamos, New Mexico 87545

[Original manuscript received 29 November 1994; in revised form 3 August 1995]

ABSTRACT *In this paper, we describe a non-hydrostatic anelastic model for simulating stratified flows in terrain-following coordinates. The model is based solely on non-oscillatory forward-in-time integration schemes, and our primary goal is to demonstrate the utility of such methods for modelling small-scale atmospheric dynamics. We use the formal similarity of the Eulerian and semi-Lagrangian equations of two-time-level approximations to construct a unified model that readily allows selection of either formulation. We apply the model to two test problems of stratified flows past isolated obstacles. We use these tests to validate the forward-in-time approach against a traditional centred-in-time-and-space Eulerian model, and to discuss the relative accuracy and efficiency of the two formulations of our model. One problem illustrates the efficacy of flux-form Eulerian methods, while the other demonstrates strengths of the semi-Lagrangian approach.*

RÉSUMÉ *On décrit un modèle non hydrostatique anélastique simulant des écoulements stratifiés dans les coordonnées épousant la forme du terrain. Il est basé entièrement sur des schémas d'intégration apériodiques à pas de temps vers l'avant, et l'on cherche principalement à démontrer l'utilité de telles méthodes pour la modélisation de la dynamique atmosphérique à petite échelle. On utilise la similarité formelle des équations eulériennes et semi-lagrangiennes discrétisées à deux niveaux temporels pour construire un modèle unifié qui permet facilement la sélection de l'une ou l'autre des formules. On applique le modèle à*

---

*The National Center for Atmospheric Research is sponsored by the National Science Foundation.
†Los Alamos National Laboratory is operated by the University of California for the U.S. Department of Energy.

deux essais d'écoulement stratifié au dessus d'obstacles isolés. Ces essais servent à valider l'approche du pas en avant par rapport au modèle eulérien traditionel avec pas temporel et spacial centrés, et à examiner la précision relative et l'efficacité des deux formules de notre modèle. Un essai montre l'efficacité des méthodes eulériennes sous forme de flux et l'autre, les avantages de l'approche semi-lagrangienne.

---

1 Introduction

Atmospheric and oceanic motions are characterized by large Reynolds' number implying that the advective transport of the fluid variables plays an important and often dominant role in the evolution of the flow. Thus the accurate treatment of transport processes in numerical models for atmospheric and oceanic fluids becomes an essential aspect of accurate flow simulations. For decades, finite-difference modelling of geophysical motions has been dominated by centred-in-time-and-space three-time-level methods (hereafter, CTS). The popularity of CTS schemes may be attributed to their conceptual simplicity, which derives from using mid-point integration rules that make at least second-order-accurate approximations to the evolution equations straightforward, regardless of the complexity of associated forcings. In geophysical applications, such complexity arises primarily from the representation of the governing fluid equations in a rotating, curvilinear system of coordinates. Notwithstanding their popularity, there are many disadvantages to CTS methods including large dispersive errors, multiple solutions and the lack of nonlinear stability. In atmospheric models, the latter two are usually mitigated by means of explicit high-frequency filters (Lilly, 1962; Smagorinsky, 1963; Robert, 1966; Robert et al., 1970; Shapiro, 1970; Asselin, 1972) and specialized conservative versions of CTS schemes (Lilly, 1965; Arakawa, 1966).

The class of forward-in-time, two-time-level methods (often referred to as one-step Lax-Wendroff, Crowley, or dissipative schemes; hereafter, FT) have been discussed by the atmospheric community since the mid-sixties (Leith, 1965; Crowley, 1968). However, FT schemes have only rarely been considered for modelling complex flows. During the last two decades, FT methods have been improved by numerous developments in the areas of multidimensionality, non-Cartesian geometries, monotonicity and/or sign preservation, as well as overall accuracy and stability (see LeVeque, 1996, for a succinct review). Among these, developments of various nonoscillatory techniques (Van Leer, 1974; Zalesak, 1979; Harten, 1983; Smolarkiewicz, 1984; Leonard, 1988) made FT methods particularly attractive for estimating the transport of sign-preserving thermodynamic fields–e.g., water vapour, salinity, potential temperature and density, and chemical constituents. As a result, many atmospheric and oceanic models now employ a hybrid approach in which the kinematic equations use CTS while the thermodynamic equations use FT methods (e.g. Smolarkiewicz and Clark, 1986; Schumann et al., 1987; Tripoli, 1992; Bleck et al., 1992; Lin et al., 1994). Although the advantages of FT methods are well-recognized in the context of modelling the transport of tracers, their utility for modelling dynamics is not widely appreciated.

During the last decade, the atmospheric community has invested substantial effort in the development of semi-Lagrangian methods. These methods differ significantly from traditional Eulerian schemes. In particular, semi-Lagrangian algorithms integrate the fluid equations along flow trajectories rather than at the grid point, i.e. in the sense of ordinary- rather than partial-differential equations. In contrast to Eulerian models, early semi-Lagrangian models for atmospheric flows (cf. Krishnamurti, 1962) employed two- rather than three-time-level discretizations. The seminal work of Robert (1981) has rejuvenated interested in semi-Lagrangian methods for numerical weather prediction (NWP) and has stimulated the rapid development of two-time-level approximations (see Staniforth and Côté, 1991, for a review). The principal advantage of two- over three-time-level schemes–twice better accuracy at a given computational expense–has been often acknowledged in the context of semi-Lagrangian approach (cf. McDonald and Bates, 1987; Temperton and Staniforth, 1987). At present, a number of NWP models (e.g. Krishnamurti et al., 1990; McDonald and Haughen, 1992; Bates et al., 1993) rely on two-time-level semi-Lagrangian schemes demonstrating their utility for simulating regional- and global-scale flows.

Recently, Smolarkiewicz and Margolin (1993) discussed the design of fully second-order-accurate, forward-in-time approximations to a prototype prognostic equation for fluids

$$\frac{\partial G\phi}{\partial t} + \nabla \cdot (\mathbf{v}\phi) = GR. \tag{1}$$

Here ϕ is a fluid variable, $G = G(\mathbf{x})$ is the Jacobian of the coordinate transformation from the Cartesian to the curvilinear framework $\mathbf{x}$, $\mathbf{v} = G\dot{\mathbf{x}}$ is the generalized advective velocity vector,

$$\nabla \equiv \left(\frac{\partial}{\partial x^1}, \ldots, \frac{\partial}{\partial x^M}\right),$$

and R combines all forcings and/or sources. In general, both $\mathbf{v}$ and R are functionals of the fluid variables. Through a rigorous truncation error analysis of the forward-in-time differencing of the left-hand-side of (1), they derived a class of finite-difference approximations

$$\phi_{\mathbf{i}}^{n+1} = \mathcal{A}_{\mathbf{i}}\left(\phi^n + \frac{\Delta t}{2}R^n,\ \boldsymbol{\alpha}^{n+1/2},\ G\right) + R_{\mathbf{i}}^{n+1}\frac{\Delta t}{2}. \tag{2}$$

Here: $\mathcal{A}$ denotes a second-order-accurate flux-form FT scheme for integrating the homogeneous transport equation

$$\frac{\partial G\psi}{\partial t} + \nabla \cdot (\mathbf{v}\psi) = 0, \tag{3}$$

for an arbitrary variable ψ; also,

$$\boldsymbol{\alpha} \equiv \left(\frac{v^1 \Delta t}{\Delta X^1}, \ldots, \frac{v^M \Delta t}{\Delta X^M} \right)$$

is a vector of modified local Courant numbers; $\mathbf{i}$ denotes the position ($\mathbf{x} = \mathbf{i} \cdot \boldsymbol{\Delta} \mathbf{X}$) on an M-dimensional, regular grid; and the Δt and n-superscripts have the usual meaning of the time step and temporal levels, respectively. The advection of the auxiliary field $\phi + R\Delta t/2$ (rather than the fluid variable alone) is important for preserving the global accuracy and stability of the FT approximations. Use of the auxiliary field compensates a first-order truncation error term proportional to the divergence of convective flux of the forcing. This particular error term appears in those "naive" approximations to (1) that simply combine (in the spirit of CTS methods) an $\mathcal{A}$ advection scheme with a second-order-accurate approximation to $R^{n+1/2}$. Ignoring this error leads to spurious $\sim O(\Delta t)$ sinks/sources of "energy" ϕ^2 (appendix A in Smolarkiewicz and Margolin, 1993). Advecting the auxiliary field in (2) has the physical interpretation of integrating the forces along a parcel trajectory rather than at the grid point, which makes (2) congruent to the two-time-level semi-Lagrangian approximation

$$\phi_{\mathbf{i}}^{n+1} J = \phi_o^n + (R_{\mathbf{i}}^{n+1} J + R_o^n) \frac{\Delta t}{2} \tag{4}$$

of the Lagrangian counterpart of (1)

$$\frac{d\phi J}{dt} = RJ. \tag{5}$$

In this equation

$$J \equiv \frac{G(\mathbf{x})}{G(\mathbf{x}_o)} det \left\{ \frac{\partial \mathbf{x}}{\partial \mathbf{x}_o} \right\}$$

is the flow Jacobian, $(\mathbf{x}_o, t_o)$ is the departure point of the trajectory arriving at $(\mathbf{x}, t)$, and the subscript "o" denotes a field value at the departure point of the trajectory arriving at the grid point $(\mathbf{i}, n+1)$.

In this paper we exploit the similarities between the Eulerian (point-wise) and the Lagrangian (trajectory-wise) discretization of the governing equations to design a unified FT model. This combination allows us to easily assess and choose between the strengths and weaknesses of both approaches. As a conceptual framework, we consider idealized, anelastic, non-hydrostatic equations of small-scale dynamics cast in terrain-following coordinates. The theory underlying the Eulerian and semi-Lagrangian numerical algorithms has been documented in earlier works, and their basic properties of computational stability, accuracy, and efficiency are already established. In this paper we focus on the overall performance of the FT approach in representative complex atmospheric problems. Our primary goal is to demonstrate that the FT approach can be used advantageously for a wide range of applications. Our secondary goal is to show that the semi-Lagrangian and Eulerian approaches are complementary rather than competitive, and that their relative accuracy depends on the physical processes modelled.

The paper is organized as follows. In the next section, we briefly summarize the governing anelastic system of equations. The finite difference approximations, in the spirit of (2) or (4), are discussed in Section 3. Section 4 contains examples of applications and comparisons of results including those of a more traditional anelastic model based on CTS. We present our conclusions in Section 5.

2 Governing equations

The anelastic approximation adopted in our model follows the formulation of Lipps (1990) and Lipps and Hemler (1982). The curvilinear coordinate transformation is based on the presentation of Gal-Chen and Somerville (1975); however, to further improve the accuracy, the equations of motion are formulated in terms of the covariant (Cartesian) velocity components (cf. Clark (1977) and the recent work of Kapitza and Eppel (1992)). In this paper, we consider an inviscid, adiabatic, density-stratified fluid whose undisturbed, geostrophically-balanced environmental state is described by the profiles of the potential temperature and velocity: $\Theta_e = \Theta_e(\mathbf{x}_c)$, and $\mathbf{v}_e = [u_e(\mathbf{x}_c), v_e(\mathbf{x}_c), 0]$, respectively. The subscript c refers to Cartesian coordinates throughout the paper. We use a nonorthogonal terrain-following system of coordinates

$$[x, y, z] = [x_c, y_c, H(z_c - h)/(H - h)], \tag{6}$$

that assumes a model depth H and an irregular (but at least twice differentiable) lower boundary $h = h(x_c, y_c)$. The coordinate transformation enters the governing equations of motion through the metric coefficients

$$G^{IJ} = \sum_{K=1}^{M} \frac{\partial x^I}{\partial x_c^K} \frac{\partial x^J}{\partial x_c^K}, \tag{7a}$$

and the Jacobian of the transformation

$$G = det\left\{\frac{\partial \mathbf{x}_c}{\partial \mathbf{x}}\right\} = (det\{G^{IJ}\})^{-1/2}. \tag{7b}$$

Given the assumptions above, the anelastic equations may be written as follows:

$$\frac{du}{dt} = -\frac{\partial \pi}{\partial x} - G^{13}\frac{\partial \pi}{\partial z} + f(v - v_e) - \alpha(u - u_e) \equiv F^u, \tag{8a}$$

$$\frac{dv}{dt} = -\frac{\partial \pi}{\partial y} - G^{23}\frac{\partial \pi}{\partial z} - f(u - u_e) - \alpha(v - v_e) \equiv F^v, \tag{8b}$$

$$\frac{dw}{dt} = -G^{-1}\frac{\partial \pi}{\partial z} + g\frac{\Theta - \Theta_e}{\bar{\Theta}} - \alpha(w - w_e) \equiv F^w, \tag{8c}$$

$$\frac{d\Theta}{dt} = -\alpha'(\Theta - \Theta_e) \equiv F^{\Theta}, \tag{8d}$$

$$\frac{\partial \bar{\rho} G u}{\partial x} + \frac{\partial \bar{\rho} G v}{\partial y} + \frac{\partial \bar{\rho} G \omega}{\partial z} = 0. \tag{8e}$$

Here π is the pressure perturbation with respect to the undisturbed environmental profile normalized by the anelastic density, (Bacmeister and Schoeberl, 1989; Lipps, 1990); Θ is the potential temperature; f is the Coriolis parameter (the examples in Section 4 are, however, for $f \equiv 0$), g is the acceleration due to gravity; and $\omega \equiv \dot{z}$ is the contravariant "vertical" velocity component related to the covariant velocity components through

$$\omega = G^{-1}w + G^{13}u + G^{23}v. \tag{9}$$

The potential temperature $\bar{\Theta} = \bar{\Theta}(z_c)$ appearing in the denominator of the buoyancy term in (8c), and the anelastic density $\bar{\rho} = \bar{\rho}(z_c)$ in the mass continuity equation (8e) refer to the hydrostatic reference state of the anelastic expansion around a constant stability profile (for discussions of reference-state profiles, see Clark and Farley (1984), Section 2b, and Bacmeister and Schoeberl (1989), Section 2a). The attenuation forcings appearing in the momentum and entropy equations (8a)–(8d) simulate wave-absorbing devices in the vicinity of the open boundaries of the numerical model (Clark, 1977; Davies, 1983; Kosloff and Kosloff, 1986).

The full model includes a time-dependent terrain-following coordinate transformation (Prusa et al., 1993), and "moist" thermodynamics with a bulk, warm-rain parametrization of water-substance phase-change processes (Grabowski and Smolarkiewicz, 1996). A fully nonlinear, incompressible Navier-Stokes' formulation of the "dry" model is also available as an option (Rotunno and Smolarkiewicz, 1996). Although these generalizations are not used in this paper, they have influenced the formulation of the model algorithm.

3 Finite-difference approximations

a *Eulerian Formulation*

In the Eulerian formulation of the model, the prognostic equations (8a)–(8d) are cast in conservation form

$$\frac{\partial \rho^* \psi}{\partial t} + \nabla \cdot (\boldsymbol{\vartheta}\, \psi) = \rho^* F^{\psi}, \tag{10}$$

where $\rho^* \equiv \bar{\rho} G$, $\boldsymbol{\vartheta} \equiv \bar{\rho} G \mathbf{V}$, and ψ and F^{ψ} represent any of the dependent variables (u, v, w, Θ) and its associated forcing $(F^u, F^v, F^w, F^{\Theta})$. The formal equivalence of

(1) and (10) implies a second-order-accurate integral of (10) in the form analogous to (2)

$$\psi_{\mathbf{i}}^{n+1} = \mathcal{A}_{\mathbf{i}}(\tilde{\psi}^n, \boldsymbol{\alpha}^{n+1/2}, \rho^*) + 0.5\Delta t F^{\Psi}|_{\mathbf{i}}^{n+1}. \tag{11}$$

Here, $\tilde{\psi} \equiv \psi + 0.5\Delta t F^{\Psi}$, and

$$\boldsymbol{\alpha} \equiv \left(\frac{\vartheta^1 \Delta t}{\Delta X^1}, \ldots, \frac{\vartheta^M \Delta t}{\Delta X^M} \right)$$

denotes the vector of modified, density-weighted local Courant numbers. In the model, the operator $\mathcal{A}$ is based on the nonoscillatory, second-order-accurate, flux-form FT algorithm MPDATA. The theory, properties, and performance of the MPDATA family of schemes have been documented in Smolarkiewicz (1984), Smolarkiewicz and Clark (1986), Margolin and Smolarkiewicz (1989) and Smolarkiewicz and Grabowski (1990). The temporal staggering of $\boldsymbol{\alpha}$ to the $n + 1/2$ time-level is approximated by linear extrapolation

$$\boldsymbol{\alpha}^{n+1/2} = 0.5(3\boldsymbol{\alpha}^n - \boldsymbol{\alpha}^{n-1}) + O(\Delta t^2), \tag{12}$$

which ensures $\nabla \cdot \boldsymbol{\alpha}^{n+1/2} = 0$ given $\nabla \cdot \boldsymbol{\alpha}^{n-1}$, $\nabla \cdot \boldsymbol{\alpha}^n = 0$. Both the temporal staggering and preservation of "anelastic incompressibility" of the advective velocities are important for achieving second-order accuracy in (11) (cf. section 2 in Smolarkiewicz and Margolin (1993)). The numerical stability of the approximation in (11) requires appropriate bounds on $\|(\bar{\rho}G)^{-1}\boldsymbol{\alpha}\|$–the magnitude of the dimensionless velocity on the grid.

b *Semi-Lagrangian Formulation*

In the semi-Lagrangian formulation of the model, all prognostic equations (8a)–(8d) are viewed in the compact form

$$\frac{d\psi}{dt} = F^{\Psi}, \tag{13}$$

which is equivalent to (5) by the virtue of the Lagrangian form of the mass continuity equation $\rho J = \rho_o$ (Truesdell, 1966). The second-order-accurate integral of (13) along a parcel trajectory employs the trapezoidal-rule approximation (4)

$$\psi_{\mathbf{i}}^{n+1} = (\psi + 0.5\Delta t F^{\Psi})_o + 0.5\Delta t F^{\Psi}|_{\mathbf{i}}^{n+1} \equiv \tilde{\psi}_o + 0.5\Delta t F^{\Psi}|_{\mathbf{i}}^{n+1}. \tag{14}$$

The mapping of a $\tilde{\psi}$ auxiliary field to the departure points $(\mathbf{x}_o, t^n)$ of the trajectories arriving at $(\mathbf{x}_{\mathbf{i}}, t^{n+1})$ is based on a second- or higher-order-accurate, nonoscillatory, interpolation algorithm that is built upon FT Eulerian advection schemes for integrating the homogeneous transport equation (3) with constant coefficients (Smolarkiewicz and Grell, 1992). The departure points $\mathbf{x}_o(\mathbf{x_i}, t^n)$ are evaluated from the trajectory equation

$$\frac{d\mathbf{x}}{dt} = \mathbf{V} \equiv [u,\ v,\ \omega], \tag{15}$$

(hereafter, $\mathbf{V}$ refers to the contravariant velocity vector) using a second-order-accurate, implicit algorithm based on Taylor series expansion (Smolarkiewicz and Pudykiewicz (1992), Section 4b and Appendix)

$$\mathbf{x}_o = \mathbf{x_i} - \Delta t(\mathbf{V} + 0.5\Delta t\mathbf{F}^{\mathbf{V}})_o \equiv \mathbf{x_i} - \Delta t\tilde{\mathbf{V}}_o, \tag{16}$$

where $\mathbf{F}^{\mathbf{V}} = [F^u, F^v, F^\omega]$.[1] With a first guess $\mathbf{x}_o = \mathbf{x_i} - \Delta t\tilde{\mathbf{V}}|_{\mathbf{i}}^n$, one iteration of (16) suffices for a second-order-accurate prediction of $\mathbf{x}_o$, provided that the Lipschitz number $\|(\partial\tilde{\mathbf{V}}/\partial\mathbf{x})\Delta t\|$ is appropriately bounded. Note that stability of the Eulerian formulation requires limiting the magnitude of the dimensionless velocity, whereas the semi-Lagrangian formulation requires limitations of its variation.

c *Eulerian/Semi-Lagrangian Model Algorithm*

The general forms of (11) and (14) are analogous, except that the semi-Lagrangian algorithm replaces the conservative Eulerian transport of the auxiliary field $\tilde{\psi}$ by the remapping of $\tilde{\psi}$ to the departure points $\mathbf{x}_o$. By virtue of the formal analogy of the two algorithms, both (11) and (14) can be explicitly written for the anelastic system (8) as

$$u = LE(\tilde{u}) - 0.5\Delta t\left(\frac{\partial\pi}{\partial x} + G^{13}\frac{\partial\pi}{\partial z}\right) + 0.5\Delta t f(v - v_e) - 0.5\Delta t\alpha(u - u_e) \tag{17a}$$

$$v = LE(\tilde{v}) - 0.5\Delta t\left(\frac{\partial\pi}{\partial y} + G^{23}\frac{\partial\pi}{\partial z}\right) - 0.5\Delta t f(u - u_e) - 0.5\Delta t\alpha(v - v_e) \tag{17b}$$

$$w = LE(\tilde{w}) - 0.5\Delta t G^{-1}\frac{\partial\pi}{\partial z} + 0.5\Delta t g\frac{\Theta - \Theta_e}{\bar{\Theta}} - 0.5\Delta t\alpha(w - w_e), \tag{17c}$$

$$\Theta = LE(\tilde{\Theta}) - 0.5\Delta t\alpha'(\Theta - \Theta_e), \tag{17d}$$

where the operator LE denotes either Eulerian or semi-Lagrangian advection procedures in (11) and (12) or (14) and (16), respectively. To simplify the notation, we have dropped all $n+1$ and $\mathbf{i}$ indices on both sides of (17)–there is no ambiguity since only the arguments of LE refer to preceding time levels.

[1] The optional procedures compute the departure points in the spirit of Adams-Bashfort schemes via $\mathbf{x}_o = \mathbf{x_i} - 0.5\Delta t(\mathbf{V}_o + 2\mathbf{V}_{\mathbf{i}}^n - \mathbf{V}_{\mathbf{i}}^{n-1})$ or $\mathbf{x}_o = \mathbf{x_i} - \Delta t(1.5\mathbf{V}^n - 0.5\mathbf{V}^{n-1})|_{05(x_\mathbf{i}+x_o)}$ in lieu of (16); (cf. Robert, 1981; Staniforth and Côté, 1991).

The relationships (17a)–(17c) already incorporate the assumption that all prognostic variables are defined at the same grid points $\mathbf{x_i}$. In the semi-Lagrangian approach, this is the most efficient and most accurate choice insofar as the advection is concerned since all prognostic variables are advanced along the same trajectory. Staggered meshes where each component of velocity is defined at a different position (e.g., the Arakawa C-grid) are cumbersome as they either lead to a complex geometric problem in lieu of (16), or require multiple evaluations of the departure points per grid box and/or averages of different velocity components to common trajectories (cf. Robert, 1993). In our model we allow two grid configurations: the unstaggered A-grid, where all variables are defined at the same positions, and the staggered B-grid, where a pressure variable is staggered one-half grid interval in all directions with respect to the other variables. An important advantage of such grids, common to both the semi-Lagrangian and Eulerian algorithms in (17), is that they allow an algebraic inversion of the implicit system (17) and, consequently, a straightforward formulation of the boundary value problem for π necessary to complete the time-step advancement of the model variables.[2]

d *Boundary Value Problem*

An elliptic pressure equation follows from the diagnostic mass-continuity constraint (8e) imposed on the model algorithm (17) (cf. Chorin, 1968; Clark, 1977; Kapitza and Eppel, 1992). Its derivation is summarized briefly as follows: Since (17d) implies

$$\Theta = \frac{LE(\tilde{\Theta}) + 0.5\Delta t\alpha'\Theta_e}{1 + 0.5\Delta t\alpha'}, \tag{18}$$

the buoyancy term in (17c) is a known part of the solution, and (17a)–(17c) can be inverted algebraically with respect to u, v, and w. The resulting expressions are inserted in the ω-equation (9), which reduces (17a)–(17c) to

$$V^I = \mathcal{E}\left(\mathcal{V}^I - \sum_{J=1}^{3} C^{IJ}\frac{\partial\pi'}{\partial x^J}\right). \tag{19}$$

Here V^I denotes the components of the contravariant velocity vector $[u, v, \omega]$, $\pi' \equiv 0.5\Delta t\pi$, and the coefficients $\mathcal{E}$, $\mathcal{V}^I$, and C^{IJ}, which vary spatially, are provided in the Appendix. Given the contravariant representation of the solution in (19), the elliptic pressure equation follows by substitution in (8e). That is, multiplying the right-hand-side of (19) by $\rho^* \equiv \bar{\rho}G$ and taking the divergence of the resulting expressions leads to

$$\sum_{I=1}^{3}\frac{\partial}{\partial x^I}\left[\rho^*\mathcal{E}\left(\mathcal{V}^I - \sum_{J=1}^{3} C^{IJ}\frac{\partial\pi'}{\partial x^J}\right)\right] = 0. \tag{20}$$

[2]This is especially convenient in problems with significant rotation, where the Coriolis force demands a centred temporal discretization of the inertial terms.

The latter is solved, subject to appropriate boundary conditions imposed on the normal components of V^I in (19), using the iterative generalized-conjugate-residual approach of Eisenstat et al. (1983). Algorithmic details and a discussion of the solver's performance can be found in Smolarkiewicz and Margolin (1994). Given a solution to (20), we next use (19) to advance u, v, and ω in time; w is then diagnosed from (9). Finally, having advanced all model variables in time, we reevaluate the F^{Ψ} forcings required in the next time step from either (11) or (14).

The conceptual design of this anelastic model is in the spirit of those in Clark (1977) and Kapitza and Eppel (1992). The distinctive aspect of our approach is the two-time-level temporal discretization employing the same integral rules (11) or (14) for all model variables.

4 Examples of applications

a *Introductory Remarks*

Here we compare results of the semi-Lagrangian and Eulerian versions of the FT model with each other, as well as those of a more traditional CTS type anelastic code. The first example illustrates the virtues of the conservative Eulerian approach. It simulates a non-Boussinesq non-hydrostatic gravity wave, which spans both linear and nonlinear flow regimes. By contrast, the second example emphasizes the advantages of the semi-Lagrangian approach. It simulates the flow of a strongly stratified Boussinesq fluid past a three-dimensional hill–a problem that is well-studied in the literature. This highly nonlinear flow is characterized by an abundance of complex, yet coherent structures.

The unified FT model outlined in the preceding section, contains a number of optional variants of the Eulerian transport and semi-Lagrangian remapping schemes denoted symbolically by *LE* in (17). These offer different compromises among the levels of accuracy, efficiency and complexity. In order to make our comparisons as fair as possible, all the calculations reported in this paper employ (unless stated otherwise) fully monotone, second-order-accurate dissipative advection schemes. Here the descriptor "monotone" is used in the sense of flux-corrected transport (FCT)–see Zalesak (1979); Smolarkiewicz and Grabowski (1990); and Smolarkiewicz and Grell (1992). Moreover, we use a special MPDATA option in which the basic positive-definite scheme is linearized with respect to a large constant background (cf. Section 4 in Smolarkiewicz and Clark, 1986). This minimizes the differences between the *LE* operators used in both versions of the model. For instance, in a one-dimensional, constant-coefficient advection problem with the Courant number smaller than 0.5, the *LE* operators become identical except for some higher-order terms that are of no relevance to the conclusions derived.

b *Breakdown of a Vertically Propagating Gravity Wave*

In this problem we simulate the development of convective instability that results from the non-Boussinesq amplification of an interval gravity wave forced by a finite-amplitude, two-dimensional mountain (see Bacmeister and Schoeberl, 1989). The mountain is bell-shaped, $h = h_o[1 + (x/L)^2]^{-1}$ with $h_o = 628.319$ m and $L = 10^3$ m. The environmental wind is uniform, $u_e(z_c) = U = 10$ m s^{-1}. We assume: constant stability $\partial(\ln \Theta_e)/\partial z_c \equiv N^2/g = 10^{-5}$ m^{-1}; $\bar{\Theta} \equiv \Theta_e$; and constant density stratification $-\partial(\ln \bar{\rho})/\partial z_c \equiv M^2 = 1.535 \cdot 10^{-4}$ m^{-1}. The model domain is $(x, z) \in [-H, H] \times [0, H]$ with $H = 6 \cdot 10^4$ m. We cover the domain with a uniform grid with spatial increment $\Delta X = \Delta Z = 5 \cdot 10^2$ m. The boundary conditions are rigid in both directions (namely, $\omega = 0$ at $z = 0, H$ and $u = u_e$ at $x = \pm H$); the gravity-wave absorbers near the lateral and upper boundaries attenuate the solution toward environmental conditions with an inverse time scale α in (17) that increases linearly from zero at the distance $2 \cdot 10^4$ m from the boundary to 600^{-1} s^{-1} at the boundary. The initial condition is the potential flow $\nabla \times (\mathbf{v} - \mathbf{v_e}) = 0$.

This choice of model parameters leads to a particular gravity-wave solution with the following characteristics. First, the dimensionless combination of the environmental wind, stability, and mountain width, $U/NL \approx 1$, implies that the horizontal and vertical wavenumbers of the standing mountain wave are about equal, and that the problem is inherently non-hydrostatic. Second, the choice of the anelastic density profile assures an amplification of the mountain-induced perturbations of about one order of magnitude at $z_c = 0.5H$. At this height the wave amplitude and length become comparable, meaning that waves at this and higher altitudes should overturn. We estimate the group velocity at 5 m s^{-1}, approximately equal in all directions, and so expect that at some time between the third and fourth hour of simulation the stable (gravity-wave type) and unstable (convective type) flow regimes should occupy about equal portions of the model depth. We have chosen our grid to assure that both flow regimes are resolved with the same spatial accuracy, i.e., both the length of the standing wave and the spatial scale of convective eddies are covered with the same number of grid intervals ($\gtrsim 12\Delta X$ in both directions).

Figure 1 shows the semi-Lagrangian solution after 3 and 3.5 hours of the simulation on the A grid with $\Delta t = 20$ s. Note the near verticality of the isentropes at the wave crests in the upper portion of the model domain in Fig. 1a, and their apparent overturning in Fig. 1b; the corresponding calculation on the B grid (not shown) exhibits a slightly faster wave development. Figures 2 and 3 show equivalent solutions generated, respectively, with the Eulerian version of the FT model and an Eulerian, hybrid model (Clark, 1977; Smolarkiewicz and Clark, 1986) that employs flux-form CTS (FT) schemes for the transport of momentum (potential temperature). The solution in Fig. 2 employs the unstaggered A-grid and $\Delta t = 15$ s; the B-grid result (not shown) is similar. The CTS solution in Fig. 3 uses a traditional staggered C-grid discretization and requires a smaller time step for stability (here $\Delta t = 10$ s was used; no attempt has been made to establish the largest Δt permitted). Although the two Eulerian solutions differ in some details, their overall similarity is

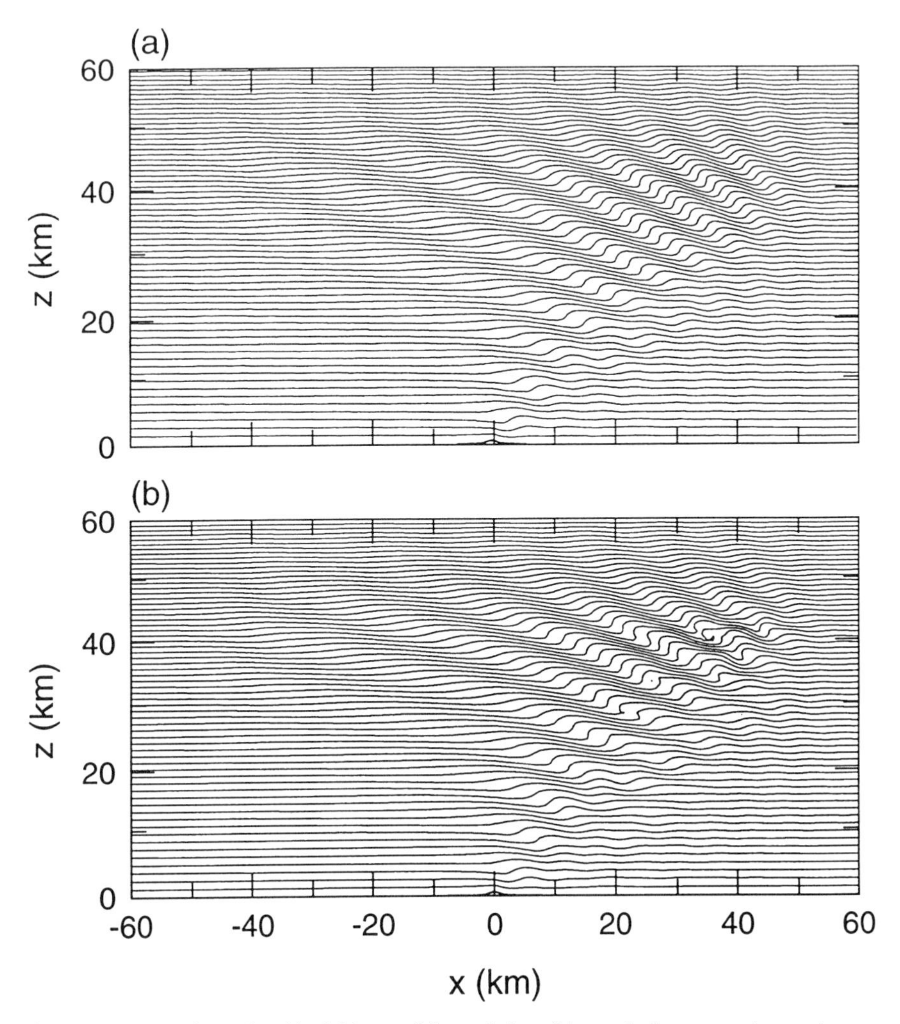

Fig. 1 Isentropes after a) 3 and b) 3.5 hours of the evolution of the vertically propagating gravity wave simulated with the semi-Lagrangian model on the A-grid. Contour interval is 4K.

remarkable, especially when the dramatic differences in their algorithmic designs[3] as well as the highly nonlinear, transient character of the developed flow are taken into account. In contrast, the differences between the Eulerian (Fig. 2 or 3) and the semi-Lagrangian (Fig. 1) approximations are much more pronounced. Although the Eulerian solutions seem to capture roughly the same stage of the wave field

[3]In addition to different temporal and spatial discretization schemes, the two models employ different "open-boundary" approximation schemes, and elliptic solvers with different residual errors.

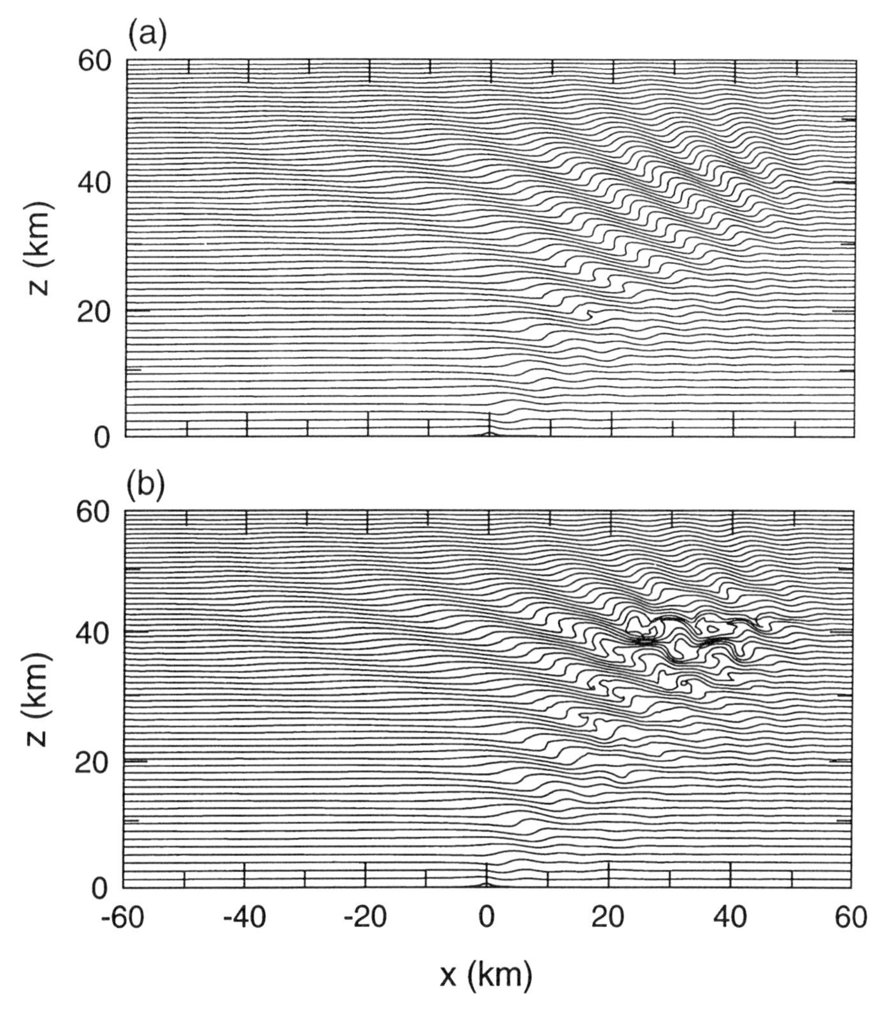

Fig. 2 As in Fig. 1 but after a) 2.5 and b) 3 hours of gravity wave evolution simulated with the Eulerian model.

evolution as that in the semi-Lagrangian results, they are shown after 2.5 and 3 h (as opposed to 3 and 3.5 h), thus predicting the onset of wave breaking about half an hour earlier in the Eulerian simulations.

Which results are closer to the true solution, and which element of the models' design is primarily responsible for the differences observed? It appears that the Eulerian models are more accurate for this problem. This conclusion is supported by double-resolution ($\Delta X = \Delta Z = 250$ m) experiments (Fig. 4), which show both semi-Lagrangian and Eulerian results after 3 h of simulation on the A-grid. In essence, both experiments capture the same stage of the wave development:

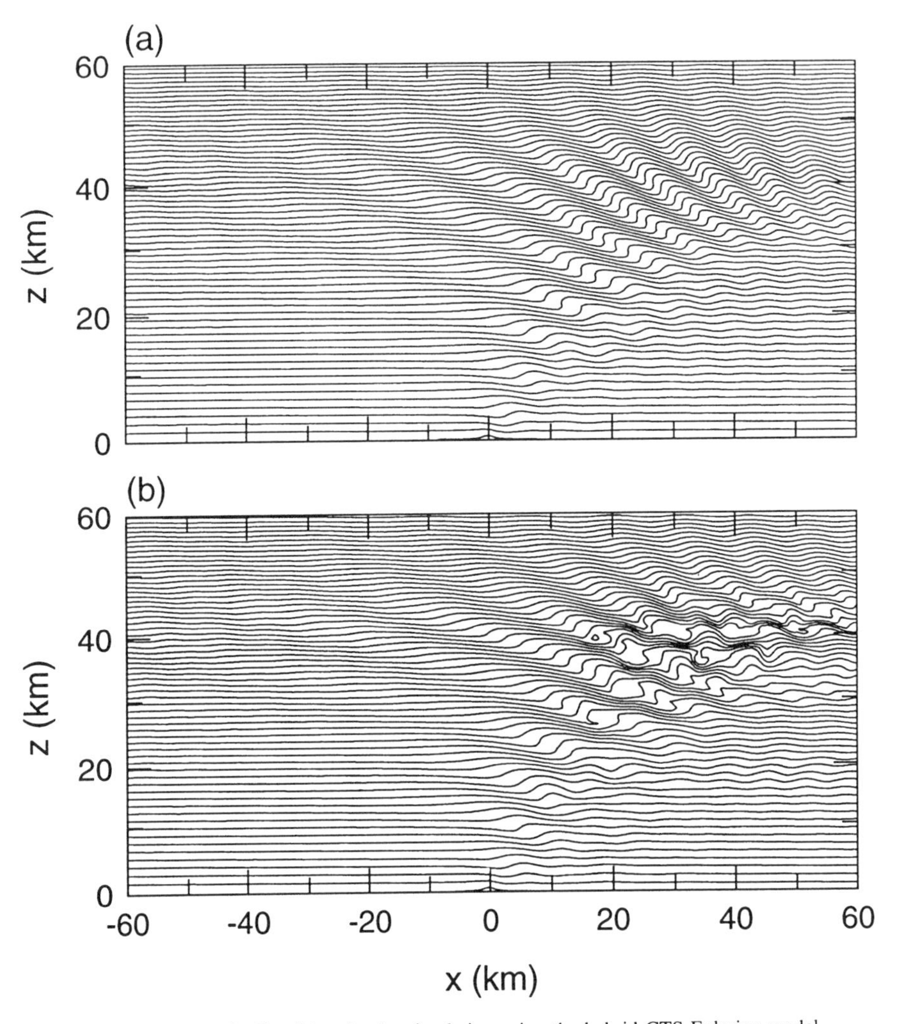

Fig. 3 As in Fig. 2 but for the simulation using the hybrid CTS Eulerian model.

comparing Fig. 4a with Fig. 1a and Fig. 4b with Fig. 2b shows that doubling the spatial resolution accelerates the wave breaking in the semi-Lagrangian model by about half an hour, but affects only fine details of the Eulerian results. The onset of wave breaking is similarly unaffected in the double resolution CTS run, and is not shown.

The overall character of the semi-Lagrangian solutions suggests that the semi-Lagrangian calculations suffer from some small implicit viscosity absent in the Eulerian model. Indeed, by introducing an explicit viscosity of constant normalized magnitude two orders smaller than the advective Courant number, we have

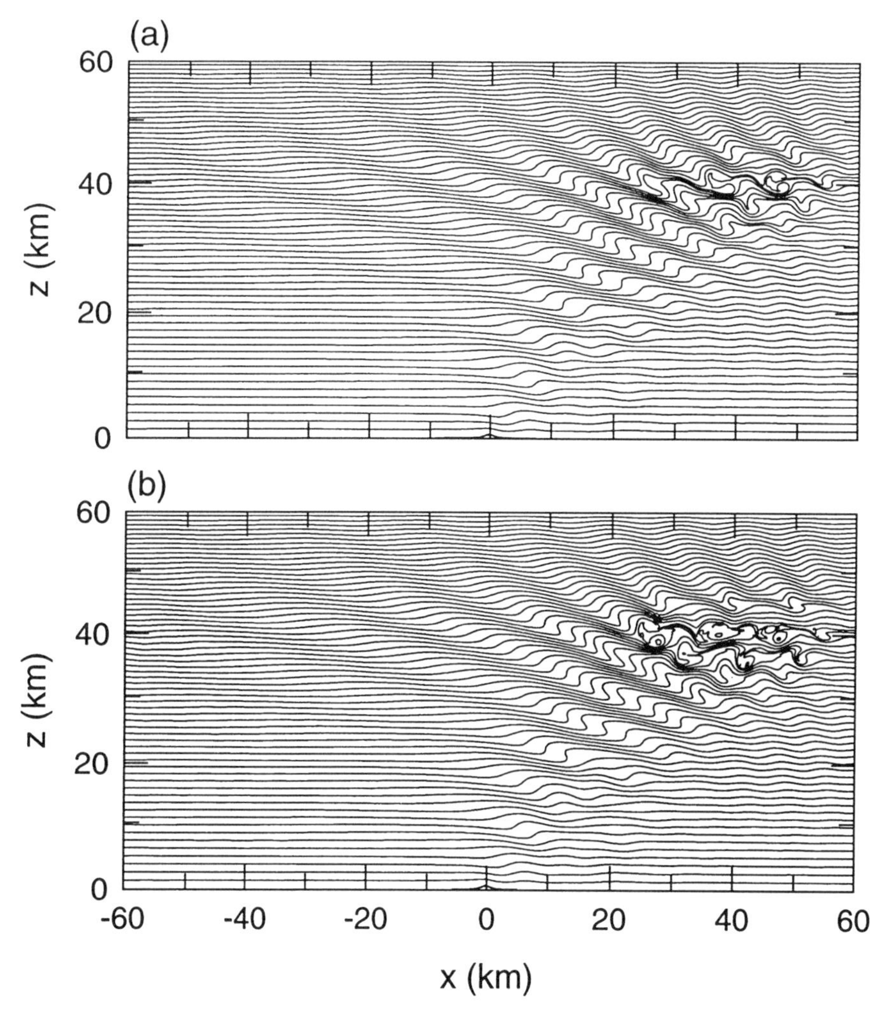

Fig. 4 Isentropes after 3 hours of the evolution of the gravity wave simulated with the semi-Lagrangian (plate a) and Eulerian (plate b) models on the double-resolution A-grid. Contour interval as in Fig. 1.

generated an Eulerian solution that does resemble the semi-Lagrangian result in Fig. 1. This hypothesis is further corroborated by an analysis of the stress profiles $\langle\bar{\rho}u'w'\rangle$ (here $\langle..\rangle$ denotes an average over the horizontal extent of the domain) resulting from "linear" calculations with a ten-times lower mountain. After 8 h of simulated time, the steady-state Eulerian solution yields a vertically propagating gravity wave of essentially constant momentum flux equal to 0.54 of the linear hydrostatic value. This is in excellent agreement with the analytic, infinitesimal-amplitude solution; (cf. Fig. 8.10, section 8.8, in Gill, 1982). By contrast, the equivalent semi-Lagrangian solution results in the momentum flux profile that decays linearly with height to reach half of the surface value at the bottom of the absorber layer ($\approx$ seven vertical wavelengths). The latter implies about 4% loss of the wave amplitude per wavelength, which is consistent with both the explicit viscosity employed in the Eulerian experiment with the original mountain and the delay of the onset of the wave breaking by about half an hour, as observed in the semi-Lagrangian calculations in Fig. 1.

In order to isolate the element of the semi-Lagrangian model algorithm primarily responsible for the apparent loss of accuracy compared to the Eulerian codes, we have performed a number of sensitivity tests. Note that the results summarized so far allow us to eliminate many potential reasons for the disparity observed. Apparently it is independent of such issues as: choice of A-, B-, or C-grid; two- versus three-time-level temporal differencing of the governing equations; dissipative versus CTS flux-form transport algorithms; residual errors in elliptic solvers; absorbing versus "radiation" lateral boundary schemes, as well as physical nonlinearity. Further tests have investigated the solutions' sensitivities to: the size of the time step; the overall accuracies of the trajectory scheme and the interpolation procedure[4]; the choice of the dependent variables[5]; global algebraic conservation of the interpolation procedure[6]; as well as imposing an anelastic mass continuity constraint on the trajectory scheme. None of those factors has significantly affected the retardation of the wave development in the semi-Lagrangian model.

Overall, the differences between the semi-Lagrangian and Eulerian results are small and of minor importance for typical tropospheric applications. Note that our spatial resolution is low–we have only two grid lengths per horizontal scale of the mountain L–and halving $\Delta X, \Delta Z$ circumvents the issue. Nevertheless, this disparity contradicts our earlier experiences with the FT schemes, which seemed to indicate the uniformly superior accuracy of the semi-Lagrangian approach (Smolarkiewicz and Rasch, 1991; Smolarkiewicz and Grell, 1992; Smolarkiewicz and Pudykiewicz, 1992; see also the following section). The experiments summarized above would seem to eliminate all but the fundamental difference between advective- and conservation-form algorithms (Roache, 1972; Section IV-C). In general, flux-form schemes have higher-order truncation errors proportional to the differentials of fluxes of the primitive variables, rather than to the differentials of the variables themselves characteristic of advective form schemes. Regardless of the

[4]Among others, we performed an experiment with an oscillatory (non-FCT), 6th-order-accurate Lagrangian-polynomial interpolation procedure.

[5]In order to assure that both models transport the same dependent variables, the semi-Lagrangian algorithm has been reformulated to solve for $\mathbf{Q} \equiv \rho^* \mathbf{v}$ and $\theta \equiv \rho^* \Theta$.

[6]Small global conservation errors ($\sim O(10^{-5})$ as measured by the normalized variations of the total entropy) were forced to zero by multiplicative adjustments of the interpolated variables; (Williamson and Rasch, 1994).

dispersive or dissipative character of the errors, this increases the overall accuracy of approximation when the fluxes of the variables exhibit a greater degree of homogeneity than the variables themselves (as, e.g., in small-amplitude gravity wave problems). Although plausible, this hypothesis may be difficult to prove.

We close this section by noting that the calculations represented by Figs 1, 2 and 3 all have Courant numbers [here $C = \max(|u|\Delta t/\Delta X + |\omega|\Delta t/\Delta Z)$] close to unity–somewhat larger/smaller than unity for the semi-Lagrangian/Eulerian runs; likewise, the Lipschitz numbers [here $L = \max(\Delta t|\partial V^I/\partial x^J|)$] are all close to but smaller than 0.5. All three experiments were run for 4 h of simulated time with respective execution times on a single processor CRAY Y-MP of 382, 405 and 1039 s.

c *Low Froude Number Flow Past a Steep Three-Dimensional Hill*

Our second example concerns the flow of a Boussinesq fluid ($\bar{\rho}$ = const.) with uniform environmental wind profile $u_e(z_c) = U = 5$ m s^{-1} ($v_e \equiv w_e \equiv 0$) and constant Brunt-Väisällä frequency $N = 10^{-2}$ s^{-1} past an axially-symmetric cosine hill $h = h_o \cos^2(\pi r/2L)$ if $r = (x^2+y^2)^{\frac{1}{2}} \leq L$ ($h = 0$ otherwise). The hill parameters are $h_o = 1.5 \cdot 10^3$ m, and $L = 3 \cdot 10^3$ m. The mesh consists of $NX \times NY \times NZ = 101 \times 81 \times 41$ grid points with the uniform grid increment $\Delta X = \Delta Y = \Delta Z = 0.15 \cdot 10^3$ m. All calculations use the same time step $\Delta t = 10$ s, which results in maximal Courant and Lipschitz numbers $\lesssim 1$ and $\lesssim 0.5$, respectively. The boundary conditions are rigid in x and z, and periodic in y; the gravity-wave absorbers near the lateral and upper rigid boundaries attenuate the solution toward environmental profiles with α in (17) increasing linearly from zero at the distance $L/2$ from the boundary to 150^{-1} s^{-1} at the boundary. The initial condition is the potential flow.

This particular choice of environmental profiles and the height of the mountain results in the interesting (in the area of stratified flows past complex terrain) fluid regime frequently referred to as a low Froude number flow. Here the Froude number ($Fr \equiv U/Nh_o$) equals 1/3. The distinguishing features of such flows include the separation and reversal of the lower upwind stream, and the formation of intense vertically-oriented vortices on the lee side of the hill (Hunt and Snyder, 1980; Smolarkiewicz and Rotunno, 1989, 1990). Recently, low Froude number flows have attracted considerable attention in the community and have been the subject of numerous theoretical, observational, and modelling (both numerical and laboratory) studies; see Reisner and Smolarkiewicz (1994) for a succinct review. Our current choice of the flow parameters and hill geometry makes the entire problem essentially non-hydrostatic and representative of laboratory experiments ($h_o/L \sim O(1)$; Hunt and Synder, 1980) rather than natural atmospheric flows ($h_o/L \sim O(0.1)$; Smolarkiewicz and Rotunno, 1989, 1990).

Figures 5 and 6 summarize the results of the low Froude number flow calculations on the A-grid using, respectively, the semi-Lagrangian and the Eulerian versions of the FT model. The surface and centre-plane flows are shown after three

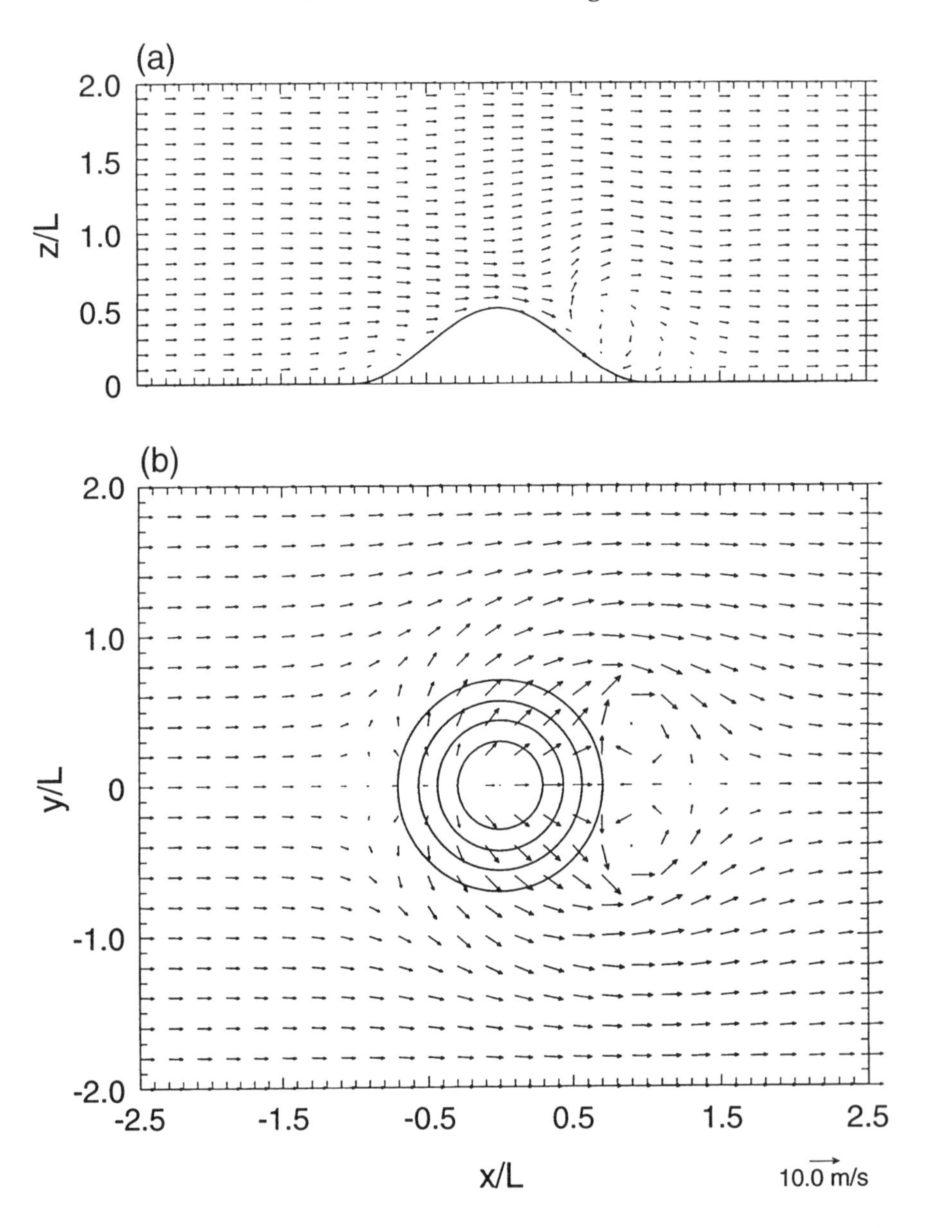

Fig. 5 Low Froude number flow ($Fr = 1/3$) past a steep three-dimensional hill simulated with the semi-Lagrangian FT model on the A-grid. The upper and lower panels show the velocity vectors at the centre plane and surface, respectively.

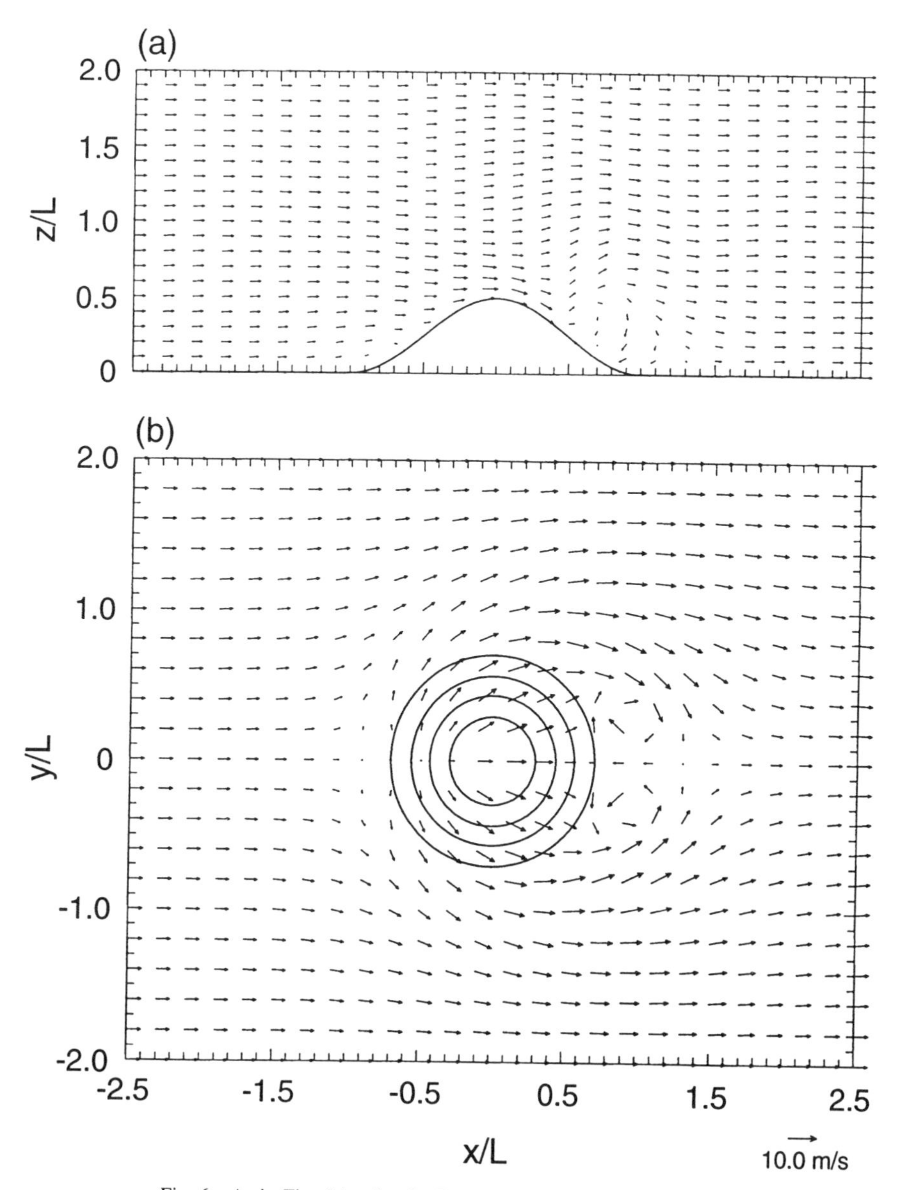

Fig. 6 As in Fig. 5 but for the Eulerian version of the FT model.

advective time scales (here $T = 0.5L/U$) when the main features of the solution have already been established (cf. Smolarkiewicz and Rotunno, 1989, 1990). Similar experiments were performed on the B-grid (not shown) as well as using the hybrid CTS model (Fig. 7) with a C-grid discretization. All calculations faithfully reproduce the topology of low Froude number flows, including the characteristic

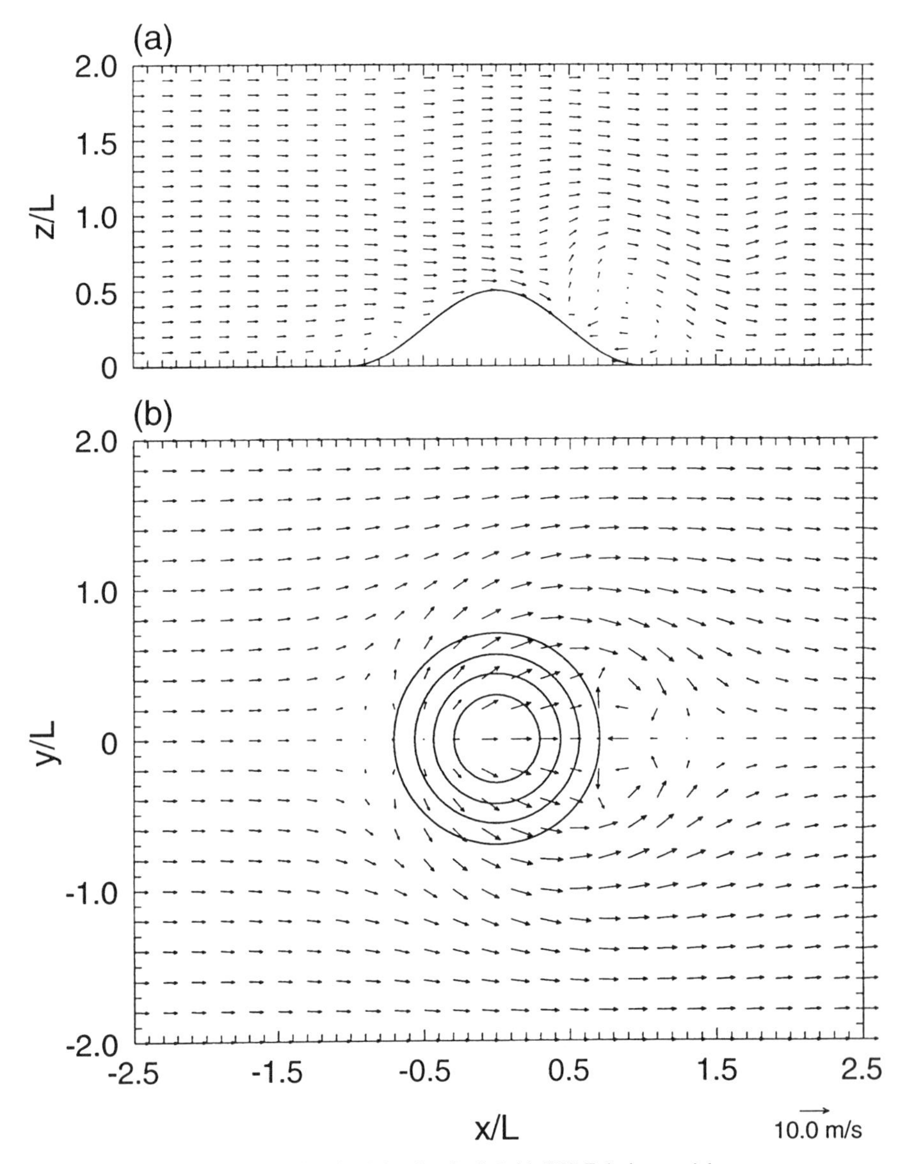

Fig. 7 As in Fig. 5 but for the hybrid CTS Eulerian model.

distribution of the singular, saddle and nodal points of the flow; (cf. Hunt and Snyder, 1980; Smolarkiewicz and Rotunno, 1989). The differences between the simulations are noticeable, but are confined to fine details. The most obvious departures are–as in the gravity wave problem discussed in the preceding section–between the semi-Lagrangian (Fig. 5) and Eulerian (Figs 6 and 7) solutions. It is

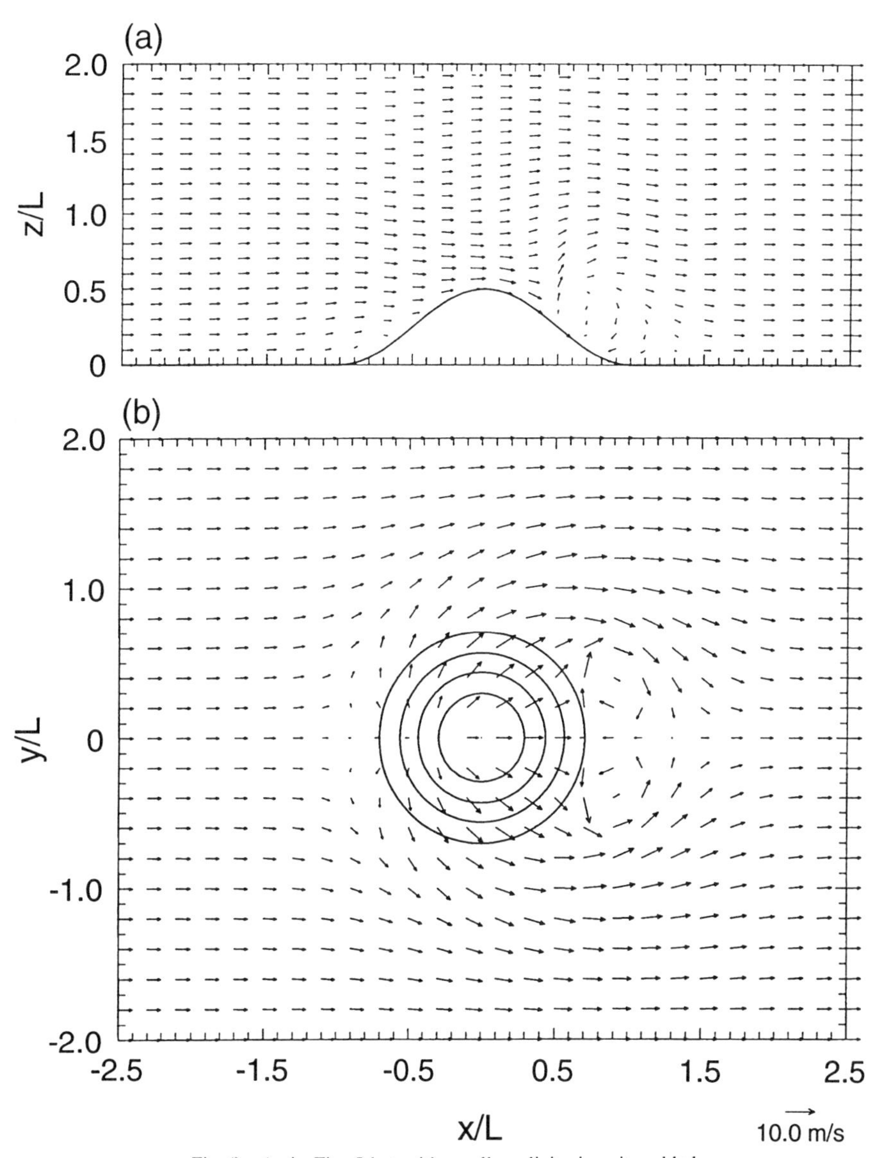

Fig. 8 As in Fig. 5 but with small explicit viscosity added.

quite apparent that both the lee eddies and the reversed, surface upwind flow are more pronounced in the semi-Lagrangian solution.

Further analysis of the results suggests that the semi-Lagrangian model better resolves some fine-scale features of the low Froude number flow. This is substantiated by the semi-Lagrangian solution in Fig. 8, similar to that in Fig. 6 but with an added explicit viscosity of constant normalized magnitude two orders smaller than the advective Courant number. In the gravity wave problem, a similar small viscosity incorporated in the Eulerian calculations had the effect of degrading the Eulerian result toward the semi-Lagrangian solution. Here, it acts in the opposite sense. When overlaying the figures, the viscous semi-Lagrangian simulation in Fig. 8 shows a weaker lateral deflection of the surface flow, a smeared interior of the lee eddies, and a somewhat less-pronounced reversed upwind flow than the explicitly inviscid results in Fig. 5. These effects tend to smooth the semi-Lagrangian solution toward the Eulerian results in Figs 6 and 7. Furthermore, adding explicit viscosity to the Eulerian FT simulation affects the solution only slightly, whereas halving the resolution of both FT experiments exhibits similar tendencies as adding viscosity to the semi-Lagrangian run. All together this suggests that the semi-Lagrangian approach is inherently less diffusive, which is more consistent with our previous experience with FT approximations.

We close this section by noting that all the experiments discussed were run for $t = 4T = 120\Delta t$. The execution times of the FT semi-Lagrangian (Fig. 5), FT Eulerian (Fig. 6) and the hybrid CTS (Fig. 7) model on a single processor CRAY Y-MP were 1765, 935, and 1108 s, respectively. Here, it is apparent that the semi-Lagrangian model is much less efficient than either of the Eulerian models. We shall return to this point in the following section.

5 Concluding remarks

Our primary goal has been to demonstrate that nonoscillatory forward-in-time (FT) methods offer sufficient accuracy and computational efficiency to be considered an attractive alternative to the centred-in-time-and-space (CTS), three-time-level approach more traditionally favoured in atmospheric models. A theoretical advantage of the nonoscillatory FT approach is the preservation of local extrema and sign of the transported variables, which assures solutions consistent with analytic properties of the modelled systems. This minimizes the need for artificial viscosity while promoting the nonlinear stability of computations. FT methods have advantages for improving computational efficiency. In theory, they require less storage and allow larger time steps. Also, they are well-suited for calculations with a variable time step, an important feature in simulations that span a variety of flow regimes.

To gain these advantages we have constructed an FT model based on the anelastic non-hydrostatic equations of motion cast in terrain-following coordinates. Although conceptually simple, this model can be applied to complex problems of small-scale atmospheric dynamics. A special feature of this model is the optional availability of either Eulerian or semi-Lagrangian formulations of FT differencing. This is made possible by the congruence of these formulations, Eqs. (2) and (4), leading to the common form of the anelastic model algorithm (17). Furthermore, the two realizations of the *LE* operator in (17) can employ the same generic FT schemes to represent either Eulerian advection or semi-Lagrangian interpolation.

We have exploited this model to compare the accuracy and efficiency of the two formulations with each other. Also, we have benchmarked it against an established CTS anelastic model. *Overall, both FT model options and CTS model produce equally meaningful results.* Finer analysis of the results exposes, however, some interesting disparities. The two test problems analyzed in Section 4 show only minor differences between FT and CTS Eulerian results compared to the differences between the semi-Lagrangian and Eulerian results. Furthermore, we have judged the Eulerian results to be more accurate in the first problem, but less accurate in the second. In light of these results, we conclude that neither formulation is inherently superior, and that the choice between these complementary model options should be problem dependent. Our unified approach facilitates this choice.

The relative computational efficiency of the Eulerian and the semi-Lagrangian FT options is also problem dependent. Both FORTRAN programs within the unified code are equally efficient and achieve roughly 200 megaflops on a single processor CRAY Y-MP, but the Eulerian algorithm performs less arithmetic operations (especially in three-dimensional problems) and so is more efficient per time step. The advantage of large time steps permitted by the semi-Lagrangian advection is useful in large-scale applications, but is less important for modelling small-scale dynamics where accuracy considerations usually dictate the Eulerian time step. Thus, whenever the accuracy permits the time step to significantly exceed the Courant condition (e.g., modelling deep atmospheres with shear), the semi-Lagrangian model becomes especially attractive. When the advantage of large time steps is permanently eschewed, the semi-Lagrangian model can be substantially accelerated by eliminating indirect addressing in FORTRAN programs. This is particularly important for massively-parallel computations.

Acknowledgements

We gratefully acknowledge the personal reviews and constructive comments of Terry Clark and Joseph Klemp. Special thanks go to Wojciech Grabowski who performed the Eulerian simulations with the hybrid CTS code. This work has been supported in part by the Department of Energy "Computer Hardware, Advanced Mathematics, Model Physics" (CHAMMP) research program.

Appendix: Coefficients of the elliptic pressure equation

The coefficients $\mathcal{E}$, $\mathcal{V}^I$ and C^{IJ} appearing in (19) and (20) follow from a somewhat tedious derivation that is outlined in Section 3d of this paper.

The multiplicative factor $\mathcal{E}$ is

$$\mathcal{E} \equiv [(1 + 0.5\Delta t\alpha)(1 + \mathcal{F}^2)]^{-1}, \tag{A1a}$$

where

$$\mathcal{F} \equiv 0.5\Delta t f(1 + 0.5\Delta t\alpha)^{-1}. \tag{A1b}$$

The auxiliary fields $\mathcal{V}^I$ can be compactly written as

$$\mathcal{V}^1 = U^* + \mathcal{F} V^*, \tag{A2a}$$

$$\mathcal{V}^2 = V^* - \mathcal{F} U^*, \tag{A2b}$$

$$\mathcal{V}^3 = \left(LE(\tilde{w}) + 0.5\Delta t g \frac{\Theta - \Theta_e}{\bar{\Theta}} \right) \frac{1 + \mathcal{F}^2}{G} + G^{13}\mathcal{V}^1 + G^{23}\mathcal{V}^2, \tag{A2c}$$

where

$$U^* \equiv LE(\tilde{u}) + 0.5\Delta t(\alpha u_e - f v_e), \tag{A3a}$$

$$V^* \equiv LE(\tilde{v}) + 0.5\Delta t(\alpha v_e + f u_e). \tag{A3b}$$

The coefficients C^{IJ} take the explicit form:

$$C^{11} = C^{22} = 1;\ C^{33} = (G^{13})^2 + (G^{23})^2 + (1 + \mathcal{F}^2)/G^2, \tag{A4a, b}$$

$$C^{12} = -C^{21} = \mathcal{F}; \tag{A4c}$$

$$C^{13} = G^{13} + \mathcal{F} G^{23};\ C^{31} = G^{13} - \mathcal{F} G^{23}. \tag{A4d, e}$$

$$C^{23} = G^{23} - \mathcal{F} G^{13};\ C^{32} = G^{23} + \mathcal{F} G^{13}. \tag{A4f, g}$$

References

ARAKAWA, A. 1966. Computational design for long-term numerical integration of the equations of fluid motions: Two-dimensional incompressible flow. *J. Comp. Phys.* **1**: 119–143

ASSELIN, R. 1972. Frequency filter for time integration. *Mon. Weather Rev.* **100**: 487–490.

BACMEISTER J.T. and M.R. SCHOEBERL. 1989. Breakdown of vertically propagating two-dimensional gravity waves forced by orography. *J. Atmos. Sci.* **46**: 2109–2134.

BATES, J.R.; S. MOORTHI and R.W. HIGGINS. 1993. A global multilevel atmospheric model using a vector semi-Lagrangian finite-difference scheme. Part 1: Adiabatic formulation. *Mon. Weather Rev.* **121**: 244–263.

BLECK, R.; C. ROOTH, D. HU and L. SMITH. 1992. Salinity-driven thermocline transients in a wind- and thermohaline-forced isopycnic coordinate model of the North Atlantic. *J. Phys. Oceanogr.* **22**: 1486–1505.

CHORIN, A.J. 1968. Numerical solution of the Navier-Stokes equations. *Math. Comp.* **22**: 742–762.

CLARK, T.L. 1977. A small-scale dynamic model using a terrain-following coordinate transformation. *J. Comp. Phys.* **24**: 186–214.

——— and R.D. FARLEY. 1984. Severe downslope windstorm calculations in two and three spatial dimensions using anelastic interactive grid nesting: A possible mechanism for gustiness. *J. Atmos. Sci* **41**: 329–350.

CROWLEY, W.P. 1968. Numerical advection experiments. *Mon. Weather Rev.* **96**: 1–11.

DAVIES, H.C. 1983. Limitations of some common lateral boundary schemes in regional NWP models. *Mon. Weather Rev.* **111**: 1002–1012.

EISENSTAT, S.C.; H.C. ELMAN and M.H. SCHULTZ. 1983.

Variational iterative methods for nonsymmetric systems of linear equations. *SIAM J. Numer. Anal.* **20**: 345–357.

GAL-CHEN, T. and R.C.J. SOMERVILLE. 1975. On the use of a coordinate transformation for the solutions of the Navier-Stokes equations. *J. Comp. Phys.* **17**: 209–228.

GILL, A.E. 1982. *Atmosphere-ocean dynamics*. International geophysics series, Vol. 30, Academic Press, San Diego, 662 pp.

GRABOWSKI, W.W. and P.K. SMOLARKIEWICZ. 1996. On two-time-level semi-Lagrangian modelling of precipitating clouds. *Mon. Weather Rev.* **124**: 487–497.

HARTEN, A. 1983. High resolution schemes for hyperbolic conservation laws. *J. Comp. Phys.* **49**: 357–393.

HUNT, C.R. and W.H. SNYDER. 1980. Experiments on stably and neutrally stratified flow over a model three-dimensional hill. *J. Fluid Mech.* **96**: 671–704.

KAPITZA, H. and D. EPPEL. 1992. The non-hydrostatic mesocale model GESIMA. Part 1: Dynamical equations and tests. *Beitr. Phys. Atmosph.* **65**: 129–146.

KOSLOFF, R. and D. KOSLOFF. 1986. Absorbing boundaries for wave propagation problems. *J. Comp. Phys.* **63**: 363–376.

KRISHNAMURTI, T.N. 1962. Numerical integration of primitive equations by quasi-Lagrangian advective scheme. *J. Appl. Meteorol.* **1**: 508–521.

———; A. KUMAR, K.S. YAP, A.P. DASTOOR, N. DAVIDSON and J. SHENG. 1990. Performance of a high-resolution mesoscale tropical prediction model. *Advances Geophys.* **32**: 133–286.

LEITH, C.E. 1965. Numerical simulation of the Earth's atmosphere. In: *Methods in Computational Physics*, Vol 4, B. Alder, S. Ferenbach and M. Rotenberg (Eds), Academic Press, New York, 1965, pp. 1–28.

LEONARD, B.P. 1988. Universal limiter for transient interpolation modelling of the advective transport equations: the ULTIMATE conservative difference scheme. NASA Technical Memorandum 100916-ICOMP-88-11, NASA Lewis, 116 pp.

LEVEQUE, R.J. 1996. High-resolution conservative algorithms for advection in incompressible flow. *SIAM J. Numer. Anal.* **33**: 627–665.

LILLY, D.K. 1962. On the numerical simulation of buoyant convection. *Tellus*, **14**: 148–172.

——— 1965. On the computational stability of numerical solutions of time-dependent non-linear geophysical fluid dynamics problems. *Mon. Weather Rev.* **93**: 11–26.

LIN, S.J.; W.C. CHAO, Y.C. SUD and G.K. WALKER. 1994. A class of the van Leer-type transport schemes and its applications to the moisture transport in a general circulation model. *Mon. Weather Rev.* **122**: 1575–1593.

LIPPS, F.B. 1990. On the anelastic approximation for deep convection. *J. Atmos. Sci.* **47**: 1794–1798.

——— and R.S. HEMLER. 1982. A scale analysis of deep moist convection and some related numerical calculations. *J. Atmos. Sci.* **39**: 2192–2210.

MARGOLIN, L.G. and P.K. SMOLARKIEWICZ. 1989. Antidiffusive velocities for multipass donor cell advection. Lawrence Livermore National Laboratory. Report # UCID-21866, 44 pp.

MCDONALD, A. and J.R. BATES. 1987. Improving the estimate of the departure point position in a two-time-level semi-Lagrangian and semi-implicit model. *Mon. Weather Rev.* **115**: 737–739.

——— and J.E. HAUGEN. 1992. A two-time-level, three-dimensional semi-Lagrangian, semi-implicit, limited-area gridpoint model of the primitive equations. *Mon. Weather Rev.* **120**: 2603–2621.

PRUSA, J.; P.K. SMOLARKIEWICZ and R. GARCIA. 1993. Application of time variable coordinate transformations and semi-Lagrangian methods to the study of deep gravity waves. *Bull. Am. Phys. Soc.* **38**: 2224.

REISNER, J.M. and P.K. SMOLARKIEWICZ. 1994. Thermally forced low Froude number flow past three-dimensional obstacles. *J. Atmos. Sci.* **51**: 117–133.

ROACHE, P.J. 1972. *Computational Fluid Dynamics*. Hermosa Publishers, Albuquerque, 446 pp.

ROBERT, A.J. 1966. The integration of a low order spectral form of the primitive meteorological equations. *J. Meteorol. Soc. Jpn.* **44**: 237–245.

———. 1981. A stable numerical integration scheme for the primitive meteorological equations. ATMOSPHERE-OCEAN, **19**: 35–46.

———. 1993. Bubble convection experiments with semi-implicit formulation of the Euler equations. *J. Atmos. Sci.* **50**: 1865–1873.

———; F.G. SHUMAN and J.P. GERRITY, JR. 1970. On partial differential equations in mathematical physics. *Mon. Weather Rev.* **98**: 1–6.

ROTUNNO, R. and P.K. SMOLARKIEWICZ. 1995. Vorticity generation in the shallow-water equations as applied to hydraulic jumps. *J. Atmos. Sci.* **52**: 320–330.

SCHUMANN, U.; T. HAUF, H. HOLLER, H. SCHMIDT and H. VOLKERT. 1987. A mesoscale model for the simulation of turbulence, clouds, and flow over mountains: Formulation and validation examples. *Beitr. Phys. Atmosph.* **60**: 413–446.

SHAPIRO, R. 1970. Smoothing, filtering, and boundary effects. *Rev. Geophys. Space Phys.* **8**: 359–387.

SMAGORINSKY, J. 1963. General circulation experiments with the primitive equations. I: The basic experiment. *Mon. Weather Rev.* **91**: 99–164.

SMOLARKIEWICZ, P.K. 1984. A fully multidimensional positive definite advection transport algorithm with small implicit diffusion. *J. Comp. Phys.* **54**: 325–362.

——— and T.L. CLARK. 1986. The multidimensional positive definite advection transport algorithm: Further development and applications. *J. Comp. Phys.* **67**: 396–438.

——— and R. ROTUNNO. 1989. Low Froude number flow past three-dimensional obstacles. Part I: Baroclinically generated lee vortices. *J. Atmos. Sci.* **46**: 1154–1164.

——— and ———. 1990. Froude number flow past three-dimensional obstacles. Part II: Upwind flow reversal zone. *J. Atmos. Sci.* **47**: 1498–1511.

——— and W.W. GRABOWSKI. 1990. The multidimensional positive definite advection transport algorithm: Nonoscillatory option. *J. Comp. Phys.* **86**: 355–375.

——— and P.J. RASCH. 1991. Monotone advection on the sphere: An Eulerian versus semi-Lagrangian approach. *J. Atmos. Sci.* **48**: 793–810.

——— and G.A. GRELL. 1992. A class of montone interpolation schemes. *J. Comp. Phys.* **101**: 431–440.

——— and J.A. PUDYKIEWICZ. 1992. A class of semi-Lagrangian approximations for fluids. *J. Atmos. Sci.* **49**: 2082–2096.

——— and L.G. MARGOLIN. 1993. On forward-in-time differencing for fluids: Extension to a curvilinear framework. *Mon. Weather Rev.* **121**: 1847–1859.

——— and ———. 1994. Variational solver for elliptic problems in atmospheric flows. *Appl. Math. Comp. Sci.* **4**: 527–551.

STANIFORTH, A. and J. CÔTÉ. 1991. Semi-Lagrangian integration schemes for atmospheric models-A review. *Mon. Weather Rev.* **119**: 2206–2223.

TEMPERTON, C. and A. STANIFORTH. 1987. An efficient two-time-level semi-Lagrangian semi-implicit integration scheme. *Q. J. R. Meteorol. Soc.* **113**: 1025–1039.

TRIPOLI, G.J. 1992. A nonhydrostatic numerical model designed to simulate scale interaction. *Mon. Weather Rev.* **120**: 1342–1359.

TRUESDELL, C. 1966. *The Mechanical Foundations of Elasticity and Fluid Dynamics*, Gordon and Breach Science Publishers, Inc., New York, 218 pp.

VAN LEER, B. 1974. Towards the ultimate conservative difference scheme. II: Monotonicity and conservation combined in a second-order scheme. *J. Comp. Phys.* **14**: 361–370.

WILLIAMSON, D.L. and P.J. RASCH. 1994. Water vapor transport in the NCAR CCM2. *Tellus*, **46A**: 34–51.

ZALESAK, S.T. 1979. Fully multidimensional flux-corrected transport algorithms for fluids. *J. Comp. Phys.* **31**: 335–362.

Semi-Lagrangian Advection on a Cubic Gnomonic Projection of the Sphere

John L. McGregor
CSIRO Division of Atmospheric Research
PB1 Mordialloc
Victoria, Australia 3195

[Original manuscript received 6 December 1994; in revised form 21 August 1995]

ABSTRACT *The cubic gnomonic projection of the sphere proposed by Sadourny (1972) is revisited. The gnomonic grid possesses six panels and has the major attraction of being quasi-uniform. Advection is performed using the semi-Lagrangian procedure of McGregor (1993). The departure points are determined accurately, even near panel boundaries. The accuracy is enhanced by a simple grid transformation which provides elements possessing more uniform area. For test cases of solid-body rotation, the semi-Lagrangian scheme performs more accurately on a gnomonic grid than on a standard Gaussian latitude-longitude grid having a similar number of grid points. The scheme is computationally efficient and can be coded without conditional jumps or calls to trigonometric functions.*

RÉSUMÉ *On réexamin la projection gnomonique cubique de la sphère proposée par Sadourny (1972). La grille gnomonique contient six panneaux et a l'avantage d'être quasi uniforme. On introduit l'advection en utilisant la procédure semi-lagrangienne de McGregor (1993). Les points de départ sont déterminés avec précision, même près des limites des panneaux. La précision est augmentée par une simple transformation de grille qui fourni des éléments ayant une surface plus uniforme. Pour les essais de rotation de corps solides, le schéma semi-lagrangien performe avec plus de précision sur une grille gnomonique que sur une grille gaussienne à latitude-longitude usuelle avec le même nombre de points de grille. Le schéma est efficace et peut être codé sans sauts conditionnels ou appels à des fonctions trigonométriques.*

1 Introduction

The primitive equations are usually solved on the globe using a latitude-longitude grid, either by means of finite differences or spectral methods. A regular latitude-longitude grid produces a clustering of points at the poles, which at first sight appears to require a significant reduction in time step to maintain Courant-Friedrichs-Levy (CFL) stability requirements. In the case of spectral models, spectral smoothing avoids the time step restriction, although it does mean that superfluous grid points are being carried near the poles. For gridpoint models Fourier filtering can be applied near the poles to avoid the time step restriction, although again this effectively means that there are superfluous grid points. Another option is to apply semi-Lagrangian time differencing (Robert, 1981) to a global latitude-longitude grid (e.g. Bates et al., 1990); although all grid points are productively used with this arrangement, one would expect some net wastage due to the much better resolution of zonal, compared to meridional, gradients in polar regions.

Several alternative grid arrangements have been proposed to provide more uniform coverage of the globe. Reduced grids with fewer points near the poles have been applied to spectral models (Hortal and Simmons, 1991) and also to gridpoint models (see Williamson, 1979); for the latter models the finite differencing becomes rather complicated. Icosahedral configurations with triangular elements have also been investigated by Sadourny et al. (1968), Williamson (1968) and Cullen (1974) with some success. Although the elements still possess significant variations in their area, those variations are much less than those of a regular latitude-longitude grid.

A different arrangement is to set up a regular rectangular grid on the panels (faces) of a cube and apply a gnomonic projection of it onto the surface of an enclosed sphere, as proposed by Sadourny (1972) and discussed by Williamson (1979). The projection is called gnomonic, meaning a projection to/from the centre of the sphere. The area elements on the surface of the sphere are quasi-uniform. A complication is that the coordinate lines lying on the sphere are not orthogonal. The non-orthogonality is not a great problem for expressing the equations of motion, and Sadourny (1972) has written the corresponding form of the shallow water equations in compact tensor notation. He also solved those equations in conservative flux form. However, his solutions were comparatively noisy, and he attributed this to increased truncation errors near the edges of the six panels. He suggested that higher order differencing could alleviate these truncation problems, although conventional higher order finite differencing does not appear to be straightforward to implement.

The present paper treats just the advection problem on a gnomonic cubic projection using the semi-Lagrangian scheme of McGregor (1993); this scheme is particularly suitable for unusual grid configurations. Semi-Lagrangian schemes have been shown to possess accurate advection properties for standard grid configurations (e.g. Staniforth and Côté, 1991) and to possess good conservation properties. For extension of the present scheme to global climate models, small a posteriori conservation corrections could be applied each time step.

2 Grid specification

a *Grid Indexing*

A cube is set up encompassing the sphere and tangent to it. Cartesian 3D coordinates are given as upper case: (X, Y, Z) for points on the sphere and (A, B, C) on the projected cube. Local 2D coordinates on the surface of a panel p are written in lower-case: (x, y, p) on the sphere and (a, b, p) on the projected cube.

On the cube each of the 6 panels has local coordinates (a, b, p) for $0 \leqslant p \leqslant 5$. The orientation of the a and b axes is shown in Fig. 1a and in spread-out form in Fig. 1b; the dotted squares show extended neighbour relationships and reveal that there are two separate families of panels in respect to neighbours: those that are odd-numbered and those that are even-numbered. Panels 3, 4 and 5 are opposite panels 0, 1 and 2 respectively. The axes A, B and C pass through the centre of panels 0, 2 and 1 respectively. The gridding for $N = 3$ is shown in Fig. 1c. Computer memory is only allocated for an $N \times N$ array for each panel, where each panel edge has $N + 1$ grid points. It can be seen from Fig. 1b that for an arbitrary variable h, the overlapping of indices at the edges is given by

$$h_{N+1,j,p} = h_{N+2-j,\,1,\,\mathrm{mod}(p+2,\,6)} \qquad \text{for } p \text{ even} \tag{1a}$$

$$h_{N+1,j,p} = h_{1,j,\,\mathrm{mod}(p+1,\,6)} \qquad \text{for } p \text{ odd} \tag{1b}$$

and

$$h_{i,\,N+1,p} = h_{i,\,1,\,\mathrm{mod}(p+1,\,6)} \qquad \text{for } p \text{ even} \tag{2a}$$

$$h_{i,\,N+1,p} = h_{1,\,N+2-i,\,\mathrm{mod}(p+2,\,6)} \qquad \text{for } p \text{ odd} \tag{2b}$$

where $1 \leqslant i \leqslant N + 1$ and $1 \leqslant j \leqslant N + 1$ on each panel. There are two extra points not covered by this indexing, marked $\otimes$ and $\odot$ in Fig. 1, respectively $h_\otimes \equiv h_{N+1,\,1,\,p}$ for even p, and $h_\odot \equiv h_{1,\,N+1,\,p}$ for odd p. There are thus a total of $6N^2 + 2$ grid points.

b *Determination of Grid Coordinates from the 3D Cartesian Position*

The departure points are to be calculated in terms of the 3D Cartesian positions (X, Y, Z) of points on a sphere of radius R; these need to be converted to units of the grid indices (i, j, p) where i and j need not be integers. First, the corresponding point on the encasing cube of side 2R has 3D Cartesian coordinates (A, B, C) given by

$$(A,\, B,\, C) = \frac{(X,\, Y,\, Z)R}{\max(|X|,\, |Y|,\, |Z|)}. \tag{3}$$

At least one of A, B, C has magnitude R which allows the panel index, p, to be determined by a sequence of conditional tests. An alternative determination which avoids conditional jumps is to calculate

$$p = \max\{A_{\mathrm{int}}(3A_{\mathrm{int}} - 3),\; B_{\mathrm{int}}(7B_{\mathrm{int}} - 3),\; C_{\mathrm{int}}(5C_{\mathrm{int}} - 3)\}/2 \tag{4}$$

where (int denotes the truncated integer value)

$$A_{\mathrm{int}} = \mathrm{int}(A/R),\; B_{\mathrm{int}} = \mathrm{int}(B/R),\; C_{\mathrm{int}} = \mathrm{int}(C/R). \tag{5}$$

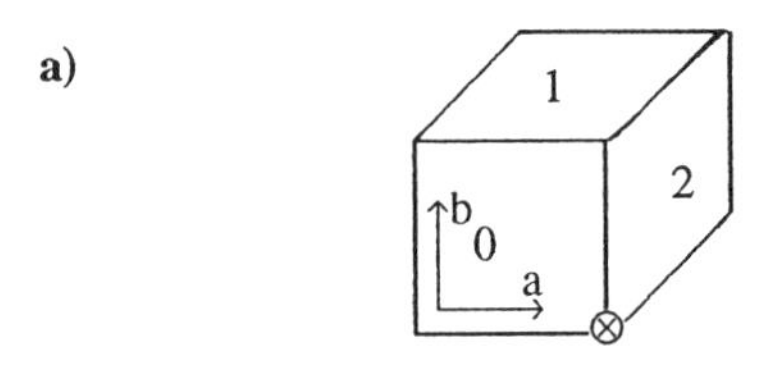

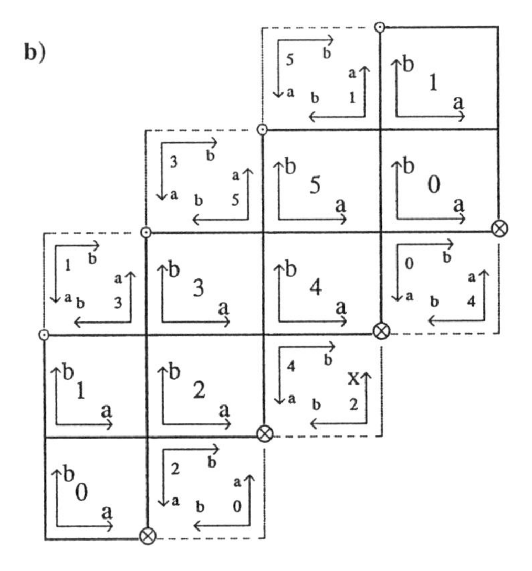

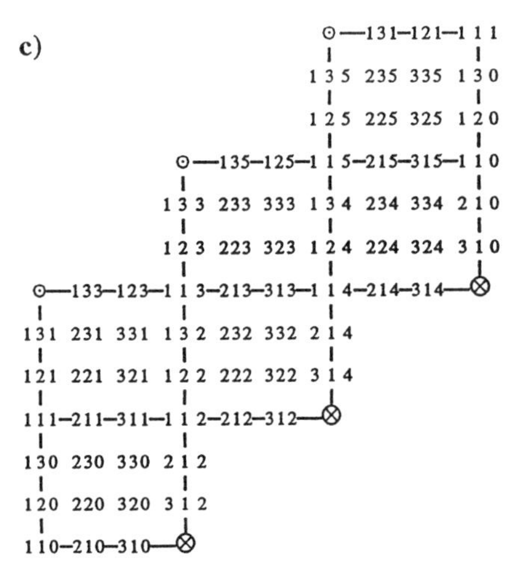

Fig. 1 Local coordinate directions on the panels of a cube and their relative orientations. A perspective view is given in (a) and unfolded views in (b) and (c). The panels in (b) with two sets of axes provide indexing information for the identification of neighbouring panels. The gridding for $N = 3$ is shown in (c). The north pole is arbitrarily assigned at the centre of panel 1, while the south pole is at the centre of panel 4.

The Cartesian coordinate position (a_c, b_c, p) on a panel of the cube is then given by

$$a_c = E_{p1}A + E_{p2}B + E_{p3}C \qquad \text{with } -R \leqslant a_c \leqslant R \tag{6}$$

$$b_c = F_{p1}A + F_{p2}B + F_{p3}C \qquad \text{with } -R \leqslant b_c \leqslant R \tag{7}$$

where the origin of each (a_c, b_c, p) is at the centre of a panel, and the non-zero members of matrices $\mathbf{E}$ and $\mathbf{F}$ are given by $E_{02} = E_{12} = E_{41} = E_{51} = 1$, $E_{23} = E_{33} = -1$, $F_{03} = F_{53} = 1$, $F_{11} = F_{21} = F_{32} = F_{42} = -1$.

c *Stretching of the Grid*

For a grid which has uniform spacing on the panels of the (a_c, b_c, p) coordinate systems on the cube, the spacing of the projection on the sphere has somewhat unequal grid spacing, with the ratio of maximum to minimum grid length being 2.1 for the case $N = 37$. A further transformation may however be performed on the panels of the cube. Let (a, b, p) denote the final coordinate system which is to be linear in grid number. An effective option is to set

$$a_c = \alpha a + \alpha' a^3/R^2 \qquad \text{where} \qquad -R \leqslant a_c \leqslant R,\ -R \leqslant a \leqslant R; \tag{8}$$

α is a constant and $\alpha' = 1 - \alpha$. There is a similar expression for b_c. The inverse calculation of a is accurately performed in just 2 Newton-Raphson iterations using

$$a_{n+1} = \frac{2\alpha' a_n^3 + R^2 a_c}{\alpha R^2 + 3\alpha' a_n^2} \qquad \text{with first guess } a_0 = a_c. \tag{9}$$

There is a similar expression for b_{n+1}. The final grid index conversion is defined by

$$(i,\ j) = \left(1 + \frac{a+R}{2R}N,\ 1 + \frac{b+R}{2R}N\right). \tag{10}$$

To determine α consider the line $b = 0$ through the centre of an arbitrary panel. It is easily shown that equal increments in a at $a = 0$ and $a = \pm R$ produce equal corresponding increments on the sphere provided that $\alpha = 0.75$. When the stretching equation (8) is applied for $\alpha = 0.75$, the ratio of maximum to minimum grid length on the sphere is 1.5 for $N = 37$. Throughout this paper only $\alpha = 1$ (the basic scheme) and $\alpha = 0.75$ will be considered; although the grid length ratio can be reduced to 1.34 for $\alpha = 0.86$, less benefit is produced in the advection tests. An example of a standard cubic gnomonic grid on the sphere for $N = 37$ is shown in Fig. 2, with an example of the $\alpha = 0.75$ stretched grid in Fig. 3.

d *Determination of the 3D Cartesian Position from the Grid Coordinates*

Given (i, j, p), the reverse specification of the Cartesian coordinates (X, Y, Z) on

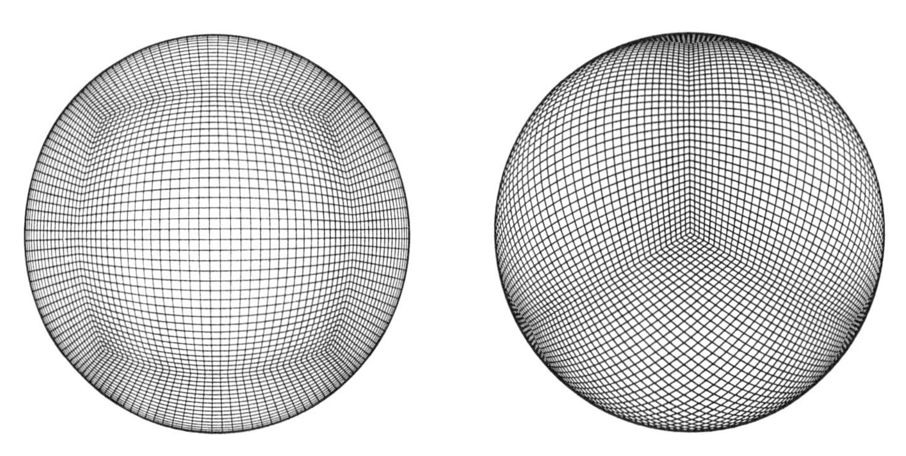

Fig. 2 Basic gnomonic cubic grid on the sphere having 6 panels each with 37 × 37 grid elements. Two alternative views from infinity are shown.

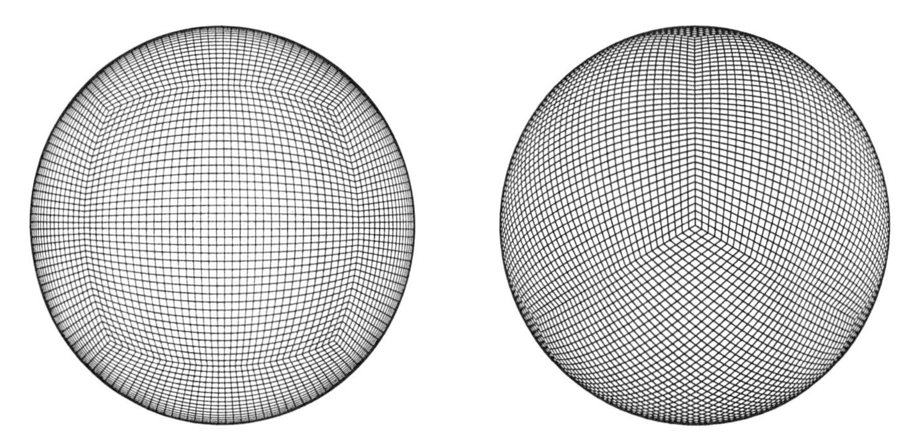

Fig. 3 Gnomonic cubic grid as in Fig. 2, but with stretching factor $\alpha = 0.75$.

the sphere is quite simple; this calculation only needs to be performed once, at the beginning of a simulation. First (10) and (8) provide respectively (a, b) and (a_c, b_c) on the cube. The 3D position (A, B, C) on the cube is then given by

$$A = R,\ B = a_c,\ C = b_c \qquad \text{for } p = 0, \tag{11a}$$

$$A = -b_c,\ B = a_c,\ C = R \qquad \text{for } p = 1, \tag{11b}$$

$$A = -b_c,\ B = R,\ C = -a_c \qquad \text{for } p = 2, \tag{11c}$$

$$A = -R,\ B = -b_c,\ C = -a_c \qquad \text{for } p = 3, \tag{11d}$$

$$A = a_c,\ B = -b_c,\ C = -R \qquad \text{for } p = 4, \tag{11e}$$

$$A = a_c,\ B = -R,\ C = b_c \qquad \text{for } p = 5. \tag{11f}$$

These expressions may be verified by inspecting Fig. 1 which shows the relative orientations of neighbouring panels of the cube. The final Cartesian position on the sphere is given by

$$(X, Y, Z) = \frac{(A, B, C)}{\sqrt{A^2 + B^2 + C^2}} R. \tag{12}$$

e *Initial Setup*

At the commencement of the simulations, the 3D coordinates for each point on the sphere are set up. The position vectors joining neighbouring points are also saved. These are used later to form requisite dot products with the 3D velocity components, which are made available at the beginning of each time step. For the B-grid version, advecting velocities are specified at the centre of each grid element. Coordinate positions are required at each of the $6N^2$ locations for each $(i + 0.5, j + 0.5, p)$. Equation (10) and the other equations are then applied exactly as above.

3 Advection on the gnomonic grid

a *Derivation of the Departure Points*

A procedure is now described for advection on a sphere in terms of the non-orthogonal coordinates (x, y) on the surface of the sphere. Unit vectors along these coordinate directions are denoted by $\hat{\mathbf{x}}$ and $\hat{\mathbf{y}}$. The projections of the velocity $\mathbf{V}$ along x and y, i.e. $\mathbf{V}\cdot\hat{\mathbf{x}}$ and $\mathbf{V}\cdot\hat{\mathbf{y}}$, are denoted by u and v respectively. The material derivative may be written as

$$\frac{d}{dt} = \frac{\partial}{\partial t} + u' \frac{\partial}{\partial x} + v' \frac{\partial}{\partial y} \tag{13}$$

where

$$u' = \frac{u - v \cos \gamma}{\sin^2 \gamma} \text{ and } v' = \frac{v - u \cos \gamma}{\sin^2 \gamma} \tag{14}$$

are the components of $\mathbf{V}$, i.e. $\mathbf{V} = u'\hat{\mathbf{x}} + v'\hat{\mathbf{y}}$, and the angle γ between the x and y axes is given by $\cos \gamma = \hat{\mathbf{x}}\cdot\hat{\mathbf{y}}$ (dot products are used in all computations instead of trigonometric function evaluation). Following McGregor (1993) the departure points of the 3D grid point located at $\mathbf{r}(\tau + \Delta t)$ can be written in terms of the truncated Taylor series of the total derivatives

$$\mathbf{r}(\tau) \cong \mathbf{r}(\tau + \Delta t) + \sum_{n=1}^{3} \frac{(-\Delta t)^n}{n!} \frac{d^n \mathbf{r}}{dt^n}(\tau + \Delta t) \tag{15}$$

where

$$\frac{d^n \mathbf{r}(t)}{dt^n} = \frac{d}{dt}\left\{\frac{d^{n-1}\mathbf{r}(t)}{dt^{n-1}}\right\} \qquad n = 2, 3. \tag{16}$$

The solution is second order accurate in time provided that d/dt is evaluated using (13) with velocities centred at time $\tau + \Delta t/2$. As in McGregor (1993) the horizontal derivatives in (13) are calculated by simple centred finite differences. Some extra economy in the computations may be achieved by first normalizing the velocity components u' and v' to units of grid point per time step in the x and y directions respectively. Equation (15) is applied separately to each Cartesian component in the form (X, Y, Z) to give corresponding departure point values; (3) through (10) give the equivalent departure points in terms of (i, j, p) where non-integer values of i and j lie between grid points. It is to be noted that the determination of departure points passes over panel boundaries in a transparent manner without loss of accuracy. In an advective calculation the values (i, j, p) then determine the value of the advected field via bicubic interpolation on a 4×4 stencil on panel p.

Note that standard bicubic interpolation can be used on each panel of the grid, despite the non-orthogonality of the axes. Near the edges, the interpolation formula is uncentred. For the normal component of advection from within one grid cell either side of an edge, it is readily shown that the amplification factor may be slightly greater than 1. This is not a problem for the present pure advection tests. Some amplification could arise in a primitive equations model with resonant effects in edge regions of orographic forcing with small scale depths, although the usual horizontal diffusion in primitive equations models should be adequate to control such amplification. If this turned out to be inadequate, quadratic rather than cubic interpolation could be applied normal and adjacent to the edges to guarantee that the amplification factor always remains less than 1. An alternative method of cubic interpolation would be to include a one grid length extension around each panel, although the vertices may then require special treatment.

The scheme is readily coded without conditional jumps by using separate loops for odd and even panels, and for edge points. The extensive use of Cartesian position vectors avoids all calls to trigonometric functions. Despite their unusual grid configuration, the cubic gnomonic advection schemes are efficient and run only 10–20% slower than the Gaussian grid version.

b *B-Grid Considerations*

The B-grid version is somewhat simpler to implement than the A-grid version. For the B-grid version the coordinate axes are rotated 45° on each panel, to more conveniently evaluate (13) at the centre of each cell; for the departure point calculations the grid may be viewed as an E-grid. After each total derivative is evaluated, values are obtained at the full grid points by a simple weighted average of the central values; higher order interpolation could be used if desired. The 8 points corresponding to the vertices of the cube use 3-point averages. The higher order total derivatives are derived by applying (13) to these interpolated full gridpoint values.

c *A-Grid Considerations*

The A-grid version avoids the above horizontal averaging, so it can be expected to maintain greater accuracy. A complication is that (13) and (14) require special treatment at the edges and vertices of the six panels. Along the edges, it is necessary to use the averages of (13) as evaluated separately for each of the two neighbouring panels; note that u' and v' have different values as used within each domain; 2-point derivatives are found to be adequate for the uncentred derivatives.

At the eight vertices, we can apply (13) and (14) within each sector for the velocity components u_q, u_r, u_s in the directions of the three adjoining grid points, taking $\gamma = 2\pi/3$, $\cos\gamma = -1/2$. Summing and averaging yields

$$\frac{d}{dt} = \frac{\partial}{\partial t} + \frac{2}{9}(4u_q + u_r + u_s)\frac{\partial}{\partial x_q} + \frac{2}{9}(u_q + 4u_r + u_s)\frac{\partial}{\partial x_r} + \frac{2}{9}(u_q + u_r + 4u_s)\frac{\partial}{\partial x_s}. \qquad (17)$$

4 Results

Test integrations are performed for a variety of solid body rotation problems using the $N = 37$ cubic gnomonic grid. This grid has 8216 grid points, similar to the 8192 points of the Gaussian grid of a T42 spectral model. First, to provide comparison with McGregor (1993), Ritchie's (1987) test problem was chosen, where a "Gaussian hill" of scale diameter 2500 km initially centred on the equator is advected over the north pole with a solid body axis of rotation passing through 45°N. Figures 4a and 4d show the initial/verifying pattern on the stretched ($\alpha = 0.75$) gnomonic and Gaussian grids respectively. Figure 4 also shows the simulated patterns after 1 rotation using 40 time steps per rotation, and the errors evaluated at the grid points; the Gaussian grid calculations are performed using the method of McGregor (1993). The gnomonic grid produces rather smaller errors.

Williamson et al. (1992) proposed that the following normalized global error estimates be plotted for each time step.

$$l_1 = I|h - h_T|/I|h_T| \qquad (18)$$

$$l_2 = \{I[h - h_T)^2]\}^{1/2}/\{I[h_T^2]\}^{1/2} \qquad (19)$$

$$l_\infty = \max|h - h_T|/\max|h_T| \qquad (20)$$

and also the normalized mean, variance, minimum and maximum. Here $I(h)$ denotes the discrete approximation to the global integral of the advected field h, and

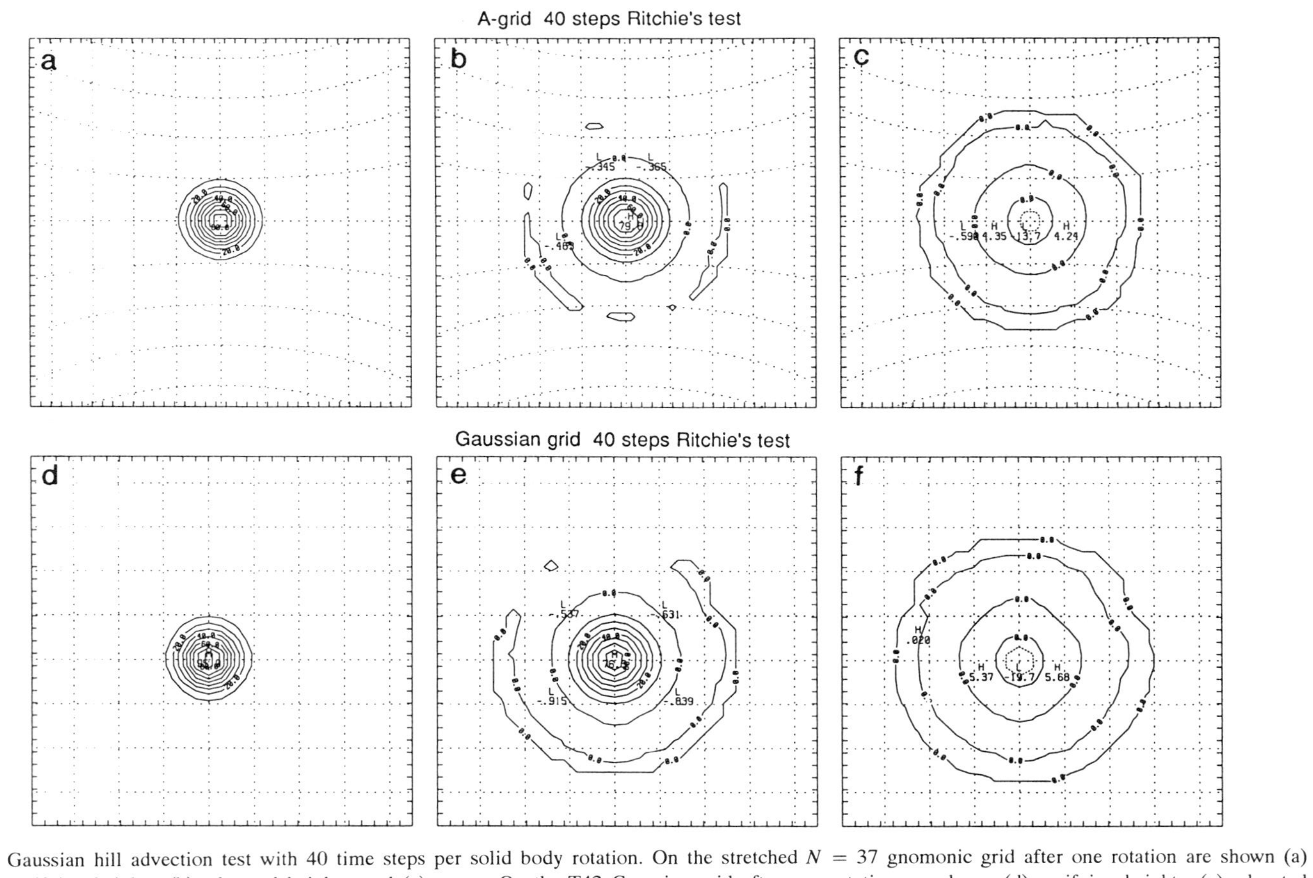

Fig. 4 Gaussian hill advection test with 40 time steps per solid body rotation. On the stretched $N = 37$ gnomonic grid after one rotation are shown (a) verifying heights, (b) advected heights, and (c) errors. On the T42 Gaussian grid after one rotation are shown (d) verifying heights, (e) advected heights, and (f) errors. Latitudes and longitudes are shown every 10°.

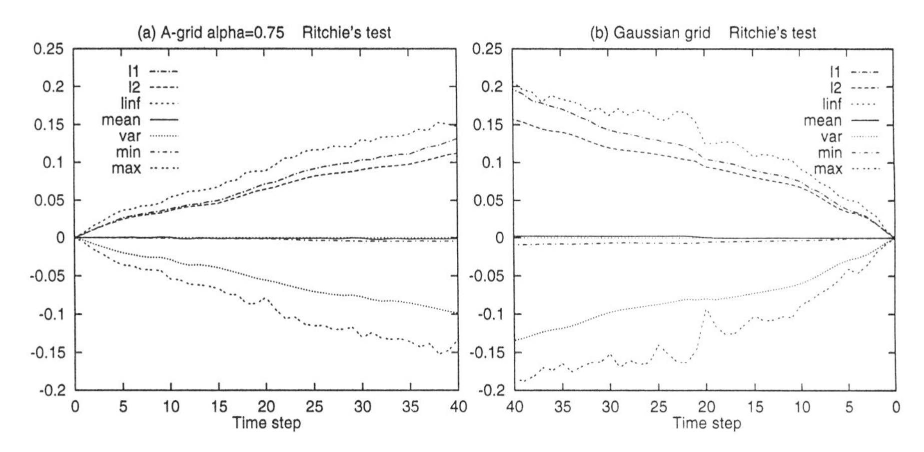

Fig. 5 Normalized errors for the Gaussian hill advection test with 40 time steps per solid body rotation for (a) the stretched $N = 37$ gnomonic grid, and (b) the T42 Gaussian grid. The second plot is plotted from right to left to assist comparisons.

TABLE 1. Normalized l_1 errors (as %) after one solid-body rotation (40 time steps per rotation) for the Gaussian hill test problem with scale diameters 10 000 km, 5000 km and 2500 km. Best results for each column are denoted by *.

	10 000 km	5000 km	2500 km
Eulerian	0.68	5.48	48.2
Ritchie I	0.70	1.90	15.2
Ritchie N	1.01	1.67	9.7*
McGregor (1993)	0.18	1.80	19.6
A $\alpha = 1$	0.17	1.60	17.2
A $\alpha = 0.75$	0.14*	1.06*	13.2
B $\alpha = 1$	0.55	1.94	17.3
B $\alpha = 0.75$	0.46	1.43	13.2

h_T denotes the true solution. The normalized errors are plotted in Fig. 5 for the Gaussian hill problem for both the stretched gnomonic grid and the Gaussian grid. The gnomonic grid shows better accuracy for all error estimates throughout the integration. No sudden changes are produced, as occurs for the normalized maximum on the Gaussian grid mid-way through the integration.

Comparisons with the l_1 errors for other methods are provided in Table 1 for the Gaussian hill with three different length scales. The Eulerian, Ritchie's interpolating and non-interpolating results are given by Ritchie (1987). A-grid and B-grid, unstretched and stretched results are included, as well as calculations performed on the T42 Gaussian grid by McGregor (1993). For the unstretched ($\alpha = 1$) gnomonic grid the maximum Courant number is about 6; for the stretched ($\alpha = 0.75$) grid

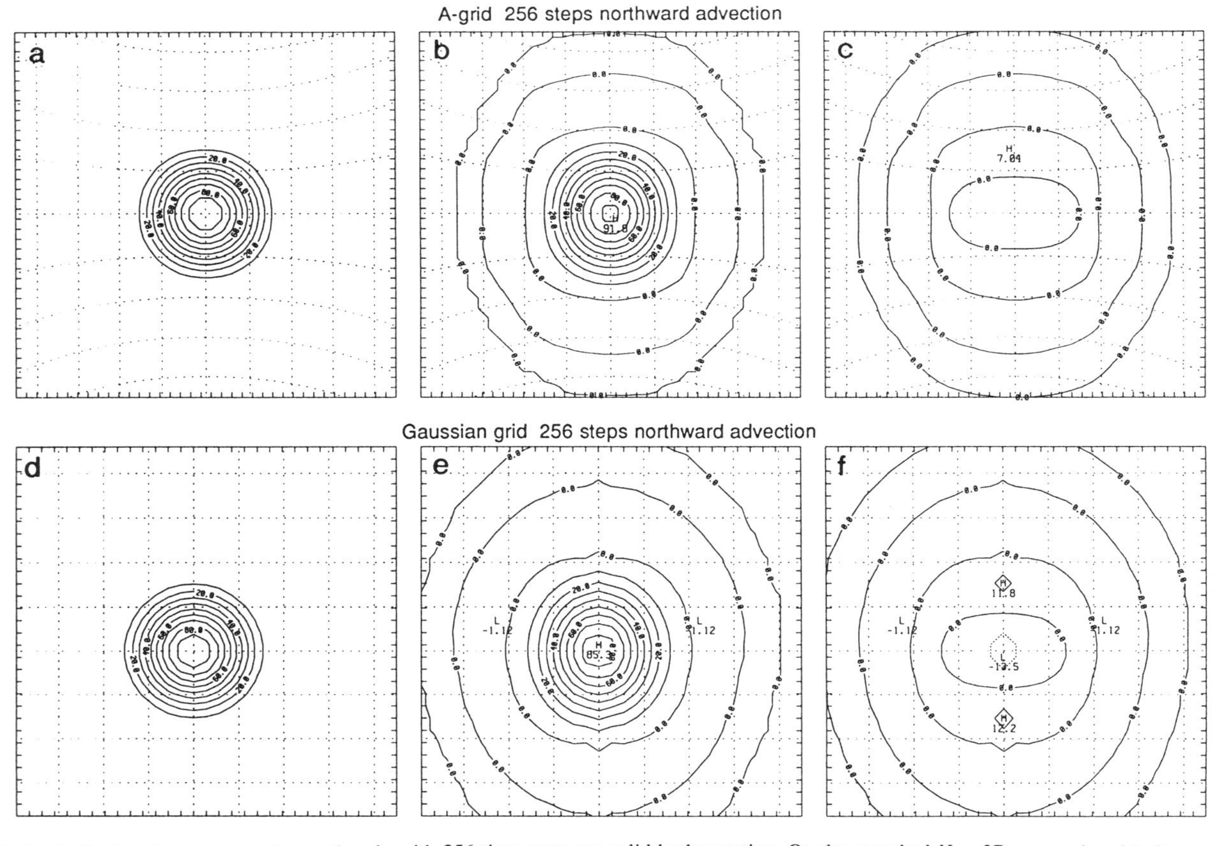

Fig. 6 Cosine bell advection test over the north pole with 256 time steps per solid body rotation. On the stretched $N = 37$ gnomonic grid after one rotation are shown (a) verifying heights, (b) advected heights, and (c) errors. On the T42 Gaussian grid after one rotation are shown (d) verifying heights, (e) advected heights, and (f) errors.

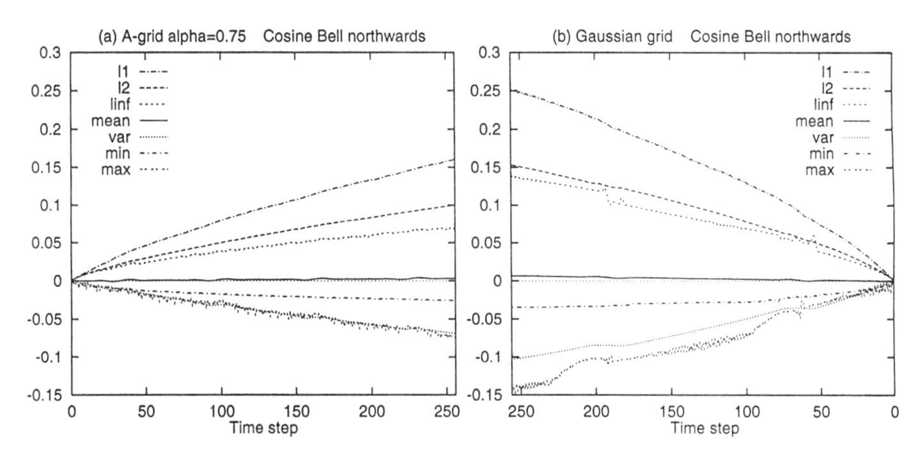

Fig. 7 Normalized errors for the cosine bell advection test with 256 time steps per solid body rotation for (a) the stretched $N = 37$ gnomonic grid and (b) the T42 Gaussian grid.

it is about 4. The stretched A-grid scheme produces the smallest l_1 errors overall, although for the 2500 km scale diameter hill Ritchie's non-interpolating scheme has smaller error.

A "cosine bell" family of solid body rotation advection tests was introduced by Williamson et al. (1992). The cosine bell resembles Ritchie's Gaussian hill having the intermediate scale diameter of 5000 km. The cosine bell starts at the equator with four alternative axes of rotation such that it moves initially either: (E) due east, (E^-) 2.9° north of due east, (N) due north, or (N^+) 2.9° east of due north.

Figure 6 presents the verifying and simulated height fields for the northward moving cosine bell after one rotation with an equatorial Courant number of 0.5; simulations are shown for the stretched gnomonic A-grid and the Gaussian T42 grid. The patterns are similar to Fig. 10 of Williamson and Rasch (1989), who compare a variety of bicubic interpolation schemes for the field interpolations. The corresponding set of normalized errors is shown in Fig. 7. Again the gnomonic grid produces consistently smaller errors than the Gaussian grid for all time steps.

Figure 8 shows the patterns for northward advection of the cosine bell with a large equatorial Courant number of about 3. The corresponding set of normalized errors is shown in Fig. 9. Once again calculations on the stretched gnomonic grid are more accurate than those on the Gaussian grid. No disruption is apparent as the pattern moves across the gnomonic panel boundaries, unlike the disruption seen on the Gaussian grid as the pattern crosses each pole. The error behaviour is clarified in Tables 2 to 6 which show a selection of the error measures for each of the cosine bell rotations and also for the Gaussian hill; A-grid and B-grid, stretched and unstretched grids are included.

As expected from the grid symmetries, it can be seen from the tables that the gnomonic E and N results are identical; likewise the E^- and N^+ results are identical.

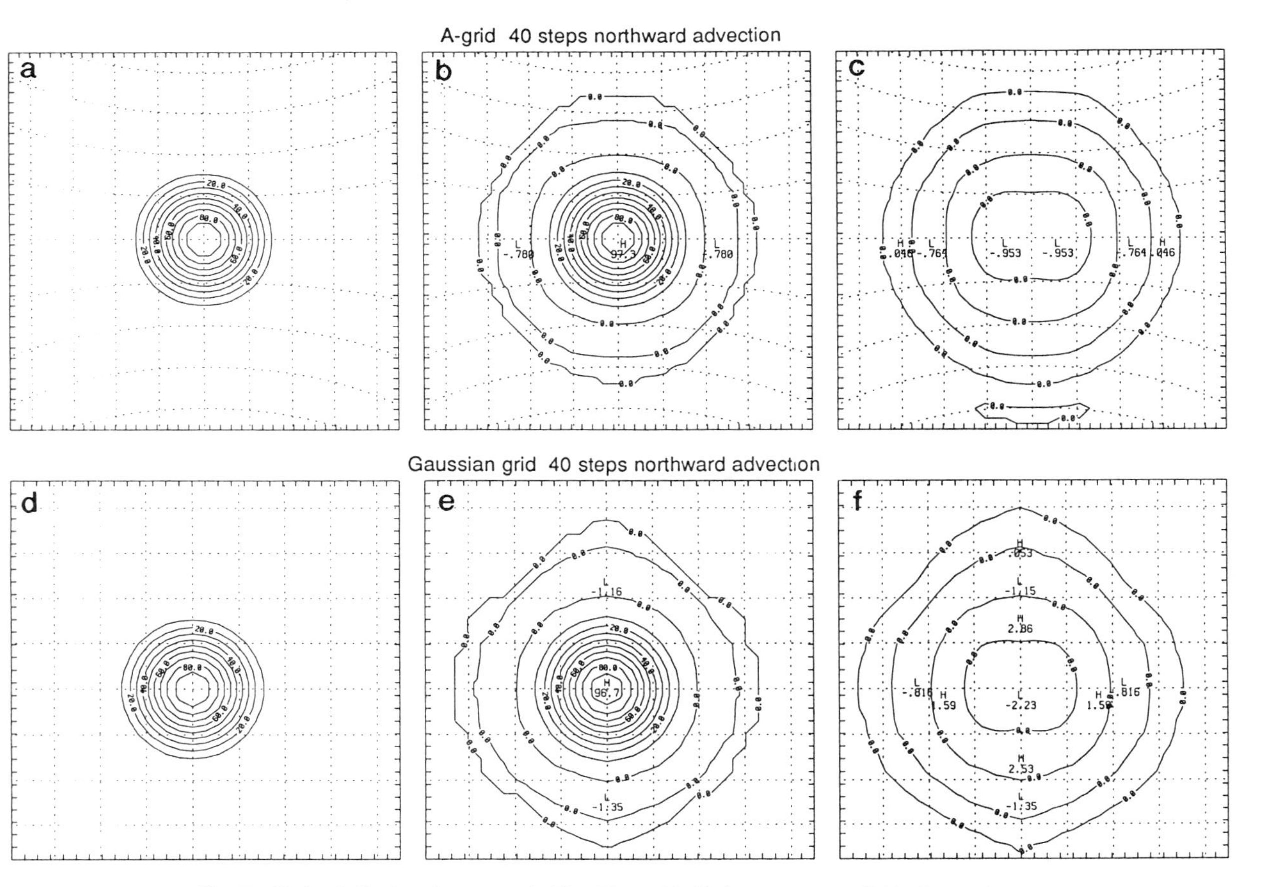

Fig. 8 Cosine bell advection test as in Fig. 6 but with 40 time steps per solid body rotation.

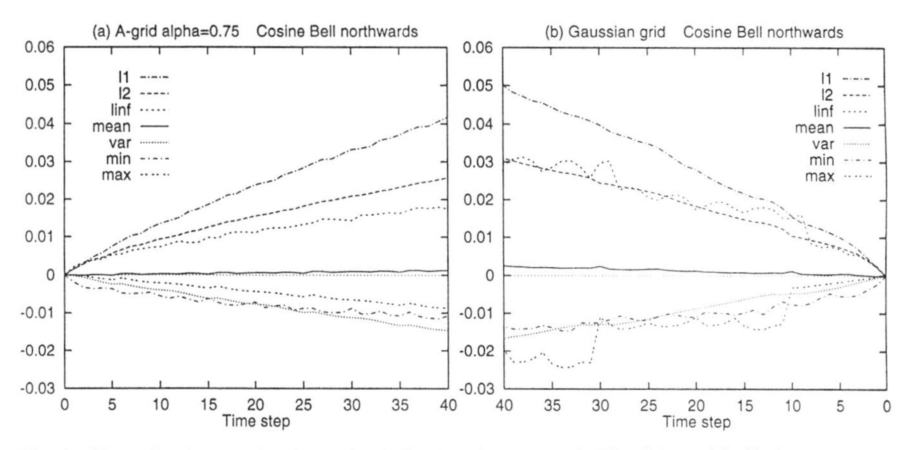

Fig. 9 Normalized errors for the cosine bell advection test as in Fig. 7 but with 40 time steps per solid body rotation.

TABLE 2. Normalized l_1 errors (as %) after one solid-body rotation (with 40 time steps per rotation). Both the Gaussian hill and cosine bell advection tests are shown.

	Gaussian grid	Cube-A $\alpha = 1$	Cube-A $\alpha = 0.75$	Cube-B $\alpha = 1$	Cube-B $\alpha = 0.75$
Gaussian hill	19.6	17.2	13.2	17.3	13.2
E	4.1	4.4	4.2	5.2	4.6
E$^-$	5.4	4.6	4.2	5.3	4.6
N	5.0	4.4	4.2	5.2	4.6
N$^+$	5.0	4.6	4.2	5.3	4.6

TABLE 3. As in TABLE 2, but for normalized error of the mean (as %).

	Gaussian grid	Cube-A $\alpha = 1$	Cube-A $\alpha = 0.75$	Cube-B $\alpha = 1$	Cube-B $\alpha = 0.75$
Gaussian hill	0.22	−0.22	−0.14	−0.22	−0.19
E	−0.08	0.55	0.13	0.55	0.13
E$^-$	−0.08	0.25	0.01	0.25	0.01
N	0.25	0.55	0.13	0.55	0.13
N$^+$	−0.02	0.25	0.01	0.25	0.01

The A-grid results are usually very similar to those for the B-grid; exceptions occur for the cosine bell for the l_1 errors where the A-grid is superior, and for the errors in the maximum where the B-grid is superior. Generally the stretched gnomonic grid performs best. As might be expected, the Gaussian grid is sometimes slightly better for the simpler eastward advection case.

The stretched grid generally produces smaller errors than the unstretched grid, especially for the mean. The stretched grid is particularly beneficial for the smaller-

TABLE 4. As in TABLE 2, but for normalized error of the variance (as %).

	Gaussian grid	Cube-A $\alpha = 1$	Cube-A $\alpha = 0.75$	Cube-B $\alpha = 1$	Cube-B $\alpha = 0.75$
Gaussian hill	−13.4	−13.4	−9.8	−13.4	−9.9
E	−1.8	−1.2	−1.5	−1.2	−1.5
E^-	−2.8	−1.9	−1.7	−1.9	−1.7
N	−1.7	−1.2	−1.5	−1.2	−1.5
N^+	−2.0	−1.9	−1.7	−1.9	−1.7

TABLE 5. As in TABLE 2, but for normalized error of the minimum (as %).

	Gaussian grid	Cube-A $\alpha = 1$	Cube-A $\alpha = 0.75$	Cube-B $\alpha = 1$	Cube-B $\alpha = 0.75$
Gaussian hill	−0.91	−0.68	−0.46	−0.69	−0.46
E	−1.42	−1.14	−1.07	−1.14	−1.04
E^-	−1.43	−1.15	−1.07	−1.15	−1.05
N	−1.35	−1.14	−1.07	−1.14	−1.04
N^+	−1.27	−1.15	−1.07	−1.15	−1.05

TABLE 6. As in TABLE 2, but for normalized error of the maximum (as %).

	Gaussian grid	Cube-A $\alpha = 1$	Cube-A $\alpha = 0.75$	Cube-B $\alpha = 1$	Cube-B $\alpha = 0.75$
Gaussian hill	−18.5	−13.8	−13.2	−13.2	−12.9
E	−1.26	−0.90	−0.87	−0.54	−0.66
E^-	−1.90	−1.36	−1.04	−0.99	−0.83
N	−1.99	−0.90	−0.87	−0.54	−0.66
N^+	−1.57	−1.36	−1.04	−0.99	−0.83

scale Gaussian hill on all error estimates except the maximum. The unstretched grid produces errors that are more like those of the Gaussian grid.

5 Concluding Comments

This study has shown that the semi-Lagrangian technique of McGregor (1993) can be applied successfully to advection on a cubic gnomonic grid, either in A-grid or B-grid form. A major attraction of the grid is its quasi-uniform resolution, especially compared to conventional latitude-longitude grids. The non-orthogonal nature of the grid does not create any problems. The scheme is straightforward to code, being somewhat simpler for the B-grid version. A stretched version of the grid was also created, producing more uniform resolution on the sphere. This form showed particular gains in accuracy over semi-Lagrangian advection on a comparable Gaussian latitude-longitude grid. These benefits were evident for a variety of Gaussian hill and cosine bell solid body rotation advection tests. Detailed normalized error measures at each time step were evaluated following Williamson et al. (1992), verifying the superior performance of the scheme throughout the integrations.

A modified cubic gnomonic grid which possesses orthogonal coordinate lines has been proposed by Rancic (personal communication, 1994). The performance of the McGregor (1993) semi-Lagrangian advection scheme is currently being evaluated on that grid. It is planned to produce a shallow water model on a gnomonic grid, and subsequently a primitive equations model. Compared to a spectral model, benefits should arise from the uniform resolution, the lack of Gibb's phenomena near orography and the related absence of negative orography over ocean areas.

References

BATES, J.R.; F.H.M. SEMAZZI, R.W. HIGGINS and S.R.M. BARROS. 1990. Integration of the shallow water equations on the sphere using a vector semi-Lagrangian scheme with a multigrid solver. *Mon. Weather Rev.* **118**: 1615–1627.

CULLEN, M.J.P. 1974. Integrations of the primitive equations on a sphere using the finite element method. *Q.J.R. Meteorol. Soc.* **100**: 555–562.

HORTAL, M. and A.J. SIMMONS. 1991. Use of reduced grids in spectral models. *Mon. Weather Rev.* **119**: 1057–1074.

MCGREGOR, J.L. 1993. Economical determination of departure points for semi-Lagrangian models. *Mon. Weather Rev.* **121**: 221–230.

RITCHIE, H. 1987. Semi-Lagrangian advection on a Gaussian grid. *Mon. Weather Rev.* **115**: 608–619.

ROBERT, A. 1981. A stable numerical integration scheme for the primitive meteorological equations. ATMOSPHERE-OCEAN, **19**: 35–46

SADOURNY, R. 1972. Conservative finite-difference approximations of the primitive equations on quasi-uniform spherical grids. *Mon. Weather Rev.* **100**: 136–144.

———; A. ARAKAWA and Y. MINTZ. 1968. Integration of the nondivergent barotropic vorticity equation with an icosahedral-hexagonal grid for the sphere. *Mon. Weather Rev.* **96**: 351–356.

STANIFORTH, A. and J. CÔTÉ. 1991. Semi-Lagrangian integration schemes for atmospheric models – a review. *Mon. Weather Rev.* **119**: 2206–2223.

WILLIAMSON, D.L. 1968. Integration of the barotropic vorticity equation on a spherical geodesic grid. *Tellus*, **20**: 642–653.

———. 1979. Difference approximations for fluid flow on a sphere. In: *Numerical Methods used in atmospheric models*, Vol. 2, GARP Publications Series No. 17. World Meteorological Organization, pp. 51–120.

——— and P.J. RASCH. 1989. Two-dimensional semi-Lagrangian transport with shape-preserving interpolation. *Mon. Weather Rev.* **117**: 102–129.

———; J.B. DRAKE, J.J. HACK, R. JAKOB and P.N. SWARTZTRAUBER. 1992. A standard test set for numerical approximations to the shallow water equations in spherical geometry. *J. Comput. Phys.* **102**: 211–224.

A Lagrangian Advection Scheme Using Tracer Points

Eigil Kaas, Annette Guldberg and Philippe Lopez
Danish Meteorological Institute,
Lyngbyvej 100, DK-2100 Copenhagen Ø

[Original manuscript received 3 February 1995; in revised form 1 September 1995]

ABSTRACT *A new quasi-Lagrangian advection scheme with similarities to the so-called Particle-In-Cell methods has been introduced with the purpose of describing advection processes in atmospheric modelling in an accurate way. The new Full Particle In Cell (FPIC) scheme has been tested and compared with other advection schemes commonly used in atmospheric modelling. The first tests deal with plane passive advection in a non-deforming as well as in a strongly deforming flow. In these cases, the FPIC scheme is compared with a traditional semi-Lagrangian advection method. The new scheme has the advantage of being exact for linear advection and in the case of non-linear advection, much less damping is seen than with the semi-Lagrangian scheme.*

In order to investigate if the FPIC scheme behaves reasonably in the atmospheric dynamical environment, it has been fully implemented in a shallow water model including orography and with semi-implicit treatment of gravity wave terms. Also in this case, the scheme behaves as well as, or better than, traditional schemes.

As it is intended to use the FPIC scheme for advection of water vapour and liquid water in a GCM, a test of advection of a positive definite, sharply varying, passive tracer in the shallow water model has been carried out. For the same reason, the FPIC scheme has also been tested for passive advection on the sphere. It turns out that the scheme also behaves excellently in these cases.

Among the advantages of FPIC are: no CFL (Courant-Friedrichs-Levy) criterion limits the length of the time step, the scheme can be formulated so that it is approximately mass conserving and high accuracy is still obtained when the resolution is decreased in order to reduce the computational cost. Furthermore, the scheme is positive definite. The most serious limitation of the scheme is that it demands more memory than traditional schemes.

RÉSUMÉ *Un nouveau schéma d'advection quasi-lagrangien comportant des similitudes avec les méthodes communément appelées "particules en cellule", et permettant de décrire avec précision les processus advectifs en modélisation atmosphérique, est présenté.*

Le nouveau schéma à "particules complètes en cellule" (FPIC) a été testé et comparé avec d'autres schémas d'advection habituellement utilisés en modélisation de l'atmosphère. Les premiers essais concernent l'advection passive dans le plan au sein de flux déformant ou non; dans ces cas, le schéma FPIC est comparé à une méthode semi-lagrangienne traditionnelle. Le nouveau schéma présente l'avantage d'être exact pour l'advection linéaire, et dans le cas de l'advection non linéaire, l'amortissement constaté est bien moindre qu'avec le schéma semi-lagrangien.

Afin de vérifier si le schéma FPIC est adéquat dans un environnement atmosphérique dynamique, il a été introduit dans un modèle en eau peu profonde incluant l'orographie, avec un traitement semi-implicite des ondes de gravité. Là encore, le schéma est aussi performant sinon meilleur que les schémas classiques.

Comme il est prévu d'appliquer le schéma FPIC pour l'advection de la vapeur d'eau et de l'eau liquide dans un MCG, un essai d'advection d'un traceur passif, défini positif et à fortes variations spatiales, a été mené à bien. Pour la même raison, le schéma FPIC a également été testé pour l'advection passive sur la sphère. Dans ces deux cas, le schéma s'avère tout aussi performant.

Parmi les avantages du schéma FPIC figurent: aucun critère CFL ne limite la longueur du pas de temps; le schéma peut être formulé de manière à conserver approximativement la masse; une précision élevée est encore possible lorsque la résolution est réduite afin de diminuer les coûts computationnels; enfin, le schéma est défini positif. La limitation la plus contraignante de ce schéma réside dans son besoin accru en mémoire, par comparaison avec les schémas traditionnels.

1 Introduction

One of the dominant processes in the atmosphere is advection. Unfortunately, the numerical techniques most commonly used to solve the equations of atmospheric motion in forecast and climate models are generally unable to describe linear passive advection exactly. This is because the classical advection schemes are dispersive or diffusive or because the diffusion techniques used to prevent spurious growth of the shortest resolved scales often destroy exactness of simple passive advection. Examples from the first group are all Eulerian formulated models based on finite difference schemes and all semi-Lagrangian schemes, and from the second group we can mention Eulerian formulated spectral models using a $K\nabla^n(\ )$ type of horizontal diffusion of the advected quantity, with K being a diffusion constant and n an arbitrary even number (e.g. 4 or 8). It is noted that for spectral models with no diffusion, linear passive advection is exact except for errors due to the initial truncation (aliasing) and errors due to the time truncation, see Machenhauer (1979). For the spectral models, however, there are often other problems with Gibbs phenomena near rapidly varying field values. An example is liquid water generated in the models parametrization packages. When this field is transformed to spectral space and back again to physical space, large "under- and overshooting" of the original values are seen because the number of degrees of freedom (for dynamical reasons) must be larger in physical than in spectral space. This will often lead to negative liquid water content, i.e. to a violation of positive definiteness, even though the problem can be somewhat reduced as mentioned by Navarra et al. (1994). Spurious under- and overshooting can also be seen for many gridpoint-based advection schemes (Eulerian or semi-Lagrangian) even though substantial effort has been put into a reduction of the problem, see e.g. Williamson and Rasch (1994).

We have tested the possibility of introducing a new scheme which is exact in the case of linear advection. The scheme, which is positive definite, is "more

Lagrangian than semi-Lagrangian" and is quite similar to the so-called Particle-In-Cell (PIC) methods introduced in the 50's (Harlow, 1957) and mainly used for simulating plasmas. See Harlow (1988) for a short review. In recent years, certain descendants of the PIC methods have been developed and used with considerable success in a wide range of hydrodynamic problems in astrophysics and plasma physics generally, see Burke (1988). The new scheme used here is of the type "Full Particle" PIC, i.e. a Lagrangian description of the fluid where all properties of the fluid are attributed to the particles. We will use the acronym FPIC for this scheme. The FPIC scheme also has certain similarities to the particle Lagrange method by Steppeler (1990), even though this scheme does not operate with field values defined at each particle. John Baumgardner (Los Alamos) has also used a PIC method for advection on an icosahedral grid.

In this paper, we will describe FPIC and compare its performance with three other well known advection schemes commonly used in atmospheric modelling. These three schemes are an Eulerian scheme using centred differences, an Eulerian spectral scheme (see Eliasen et al., 1970) and an (cubic interpolating) semi-Lagrangian scheme (see Robert, 1982). After a short description in Section 2 of the basics of FPIC, four different tests will be applied to the advection schemes:

- passive advection by non-deforming velocity;
- passive advection by a strongly deforming velocity which is constant in time;
- performance in a shallow water channel model with orography and with semi-implicit treatment of the gravity wave terms;
- passive advection by the flow in the shallow water model.

These tests will be described in Sections 3 through 6. Furthermore Sections 3 and 5 include descriptions of the implementation of FPIC in the case of passive advection and in the case of the shallow water model, respectively. Section 7 includes a discussion of problems related to its use in atmospheric modelling and some computational aspects while a summary of the findings can be found in Section 8.

2 The basics of FPIC

The idea in the FPIC scheme is to let a number of tracer points describe the field values to be advected. In this way, one obtains an advection scheme which is free of the damping seen in traditional semi-Lagrangian schemes or in many Eulerian finite difference schemes with a suitable horizontal diffusion needed to control unphysical numerical dispersion and non-linear instabilities.

The basic principle is to move tracer points holding the field values of all prognostic variables along trajectories corresponding to a time step and then obtain the forecasted gridpoint values by interpolation from the tracer points. As an example, Fig. 1 explains the difference between passive advection by constant flow in one dimension with Courant number 0.5 using a linearly interpolating semi-Lagrangian

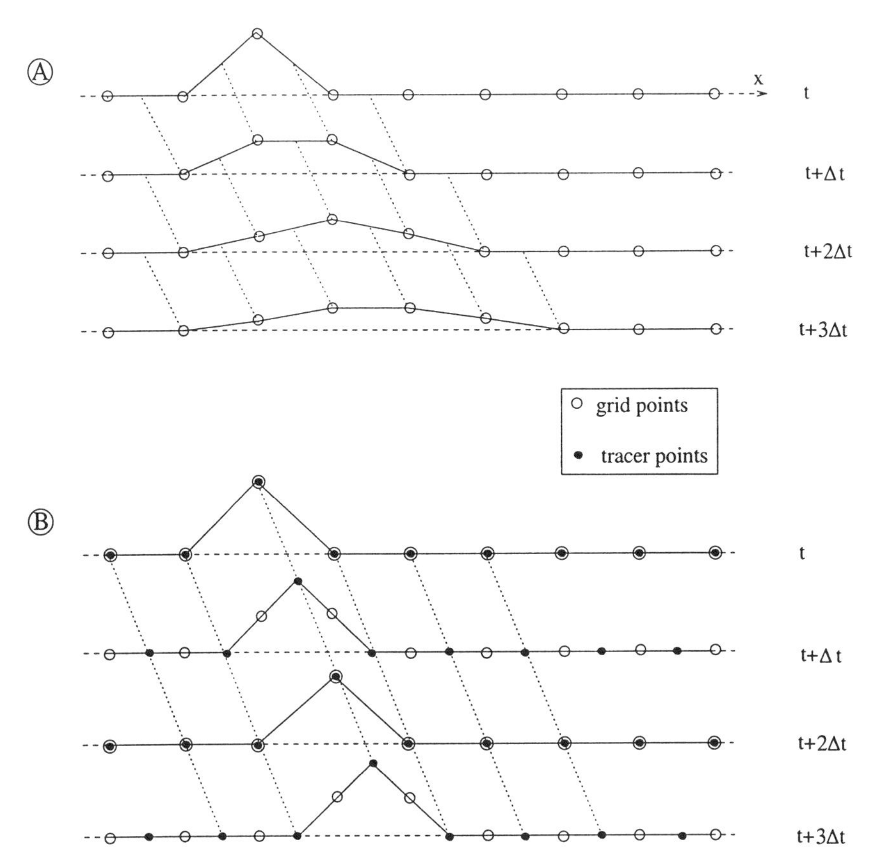

Fig. 1 Illustration of the difference between a traditional semi-Lagrangian scheme (A) and the FPIC method (B) in the case of the pure advection of a triangular structure by a one dimensional flow, constant in time. All the interpolations are supposed to be linear. While a damping due to the interpolations is observed when using the first scheme, the structure is kept unchanged in the second case.

scheme (A) and the FPIC scheme (B). The semi-Lagrangian forecast is based on upstream interpolations of trajectories ending in the grid points to be forecast. When these interpolations are repeated from time step to time step, a damping of the field takes place. In the FPIC, however, a number of tracer points are kept for a long time and the forecast at a given time simply consists of an interpolation to the grid points from the tracer points. There is also, of course, an interpolation error from this interpolation which in Fig. 1 is shown to be linear. However, contrary to the semi-Lagrangian scheme the interpolation error is not "remembered" by the model and there is then no overall damping. This is because the tracer points hold the history of the model and since the only process in this simple example is advection, no other processes can modify the field value at the tracer points.

The flow in the real atmosphere is of course not constant as in this simple example. It is therefore necessary to add and remove tracer points during a time integration of a model of the atmosphere. Otherwise a divergent and deforming flow would cause the density of tracer points to be very high or too low to allow reasonably accurate interpolations from tracer to grid points. In the real atmosphere, there are also many processes other than advection. At each time step, the tendencies due to these processes are calculated in the gridpoint space and transferred to the tracer points via an interpolation.

The details of FPIC will be explained in the following sections.

3 Plane passive advection tests in a constant flow

In this section, we describe two-dimensional plane advection tests both in a flow which is constant in time and space and for steady solid body rotation. The general prognostic equation for plane passive advection of a variable Ψ is:

$$\frac{d\Psi}{dt} = \frac{\partial\Psi}{\partial t} + \mathbf{V}\cdot\nabla\Psi = \frac{\partial\Psi}{\partial t} + u\frac{\partial\Psi}{\partial x} + v\frac{\partial\Psi}{\partial y} = 0 \tag{1}$$

where $\mathbf{V} = (u, v)$ is the two-dimensional velocity vector and x and y are Cartesian coordinates. We have used the FPIC scheme to solve this equation and compared the results with a bicubic Lagrange interpolating semi-Lagrangian advection method.

a *Description of the FPIC method in the case of passive advection*

The FPIC scheme is similar to a three time level semi-Lagrangian advection scheme for which the discretized prognostic equation reads:

$$\{\Psi^{n+1}\}_F = \{\Psi^{n-1}\}_O \tag{2}$$

Here the subscript F denotes the grid point to be forecast, while the subscript O denotes the trajectory departure point, upstream relative to F, and n is a time step counter. Generally the point O lies between grid points and the value there has to be evaluated by an interpolation which, in this paper, is always bicubic Lagrange. The trajectory from O to F is calculated by an iterative procedure using two iterations of the type:

$$\{x\}_O = \{x\}_F - 2\Delta t\{u\}_M \tag{3}$$

$$\{y\}_O = \{y\}_F - 2\Delta t\{v\}_M \tag{4}$$

where the subscript M denotes the mid-point between points F and O. The first guess of M is:

$$\{x\}_M = \{x\}_F - \Delta t\{u\}_F \tag{5}$$

$$\{y\}_M = \{y\}_F - \Delta t\{v\}_F \tag{6}$$

The velocities at the point M have to be interpolated from surrounding gridpoint values where they are known. This interpolation is bilinear. This trajectory scheme is a rather low-order scheme and of course higher order schemes could have been used.

In the new scheme, we formally use the same prognostic equation (2) as for the semi-Lagrangian scheme. Here however, the forecast of Ψ is made at tracer points, meaning that both points F and O lie between the grid points. Assuming that, at a certain time, we know all tracer point positions $\{x\}_O, \{y\}_O$ and the field value $\{\Psi\}_O$ in these points, the procedure in each time step is as follows:

1. For each tracer point calculate its trajectory corresponding to two time steps (three time level scheme). This is done in exactly the same way as for the traditional semi-Lagrangian scheme using two iterations. The trajectory we calculate with the FPIC scheme is, however, a downstream trajectory meaning that we find F (the unknown arrival point of tracer points) instead of O. The first guess of M is simply:

$$\{x\}_M = \{x\}_O + \Delta t\{u\}_O \tag{7}$$

$$\{y\}_M = \{y\}_O + \Delta t\{v\}_O \tag{8}$$

   The arrival point F is determined by:

$$\{x\}_F = \{x\}_O + 2\Delta t\{u\}_M \tag{9}$$

$$\{y\}_F = \{y\}_O + 2\Delta t\{v\}_M \tag{10}$$

2. Advect (re-position) the tracer points according to their individual trajectories. Each tracer point is associated with the grid box where it arrives.
3. Add tracer points where their density is "too" low, namely in grid boxes where there are no tracer points. The number of tracer points to be added in this case is set by a model parameter (*maxtr*), which is also the upper limit on the number of tracer points allowed in a grid box. The new tracer points that are added are evenly distributed in the actual grid box. The field values in the new points are interpolated from the tracer point values in the surrounding grid boxes, using a simple bilinear interpolation (see below).
4. Interpolate tracer point values to grid points. This is done using a bilinear interpolation (see below).
5. Remove tracer points where they are "too" dense. By too dense we mean that a grid box includes more than *maxtr* tracer points (see item 3). In this paper, the value of *maxtr* has been set to four, which equalizes the number of tracer points that are generated inside each grid box at time $t = 0$. The practical removal is carried out as a successive merging of the two closest tracer points, using a Euclidian distance.

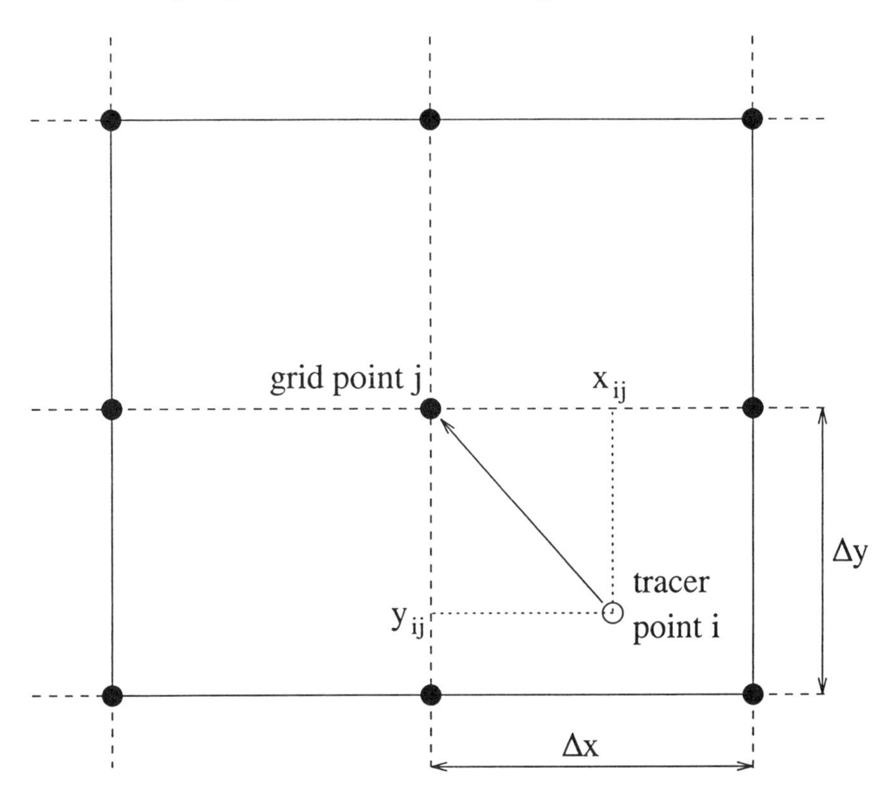

Fig. 2 Illustration of the bilinear interpolation from tracer to grid points in the FPIC method. The plot defines the coordinates x_{ij} and y_{ij} which are used in the computation of the weight of the tracer point i relative to the gridpoint j.

To interpolate from tracer to grid points on a regular mesh, we define a bilinear interpolation as follows.

Let us first notice that each grid point j is common to four grid boxes as shown on Fig. 2. To each tracer point i located inside one of these four adjacent grid boxes, we can attribute a weight w_{ij} relative to gridpoint j. This weight is defined by:

$$w_{ij} = \left(1 - \frac{x_{ij}}{\Delta x}\right)\left(1 - \frac{y_{ij}}{\Delta y}\right) \tag{11}$$

where x_{ij} and y_{ij} are the coordinates of the tracer point i relative to the gridpoint j, and where Δx and Δy are the resolutions of the grid in the x and y directions, respectively.

If f_i denotes the value at the tracer point i of the field to be interpolated, $w_{ij}f_i$ represents the contribution of this tracer point to the field value at gridpoint j.

Finally, we can determine the interpolated value of the field F_j at the gridpoint j by:

$$F_j = \frac{\sum_{i=1}^{N_j} w_{ij} f_i}{\sum_{i=1}^{N_j} w_{ij}} \tag{12}$$

where N_j is the number of tracer points cumulated over the four grid boxes surrounding the gridpoint j.

In the FPIC method, we occasionally need to add some tracer points in empty grid boxes (see item 3). In such boxes, we have chosen to define four new tracer points using a bilinear interpolation from other tracer points contained in the four surrounding grid boxes. The weighting function is similar to the one that we defined for interpolating from tracer to grid points.

b *Results*

Two different experiments have been run in the case of plane passive advection. In both tests, the initial field values vary between 0 and 1 arbitrary unit. The 64×64 point domain we consider here is a square with coordinates x and y between 0 and 1.

In the first experiment (see Fig. 3), a two-dimensional step function structure is initially located at the centre of the square domain and is then linearly advected along a diagonal using periodic boundary conditions, until it comes back to its departure point. The Courant number is arbitrarily set to 0.64. The comparison of the performances of the traditional semi-Lagrangian scheme and the FPIC method confirms that the latter gives the best results. Indeed, the semi-Lagrangian method suffers from a damping that grows with the integration time whereas the advected structure given by the FPIC remains close to the analytic solution with an error which never exceeds the initial error due to the way in which the initial tracer points are generated.

These verifications are repeated in a second experiment (see Fig. 4), in which a slotted cylindrical structure is initially placed close to the left edge of the domain and is then advected through solid body rotation around the centre of the domain. The maximum Courant number in the area is set to 1. The semi-Lagrangian method still causes a damping which remains reasonable only if a small number of complete rotations are carried out. Once more, we can notice that the FPIC scheme has been formulated so as to be almost exact in the case of pure advection. In this case, however, a small damping is seen because a few tracer points are merged (see item 5 above).

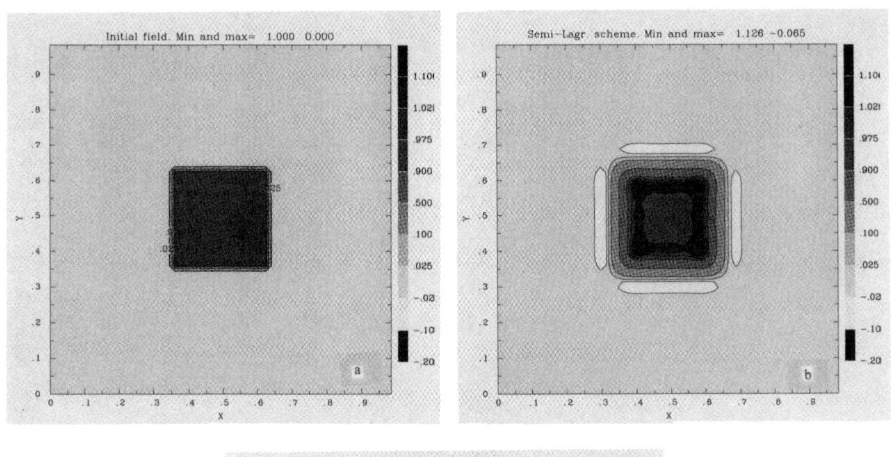

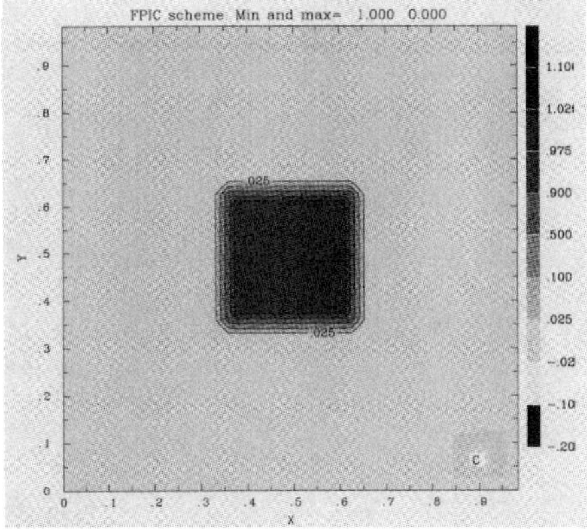

Fig. 3 An initial two-dimensional step function a) undergoes linear plane passive advection along a diagonal of a 64 × 64 point square domain, using periodic boundary conditions. The advected field is displayed when it is back to its initial central position for b) the semi-Lagrangian scheme and c) the FPIC method.

4 Plane passive advection tests in a strongly non-linear flow

a *Description of the experiment*

This section will be devoted to the description of strongly non-linear advection tests in a flow which is constant in time. The equations and the methods used in this part are the same as those presented in the previous section. However, the velocity field is now defined by its two components u and v:

$$u(x, y) = -\sin(2\pi x)\cos(2\pi y) \tag{13}$$

$$v(x, y) = \cos(2\pi x)\sin(2\pi y) \tag{14}$$

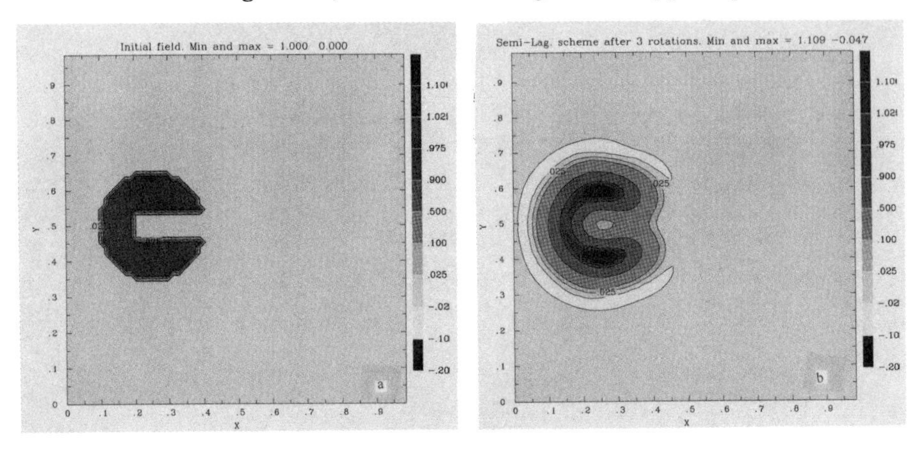

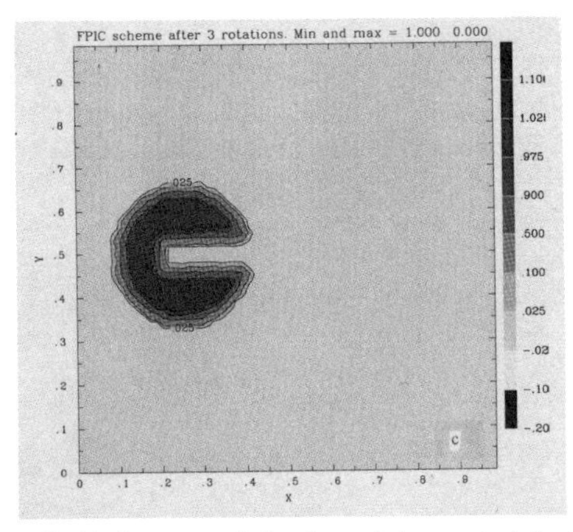

Fig. 4 An initial slotted cylindrical structure a), first located close to the left edge of the square domain, is advected through solid body rotation around the centre of the domain. The advected field is plotted after three complete rotations for b) the semi-Lagrangian scheme and c) the FPIC method.

where x and y are the coordinates in the square domain and both vary between 0 and 1. The corresponding non-linear flow is non-divergent and the initial two-dimensional step function structure to be advected is identical to the one shown in Fig. 3a. The advection scheme is first run for an integration time equal to 1 – i.e. until an "intermediate" time – and then the velocity orientation is reversed and the scheme is performed for another unit integration time so that the structure turns back to its initial position at the final time step. The maximum Courant number is set to 1.

b *Results*

In the current experiment, we have compared the results given by the traditional

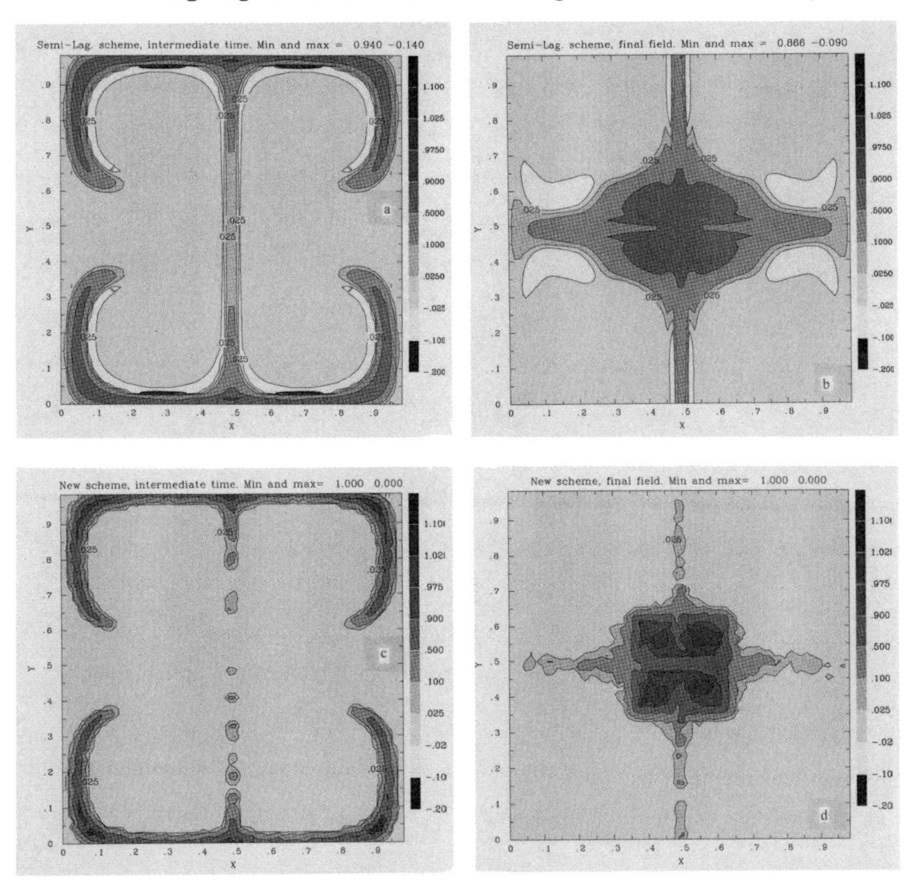

Fig. 5 An initial two-dimensional step function structure, the same as the one defined in Fig. 3a, first located at the centre of the square domain, is advected by a non-linear and non-divergent flow, defined in the text. The advected field is plotted at an intermediate time when the structure is put through the strongest deformation and at the final time when it turned back to its initial position, using a) and b) the semi-Lagrangian scheme and c) and d) the FPIC method, respectively.

semi-Lagrangian scheme and the FPIC method at both the intermediate and final times. It clearly appears in Fig. 5 that the FPIC is performing better than the former scheme, especially as regards the restoring of the two-dimensional step function structure at the final time step. Indeed, the semi-Lagrangian method is penalized by a damping that is really much stronger than in the case of the new scheme. The damping observed in the FPIC final results is mainly caused by the addition and removal of tracer points along the run and to a lesser extent by inaccuracies in the calculation of the trajectories.

5 Implementation of FPIC into a shallow water model

In this section, we describe the implementation of FPIC in a shallow water model

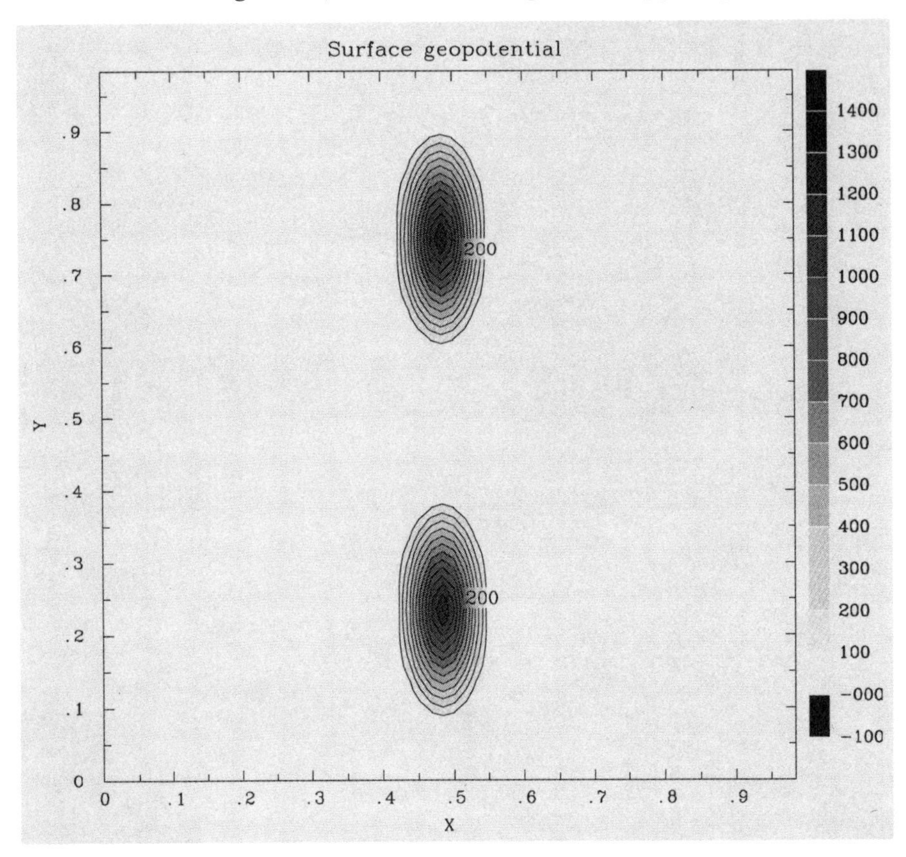

Fig. 6 Geopotential of the orography ϕ_s in the shallow water model. The two mountains are symmetrical about the equator ($y = 0.5$).

including orography and with semi-implicit treatment of the gravity wave terms. For comparison we also describe the implementation of the three traditional schemes. The FPIC is compared with a semi-Lagrangian gridpoint scheme (see Robert, 1982), an Eulerian gridpoint scheme and an Eulerian spectral scheme (see Eliasen et al., 1970).

The model is formulated as a "global channel" model permitting a regular grid and omitting all terms (except the Coriolis force) in the governing equations associated with the spherical geometry of the Earth. The orography of the model is shown in Fig. 6, which also shows the general geometry. The time development of the velocity components u and v and the geopotential ϕ (depth of the fluid) is determined by the usual shallow water equations:

$$\frac{du}{dt} = \frac{\partial u}{\partial t} + u\frac{\partial u}{\partial x} + v\frac{\partial u}{\partial y} = fv - \frac{\partial(\phi + \phi_s)}{\partial x} \tag{15}$$

$$\frac{dv}{dt} = \frac{\partial v}{\partial t} + u\frac{\partial v}{\partial x} + v\frac{\partial v}{\partial y} = -fu - \frac{\partial(\phi + \phi_s)}{\partial y} \tag{16}$$

$$\frac{d\phi}{dt} = \frac{\partial \phi}{\partial t} + u\frac{\partial \phi}{\partial x} + v\frac{\partial \phi}{\partial y} = -\phi D + \mathcal{F}_\phi \tag{17}$$

where f is the Coriolis parameter, ϕ_s the geopotential of the orography, $D = (\frac{\partial u}{\partial x} + \frac{\partial v}{\partial y})$ the divergence. All the variables are defined as periodic in x and y. Furthermore, u, ϕ and ϕ_s are symmetrical about the equator ($y = 0.5$) whereas v and f are asymmetrical about the equator. $\mathcal{F}_\phi$ is a zonally symmetrical Newtonian "cooling" type driving of the form not affecting the total mass:

$$\mathcal{F}_\phi = \frac{[\phi - \bar{\phi}]^0 - [\phi - \bar{\phi}]}{\tau} \tag{18}$$

where $\bar{\phi}$ is the average ϕ value over the whole grid. [ ] denote the zonal average of the variable and τ is a time constant equal to 172800 seconds (48 hours). The x and y extension of the domain is 20000 × 20000 kilometres.

The reason for this simple choice of model is that possible problems with spherical geometry are discarded while basic shallow water dynamics including orographic effects are described.

For each of the four advection schemes presented here, a three-time-level time stepping is used and a stable semi-implicit treatment of the gravity wave terms (see Robert et al., 1972) damps the speed of the fastest modes of the model and thus permits long time steps. For the two Eulerian advection schemes the maximum possible time step is then given by the usual Courant-Friedrichs-Levy criterion, while the maximum time step for the semi-Lagrangian and the FPIC schemes are much less restrictive, since the time step is only limited by the speed of inertial waves and the gradient of the wind (see Pudykiewicz et al., 1985).

For all schemes, a Robert time filter is used with a constant of 0.05. The formulation of each scheme will be described in the following subsections.

a *The Eulerian schemes*

The time discretization for the two Eulerian schemes can be written:

$$u^{n+1} = \left(u^{n-1} - 2\Delta t\left(u^n\frac{\partial u^n}{\partial x} + v^n\frac{\partial u^n}{\partial y}\right)\right) + 2\Delta t f v^n - 2\Delta t\left(\frac{\partial(\phi + \phi_s)}{\partial x}\right)_{tt} \tag{19}$$

$$v^{n+1} = \left(v^{n-1} - 2\Delta t\left(u^n\frac{\partial v^n}{\partial x} + v^n\frac{\partial v^n}{\partial y}\right)\right) - 2\Delta t f u^n - 2\Delta t\left(\frac{\partial(\phi + \phi_s)}{\partial y}\right)_{tt} \tag{20}$$

$$\phi^{n+1} = \left(\phi^{n-1} - 2\Delta t\left(u^n\frac{\partial \phi^n}{\partial x} + v^n\frac{\partial \phi^n}{\partial y}\right)\right) - 2\Delta t\phi_0 D_{tt} - 2\Delta t(\phi^n - \phi_0)D^n$$

$$+ 2\Delta t\, \mathcal{F}_\phi^{n-1} \tag{21}$$

where $\phi_0 = 5000g$ and $(\ )_{tt} = \alpha(\ )^{n+1} + \beta(\ )^n + \gamma(\ )^{n-1}$. In this last equation, α, β and γ are parameters fulfilling $\alpha+\beta+\gamma = 1$, and are meant to allow an off-centring of semi-implicit terms (see Rivest et al., 1994).

In the procedure for solving these equations, ϕ^{n+1} is isolated resulting in a Helmholtz equation in ϕ^{n+1}.

In the gridpoint version, all space derivatives are calculated using centred differences and the Helmholtz equation is solved using a standard five point finite difference approximation (see Swartztrauber and Sweet, 1975). To ease comparison with the other schemes, no variable staggering is introduced; meaning an Arakawa A-grid is used.

The spectral model uses double Fourier series as expansion functions. It is formulated as a usual transform model where all prognostic variables are transferred back and forth between the spectral and gridpoint spaces using computationally efficient Fast Fourier Transforms. The prognostic variables are truncated at a given model resolution in spectral space and then transformed to and used in gridpoint space.

The Helmholtz equation is easily solved in spectral space.

In both the Eulerian schemes, a weak fourth order horizontal diffusion is applied to all prognostic variables and the diffusion constant is set to 10^{-13} (the unit is $s^{-1}\ m^{-4}$). In the spectral model, this diffusion is as usual formulated in an implicit way, whereas a second order Shapiro filter is used in the gridpoint model (a second order Shapiro filter is a fourth order scheme).

b *The traditional semi-Lagrangian scheme*

For the traditional semi-Lagrangian scheme, the shallow water model equations can be discretized into:

$$u^{n+1} = u^{n-1} + 2\Delta t f v^n - 2\Delta t \left(\frac{\partial(\phi + \phi_s)}{\partial x}\right)_{tt} \tag{22}$$

$$v^{n+1} = v^{n-1} - 2\Delta t f u^n - 2\Delta t \left(\frac{\partial(\phi + \phi_s)}{\partial y}\right)_{tt} \tag{23}$$

$$\phi^{n+1} = \phi^{n-1} - 2\Delta t \phi_0 D_{tt} - 2\Delta t(\phi^n - \phi_0)D^n + 2\Delta t\, \mathcal{F}_\phi^{n-1} \tag{24}$$

Terms at time $n-1$ are calculated at the departure point of the trajectory and terms at time n are calculated as the mean between the values at the departure point and at the arrival point. Field values at the departure point are computed using a bicubic Lagrange interpolation. Departure points are determined in the same way as described in Section 3, that is using bilinear interpolations of the wind and two iterations.

The resulting Helmholtz equation is again solved using the standard five point difference approximation. Spatial derivatives are calculated using centred differences.

Note that in the case of the semi-Lagrangian scheme, no horizontal diffusion is required, since the upstream interpolations play the role of diffusion.

c *The FPIC scheme*

The equations can be written as for the traditional semi-Lagrangian scheme:

$$\begin{aligned}\left\{u^{n+1} + 2\Delta t\alpha\left(\frac{\partial(\phi+\phi_s)}{\partial x}\right)^{n+1}\right\}_F = &\left\{u^{n-1} + \Delta t\left(fv^n - \beta\left(\frac{\partial(\phi+\phi_s)}{\partial x}\right)^n\right)\right. \\ &\left. + 2\Delta t\gamma\left(\frac{\partial(\phi+\phi_s)}{\partial x}\right)^{n-1}\right\}_O \\ &+ \left\{\Delta t\left(fv^n - \beta\left(\frac{\partial(\phi+\phi_s)}{\partial x}\right)^n\right)\right\}_F \end{aligned} \quad (25)$$

$$\begin{aligned}\left\{v^{n+1} + 2\Delta t\alpha\left(\frac{\partial(\phi+\phi_s)}{\partial y}\right)^{n+1}\right\}_F = &\left\{v^{n-1} + \Delta t\left(-fu^n - \beta\left(\frac{\partial(\phi+\phi_s)}{\partial y}\right)^n\right)\right. \\ &\left. + 2\Delta t\gamma\left(\frac{\partial(\phi+\phi_s)}{\partial y}\right)^{n-1}\right\}_O \\ &+ \left\{\Delta t\left(-fu^n - \beta\left(\frac{\partial(\phi+\phi_s)}{\partial y}\right)^n\right)\right\}_F \end{aligned} \quad (26)$$

$$\begin{aligned}\{\phi^{n+1} + 2\Delta t\alpha\phi_0 D^{n+1}\}_F = &\{\phi^{n-1} + \Delta t((1-\beta)\phi_0 - \phi^n)D^n) \\ &+ 2\Delta t(-\gamma\phi_0 D^{n-1} + \mathcal{F}_\phi^{n-1}\}_O \\ &+ \{\Delta t((1-\beta)\phi_0 - \phi^n)D^n\}_F \end{aligned} \quad (27)$$

where subscripts O and F respectively denote the origin and final points of a trajectory, as in Section 3.

Referring to Section 3a where the FPIC was described in the case of passive advection, the steps for solving (25), (26) and (27) using the FPIC scheme are:

1. At grid points, calculate the terms in (25), (26) and (27) marked $\{\ \}_O$ except the full field values u^{n-1}, v^{n-1} and ϕ^{n-1}. For these calculations, simple centred differences are used to calculate spatial derivatives.
2. Using a bicubic Lagrange interpolation, terms from 1. are added to the tracer point values u^{n-1}, v^{n-1} and ϕ^{n-1}. The interpolation is of the same type as the one used to find the departure point value in the semi-Lagrangian scheme. Note that tracer points are still located at their departure points.

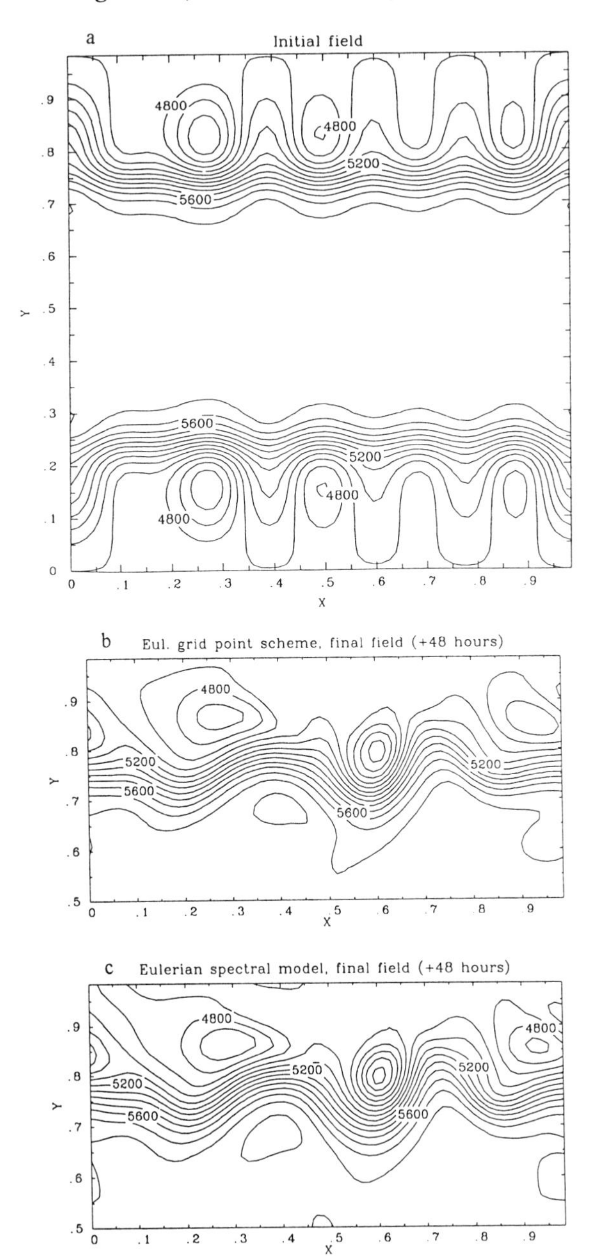

Fig. 7 The initial height ($\phi + \phi_s$) (in metres) field used in the shallow water model is drawn in Fig. 7a. In this case, the horizontal grid has 64 × 64 points and a square truncation at wavenumber 21 is used in the Eulerian spectral model. The final height field is plotted for the "northern" half of the domain and for the four advection schemes described in the text: b) the Eulerian gridpoint scheme, c) the Eulerian spectral method, d) the traditional semi-Lagrangian scheme and e) the FPIC.

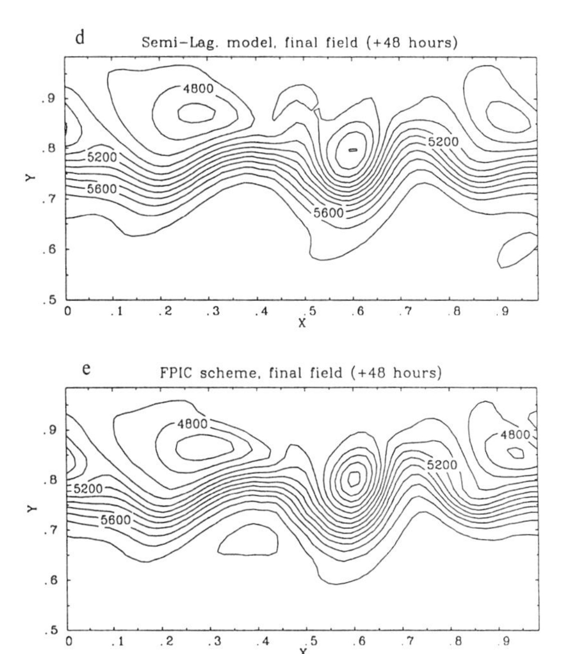

Fig. 7 *(concluded).*

3. For each tracer point calculate its trajectory corresponding to two time steps (three time level scheme). See item 1. in Section 3a.
4. Advect (re-position) the tracer points according to their individual trajectories. Each tracer point is associated with the grid box where it arrives.
5. Add tracer points where their density is "too" low. Same as item 3. in section 3a.
6. Interpolate tracer point values to grid points as in item 4. in Section 3a. At the grid points we now know the terms on the right hand side of (25), (26) and (27), except for the $\{\ \}_F$ terms.
7. In gridpoint space, calculate the $\{\ \}_F$ terms on the right hand side of (25), (26) and (27) and add them to the values obtained in 6. Now we have the full right hand side of the equations at each grid point.
8. Solve the system of equations in grid points at time level $n + 1$ in the same way as it was done for the semi-Lagrangian scheme. Now we have forecasted values of u^{n+1}, v^{n+1} and ϕ^{n+1} in gridpoint space.
9. In order to finish the forecast at the tracer points, we need to interpolate the terms, except u^{n+1}, v^{n+1} and ϕ^{n+1}, on the left hand side of (25), (26) and (27) to the tracer point positions and subtract the results from the tracer point values. Similarly, the explicit $\{\ \}_F$ terms on the right hand side must be interpolated and added to the tracer point values.
10. Same as item 5. in Section 3a.

We note that item 9. can be postponed two time steps and then merged with item 2. which means that only one computationally expensive interpolation must be performed from grid points to tracer points at each time step.

The addition and removal of tracer points plays the role of a subgrid scale diffusion. Since additions/removals mainly take place where there is divergence/ convergence, this diffusion is similar to a diffusion based on the divergence of the flow. In addition the FPIC scheme has some numerical diffusion from the interpolations of the Coriolis and pressure gradient terms.

d *Forecasted height field in the shallow water model*

In all the experiments we carried out, the values of the coefficients α, β and γ were set to 0.5, 0.0 and 0.5 respectively, i.e. a centred semi-implicit scheme was used.

In a first set of tests, we have run the shallow water model using a resolution of 64×64 grid points for all schemes and a truncation at wavenumber 21 for the Eulerian spectral model. For the spectral model a Fourier decomposition in two dimensions is used. The model parameter *maxtr* was set equal to 4.

The synthetic initial field of the height $(\phi + \phi_s)$ which fulfills the geostrophic balance is plotted in Fig. 7a. This initial field was chosen by "trial and error" until a strong development was seen.

The other plots in Fig. 7 show the 48-hour forecast height field when using the four different previously described advection schemes. It turns out that the field patterns are more developed in the FPIC scheme than in the Eulerian grid-point scheme, and the semi-Lagrangian method apparently suffers from a certain damping. We can also notice that in the current case, the FPIC leads to a final height field which looks very similar to the one given by the Eulerian spectral model and which exhibits well developed centres of high and low values of ϕ.

A second set of tests was run using a doubled resolution of 128×128 points and a truncation at wavenumber 42 in spectral mode.

The corresponding results which are plotted in Fig. 8, suggest that the Eulerian gridpoint and the semi-Lagrangian schemes behave in a better way than with the lower resolution. On the contrary, the increase in the resolution does not influence the results of the Eulerian spectral model much. The FPIC still leads to more developed field patterns than the other schemes. However, weak spurious resonance can be observed especially at the centre of Fig. 8d. This problem can, however, be easily overcome by using an off-centred semi-implicit scheme, as was shown by Rivest et al. (1994). (The Courant number is approximately 1.5 and reducing this number causes the noise to disappear.)

We also notice that the FPIC and the spectral Eulerian model with the low resolution (see Fig. 7e) perform as well as the other schemes with the high resolution.

It may be speculated that the horizontal diffusion leads to a damping in the two Eulerian models and if this damping was not present, they would outperform the other two models. This is, however, not the case since the diffusion is small and only affects the forecasts marginally, except that the q-field in the Eulerian forecast is very noisy without diffusion.

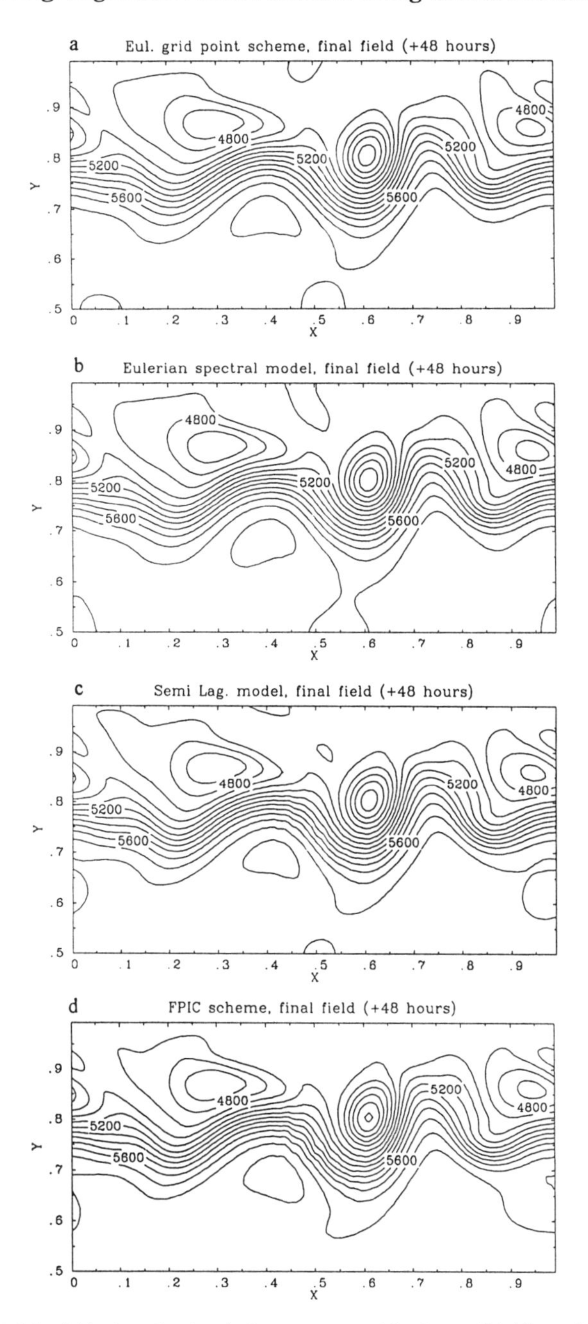

Fig. 8 Final height field given by the shallow water model when a "high" resolution of 128 × 128 points (and a square truncation at wave number 42 in spectral mode) is used for a) the Eulerian gridpoint scheme, b) the Eulerian spectral method, c) the semi-Lagrangian scheme and d) the FPIC.

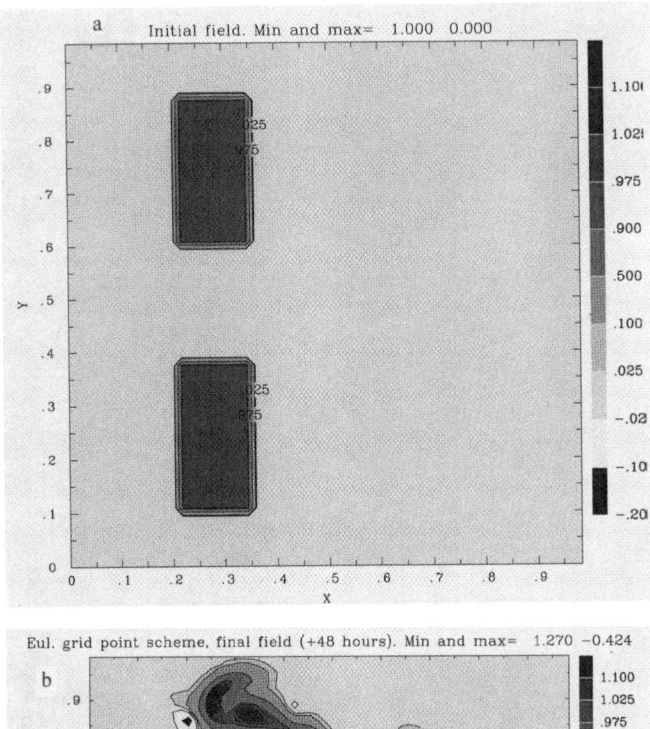

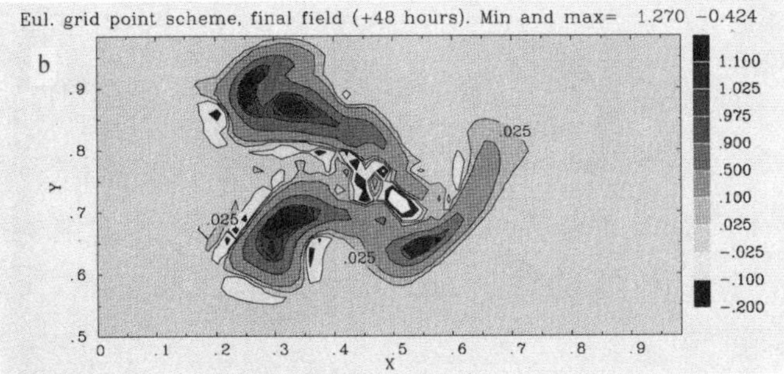

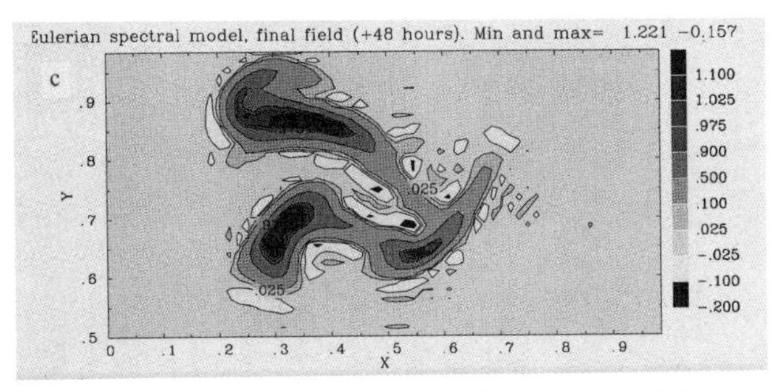

Fig. 9 The initial field of the tracer quantity used in the shallow water model is shown by Fig. 9a. The final tracer field is plotted for the four advection schemes described in the text: b) the Eulerian gridpoint scheme, c) the Eulerian spectral method, d) the traditional semi-Lagrangian scheme and e) the FPIC.

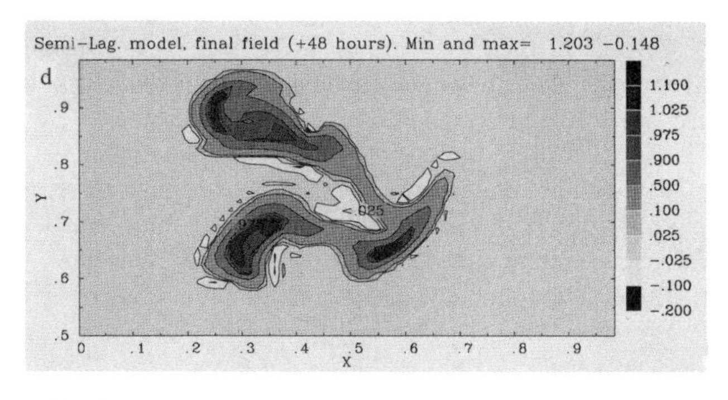

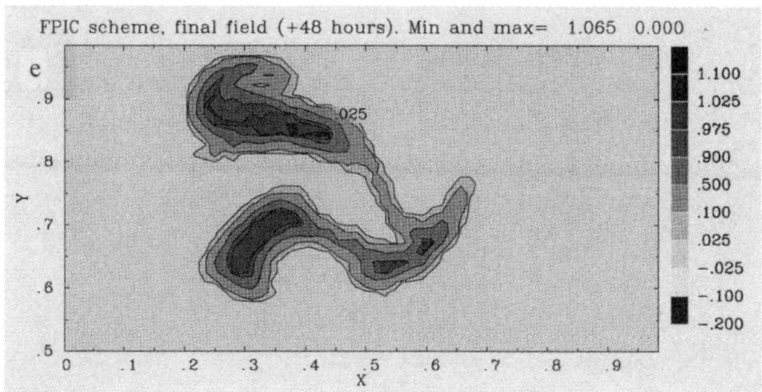

Fig. 9 *(concluded).*

6 Passive advection of a tracer quantity by the flow in the shallow water model

In addition to solving the shallow water equations for u, v and ϕ, the advection of a (positive definite) tracer q is studied. The prognostic equation per unit area is:

$$\frac{dq}{dt} = \frac{\partial q}{\partial t} + u\,\frac{\partial q}{\partial x} + v\,\frac{\partial q}{\partial y} = -q\left(\frac{\partial u}{\partial x} + \frac{\partial v}{\partial y}\right) = -qD \tag{28}$$

In the numerical treatment, the procedure is similar to that described above for the geopotential, but with the right hand side treated explicitly, i.e. with $\alpha = 0$, $\beta = 1$ and $\gamma = 0$. The same four advection schemes as in Section 5 are tested and the advective flow (u, v) in (28) is taken from the corresponding shallow water models, i.e. the flow is slightly different for each advection scheme. As above, we use periodic boundary conditions and symmetry about the "equator". As already mentioned in Section 5, a diffusion is used for the two Eulerian schemes with the same diffusion constant $(= 10^{-13})$. For the FPIC scheme, the parameter *maxtr* is set to 4.

The arbitrarily chosen initial q-field is shown in Fig. 9a. Since it is positive or zero everywhere it should ideally remain so throughout the integration. By comparing the results after 48 hours of integration (see Fig. 9) it is seen that only the FPIC scheme has non-negative values of q. For this reason it is suggested that the FPIC scheme may be very suitable for integration of the variables describing the positive definite humidity variables in atmospheric models. In traditional models one usually has to perform an ad hoc zero setting of negative humidity and liquid water values in order to obtain reasonable behaviour of the physical parametrization packages.

Even though it cannot be seen in Fig. 9 the minimum value of q for the FPIC scheme is slightly negative. This is due to the use of a three-time-level integration scheme and inaccuracies in the interpolations of tendencies from grid points to tracer points. The problem can, however, be solved by introducing an approximately mass conserving version of the FPIC scheme (see Section 7).

7 Discussion

We have seen above that the FPIC scheme behaves very well in a shallow water model. There are, however, some important questions which need to be answered before it can be stated whether the FPIC scheme is of potential importance for use in atmospheric modelling. From a climate modelling point of view it is important that the numerical integration techniques are mass conserving, energy conserving, enstrophy conserving, etc. It is possible to obtain a FPIC scheme which is practically mass conserving – the details of the formulation will be given in a forthcoming paper. As it is our main purpose to implement the FPIC scheme in a global atmospheric model the next step is to test the method for passive advection on a spherical grid. The results of this test will also be presented in a forthcoming paper. It is also of importance to all possible applications that the scheme be computationally efficient.

a *Computational efficiency*

In order to estimate the computational efficiency of the FPIC method, the CPU usage of the FPIC scheme has been compared with the CPU usage of the three other schemes used for comparison in this paper.

A 48-hour forecast was made with the shallow water model with three different resolutions: 32 × 32, 64 × 64 and 128 × 128 grid points. The runs were done on a Silicon Graphics Indy machine and in Table 1 the CPU seconds used by the four different schemes are shown for each of the three resolutions. The numbers in parentheses are the length of the time step used in each case.

It is seen that when the resolution is changed from 32 × 32 to 64 × 64, the CPU times are raised by approximately a factor of 4 consistent with the fact that the number of grid points is raised by a factor of 4. Similarly when the resolution is changed from 64 × 64 to 128 × 128 the CPU times are approximately 8 times larger (the number of grid points is four times larger and the time step is divided by two).

The FPIC method is the most time consuming when the pure CPU times are regarded, but it should be noticed that with the Eulerian grid point and the semi-Lagrangian methods, one will have to use a higher resolution in order to obtain results as good as those obtained with the FPIC method. Using the same resolution, similar results are obtained for the height field using the Eulerian spectral and the

TABLE 1. The CPU usage in seconds by the four different advection schemes: Eulerian grid point, Eulerian spectral, semi-Lagrangian and FPIC in three different resolutions: 32 × 32, 64 × 64 and 128 × 128 grid points, for a 48-hour forecast with the shallow water model. The numbers in parentheses are the length of the time step used in each case.

	Resolution		
	32 × 32	64 × 64	128 × 128
Eulerian grid point	8 ($\frac{1}{2}$ h)	30 ($\frac{1}{2}$ h)	235 ($\frac{1}{4}$ h)
Eulerian spectral	13 ($\frac{1}{2}$ h)	64 ($\frac{1}{2}$ h)	557 ($\frac{1}{4}$ h)
Semi-Lagrangian	6 (1h)	24 (1h)	210 ($\frac{1}{2}$ h)
FPIC	35 (1h)	145 (1h)	1015 ($\frac{1}{2}$ h)

FPIC methods, but for the tracer field a more realistic positive definite forecast is produced with the FPIC scheme.

The timings in Table 1 are given for the approximate mass conserving FPIC scheme which is some 30% more costly than the basic version described in Subsection 5c, and it should also be mentioned that no special effort has been made to optimize the code.

It is obvious that the FPIC scheme demands more memory than the other schemes tested in this paper. This is because tracer point positions as well as field values are used as prognostic variables.

8 Summary and conclusion

In this paper a new advection scheme (FPIC) has been tested. It is similar to the so-called particle-in-cell methods where both moving particles and a regular grid are used and it has here been demonstrated that FPIC behaves very well in the case of passive advection. We have furthermore combined the use of FPIC and the semi-implicit scheme by Robert et al. (1972) to solve the shallow water equations. The results so obtained have been compared with those obtained using a few other commonly used advection schemes in atmospheric models, i.e. a spectral and a gridpoint Eulerian scheme and a semi-Lagrangian gridpoint scheme. Also a passive tracer advected by the shallow water model flow has been investigated.

The findings are summarized in the following list:

- Except for initial interpolation error from grid points to tracer points the FPIC scheme is exact in the case of plane passive advection by constant flow.
- As with a semi-Lagrangian advection scheme there is no CFL criterion for advection which limits the time step for the FPIC scheme.
- The FPIC scheme can be combined with the semi-implicit scheme.
- It is possible to formulate FPIC to be approximately mass conserving.

- In a model including divergence, it is possible to formulate FPIC to be formally positive definite for a passive tracer.
- The scheme appears to consume more CPU-time than the other schemes tested, but when accuracy is also considered, this picture is less certain.
- The FPIC scheme demands considerably more memory than the other schemes tested.

We conclude that the FPIC scheme is of potential use in atmospheric modelling even though further developments are needed. Future work will include tests of FPIC in a global shallow water model and later in a full baroclinic model. From a computational point of view substantial work has to be done in order to optimize the computer code.

Acknowledgements

The work presented in this paper was carried out within a joint NOrdic CLimate Modelling Project, NOCLIMP, funded by the Nordic Council of Ministers (project number FS/ULF/93002). The authors are grateful to John Dukowicz from the Los Alamos National Laboratory, NM, who drew our attention to the similarity between the new scheme described here and the Particle-In-Cell methods.

References

BURKE, P.G. 1988. Computer Physics Communications. P.G. Burke (Honorary Ed.), *Europhys. J.* **48**: 174 pp.

ELIASEN, E.; B. MACHENHAUER and E. RASMUSSEN. 1970. On a numerical method for integration of the hydrodynamical equations with a spectral representation of the horizontal fields. Report No. 2, Institut for teoretisk meteorologi, University of Copenhagen, 37 pp.

HARLOW, F.H. 1957. Hydrodynamic problems involving large fluid distortions. *J. Assoc. Comput. Mach.* **4**: 137–142.

———. 1988. PIC and its progeny, Computer Physics Communications. *Europhys. J.* **48**: 1–10.

MACHENHAUER, B. 1979. The spectral method. Global Atmos. Res. Prog. (GARP) publication series No. 17, WMO-ICSU Joint Organizing Committee, pp. 121–275.

NAVARRA, A.; W.F. STERN and K. MIYAKODA. 1994. Reduction of the Gibbs oscillation in spectral model simulations. *J. Clim.* **7**: 1169–1183.

PUDYKIEWICZ, J.; R. BENOIT and A. STANIFORTH. 1985. Preliminary results from a partial LRTAP model based on an existing meteorological forecast model. ATMOSPHERE-OCEAN, **23**: 267–303.

RIVEST, C.; A. STANIFORTH and A. ROBERT. 1994. Spurious resonant response of semi-Lagrangian discretizations to orographic forcing: Diagnosis and solution. *Mon. Weather Rev.* **122**: 366–376.

ROBERT, A. 1982. A semi-Lagrangian and semi-implicit numerical integration scheme for the primitive meteorological equations., *J. Meteorol. Soc. Jpn.* **60**: 319–325.

———; J. HENDERSON and C. TURNBULL. 1972. An implicit time integration scheme for baroclinic models of the atmosphere. *Mon. Weather Rev.* **100**: 329–335.

STEPPELER, J. 1990. The concept of particle Lagrange methods. *Meteorol. Rdsch.* **43**: 23–31.

SWARTZTRAUBER, P.N. and R.A. SWEET. 1975. Efficient Fortran subprograms for the solution of elliptic partial differential equations. NCAR Technical Note IN/IA-109, 139 pp.

WILLIAMSON, D.L. and P.J. RASCH. 1994. Water vapor transport in the NCAR CCM2. *Tellus,* **46A**: 34–51.

The Formulation of the André Robert MC² (Mesoscale Compressible Community) Model

René Laprise,[1] Daniel Caya, Guy Bergeron and Michel Giguère
Cooperative Centre for Research in Mesometeorology (CCRM),
and
Atmospheric Sciences, Department of Earth Sciences, UQAM

[Original manuscript received 4 January 1995; in revised form 3 October 1995]

ABSTRACT *A description of the numerical formulation of the dynamics module of the Mesoscale Compressible Community (MC^2) model is presented. This model is based on the fully elastic, semi-implicit semi-Lagrangian model developed by Tanguay et al. (1990). This version was extended to incorporate topography by Denis (1990), and later variable vertical resolution was added as an option. This article is a condensed version of an extensive report by Bergeron et al. (1994) that documents all the numerical aspects of the MC^2 model. The performance of the model is illustrated through a sample of results obtained on a wide range of physical problems.*

RÉSUMÉ *Cet article présente une description de la formulation numérique du module de la dynamique du modèle de mésoéchelle compressible communautaire (MC^2). Ce modèle est basé sur les équations pleinement élastiques qui sont solutionnées par les schémas semi-implicite et semi-lagrangien comme dans Tanguay et al. (1990). La présente version incorpore la topographie comme dans Denis (1990), et elle permet l'utilisation d'un étirement de la maille verticale. Cet article constitue un condensé du volumineux rapport de Bergeron et al. (1994) qui décrit tous les divers aspects numériques du modèle MC^2. La performance du modèle est illustrée avec un échantillon de résultats obtenus sur une grande variété de problèmes physiques.*

1 Introduction

A description of the dynamical framework and time discretization of the MC^2 (Mesoscale Compressible Community) model originally developed by the late André Robert and his colleagues is presented. This non-hydrostatic model evolved

[1]Corresponding author's address: Prof. René Laprise, Department of Earth Sciences, Université du Québec à Montréal, 515 Sainte-Catherine St. West, P.O. Box 8888, Stn "Downtown", Montréal (Québec), Canada H3C 3P8

from an earlier version developed by Robert (1981, 1982) and Robert et al. (1985) to demonstrate the advantages of using semi-Lagrangian (SL) advection in a semi-implicit (SI) limited-area hydrostatic model. The model was later extended to the non-hydrostatic framework by replacing the primitive equations with fully elastic Euler equations by Tanguay et al. (1990). It was shown that SI-SL techniques made possible the integration of Euler equations at large scale with little computational overhead compared to more traditional hydrostatic equations.

The MC^2 model uses the lateral boundary nesting strategy developed by Yakimiw and Robert (1990). Topography was later incorporated in the model by Denis (1990) through the use of the Gal-Chen and Somerville (1975) terrain-following vertical coordinate transformation. Variable vertical resolution formalism was developed (Trudel and Robert, personal communication) and later implemented as an option.

The MC^2 model was developed with the aim of becoming a "universal" dynamical-numerical framework applicable to a wide range of meteorological and fluid dynamics problems, ranging in scale from the hemispheric through the mesoscale and the convective scale, down to the micrometre scale. A variety of subgrid-scale parametrization packages are currently being developed and tested within the MC^2 model by a number of mesoscale modelling groups in Canada and abroad.

In this article, a description of the dynamical and numerical formulation of MC^2 will be presented for the first time in the refereed literature. In the next Section, the formulation of the Euler equations in terrain-following coordinates will be presented, and the time discretization of the resulting equations will be described in the following Section. Some of the applications of MC^2 that have recently been performed or that are currently taking place will be briefly reviewed in the last Section of this article. Finally, the main features and properties of the MC^2 model will be summarized in the Conclusion.

2 The Euler equations in generalized coordinates

The Euler equations describing gaseous flow at the surface of the rotating earth, taking account of the traditional approximations in meteorology (Phillips, 1966), assume the following form when expressed in a conformal (X, Y) projection (e.g. Tanguay et al., 1990):

$$\frac{dU}{dt} = f\,V - K\frac{\partial S}{\partial X} - RT\frac{\partial q}{\partial X} + F_X$$

$$\frac{dV}{dt} = -f\,U - K\frac{\partial S}{\partial Y} - RT\frac{\partial q}{\partial Y} + F_Y$$

$$\frac{dw}{dt} = -g - RT\frac{\partial q}{\partial z} + F_z$$

$$(1-\alpha)\frac{dq}{dt} = -S\left(\frac{\partial U}{\partial X} + \frac{\partial V}{\partial Y}\right) - \frac{\partial w}{\partial z} + \frac{L}{T}$$

$$\frac{dT}{dt} = \alpha T \frac{dq}{dt} + L$$

$$\frac{dM}{dt} = E$$

$$\frac{dC}{dt} = B.$$

The variables used in these equations are: $q = \ln(p/p_o)$ with p for pressure and p_o a constant; U, V and w are the wind components according to the X, Y and z coordinates, respectively; M and C are the water vapour and liquid water specific quantities, respectively; f is the Coriolis parameter; K is the pseudo kinetic energy per unit mass $K = (U^2 + V^2)/2$; $S = m^2$ is the conformal projection metric term ($m = (1 + \sin\varphi_0)/(1 + \sin\varphi)$ for a polar stereographic projection that is used for meteorological applications); L represents the heat sources or sinks affecting temperature T; F_X, F_Y and F_z correspond to sources or sinks of momentum in each direction; E and B are the source or sink terms for moisture and liquid water content, respectively. The constants are:

g: acceleration due to gravity
R: gas constant for air
α: $= R/C_p$
C_p: heat capacity of air at constant pressure
φ: latitude
φ_o: reference latitude for the polar stereographic conformal transformation.

The total derivative in conformal coordinates is to be interpreted as follows (Haltiner and Williams, 1980; Chap. 1):

$$\frac{d}{dt} = \frac{\partial}{\partial t} + S\left(U\frac{\partial}{\partial X} + V\frac{\partial}{\partial Y}\right) + w\frac{\partial}{\partial z}$$

where the (U, V) components of the "image" wind velocity are defined in terms of the (u, v) components of the true horizontal velocity in the (x, y, z) coordinates:

$$\begin{pmatrix} U \\ V \end{pmatrix} = \frac{1}{m}\begin{pmatrix} -\sin\lambda & -\cos\lambda \\ \cos\lambda & -\sin\lambda \end{pmatrix}\begin{pmatrix} u \\ v \end{pmatrix}$$

For meteorological applications, the (X, Y) components of the polar stereographic coordinates are scaled and rotated functions of longitude λ:

$$\begin{pmatrix} dX \\ dY \end{pmatrix} = m\begin{pmatrix} -\sin\lambda & -\cos\lambda \\ \cos\lambda & -\sin\lambda \end{pmatrix}\begin{pmatrix} dx \\ dy \end{pmatrix}$$

a *Transformed vertical coordinate*

It is convenient to perform a transformation of vertical coordinate in order to simplify the implementation of topography in the model. A new vertical variable coordinate Z is introduced, which initially will be defined rather loosely by simply demanding that it be a monotonic function of the original coordinate z; the precise definition of the new vertical coordinate Z will be provided later. Given a field $A(X, Y, z)$, partial derivatives will take the following form in accordance with the chain rule of differentiation (e.g., Kasahara, 1974):

$$\left(\frac{\partial A}{\partial c}\right)_z = \left(\frac{\partial A}{\partial c}\right)_Z + \left(\frac{\partial A}{\partial Z}\right)_c \left(\frac{\partial Z}{\partial c}\right)_z$$

$$\left(\frac{\partial A}{\partial z}\right) = \frac{\partial A}{\partial Z}\left(\frac{\partial Z}{\partial z}\right)$$

where c represents either the X or Y coordinates or time t. The index z or Z besides a partial derivative indicates which variable is maintained constant during differentiation.

The Euler equations take the following form in (X, Y, Z) coordinates:

$$\frac{dU}{dt} = f\,V - K\left(\frac{\partial S}{\partial X}\right)_Z - RT\left\{\left(\frac{\partial q}{\partial X}\right)_Z + \left(\frac{\partial q}{\partial Z}\right)\left(\frac{\partial Z}{\partial X}\right)_z\right\} + F_X \tag{1}$$

$$\frac{dV}{dt} = -f\,U - K\left(\frac{\partial S}{\partial Y}\right)_Z - RT\left\{\left(\frac{\partial q}{\partial Y}\right)_Z + \left(\frac{\partial q}{\partial Z}\right)\left(\frac{\partial Z}{\partial Y}\right)_z\right\} + F_Y \tag{2}$$

$$\frac{dw}{dt} = -g - RT\frac{\partial q}{\partial Z}\left(\frac{\partial Z}{\partial z}\right) + F_Z \tag{3}$$

$$(1-\alpha)\frac{dq}{dt} = -\left(\frac{\partial z}{\partial Z}\right)^{-1}\frac{d}{dt}\left(\frac{\partial z}{\partial Z}\right) - S\left\{\left(\frac{\partial U}{\partial X}\right)_Z + \left(\frac{\partial V}{\partial Y}\right)_Z\right\} - \frac{\partial W}{\partial Z} + \frac{L}{T} \tag{4}$$

$$\frac{dT}{dt} = \alpha T\frac{dq}{dt} + L \tag{5}$$

$$\frac{dM}{dt} = E \tag{6}$$

$$\frac{dC}{dt} = B. \tag{7}$$

The details of the transformation of the vertical coordinate for the continuity equation (4) can be found in Bergeron et al. (1994). The total derivative must now be interpreted as follows:

$$\frac{d}{dt} = \left(\frac{\partial}{\partial t}\right)_Z + S\left\{U\left(\frac{\partial}{\partial X}\right)_Z + V\left(\frac{\partial}{\partial Y}\right)_Z\right\} + W\frac{\partial}{\partial Z} \tag{8}$$

and the generalized vertical velocity in the Z coordinate system is interpreted as follows:

$$W = \left(\frac{\partial Z}{\partial t}\right)_z + S\left\{U\left(\frac{\partial Z}{\partial X}\right)_z + V\left(\frac{\partial Z}{\partial Y}\right)_z\right\} + w\,\frac{\partial Z}{\partial z}. \quad (9)$$

The change of the vertical coordinate has brought terms such as $(\partial Z/\partial X)_z$, $(\partial Z/\partial Y)_z$ and $(\partial Z/\partial z)$ in the field equations, and these terms will be referred to as vertical coordinate transformation metrics.

The vertical coordinate Z is now defined such as to possess the following properties:

- The $Z = 0$ surface is following the terrain features and thus corresponds to the bottom of the model atmosphere;
- The $Z = H$ surface corresponds to a constant $z = H$ surface, and it defines the top of the model atmosphere;
- The $Z(z)$ transformation is monotonic;
- It permits variable vertical resolution despite the fact that the vertical discretization of the model assumes that ΔZ is constant.

The Gal-Chen terrain-following coordinate (Gal-Chen and Somerville, 1975), henceforth noted ζ, possesses the first 3 properties; it is defined by the following relation:

$$\zeta(X, Y, z) = \left[\frac{z - h_0(X, Y)}{H - h_0(X, Y)}\right] H$$

where H is the top of the model atmosphere, ζ is the Gal-Chen scaled-height with units of length, and $h_0(X, Y)$ is the topographic height.

This type of terrain-following coordinate has the advantage that the kinematic surface boundary condition, which is expressed by the following relation in z coordinate

$$w(X, Y, z = h_o, t) = \mathbf{V}(X, Y, z = h_o, t) \cdot \mathbf{V} h_o,$$

takes the form of a homogeneous condition in the ζ coordinate:

$$\dot{\zeta}(X, Y, \zeta = 0, t) = \left.\frac{d\zeta}{dt}\right|_{(X, Y, \zeta=0, t)} = 0.$$

Various metric terms will appear in the Euler equations when expressed in the ζ coordinate; these metrics may be written in compact form as follows:

$$\left(\frac{\partial \zeta}{\partial X}\right)_z = -\frac{(H - \zeta)}{(H - h_0)}\left(\frac{\partial h_0}{\partial X}\right) = \frac{g_1}{g_0}$$

$$\left(\frac{\partial \zeta}{\partial Y}\right)_z = -\frac{(H - \zeta)}{(H - h_0)}\left(\frac{\partial h_0}{\partial Y}\right) = \frac{g_2}{g_0}$$

$$\left(\frac{\partial \zeta}{\partial z}\right) = \frac{H}{(H - h_0)} = \frac{1}{g_0}$$

with

$$g_0 = \frac{(H - h_0)}{H}$$

$$g_1 = -\frac{(H - \zeta)}{H}\left(\frac{\partial h_0}{\partial X}\right)$$

$$g_2 = -\frac{(H - \zeta)}{H}\left(\frac{\partial h_0}{\partial Y}\right)$$

As previously noted, the vertical discretization of MC^2 assumes uniform spacing of the levels in the model coordinate, i.e. constant ΔZ. Variable vertical resolution is introduced by a stretching transformation, $Z(\zeta)$. The vertical coordinate transformation metrics will be expressed in terms of ζ as follows:

$$\left(\frac{\partial Z}{\partial X}\right)_z = \left(\frac{\partial Z}{\partial X}\right)_\zeta + \left(\frac{\partial Z}{\partial \zeta}\right)\left(\frac{\partial \zeta}{\partial X}\right)_z = \left(\frac{\partial Z}{\partial \zeta}\right)\left(\frac{\partial \zeta}{\partial X}\right)_z$$

$$\left(\frac{\partial Z}{\partial Y}\right)_z = \left(\frac{\partial Z}{\partial Y}\right)_\zeta + \left(\frac{\partial Z}{\partial \zeta}\right)\left(\frac{\partial \zeta}{\partial Y}\right)_z = \left(\frac{\partial Z}{\partial \zeta}\right)\left(\frac{\partial \zeta}{\partial Y}\right)_z$$

$$\left(\frac{\partial Z}{\partial z}\right) = \left(\frac{\partial Z}{\partial \zeta}\right)\left(\frac{\partial \zeta}{\partial z}\right).$$

Since Z is only a function of ζ, the horizontal derivatives of Z on constant ζ surfaces vanish identically. The derivatives of ζ with respect to X, Y and z are easily obtained from the definition of the Gal-Chen coordinate. In practice the new coordinate needs only to be defined in discrete form by specifying the number and position of the Z levels in the ζ domain. This determines the thickness ratio between the model levels (ΔZ) and Gal-Chen coordinate $\Delta\zeta$, i.e. in discrete form:

$$\left(\frac{\partial Z}{\partial \zeta}\right) \cong \left(\frac{\Delta Z}{\Delta \zeta}\right).$$

It is interesting to note here that the modified Gal-Chen coordinate still offers the benefit of a homogeneous kinematic condition at the surface, i.e. $W(X, Y, Z = 0, t) = 0$, as well as at the top of the model atmosphere. In the Z coordinate, the metric terms become:

$$\left(\frac{\partial Z}{\partial X}\right)_z = \left(\frac{\partial Z}{\partial \zeta}\right)\left(\frac{\partial \zeta}{\partial X}\right)_z = \frac{g_1}{g_0}\left(\frac{\partial Z}{\partial \zeta}\right) = \frac{G_1}{G_0}$$

$$\left(\frac{\partial Z}{\partial Y}\right)_z = \left(\frac{\partial Z}{\partial \zeta}\right)\left(\frac{\partial \zeta}{\partial Y}\right)_z = \frac{g_2}{g_0}\left(\frac{\partial Z}{\partial \zeta}\right) = \frac{G_2}{G_0}$$

$$\left(\frac{\partial Z}{\partial z}\right) = \left(\frac{\partial Z}{\partial \zeta}\right)\left(\frac{\partial \zeta}{\partial z}\right) = \frac{1}{g_0}\left(\frac{\partial Z}{\partial \zeta}\right) = \frac{1}{G_0}$$

where

$$G_0 = g_0\left(\frac{\partial \zeta}{\partial Z}\right)$$

$$G_1 = g_1$$

$$G_2 = g_2.$$

The Euler equations then take the following form in the vertically stretched terrain-following $Z(\zeta(z))$ coordinate system:

$$\frac{dU}{dt} = fV - K\left(\frac{\partial S}{\partial X}\right)_Z - RT\left\{\left(\frac{\partial q}{\partial X}\right)_Z + \frac{G_1}{G_0}\left(\frac{\partial q}{\partial Z}\right)\right\} + F_X \tag{10}$$

$$\frac{dV}{dt} = -fU - K\left(\frac{\partial S}{\partial Y}\right)_Z - RT\left\{\left(\frac{\partial q}{\partial Y}\right)_Z + \frac{G_2}{G_0}\left(\frac{\partial q}{\partial Z}\right)\right\} + F_Y \tag{11}$$

$$\frac{dw}{dt} = -g - \frac{RT}{G_0}\left(\frac{\partial q}{\partial Z}\right) + F_Z \tag{12}$$

$$(1-\alpha)\frac{dq}{dt} = S(F_1U + F_2V) - S\left\{\left(\frac{\partial U}{\partial X}\right)_Z + \left(\frac{\partial V}{\partial Y}\right)_Z\right\} - \frac{1}{G_0}\left(\frac{\partial G_0 W}{\partial Z}\right) + \frac{L}{T} \tag{13}$$

$$\frac{dT}{dt} = \alpha T\frac{dq}{dt} + L \tag{14}$$

$$W = \frac{S(G_1U + G_2V) + w}{G_0} \tag{15}$$

$$\frac{dM}{dt} = E \tag{16}$$

$$\frac{dC}{dt} = B \tag{17}$$

where

$$F_1 = \frac{1}{g_0 H}\left(\frac{\partial h_0}{\partial X}\right)$$

$$F_2 = \frac{1}{g_0 H}\left(\frac{\partial h_0}{\partial Y}\right).$$

The details of the algebraic manipulations required to arrive at these equations may be found in Bergeron et al. (1994). Since all the terms of the equations are now calculated in (X, Y, Z) coordinates, the subscript Z will be henceforth omitted to simplify the notation.

3 Time discretization

This section describes the time discretization of the Euler equations in their new vertical coordinate. Time integration is done by means of a semi-Lagrangian and semi-implicit scheme similar to that presented by Tanguay et al. (1990). For purposes of applying the semi-implicit scheme, the T and q fields are decomposed in two parts consisting of a basic state (ψ^*) and a perturbation (ψ'). An isothermal atmosphere in hydrostatic equilibrium is used for the basic state so that:

$$T(X, Y, Z, t) = T^* + T'(X, Y, Z, t)$$

$$q(X, Y, Z, t) = q^*(z(X, Y, Z)) + q'(X, Y, Z, t)$$

with

$$\frac{dq^*}{dz} = -\frac{g}{RT^*}$$

$$q^*(z(X, Y, Z)) = q_0 - \frac{gz(X, Y, Z)}{RT^*}$$

where T^* and q_o are constant.

Once these expressions for T and q are inserted into the Euler equations, these become:

$$\frac{dU}{dt} + RT^*\frac{\partial q'}{\partial X} = fV - K\frac{\partial S}{\partial X} - RT'\frac{\partial q'}{\partial X} - RT\frac{G_1}{G_0}\frac{\partial q'}{\partial Z} + F_X$$

$$\frac{dV}{dt} + RT^*\frac{\partial q'}{\partial Y} = -fU - K\frac{\partial S}{\partial Y} - RT'\frac{\partial q'}{\partial Y} - RT\frac{G_2}{G_0}\frac{\partial q'}{\partial Z} + F_Y$$

$$\frac{dw}{dt} + \frac{RT^*}{G_0}\frac{\partial q'}{\partial Z} - g\frac{T'}{T^*} = -\frac{RT'}{G_0}\frac{\partial q'}{\partial Z} + F_Z$$

$$(1-\alpha)\left[\frac{dq'}{dt}-\frac{gw}{RT^*}\right]+S\left[\frac{\partial U}{\partial X}+\frac{\partial V}{\partial Y}\right]+\frac{g_0}{G_0}\frac{\partial G_0 W}{\partial Z}=S(F_1U+F_2V)$$

$$-\frac{(1-g_0)}{G_0}\frac{\partial G_0 W}{\partial Z}+\frac{L}{T}$$

$$\frac{dT'}{dt}-\alpha T^*\frac{dq'}{dt}+\frac{\alpha g}{R}w=\frac{\alpha T'}{(1-\alpha)}$$

$$\times\left\{S(F_1U+F_2V)-S\left[\frac{\partial U}{\partial X}+\frac{\partial V}{\partial Y}\right]-\frac{1}{G_0}\frac{\partial G_0 W}{\partial Z}+\frac{L}{T}\right\}+L$$

$$w-G_0W=-S(G_1U+G_2V)$$

$$\frac{dM}{dt}=E$$

$$\frac{dC}{dt}=B.$$

Note that the linear terms responsible for elastic and gravity waves have been grouped to the left-hand sides of the equations for convenience. From this point on, the diabatic terms will be omitted, as these are incorporated by "process splitting" through a correction step once the inviscid adiabatic time step is completed.

In order to allow for a three-dimensional or two-dimensional (horizontal) semi-Lagrangian scheme, the following Lagrangian derivative (D/Dt) is introduced:

$$\frac{D}{Dt}=\frac{\partial}{\partial t}+S\left\{U\frac{\partial}{\partial X}+V\frac{\partial}{\partial Y}+\mu W\frac{\partial}{\partial Y}\right\}.$$

The μ coefficient is a switch that allows us to have a three-dimensional ($\mu=1$) or a two-dimensional Lagrangian derivative ($\mu=0$). In the two-dimensional case, vertical advection is incorporated with the explicit terms on the right-hand sides of the equations.

The terms responsible for elastic and gravity waves are computed implicitly as time averages along Lagrangian displacement $(-t)$, leading to the following time-discretized system:

$$\frac{DU}{Dt}+RT^*\overline{\frac{\partial q'}{\partial X}}^t=R_U$$

$$\frac{DV}{Dt}+RT^*\overline{\frac{\partial q'}{\partial Y}}^t=R_V$$

$$\frac{Dw}{Dt} + RT^* \frac{g_0}{G_0} \overline{\frac{\partial q'}{\partial Z}}^t - g \frac{\overline{T'}^t}{T^*} = R_w$$

$$(1-\alpha)\left[\frac{Dq'}{Dt} - \frac{\overline{gw}^t}{RT^*}\right] + S\left[\overline{\frac{\partial U}{\partial X}}^t + \overline{\frac{\partial V}{\partial Y}}^t\right] + \frac{g_0}{G_0} \overline{\frac{\partial G_0 W}{\partial Z}}^t = R_q$$

$$\frac{DT'}{Dt} - \alpha T^* \frac{Dq'}{Dt} + \frac{\alpha g}{R} \bar{w}' = R_T$$

$$\bar{w}' - G_0 \overline{W}^t = R_W$$

$$\frac{DM}{Dt} = R_M$$

$$\frac{DC}{Dt} = R_C$$

where the right-hand terms have the following meaning:

$$R_U = f\,V - K \frac{\partial S}{\partial X} - RT' \frac{\partial q'}{\partial X} - RT \frac{G_1}{G_0} \frac{\partial q'}{\partial Z} - (1-\mu) W \frac{\partial U}{\partial Z}$$

$$R_V = -f\,U - K \frac{\partial S}{\partial Y} - RT' \frac{\partial q'}{\partial Y} - RT \frac{G_2}{G_0} \frac{\partial q'}{\partial Z} - (1-\mu) W \frac{\partial V}{\partial Z}$$

$$R_w = RT^* \frac{g_0}{G_0} \frac{\partial q'}{\partial Z} - \frac{RT}{G_0} \frac{\partial q'}{\partial Z} - (1-\mu) W \frac{\partial w}{\partial Z}$$

$$R_q = S(F_1 U + F_2 V) - \frac{(1-g_0)}{G_0} \frac{\partial G_0 W}{\partial Z} - (1-\alpha)(1-\mu) W \frac{\partial q'}{\partial Z}$$

$$R_T = \frac{\alpha T'}{(1-\alpha)} \left\{ S(F_1 U + F_2 V) - S\left[\frac{\partial U}{\partial X} + \frac{\partial V}{\partial Y}\right] - \frac{1}{G_0} \frac{\partial G_0 W}{\partial Z} \right\}$$
$$- (1-\mu) W \frac{\partial (T' - \alpha T^* q')}{\partial Z}$$

$$R_W = -S(G_1 U + G_2 V)$$

$$R_M = -(1-\mu) W \frac{\partial M}{\partial Z}$$

$$R_C = -(1-\mu) W \frac{\partial C}{\partial Z}$$

The Lagrangian time derivative is evaluated as a second-order centred time difference along the Lagrangian trajectory

$$\frac{D\psi}{Dt} = \frac{\psi(X,\, Y,\, Z,\, t+\Delta t) - \psi(X-2\alpha,\, Y-2\beta,\, Z-2\gamma\mu,\, t-\Delta t)}{2\Delta t}$$

and the implicit terms are treated as off-centred time averages

$$\bar{\psi}^t = \frac{(1+\varepsilon)\psi(X,\, Y,\, Z,\, t+\Delta t) + (1-\varepsilon)\psi(X-2\alpha,\, Y-2\beta,\, Z-2\gamma\mu,\, t-\Delta t)}{2}$$

The ε coefficient represents the degree to which the average is off-centred. This approach is used to reduce the singularity problem of stationary forcings with long time steps (Tanguay et al., 1992) (see also Rivest et al. (1994) for a discussion of the problem in shallow water, and Héreil and Laprise (1995) in baroclinic atmosphere). The $\psi(X, Y, Z)$ functions correspond to the fields on grid points; the upstream positions $(X-2\alpha, Y-2\beta, Z-2\gamma\mu)$ usually do not correspond to grid points, however, and the $\psi(X-2\alpha, Y-2\beta, Z-2\gamma\mu)$ terms are obtained by multi-dimensional cubic spline interpolation in $(2+\mu)$ dimensions (see Bergeron et al., 1994). The Lagrangian displacements (α, β, γ) are calculated iteratively, as follow (Robert, 1985):

$$\alpha(X,\, Y,\, Z,\, t) = \Delta t S U(X-\alpha,\, Y-\beta,\, Z-\gamma\mu,\, t)$$

$$\beta(X,\, Y,\, Z,\, t) = \Delta t S V(X-\alpha,\, Y-\beta,\, Z-\gamma\mu,\, t)$$

$$\gamma(X,\, Y,\, Z,\, t) = \mu \Delta t W(X-\alpha,\, Y-\beta,\, Z-\gamma\mu,\, t)$$

The nonlinear R_ψ terms on the right-hand side of the equations are evaluated as uncentred averages of their values at time t at the position of the two end points of the Lagrangian trajectory (Tanguay et al., 1992):

$$\bar{R}_\psi^{traj} = \frac{(1+\varepsilon)R_\psi(X,\, Y,\, Z,\, t) + (1-\varepsilon)R_\psi(X-2\alpha,\, Y-2\beta,\, Z-2\gamma\mu,\, t)}{2}$$

Grouping on the left-hand side the terms that are functions of variables at time $t+\Delta t$, the field equations become:

$$U + (1+\varepsilon)\Delta t R T^* \frac{\partial q'}{\partial X} = Q_U \qquad (18)$$

$$V + (1+\varepsilon)\Delta t R T^* \frac{\partial q'}{\partial Y} = Q_V \qquad (19)$$

$$w + (1+\varepsilon)\Delta t R T^* \frac{g_0}{G_0}\frac{\partial q'}{\partial Z} - (1+\varepsilon)\Delta t g \frac{T'}{T^*} = Q_w \qquad (20)$$

$$(1-\alpha)\left[q' - (1+\varepsilon)\Delta t \frac{gw}{RT^*}\right] + (1+\varepsilon)\Delta t S \left[\frac{\partial U}{\partial X} + \frac{\partial V}{\partial Y}\right]$$

$$+ (1+\varepsilon)\Delta t \frac{g_0}{G_0} \frac{\partial G_0 W}{\partial Z} = Q_q \quad (21)$$

$$T' - \alpha T^* q' + (1+\varepsilon)\Delta t \frac{\alpha g}{R} w = Q_T \quad (22)$$

$$G_0 W - w = -\frac{Q_W}{(1+\varepsilon)\Delta t} \quad (23)$$

$$M = Q_M \quad (24)$$

$$C = Q_C \quad (25)$$

where the Q_ψ terms are defined as follows

$$Q_\psi = P_\psi + 2\Delta t \bar{R}_\psi^{traj}$$

The R_ψ terms correspond to expressions that may be evaluated based on dependent variables at time t:

$$R_U = f V - K \frac{\partial S}{\partial X} - RT' \frac{\partial q'}{\partial X} - RT \frac{G_1}{G_0} \frac{\partial q'}{\partial Z} - (1-\mu) W \frac{\partial U}{\partial Z} \quad (26)$$

$$R_V = -f U - K \frac{\partial S}{\partial Y} - RT' \frac{\partial q'}{\partial Y} - RT \frac{G_2}{G_0} \frac{\partial q'}{\partial Z} - (1-\mu) W \frac{\partial V}{\partial Z} \quad (27)$$

$$R_w = RT^* \frac{g_0}{G_0} \frac{\partial q'}{\partial Z} - \frac{RT}{G_0} \frac{\partial q'}{\partial Z} - (1-\mu) W \frac{\partial w}{\partial Z} \quad (28)$$

$$R_q = S(F_1 U + F_2 V) - \frac{(1-g_0)}{G_0} \frac{\partial G_0 W}{\partial Z} - (1-\alpha)(1-\mu) W \frac{\partial q'}{\partial Z} \quad (29)$$

$$R_T = \frac{\alpha T'}{(1-\alpha)} \left\{ S(F_1 U + F_2 V) - S \left[\frac{\partial U}{\partial X} + \frac{\partial V}{\partial Y}\right] - \frac{1}{G_0} \frac{\partial G_0 W}{\partial Z} \right\}$$

$$- (1-\mu) W \frac{\partial (T' - \alpha T^* q')}{\partial Z} \quad (30)$$

$$R_W = -S(G_1 U + G_2 V) \quad (31)$$

$$R_M = -(1-\mu) W \frac{\partial M}{\partial Z} \quad (32)$$

$$R_C = -(1-\mu)W\frac{\partial C}{\partial Z} \tag{33}$$

and the P_ψ terms correspond to expressions that may be evaluated based on dependent variables at time $t - \Delta t$:

$$P_U = U - (1-\varepsilon)\Delta t RT^*\frac{\partial q'}{\partial X} \tag{34}$$

$$P_V = V - (1-\varepsilon)\Delta t RT^*\frac{\partial q'}{\partial Y} \tag{35}$$

$$P_w = w - (1-\varepsilon)\Delta t RT^*\frac{g_0}{G_0}\frac{\partial q'}{\partial Z} + (1-\varepsilon)\Delta t g\frac{T'}{T^*} \tag{36}$$

$$P_q = (1-\alpha)\left[q' + (1-\epsilon)\Delta t\frac{gw}{RT^*}\right] - (1-\varepsilon)\Delta t S\left[\frac{\partial U}{\partial X} + \frac{\partial V}{\partial Y}\right] - (1-\varepsilon)\Delta t\frac{g_0}{G_0}\frac{\partial G_0 W}{\partial Z} \tag{37}$$

$$P_T = T' - \alpha T^* q' - (1-\varepsilon)\Delta t\frac{\alpha g}{R}w \tag{38}$$

$$P_W = -(1-\varepsilon)\Delta t(w - G_0 W) \tag{39}$$

$$P_M = M \tag{40}$$

$$P_C = C \tag{41}$$

These equations form a system of coupled equations for the dependent variables at time $t + \Delta t$ in terms of variables known at times t and $t - \Delta t$. This system is solved by successively eliminating all dependent variables but one (q' in this case), leading to an elliptic equation for that variable at time $t + \Delta t$:

$$C_1[(1-\alpha) - (1+\varepsilon)^2(\Delta t)^2 RT^* S\nabla^2]q' - (1+\varepsilon)^2(\Delta t)^2 RT^* D_2[D_1(q')] = A_2 \tag{42}$$

where

$$A_1 = A_5 - \frac{Q_W}{(1+\varepsilon)\Delta t}C_1$$

$$A_2 = C_1 A_3 - (1+\varepsilon)\Delta t D_2(A_1)$$

$$A_3 = A_4 + Q_W\frac{(1-\alpha)g}{RT^*}$$

$$A_4 = Q_q - (1+\varepsilon)\Delta t S \left[\left(\frac{\partial Q_U}{\partial X} \right)_Z + \left(\frac{\partial Q_V}{\partial Y} \right)_Z \right]$$

$$A_5 = Q_w + (1+\varepsilon)\Delta t \frac{g}{T^*} Q_T$$

$$C_1 = 1 + (1+\varepsilon)^2 (\Delta t)^2 \frac{\alpha g^2}{RT^*}$$

and the vertical operators D_1 and D_2 are defined as follow:

$$D_1(\psi) = \frac{g_0}{G_0} \left(\frac{\partial \psi}{\partial Z} \right) - \frac{\alpha g}{RT^*} \psi$$

$$D_2(\psi) = \left(\frac{g_0}{G_0} \right) \frac{\partial \psi}{\partial Z} - \frac{g(1-\alpha)}{RT^*} \psi$$

This Helmholtz equation is solved by a variant of the alternating-direction implicit method of Peaceman and Rachford (1955) using the lateral boundary conditions provided through the nesting scheme to obtain q' at time $t + \Delta t$. Finally, the W, w, T', U and V fields at time $t + \Delta t$ are obtained by back-substitution using the following relations:

$$W = \frac{1}{C_1 G_0} [A_1 - (1+\varepsilon)\Delta t RT^* D_1(q')] \tag{43}$$

$$w = \frac{Q_W}{(1+\varepsilon)\Delta t} + G_0 W \tag{44}$$

$$T' = Q_T + \alpha T^* q' - (1+\varepsilon)\Delta t \frac{\alpha g}{R} w \tag{45}$$

$$U = Q_U - (1+\varepsilon)\Delta t RT^* \left(\frac{\partial q'}{\partial X} \right)_Z \tag{46}$$

$$V = Q_V - (1+\varepsilon)\Delta t RT^* \left(\frac{\partial q'}{\partial Y} \right)_Z \tag{47}$$

$$M = Q_M \tag{48}$$

$$C = Q_C \tag{49}$$

The algebraic details of this back-substitution process may be found in Bergeron et al. (1994).

The spatial derivative in the above and preceding equations are calculated by

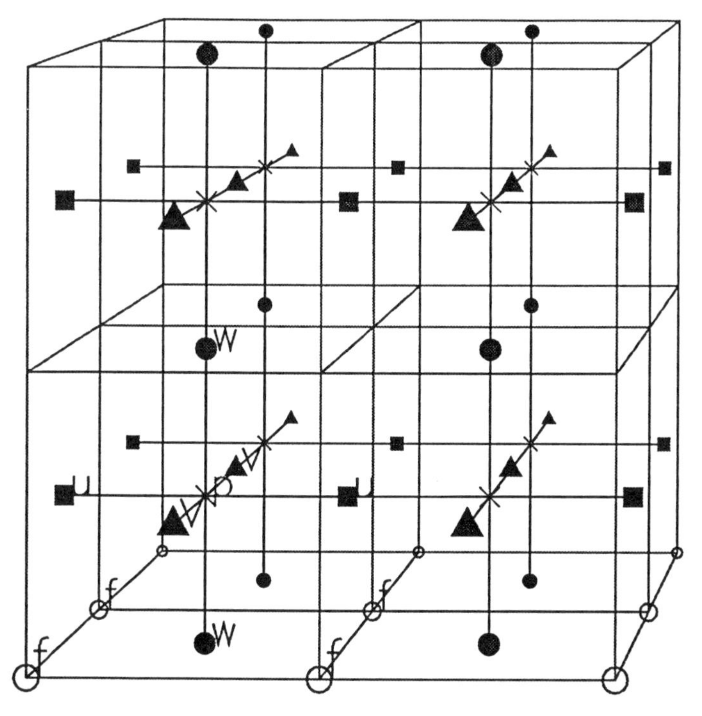

Fig. 1 A perspective view of the positioning of variables in MC² model. Squares and triangles denote the position of the U and V winds, respectively; closed circles that of the W, w and T variables; and asterisks that of the $q = \ln p$ variables. The Coriolis parameter f is defined on open circles points.

second-order finite differences on an Arakawa-C staggered grid shown in Fig. 1. Lateral boundary forcing is further blended with the regional model forecast values through a variant of the Davies (1976) sponge procedure developed by Yakimiw and Robert (1986). Once an adiabatic inviscid step has been accomplished, the physical forcings are incorporated through a correction step. The forecast is completed by applying a Robert's running time filter (Asselin, 1972).

4 Some applications of the MC² model

The three-dimensional nested MC² model and a closed-domain two-dimensional variant of it known as SISLAM (Semi-Implicit Semi-Lagrangian Atmospheric Model) have been applied on a number of physical problems by André Robert, his students and collaborators. Some of these applications will be reviewed briefly below to show the versatility of the model.

Coarse-resolution (127 km) short-term (24 h) regional forecasts based on the fully elastic Euler equations were shown by Tanguay et al. (1990) to be computationally feasible by using semi-implicit semi-Lagrangian algorithms. At this scale, the resulting model proved to be comparable to the hydrostatic equations in terms of computational cost and forecast quality. The time step used for these forecasts was 30 minutes, which is 800 times larger than the time step required to integrate such a fully elastic model with an explicit marching scheme. A variable-mesh version of this model was later tested by Ben Hadj Tahar (1992).

In May 1992, the modelling group of CCRM (Cooperative Centre for Research in Mesometeorology) chose the MC^2 model for its modelling research activities. At that point, Dr. Michel Béland, then Chief of the Division de Recherches en Prévision Numérique (RPN) at the Atmospheric Environment Service, agreed to devote some human resources from his group (Benoit et al., unpublished manuscript) to develop an infrastructure around the model and to service community users.

Benoit et al. (unpublished manuscript) have implemented RPN's forecast physics parametrization (Mailhot et al., 1989) in the MC^2 model and have carried out a number of experiments, four of which will be reviewed briefly here. (i) Benoit et al. (1994a) used MC^2 with telescoping grids, from 50 km to 5 km and then down to 2 km, to forecast a few severe spring-time precipitation events over the very rugged terrain of the Columbia River watershed in British Columbia (Canada). The model hourly forecast rain field was fed into physically-based hydrological models which in turn simulated hydrographs for the tributaries of the Columbia River. The quality of these synthetic hydrographs was shown to be superior to those derived solely from surface observations that suffer from elevation biases. (ii) The MC^2 model also participated in the WMO-sponsored COMPARE (Comparison of Mesoscale Prediction and Research Experiments) exercise in which the forecasts produced by more than 10 mesoscale models were intercompared at various resolutions culminating at 25 km and 54 levels. MC^2 was the only non-hydrostatic model to be run on the IOP-14 (Intensive Operating Period-14) rapid East Coast cyclogenesis event of CASP I (Canadian Atlantic Storm Program-I) (Mailhot et al., 1989); MC^2 performed in a very similar fashion to the Canadian RFE (Regional Finite Element) model (Mailhot et al., 1989), with excellent 24- to 36-hour prognoses of the coastal low (Gyakum et al., 1996). (iii) MC^2 was run daily by the Canadian Meteorological Centre, in real-time support to aircraft missions of the Canadian BASE (Beaufort and Arctic Storm Experiment) field experiment, during a six-week period in the fall of 1994 (Benoit et al., 1994b); the grid resolution was 15 km on a $(1500\ \text{km})^2$ domain, with initial and lateral boundary conditions supplied by the operational 50-km RFE model. The MC^2 model forecasts showed enhanced mesoscale variations in the low-level circulation and precipitation due to the improved definition of topography and of complex land, ocean and sea-ice distribution in the Mackenzie Delta region. (iv) In the examination of a case of downslope windstorms called SUÈTES winds over Cape Breton (Nova Scotia, Canada), MC^2 was applied on a cascade of grids down to a resolution of 2 km which allows the resolution of the 700-m highlands interior and their sharp slopes. In a south-easterly flow due to an approaching depression, the model generated a surface wind speedup factor of two, reaching about 60 kts near Chéticamp, in agreement with the hourly observations (Benoit et al., unpublished manuscript). The vertical structure generated by the model over the Cape resembles a classical mountain wave with a critical level near 600 hPa, due to the turning of the synoptic flow; non-dimensional parameters indicate rather non-hydrostatic conditions.

Also in the mesoscale domain, Tremblay (1994) used the MC^2 model at 30 and 10 km for modelling a particularly severe case of flash flooding due to a very slow moving, long lasting large mesoscale precipitating system over Montréal (Québec, Canada) on 14 July 1987. Tremblay, Szyrmer and Zawadzki (personal communication) are also running a version of MC^2 including a bulk parametrization of hydrometeors using moments methods developed by Zawadzki et al. (1993) and further refined by Szyrmer (1996, unpublished manuscript) to study the problem of icing in cloud environments.

The Regional Climate Modelling group at UQAM has implemented the Canadian Climate Centre second-generation General Circulation Model physics parametrization (McFarlane et al., 1992) in the MC^2 model. The resulting Regional Climate Model (RCM) is developed as a tool to regionalize the climate simulation of the Canadian GCM (Caya et al., 1995a, 1995b). The efficient numerics of MC^2 allow the use of the same 20-min time step at 45 km that is used to integrate the T32 GCM whose effective resolution is of the order of 350 km; this results in a factor five computational saving compared to other RCMs that do not use semi-implicit semi-Lagrangian techniques.

The MC^2 model has been validated for the classical mountain wave problem by Héreil (1993), Rivals (1993) and Pinty et al. (1995). These experiments show that it is possible to obtain accurate mountain wave solutions with SI-SL numerics by using "reasonable" values of the time step. Despite problems pointed out by Tanguay et al. (1992), Rivest et al. (1994) and Héreil and Laprise (1995) related to exploiting the long time steps permitted by the semi-implicit semi-Lagrangian algorithm, when stationary forcings are present, it is noteworthy that the SI-SL approach still allows time steps that are 3 to 5 times longer than those used by more traditional Eulerian anelastic models.

A two-dimensional version of the model was validated on the classical dry bubble convection problem by Robert (1993) where it was shown that the semi-implicit semi-Lagrangian approach was viable even for convection modelling. These tests answered some concerns that had been raised about the suitability of using a statically stable basic state for the SI decomposition in situations when the environment is neutrally stratified or even statically unstable. The SL treatment of advection resulted in greater accuracy and less numerical noise than commonly used Eulerian models.

Pellerin (1992) and Pellerin et al. (1995a, b) have expanded these tests to moist convection. Figure 2 displays the results of a simulation of downward convection caused by evaporative cooling of water droplets released at the top of the domain and falling in a dry environment. The resolution for this experiment was 1.5 mm and the time step 25 ms. The two-dimensional domain is 90 cm high by 60 cm wide,

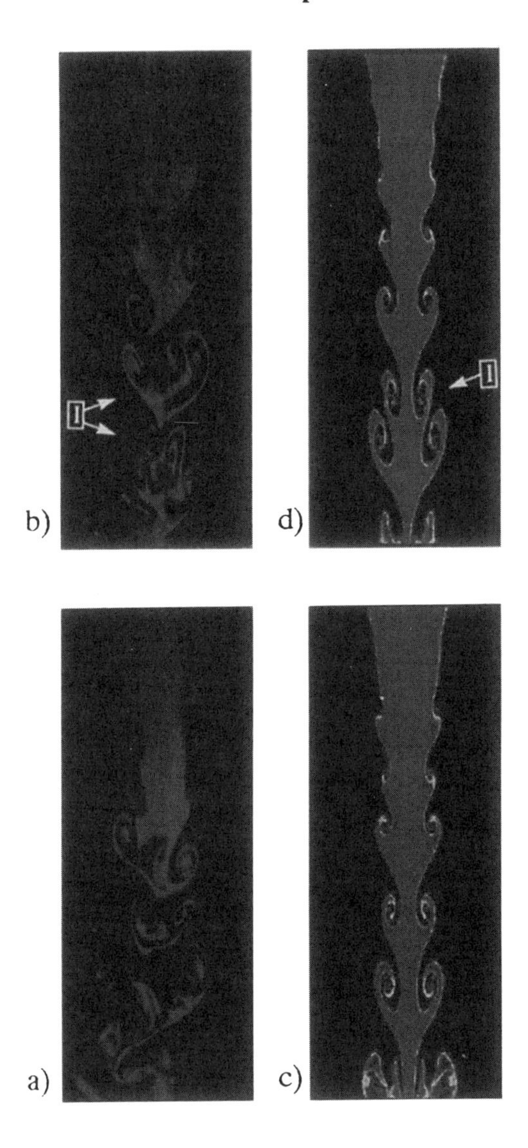

Fig. 2 Comparison of laboratory experiment (left) and numerical model simulation (right) of cloud water distribution in a downdraft induced by cooling due to the evaporation of cloud droplets as they fall in a dry environment after their release at the top of the tube. The portion of the apparatus that is shown is roughly 40 cm high, and the model results correspond to two instants near the end of a 7 s simulated period. (Reproduced with permission from Pellerin et al. (1995b).)

thus the computational grid has 600 by 400 points. Figures 2c and d display the liquid water concentration at two instants near the end of the 7 s long integration. For comparison, the results of a laboratory experiment are shown in Figs 2a and b: the details of this experiment may be found in Pellerin (1992) or Pellerin et al. (1995b). Near the top of the domain, the cloud water field reflects the laminar nature of the downward flow there. A little farther down, the jet undergoes a transition to turbulence, with undulations developing at the interface of the cloudy region. The rolls simulated by the model bear a striking resemblance to those observed in the laboratory experiment: their dimension and growth rate (as visualized by the increasing number of spiral threads in the rolls as they move in the lower part of the domain) are similar. Vortex collision process is observed in the model results and in the laboratory experiment; an example of such an event is marked by the "I" on panels b and d. The two-dimensional nature of the model experiment which results in a greater symmetry than in the laboratory experiment, does not allow a detailed comparison with observations. This comparison nevertheless shows that the SI-SL model adequately resolves the fine scales that develop under realistic experimental conditions.

The model has also been applied to simulate liquids through a suitable scaling of the background temperature and pressure of the "gas" to correspond to the thermal expansion of water. This version of the model has been validated on the classical experiment of a differentially heated rotating annulus by Ménard (1994). Figure 3 shows an example of his results. For these simulations, no-slip wall boundary conditions were imposed on the (2-cm radius) inner and (5-cm radius) outer cylinders where a constant 5°C temperature difference was maintained. At the top and bottom, flat insulated boundaries were imposed, with no-slip condition at the bottom and free-slip at the 3 cm high top. The rotation rate was 0.8 s^{-1} for this experiment, and the time step was 0.03 s. Relatively coarse resolution was used for this experiment, with 83 points in the angular direction, 13 in the radial direction and 10 layers. At such resolution, the details of the thin boundary layers that develop near the solid boundaries are lost, and only the "outer" flow is reproduced with some fidelity. Figure 3 displays the fields after 8 min of integration, once a quasi-equilibrium has been achieved. A wavenumber 5 pattern is simulated, and the phase difference between the thermal and flow fields indicates that positive baroclinic energy conversion plays a major role in the maintenance of these eddies. Again the SI-SL treatment of the Navier-Stokes or incompressible equations have been shown to be more efficient than models that have thus far been used in the published literature on this topic.

The SI-SL approach has also been shown to be applicable to supersonic flow simulations. Robert and Ben Hadj Tahar (personal communication) have computed the time evolution of supersonic flow around a solid object at Mach number 1.5 on a mesh size of 1 mm with a time step of 4 μs, for a Courant number of 2, in a two-dimensional domain with 750 by 250 grid points (results not shown here).

Classical von Karman vortex streets generated by flow around a solid object at

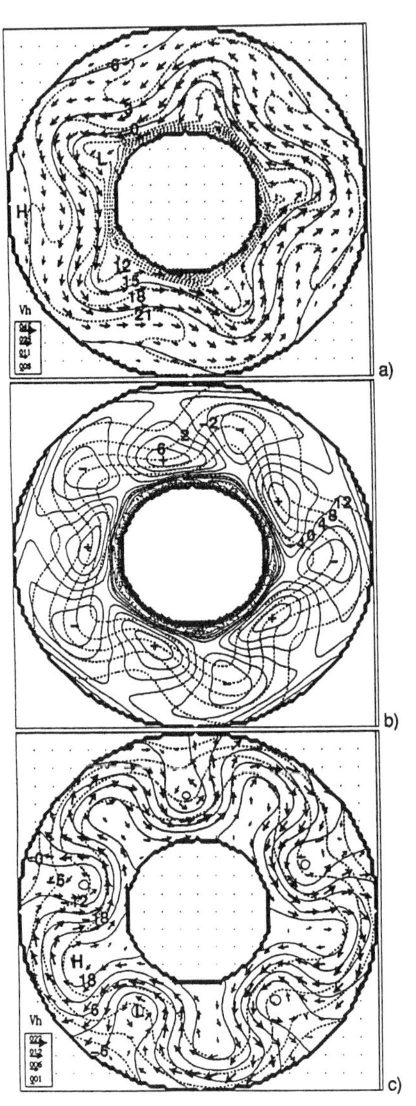

Fig. 3 Differentially heated rotating annulus experiment for Rossby regime wavenumber five. Panels (a) and (c) present the results at the top (z = 3 cm) and bottom (z = 0 cm) of the annulus, respectively, while panel (b) is at mid-level height. For (a) and (c), solid and dotted curves correspond to pressure and temperature perturbations, respectively, while arrows show the horizontal flow. The arrow length legend is shown in the left-hand corner: 045 means 0.45 cm s^{-1}. High and low pressure centres are indicated by H and L, respectively. In (a), the contour intervals are 3×10^{-2} Pa for pressure and 3×10^{-1} °C for temperature, while in (c), the contour interval is 6×10^{-2} Pa for pressure. For (b), solid and dotted curves show respectively the vertical motion and temperature perturbations at mid-level (z = 1.5 cm), with + and − indicating upward and downward motion, respectively. The contour intervals are 2×10^{-2} cm s^{-1} for vertical velocity and 4×10^{-1} °C for temperature. (Reproduced with permission from Ménard (1994).)

Reynolds number 100 have been successfully simulated with the SI-SL approach by Larocque (1995), using a two-dimensional computational grid of 250 by 250 points with mesh spacing of 1 mm and a time step of 0.01 s for a flow speed of 100 mm s^{-1} around a 15 mm circular object. No-slip boundary condition is used on the solid object and free-slip on the domain limits. The flow field after 8 s is displayed in Fig. 4. The alternating vortices in the trail of the obstacle are in excellent agreement with other published results (see Larocque (1995) for the details of the comparison).

In the same study, Larocque (1995) has simulated an internal flow in a driven lid cavity. In this case, a fluid fills a 1 m^2 closed two-dimensional cavity whose lid is moving. No-slip boundary condition is used on all sides. Figure 5 shows the flow in the cavity after 2,067 s at Reynolds number 1000. The lid is moving from left to right, with a velocity that is maximum at the centre (15 mm s^{-1}) and decreases to zero at the edge of the side walls. The time step for this simulation was 0.52 s. The details of the central and secondary vortices were shown by Larocque (1995) to agree very well with other published results despite the fact that a relatively coarse mesh of 128 by 128 grid points was used in this simulation.

Recently, Szyrmer and Zawadzki (personal communication) have used the same model to analyze the response of a thermal sensor designed to be mounted on an aircraft to measure in-cloud temperature. For this experiment, a two-dimensional mesh with 1600 $\times$ 600 grid points with mesh size of 10 μ and a time step of 100 ns has been used. A deflector is inserted upstream of the thermal sensor to shield the sensor from hydrometeors without interfering with its proper ventilation. The simulations have shown that eddies shed from the edges of the shield can substantially affect the measurement of temperature on the sensor (results not shown here).

5 Conclusion

A formal description of the MC^2 model developed by André Robert, his colleagues and students has been presented. This model is based on the fully elastic non-hydrostatic Euler equations, and hence the dynamical framework of the model is suitable at all scales of fluid mechanics. Even though the model is cast in a form suitable for a gas, it is possible to scale the temperature in order to accurately model the thermal expansion of a liquid (Ménard, 1994). The semi-implicit semi-Lagrangian (SI-SL) time discretization of MC^2 allows for an efficient integration of the model. It has been shown by Tanguay et al. (1990) that this non-hydrostatic model could be integrated at large-scale without any computational penalty compared to more traditional hydrostatic models.

The semi-Lagrangian treatment of advection implies that the time step is not limited by the wind speed. The semi-implicit treatment of linear waves implies that the model time step is not limited by the phase speed of fast moving waves either. Hence, the SI-SL approach allows the choice of the time step to be based on accuracy rather than stability considerations, unlike the explicit Eulerian approach

Fig. 4 Streamfunction (upper) and vorticity (lower) fields of two-dimensional flow around a solid at Reynolds number 100 after 800 time steps (8 s of simulated time). (Reproduced with permission from Larocque (1995).)

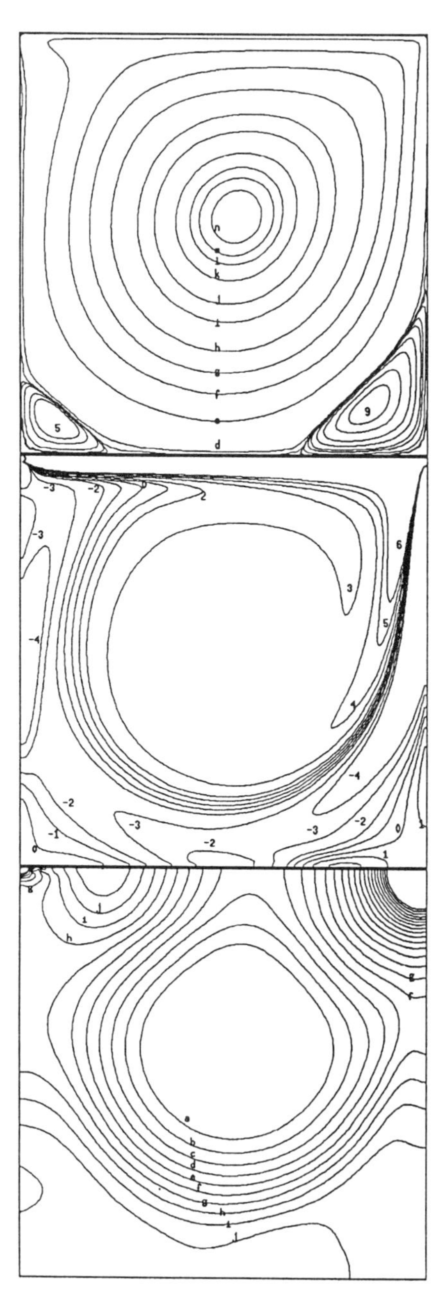

Fig. 5 Lid driven internal flow in a cavity at Reynolds number 1000. The lid is moving from left to right in this simulation. Panels show the streamfunction (upper), the vorticity (middle) and the pressure field (lower) after 4,000 time steps (2,067 s of simulated time). (Reproduced with permission from Larocque (1995).)

in which numerical stability dictates the maximum time step. In most practical situations, accuracy considerations are far less stringent than stability ones. Computational savings can result since fewer time steps are required to achieve a given integration length.

The model has been validated on a number of classical problems for which non-hydrostatic effects are important. Robert (1993) has shown that the SI approach is valid even when the environmental stability differs substantially from the stable reference profile used for the SI decomposition. The interpolative nature of SL makes it considerably less dispersive than usual finite-difference Eulerian methods. This results in fields that are more coherent and less prone to false ripples than unconstrained Eulerian schemes.

The approach pioneered by André Robert has therefore been shown to provide a computationally efficient unified framework applicable to a wide range of atmospheric and, more generally, geophysical fluid dynamics problems: it hence constitutes the basis of a "universal" model.

Acknowledgements

Parts of this article were taken from sections of an unpublished CCRM Report (Bergeron et al., 1994). The following people have contributed in various ways to the development, documentation and/or validation of the MC^2 model, and/or to the results that were shown in RL's talk at the André J. Robert Symposium held at UQAM on 5–7 October 1994: Dr Robert Benoit and his collaborators, Michel Desgagné and Yves Chartier (Division de Recherches en Prévision Numérique (RPN), Atmospheric Environment Service); Pierre Pellerin, Sylvain Ménard and Yvan Larocque, former M.Sc. Students, and Néji Ben Hadj Tahar, Research Associate, under the direction of André Robert; Prof. Isztar Zawadzki (now at Department of Atmospheric and Oceanic Sciences, McGill University) and his Ph.D. Student at UQAM, Wanda Szyrmer; Drs Jean-Pierre Pinty and Évelyne Richard (CNRS, Université Paul-Sabatier (UPS), Toulouse, France); Philippe Héreil and Hélène Rivals, D.E.A. Stagiaires at UQAM with RL in 1993; and finally, André Robert himself who developed the initial version of MC^2 and carried out validation tests covering a wide range of scales. This research was supported by the Natural Sciences and Engineering Research Council (NSERC) (Operating and RCM Strategic Grants), the Atmospheric Environment Service, the Canadian Green Plan funds of the Canadian Climate Centre and Forestry Canada (Ontario Region), and internal grants from UQAM.

References

ASSELIN, R.A. 1972. Frequency filter for time integrations. *Mon. Weather Rev.* **100**: 487–490.

BEN HADJ TAHAR, N. 1992. Introduction d'une grille variable dans un modèle atmosphérique non hydrostatique. M. Sc. Thesis, Atmos. Sci., Physics Dept., UQAM, 100 pp.

BENOIT, R.; P. PELLERIN, J. MAILHOT and V. LEE. 1994a. Modelling of late spring intense orographic precipitation over the Canadian portion of the Columbia River basin. 6th AMS Conf. on mesoscale processes, 18–22 July 1994, Portland, Oregon, pp. 583–586.

———; ———, ———, M. DESGAGNÉ and Y. CHARTIER. 1994b. A real-time very-high resolution forecast model for the field phase of the Beaufort and Arctic Storms Experiment (BASE). 10th AMS Conference on Numerical Weather Prediction, 18–22 July 1994, Portland, Oregon, pp. 451–453.

BERGERON, G.; R. LAPRISE and D. CAYA. 1994. The numerical formulation of MC², the mesoscale community compressible model, 165 pp. (CCRM report, available at the corresponding author's address).

CAYA, D.; R. LAPRISE, M. GIGUÈRE, G. BERGERON, J.-P. BLANCHET, B.J. STOCKS, G.J. BOER and N.A. MCFARLANE. 1995a. Description of the Canadian RCM. IBFRA Report, Saskatoon, 1994. Water, Air and Soil Pollution, 82, pp. 477–482.

———; ———, ———, ———. 1995b. Preliminary results of a two-month simulation of the Canadian RCM. *In*: Research Activities in Atmospheric and Oceanic Modelling, WMO/TD - No. 665, Report No. 21, 7.3–7.5.

DAVIES, H.C. 1976. A lateral boundary formulation for multi-level prediction models. *Q. J. R. Meteorol. Soc.* **102**: 405–418.

DENIS, B. 1990. Introduction de la topographie dans un modèle atmosphérique non hydrostatique. M. Sc. Thesis, Atmos. Sci., Physics Dept., UQAM, 121 pp., (available at the corresponding author's address).

GAL-CHEN, T. and R. SOMERVILLE. 1975. On the use of a coordinate transformation for the solution of the Navier-Stokes equations. *J. Comput. Phys.* **17(2)**: 209–228.

GYAKUM, J.R.; M. CARRERA, D.-L. ZHANG, S. MILLER, J. CAVEEN, R. BENOIT, T. BLACK, A. BUZZI, C. CHOUIMARO, M. FANTINI, C. FOLLONI, J.J. KATZFEV, Y.-H. KUO, F. LALAURETTE, S. LOW-HAM, J. MAILHOT, P. MALGUZZI, J.L. MCGREGOR, M. NAKAMURA, G. TRIPOLI and C. WILSON. 1996. A regional model intercomparison using a case of explosive oceanic cyclogenesis. *Weather Forecast.* **11**: 521–543.

HALTINER, G.J. and R.T. WILLIAMS. 1980. *Numerical Prediction and Dynamic Meteorology*. John Wiley and Sons, 477 pp.

HÉREIL, P. 1993. Sensibilité de la solution d'ondes de relief au choix du pas de temps dans un modèle non hydrostatique semi-implicite semi-lagrangien. D.E.A. Report directed by R. Laprise, Université Paul-Sabatier, 39 pp., (available at the corresponding author's address).

——— and R. LAPRISE. 1995. Sensitivity of internal gravity wave solutions to the timestep of a semi-implicit semi-Lagrangian non-hydrostatic model. *Mon. Weather Rev.* **124(4)**: 972–999.

KASAHARA, A. 1974. Various vertical coordinate systems used for numerical weather prediction. *Mon. Weather Rev.* **102**: 507–522.

LAROCQUE, Y. 1995. Étude d'un écoulement d'un fluide autour d'un cylindre circulaire à l'aide d'un modèle numérique aux équations de Navier-Stokes. M. Sc. Thesis, Atmos. Sci., Physics Dept., UQAM, 84 pp.

MAILHOT, J.; C. CHOUINARD, R. BENOIT, M. ROCH, G. VERNER, J. CÔTÉ and J. PUDYKIEWICZ. 1989. Numerical forecasting of winter coastal storms during CASP: Evaluation of the Regional Finite-Element Model. ATMOSPHERE-OCEAN, **27(1)**: 24–58.

MCFARLANE, N.A.; G.J. BOER, J.-P. BLANCHET and M. LAZARE. 1992. The Canadian Climate Centre second generation General Circulation Model and its equilibrium climate. *J. Clim.* **5(10)**: 1013–1044.

MÉNARD, S. 1994. Étude de l'écoulement dans le bassin annulaire de laboratoire à l'aide du nouveau modèle numérique UQAM. M. Sc. Thesis, Atmos. Sci., Physics Dept., UQAM, 80 pp.

PEACEMAN, D.W. and H.H. RACHFORD. 1955. The numerical solution of parabolic and elliptic differential equations. *SIAM*, **3(1)**: 28–41.

PELLERIN, P. 1992. Validation d'un nouveau modèle numérique basé sur les équations d'Euler (modèle UQAM de convection). M. Sc. Thesis, Atmos. Sci., Physics Dept., UQAM, 79 pp.

———; R. LAPRISE and I. ZAWADZKI. 1995a. Semi-Lagrangian solutions to the advection-condensation problem. *Mon. Weather Rev.* **123(11)**: 3318–3330.

———; I. ZAWADZKI and R. LAPRISE. 1995b. Comparison between downward convection produced in laboratory and that simulated by a numerical model. *Bull. Can. Meteorol. Oceanogr. Soc.* **23(2/3)**: 3–6.

PHILLIPS, N.A. 1966. The equations of motion for a shallow rotating atmosphere and the "traditional approximation." *Mon. Weather Rev.* **23**: 626–628.

PINTY, J.-P.; R. BENOIT, E. RICHARD and R. LAPRISE.

1995. Simple tests of a semi-implicit semi-Lagrangian model on 2D mountain wave problems. *Mon. Weather Rev.* **123(10)**: 3042–3058.

RIVALS, H. 1993. Validation d'un modèle non hydrostatique semi-implicite semi-lagrangien par des simulations d'ondes de relief. D.E.A. Report directed by R. Laprise, Université Paul-Sabatier, 32 pp. (available at the corresponding author's address).

RIVEST, C.; A. STANIFORTH and A. ROBERT. 1994. Spurious resonant response of semi-Lagrangian discretizations to orographic forcing: diagnosis and solution. *Mon. Weather Rev.* **122**: 366–376.

ROBERT, A. 1993. Bubble convection experiments with a semi-implicit formulation of the Euler equations. *J. Atmos. Sci.* **50(13)**: 1865–1873.

———. 1981. A stable numerical integration scheme for the primitive meteorological equations. ATMOSPHERE-OCEAN, **19**: 319–325.

———. 1982. A semi-implicit and semi-Lagrangian numerical integration scheme for the primitive meteorological equations. *J. Meteorol. Soc. Jpn.* **60**: 319–325.

———; T.L. YEE and H. RITCHIE. 1985. A semi-Lagrangian and semi-implicit numerical integration scheme for multilevel atmospheric models. *Mon. Weather Rev.* **113**: 388–394.

TANGUAY, M.; A. ROBERT and R. LAPRISE. 1990. A semi-implicit semi-Lagrangian fully compressible regional forecast model. *Mon. Weather Rev.* **118(10)**: 1970–1980.

———; E. YAKIMIV, H. RITCHIE and A. ROBERT. 1992. Advantages of spatial averaging in semi-implicit semi-Lagrangian schemes. *Mon. Weather Rev.* **120(1)**: 115–123.

TREMBLAY, A. 1994. Simulations of the 14 July 1987 squall line using a fully compressible model. ATMOSPHERE-OCEAN, **32(3)**: 567–603.

YAKIMIW, E. and A. ROBERT. 1990. Validation experiments for a nested grid-point regional forecast model. ATMOSPHERE-OCEAN, **28(4)**: 466–472.

ZAWADZKI, I.; L. OSTIGUY and R. LAPRISE. 1993. Retrieval of the Microphysical Properties in a CASP Storm by Integration of a Numerical Kinematic Model. ATMOSPHERE-OCEAN, **31(2)**: 201–233.

MC2 Model Performance during the Beaufort and Arctic Storm Experiment

Robert Benoit[1], Simon Pellerin[1] and Wei Yu[2]*
[1] *Recherche en Prévision Numérique*
[2] *Data Assimilation and Satellite Meteorology Division*
Service de l'environnement atmosphérique
Dorval, Québec, Canada

[Original manuscript received 20 May 1995; in revised form 25 September 1995]

ABSTRACT *A special operational forecasting guidance was produced during the field phase of the Beaufort and Arctic Storms Experiment (BASE) with the mesoscale compressible community model (MC2). After a brief description of the model, its performance during BASE is compared to observations. The in-house numerical database assembled at RPN especially for the BASE model is described briefly. Based on the 26 September 1994 precipitation bands event at Tuktoyaktuk, a case study shows the model sensitivity to the ice extent in the Beaufort Sea. Some applications of the MC2 model and its two-dimensional variant are mentioned to show the versatility of the model.*

RÉSUMÉ *On a fourni, durant la phase sur le terrain de BASE (Expérience des tempêtes sur la mer de Beaufort et l'océan Arctique), une assistance spéciale pour la prévision en temps réel basée sur le modèle de mésoéchelle compressible communautaire (MC2). On décrit brièvement le modèle et compare ses résultats aux observations durant BASE. La base de données numériques maison assemblée au RPN spécialement pour le modèle BASE est aussi décrite. Une étude de cas, utilisant les bandes de précipitation du 24 septembre 1994 sur Tuktoyaktuk, montre que le modèle est sensible à l'étendue de glace sur la mer de Beaufort. Quelques applications du modèle MC2 et sa variante à deux dimensions sont mentionnées pour en montrer la flexibilité.*

1 Introduction

The André J. Robert Memorial Symposium on Numerical Methods in Atmospheric and Oceanic Sciences was held in Montréal at the Université du Québec à Montréal (UQAM) in October 1994. The symposium was dedicated to Dr. André Robert and emphasized his contributions to meteorological science, particularly in the field of numerical weather prediction. The CCRM's (Cooperative Centre for Research in

*Current affiliation: Centre de recherche en calcul appliqué, Montréal, Québec, Canada.

Mesometeorology) MC2 (mesoscale compressible community) model is the ultimate model inherited from Robert before his death in November 1993. It is among the few currently available 3D fully-compressible models that use a semi-implicit/semi-Lagrangian time scheme.

Starting with the fully-elastic non-hydrostatic model developed recently by Tanguay, Robert and Laprise (1990; denoted herein by TRL), a team of scientists from Recherche en Prévision Numérique (RPN) and UQAM has worked actively since May 1992 to document, validate and improve upon this versatile modelling tool which is applicable to a variety of needs of the research community both in Canada and abroad. Coarse-resolution (127 km) short-term (24 h) regional forecasts were shown by TRL to be feasible by a semi-implicit semi-Lagrangian discretization of the fully-elastic equations. The resulting model proved to be comparable to the hydrostatic equations in terms of computational cost. The time step used for these forecasts was 30 minutes, which is 800 times larger than the time step that would be required to integrate the elastic equations with an explicit marching scheme. Orography was later implemented through the use of a Gal-Chen terrain-following coordinate transformation by Denis (1990).

The team's objectives have been to adapt the model coding to current RPN standards, to provide an extensive Users' Guide and documentation, to include all necessary features (such as nesting, boundary layer, convection, radiation) needed for the current applications and to maintain a central model library (with modules contributed from the community) that allows easy porting to various computers. By January 1994, the objectives were reached and the delivery of the model to research groups was started; the MC2 community soon became international. Support to remote users as well as contributions to the central community model library now occurs.

A few subgrid-scale physical parametrization packages have already been implemented in MC2: the RPN Physics for application to fine-scale short-range forecasting by Benoit et al. (1997) at RPN; the Canadian Climate Centre second-generation General Circulation Model Physics for application to regional climate modelling by Laprise (this volume) at UQAM; a cloud micro-physics package (Zawadzki et al., 1993) for mesoscale modelling implemented by Tremblay et al. (1994).

A brief overview of several recent applications of the MC2 model is presented in the companion paper by Laprise (this volume). Suffice to say here that these applications cover regional climate, intense orographic precipitation, east coast cyclogenesis, downslope windstorm over Canadian landscape, flash flood and large amplitude gravity wave, plus several micro-scale simulations of turbulent and stratified flows.

As part of the Beaufort and Arctic Storms Experiment (BASE), RPN configured a special version of the MC2 for daily short-term forecasting (30 hours) of the few hundreds of kilometres around the Mackenzie River Delta. This version has been successfully run in real-time for a period of 6 1/2 weeks over the BASE area. The highly stable behaviour of the model and its good performance during BASE is an impressive demonstration of its reliability as a real-time forecasting tool. The BASE was conducted over the southern Beaufort Sea and Mackenzie River Delta from 1 September 1994 to 14 October 1994. It provided a unique source of information on Arctic weather systems as two X-band Doppler radars, a research aircraft, enhanced rawinsondes and satellite images were available.

The project had 12 Intensive Observational Periods (IOPs). Sixty aircraft-hours were flown and 88 dropsondes launched during the six and a half weeks. There were 60 radiosonde flights at Tuktoyaktuk and 37 extra flights at Inuvik in addition to their regular launches. Doppler radar observations were essentially continuous at Inuvik throughout the experiment and from 7 September 1994 onwards at Tuktoyaktuk. Each radar monitored a reliable ring of 60 km radius in Doppler mode which almost touched the other ring (Tuktoyaktuk-Inuvik distance being 130 km).

The requirement of a daily forecast prepared with real-time data was an opportunity to test the MC2 in a quasi-operational context within the production environment of the Canadian Meteorological Centre (CMC). The main feature of this special version included a limited-area domain centred on Inuvik with a horizontal resolution of 15 km. Parametrization of physical processes was modified compared to the RPN/CMC physics library to include a more sophisticated treatment of cloud water and precipitation, and to allow interaction between clouds and radiation. Some of the special observations from the BASE region were used in order to improve the initial conditions.

The Inuvik airport terminal building, located about 10 km from the main town, housed the field project forecast office and main operations centre. During BASE, the special model contributed to the daily detailed weather briefing at the main operations centre for the planning of the experimental weather observation operations. MC2 was run on the NEC SX-3 computer at CMC in Dorval. 3D model grid output of the basic fields was archived every 3 hours and was transmitted to Inuvik via a dedicated satellite uplink. About 50 MBytes per run were thus transmitted to the operations centre and then processed locally at pressure levels before visualisation on workstations.

In 1986, as part of the field phase of CASP I on the Canadian east coast, RPN had also developed an experimental meso-alpha scale version (100 km) of the Canadian Regional Finite-Element (RFE) model (Mailhot et al., 1989). Using the operational 190 km resolution version of the model as a starting point, the special RFE version for CASP I included modifications to the physics package and improvements in the geophysical and surface field analysis. A post-analysis of the model performance has led to overall better meteorological performance of the special guidance as a result of increased horizontal resolution and improvements in the surface analysis and physics (Mailhot and Chouinard, 1989).

None of the two special models designed for BASE and CASP I field experiments have taken advantage of an initial mesoscale analysis on their computational grid. However the CASP I set up included an improvement in sea surface temperature (SST) compared to the existing operational environment. According to Mailhot et al. (1989) the higher resolution SST analysis and the improvement in the mountain field representation were crucial factors in the overall performance of the special model compared to its operational version.

The main purpose of this paper is to present a preliminary evaluation of the BASE special numerical forecasting guidance. This will be based on a comparison of BASE model surface field forecasts against observations and forecasts from the Canadian 50-km resolution operational RFE model (Mailhot et al., 1995). It is important to keep in mind in comparing the forecast performance of the two models that in spite of having a higher horizontal resolution, an improved ice analysis and mountain field representation, the MC2 initial conditions resulted from an assimilation cycle designed to provide the RFE model with more detailed analyses in a dynamically consistent manner. The consistent mesoscale structures developed and supported by the MC2 model were not always anchored with an analysis designed on the computational grid of the model.

During the special model selection phase of the project (winter 1993) four model configurations were examined (Benoit et al., 1994). Results of two 25-km resolution versions (with different central windows) of the RFE model were then compared to those of 15-km and 25-km resolution versions of the MC2 model. The four candidates were tested for their performances in 5 storm event integrations typical of the BASE time period. The few observational data available to estimate model accuracy indicated an overall better performance of the 15-km MC2 model. The mesoscale structures supported by this later version, sometimes showing plausible non-hydrostatic signatures, made it very attractive as a special guidance tool intended for a research experiment. Another argument in favour of choosing the 15-km MC2 version was its reduced computational time compared to that required by either of the two 25-km RFE versions to perform a similar integration. A more detailed appraisal of the special model performance during BASE will be performed later as more information collected during the field phase becomes available.

The primary features of the BASE model, its initialisation procedure and timing strategy for the data transfer are covered in Section 2. The value of the special model relative to the operational RFE model is then examined in Section 3 in terms of exactness of their precipitation, surface temperature and wind (direction and speed), and sea-level pressure forecasts. The in-house numerical database assembled at RPN especially for the BASE model will be described briefly in Section 4. A set of high-resolution runs described in Section 5 reveals the model sensitivity to the Beaufort Sea ice extent based on the 26 September 1994 precipitation bands event at Tuktoyaktuk.

2 Model description

a *Dynamics and Physical Processes*

The MC2 model was the special real-time very-high resolution forecasting guidance in support of the field phase and it was used in the planning of the logistics of the BASE experiment, mostly for the operation of the research aircraft. MC2 is a finite-difference semi-implicit semi-Lagrangian model solving the non-hydrostatic elastic equations on a polar-stereographic limited-area grid nested in time-dependent boundaries. A comprehensive description of the dynamical framework of MC2 is given by Laprise (this volume) while the inclusion of the full physics parametrizations in MC2 and the extension needed to make MC2 a full mesoscale weather forecasting model is found in Benoit et al. (1995).

The special model was run on a polar-stereographic grid with a horizontal resolution of 15 km for 30 hours, once per day, at night, using the lateral boundary conditions every 3 hours from the operational RFE model and the initial conditions from the 0000 UTC 50-km Regional Data Assimilation System (RDAS) of CMC (Chouinard et al., 1994). Figure 1 illustrates the limited-area (1500 km × 1500 km) covered by the 101 × 101 uniform resolution grid used in the BASE version. The outer edge of the domain, delimited by a square box, depicts the 12-point thick time-dependent boundary zone. The initialisation included parts of the BASE special observations (extra radiosondes and surface stations) and 25-km resolution analyses for sea-ice based on Special Sensor Microwave Imager (SSMI) satellite data.

Two different configurations of parametrization schemes for physical processes were used during the experiment. From 1 September to 11 September, a Kuo-type (Kuo 1965, 1974) deep convection scheme combined with a grid-scale condensation in supersaturated layers scheme was used (identical with the one used by the RFE). For the rest of the period, the MC2 model was integrated with a new cloud water scheme (Sundqvist et al., 1989) in which the combined liquid and solid cloud water is a single predictive variable of the model.

This scheme produces both deep convective and stratiform clouds. The convective condensation is based on the check of atmospheric instability. It is activated in the presence of water vapour convergence and conditional instability. The stratiform clouds are generated if the relative humidity, U, exceeds a critical condensation threshold value, U_0. The micro-physical processes such as the auto conversion, coalescence and Bergeron-Findeisen mechanisms are taken into account in the cloud scheme. The ice existence in clouds and precipitation is parametrized as a function of temperature (Matveev, 1984; Sundqvist, 1993). The scheme also generates liquid and solid precipitation fluxes across the model atmosphere.

Another change in this new physics package compared with the RPN physics library is the treatment of the interaction between clouds and radiation. The cloud water predicted by the cloud scheme is directly introduced into the radiation calculation. The interaction between clouds and radiation is more consistent in the new configuration than in the operational model where a diagnostic cloud water independent of the cloud scheme is used in the solar radiation calculation. The 25 computational levels of the MC2 model were chosen to match the same pressure values as the sigma levels of the operational RFE for a standard atmosphere. Figure 2 shows the vertical distribution of levels.

The Stevenson screen temperature level was assumed to be 1.5 m while the wind screen height was taken as 10 m (same as the RFE). These variables are

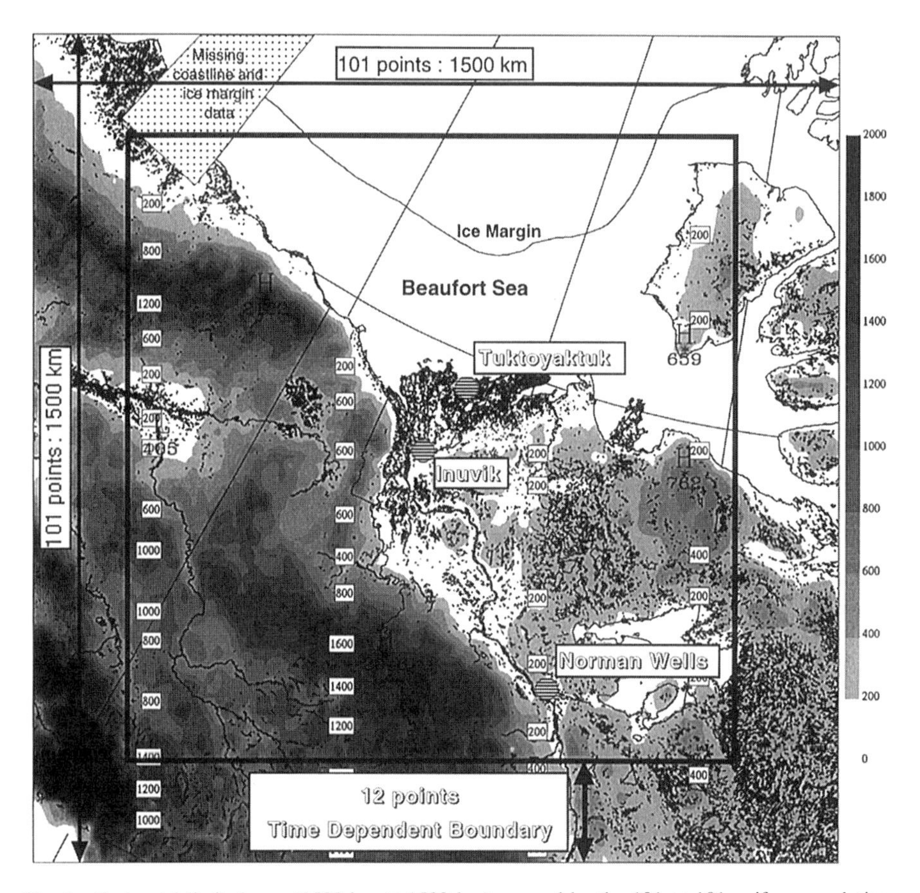

Fig. 1 Horizontal limited-area (1500 km × 1500 km) covered by the 101 × 101 uniform resolution grid used in the MC2-BASE version. The outer edge of the domain delimited by a square box depicts the 12 points time dependant boundary zone. Background shaded field illustrates the topography representation of the BASE area used by the MC2 model (shade interval: 200 metres). The distribution of rivers and lakes is also depicted. A climatological ice margin is shown over the Beaufort Sea.

used as physics processes output on surface levels and then treated by the dynamic component of the model. In contrast with the RFE, winds are not represented on the MC2 surface level but on staggered "momentum" levels. This means that the diagnostic wind from the physics is not used in MC2 (Benoit et al., 1997). The lowest level where wind components are represented in the BASE version of MC2 is located at 36.6 metres.

b *Analysis and Surface Descriptors*

The RDAS operates in a 12-h spin-up mode, initiating each cycle from the global

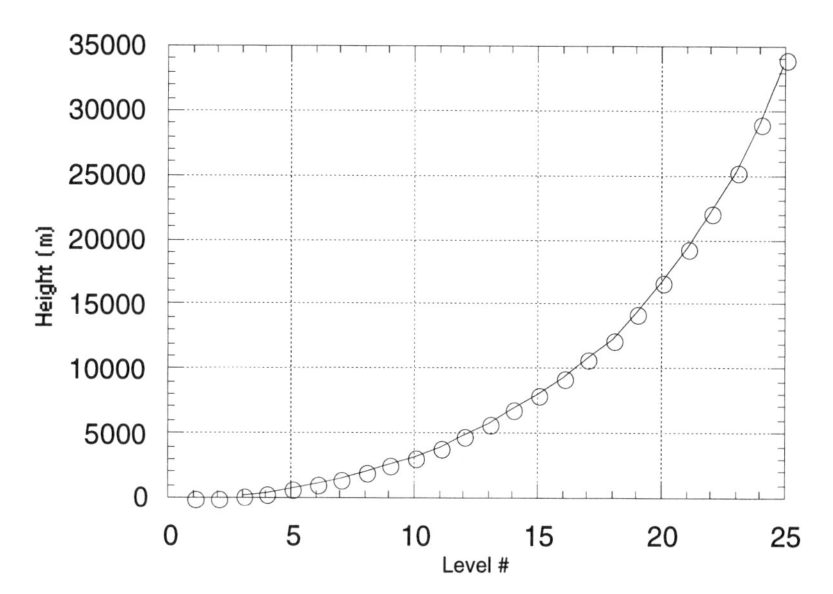

Fig. 2 Vertical distribution of the BASE-MC2 25 computational levels. Levels were chosen to match the same pressure values as the sigma levels of the operational RFE for a standard atmosphere.

cycle to redefine the large-scale flow (Chouinard et al., 1994). The spin-up cycle uses two 6-h forecast trial fields from the RFE model to achieve a good dynamical balance by maximizing the coherence between the regional forecast model and the data assimilation cycle. The 25 levels of the 50-km horizontal resolution analysis resulting from RDAS were then used to initiate the operational RFE model.

The BASE model was initialized with surface pressure, geopotential height, temperature, specific humidity, vertical velocity and wind components from the 25 σ levels of the 0-h regional model forecast (after the RFE normal-mode initialization). In addition to these dynamical fields, some other initial geophysical surface fields are required to start the MC2 model. During BASE, ocean-land mask, launching height (for use in the gravity-wave drag scheme) and roughness fields originated from interpolation of the U.S. Navy dataset. The 1.875° resolution surface soil moisture monthly climatology used during BASE was obtained from ECMWF (European Centre for Medium-range Weather Forecasts) and is described in Louis (1981). Surface air temperature analysis resulted from a 6-h interval optimum interpolation of air temperature reported at the Stevenson screen level. The trial field is taken from the σ=1 forecast of temperature from the Canadian global spectral model. The deep soil temperature field is inferred from the surface air temperature analysis according to the following algorithm:

$$TP_{t=0} = 0.8TP_{t=-6} + 0.2TS_{t=0}$$

where *TP* and *TS* represent deep-soil and surface air temperature analyses respectively and the subscripts refer to the time of validity. Sea-surface temperature is derived every 24 h from ship, buoy and drifter reports using an optimal interpolation with the previous analysis as trial field. Snow-cover analyses were obtained by blending snow depth reports with previous analyses and monthly climatology (0.5° resolution from U.S. Air Force; Foster and Davy 1988) using a statistical interpolation technique.

In order to get a better spatial definition of the physical forcing, terrain orography and sea-ice fields were improved compared to those currently used by the RFE model. The daily initial ice field is derived from an interpolation of the 25-km CMC analysis based on SSMI satellite data. In atmospheric models, as in many other types of geophysical models, the notion of terrain elevation is distinct from the definition of actual pointwise elevation of terrain and is now frequently given the name of DEM (digital elevation model). The DEM is a two-dimensional array or continuous function of horizontal position giving the modelled elevation of a fictitious surface which the atmospheric model will consider to be the surface of the Earth. The DEM surface is obtained by some form of averaging of actual spot elevations but locally can be very different from the real terrain elevation. To define the 15-km DEM for the MC2, an interpolation of the 10-arc-minutes resolution U.S. Navy DEM was used as a starting point for a Cressman-type correction with the 20-arc-seconds (or 500-metre) resolution DEM from the Communications Research Centre (CRC, from Canadian Department Of Communications, Whitteker,1992). This correction is applied to remove an error of the U.S. Navy DEM for the hills (much too high) located around 69°, North 128°, West and to slightly sharpen the various mountain ranges in the BASE area, particularly the Richardson and Brooks ranges. Unfortunately, the Arctic coast portion of the dataset covers only 3 degrees of latitude along the ocean-land interface in the Beaufort-Mackenzie Delta area. The resulting topographic representation of the BASE area used by the special model is illustrated in Fig. 1. One should also note that in contrast to the RFE model, the MC2 does not explicitly filter topography to remove two grid length waves. The only averaging is due to the staggering required to project input arrays to the Arakawa-C staggered grid.

Background contours in Fig. 3 represent differences in terrain elevation seen by both model during BASE with RFE mountain field taken as reference. Contour configuration in Fig. 3 reveals higher MC2 terrain elevation compared to the RFE topography field in mountainous areas and lower terrain height representation in valley areas. Important differences are present not only in the first 3 degrees of latitude along the coast line where the CRC DEM has been used to correct the U.S. Navy DEM, but also exist in the south-west corner of the domain where differences in excess of 1000 m prevail. Departures of RFE and MC2 DEM elevations from the actual height for several stations are plotted on Fig. 3. The maximum elevation difference at a station location inside the MC2 *free* domain is observed at Burwash station where the RFE DEM exceeds, by 744 m, the actual 807 m ASL Burwash elevation. The largest bias seen by the MC2 model, excluding its *nesting* zone, is at

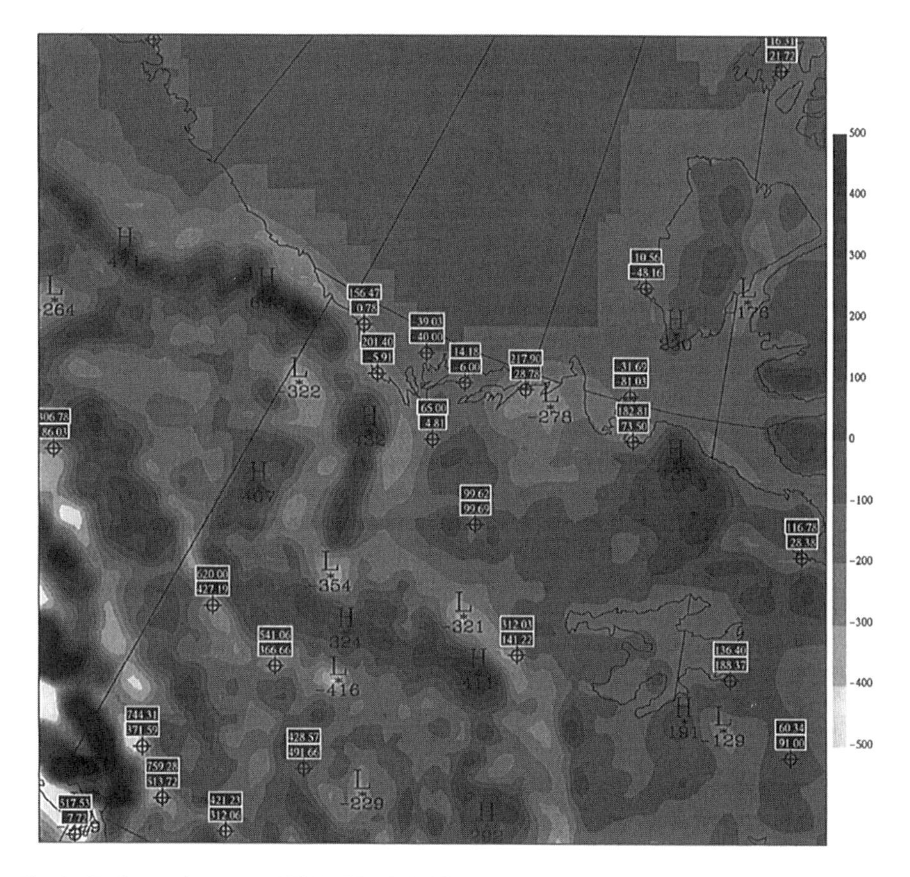

Fig. 3 Background contours: MC2-RFE DEM difference with RFE taken as reference (contour interval: 100 metres). DEM bias at observation stations coordinates (indicated by circles) are shown in black boxes on top of the station locations (MC2 DEM bias: bottom numbers; RFE DEM bias: top numbers).

the Dawson station where a difference of 427 m is noticed for an actual elevation of 370 m above sea level. The average MC2 DEM deviation is +128 m while the RFE DEM average deviation is +244 m.

3 Observations and Model Performances

a *Precipitation*

Precipitation results from a complex interaction between dynamical and physical processes and for this reason it is an important meteorological element in the perspective of numerical weather modelling. An important contribution of the modelling effort for BASE has been to test and incorporate a new clouds/precipitation scheme into the existing RPN physics library. This is believed to be the main factor responsible for the improvement of the precipitation forecasts of the BASE model compared to the operational model performance.

Four synoptic stations were used to validate MC2 precipitation forecasts. Inuvik is located at the middle grid point of the MC2 domain and has an elevation of 68 m ASL (DEM: 73 m). One hundred and fifty kilometres west of Great Bear Lake, at the foot of the Mackenzie Mountains, Norman Wells station is 74 m ASL (DEM: 215 m). With an elevation of 370 m (DEM: 797 m), Dawson station is located in the southwest portion of the domain in the Yukon River valley of the Ogilvie mountains. The fourth station is Burwash station, 807 m high (DEM: 1179 m) and located at the extreme southwest of the domain, a couple of points away from the nesting zone of the model.

Inuvik experienced a slightly above normal rainfall and snowfall during BASE (Hudson and Crawford, 1995). Inuvik received 42 mm of precipitation during the 44 days of the experiment. Summing the 24-hour point forecasts for Inuvik station over the period of the experiment, the MC2 model showed an accumulation of 44 mm while the operational RFE model totalled 37 mm of precipitation, underestimating the observed amount of precipitation by 5 mm (see Fig. 4).

For the three other locations, which are all strongly influenced by high topography, both models overestimate precipitation amounts and this overestimation increases with the elevation of the station. The MC2 model showed forecast accumulations closer to the observations than the RFE. For example, Burwash received 53 mm of precipitation during BASE while 78 mm and 152 mm were forecast by the MC2 and the RFE models, respectively.

Averaging over the four stations, about 45 mm of precipitation were observed, 55 mm were predicted by the MC2 model and over 80 mm by the RFE. Maxima in the MC2 forecast accumulation of precipitation during BASE were highly correlated to the mountain field. Figure 4 shows several maxima of about 195 mm along the Brooks range. A series of maxima coincides with the Richardson mountains with a maximum of 205 mm in its northern and highest portion. Another range of maxima, extending in an almost west-east line in the southern portion of the MC2 domain, coincides with the location of the Mackenzie mountains. An important maximum of 203 mm appears just east of the Yukon/Alaska border associated with the north-south ridges of the Ogilvie mountains. Eastward from the Mackenzie River the cumulative precipitation field shows a quasi-uniform area of about 40 mm of precipitation. This area is lake-strewn and almost flat, with an elevation less than 400 metres.

Examination of weekly and daily accumulation of precipitation shows that there is an improvement in the higher resolution MC2 forecasts compared to the lower resolution RFE. This improvement is weaker as the elevation of the station is lower but remains noticeable.

Figure 5 shows the cumulative forecast frequencies, based on 44 daily samples, as a function of the precipitation absolute error for both models. About 50% of

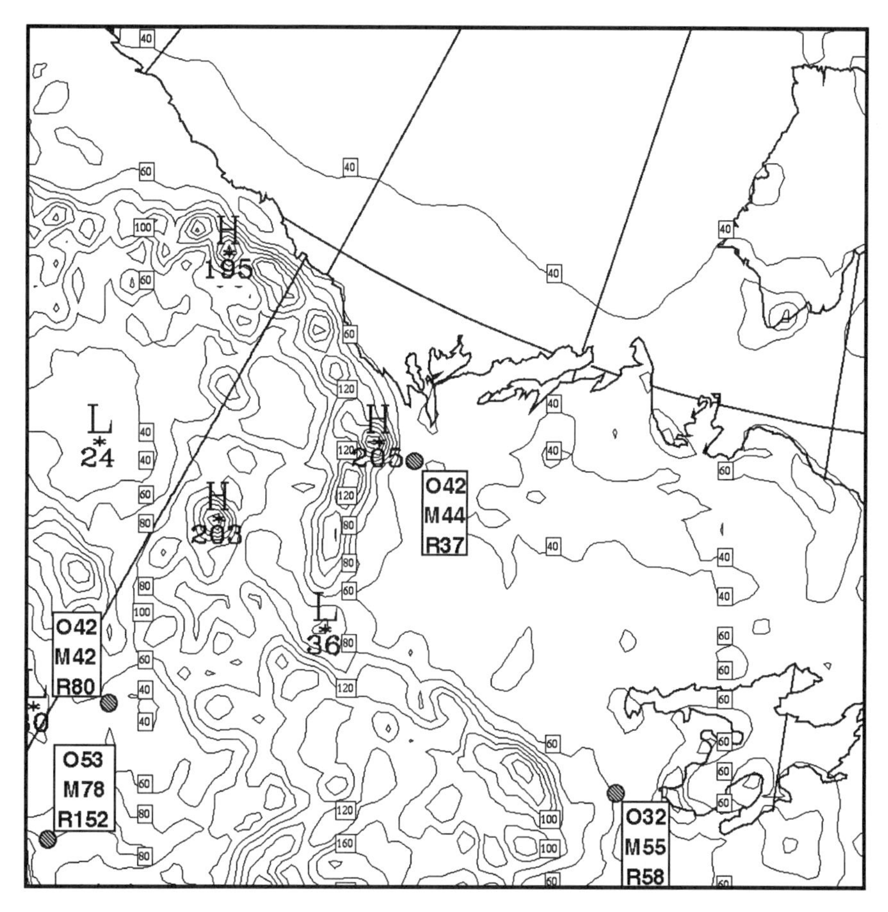

Fig. 4 Background contours: Total MC2 forecast accumulation of precipitation during BASE (44 days) based on 24-h forecasts (contour interval: 20 mm). Hatched circles represent the location of the four stations of Inuvik, Norman Wells, Dawson and Burwash. The three numbers next to every one of the four stations and preceded by O, M and R represent the observed, MC2 forecast and RFE forecast total accumulations of precipitation (in mm) respectively.

the forecasts of the two models produced point forecasts with errors less than 1 mm. This graph shows that for all cases having a forecast error greater than 1 mm, which covers 50% of the cases, the MC2 always had a lesser error. The weaker MC2 DEM bias to the real elevation of observation stations compared to the RFE DEM bias is not believed to play a crucial role in the increased performance of the MC2 model relative to the RFE for precipitation forecast. The precipitation amount is known to be generally related to the elevation gradient: the stronger the elevation gradient, the higher the expected precipitation amount. Due to the strong

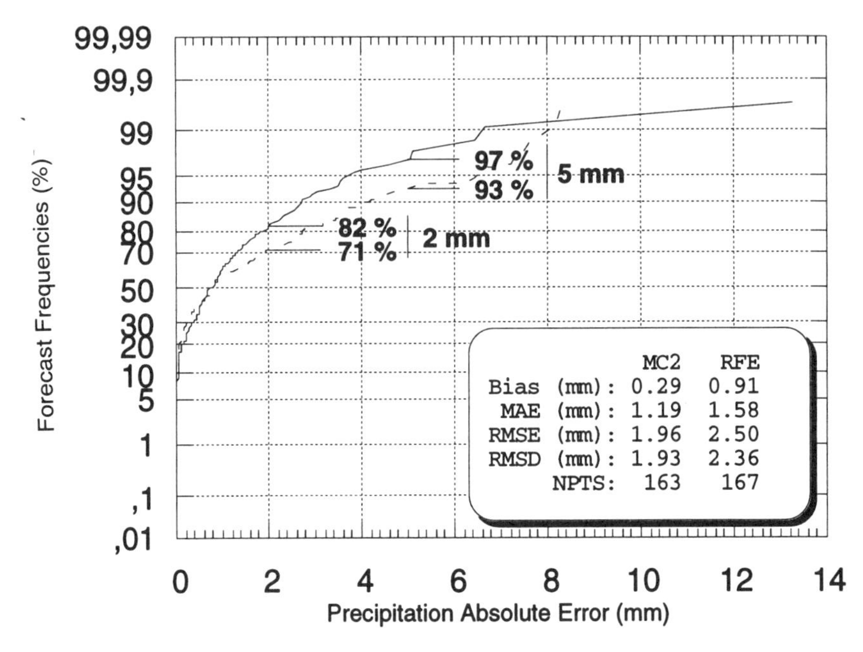

Fig. 5 Precipitation forecast frequencies vs. precipitation accumulation errors for the MC2 model (solid curve) and RFE model (dashed line). Statistics cover the 44 days of BASE with one 24-h forecast per day. Bias, mean absolute error (MAE), root mean square error (RMSE), unbiased root mean square error (RMSD) and the number of data points (NPTS) are indicated in the inset for both models.

topographic filtering in the RFE model, the resulting DEM gradient is significantly weaker compared to the DEM gradient seen by the MC2.

Based on the above argument, we would expect a lower amount of precipitation forecast by the RFE compared to the amount forecast by MC2. Consequently, the increased resolution and the improvement in the clouds/precipitation parametrization scheme are believed to be the most important ingredients to the higher accuracy of the MC2 precipitation forecast. In the tested condensation scheme, only part of the condensed cloud water is precipitated, while in a diagnostic cloud scheme such as that in the RFE model, all condensed cloud water is precipitated at the same time step. In reality, the condensed cloud water cannot totally fall in one model time step (5 minutes for MC2 and 12 minutes for 50 km RFE). The present model intercomparison exercise may imply that the cloud water storage mechanism, as it is considered in the prognostic cloud scheme, has an important impact on the simulated precipitation as well as on the associated water cycle, especially in a high-resolution model with a short time step. Preliminary case studies of Beaufort-Mackenzie delta storms and experimentation with the Canadian Global Spectral model have also shown clearly that the use of the Sundqvist cloud water scheme reduced the *fall* season amount of precipitation forecast compared to forecasts performed with the operational grid-scale condensation parametrization. However, we should note here that experimentation with the prognostic cloud scheme in the MC2 model is just in the preliminary stage. Detailed validation and refined parametrization of different physical processes involved in condensation are still needed, considering the remaining MC2 model precipitation error.

b *Temperature*

The first eight days of the BASE experiment were characterized by above-normal temperatures in Inuvik with maximum temperatures observed in the low teens. On 9 September, the temperature regime shifted to cold and remained between 0 to 5°C until the end of the month, causing its departure from the normal to decrease gradually. The maximum temperature during the first 14 days of October remained slightly above normal at Inuvik, dropping gradually from 4°C to −1°C by the end of the experiment.

Twenty-four hour screen-level temperature forecasts have been compared to synoptic surface observations during the 44 days of the experiments for 10 to 11 stations every day, giving a sample of 482 data points for the MC2 model and 492 points for the RFE (Fig. 6a). Model temperature is taken as is, from the RFE $\sigma = 1$ level and from the zero-elevation level of the MC2, without any adjustment. Both model forecasts are reasonably good and show forecast versus observation least squares regression line slope coefficients of 0.81 and 0.77 for MC2 and RFE, respectively. MC2 temperature forecasts present a weak cold bias of −0.38°C while the RFE reveals a stronger cold bias of −1.05°C. On average, the magnitude of the MC2 temperature error is slightly less than the RFE mean absolute error (2.36 and 2.62°C respectively). Both models have good skill at forecasting temperature: 70% of the forecasts are within 3°C of the observations (see Fig. 6a). 89% of the MC2 forecasts remain within 5°C of the observations as compared to 84% for the RFE.

For the highest of the 11 stations plotted on Fig. 6b, the average improvement over the 44 days of the high-resolution model is 1.6°C. For the lower elevation stations, the advantage of the MC2 model over the RFE reduces gradually. The temperature mean absolute errors of the two models for stations close to sea level become identical and less than one degree on average.

The overall differences in the surface temperature forecast can be explained by the 116 m mean spread between the DEM bias of the MC2 and the RFE models. In a standard atmosphere (with a lapse rate of 6°/km), this would lead to a 0.7° difference between the surface temperature bias of the two models. Accordingly, the inset of Fig. 6a shows a difference of 0.67° between the surface temperature bias of the two models.

c *Wind*

Examination of time series of observations and 24-h surface wind forecasts shows that for wind speeds higher than 1.5 m/s, wind direction is slightly more accurate

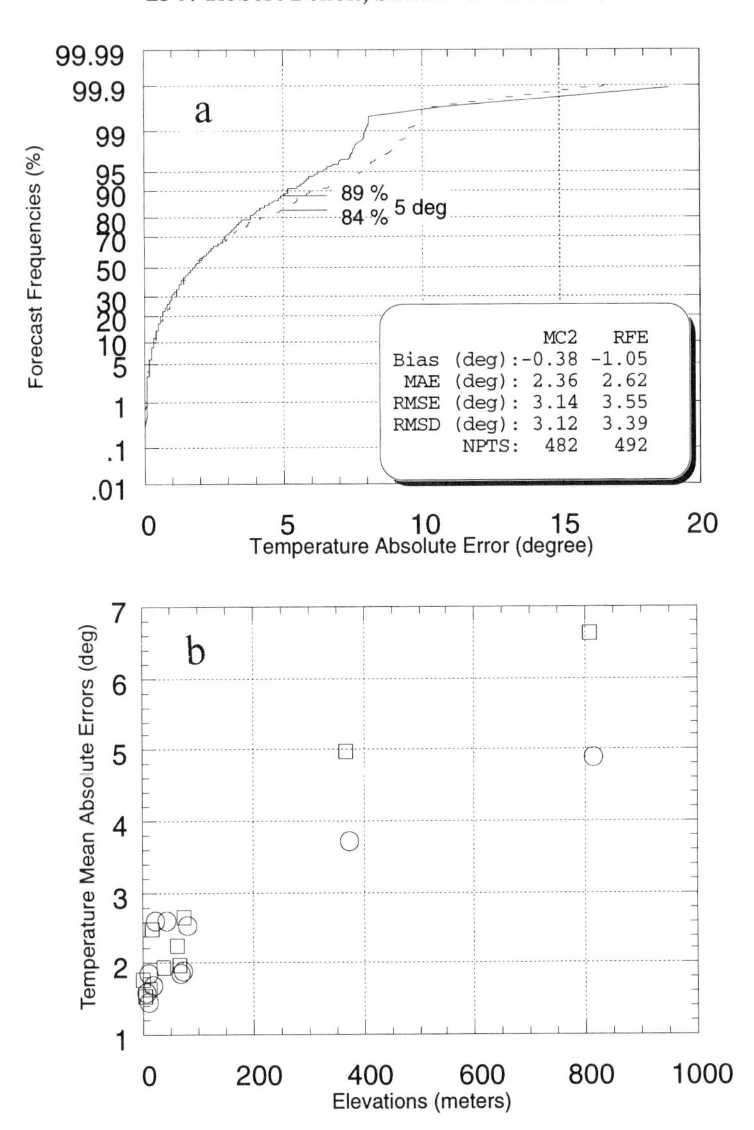

Fig. 6 a) Surface temperature forecast frequencies vs. surface temperature errors for MC2 model (solid curve) and RFE model (dashed line). Statistics cover the 44 days of BASE with one 24-h forecast per day. Inset for both models as in Fig. 5. b) Temperature mean absolute errors vs. real station elevations. For every one of the 11 stations, statistics cover the 44 days of BASE with one 24-h forecast per day. Circles are for MC2 and squares for the RFE.

in the 15-km MC2 model than in the 50-km RFE model. The small improvement is partly caused by the finer representation of the mountain field in the high-resolution model.

The compared prognostic winds are taken for both models at their first computational wind level. For the RFE model, this level is at the surface while for the MC2, the first momentum level is located at 36.6 m above the ground: those winds are taken as surface wind values. This may cause a stronger positive bias in the MC2 prognostic wind speed at the surface compared to the RFE.

The two forecast frequency graphs of Fig. 7 summarize the performance of the RFE and MC2 models for wind speed and wind direction forecasts during BASE. Figure 7a shows a similar 25% forecast frequency of the two models for wind speed absolute errors less than 1 m/s. The same plot indicates that 56% of the RFE surface wind speed forecasts were within 2 m/s of the observations, compared to 47% (worse) for the MC2. The MC2 wind speed forecast bias to the observation is greater by 0.62 m/s when compared to the RFE forecast. However, its mean absolute error and root mean square error are only slightly increased by 0.35 and 0.39 m/s respectively above the RFE values of 2.16 and 2.78 m/s.

Figure 7b indicates that 30% of the surface wind direction forecasts were within 17° of the observations for the two models. It shows also that a slightly higher fraction of the MC2 forecasts (67%) were within 45° of the observed wind direction, compared to the RFE frequency (58%) for the same error. Computation of the mean absolute and root mean square errors for the two surface wind direction forecasts shows that the RFE errors are slightly greater, by 5°, than those of MC2.

d *Sea-level pressure*

BASE took place during the stormiest period of the year, according to climatology. In September, typical storms reaching the Mackenzie Delta and the southern Beaufort Sea originate from the Pacific or track from west to east along 70°N. During that month, low pressure systems from the Pacific move north-east across Alaska and the Yukon into the southern Beaufort Sea where they normally redevelop and intensify. During BASE however, low pressure centres of Pacific origin reformed and deepened in the Mackenzie Delta and southern Beaufort Sea area without pressure-gradient intensification (Hudson and Crawford, 1995).

From September through October, the origin of the general circulation over the BASE area changes from the Pacific to Arctic Oceans and a secondary storm track gradually develops north of 70°N. During October, disturbances of Pacific origin decrease in frequency, while storms moving in a west to east track become more frequent. As Arctic storms become more frequent, cold air interacts with the open water resulting in a high incidence of stratocumulus clouds and snow showers (Hudak, 1994; Hudak et al., 1995).

Atmospheric pressure is controlled principally by large-scale forcing. It results in very weak differences in the 24-h forecast of the sea-level pressure for the two

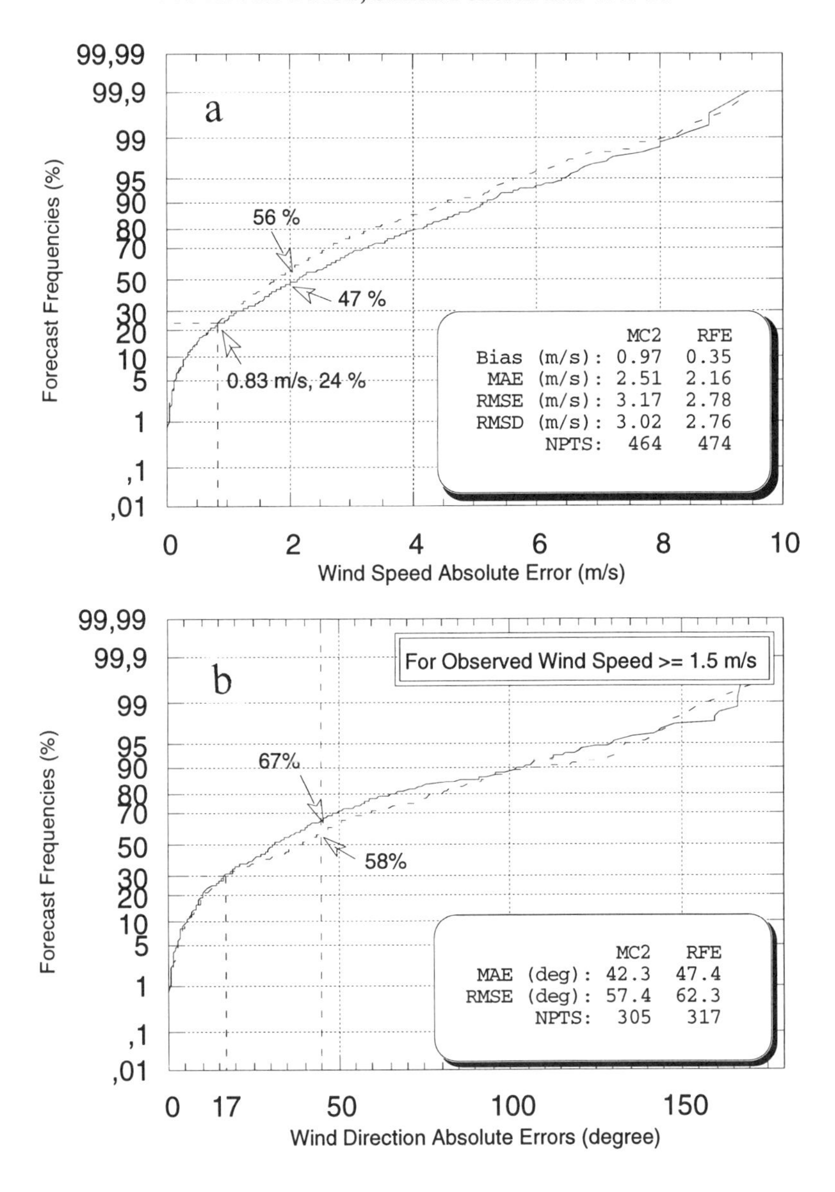

Fig. 7 Forecast frequencies for surface wind speed (7a) and surface wind directions (7b) vs. surface wind speed and direction errors for MC2 model (solid curve) and RFE model (dashed line). Statistics cover the 44 days of BASE with one 24-h forecast per day. Wind directions for which wind speed were less than or equal to 1.5 m/s were not considered in 7b. Inset for both models as in Fig. 5.

models considered. RFE and MC2 sea-level pressure forecasts are highly correlated to the observations. More than 75% of the forecasts were within 5 hPa of the station observations. This is illustrated in Fig. 8a where curves of forecast frequencies versus sea-level pressure absolute errors are superposed.

Pressure tendencies over 24 h are equally well predicted by the two models. Figure 8b shows the time series of the 24-h variation of the observed, RFE and MC2 forecast sea-level pressure in Inuvik. This latter plot reveals that both models agree closely with the observed sea-level pressure tendencies and that the differences in the two forecasts are negligible. Bias, mean absolute error and root mean square error, shown in Fig. 8a, also reflect the similarity in accuracy of the two models to forecast sea-level pressure.

4 Data base

In order to document and make easily accessible the numerical weather prediction activity for BASE, a special archive was assembled by the authors which resides on a robotic mass storage system at the CMC/RPN site. The database is divided into a tree structure with branches containing sets of files of the same type.

One branch contains the evolution of the boundary conditions that were imposed on MC2 for each run of the experiment. This branch of the archive is also a copy of part of the operational RFE forecast during BASE. Another branch includes a subset of MC2 forecasts on pressure levels. Primary fields of MC2 forecasts on Gal-Chen (computational) levels (Gal-Chen and Somerville, 1975) are also archived. To give an overview of the different synoptic situations that prevailed during BASE, a limited portion of the Regional Data Analysis grid covering the Beaufort Sea, the Chukchi Sea, the Bering Sea, the Gulf of Alaska and the Northern Pacific Ocean down to Vancouver Island has been archived in another branch. MC2 model area coverage for the BASE archive is illustrated in Fig. 1. A complete description of the database can be found in Hudson and Crawford (1995).

5 Case study

To illustrate the type of mesoscale structures that the 15-km MC2 generated in real-time during BASE, we will examine, in this section, one event that was documented with the BASE special observations and for which the model demonstrated a certain skill. The resolution was further enhanced beyond the 15-km value down to 2 km to illustrate what can be achieved during subsequent modelling studies of the Beaufort/Mackenzie area by the research community using the MC2 model.

Mesoscale disturbances of a convective nature in polar air-masses are often accompanied by severe weather and affect the large-scale weather in different ways. Polar lows described by Kellogg and Twitchell (1986) as “small vortices that form in very cold air near the ice edge and intensify as they move over the open ocean” are one type of these small-scale Arctic disturbances. Although no polar low was observed in the southern Beaufort Sea during BASE, streamers known to be polar low precursors were observed several times. When strong vertically-

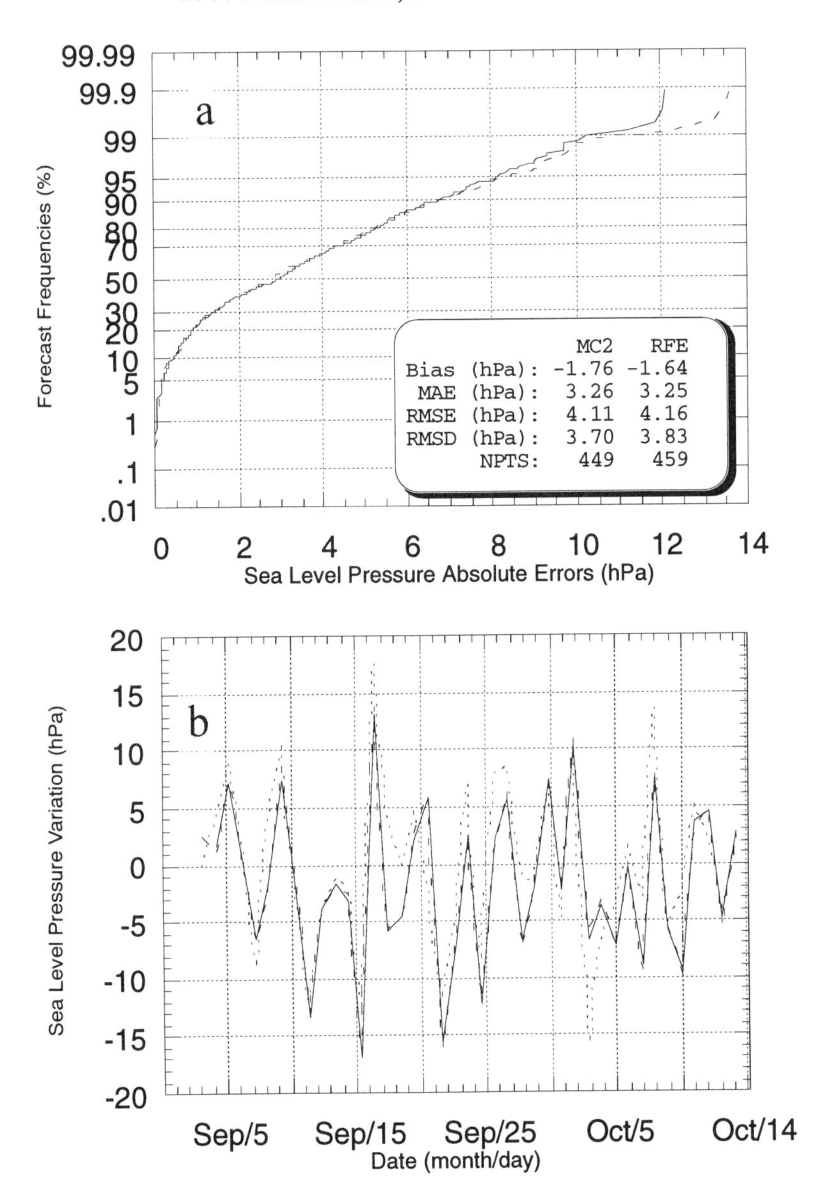

Fig. 8 a) Sea-level pressure forecast frequencies vs. sea-level pressure errors for MC2 model (solid curve) and RFE model (dashed line). Statistics cover the 44 days of BASE with one 24-h forecast per day. Inset for both models as in Fig. 5. b) Time series of the 24-h variation of the sea-level pressure in Inuvik observed (dotted line) and forecast (24-h forecasts) by the RFE (dashed line) and the MC2 (solid line) models.

sheared northerly winds push cold air from the ice and/or snow cover to the open water surface and when substantial heat and moisture fluxes originate from the underlying ocean, destabilization of the boundary layer air mass occurs and cloud streets or streamers may form over the open sea. After their formation, they may evolve into a polar low if appropriate support is present (Parker et al., 1990). The open water portion of the Beaufort Sea is an important source of heat and moisture leading, under certain conditions, to the formation of streamers and polar lows.

The year-to-year variability in sea-ice extent is very large. Beaufort Sea climatology shows that ice begins to form in late September or early October. The amount of open water is climatologically at its maximum in mid-September. Freeze-over begins in the north as first-year ice forms along the outer fringes of the permanent pack and in the shallow coastal areas. Typically the freeze-over near the coast down to the vicinity of Tuktoyaktuk occurs around 15 October. A complete freeze-over of the Beaufort Sea down to Cape Parry in the southern Amundsen Gulf coast typically takes place around 15 November (Hudak, 1994).

There was a normal amount of open water during BASE 1994 but there was an abnormal amount of ice in Amundsen Gulf, and the ice motion across the entire southern Beaufort Sea showed anti-clockwise circulation rather than the normal clockwise Beaufort gyre (Hudson and Crawford, 1995).

A deep low pressure system from the Gulf of Alaska went through the BASE area on 25 September 1994, leaving a cold air outbreak behind its cyclone on 26 September. Snow streamer bands formed in the north-westerly winds in the vicinity of Tuktoyaktuk as a result of the cold air interacting with the open water of the Beaufort Sea.

On 26 September, an organized banded-structure was observed just to the west of the Tuktoyaktuk radar. It appears that vortex structures were especially pronounced along both edges of the band. Between 0400 and 0500 UTC, the Tuktoyaktuk radar showed streamers beginning to form in WNW-ESE alignment associated with moderate snow showers. At 1900 UTC, streamers were organized in W-E lines from the Tuktoyaktuk radar to 30 km west of it. An aircraft flight was made into and also in the vicinity of these features with most of the observations made within 30 km of the radar. Additionally, between 2200 UTC 26 September and 0100 UTC 27 September, multiple vortices formed in the cloud bands at the land/water interface. The precipitation bands event of 26 September was particularly interesting as it was not fully resolved on either the prognostic charts or the analyses (Hudson and Crawford, 1995).

Four high-resolution integrations on the 26 September case have been realized with MC2. The model was first run with a horizontal resolution of 15 km with initial conditions from the 0000 UTC 26 September 50-km RDAS of the CMC. This first run was initiated with a 25-km resolution ice analysis based on SSMI data. The analysis shows a last area to be covered by ice some 15 to 30 km offshore. A second run was made with the same MC2 configuration and initial conditions except for an ice cover typical of the freeze-over of the Beaufort Sea.

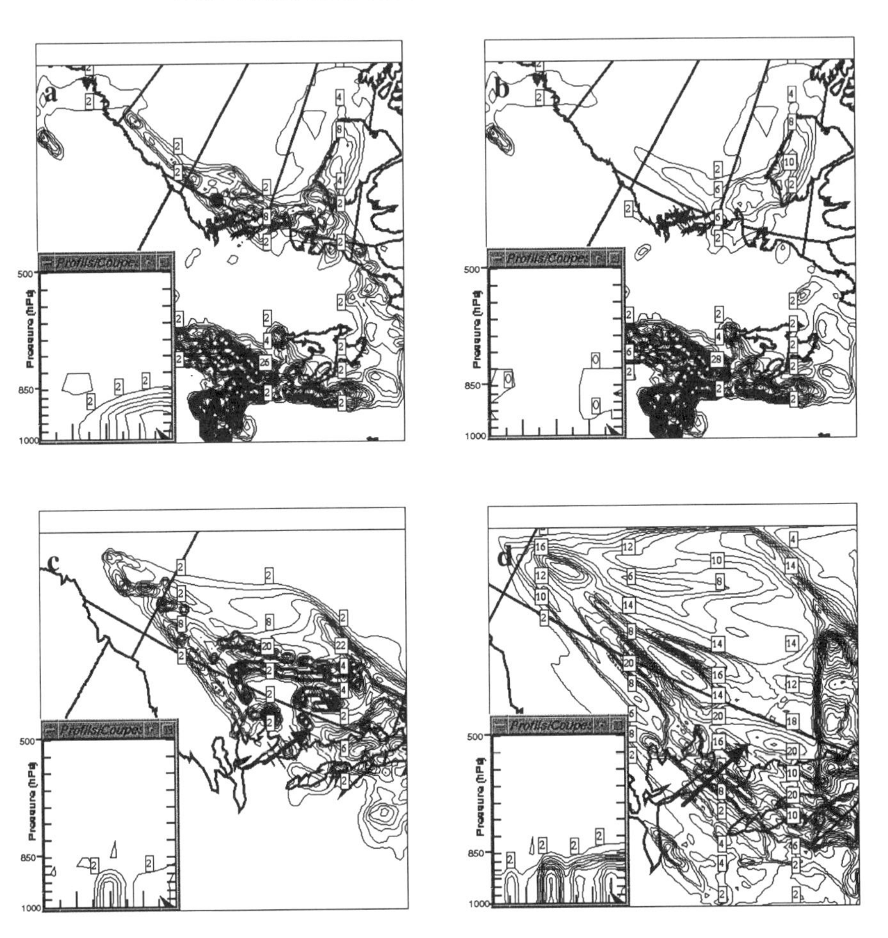

Fig. 9 MC2 simulation of the 950-hPa solid phase precipitation flux (mm/h) on 26 September 1994 at 1500 UTC (panels a and b) or 1600 UTC (panels c and d). Inset on each of the four panels is the vertical cross-section of vertical motion across Kugmallit Bay (common baseline given by arrow visible on panels c, d). Contours are every 0.2 mm/h and labels must be scaled by 0.1. a) 15-km MC2. Ice: SSMI-based analysis. b) 15-km MC2. Ice: Freeze-over scenario of the Beaufort Sea. c) 5-km MC2. Ice: SSMI-based analysis. d) 2-km MC2. Ice: SSMI-based analysis.

The results of the first 15-km resolution run were used as the boundary/initial conditions for a very-high resolution 12-h cascade run of the MC2 at 5 km around Tuktoyaktuk starting at 1200 UTC 26 September. Once again, output from the third run was used to initiate a fourth cascade run with a resolution of 2-km for the same period of time. The three windows used for those cascaded runs are shown on Fig. 9. A finer vertical resolution in the lower atmosphere has been used for the 2-km run to better resolve the shallow convective layer over the open water. For the three other experiments, the vertical distribution of the computational levels was identical to that of the BASE model configuration levels.

For all but the 2-km run, MC2 was run with the same cloud water scheme (Sundqvist et al., 1989) and Kuo convection scheme that were used during BASE. For the 2-km experiment, no subgrid-scale convective scheme was used to see how much of the convection would be resolved explicitly by the model dynamics.

Comparisons between the two 15-km runs, initiated with different ice representations, show the model sensitivity of the freeze-over of the Beaufort Sea in the presence of a strong north-westerly airflow. Figures 9a and 9b show that the intensity and the variability in the solid precipitation flux pattern is very sensitive to the presence of open water near the coast. Weaker coastal precipitation rates result from the freeze-over Beaufort Sea run, and distribution of the precipitation was kept only in the vicinity of the main cold front located over Banks Island and Amundsen Gulf. A vertical cross-section of the precipitation flux at 1500 UTC over Kugmallit Bay taken perpendicular to the prevailing air flow (shown in the lower left corner Fig. 9a) reveals that the 15-km BASE model was unable to resolve the observed snow band structures.

On the other hand, higher resolution runs (Figs 9c and d) display enhanced precipitation rates and banded-like precipitation patterns in the vicinity of the coast. We can see in Fig. 9d that the banded configuration of the precipitation flux is much better resolved by the 2-km version of the model, especially in the immediate vicinity of the coast of Kugmallit Bay. The vertical cross-section of the solid precipitation flux for the 2-km run (lower left corner of Fig. 9d), along the same baseline as for the lower resolution runs, displays a series of shallow (1000–850 hPa) maxima corresponding to the individual streamers. The 5-km run cross-section in Fig. 9c does not show such variability. It is concluded that the shallow convective circulation of the streamer clouds begins to be resolved explicitly by the MC2 dynamics on the 2-km mesh. A vertical cross-section of horizontal and vertical velocities through the streamers showed structures of mesoscale vortices very similar to the one described in observational studies.

6 Conclusion

An overall appraisal of the MC2 model version for BASE has been presented with the objective of investigating the model forecast ability for significant weather elements such as precipitation, surface winds, surface temperature and sea-level pressure. During the model selection activities and the field phase of the experiment the model has proven its skill in developing and sustaining mesoscale features even when initialized with synoptic-scale analyses only. Details in the simulated mesoscale circulation and three dimensional structures of mesoscale vortices were very similar to their characteristic features obtained from observational studies.

Comparisons of point model forecasts with observations have shown that a modest improvement is achieved in the accuracy of the forecast by using a high-resolution model when this model is initialized with a low-resolution analysis. The presence of a signal at the scale of the forecast model in the initial state of the atmosphere is believed to be very important to anchor the evolution of the mesoscale structures. The RDAS is based on RFE trial fields and statistics of the Canadian Global Model errors, and this naturally results in an analysis that is dynamically less coherent with the MC2 model scale than with the RFE model mesh.

Due to the shallow nature of the polar air masses, changes in the physical parametrization schemes are believed to be the prime factor of the increased performance of the BASE high-resolution model precipitation forecast. Higher resolution integrations, refined DEM and the absence of explicit mountain field smoothing result in an improvement of surface temperature forecast. Bias of the surface temperature forecast compared to the observed Stevenson screen temperature was shown to be directly proportional, by a factor equivalent to the lapse rate of a standard atmosphere, to the bias of the DEM to the actual terrain elevation.

Surface wind forecast statistical performance evaluation shows very little change with the use of a higher resolution model. Comparisons with the operational RFE model forecast have shown that the BASE model forecast results in a weak improvement in the wind direction forecast and a slight degradation of the overall performances of the 24-h wind speed forecast. However, the MC2's remarkably higher variability of forecast small-scale surface wind structures reinforces the argument in favour of the necessity of a higher resolution initial analysis at the scale of the model.

The RFE and MC2 models have performed equally well in their 24-h forecasts of the sea-level pressure and its 24-h tendency. This results again in an overall performance of point forecasts during the whole BASE experiment. Examination of sea-level pressure patterns (such as would be associated with the case studied in Section 5) have shown very realistic small scale features developed by the MC2 model that were absent from the RFE model forecasts. Even though these mesoscale structures are believed to be real (they have not been compared to high-resolution analyses so far), their high variability and amplitudes could lead to a greater departure from the observed values just because of their phase errors.

During the BASE runs, MC2 played one of its roles as a laboratory at RPN for the development of high-resolution physics by the testing of the cloud water scheme and with explicit simulation of small-scale convection. It is also expected that a cloud-resolving model will soon be developed with MC2.

The problem of data assimilation at the mesoscale, which ultimately is to be treated by the 4-dimensional variational technique, will be examined, in its simpler avenues, with the MC2. In view of the present paper, we consider that it is very important to experiment with existing analysis techniques in order to cover the great lack of finely-gridded initial conditions needed to reap the dynamical skill of the MC2.

These actions, based on the excellent model inherited from André Robert will contribute positively to the advancement of solving the problem of forecasting precipitation and other weather elements in Canada.

Acknowledgements

Special thanks are due to the Québec Weather Centre for giving S. Pellerin the opportunity to work on the development of the special numerical forecasting guidance for BASE. Also thanked is E. Hudson for the tremendous amount of data and information provided during the model post-analysis. Discussions with other personnel at RPN and CMC (Development and Operations) are gratefully acknowledged. The involvement of the BASE lead scientists and their appreciation of the modelling support is noted. The final version of the figures benefited from useful criticisms and patient support from Y. Chartier at RPN. Funding for this work comes principally from the Canadian GEWEX (Global Energy and Water Cycle Experiment).

References

BENOIT, R., M. DESGAGNÉ, P. PELLERIN, S. PELLERIN, Y. CHARTIER and S. DESJARDINS. 1997. The Canadian MC2: a semi-Lagrangian, semi-implicit wide-band atmospheric model suited for fine-scale process studies and simulation. *Mon. Weather Rev.* **125**

———; S. PELLERIN, J. MAILHOT, M. DESGAGNÉ and Y. CHARTIER. 1994. A real-time very-high-resolution forecast model for the field phase of the Beaufort and Arctic Storms Experiment (BASE). Paper 12A.5, 10th Conf. on Numerical Weather Prediction, July 1994, Portland, Oregon. Am. Meteorol. Soc., Boston, MA, pp. 451–453.

CHOUINARD, C.; J. MAILHOT, H. L. MITCHELL, A. STANIFORTH and R. HOGUE. 1994. The Canadian regional data assimilation system: Operational and research applications. *Mon. Weather Rev.* **122**: 1306–1325.

CRAWFORD, R.W. cited 1995. BASE Home Page. World Wide Web HTML document. Available on line from http://www.dow.on.doe.ca/BASE.

DENIS, B. 1990. Introduction de la topographie dans un modèle atmosphérique non hydrostatique. M. Sc. Thesis, Atmos. Sci., Physics Dept., UQAM, 121 pp, available from Département de physique, Université du Québec à Montréal, 1193 Phillips Square, Room C-3360, P. O. Box 8888, Stn "Downtown", Montréal (Québec), Canada H3C 3P8.

FOSTER, D.J. and R.D. DAVY. 1988. Global snow depth climatology, USAF Environmental Technical Applications Center, Note TN-88-006, 49 pp.

GAL-CHEN, T. and R. SOMERVILLE. 1975. On the use of a coordinate transformation for the solution of the Navier-Stokes equations. *J. Comput. Phys.* **17(2)**: 209–228.

HUDAK, D.R. 1994. A Storm Climatology of the Southern Beaufort Sea and the Mackenzie River Delta. Internal Report available from AES Canada, 4905 Dufferin St. Downsview, Ontario, Canada M3H 5T4.

———; R.E. STEWART, G.W.K. MOORE and E.T. HUDSON. 1995. Synoptic considerations of storms in the southern Beaufort Sea: Expectations For BASE. *In*: Proc. 4th Conf. Polar Meteorol. and Oceanog. 15–20 January 1995, Dallas, Texas. Am. Meteorol. Soc., pp. 234–237.

HUDSON, E. and R.W. CRAWFORD. 1995. Beaufort & Arctic Storms Experiment (BASE) Meteorology and field data summary. Cloud Physics Research Division, Environment Canada. 180 pp (approx.). Available form AES, 4905 Dufferin St. Downsview, Ontario, Canada M3H 5T4.

KELLOGG, W. and P. TWITCHELL. 1986. Summary of the Workshop on Arctic Lows 9-10 May 1985, Boulder, Colorado. Bull. Am. Meterol. Soc. **67(2)**: 186–193.

KUO, H.L. 1965. On formation and intensification of tropical cyclones through latent heat releases by cumulus convection. *J. Atmos. Sci.* **22**: 40–63.

———. 1974. Further studies of the parametrization of the influence of cumulus convection on large-scale flow. *J. Atmos. Sci.* **31**: 1232–1240.

LOUIS, J.F. 1981. ECMWF Forecast model documentation manual, Vol 1, Appendix 1, pp. A1.21-A1.33, ECMWF.

MAILHOT, J. 1994. The Regional Finite-Element (RFE) Model Scientific Description. Part 2: Physics. Documentation manual available from RPN.

——— and C. CHOUINARD. 1989. Numerical forecasts of explosive winter storms: Sensitivity experiments with a meso-alpha-scale model. *Mon. Weather Rev.* **117**: 1311–1343.

———; ———, R. BENOIT, M. ROCH, G. VERNER, J. CÔTÉ and J. PUDYKIEWICZ. 1989. Numerical forecasting of winter coastal storms during CASP: Evaluation of the regional finite-element model. ATMOSPHERE-OCEAN, **27**: 27–58.

———; R. SARRAZIN, B. BILODEAU, N. BRUNET, A. MÉTHOT, G. PELLERIN, C. CHOUINARD, L. GARAND, C. GIRARD and R. HOGUE. 1995. Changes to the Canadian regional forecast system: Description and evaluation of the 50-km version. ATMOSPHERE-OCEAN, **33**: 55 80.

MATVEEV, L.T. 1984. *Cloud dynamics*. D. Reidel Publishing Co., 340 pp.

PARKER, N.; E. HUDSON, R. BENOIT and M. ROCH. 1990. Polar low handbook for Canadian waters. Atmospheric Environment Service Report, 106 pp.

SUNDQVIST, H. 1993. Inclusion of ice phase of hydrometeors in cloud parametrization for mesoscale and large scale models. *Beitr. Phys. Atmos.* **66**: 137–147.

———; E. BERGE and J.E. KRISTJANSSON. 1989. Condensation and cloud parametrization studies with a mesoscale numerical weather prediction model. *Mon. Weather Rev.* **117**: 1641–1657.

TANGUAY, M.; A. ROBERT and R. LAPRISE. 1990. A semi-implicit semi-Lagrangian fully compressible regional forecast model. *Mon. Weather Rev.* **118**: 1970–1980.

TREMBLAY, A.; A. GLAZER, W. SZYRMER and W. YU. 1994. Mesoscale simulations of cloud microphysics within a winter storm: some implications for large-scale ice phase schemes. *In*: Proc. ECMWF IGCSS Workshop on modelling and assimilation of clouds. 31 October – 4 November 1994, Reading, U.K. pp. 337–346.

TWITCHELL, P.F.; E.A. RASMUSSEN and K.L. DAVIDSON. 1989. *Polar and Arctic Lows*. A. Deepak Publ., Hampton, Virginia, 421 pp.

WHITTEKER, J. 1992. Storage specifications for the 500-metre-grid terrain data. Communications Research Centre Report RP-92-002. Ottawa, 16 pp.

ZAWADZKI, I.; L. OSTIGUY and R. LAPRISE. 1993. Retrieval of the microphysical properties in a CASP storm by integration of numerical kinematical model. ATMOSPHERE-OCEAN, **31**: 201–233.

Preliminary Results from a Dry Global Variable-Resolution Primitive Equations Model

Jean Côté*, Sylvie Gravel*, André Méthot†,
Alain Patoine†, Michel Roch* and Andrew Staniforth*
*Recherche en prévision numérique et †Centre météorologique canadien
Service de l'environnement atmosphérique, 2121 Route Transcanadienne,
porte 500 — Dorval, Québec, Canada H9P 1J3

[Original manuscript received 4 February 1995; in revised form 17 July 1995]

ABSTRACT *The viability of a proposed global variable-grid strategy was previously tested (Côté et al., 1993) using a shallow-water-equations prototype. It was demonstrated that by using a global variable-resolution mesh, a high-resolution short-term forecast can be obtained for a region of interest at a fraction of the cost of using uniformly-high resolution everywhere. The prototype is generalized here to use the hydrostatic primitive equations. Preliminary results obtained with an adiabatic version of this baroclinic model are presented. They confirm the potential of the proposed strategy: differences between the 48-h forecasts for the 500 hPa geopotential height and mean-sea-level pressure fields obtained from a uniform 1.2° model, and those obtained from a variable-mesh model with equivalent resolution on a 81.6° × 60° sub-domain, are small.*

RÉSUMÉ *Précédemment (Côté et al., 1993), un modèle modial des équations de St-Venant a permis de tester la viabilité d'une stratégie à résolution variable et démontrer qu'une prévision de courte échéance à haute résolution peut être obtenue pour une région donnée à une fraction du coût de la prévision effectuée à résolution uniforme équivalente. Ici, ce prototype est généralisé aux équations primitives hydrostatiques. Les résultats préliminaires de la version adiabatique de ce modèle barocline sont présentés. Ils confirment le potentiel de la stratégie que nous proposons: les différences entre des prévisions de 48 heures du géopotentiel à 500 hPa et de la pression au niveau moyen de la mer obtenues avec le modèle à une résolution uniforme de 1,2°, et celles obtenues avec le modèle à résolution variable ayant une résolution équivalente sur une fenêtre de 81,6° × 60°, sont petites.*

1 Introduction

A new model is being developed to meet the operational weather forecasting needs of Canada for the coming years. These presently include short-range regional forecasting, medium-range global forecasting and data assimilation. In the future they will probably include nowcasting at the mesoscale and dynamic extended-range forecasting on monthly to seasonal time-scales.

The Canadian Meteorological Centre currently runs two operational data assimilation and forecasting cycles, as do a number of other national weather forecasting centres. The first of these addresses medium-range needs by using a global data assimilation system (Mitchell et al., 1993) and a global spectral forecast model (Ritchie and Beaudoin, 1994). It also initializes the 12-h regional data assimilation spin-up system (Chouinard et al., 1994) of the second cycle, which provides the higher-resolution regional analyses used by the variable resolution finite-element model (Mailhot et al., 1994) to produce more detailed short-range (<2 days) forecasts over N. America and some of its adjacent waters. This two-cycle strategy requires the maintenance, improvement and optimization of two sets of libraries and procedures. It has motivated the definition of a new strategy to consolidate these two complementary aspects of weather forecasting within a single model framework, and thereby rationalize resource utilization.

The essence of the approach (Côté et al., 1993) is to develop a single highly efficient model that can be reconfigured at run time to *either* run globally at uniform resolution (with possibly degraded resolution in the southern hemisphere), *or* to run with variable resolution over a global domain such that high resolution is focused over an area of interest. To demonstrate the potential of this strategy, a shallow-water global variable-resolution prototype has been developed (Côté et al., 1993) that uses an arbitrarily rotated latitude-longitude mesh. In this way a high-resolution subdomain can be focused on any geographical area of interest, making it a more flexible strategy than the operational finite-element regional model, which is best suited to extra-tropical applications. This shallow water model has been successfully integrated on a variety of meshes right down to the meso-gamma scale (with 250 m resolution over a 100 km × 100 km subdomain). An important conclusion is that the overhead of using a model of global extent for short-range forecasting over an area of interest is no worse than that associated with the sponge regions of traditional limited-area models.

The goal of this paper is to briefly describe a generalization of the shallow water formulation to the dry hydrostatic primitive equations, and to give some preliminary results from this baroclinic prototype. This is an important step towards the ultimate goal, a full-physics non-hydrostatic global Euler equations model using a pressure-type hybrid vertical coordinate (Laprise, 1992).

2 Model formulation

a *Governing Equations*

The governing equations (Kasahara, 1974; Simmons and Burridge, 1981) are the hydrostatic primitive equations for frictionless adiabatic flow on a rotating sphere:

$$\frac{d\mathbf{V}}{dt} + RT\,\nabla \ln p + \nabla \phi + f(\mathbf{k} \times \mathbf{V}) = 0, \tag{1}$$

$$\frac{d}{dt}\ln\left|\frac{\partial p}{\partial \eta}\right| + \nabla \cdot \mathbf{V} + \frac{\partial \dot{\eta}}{\partial \eta} = 0, \tag{2}$$

$$\frac{d}{dt}[\ln(p^{-\kappa}T)] = 0, \tag{3}$$

$$\frac{\partial \phi}{\partial \eta} = -RT\frac{\partial \ln p}{\partial \eta}, \tag{4}$$

where

$$\frac{d}{dt} = \frac{\partial}{\partial t} + \mathbf{V} \cdot \nabla + \dot{\eta}\frac{\partial}{\partial \eta}, \tag{5}$$

and

$$\eta = \frac{p - p_T}{p_S - p_T}, \tag{6}$$

is the terrain-following vertical coordinate. In the above, (1)–(4) are respectively the horizontal momentum, continuity, thermodynamic and hydrostatic equations. Also: $\mathbf{V}$ is the horizontal velocity, T is temperature, p is pressure, p_S and p_T are its respective values at the surface and at the top of the model atmosphere, $\phi = gz$ is the geopotential height, f is the Coriolis parameter, $\kappa = R/c_p$, and R and c_p are respectively the gas constant and specific heat of dry air.

The boundary conditions are periodicity in the horizontal; and no motion across the top and bottom of the atmosphere, where the top is at constant pressure p_T. Thus

$$\dot{\eta} \equiv \frac{d\eta}{dt} = 0 \text{ at } \eta = 0, 1. \tag{7}$$

b *Temporal Discretization*

As a preparatory step for the time discretization, the thermodynamic equation (3) is rewritten as

$$\frac{d}{dt}\left[\ln\left(\frac{T}{T^*}\right) - \kappa \ln\left(\frac{p}{p^*}\right)\right] - \kappa\dot{\eta}\frac{d}{d\eta}(\ln p^*) = 0, \tag{8}$$

where

$$p^* = p_T^* + (p_S^* - p_T^*)\eta, \tag{9}$$

and p_S^* and $p_T^* \equiv p_T$ are the respective bottom and top constant pressures of a motionless isothermal ($T^* \equiv$ constant) reference atmosphere. Equation (9) is a direct consequence of the definition (6) of the vertical coordinate, and (8) is written this way to ensure the computational stability of the gravitational oscillations.

The time discretization is fully-implicit/semi-Lagrangian. This was inspired by the introduction by André Robert of the semi-implicit semi-Lagrangian scheme (Robert, 1981, 1982) and its further development (see Staniforth and Côté, 1991; Staniforth, this volume). Indeed, half the papers in this André J. Robert memorial volume were also inspired in part by his work on the subject (Benoit et al.; Cullen et al.; Daley; Geleyn and Bubnova, Kaas et al.; Laprise et al.; Leslie and Purser; Machenhauer and Olk; McGregor; Ritchie; Smolarkiewicz and Margolin; Temperton; Williamson).

Consider a prognostic equation of the form

$$\frac{dF}{dt} + G = 0, \tag{10}$$

where F represents one of the prognostic quantities $\{\mathbf{V}, \ln|\partial p/\partial\eta|, \ln(T/T^*) - \kappa \ln(p/p^*)\}$, and G represents the remaining terms, some of which are nonlinear. Such an equation is approximated by time differences and weighted averages along a trajectory determined by an approximate solution to

$$\frac{d\mathbf{x}_3}{dt} = \mathbf{V}_3(\mathbf{x}_3, t), \tag{11}$$

where $\mathbf{x}_3$ and $\mathbf{V}_3$ are the three-dimensional position and velocity vectors respectively. Thus

$$\frac{(F^n - F^{n-1})}{\Delta t} + \left(\frac{3}{4}G^n + \frac{1}{4}G^{n-2}\right) = 0, \tag{12}$$

where $\psi^n = \psi(\mathbf{x}_3, t)$, $\psi^{n-m} = \psi[\mathbf{x}_3(t - m\Delta t), t - m\Delta t]$, $\psi = \{F, G\}$, $t = n\Delta t$. The upstream position of the trajectory at time $t - \Delta t$ is computed (Temperton and Staniforth, 1987; McDonald and Bates, 1987) using winds extrapolated at meshpoints from times $t - \Delta t$ and $t - 2\Delta t$ to time $t - \Delta t/2$. The upstream position of the trajectory at time $t - 2\Delta t$ is then obtained (Rivest et al., 1994) by extending the trajectory back along a great circle.

Note that this scheme is decentred along the trajectory as in Rivest et al. (1994), in anticipation of the introduction of orography (not included in the results of the present paper). In this way the spurious resonant response arising from a centred approximation in the presence of orography will be avoided. This decentred scheme nevertheless shares the $O(\Delta t^2)$ accuracy of the more usual centred schemes: an $O(\Delta t)$ decentred scheme would also address the spurious resonance problem but would unduly limit the maximum permissible time step for reasons of accuracy (Rivest et al., 1994). Cubic interpolation is used everywhere for upstream evaluations (c.f. 12) except for the trajectory computations (c.f. 11), where linear interpolation is used with no visible degradation in the results.

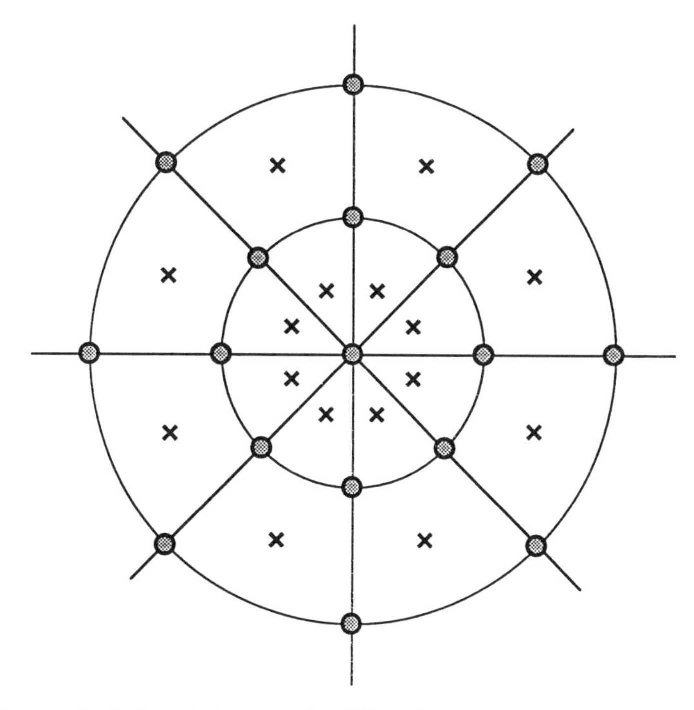

Fig. 1 Schematic for horizontal placement of variables. Circled points: p, ϕ. Crossed points: $\mathbf{V}$, T, $\dot{\eta}$.

Grouping terms at the new time on the left-hand side and known quantities on the right-hand side, (12) may be rewritten as

$$\left(F + \frac{3}{4}\Delta t G\right)^n = F^{n-1} - \frac{1}{4}\Delta t G^{n-2}. \tag{13}$$

This yields a set of coupled nonlinear equations for the unknown quantities at the meshpoints of a regular grid at the new time t, the efficient solution of which is discussed below. A fully-implicit time treatment, such as that adopted here, of the nonlinear terms has the useful property of being inherently computationally more stable than an explicit one (e.g. those of Bates et al., 1993 and McDonald and Haugen, 1992, whose computational stability is analysed in Gravel et al., 1993).

c *Spatial Discretization*

A variable-resolution finite-element discretization, based on that described in Côté et al. (1993), is used in the horizontal with a placement of variables as shown schematically in Fig. 1. It has the advantage that only one set of trajectories is required since all of the right-hand sides of the equations are evaluated at cell centres before interpolation. Other placements are, however, possible and are currently

TABLE 1. Experimental configurations.

Experiment	Mesh Uniform/Variable	Coordinate System Rotated?
1	Uniform	No
2	Uniform	Yes
3	Variable	Yes

being examined. The vertical discretization is modelled after that of Tanguay et al. (1989).

d *Solving the Coupled Nonlinear Set of Discretized Equations*
After spatial discretization the coupled set of nonlinear equations still has the form of (13). Terms on the right-hand side, which involve upstream interpolation, are evaluated once and for all. The coupled set is rewritten as a linear one (where the coefficients depend on the basic state) plus a perturbation which is placed on the right-hand side and which is relatively cheap to evaluate. This set is then solved iteratively using the linear terms as a kernel, and the nonlinear terms on the right-hand sides are re-evaluated at each iteration using the most-recent values. The linear set can be algebraically reduced to the solution of a three-dimensional elliptic-boundary-value problem. In practice the cost of solving the coupled set of nonlinear equations is only marginally more expensive than the iterative solution of the variable-coefficient linear set. The most significant contribution to the present cost of a time step is that of interpolation, which is the same regardless of whether the coupled set of equations is linear or nonlinear.

3 Preliminary results

a *Methodology*
To make a preliminary assessment of the global variable-mesh strategy described above, we performed three 48-h integrations starting from the same initial data, the Canadian Meteorological Centre analysis valid at 12 UTC 12 February 1993. For all integrations, the (adiabatic) model was run with 23 vertical levels ($p_T = 10$ hPa) using a 30-min time step and a Laplacian diffusion with a coefficient of 1.5×10^5 m^2 s^{-1}. There was no topography and there were no heat or momentum fluxes.

The configurations of the three experiments are summarized in Table 1. The purpose of the uniform-resolution experiments is to validate a uniform-resolution ground truth (Expt. 2) against which the variable-resolution forecast of Expt. 3 can be compared and evaluated. Both uniform-resolution experiments were performed at 1.2° resolution (in both longitude and latitude), using either a mesh with the poles of the coordinate system coincident with respect to the geographical ones (Fig. 2a), or rotated (i.e., oriented as in Fig. 2b, but with resolution uniform everywhere). Note that for pictorial clarity, only every second latitude and longitude is plotted in Fig. 2. Ideally it would have been preferable to run these experiments at the

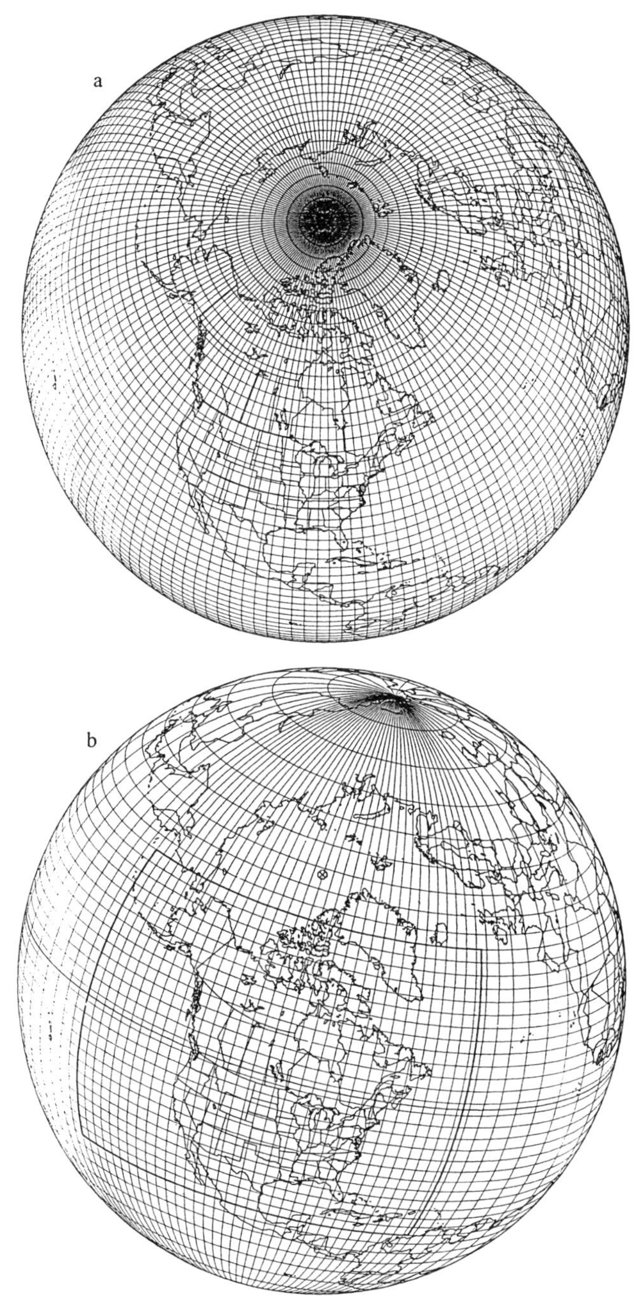

Fig. 2 (a) The uniform 1.2° resolution 300 × 151 mesh used for Expt. 1 and (after rotation) Expt. 2: for clarity only every 2nd point in each direction is plotted. (b) A variable-resolution 119 × 87 mesh having an 81.6° × 60° window of uniform 1.2° resolution, centred on (103°W, 51°N), and used for Expt. 3: for clarity only every 2nd point in each direction is plotted.

same 0.5° resolution as for the analogous ones presented in Côté et al. (1993) using the shallow-water prototype. However, at the present stage of development the use of computer memory by the model has not yet been sufficiently optimized to permit this for the uniform grid. This limits the agreement between the results of the experiments due to the more rapid growth of the spatial truncation errors outside the window of uniform resolution, and their advection into the region of interest.

The variable-resolution Expt. 3 was performed on the mesh depicted in Fig. 2b, where the poles of the coordinate system are rotated with respect to the geographical ones. Its purpose is to demonstrate our thesis that we can reasonably well (given the above-mentioned current limit on resolution) reproduce the forecast over the 1.2° uniform resolution window at a fraction of the cost of using 1.2° uniform resolution everywhere. The resolution of the variable mesh degrades smoothly away in each direction (each successive meshlength is approximately 11% larger than its predecessor) from an 81.6° × 60° uniform-resolution (1.2°) window centred on a point of the equator of a rotated coordinate system, located at (103°W, 51°N) in geographical coordinates. Uniform resolution again refers to uniform spacing in latitude and longitude: however, the meshpoints of the window are also almost uniformly spaced over the sphere with a meshlength that varies between approximately 114 and 132 km.

b *Uniform Resolution Experiments*

The 500-hPa height and mean sea level pressure (mslp) fields of the initial analysis used for the experiments are shown in Fig. 3. The resulting 2-day forecasts of these fields for Expts 1 and 2 (i.e., using uniform 1.2° resolution on the unrotated and rotated meshes) are shown in Figs 4 and 5 respectively. The global r.m.s. forecast differences (computed after interpolating the forecast of Expt. 1 to the rotated mesh of Expt. 2) are small (4.9 m and 0.6 hPa respectively), showing that the effect of rotating the mesh while keeping its resolution uniform is small.

c *Variable Resolution Experiment*

The integration of Expt. 2 (i.e., uniform resolution everywhere in the rotated coordinate system) is considered to be the ground truth for the purposes of validating the 48-h forecast of the variable resolution integration (Expt. 3): the meshes of both integrations are identical over the uniform resolution window of Fig. 2b. The 2-day variable resolution forecast is shown in Fig. 6 and may be compared to that of the control (Fig. 5). The two forecasts (Expts 2 vs. 3) are quite close over the uniform resolution area of interest (defined by the curvilinear rectangle of Fig. 6a). This confirms the thesis that the forecast over the 1.2° uniform resolution window can be well reproduced at a fraction of the cost of using 1.2° uniform resolution everywhere. However, they are significantly different over areas of low resolution, as indeed they should be. Quantifying this, the global r.m.s. differences between the forecasts of Expts 2 and 3 are 26.8 m and 3.4 hPa for the 500-hPa height

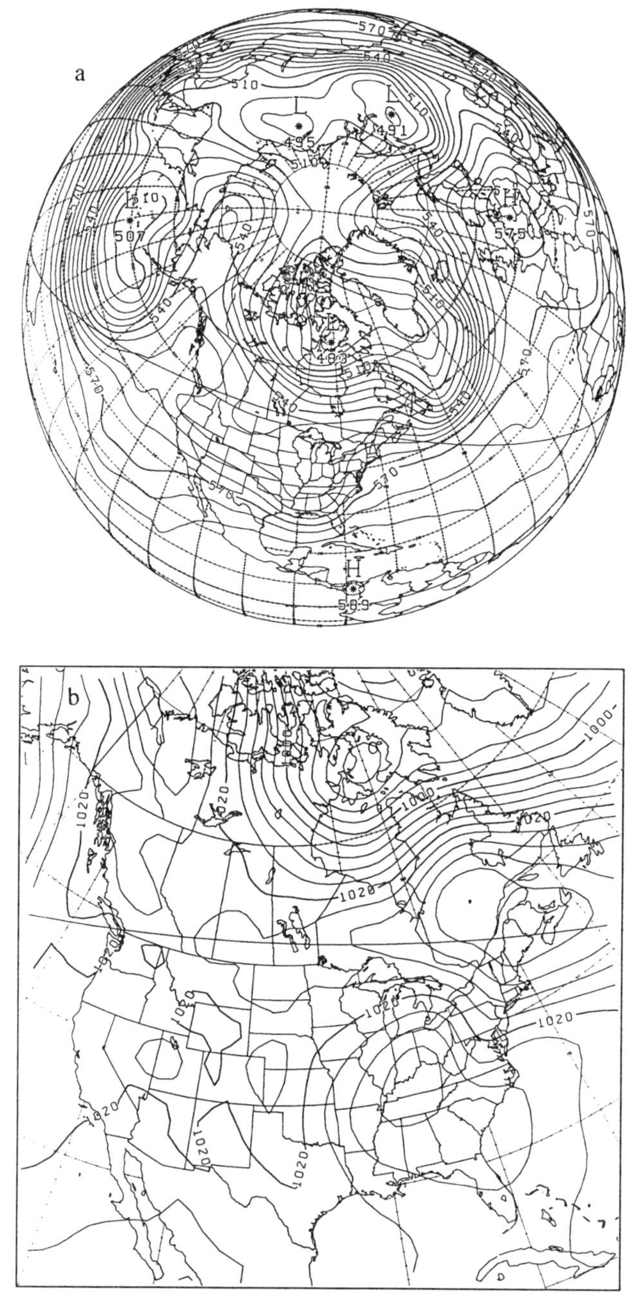

Fig. 3 (a) Initial geopotential height at 500 hPa in dam on an orthographic projection; contour interval = 6 dam. (b) Initial mslp in hPa for a N. American window on a polar sterographic projection; contour interval = 4 hPa.

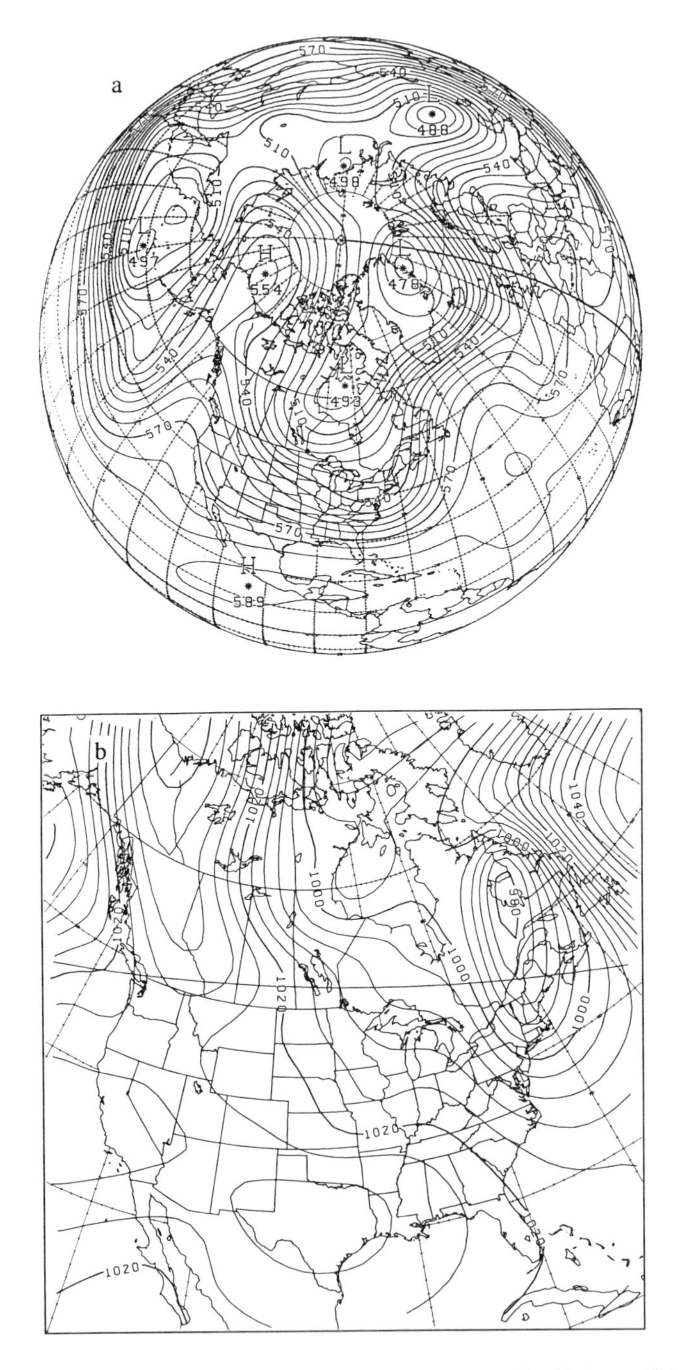

Fig. 4 (a) Same as in Fig. 3a, but at 48-h for Expt. 1. (b) Same as in Fig. 3b, but at 48-h for Expt. 1.

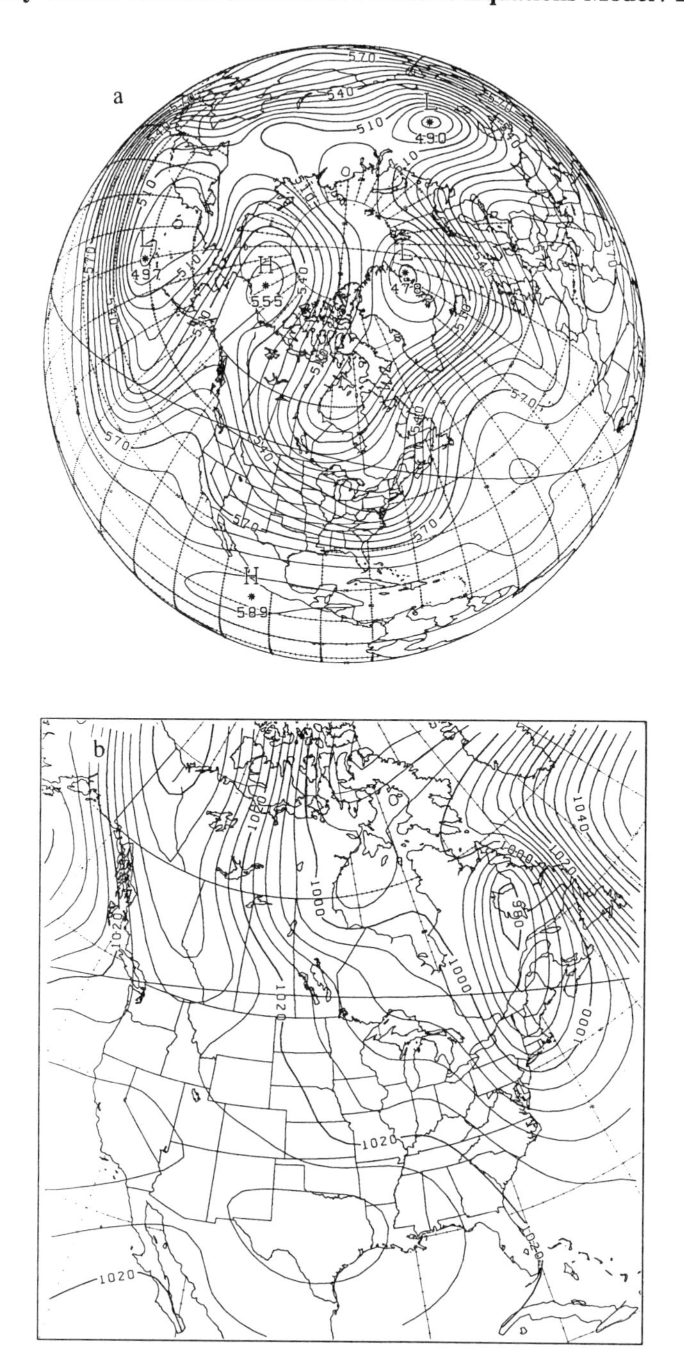

Fig. 5 (a) Same as in Fig. 3a, but at 48-h for Expt. 2. (b) Same as in Fig. 3b, but at 48-h for Expt. 2.

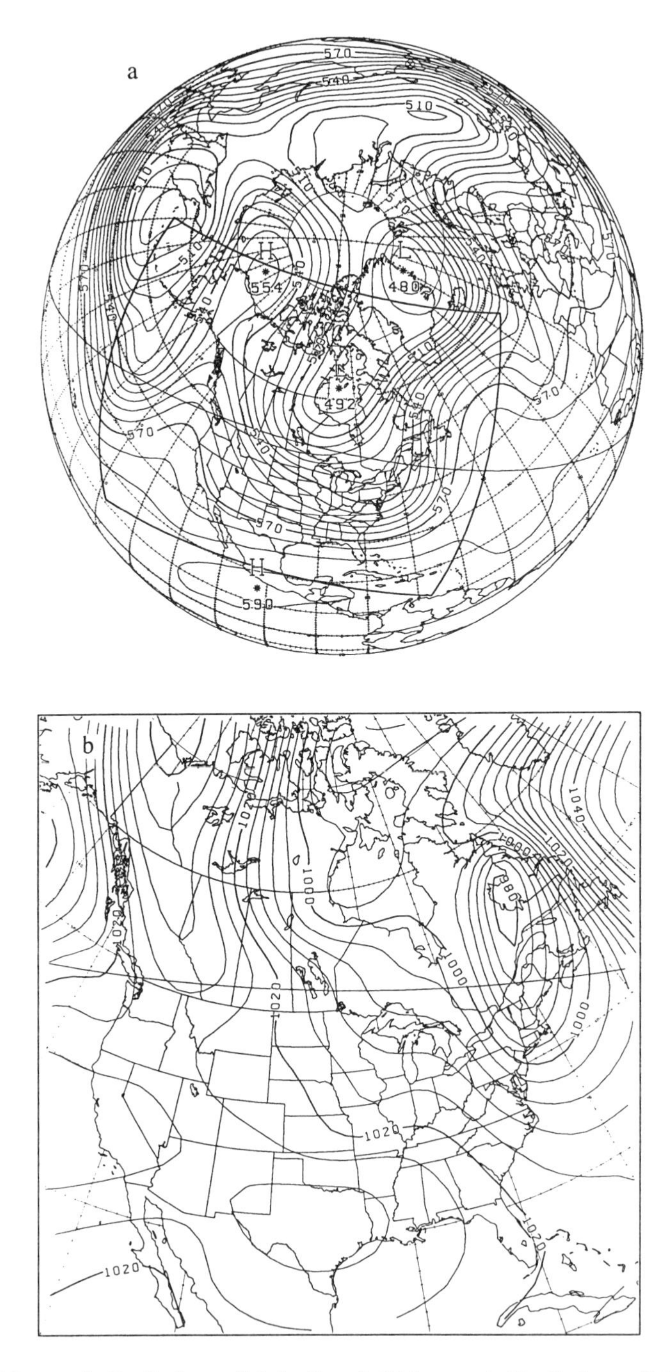

Fig. 6 (a) Same as in Fig. 3a, but at 48-h for Expt. 3. (b) Same as in Fig. 3b, but at 48-h for Expt. 3.

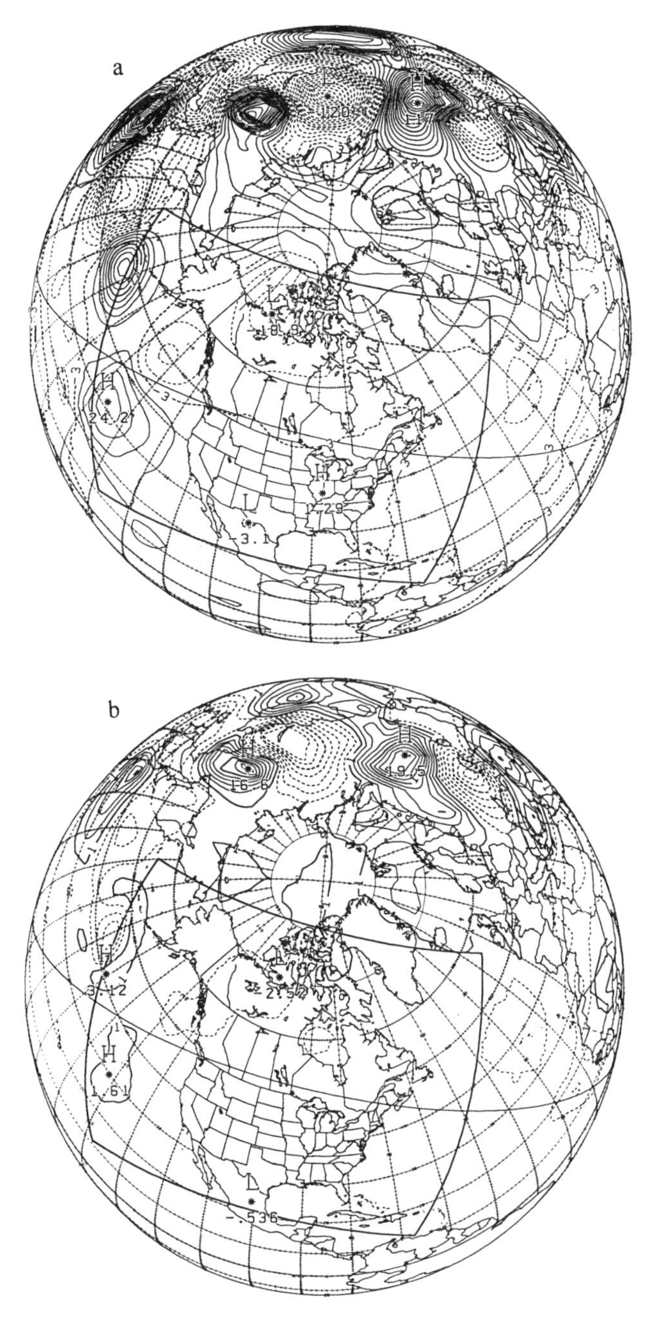

Fig. 7 Difference between 48-h forecasts of Expt. 2 and Expt. 3 on an orthographic projection for: (a) 500 hPa geopotential height; contour interval = 6 m. (b) mslp; contour interval = 2 hPa.

and mslp fields respectively, whereas they are only 7.0 m and 0.6 hPa over the curvilinear rectangle, where the meshpoints of the two grids are coincident.

Note that the spatial truncation errors associated with the variable resolution portion of the model's grid propagate with the speed of the local wind. This has to be taken into account when defining a uniform resolution region of interest for the model. It has to be sufficiently large, so that the entire region is not unduly contaminated by the error advected from the variable portion of the grid during the time of integration. It is therefore a compromise between the width of this region and the length of the run. The differences between the forecasts of Expts 2 and 3 increase as a function of the proximity to the upstream boundaries of the uniform resolution window, due to the inflow from the coarser resolution outer domain (see Fig. 7). They would have been substantially smaller had it been possible to run the experiments at 0.5° resolution instead of 1.2°.

4 Future work and conclusion

All of the attributes (output program, postprocessors, time series, zonal diagnostic extractors, etc.) of a modern operational model are being added to the dynamical core of the described model, including topography. The physical parameterizations currently used by both Canadian operational models are being introduced into the new model using a plug-compatible approach, and each physical process will be individually tested. Once this has been successfully accomplished the model could then be considered as a replacement model at the Canadian Meteorological Centre for both operational models, the global spectral and the regional finite-element models. The next stage of development would then be to complete the coding of a non-hydrostatic version and to fully evaluate it for mesoscale applications where the hydrostatic assumption breaks down.

The preliminary results shown in the present work are encouraging. The strategy that was developed in a 2D prototype and recently extended to 3D shows promising results in progressing towards a unified data assimilation and forecast system, at the heart of which lies a single multipurpose and multiscale numerical model.

Acknowledgements

We gratefully acknowledge the helpful comments of the editorial committee and an anonymous reviewer.

References

BATES, J.R.; S. MOORTHI and R.W HIGGINS. 1993. A global multilevel atmospheric model using a vector semi-Lagrangian finite-difference scheme. Part I: Adiabatic formulation. *Mon. Weather Rev.* **121**: 244–263.

CHOUINARD, C.; J. MAILHOT, H.L. MITCHELL, A. STANIFORTH and R. HOGUE. 1994. The Canadian regional data assimilation system: operational and research applications. *Mon. Weather Rev.* **122**: 1306–1325.

CÔTÉ, J.; M. ROCH, A. STANIFORTH and L. FILLION. 1993. A variable-resolution semi-Lagrangian finite-element global model of the shallow-water equations. *Mon. Weather Rev.* **121**: 231–243.

GRAVEL, S.; A. STANIFORTH and J. CÔTÉ. 1993. A stability analysis of a family of baroclinic semi-Lagrangian forecast models. *Mon. Weather Rev.* **121**: 815–824.

KASAHARA, A. 1974. Various vertical coordinate systems used for numerical weather prediction. *Mon. Weather Rev.* **102**: 509–522.

LAPRISE, R. 1992. The Euler equations of motion with hydrostatic pressure as independent variable. *Mon. Weather Rev.* **120**: 197–207.

MAILHOT, J.; R. SARRAZIN, B. BILODEAU, N. BRUNET, A. MÉTHOT, G. PELLERIN, C. CHOUINARD, L. GARAND, C. GIRARD and R. HOGUE. 1994. Changes to the Canadian regional forecast system – description and evaluation of the 50 km version. ATMOSPHERE-OCEAN, **33**: 55–80.

MCDONALD, A. and J.R. BATES. 1987. Improving the estimate of the departure point position in a two-time-level semi-Lagrangian and semi-implicit model. *Mon. Weather Rev.* **115**: 737–739.

——— and J.E. HAUGEN. 1992. A two-time-level, three-dimensional semi-Lagrangian, semi-implicit, limited-area gridpoint model of the primitive equations. *Mon. Weather Rev.* **120**: 2603–2621.

MITCHELL, H.L.; C. CHARETTE, S.J. LAMBERT, J. HALLÉ and C. CHOUINARD. 1993. The Canadian global data assimilation system: Description and evaluation. *Mon. Weather Rev.* **121**: 1467–1492.

RITCHIE, H. and C. BEAUDOIN. 1994. Approximations and sensitivity experiments with a baroclinic semi-Lagrangian spectral model. *Mon. Weather Rev.* **122**: 2391–2399.

RIVEST, C.; A. STANIFORTH and A. ROBERT. 1994. Spurious resonant response of semi-Lagrangian discretizations to orographic forcing: Diagnosis and solution. *Mon. Weather Rev.* **122**: 366–376.

ROBERT, A. 1981. A stable numerical integration scheme for the primitive meteorological equations. ATMOSPHERE-OCEAN, **19**: 35–46.

———. 1982. A semi-Lagrangian and semi-implicit numerical integration scheme for the primitive meteorological equations. *J. Meteorol. Soc. Jpn.* **60**: 319–325.

SIMMONS, A.J. and D.M. BURRIDGE. 1981. An energy and angular-momentum conserving vertical finite-difference scheme and hybrid vertical coordinate. *Mon. Weather Rev.* **109**: 758–766.

STANIFORTH, A. and J. CÔTÉ. 1991. Semi-Lagrangian integration schemes for atmospheric models – a review. *Mon. Weather Rev.* **119**: 2206–2223.

TANGUAY, M.; A. SIMARD and A. STANIFORTH. 1989. A three-dimensional semi-Lagrangian scheme for the Canadian regional finite-element forecast model. *Mon. Weather Rev.* **117**: 1861–1871.

TEMPERTON, C. and A. STANIFORTH. 1987. An efficient two-time-level semi-Lagrangian semi-implicit integration scheme. *Q. J. R. Meteorol. Soc.* **113**: 1025–1039.

The Fully-Elastic Equations Cast in Hydrostatic-Pressure Coordinate: Accuracy and Stability Aspects of the Scheme as Implemented in ARPEGE/ALADIN

Jean-François Geleyn
Centre National de Recherches Météorologiques
Météo-France, Toulouse, France

and

Radmila Bubnová[1]
Czech Hydrometeorological Institute
Prague, Czech Republic

[Original manuscript received 9 December 1994; in revised form 21 August 1995]

ABSTRACT *A novel form of the fully elastic equations cast in hydrostatic-pressure terrain-following coordinate has been used to develop a non-hydrostatic version of the already existing NWP model ARPEGE/ALADIN. While the main features of the implemented elastic dynamics and first validation results have already been published (Bubnová et al., 1995), some specific details of the numerical application are emphasized in this paper. The complexity of the elastic dynamics puts additional constraints on the vertical discretization scheme which, together with the semi-implicit scheme design, leads to a particular choice of "non-hydrostatic" prognostic variables. Further, the necessary nonlinear implicit correction, introduced via iterations of the semi-implicit step to ensure the stability, is outlined. Problems related to the surface boundary condition are discussed and some experiments on idealized nonlinear non-hydrostatic mountain flow have been done to test the precision of the scheme. Results of a real data simulation of lee waves in a pseudo-3D form are also presented.*

RÉSUMÉ *On utilise une forme originale des équations entièrement élastiques exprimées dans une coordonnée de pression hydrostatique épousant le terrain pour élaborer une version non hydrostatique du modèle de prévision courant ARPEGE/ALADIN. Bien que les principaux éléments de la dynamique élastique appliquée et les premiers résultats de validation aient déjà été publiés (Bubnovà et al., 1995), on souligne ici des détails spécifiques de l'application numérique. La complexité de la dynamique élastique entraîne des contraintes*

[1]Currently on partial leave of absence to CNRM, Météo-France, Toulouse

supplémentaires pour le schéma de discrétisation verticale, qui, avec le calcul du schéma semi-implicite, amène à un choix particulier de variables pronostiques non hydrostatiques. De plus, on présente une correction non linéaire implicite nécesssaire, effectuée au moyen d'itérations du pas semi-implicite pour assurer la stabilité. Les problèmes associés à la condition à la limite inférieure sont examinés; quelques essais sur des écoulements de montagne non linéaires non hydrostatiques ont été effectués pour vérifier la précision du schéma. On présente aussi les résultats d'une simulation réelle des ondes de sillage dans une forme quasi-tridimensonnelle.

---

1 Introduction

The present paper is not aimed at giving a full description of the work accomplished when implementing, in the ARPEGE/ALADIN (Action de Recherche Petite Echelle Grande Echelle/Aire Limitée Adaption Dynamique développement International) code, the fully-elastic Euler equations in hydrostatic-pressure terrain-following coordinate suggested by Laprise (1992), hereafter L92. For that purpose the reader is referred to Bubnová et al. (1995), hereafter BHBG95. Owing to the topic of this special issue, our goal will rather be to identify and study in detail the areas where specific choices in numerical algorithms have had a clear impact on the accuracy and/or the stability of the solution obtained. Since the development is still a rather fresh one, some minor aspects already differ from what was reported in the basic paper BHBG95 and may still evolve slightly in the future. Nevertheless it is our belief that the validation, also partly reported here, is now sufficiently advanced to make a few statements with confidence. Yes, the solution advocated in L92, when Laprise translated the lessons of the "MC2" (Mesoscale Compressible Community (model)) development (Tanguay et al., 1990; Denis, 1990) to the more classical "pressure-type vertical coordinate" models, is workable and can be implemented at reasonable development and computing costs. No, it is not so simple as was suggested by L92 when the following two constraints have to be obeyed: firstly, leaving untouched all or most of the characteristics of an already existing classical "pressure-type terrain-following vertical coordinate" hydrostatic model; secondly, introducing the elastic version as a fully switchable option, all things being otherwise equal. The consequences of those constraints and the numerical framework in which this study was performed will be described in Section 2. Section 3 will deal with the above-mentioned central numerical aspects. Section 4 will be dedicated to the present stage of progress of the validation, leading to most of the conclusions and indications for future work outlined in Section 5.

2 Specific scope of the reported work

For the reader who is unaware of Laprise's (1992) suggestion, let us recall its three main points: (i) using as vertical coordinate a suitable monotonous function of the weight of air above the point being considered (hydrostatic pressure), it is possible to cast the fully-elastic Euler equations in a form that degenerates back to the so-called primitive equations when cancelling the terms associated with the propagation of sound waves; (ii) despite the associated introduction of two new prognostic variables (for instance real pressure and vertical velocity), fast propagating waves can still be controlled by the same type of semi-implicit (or semi-implicit semi-Lagrangian) algorithm as the one introduced by Robert (1969) for primitive equations; (iii) the vertical/horizontal separability implicit to this algorithm allows an elegant introduction of the orographic effects via the choice of a terrain-following (e.g. sigma- or hybrid-) type for the above-mentioned monotonous function of the hydrostatic pressure, leading to an unambiguous definition of the surface boundary condition.

We shall see in the following section that each of these three items was at the same time a solid guideline and a source of theoretical and practical problems when implementing the scheme. Those problems may have arisen from unwanted effects of the combination of our separate choices (a quite unlikely, but nevertheless possible, situation) or from two basic constraints imposed on us by the framework in which the development had to take place. The fact that the ALADIN limited-area model belongs to the "family" of the ARPEGE/IFS (Integrated Forecasting System) code (Courtier et al., 1991) led us to formulate the vertical discretization in one of the two most used Lorenz-type vertical coordinate (Simmons and Burridge, 1981; hereafter SB81) while things may (or may not) have been easier with the formulation of Arakawa and Suarez (1983) or in a Charney-Phillips type of coordinate. The spectral character of the ALADIN model, following Machenhauer and Haugen (1987) in order to keep full compatibility with ARPEGE/IFS, requires the inversion of the Helmholtz operator to happen in spectral space where it becomes trivially diagonal. This requires, however, that all fast waves be solutions of the associated linearized set of equations. It is as yet unknown whether or not a finite-difference horizontal discretization would have simplified the solution of our second algorithmic problem or not.

These reservations notwithstanding (perhaps they will form the incentive for new trials of some of the ideas reported in BHBG95) we shall now concentrate on how to minimize the changes to the existing code when implementing the suggestion of L92.

The first goal is to leave untouched the semi-implicit algorithm tailored to the SB81 vertical discretization, whose main characteristic is the absence of any intrinsic definition of the pressures at full levels. It has to be reflected in the choice of the new prognostic variables. Starting from there and forgetting, for the purpose of this paper, some minor consequences on the nonlinear part of the equations, the elimination procedure towards the structure equation creates a surprise. There is a property of the continuous vertical set of semi-implicit operators that is not reflected in the discretized one proposed by SB81 and that matters only for the elastic case! Fortunately, for some deeper reasons that we did not investigate, the two consequences of this new constraint are internally compatible, the only practical consequence being a small vertical smoothing of the "Brunt-Väisälä" time independent term of the structure equation. This will be detailed in Section 3a.

After dealing with the consequences of the last remark on the computation of the pressure gradient force term (not reported here), the design of the full non-linear scheme is trivial except for one small but crucial point: with time-dependent geometrical shapes of the model's grid-boxes in pressure-type coordinate, the 3D divergence term of Euler's equations is nonlinear whereas the corresponding (2D) term of the hydrostatic equations is linear, a fact overlooked in the L92 analysis. This imposes an iterative solution for the Helmholtz equation, the consequences of which on coding and computational price have to be kept minimal, as will be reported in Section 3b.

The choice of new prognostic variables to be outlined in Section 3a leads to a common vertical staggering for vertical velocities in z and in the generalized vertical coordinate, another apparently desirable feature. But the first one being of the prognostic type and the second one purely diagnostic, the choice of the horizontal wind at the earth's surface now has more consequences for the equations. One of them, linked to the above-mentioned nonlinearity of the 3D divergence, if left untreated, leads to the unstable behaviour of the full model. Our solution, reported in Section 3c, has some heuristic aspects which may disappear if we use a more sophisticated approximation than up to now to discretize the so-called free-slip boundary condition for determining the horizontal wind at the surface (Laprise, personal communication).

3 Details of the numerical problems encountered

a *Vertical Discretization*

We start from the linearized version of Laprise's set of equations, using, for the time being, as additional variables the vertical velocity w and the pressure departure from the hydrostatic pressure $\tilde{p} = p - \pi$ (as advocated in L92). The basic state values of these two new variables are zero everywhere so that the linearized equations are

$$\begin{aligned}
&\mathbf{V} = \mathbf{V}^* + \mathbf{V}'; \quad && \mathbf{V}^* = 0\\
&w = w^* + w'; && w^* = 0\\
&T = T^* + T'; && T^* = \text{const.}\\
&p = \pi^* + \pi' + \tilde{p}; && \pi^* = A(\eta) + B(\eta)\pi_s^*,\ \pi_s^* = \text{const.}\\
&\phi = \phi^* + \phi'; && \partial\phi^*/\partial\eta = -m^*/\rho^*,\ \phi_s^* = \text{const.}
\end{aligned}$$

where $\mathbf{V}$ is the horizontal wind component, w is the vertical wind component, T is the temperature, p is the true pressure, π is the "hydrostatic pressure" as defined in L92, ϕ is the geopotential and η is the hybrid vertical coordinate, introduced implicitly as usual but this time based on pressure π. Further, m is a short hand notation for $\partial\pi/\partial\eta$. The basic state is denoted by $*$ and the perturbations are primed.

Thus the linearized set of equations are as follow:

$$\frac{\partial \mathbf{V}'}{\partial t} = -\nabla\phi' - \frac{R_a T^*}{\pi^*}(\nabla\pi' + \nabla\tilde{p}) \tag{1}$$

$$\frac{\partial w'}{\partial t} = \frac{g}{m^*}\frac{\partial \tilde{p}}{\partial \eta} \tag{2}$$

$$\frac{\partial T'}{\partial t} = \frac{R_a T^*}{C_{pa}\pi^*}\left(\omega' + \frac{\partial \tilde{p}}{\partial t}\right) \tag{3}$$

$$\frac{\partial \tilde{p}}{\partial t} = -\pi^* \frac{C_{pa}}{C_{va}}\left(D' - g\,\frac{\rho^*}{m^*}\frac{\partial w'}{\partial \eta}\right) - \omega' \tag{4}$$

$$\frac{\partial \pi'_s}{\partial t} = -\int_0^1 D' m^* d\eta \tag{5}$$

with

$$\omega' \equiv \frac{\partial \pi'}{\partial t}. \tag{6}$$

The first obvious property of this set of equations is that equations (2) and (4) would allow a far better vertical discretization if the variables w and $\tilde{p}$ are not forecast at the same level. Furthermore, the horizontal divergence D being naturally defined on momentum levels (called the full levels), it appears natural, for the consistency of the right hand side of equation (4), to have a vertical staggering similar to that of the generalized coordinate vertical velocity $\dot{\eta}(\partial\pi/\partial\eta)$, w being defined on the half levels. Rather than having to advect variables situated on two types of levels, the last logical step in this reasoning leads us to choose as a new prognostic variable the layers' scaled increment in vertical velocity (a pseudo vertical divergence whose self-identical linearization gives us back the complement term to D' in equation (4)) :

$$\hat{d} = -g\,\frac{\rho^*}{m^*}\frac{\partial w}{\partial \eta} \tag{7}$$

The transformation of equation (2) leads to the appearance of the vertical Laplacian operator. The ω' term at the end of equation (4) needs a type of vertical integral evaluation similar to the so-called conversion term in a primitive equation set (the latter disappears from equation (3) by combination with (4)). Finally, the evaluation of vertical velocities at all $L+1$ half levels of an L-layer model will require, in addition to the forecast values of the L pseudo vertical divergences, the diagnostic knowledge of the vertical velocity at the surface and hence of the horizontal wind there. In that sense, the lower boundary condition starts to play a quasi-prognostic role in our set of equations. In the following we shall use the same notation for full and half levels as in BHBG95: index l denotes full levels ($l = 1, \ldots, L$), index $\tilde{l}$ denotes half levels ($\tilde{l} = 0, \ldots, L$).

Returning to the set of linearized equations, we can see that, in the new form of equation (4)

$$\frac{\partial \tilde{p}}{\partial t} = -\pi^* \frac{C_{pa}}{C_{va}}(D' + \hat{d}) - \omega' \tag{8}$$

we cannot escape evaluating the pressure π^* at full levels. As already mentioned, one of the essential features of the SB81 set of equations is to avoid such an evaluation. Here also, a change of variable will help to solve this problem. Setting

$$\hat{\mathcal{P}} = \frac{p - \pi}{\pi^*} \tag{9}$$

to replace $\tilde{p}$ as a prognostic variable ($\hat{\mathcal{P}}$ is again identical to its linearized value), one can see that the above-mentioned problem disappears and a careful inspection of the full set of equations indicates that it does not reappear elsewhere.

When we now want to reduce the set of equations to the single structure equation for fast propagating waves, a new difficulty appears that did not exist in the hydrostatic case. Interaction between the vertical integral at the end of the new form of equation (4)

$$\frac{\partial \hat{\mathcal{P}}}{\partial t} = -\frac{C_{pa}}{C_{va}}(D + \hat{d}) + \frac{1}{\pi^*} \int_0^{\eta} m^* D d\eta \tag{10}$$

and the pressure gradient force term of equation (1) will create an isolated double integral that can only be eliminated if we make use of the per-parts integration rule:

$$\int_{\pi}^{\pi_s} \frac{1}{\pi^2} \int_0^{\pi} (\ ) d\pi' d\pi'' = \int_{\pi}^{\pi_s} \frac{1}{\pi}(\ ) d\pi' + \frac{1}{\pi} \int_0^{\pi} (\ ) d\pi' - \frac{1}{\pi_s} \int_0^{\pi_s} (\ ) d\pi'. \tag{11}$$

This does not create any problem in the continuous case but, when going to the vertically discretized linear operators, some simple algebra indicates that the "natural" operators chosen in SB81 are not fulfilling the discretized equivalent of identity (11).

Some more cumbersome algebra shows that this handicap can be eliminated by replacing, for the computation of the logarithmic thickness of the layers δ_l,

$$(\delta_l)_{A1} = \ln \frac{\pi_{\tilde{l}}}{\pi_{\tilde{l}-1}} \tag{12}$$

with

$$(\delta_l)_{A2} = \frac{\pi_{\tilde{l}} - \pi_{\tilde{l}-1}}{\sqrt{\pi_{\tilde{l}-1}\pi_{\tilde{l}}}}. \tag{13}$$

There may, however, be some detrimental effect of this new choice if the energy conservation property of the SB81 set is not maintained. A priori, this should be

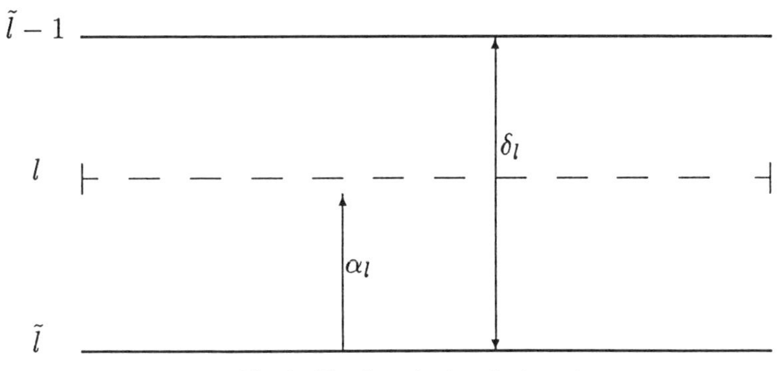

Fig. 1 The discretization of a layer l.

the case since the system used for computing δ_l, α_l (counterpart of δ_l between half and full levels, see Fig. 1) and the implicit hydrostatic pressure π at full levels that links them, is now over-determined by one equation. Fortunately one can prove that our new choice for δ_l suppresses this over-determination. One obtains:

$$\alpha_l = 1 - \frac{\pi_l}{\pi_{\tilde{l}}} \tag{14}$$

$$\pi_l = \frac{\pi_{\tilde{l}} - \pi_{\tilde{l}-1}}{\delta_l} = \sqrt{\pi_{\tilde{l}-1}\pi_{\tilde{l}}}. \tag{15}$$

We did not investigate the reasons why the fulfilment of the discrete version of the per-parts rule should diminish the number of conditions required for energy conservation in the SB81 formalism. This is surely not fortuitous, but, in retrospect, it appears that it was that very property which allowed us to keep the elastic version of the model as an enhancement of the primitive equation one. The latter step obviously also requires that the hydrostatic version be run with the second option for the computation of δ_l, something that can be done without any apparent disadvantage, once some of the consequences on the pressure gradient force term in the full nonlinear set of equations have been worked out. Conversely we verified that trying to run the full elastic model, once finalized, when substituting back the original form for δ_l, leads to immediate numerical instability.

The story is, however, not completely finished at this point. Indeed the above-mentioned additional terms that appear in both the p and ϕ parts of the pressure gradient force term are multiples of $\delta_l - 2\alpha_l$, the same term that we find in the deviation between the structure equation of the discretized set of linear equations and the discretized version of the continuous structure equation. The latter difference is again the consequence of the discretization of a double vertical integral (see details in BHBG95) but we are apparently not in the same fortunate position this time. Our colleague P. Bénard (personal communication) showed that there indeed exists no way to overcome this problem. The similarity between the correction terms in the linear and nonlinear parts of the set of equations, however, strongly suggests that the two consequences of all the preceding choices may be interdependent. We did not investigate this point further, but once again we verified that setting back a posteriori both sets of correction terms to zero creates numerical instability. Given the fact that, for most of the atmospheric levels, $\delta_l - 2\alpha_l$ is very small (see equations (13) and (14)), we thus have strong reasons to believe that our set of vertical discretization choices is the most consistent one possible, given the initial constraints indicated in the previous section. It should finally be mentioned that, despite the cumbersome method needed to obtain their formulation, the overall changes to the code associated with this part of the work are well isolated and relatively small.

b *Stability of the Semi-Implicit Scheme*

As a consequence of the changes of variable explained in the previous section, the full system of nonlinear equations now takes the form:

$$\frac{d\mathbf{V}}{dt} + \frac{RT}{p}\nabla p + \frac{1}{m}\frac{\partial p}{\partial \eta}\nabla \phi = \mathcal{V} \tag{16}$$

$$\begin{aligned} &\frac{d(\hat{d})}{dt} + g^2 \frac{\rho^*}{m^*}\frac{\partial}{\partial \eta}\left(\frac{1}{m}\frac{\partial \pi^* \hat{\mathcal{P}}}{\partial \eta}\right) - g\frac{\rho^*}{m^*}\frac{\partial \mathbf{V}}{\partial \eta}\cdot \nabla w \\ &\quad + \hat{d}\frac{1}{m}\frac{\partial}{\partial \eta}(\dot{\eta} m) - \hat{d}\dot{\eta}\frac{\partial(\ln \rho^*)}{\partial \eta} \\ &- g\rho^*(\dot{\eta} m)\frac{1}{m^*}\frac{\partial}{\partial \eta}\left(\frac{1}{m}\frac{\partial w}{\partial \eta}\right) + g\rho^*\dot{\eta}\frac{\partial}{\partial \eta}\left(\frac{1}{m^*}\frac{\partial w}{\partial \eta}\right) = -g\frac{\rho^*}{m^*}\frac{\partial \mathcal{W}}{\partial \eta} \end{aligned} \tag{17}$$

$$\frac{dT}{dt} + \frac{RT}{C_v} D_3 = \frac{Q}{C_v} \tag{18}$$

$$\frac{d\hat{\mathcal{P}}}{dt} + \frac{C_p}{C_v}\frac{p}{\pi^*} D_3 + \frac{\dot{\pi}}{\pi^*} + \frac{\hat{\mathcal{P}}}{\pi^*}\dot{\eta}\frac{d\pi^*}{d\eta} = \frac{1}{\pi^*}\frac{Qp}{C_v T} \tag{19}$$

$$\frac{\partial \pi_s}{\partial t} + \nabla \cdot \int_0^1 m\mathbf{V} d\eta = 0 \tag{20}$$

$$\begin{aligned} m\dot{\eta} &= \frac{\partial \pi}{\partial \pi_s}\int_0^1 \nabla \cdot (m\mathbf{V}) d\eta \\ &\quad - \int_0^\eta \nabla \cdot (m\mathbf{V}) d\eta \end{aligned} \tag{21}$$

$$\frac{d\phi}{dt} = gw \tag{22}$$

$$\frac{\partial \phi}{\partial \eta} = -m\frac{RT}{\pi + \pi^* \hat{\mathcal{P}}} \tag{23}$$

$$\begin{aligned} D_3 &= \nabla \cdot \mathbf{V} + \frac{1}{m}\frac{p}{RT}\nabla \phi \\ &\quad \cdot \left(\frac{\partial \mathbf{V}}{\partial \eta}\right) - \frac{g}{m}\frac{p}{RT}\frac{\partial w}{\partial \eta} \end{aligned} \tag{24}$$

Equation (22) is only used at the surface to derive the vertical acceleration needed in the second term of equation (17). The direct link between $\partial\hat{\mathcal{P}}/\partial t$ and D_3 in equation (19) reflects the three-dimensional character of the propagation of sound waves in the Eulerian equations. According to the analysis of L92, the use of the semi-implicit algorithm should indeed slow down those waves sufficiently to avoid any violation of the CFL stability condition, as is the case with gravity waves in the primitive equation set. However, inspection of equation (24) indicates that this is not the case. The middle term of the right-hand-side, involving the product of the vertical wind shear by the horizontal geopotential gradient, cannot be linearized. Neither can the difference between the true vertical divergence (third term of the right-hand-side) and our equivalent prognostic variable. Hence, as already noticed by Ikawa (1988) for other types of non-hydrostatic models, some part of the sound wave propagation will escape the control of the semi-implicit procedure and, provided conditions are "favourable", this can lead to numerical instability.

Indeed this numerical instability always happened in our tests and a solution had to be sought for it. It was rather easy to identify the similarity between our problem of nonlinearity inside the time-averaging process central to the semi-implicit treatment and the one related to the $\ln\phi$ term in the model of Côté and Staniforth (1988). Indeed it is sufficient to implement their solution based on the iteration of the time-averaging procedure of the semi-implicit scheme only for the term:

$$Y = \frac{1}{m}\frac{p}{RT}\nabla\phi \cdot \left(\frac{\partial \mathbf{V}}{\partial \eta}\right) + \hat{d}\left(\frac{\Delta\phi^*}{\Delta\phi} - 1\right) \tag{25}$$

Where Y denotes the full non-linear part of D_3, $\Delta\phi$ is the geopotential thickness of the layer and $\Delta\phi^*$ its basic-state counterpart. The semi-implicit scheme is extended by an iterative algorithm, involving additional corrections to the right-hand-sides of the temperature equation:

$$\widetilde{T_i^+} = \widetilde{T_{i-1}^+} - \frac{R_a T^*}{C_{va}}\left(\frac{Y^{t+\Delta t} + Y^{t-\Delta t}}{2} - Y^t\right)_{i-1} \Delta t \tag{26}$$

and of the reduced non-hydrostatic pressure departure equation:

$$\widetilde{\hat{\mathcal{P}}_i^+} = \widetilde{\hat{\mathcal{P}}_{i-1}^+} - \frac{C_{pa}}{C_{va}}\left(\frac{Y^{t+\Delta t} + Y^{t-\Delta t}}{2} - Y^t\right)_{i-1} \Delta t \tag{27}$$

where i is the index of an iteration step and $\widetilde{T_0^+}$ and $\widetilde{\hat{\mathcal{P}}_0^+}$ denote the right-hand-side resulting from the original "explicit" part of the time step calculations. The full recomputing of the Y term at each iteration is however very expensive, owing to the need to perform the nonlinear calculations in gridpoint space while the semi-implicit operator is solved in spectral space. Hence we looked for a mixed solution in which only essential parts of the Y term would have to be re-evaluated at $t + \Delta t$ for each iteration (via direct and inverse fast Fourier transforms after and before the actual recomputation). It turned out that only the surface hydrostatic pressure, the horizontal wind, the temperature and the pseudo vertical divergence have to be reevaluated, which leads to a substantial reduction of the computational overhead associated with each iteration without any loss of accuracy (see next Section).

Thus we may replace $Y^{t\pm\Delta t}$ in equations (26) and (27) by $\widehat{Y^{\pm}}$:

$$\widehat{Y^{\pm}} = \frac{1}{m^{\pm}} \frac{\pi^{\pm}(1 + \hat{\mathcal{P}}^{0}\pi^{*}/\pi^{\pm})}{R^{0}T^{\pm}} (\nabla\phi)^{0} \cdot \left(\frac{\partial \mathbf{V}^{\pm}}{\partial \eta}\right) + \hat{d}^{\pm}\left(\frac{\Delta\phi^{*}}{\Delta\phi^{\pm}} - 1\right) \tag{28}$$

where (+, 0, −) correspond to the time levels ($t + \Delta t$, t, $t - \Delta t$). Furthermore, as in Côté and Staniforth (1988), it seems that, for most meteorological applications, one single iteration is sufficient to obtain both stability and a very acceptable level of accuracy (again see next Section). However, the implications for the code implementation of this solution are not straightforward and indeed they create the major technical hurdle on the way to a fully switchable elastic version, at least in our spectral model.

One may indeed speculate that the "Y iterations" could be merged with the iterations that are anyhow necessary to solve the Helmholtz equation in gridpoint space for a finite-difference model, in which case one of the major advantages of spectral modelling would disappear for our special application. This remains however to be verified and, should it be wrong, the additional problem brought by the nonlinearity of the three-dimensional divergence would be the same with spectral and finite-difference models.

Laprise (personal communication) suggested that one may well go around the problem of the nonlinear terms in equation (24) by combining their critical parts to obtain better behaving terms plus the time derivative of the vertical gradient of ϕ, a solution similar to the one employed in the "MC2" model (Tanguay et al., 1990). We did not investigate this possibility further since the new version of equation (18) would then have totally lost its compatibility with its counterpart in the hydrostatic version of ARPEGE/ALADIN, but this issue remains a possible one for someone trying to implement the L92 idea "from scratch".

c *Surface Boundary Condition*

As already mentioned at the end of Section 2, the choice of the surface horizontal wind is going to play a crucial role in our algorithm. Owing to the complexity of some of the consequences that we shall now demonstrate, it appeared necessary to first have a simple choice, e.g.

$$\mathbf{V}_{\tilde{L}} = \mathbf{V}_{L} \tag{29}$$

the very first-order simplification of the so-called free-slip boundary condition. As an immediate consequence the surface vertical velocity becomes

$$w_s = \frac{1}{g}\mathbf{V}_L \cdot \nabla\phi_s \tag{30}$$

but this will not be the only use of our boundary condition for $\mathbf{V}_{\tilde{L}}$.

Indeed, looking at equation (17), one sees that we need an estimation of ∇w on full levels. This will be obtained through a vertical integration of $\nabla \hat{d}$ similar to that of temperature for geopotential, starting at the bottom from:

$$\nabla w_s = \frac{1}{g} \nabla (\mathbf{V}_L \cdot \nabla \phi_s). \tag{31}$$

Equation (17) contains a more hidden use of the surface boundary condition: the computation of the second term on the left-hand-side requires the knowledge of $g/m\, \partial\tilde{p}/\partial\eta$ at the surface, i.e., according to the nonlinear equivalent of equation (2), dw_s/dt, in other words the surface vertical acceleration. Using our boundary condition again for the surface horizontal wind and its horizontal derivatives, one gets:

$$g \left(\frac{dw}{dt}\right)_{\tilde{L}} = g^2 \left(\frac{1}{m} \frac{\partial(p-\pi)}{\partial\eta}\right)_{\tilde{L}} = u_L^2 \frac{\partial^2 \phi_{\tilde{L}}}{\partial x^2} + v_L^2 \frac{\partial^2 \phi_{\tilde{L}}}{\partial y^2} + 2u_L v_L \frac{\partial^2 \phi_{\tilde{L}}}{\partial x \partial y}$$
$$+ \left(\mathcal{V} - \frac{RT}{p} \nabla p - \frac{1}{m} \frac{\partial p}{\partial \eta} \nabla \phi\right)_L \cdot \nabla \phi_{\tilde{L}} \tag{32}$$

where $\mathcal{V}$ contains the Coriolis and curvature terms. Finally the most difficult use of the surface wind occurred in the evaluation of the first term of the right-hand-side from equation (25). It appeared that the "zero vertical derivative" approach had to be extended not only to the next full level, but in fact to the next half level. In order to avoid only displacing the problem of vertical consistency in the evaluation of $\partial\mathbf{V}/\partial\eta$ one layer up, we elected to "absorb" the discrepancy across four layers (see Fig. 2).

The choice of four layers to proceed with the adjustment was dictated by the earlier results reported in BHBG95, where the numerical instability associated with the use of the "natural" values for all $\mathbf{V}_{\tilde{L}}$ was treated by an unnecessarily heavy filtering (0.25, 0.25, 0.25, 0.25) throughout the vertical.

As mentioned at the very end of Section 2, it might well be that computing $\mathbf{V}_{\tilde{L}}$ in order to have a zero derivative of the wind parallel to the surface along the normal direction to the latter would help solve this problem for which our current solution is rather heuristic. It may well be, however, that this improvement would be offset by a deterioration due to the practical impossibility of extending the computation corresponding to equation (32) to the new formulation of $\mathbf{V}_{\tilde{L}}$. Intermediate complexity with some simplifying assumptions might offer an acceptable compromise, but we have not yet studied this issue. The top boundary condition is symmetric, for reasons of simplicity.

Let us finally state that studying alternative formulations for this crucial choice of $\mathbf{V}_{\tilde{L}}$ is very easy since additions to the hydrostatic code are small and easily isolated from the rest of the algorithm.

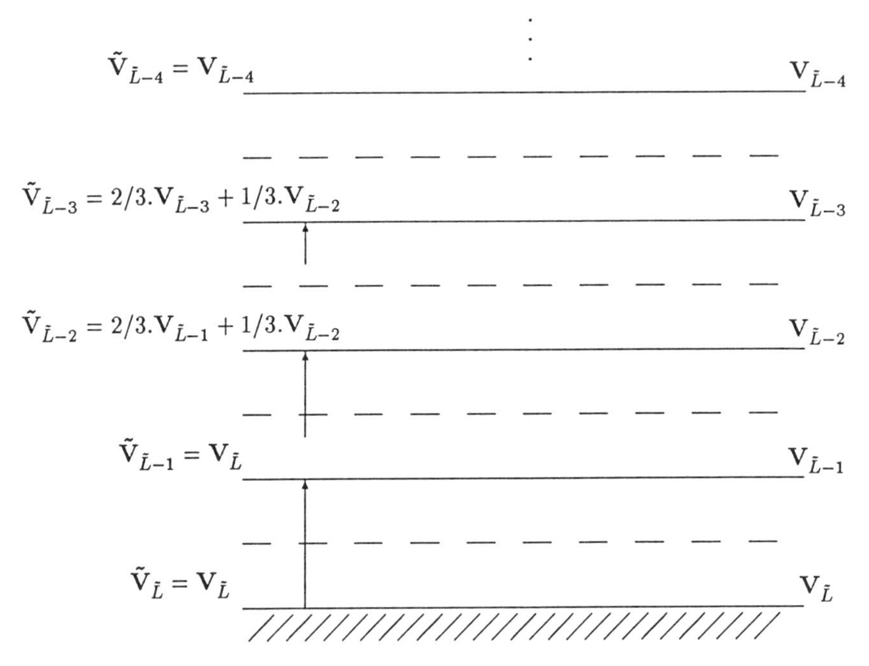

Fig. 2 Bottom boundary condition for the first part of the Y term: redefinition of the half level horizontal wind values (denoted $\tilde{\mathbf{V}}$ on the left side) from those calculated originally (denoted $\mathbf{V}$ on the right side). The arrows symbolize the partial influence of the surface level.

4 First validation tests for purely non-hydrostatic flows

a *Nonlinear Non-Hydrostatic 2D Regime*

The fully elastic dynamics approach was first validated on 2D idealized flows over a bell-shaped mountain. Four different types of regimes were simulated and the basic results are reported in BHBG95. Among these regimes the nonlinear non-hydrostatic flow represents the most interesting one and it has been chosen for testing the precision and stability of the vertical discretization of the important nonlinear terms specific to the elastic dynamics (see previous Section).

The specifications of the nonlinear non-hydrostatic flow are as follow: dry atmosphere with a constant temperature gradient imposed by the constant value of the Brunt-Väisälä frequency $N = 0.01\ \mathrm{s}^{-1}$ with a reference flat surface temperature $T_0 = 285$ K; uniform flow $U = 4\ \mathrm{m\ s}^{-1}$; half-width of the mountain $a = 400\ \mathrm{m} = 5\Delta x$ and height $h = 400$ m. The most important parameters are the critical values of:

– adimensional width of the mountain $Na/U = 1$ for the non-hydrostatic effects;
– adimensional height of the mountain $Nh/U = 1$ for the nonlinearity of the flow.

Hence, this regime is difficult to tune and it is likely to show most of the weaknesses

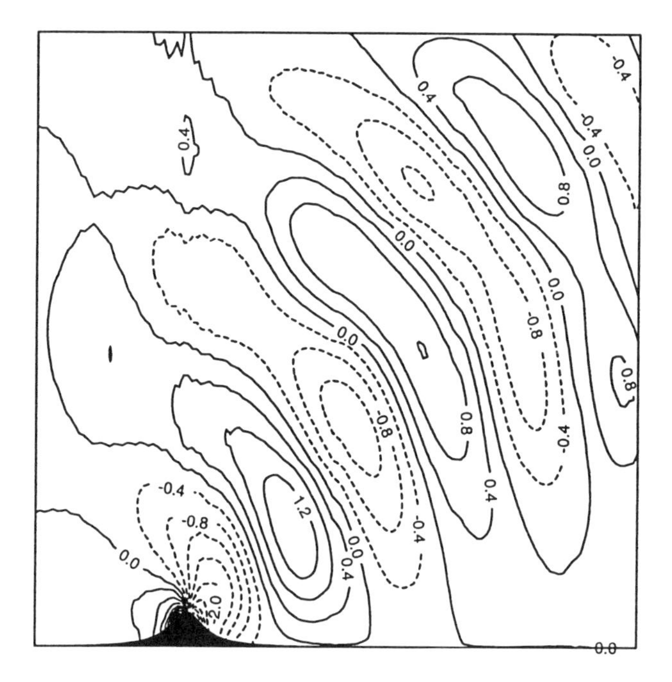

Fig. 3 Nonlinear non-hydrostatic flow: vertical velocity w (m/s).

of the discretization and of the formulation of boundary conditions. Simulations are compared against a pseudo-analytically obtained solution (Bacmeister, 1987; Long, 1954).

One of the most crucial tasks in the formulation of the nonlinear elastic dynamics is the vertical discretization of the nonlinear part of the total 3D divergence, which contains mixed horizontal and vertical derivatives. After changing the top and bottom boundary conditions of the vertical derivative of horizontal wind from those in BHBG95 to those indicated in Section 3c, the simulation has provided results closer to the pseudo-analytical solution. Normalized values of momentum flux (0.743) and drag (0.715) match quite well the expected value of 0.764. Also, compared to Fig. 6 in BHBG95, the wave pattern of vertical velocity w (Fig. 3) no longer gets a spurious secondary wave train above the main one. The wave pattern of the adimensional zonal wind perturbation is shown in Fig. 4 together with the pseudo-analytical solution. The agreement is quite satisfactory.

The other tests have been devoted to the iterations of the semi-implicit scheme. Previous experiments have already shown that from the stability point of view it is sufficient to make just one iteration of the semi-implicit operator. It has also turned out that it is unnecessary to iterate the full part of the nonlinear D_3 term: the scheme is stable even if the horizontal derivative of geopotential $\nabla_\eta \phi$ and the variable $\hat{\mathcal{P}}$ are evaluated only at the current time level t.

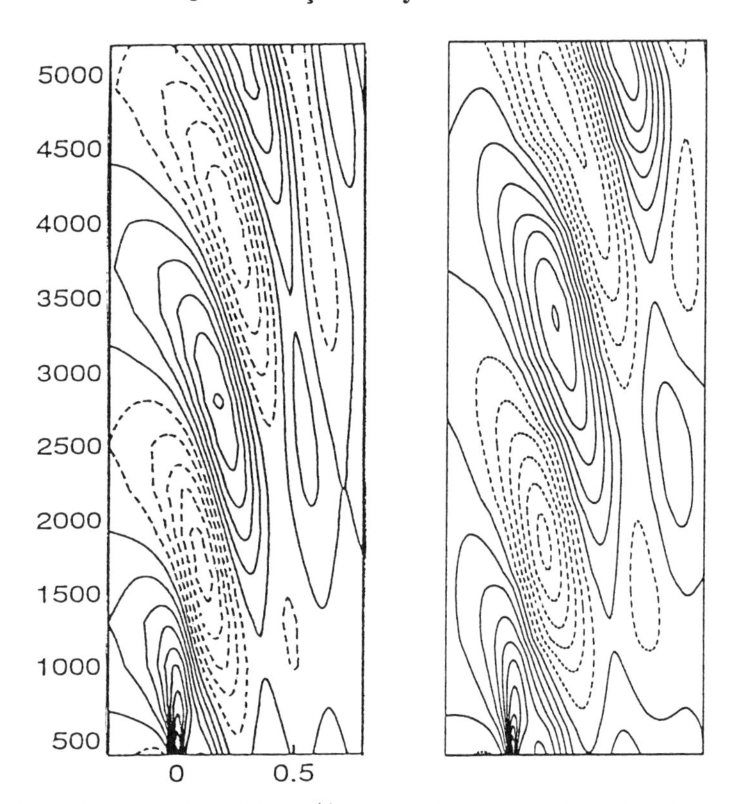

Fig. 4 Nonlinear nonhydrostatic flow: u'/U. The pseudo-analytical solution is on the left, the ALADIN simulation is on the right. Isolines are drawn with an interval of 0.1.

Since the simplification of the iterative algorithm is very favourable for an important reduction of computational costs, both above-mentioned issues were tested with respect to the quality of the 2D nonlinear non-hydrostatic simulation. As demonstrated in Fig. 5, the differences in solutions are really negligible. This is also confirmed by values of diagnosed momentum flux and drag, where the maximum difference only reaches a few thousandths of a percent.

b *PYREX'90 Pseudo-3D Simulation*

The vertical plane version of the model was also used to simulate the phenomenon of real lee waves observed during the PYREX'90 field measurement experiment (Bougeault et al., 1993). This choice was made since it allows the use of a fine resolution in the required horizontal domain length while still maintaining quite a low computational cost. The 2D dynamics were slightly modified in order to introduce a non-zero Coriolis parameter and to include the geostrophic adjustment.

The initial and lateral boundary conditions were obtained from data analyzed

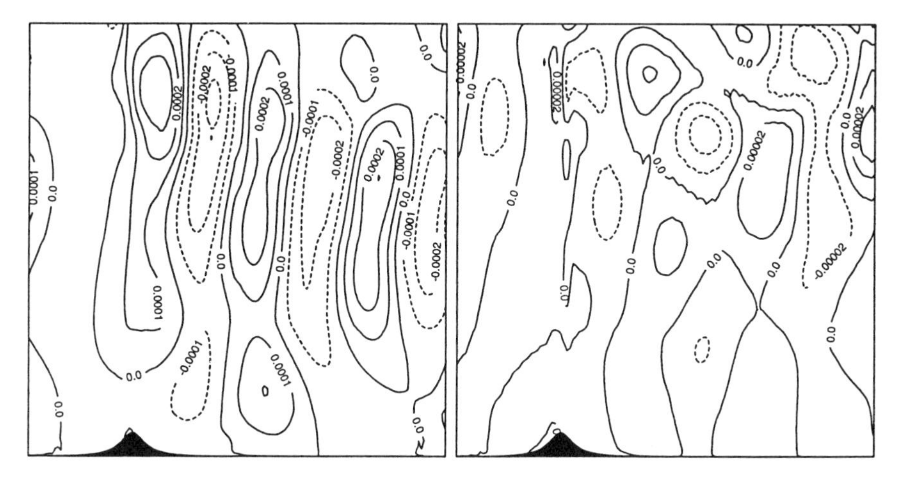

Fig. 5 Left panel: difference in vertical velocity w (m/s) between the simulation with 2 iterations and the reference one. The plotting interval is 10^{-4} (m/s). Right panel: difference in vertical velocity w (m/s), between the simulation with simplified iterations and the reference one. The plotting interval is 2.0×10^{-5} (m/s).

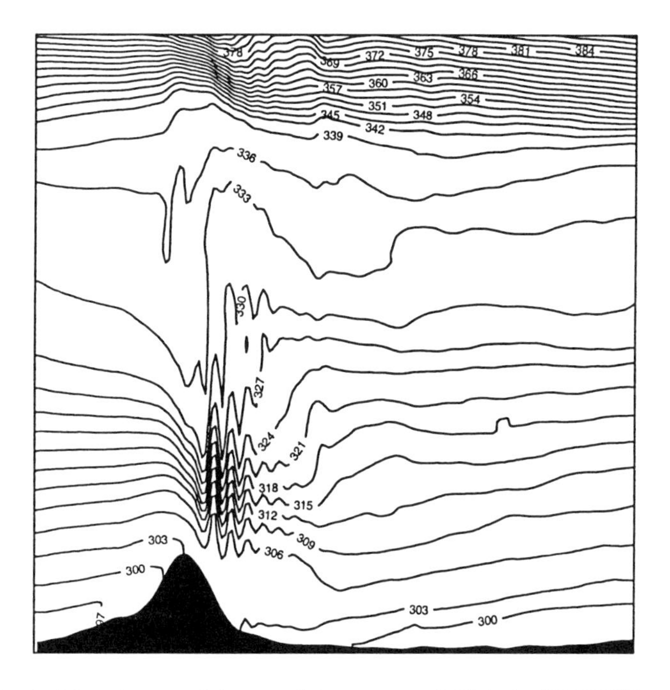

Fig. 6 6-hour forecast of the potential temperature θ (K) in the PYREX'90 case.

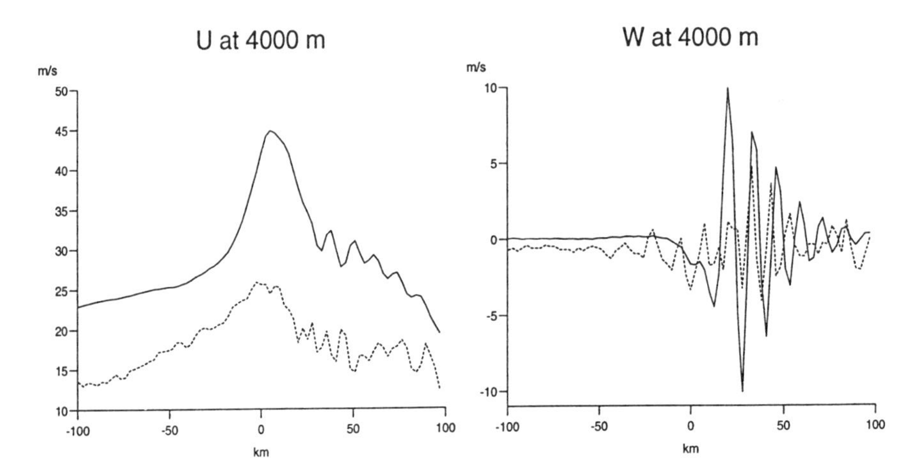

Fig. 7 PYREX'90: ALADIN simulation (solid lines) vs measurement at flight level 4000 m (dashed lines). Left panel: wind component u (m/s); right panel: vertical velocity w (m/s).

by ECMWF (T213, 31 levels). The simulation started on 15/10/1990 at 0000 UTC and the forecast carried out to 6 h. The vertical plane, perpendicular to the main mountain ridge, was 581 km long and about 30 km high, the resolution being 2.55 km in the horizontal and 150 levels in the vertical. The lee wave pattern in the potential temperature field at 6 h of the forecast is shown in Fig. 6.

However, the modified 2D dynamics do not allow the air to flow around the ridge of the Pyrénées; hence the simulated upslope winds get too strong with respect to the observations. This effect is well expressed in the wind field in Fig. 7, where the simulation is presented together with the aircraft observations taken between 0602 UTC and 0645 UTC. The simulated wavelength is quite realistic, even if still a bit too long. This is probably due to the coarseness of our grid which does not allow an accurate simulation of waves with a wavelength of about 10 km.

5 Conclusions

The numerical application of this new form of the fully elastic dynamics has highlighted some specific problems, which do not exist either in other non-hydrostatic formulations or in the hydrostatic dynamics one obtains by degeneration of the present one. One may say that some issues have to be solved in any elastic model as, for instance, control of the acoustic waves, bottom boundary condition formulation, etc. However, a concrete method of discretization, imposed by the vertical coordinate and constraints given by the already existing hydrostatic scheme (which we wish to keep as a switchable option), may be quite different and may have other specific crucial points.

In the case of the ARPEGE/ALADIN model which we described, the initial constraints were given by the SB81 vertical discretization, which fortunately allowed the required extensions towards the fully elastic dynamics in a logical way. The less pleasant feature (at least in our case of a spectral model) is the necessity of iterating the semi-implicit scheme in order to maintain control of the acoustic waves propagation. The other difficulty is the formulation of the best possible surface boundary condition for some nonlinear terms of the fully elastic equations, which do not break down the stability of the scheme and at the same time ensure the accuracy of the solution.

We may conclude that the present implementation of the fully elastic dynamics in the ARPEGE/ALADIN model is satisfying from the accuracy and stability point of view. Nevertheless, some future improvements are a priori not excluded, namely those concerning the formulation of the surface boundary condition. Otherwise we have already demonstrated that the new formulation of the elastic dynamics works well provided it has a very exact vertical discretization, in which all choices need to be carefuly examined.

Acknowledgments

Our co-authors of the BHBG95 paper, Gwenaëlle Hello and Pierre Bénard, through all their work but especially that on the properties of the pure linear model and on the new specificity of the vertical operators, have provided much needed help for the specific issues reported here. We also thank René Laprise for his very constructive comments and suggestions and for following the progress of our work with deep interest. We are, of course, indebted to all the people, inside the ARPEGE/IFS and ARPEGE/ALADIN project, who made it possible to add this final touch of an elastic variant to an already quite huge and multi-purpose code. This work has been made possible by the scholarship provided to the second author by the French "Ministère de l'Enseignement Supérieur et de la Recherche".

References

ARAKAWA, A. and M.J. SUAREZ. 1983. Vertical differencing of the primitive equations in sigma coordinates. *Mon. Weather Rev.* **111**: 34–45.

BACMEISTER, J. 1987. Nonlinearity in transient two dimensional flow over topography. PhD thesis, Princeton University, 187 pp.

BOUGEAULT, P.; A. JANSA, J.L. ATTI, I. BEAU, B. BENECH, R. BENOIT, P. BESSEMOULIN, J.L. CACCIA, J. CAMPINS, B. CARRISIMO, J.L. CHAMPEAUX, M. CROCHET, A. DRUILHET, P. DURAND, A. ELKHALFI, P. FLAMANT, A. GENOVES, M. GEORGELIN, K.P. HOINKA, V. KLAUS, E. KOFFI, V. KOTRONI, C. MAZAUDIER, J. PELON, M. PETITDIDIER, Y. POINTIN, D. PUECH, E. RICHARD, T. SATOMURA, J. STEIN and D. TANNHAUSER. 1993. The atmospheric momentum budget over a major mountain range: First results of the PYREX field program. *Annales geophysicae*, **11**: 395–418.

BUBNOVÁ, R.; G. HELLO, P. BÉNARD and J.-F. GELEYN. 1995. Integration of the fully elastic equations cast in the hydrostatic pressure terrain-following coordinate in the framework of the ARPEGE/Aladin NWP system. *Mon. Weather Rev.* **123**: 515–535.

CÔTÉ, J. and A. STANIFORTH. 1988. A two-time-level semi-Lagrangian semi-implicit scheme for spectral models. *Mon. Weather Rev.* **116**: 2003–2012.

COURTIER, P.; C. FREYDIER, J.-F. GELEYN, F. RABIER and M. ROCHAS. 1991. The Arpege project at Météo-France. *In*: ECMWF 1991 Seminar Proceedings: Numerical methods in atmospheric models; 9–13 Sept. 1991, ECMWF, Reading, U.K. vol. II, pp. 193–231.

DENIS, B. 1990. Introduction de la topographie dans un modèle atmosphérique nonhydrostatique. M.Sc. thesis, Physics Department, Université du Québec à Montréal, 121pp, (Available from P.O. Box 8888 station A Montreal Quebec, H3C 3P8).

IKAWA, M. 1988. Comparison of some schemes for nonhydrostatic models with orography. *J. Meteorol. Soc. Jpn.* **66**: 753–766.

LAPRISE, R. 1992. The Euler equations of motion with hydrostatic pressure as an independent variable. *Mon. Weather Rev.* **120**: 197–207.

LONG, R.R. 1954. Some aspects of the flow of stratified fluids. II. Experiments with a two fluids system. *Tellus*, **6**: 97–115.

MACHENHAUER, B. and J.E. HAUGEN. 1987. Test of a spectral limited area shallow water model with time dependent lateral boundaries conditions and combined normal mode/semi-Lagrangian time integration schemes. *In*: Workshop Proceedings: Techniques for horizontal discretization in numerical weather prediction models; 2–4 Nov. 1987, ECMWF, Reading, U.K., pp. 361–377.

ROBERT, A.J. 1969. The integration of a spectral model of the atmosphere by the implicit method. *In*: Proc. of the WMO/IUGG Symp. on NWP, 26 Nov. – 4 Dec. 1968, Tokyo, Japan Meteorol. Agency, VII, pp. 19–24.

SIMMONS, A. and D. BURRIDGE. 1981. An energy and angular momentum conserving vertical finite-difference scheme and hybrid vertical coordinates. *Mon. Weather Rev.* **109**: 2003–2012.

TANGUAY, M.; A. ROBERT and R. LAPRISE. 1990. A semi-implicit semi-Lagrangian fully compressible regional forecast model. *Mon. Weather Rev.* **118**: 1970–1980.

Climate Simulations with a Spectral, semi-Lagrangian Model with Linear Grids

David L. Williamson
National Center for Atmospheric Research,
Boulder, CO 80307

[Original manuscript received 4 January 1995; in revised form 30 August 1995]

ABSTRACT *Semi-Lagrangian, semi-implicit, spectral transform atmospheric models do not require conventional quadratic unaliased Gaussian grids because advection is not expressed as an Eulerian quadratic product in their formulation. The conventional T42 quadratically unaliased grid with 128 (64) points in longitude (latitude) supports a T63 spectral truncation, and similarly for other resolutions. We refer to these types of truncation/grid combinations as aliased or linear grids, since they can be considered as being aliased for quadratic products or unaliased for linear terms. In this paper, climate simulations based on linear grids are compared with ones based on quadratic unaliased grids. It is shown that for spectral resolutions above T42, the spectral truncation, rather than the transform grid on which the physical parametrizations are calculated, dominates the accuracy of simulations. Thus the linear grid provides a 50 percent increase in resolution of both horizontal dimensions at a negligible additional cost.*

RÉSUMÉ *L'advection n'étant pas représentée comme un produit eulérien quadratique dans la formulation des modèles atmosphériques à transformation spectrale semi-lagrangiens et semi-implicites, ils ne nécessitent donc pas la grille gaussienne usuelle sans repliement quadratique. La grille sans repliement quadratique classique T42 de 128 (64) points de longitude (latitude) supporte une troncature spectrale T63 pour les opérations linéaires et il en est de même pour d'autres résolutions. Ces types de combinaisons troncature/grille sont appelées des grilles à repliement ou linéaires car on peut les considérer comme causant des repliements pour les produits quadratiques mais pas pour les termes linéaires. On compare des simulations du climat basées sur des grilles linéaires à d'autres basées sur des grilles quadratiques sans repliement. On montre que pour les résolutions spectrales au-dessus de T42, la troncature spectrale, plutôt que la grille de transformation sur laquelle les paramétrages physiques sont calculés, domine la précision des simulations. La grille linéaire fournit donc une augmentation de 50% de la résolution des deux dimensions horizontales à un faible coût supplémentaire.*

1 Introduction

Eulerian, global, baroclinic spectral transform models generally adopt quadratic unaliased computational Gaussian grids, even though higher order nonlinearities arise in the governing equations, especially when a complete suite of physical parametrizations is included. The *quadratic grid* is the minimum (tensor product) latitude-longitude Gaussian grid for which the transform of the product of two fields, each of which can be represented in spectral space, has no aliasing onto the waves of the underlying spectral representation (see Machenhauer, 1979, for a review of the spectral transform method). Early studies by Bourke et al. (1977) indicated that grids finer than quadratic were not required. Finer grids introduced changes in short forecasts that Bourke et al. (1977) considered insignificant. General experience since their study has verified this finding, and all Eulerian spectral transform models adopt quadratic unaliased grids. Grids coarser than quadratic, on the other hand, were unsatisfactory in the Bourke et al. (1977) experiments, and led to instability. Once again, general experience confirms this conclusion. Many Eulerian spectral models, such as the National Center for Atmospheric Research (NCAR) Community Climate Model, CCM2 (Hack et al., 1993), continue to require a quadratic unaliased grid. The CCM2, in its standard configuration, exhibits instability with a grid that is coarser than quadratic. Of course, the model could undoubtedly be stabilized with a judicious choice of diffusion or spatial filter; the obvious trivial examples being a filter with a sharp cutoff at 2/3 the smallest resolved wave or diffusion which very strongly damps the upper third of the spectrum. The stability associated with quadratic unaliased calculations in Eulerian spectral transform models is also consistent with experience with finite-difference based Eulerian models (Arakawa, 1966). Quadratic conserving approximations are adequate to prevent nonlinear instability in baroclinic models having higher order nonlinearities (Arakawa and Lamb, 1977)

Côté and Staniforth (1988) suggested that a linear grid might be suitable for semi-Lagrangian spectral transform models. The *linear grid* is defined to be the minimum grid required for transformations of a field from spectral space to grid point space and back again to spectral without loss of information. With such a grid, only linear terms are unaliased. Côté and Staniforth pointed out that, in the semi-Lagrangian method, the nonlinearities associated with advection are hidden in the trajectory calculation and interpolations. The latter implicitly contain some damping. In addition, semi-Lagrangian finite-difference and finite-element schemes are computationally stable. Côté and Staniforth (1988) also point out that, after the semi-Lagrangian discretization, and ignoring the physical parametrizations, the dynamical governing equations are only weakly nonlinear. Côté and Staniforth explored the use of a linear grid experimentally in the shallow water equation framework. They carried out 5-day forecasts initialized from atmospheric analyses with a semi-Lagrangian model with linear and quadratic grids. The linear grid gave forecasts of comparable quality to ones with a quadratic grid. Côté and Staniforth also considered the energy conservation characteristics of 20-day runs. Although energy conservation was not quite as good with the linear grid as with the quadratic, in both cases it was better than in their Eulerian benchmark integration.

Côté and Staniforth's (1988) results are from an unforced, barotropic model, and the lengths of the integrations are short from a climate perspective. The results might be different for a baroclinic model with a complete suite of physical parametrizations, run in a climate simulation mode. In multi-year climate applications, any slow systematic accumulation of error might ultimately lead to catastrophic results. Current climate model physical parametrizations, with interactive diagnosed or predicted clouds, tend to force models strongly on the smallest resolved scales. The parametrizations are very complex and result in aliasing which, presumably, is of some very high order. With the quadratic grid, the upper third of the scales generated by the parametrizations are automatically filtered out. Therefore, the direct results on those scales of these rather expensive calculations are lost. The issue is whether aliasing significantly affects the larger, retained scales differently when the calculation is done on a linear or on a quadratic grid. The affect is difficult to isolate directly, without carrying out parallel diagnostic calculations on the two grids and truncating them to the same spectral resolution in the same model simulation. In the following, we study the effect indirectly by examining a series of climate simulations with the semi-Lagrangian version of the NCAR CCM2 at different horizontal spectral resolutions with linear and quadratic grids. We will show that linear grids pose no problem for climate simulations with semi-Lagrangian, spectral transform models. The spectral truncation, rather than the transform grid on which the physical parametrizations are calculated, dominates the accuracy of simulations. Thus, linear grids provide a 50 percent increase in the resolution of both horizontal dimensions at negligible additional cost.

2 Model

The model used for the simulations described in this paper is an experimental, semi-Lagrangian version of the NCAR Community Climate Model, CCM2 (Williamson and Olson, 1994). The basic CCM2 is built around an Eulerian, spectral transform dynamical component and was developed at T42 horizontal spectral resolution with 18 vertical hybrid levels. Water vapour transport is treated in a monotonic semi-Lagrangian manner (Williamson and Rasch, 1994). Fourth order horizontal diffusion is applied to the dynamical variables, but not the water vapour. Details are provided in Hack et al. (1993). The model contains a complete complement of physical parametrizations. The planetary boundary layer is parametrized by a nonlocal scheme described in Holtslag and Boville (1993). McFarlane's (1987) parametrization of momentum flux divergence by stationary gravity waves is included. A mass flux scheme developed by Hack (1994) is used to represent all types of moist convection. The diagnosed cloud fraction depends on relative humidity, vertical motion, static stability and convective precipitation rate (Kiehl et al., 1994). The solar radiative heating is computed using a delta-Eddington parametrization with 18 spectral bands (Briegleb, 1992). Both diurnal and annual cycles are included. A Voigt correction has been incorporated in the longwave parametrization (Kiehl and Briegleb, 1991). The land temperature is calculated with soil heat capacity represented by a four layer diffusion model with heat capacities specified for each layer to capture the major observed climatological cycles. The land has specified soil hydrologic properties. Sea-surface temperatures are specified by linear interpolation between the climatological monthly mean values of Shea et al. (1990). Sea-ice extent is also specified from climatological monthly mean values. The climate simulated by the standard model at T42 spectral resolution with 18 levels is summarized in Hack et al. (1994). The effect on the climate of changes in the horizontal resolution from R15 to T106 spectral truncations is discussed in Williamson et al. (1995).

As mentioned above, an experimental version of the CCM2 with semi-Lagrangian dynamics was developed by Williamson and Olson (1994). Other than the dynamical component, this model is identical to the CCM2. Williamson and Olson describe some aspects of the simulated climate that are affected by the change from Eulerian to semi-Lagrangian dynamics. Overall, the climates are rather similar, the differences between the Eulerian and semi-Lagrangian versions with the same spectral resolution and quadratic grids being less than found between the Eulerian CCM2 at different spectral resolutions. This experimental semi-Lagrangian version was used for all simulations discussed in this paper. The physical parametrizations were not retuned to compensate for some of the differences introduced by the change from Eulerian to semi-Lagrangian. It will be seen that the resolution signal is strong enough that these differences are likely to be unimportant.

3 Simulations

In this section we compare selected climate statistics from semi-Lagrangian simulations at spectral truncations of T31, T42 and T63 on quadratic grids, and T42 and T63 on linear grids. The spectral truncations were chosen to be a subset of those included in the resolution study of Williamson et al. (1995) involving the Eulerian CCM2. This choice was made to establish that the semi-Lagrangian version shows the same resolution signal as CCM2 even though the semi-Lagrangian version was not tuned to produce its optimal climate. Instead of exact linear grids, the quadratic Gaussian grids from the next lower spectral truncation are used, even though these grids are not quite as coarse as the linear ones. This strategy was chosen so that the specified lower boundary datasets would be common and would influence the surface fluxes and other physical parametrizations in the same way. (As an aside, we note that the T64 linear grid actually requires 128 (66) longitude (latitude) points.) Each simulation with a "linear" grid can be compared to one with the same spectral truncation but different (quadratic) grid, or to one with the same grid but different spectral truncation. For convenience the grids are referred to here as quadratic and linear, recognizing that the term "linear" is not quite correct. The simulations are referenced by spectral truncation and number of longitudinal grid points, including L and Q as a reminder for which type of grid is employed. Thus the simulations on quadratic grids will be referred to as T31/Q96, T42/Q128 and T63/Q192. Those on linear grids will be referred to as T42/L96 and T63/L128. All simulations started on 1 September of an Eulerian CCM2 simulation and ran 17 months through two Januaries, except the T63/Q192 which only ran five months through one January

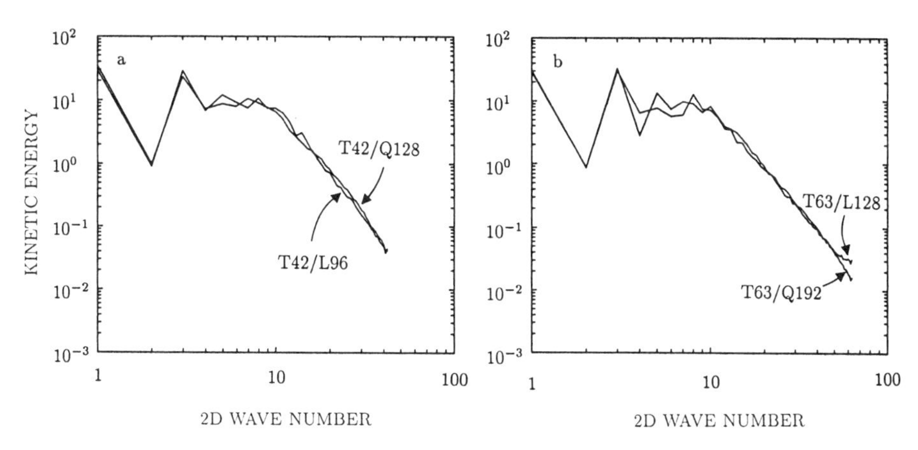

Fig. 1 January average kinetic energy spectra (total kinetic energy versus spherical wavenumber) at 500 mb (a) from T42/Q128 and T42/L96 simulations with horizontal diffusion coefficient of 1×10^{16} m^4 s^{-1}, and (b) from T63/Q192 and T63/L128 simulations with diffusion coefficient of 2×10^{15} m^4 s^{-1}.

and the T42/Q128 which ran 41 months through four Januaries. Unless otherwise stated, averages are over all available Januaries. A time step of 20 minutes was adopted for all simulations (in rare instances, the surface exchange parametrization in the frozen CCM2 has difficulty with a longer time step). All coefficients in the parametrizations are the same for all simulations, except the horizontal diffusion coefficient. For each spectral truncation, these are taken to be the values used in the Eulerian version (Williamson et al., 1995). Thus T31/Q96 uses 2×10^{16} m^4 s^{-1}, T42/Q128 and T42/L96 use 1×10^{16} m^4 s^{-1}, and T63/Q192 and T63/L128 use 2×10^{15} m^4 s^{-1}. The surface geopotential for a given spectral resolution is taken from the Eulerian model with the same spectral resolution, but synthesized to the appropriate quadratic or linear grid. In other words, T42/Q128 and T42/L96 have the same surface geopotential in spectral space, and T63/Q192 and T63/L128 have the same.

We first consider whether small scale, spatial noise accumulates to create a problem with the linear grids. Figure 1 shows the 500 mb kinetic energy spectra as a function of two-dimensional wavenumber for the T42/L96 simulation compared to the T42/Q128, and for the T63/L128 simulation compared to the T63/Q192, all averaged for a single January only. The T42/L96 and T42/Q128 are extremely close, with the T42/Q128 dropping off slightly faster than the T42/L96 at the tail of the spectra. Compared to the T63/Q192, the T63/L128 shows an increase in energy for wavenumbers above 50. This difference is comparable to that found between the Eulerian and semi-Lagrangian versions, and to that which occurs when the diffusion coefficient is halved or doubled (both examples are shown in Fig. 2 of Williamson and Olson, 1994). The increase in kinetic energy above wavenumber 50 in the T63/L128 simulation occurs gradually throughout the first four months

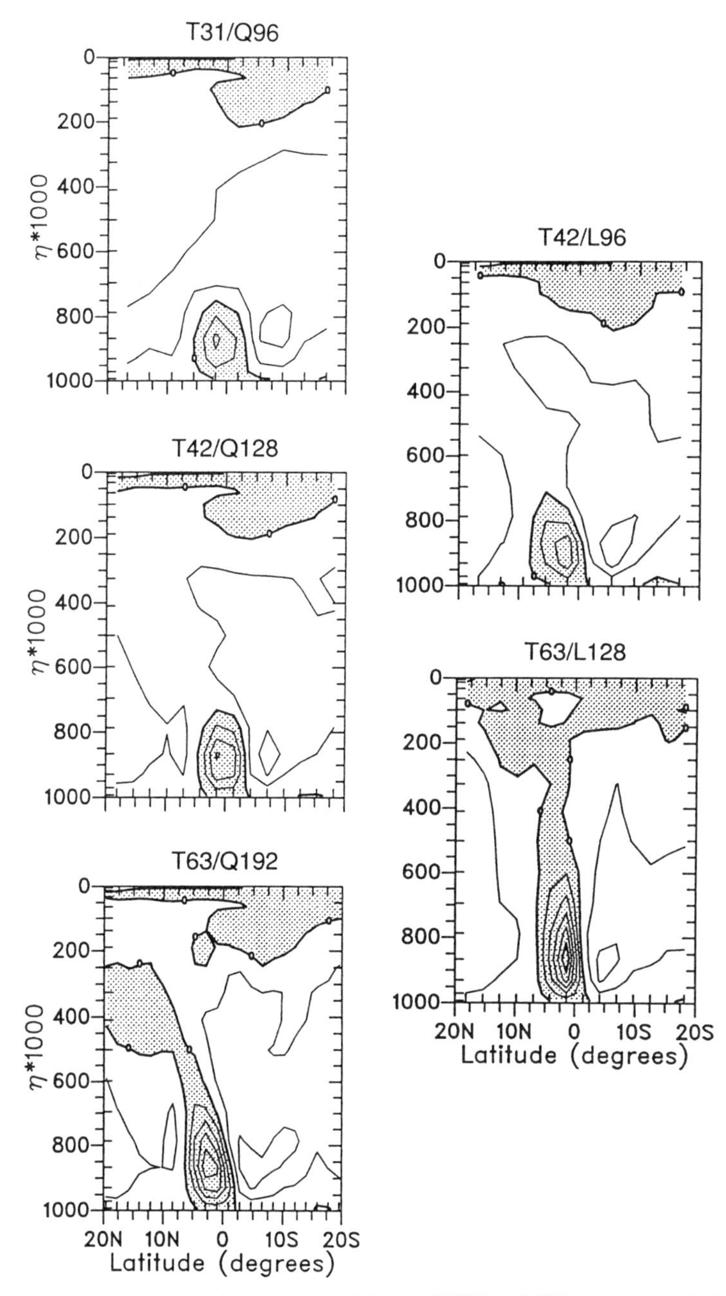

Fig. 2 January average, zonal average over Atlantic (30°W to 7.5°E) pressure vertical velocity (ω) for T31/Q96, T42/Q128, T63/Q192, T42/L96 and T63/L128 simulations. Contour interval is 20 mb day^{-1}, negative (upward) regions stippled. Averages are taken on model (η) surfaces and η*1000 corresponds approximately to pressure in mb.

of the simulation, after which the spectrum does not change. Above wavenumber 10, the spectrum averaged for the second January overlays that from the first. The differences below wavenumber 10, seen in Fig. 1, are comparable to the year-to-year differences that occur in the Eulerian CCM2 in long control integrations.

We now present a collection of fields which showed a fairly clear resolution signal in the Eulerian resolution study mentioned above (Williamson et al., 1995, hereafter referred to as WKH95), and make comparisons with results from that study. Overall, the semi-Lagrangian integrations with the quadratic grid show the same resolution signal as the Eulerian simulations, with similar structures at like resolutions. The effect of horizontal resolution was clearly seen in WKH95 in the tropics in zonal averages taken over individual ocean basins. In each tropical basin, the phenomena tend to be zonally aligned. Separation into individual basins eliminates cancellation that occurs when, for example, the upward branch of the local Hadley circulation forms at different latitudes in different basins. Figure 2 shows the pressure vertical velocity, ω, in the tropics, zonally averaged over the Atlantic ocean (30°W longitude to 5°E longitude). This figure can be compared with the corresponding Fig. 6 in WKH95. The semi-Lagrangian simulations with the quadratic grids (left column, Fig. 2) show the same resolution signal as the Eulerian model. The strength, width, and depth of the upward cell ($\omega < 0$) all vary with resolution. The accompanying subsidence regions also show dependence on resolution. The T31/Q96 simulation shows a weak, shallow cell of upward motion, while the T42/Q128 shows a somewhat stronger, slightly narrower vertical motion cell. Although the vertical extent has not increased significantly, the neighbouring subsidence is deeper and more focused, particularly on the southern flank. The T63/Q192 simulation shows a significant increase in vertical extent and strength of the upward cell. In the T63/Q192 simulation, the central values are almost double those in the T42/Q128. The T42/L96 simulation falls between the T31/Q96 and T42/Q128, but is closer to the T42/Q128 in both the strength of the upward branch and depth of the subsidence. The T63/L128 has the characteristics of the T63/Q192 rather than the T42/Q128. The upward branch is stronger and extends throughout the troposphere.

Figure 3 shows the zonal average specific humidity tendency from the convective parametrization and stable condensation in the equatorial Atlantic, and is comparable to Fig. 10 of WKH95. As in Fig. 2, the zonal average is taken from 30°W to 5°E. Both T31/Q96 and T42/Q128 simulations show detrainment from the convective parametrization above 800 mb throughout much of the tropical Atlantic region including over the upward branch of the Hadley circulation. The T63/Q192 simulation shows such detrainment associated with the trade cumulus on the subsiding flanks of the Hadley cell, but it also shows condensation throughout the column of the ascending branch, where stronger ascending motion is producing more vertical transport by the resolved scales. Once again, the overall impression is that the T42/L96 simulation falls between the T31/Q96 and T42/Q128, but the T63/L128 simulation is clearly more like the T63/Q192 than the T42/Q128.

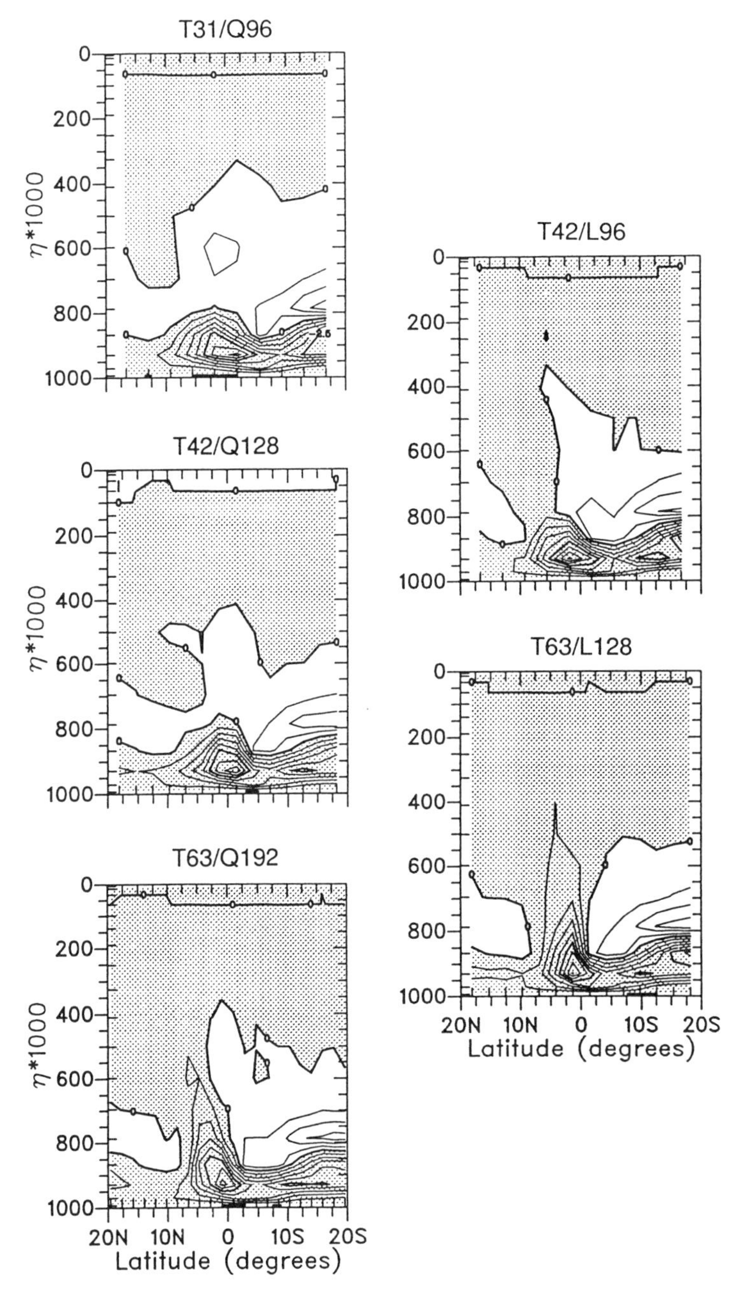

Fig. 3 January average, zonal average over Atlantic (30°W to 7.5°E) specific humidity tendency due to convective parametrization and stable condensation for T31/Q96, T42/Q128, T63/Q192, T42/L96 and T63/L128 simulations. Contour interval is 0.5 g kg^{-1} day^{-1}, negative regions stippled.

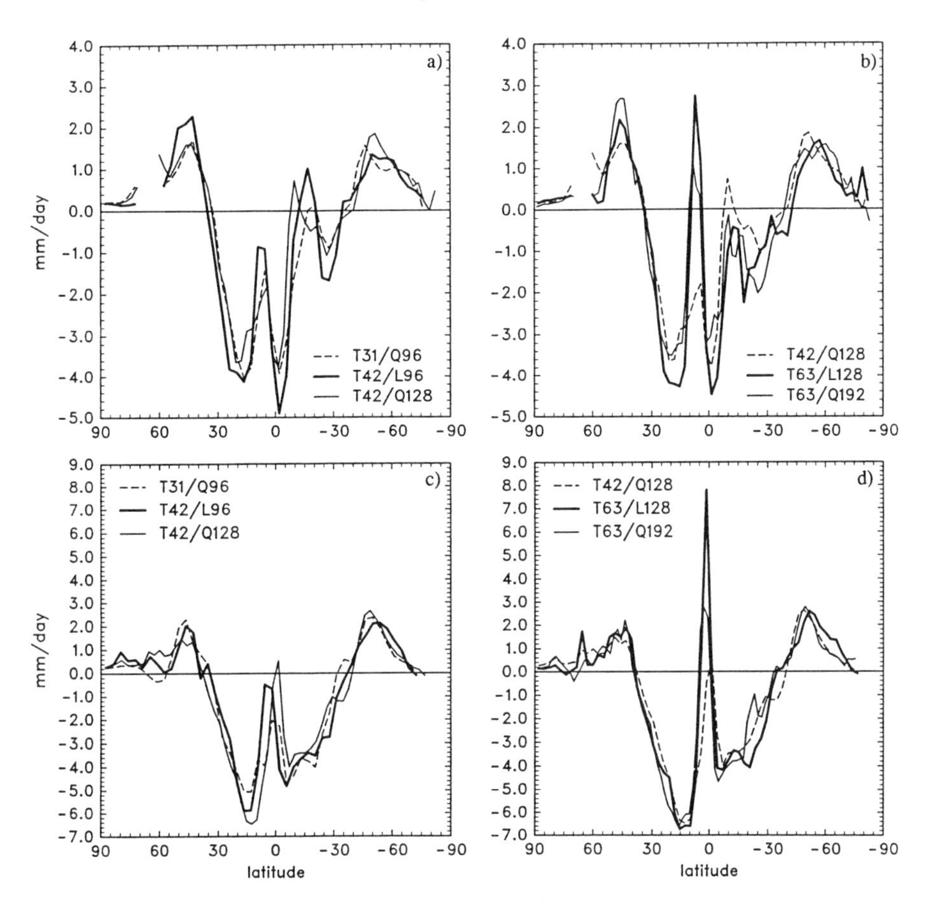

Fig. 4 January average, zonal average over (a,b) the Pacific, and (c,d) the Atlantic of precipitation minus evaporation for T31/Q96, T42/Q128, T63/Q192, T42/L96 and T63/L128 simulations.

Figure 4 shows the precipitation minus evaporation (P – E), zonally averaged over the Atlantic and Pacific ocean basins (including sea-ice). The major differences with resolution are associated with the ascending branch of the Hadley circulation, and attributable primarily to the precipitation. The lower resolution triplet (T31/Q96, T42/L96 and T42/Q128; left column) shows little systematic variation between resolutions. The higher resolution triplet, however, (T42/Q128, T63/L128 and T63/Q192; right column) exhibits the principal signal seen in the Eulerian experiments (Fig. 12 of WKH95), with an increase in P – E with increasing resolution in the ascending branch of the Hadley cell. The T63/L128 simulation is stronger even than the T63/Q192, and is approaching the values from the T106 Eulerian simulation in WKH95.

Figure 5 (top row) shows the zonal wind stress zonally averaged over the Pacific (including sea-ice points). The figure also includes atmospheric estimates calculated

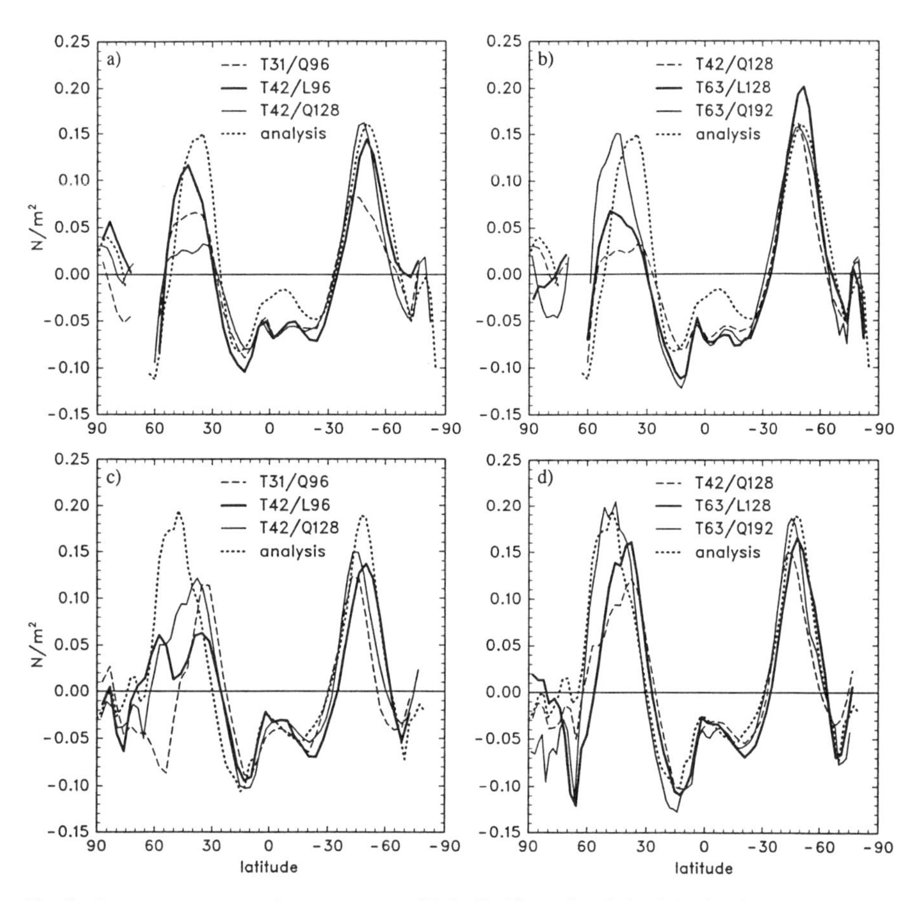

Fig. 5 January average, zonal average over (a,b) the Pacific, and (c,d) the Atlantic of zonal component of surface stress for T31/Q96, T42/Q128, T63/Q192, T42/L96 and T63/L128 simulations and estimates from analyses.

from the European Centre for Medium Range Weather Forecasts (ECMWF) wind analyses (Trenberth et al., 1989), labelled analysis in the figure. The most noticeable changes with resolution are in the Northern and Southern Hemisphere storm tracks from the lower resolution T31/Q96 simulation to the T42/Q128. In the Southern Hemisphere, the T42/L96 values match the T42/Q128. There is no change from T42/Q128 to T63/Q192, but the T63/L128 exceeds both and matches the T106 Eulerian values in WKH95, which exceed the atmospheric estimates. The Northern hemisphere does not show a consistent signal with resolution and is much more variable than the Eulerian simulations (Fig. 14 of WKH95). The same can be said about the averages over the Atlantic shown in the bottom row of Fig. 5. The variability seen with the semi-Lagrangian simulations is likely the result of the small sample in these experiments. Each simulation is for only two years, a relatively

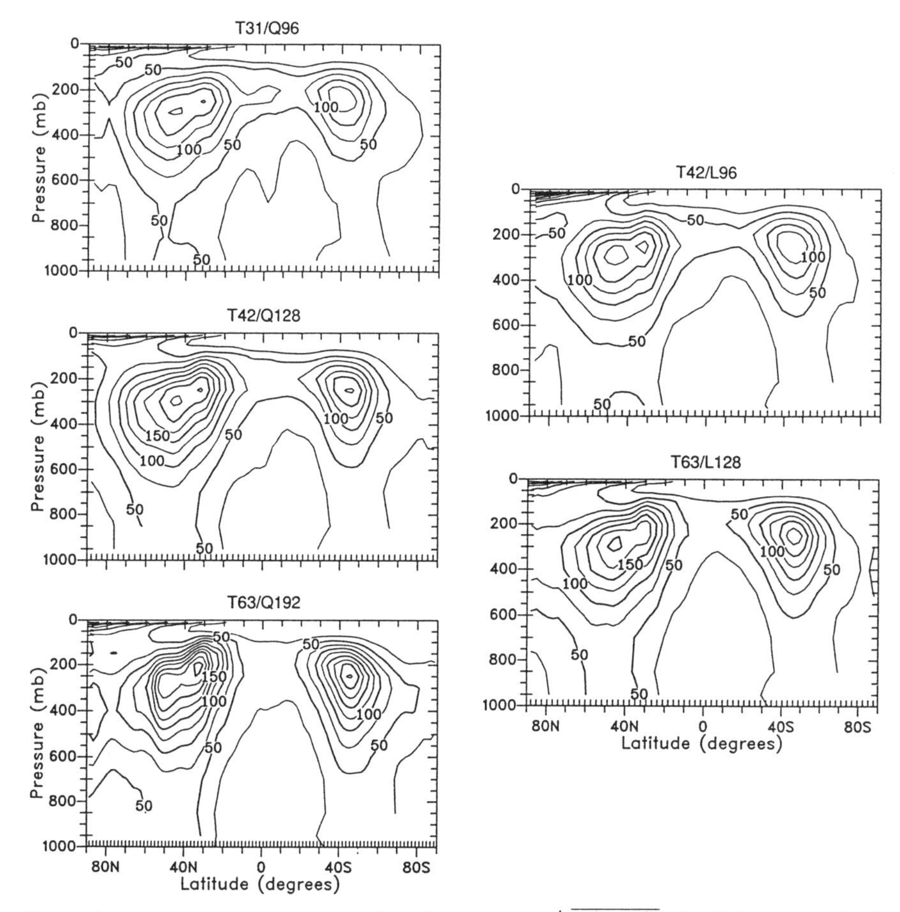

Fig. 6 January average, zonal average eddy kinetic energy, $[\frac{1}{2}\,\overline{(u'^2 + v'^2)}]$, for T31/Q96, T42/Q128, T63/Q192, T42/L96 and T63/L128 simulations. Contour interval is 25 $m^2\ s^{-2}$. Calculation was done with data interpolated to pressure surfaces.

short period for stable statistics in the mid-Latitudes. The only conclusion to be drawn from Fig. 5 is that there is nothing to indicate that the linear grids cause a problem.

Figure 6 shows the zonal average eddy kinetic energy, $[\frac{1}{2}\overline{(u'^2 + v'^2)}]$. The overbar denotes the time average, the prime denotes the deviation from the time average, and the square brackets denote the zonal average. In the simulations with the quadratic grids (left column), the upper tropical tropospheric values decrease with increasing resolution as in the Eulerian simulations (Fig. 5 of WKH95). For this feature, the T42/L96 values look more like those from the T42/Q128 simulation than those from the T31/Q96. Similarly, the T63/L128 values look more like those from the T63/Q192 than from the T42/Q128. In mid-latitudes, the amplitude of the cells increases with increasing resolution using the quadratic grids. The Eulerian simulations showed the same feature (Fig. 5 of WKH95) in the Southern Hemisphere, but the Northern Hemisphere appeared to converge around T42. The Northern Hemisphere semi-Lagrangian values reach the levels calculated from atmospheric analyses (Fig. A4 of Trenberth, 1992) while the Eulerian values fall short of analyzed levels. In both hemispheres, the T42/L96 values match the T31/Q96 rather than the T42/Q128, and the T63/L128 match the T42/Q128 rather than the T63/Q192. In the mid latitudes, the zonal average eddy kinetic energy does not show an advantage from the additional spectral resolution.

4 Conclusions

Many Eulerian spectral transform models require a quadratic unaliased grid for numerical stability. For example, the NCAR CCM2 is numerically unstable with coarser grids. Semi-Lagrangian models can be integrated stably with linear grids, without extra diffusion or smoothing. The linear grid is defined to be the minimum grid required for transformations of a field from spectral space to gridpoint space and back again to spectral without loss of information. The linear grid offers significant computational savings as it contains only $2/3$ the number of grid points of the quadratic grid in each horizontal dimension.

With a fixed spectral truncation, compared to the quadratic grid, the linear grid results in a slight increase in energy in the wavenumbers close to the spectral truncation; however, this increase is no larger than the changes introduced by arbitrary aspects of the model design. For example, the increase is comparable to that found when changing from semi-Lagrangian to Eulerian versions, or to that caused by halving the diffusion coefficient.

In Eulerian simulations with the CCM2, there is a consistent resolution signal seen for T42 truncation and above (Williamson et al., 1995). Below T42, the picture is less clear. Experiments with a variety of models show large differences between low resolution (T21 or R15 truncations) and medium resolution (T42 truncation) simulations (Boer and Lazare, 1988; Tibaldi et al., 1990; Boville, 1991; Kiehl and Williamson, 1991). Many groups conclude that a model at resolution lower than T42 seems to represent a "different" model and behaves very differently than the same model at T42 and above. The semi-Lagrangian simulations with quadratic grids show similar behaviour. The linear grid shows a clear advantage in the tropics at T63. The T63 simulation on the linear grid is comparable to the T63 simulation on the quadratic grid, but at the expense of a T42 quadratic grid simulation. The advantage of the linear grid at lower resolution (T42) is less clear. This may be because the underlying grid is too coarse to adequately represent the surface characteristics. The advantage of the linear grid is less clear in mid-latitudes for both T42 and T63 truncations, but the issue is clouded by the small sample size of two years only in the current experiments. At the very least, there is no indication that the linear grids introduce problems.

A possible additional advantage not considered in this study, is that the linear grid and spectral space are more commensurate, and problems associated with spectral ringing may be minimized. This is certainly the case in longitude, where the spectral and grid spaces have the same number of degrees of freedom, and thus spectral ringing is not introduced on the grid. Unfortunately, this is not the case in latitude, and the latitudinal transform and inverse will continue to introduce oscillations on the grid, but perhaps, their amplitudes will be less than with the quadratic grid, an aspect we have not yet examined.

Acknowledgments

I would like to thank Jerry Olson for developing the semi-Lagrangian codes and running the experiments described in this paper, James Hack for helpful comments on the original draft, and Paula Drager for editorial improvements. This work was partially supported by the Computer Hardware, Advanced Mathematics and Model Physics (CHAMMP) Program which is administered by the Office of Energy Research under the Office of Health and Environmental Research in the U.S. Department of Energy, Environmental Sciences Division.

References

ARAKAWA, A. 1966. Computational design for long-term numerical integration of the equations of fluid motion: Two-dimensional incompressible flow. *J. Comp. Phys.* **1**: 119–143.

——— and V.R. LAMB. 1977. Computational design of the basic dynamical processes of the UCLA general circulation model. In: *Methods in Computational Physics*, Vol. 17, J. Chang (Ed.), Academic Press, New York, pp. 173–265.

BOER, G.J. and M. LAZARE. 1988. Some results concerning the effect of horizontal resolution and gravity wave drag on simulated climate. *J. Clim.* **1**: 789–806.

BOURKE, W.; B. MCAVANEY, K. PURI and R. THURLING. 1977. Global modeling of atmospheric flow by spectral methods. In: *Methods in Computational Physics*, Vol. 17, J. Chang (Ed.), Academic Press, New York, pp. 267–324.

BOVILLE, B.A. 1991. Sensitivity of simulated climate to model resolution. *J. Clim.* **4**: 469–485.

BRIEGLEB, B.P. 1992. Delta-Eddington approximation for solar radiation in the NCAR Community Climate Model. *J. Geophys. Res.* **97**: 7603–7612.

CÔTÉ, J. and A. STANIFORTH. 1988. A Two-time-level semi-Lagrangian semi-implicit scheme for spectral models. *Mon. Weather Rev.* **116**: 2003–2012.

HACK, J.J. 1994. Parameterization of moist convection in the NCAR Community Climate Model, CCM2. *J. Geophys. Res.* **99**: 5551–5568.

———; B.A. BOVILLE, B.P. BRIEGLEB, J.T. KIEHL, P.J. RASCH and D.L. WILLIAMSON. 1993. Description of the NCAR Community Climate Model (CCM2). NCAR Technical Note NCAR/TN–382+STR, 108 pp.

———; ———, J.T. KIEHL, P.J. RASCH and D.L. WILLIAMSON. 1994. Climate statistics of the NCAR Community Climate Model (CCM2). *J. Geophys. Res.* **99**: 20785–20813.

HOLTSLAG, A.A. and B.A. BOVILLE. 1993. Local versus nonlocal boundary-layer diffusion in a global climate model. *J. Clim.* **6**: 1825–1842.

KIEHL, J.T. and B.P. BRIEGLEB. 1991. A new parameterization of the absorptance due to the 15 μm band system of carbon dioxide. *J. Geophys. Res.* **96**: 9013–9019.

——— and D.L. WILLIAMSON. 1991. Dependence of cloud amount on horizontal resolution in the National Center for Atmospheric Research Community Climate Model. *J. Geophys. Res.* **96**: 10955–10980

———; J. HACK and B.P. BRIEGLEB. 1994. The simulated earth radiation budget of the NCAR CCM2 and comparisons with the earth radiation budget experiment. *J. Geophys. Res.* **99**: 20815–20827.

MACHENHAUER, B. 1979. The spectral method. *In:* Numerical Methods used in Atmospheric Models, Vol. II, Global Atmospheric Research Programme, WMO-ICSU Joint Organizing Committee, pp. 121–275.

MCFARLANE, N.A. 1987. The effect of orographically excited wave drag on the general circulation of the lower stratosphere and troposphere. *J. Atmos. Sci.* **44**: 1775–1800.

SHEA, D.J.; K.E. TRENBERTH and R.W. REYNOLDS. 1990. A global monthly sea surface temperature climatology. NCAR Technical Note, NCAR/TN–345+STR, Boulder, CO, 167 pp.

TIBALDI, S.; T. PALMER, C. BRANKOVIC and U. CUBASCH. 1990. Extended range predictions with ECMWF models: II Influence of horizontal resolution on systematic error and forecast skill. Q.J.R. Meteorol. Soc. **114**: 639–664.

TRENBERTH, K.E. 1992. Global analyses from ECMWF and atlas of 1000 to 10 mb circulation statistics. NCAR/TN–373+STR, Boulder, CO 191 pp.

———; J.G. OLSON and W.G. LARGE. 1989. A global ocean wind stress climatology based on ECMWF analyses. NCAR/TN–338+STR, Boulder, CO 93 pp.

WILLIAMSON, D.L. and J.G. OLSON. 1994. Climate simulations with a semi-Lagrangian version of the NCAR Community Climate Model. *Mon. Weather Rev.* **122**: 1594–1610.

——— and P.J. RASCH. 1994. Water vapor transport in the NCAR CCM2. *Tellus*, **46A**: 31–51.

———; J.T. KIEHL and J.J. HACK. 1995. Climate sensitivity of the NCAR Community Climate Model (CCM2) to horizontal resolution. *Clim. Dyn.* **11**: 377–397.

Treatment of the Coriolis Terms in Semi-Lagrangian Spectral Models

Clive Temperton
European Centre for Medium-Range Weather Forecasts
Shinfield Park, Reading, Berks, U.K.

[Original manuscript received 1 March 1995; in revised form 25 August 1995]

ABSTRACT *In a conventional three-time-level semi-Lagrangian scheme, the Coriolis terms are treated in an explicit centred fashion. This option is not available in a two-time-level scheme, and an alternative treatment must be sought. Two possible alternatives are tested here in the framework of a three-time-level scheme. Both are stable and accurate, but only one generalizes easily to a rotated coordinate system; this alternative is based on absorbing the Coriolis terms into the semi-Lagrangian advection.*

RÉSUMÉ *Dans un schéma semi-lagrangien à trois niveaux temporels les termes de Coriolis sont traités de façon explicite et centrée. Cette option n'étant pas disponible dans un schéma à deux niveaux temporels, on doit donc chercher une solution de remplacement. On essaie deux solutions possibles dans le cadre d'un schéma à trois niveaux temporels. Les deux sont stables et précises mais une seule se généralise facilement à un système de coordonnées tournées. Cette dernière fonctionne en absorbant les termes de Coriolis dans l'advection semi-lagrangienne.*

1 Introduction

In September 1991, a new high-resolution (T213, 31-level) spectral model became operational at the European Centre for Medium-Range Weather Forecasts (ECMWF). A considerable gain in the computational efficiency of the model was required to produce operational forecasts at this resolution with the available computer resources, and to this end the new model used a three-time-level semi-Lagrangian semi-implicit integration scheme as pioneered by Robert (1981, 1982). The formulation of this model and details of its performance are described in a recent paper by Ritchie et al. (1995).

In March 1994 a new version of the model was implemented operationally. While the scientific details of the forecast model itself remained essentially the same, the new code included many additional features required for three- and four-dimensional variational data assimilation (Thépaut and Courtier, 1991; Andersson et al., 1994) and for determining optimal unstable perturbations for ensemble prediction (Buizza et al., 1993). This new code was developed jointly by ECMWF (where it is known as the Integrated Forecast System, IFS) and Météo-France (where it is known as ARPEGE). In the present context, one aspect of note is the operational use of this model by Météo-France in rotated and stretched mode (Courtier and Geleyn, 1988).

The model is currently being adapted to make use of a *two*-time-level semi-Lagrangian scheme, both to take advantage of greater computational efficiency (Staniforth and Côté, 1991) and in the hope of reducing storage requirements when the model is used for four-dimensional data assimilation. Since the Coriolis terms can then no longer be treated in a simple explicit manner as in Ritchie et al. (1995), it is necessary to seek an alternative treatment. There are at least two possibilities, both of which can be evaluated within the framework of the well-established three-time-level scheme. These investigations form the subject of the present paper.

2 Semi-implicit treatment of Coriolis terms

The first and more obvious alternative is to treat the Coriolis terms semi-implicitly (i.e., to average them in time and space along the trajectory). As in Ritchie et al. (1995), the resulting set of coupled equations to be solved for the variables at the new time level $(t + \Delta t)$ has the same form whether the underlying scheme is Eulerian or semi-Lagrangian. For simplicity, the details are set out here for the corresponding shallow-water equation model. The equations take the form:

$$U^{+} - f\Delta t V^{+} + \frac{\Delta t}{a}\frac{\partial \phi^{+}}{\partial \lambda} = R_1 \tag{1}$$

$$V^{+} + f\Delta t U^{+} + \frac{\Delta t}{a}\cos\theta\,\frac{\partial \phi^{+}}{\partial \theta} = R_2 \tag{2}$$

$$\phi^{+} + \Phi\Delta t D^{+} = S_3 \tag{3}$$

where the superscript + indicates unknown values at $(t + \Delta t)$, while the right-hand sides contain all the known terms. In (1)–(3) U and V are the wind components multiplied by $\cos\theta$, ϕ is geopotential, D is divergence, and Φ is the mean geopotential.

Transforming (1)–(3) to spectral space and taking the curl and divergence of the equations for the wind components yields, for each zonal wavenumber m and for $m \leqslant n \leqslant N$ where N is the maximum total wavenumber:

$$(1 - i\alpha_n^m)\zeta_n^{+} + 2\Omega\Delta t\left[\frac{(n+1)}{n}\,\varepsilon_n^m D_{n-1}^{+} + \frac{n}{(n+1)}\,\varepsilon_{n+1}^m D_{n+1}^{+}\right] = (S_1)_n \tag{4}$$

$$(1 - i\alpha_n^m)D_n^+ - 2\Omega\Delta t\left[\frac{(n+1)}{n}\,\varepsilon_n^m\zeta_{n-1}^+ + \frac{n}{(n+1)}\,\varepsilon_{n+1}^m\zeta_{n+1}^+\right]$$

$$- \frac{\Delta t n(n+1)}{a^2}\,\phi_n = (S_2)_n \tag{5}$$

$$\phi_n^+ + \Phi\Delta t D_n^+ = (S_3)_n \tag{6}$$

where

$$\alpha_n^m = \frac{2\Omega m\Delta t}{n(n+1)}, \qquad \varepsilon_n^m = \left(\frac{n^2 - m^2}{4n^2 - 1}\right)^{1/2},$$

$$S_1 = \mathrm{curl}(R_1, R_2) \quad \text{and} \quad S_2 = \mathrm{div}(R_1, R_2).$$

In (4)–(6) the zonal wavenumber index m has for clarity been omitted from the spectral coefficients ζ_n^+, D_n^+, ϕ_n^+. Equations (4)–(6) are derived from (1)–(3) using standard recurrence relations and orthogonality properties of the Legendre polynomials, and in practice may be obtained by adapting Bourke's (1972) derivation of the spectral shallow-water equations.

These equations differ in two important ways from those obtained without the additional Coriolis terms. First, the vorticity and divergence equations are coupled together. Second, although an independent set of equations is obtained for each zonal wavenumber m, the vorticity components at total wavenumber n are coupled to the divergence components at total wavenumbers $(n-1)$ and $(n+1)$, and vice versa.

To solve the equations, all the spectral components ($m \leqslant n \leqslant N$) for each variable are assembled into a vector for each wavenumber m. Thus, (4)–(6) become:

$$\mathbf{J}\boldsymbol{\zeta}^+ + \mathbf{F}\mathbf{D}^+ = \mathbf{S}_1 \tag{7}$$

$$\mathbf{J}\mathbf{D}^+ - \mathbf{F}\boldsymbol{\zeta}^+ - \mathbf{L}\boldsymbol{\phi}^+ = \mathbf{S}_2 \tag{8}$$

$$\boldsymbol{\phi}^+ + \Phi\Delta t\mathbf{D}^+ = \mathbf{S}_3 \tag{9}$$

where

$$\mathbf{J} = \mathrm{diag}\{1 - i\alpha_n^m\},$$

$$\mathbf{F} = 2\Omega\Delta t\begin{bmatrix} 0 & f_m^+ & & & & \\ f_{m+1}^- & 0 & f_{m+1}^+ & & & \\ & f_{m+2}^- & 0 & f_{m+2}^+ & & \\ & & \ddots & \ddots & \ddots & \\ & & & f_{N-1}^- & 0 & f_{N-1}^+ \\ & & & & f_N^- & 0 \end{bmatrix},$$

$$f_n^+ = \frac{n}{(n+1)} \varepsilon_{n+1}^m,$$

$$f_n^- = \frac{(n+1)}{n} \varepsilon_n^m,$$

$$\mathbf{L} = \frac{\Delta t}{a^2} \operatorname{diag}\{n(n+1)\}.$$

Equations (7)–(9) can then be combined to give:

$$(\mathbf{J} + \mathbf{F}\mathbf{J}^{-1}\mathbf{F} + \Phi\Delta t\mathbf{L})\mathbf{D}^+ = \mathbf{S}_2 + \mathbf{F}\mathbf{J}^{-1}\mathbf{S}_1 + \mathbf{L}\mathbf{S}_3. \quad (10)$$

The matrix on the left-hand side of (10) is pentadiagonal with zeros on the sub- and superdiagonals; thus (10) decouples into two tridiagonal systems, one for the "even" components and one for the "odd" components. The tridiagonal matrices have *complex* entries, but they are diagonally dominant and the usual solution algorithm (equivalent to Gaussian elimination without pivoting) is stable. Equation (10) is essentially the same as that obtained by Côté and Staniforth (1988) for a two-time-level semi-Lagrangian spectral model.

Proceeding to the case of a multi-level model, the semi-implicit equations including Coriolis terms may be solved by diagonalizing the vertical operators which couple the unknowns at different model levels, resulting in a set of equations of the form (7)–(9) for each vertical mode, with the mean geopotential height $\mathbf{\Phi}$ in (9) being replaced by the equivalent geopotential depth for each mode.

The semi-implicit treatment of the Coriolis terms was tested in the three-time-level semi-Lagrangian spectral model, using the "vertically non-interpolating" version (Ritchie et al., 1995) at resolution T106, 31 levels, and compared with the standard treatment of the Coriolis terms over a set of six independent cases evenly spaced throughout the year. In terms of verification scores, the choice of treatment of the Coriolis terms has very little impact on the results. A typical example is shown in Fig. 1, for the 500-hPa height field over the Northern Hemisphere. Visual comparison of charts confirms that the forecast fields are very similar.

These results suggest that a semi-implicit treatment of the Coriolis terms would also be a viable option in a *two*-time-level semi-Lagrangian integration scheme. A problem would, however, arise in the case of rotated and stretched coordinates: the rotation would destroy the horizontal separability of the equations to be solved in spectral space (essentially because the Coriolis parameter becomes a function of longitude as well as latitude in the transformed coordinate system). Although it should be possible to solve the resulting set of equations in spectral space, it would certainly be more complicated (and more expensive) than in the non-rotated case.

3 Advective treatment of Coriolis terms

A less obvious alternative was proposed by Rochas (1990). Again for simplicity, we

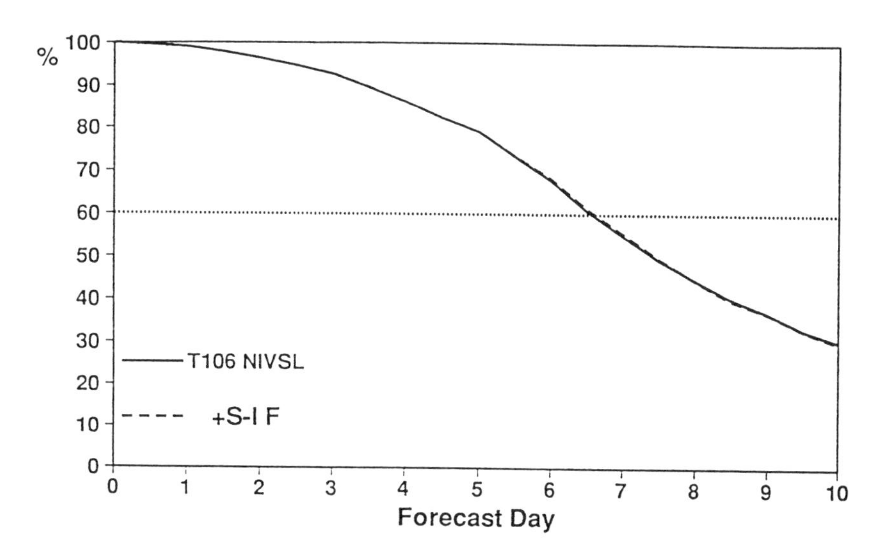

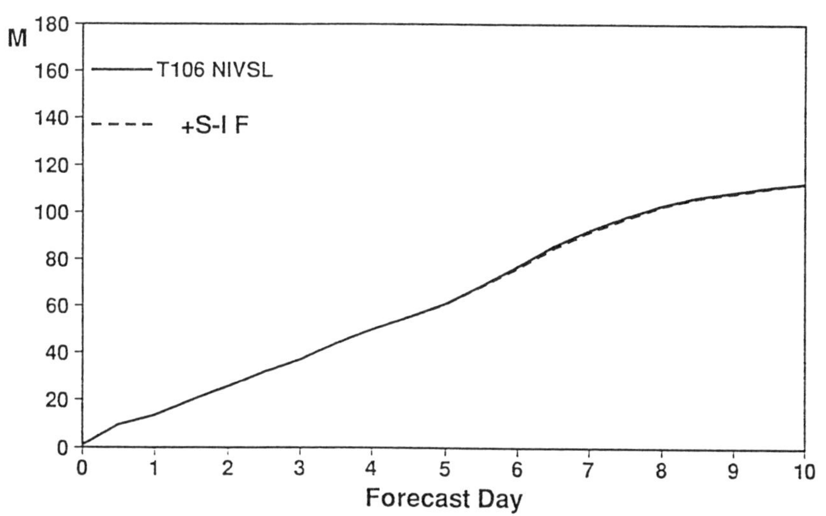

Fig. 1 Anomaly correlation and rms error scores for the 500-hPa height field over the Northern Hemisphere, averaged over 6 cases. Solid line, standard formulation; dashed line, with semi-implicit Coriolis terms.

consider the treatment of the Coriolis terms in the shallow-water equations. Recall that semi-Lagrangian schemes on the sphere handle the momentum equations in *vector* form to avoid an instability due to the metric term (Ritchie, 1988). Thus, the momentum equation is written as

$$\frac{d\boldsymbol{v}}{dt} + 2\boldsymbol{\Omega} \times \mathbf{v} + \nabla\phi = 0 \tag{11}$$

where the total derivative operator

$$\frac{d}{dt} \equiv \frac{\partial}{\partial t} + \mathbf{v} \cdot \nabla$$

is discretized in a semi-Lagrangian fashion. In the case of the advection of a *vector* quantity such as $\mathbf{v} = (u, v)$ on the sphere, it has to be borne in mind that the orientation of the local coordinate system changes as we move along the trajectory (Ritchie, 1988).

Rochas (1990) pointed out that, since $\mathbf{v} = d\mathbf{r}/dt$ where $\mathbf{r}$ is the radial position vector, (11) can be rewritten as

$$\frac{d}{dt}(\boldsymbol{v} + 2\boldsymbol{\Omega} \times \mathbf{r}) + \nabla \phi = 0. \tag{12}$$

In a semi-Lagrangian discretization of (12), the Coriolis terms are absorbed into the advection. The change does not affect the trajectory itself, but only the quantity being advected. Since the term $2\boldsymbol{\Omega} \times \mathbf{r}$ is known everywhere (in component form it is just $(2\Omega\, a \cos\theta, 0)$), the extra term is simply added to $\mathbf{v}$ at the departure point of the trajectory and subtracted again at the arrival gridpoint. The equations to be solved for the variables at the new time-level $(t + \Delta t)$ have exactly the same form as in the case of a simple explicit treatment of the Coriolis terms; the only change is that the right-hand sides have been computed in a different way.

Thus, the idea proposed by Rochas (1990) has a clear advantage over the semi-implicit treatment of the Coriolis terms in the rotated and stretched configuration of the model, since the equations resulting from the semi-implicit scheme remain horizontally separable and easy to solve (the Laplacian operator is invariant with respect to a rotation of the coordinate system).

Preliminary tests of this option within the three-time-level scheme, using the "old" (pre-IFS) ECMWF model at T106 resolution, were disappointing. Figure 2 shows the verification scores for the 500-hPa height field over the Southern Hemisphere, averaged over a sample of four independent cases. The solid line is for the standard semi-Lagrangian scheme, while the dashed line is for the alternative treatment and shows a clear degradation of the results. It was then realized that in order to speed up the calculation, the treatment of spherical geometry in the determination of the trajectory and handling of the advection of the wind vector (Ritchie, 1988) had been replaced by approximations as described by Ritchie and Beaudoin (1994), moreover leaving out some higher-order terms. Although these approximations were perfectly adequate for the conventional formulation of the semi-Lagrangian scheme, it was suspected that the accurate treatment of spherical geometry might be more important for the new formulation. This was confirmed by removing the approximations and reverting to a more accurate treatment of the geometry, resulting in the dotted line in Fig. 2.

The option based on a semi-Lagrangian discretization of (12) has more recently

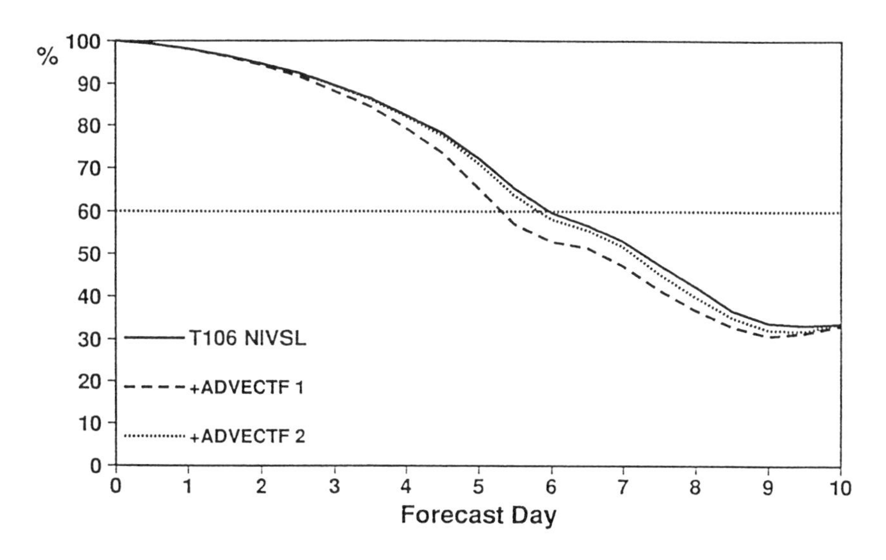

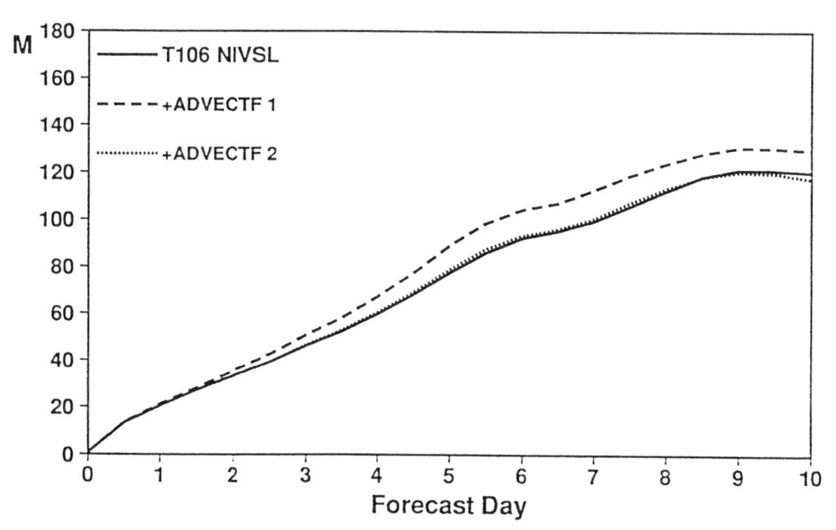

Fig. 2 Anomaly correlation and rms error scores for the 500-hPa height field over the Southern Hemisphere, averaged over 4 cases, using the "old" (pre-IFS) version of the model. Solid line, standard formulation; dashed line, with Coriolis terms absorbed into the advection; dotted line, the same but with more accurate spherical geometry.

been evaluated in the current (IFS) version of the ECMWF model, which retains an accurate treatment of spherical geometry. Figure 3 shows results for the 500-hPa height field over the Northern Hemisphere, averaged over six independent cases. These results indicate, if anything, a slight advantage of the new formulation over the conventional treatment of the Coriolis terms. As a result of these experiments,

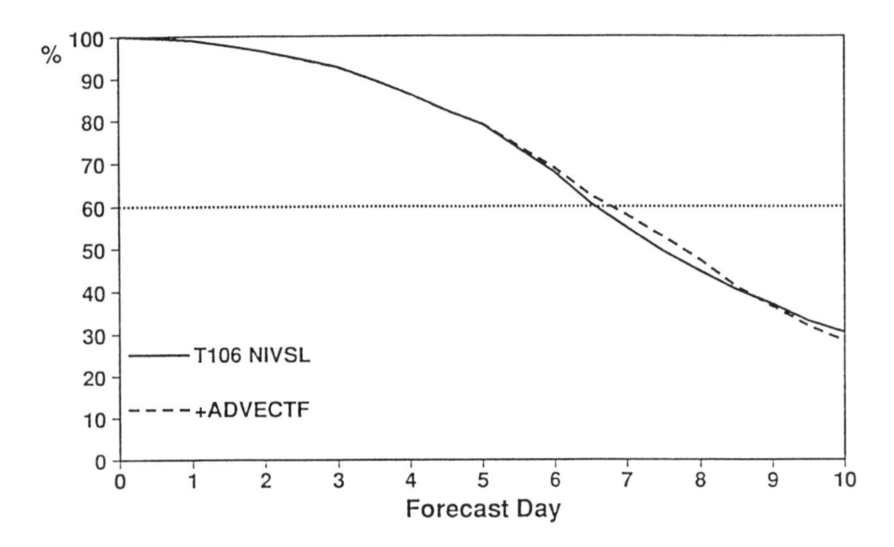

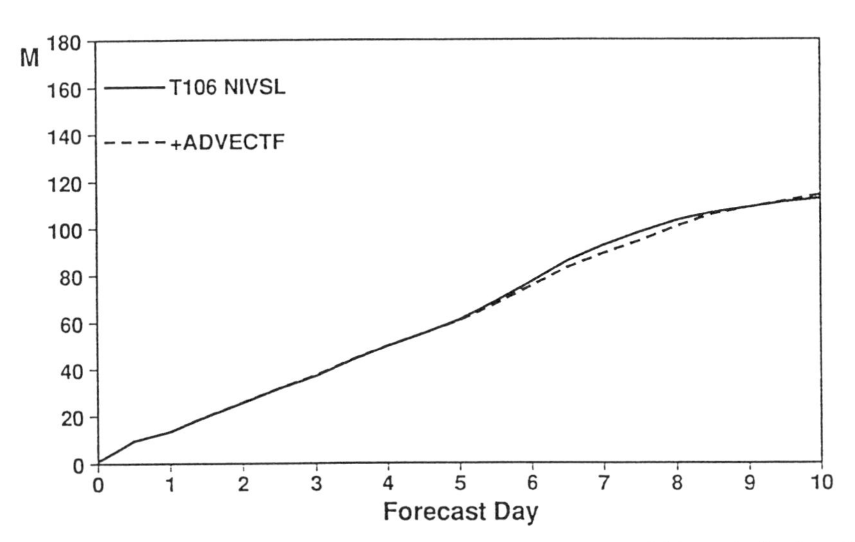

Fig. 3 Anomaly correlation and rms error scores for the 500-hPa height field over the Northern Hemisphere, averaged over 6 cases, using the IFS version of the model. Solid line, standard formulation; dashed line, with Coriolis terms absorbed into the advection.

the new formulation will shortly become the default option in the operational three-time-level model.

The "advective" treatment of the Coriolis terms in a two-time-level semi-Lagrangian scheme is exactly analogous to that in the three-time-level scheme. There is one potential drawback: since the determination of the trajectory in a two-time-level scheme requires a wind field extrapolated forward in time (Staniforth and Côté, 1991), it is conceivable that incorporating the Coriolis terms in the advection might be unstable. This possibility is hinted at by the analysis of Bates et al. (1995) for a two-time-level scheme based on the advection of potential vorticity, though the scheme of Bates et al. is by no means equivalent to that proposed here. Fortunately, we now have enough experience with a preliminary two-time-level version of the ECMWF semi-Lagrangian spectral model to be sure that the treatment of the Coriolis terms proposed by Rochas (1990) does remain stable when the advection is based on a time-extrapolated wind field.

4 Conclusions

Two possible treatments of the Coriolis terms in a two-time-level semi-Lagrangian model have been tested by modifying the ECMWF three-time-level semi-Lagrangian spectral model. Including the Coriolis terms in the semi-implicit scheme is stable and accurate, but results in a difficult problem to be solved in spectral space when the model is used with a rotated coordinate system. An alternative scheme proposed by Rochas (1990), in which the Coriolis terms are incorporated in the semi-Lagrangian advection, is also stable and accurate provided that the spherical geometry of the problem is handled accurately. This alternative scheme has the additional advantage that the problem to be solved in spectral space has just the same simple form in a rotated coordinate system as in the corresponding unrotated case.

References

ANDERSSON, E.; J. PAILLEUX, J-N. THÉPAUT, J.R. EYRE, A.P. MCNALLY, G.A. KELLY and P. COURTIER. 1994. Use of cloud-cleared radiances in three/four-dimensional variational data assimilation. *Q. J. R. Meteorol. Soc.* **120**: 627–653.

BATES, J.R.; Y. LI, A. BRANDT, S.F. MCCORMICK and J. RUGE. 1995. A global barotropic primitive equation model based on the semi-Lagrangian advection of potential vorticity. *Q. J. R. Meteorol. Soc.* **121**: 1981–2005.

BOURKE, W. 1972. An efficient, one-level, primitive equation spectral model. *Mon. Weather Rev.* **100**: 683–689.

BUIZZA, R.; J. TRIBBIA, F. MOLTENI and T. PALMER. 1993. Computation of optimal unstable structures for a numerical weather prediction model. *Tellus*, **45A**: 388–407.

CÔTÉ, J. and A. STANIFORTH. 1988. A two-time-level semi-Lagrangian semi-implicit scheme for spectral models. *Mon. Weather Rev.* **116**: 2003–2012.

COURTIER, P. and J-F. GELEYN. 1988. A global weather prediction model with variable resolution: application to shallow-water equations. *Q. J. R. Meteorol. Soc.* **114**: 1321–1346.

RITCHIE, H. 1988. Application of the semi-Lagrangian method to a spectral model of the shallow water equations. *Mon. Weather Rev.* **116**: 1587–1598.

——— and C. BEAUDOIN. 1994. Approximations and sensitivity experiments with a baroclinic semi-Lagrangian spectral model. *Mon. Weather Rev.* **122**: 2391–2399.

———; C. TEMPERTON, A. SIMMONS, M. HORTAL, T. DAVIES, D. DENT and M. HAMRUD. 1995. Implementation of the semi-Lagrangian method in a high-resolution version of the ECMWF forecast model. *Mon. Weather Rev.* **123**: 489–514.

ROBERT, A. 1981. A stable numerical integration scheme for the primitive meteorological equations. ATMOSPHERE-OCEAN, **19**: 35–46.

———. 1982. A semi-Lagrangian and semi-implicit numerical integration scheme for the primitive meteorological equations. J. Meteorol. Soc. Jpn. **60**: 319–325.

ROCHAS, M. 1990. ARPEGE Documentation, Part 2, Ch. 6 (available from Météo-France).

STANIFORTH, A. and J. CÔTÉ. 1991. Semi-Lagrangian integration schemes for atmospheric models – a review. *Mon. Weather Rev.* **119**: 2206–2223.

THÉPAUT, J-N. and P. COURTIER. 1991. Four-dimensional variational data assimilation using the adjoint of a multilevel primitive-equation model. *Q. J. R. Meteorol. Soc.* **117**: 1225–1254.

The Use of Adjoint Equations in Numerical Weather Prediction

P. Courtier and F. Rabier
ECMWF, Shinfield Park, Reading Berkshire RG2 9AX, U.K.

[Original manuscript received 23 January 1995; in revised form 28 September 1995]

ABSTRACT *The adjoint equations allow computation of the sensitivity of one output parameter of a model to all input parameters. After a brief introduction of this technique, its main applications to numerical weather prediction are described.*

The most trivial application of the adjoint model is to investigate the sensitivity to initial conditions or model parameters. The fact that the adjoint method allows computation of the gradient of a cost function with respect to some parameters in an efficient way makes the minimization of a cost function using descent algorithms possible. This may be applied to estimation problems like variational assimilation.

Another use of the adjoint model is the evaluation of the covariances of forecast error in the Kalman filtering context. Finally, the estimation of the singular vectors of a linearized model using its adjoint is relevant for predictability studies.

RÉSUMÉ *Les équations adjointes permettent le calcul de la sensibilité d'un paramètre de sortie d'un modèle à tous les paramètres d'entrée. Une brève explication de la technique est suivie de ses principales applications à la prévision météorologique numérique.*

L'application la plus simple du modèle adjoint est l'examen de la sensibilité aux conditions initiales ou aux paramètres du modèle. Le fait que la méthode adjointe permette le calcul du gradient d'une fonction de coût par rapport à certains paramètres de façon efficace, rend possible la minimisation d'une fonction de coût en utilisant des algorithmes descente. On peut appliquer cette technique aux problèmes d'estimation comme l'assimilation variationnelle.

Un autre usage du modèle adjoint est l'évaluation des covariances des erreurs de prévision dans le contexte du filtrage de Kalman. Enfin, l'estimation des vecteurs singuliers d'un modèle linéaire à l'aide de son adjoint est pertinente pour les études de prévisibilité.

1 Introduction

A wide range of problems pertain to evaluating the impact of some changes to the input parameters of a model on the output parameters. Mathematically, this can be interpreted in the linear context as the computation of the Jacobian matrix of the model (the matrix of the partial derivatives of the output parameters with respect to the input parameters).

In numerical experimentation one often performs two experiments which differ only in the specification of one input parameter of a model and then studies the impact on the output parameters. A typical example is the study of the impact of a doubling of the quantity of carbon dioxide on atmospheric circulation. This is approximately equivalent to computing one column of the Jacobian matrix, by means of finite differences.

In other types of problems one may need to determine the sensitivity of one output parameter with respect to all of the input parameters. For example, if the forecast of a low is not deep enough, one may want to evaluate the sensitivity of this forecast error to the initial conditions, and in operational practice this is done by the bench forecasters tracing back the forecast error. Mathematically, the problem is to compute a row of the Jacobian matrix.

We have seen above that a column of the Jacobian matrix could be evaluated at the cost of a few experiments (actually two with the finite difference algorithm). Is it possible to evaluate a row of the Jacobian matrix at a similar cost? This question may be rephrased as: Is it possible to devise an algorithm for applying the transpose of the Jacobian matrix to a vector at a reasonable cost?

Such an algorithm can be provided by adjoint equations, introduced in meteorology by Marchuk (1974). A row of the Jacobian matrix can then be computed for large-scale problems at a reasonable cost, and applications using such computations have been implemented in practice. Since then, the adjoint of a large number of 3D primitive equation models has been developed.

The purpose of this paper is to introduce the main applications of the adjoint equations which have been performed in numerical weather prediction so far. After a brief (and non mathematical) introduction of the adjoint equations (Section 2) we present an overview of their applications by considering their application to sensitivity experiments (Section 3); in Section 4, we introduce the most widely known application of variational assimilation which is a particular case of estimation, Section 5 is devoted to Kalman filtering and Section 6 to predictability with the estimation of the singular vectors of a linearized model. The examples used have been taken, for the most part, from achievements of the ARPEGE/IFS (Action de Recherche Petite Echelle Grande Echelle/Integrated Forecasting System) collaboration between ECMWF (European Centre for Medium-range Forecasts) and Météo-France. The purpose of the collaboration was to develop variational assimilation and therefore the adjoint of a 3D primitive equation model. For a more exhaustive list of references (up to 1992), the reader is referred to Courtier et al. (1993b). The adjoint equations have been used for a long time by mathematicians as a theoretical tool; this is not discussed in the present paper.

2 The Adjoint Equations

Let $\mathbf{G}$ be an operator, differentiable almost everywhere, which computes the output parameters v from the input parameters $\mathbf{u}$:

$$\mathbf{v} = \mathsf{G}(\mathbf{u}) \tag{1}$$

G may be available in practice as a computer program like a numerical weather prediction model. For perturbations $\delta\mathbf{u}$ and $\delta\mathbf{v}$ of $\mathbf{u}$ and $\mathbf{v}$, the Taylor formula provides the following equality, valid to the first order:

$$\delta\mathbf{v} = \mathsf{G}'_{\mathbf{u}}\delta\mathbf{u} \tag{2}$$

where $\mathsf{G}'_{\mathbf{u}}$ is the tangent-linear operator of G linearized in the vicinity of $\mathbf{u}$. The matrix representing $\mathsf{G}'_{\mathbf{u}}$ is the Jacobian matrix of G. In most meteorological applications, $\delta\mathbf{u}$ and $\delta\mathbf{v}$ are vectors of size 10^5 to 10^7 (the latter being valid for the 31 level, spectral with triangular truncation at total wavenumber 213 (T213L31) ECMWF operational model and the former for a T42L19 version). The Jacobian matrix cannot then be explicitly computed (and even stored). However, the following two results make it possible to apply the operator $\mathsf{G}'_{\mathbf{u}}$ to an input vector $\delta\mathbf{u}$ and its transpose to an input vector $\delta\mathbf{v}$.

(a) It is possible to write a computer program which solves (2) once the computer program for (1) is available. Its cost is of a similar order of magnitude to (1).
(b) It is possible to write a computer program of similar cost to (1) which computes an output vector $\delta\mathbf{u}$ for a given input vector $\delta\mathbf{v}$:

$$\delta_{\mathbf{u}} = \mathsf{G}'^{t}_{\mathbf{u}}\delta_{\mathbf{v}} \tag{3}$$

where $\mathsf{G}'^{t}_{\mathbf{u}}$ is the transpose of $\mathsf{G}'_{\mathbf{u}}$. This can be generalized to the adjoint $\mathsf{G}'^{*}_{\mathbf{u}}$ of $\mathsf{G}'_{\mathbf{u}}$ through appropriate metric (inner product) changes.

Result (a) is trivial in practice since the result of (2) could, for example, be obtained to the first order by finite differences in the direction $\delta\mathbf{u}$ from two integrations of (1): this is the methodology generally followed in most sensitivity experiments. For an exact solution to (2), Morgenstern (1973) discusses the complexity of tangent-linear algorithms. However, result (b) is not trivial and relies on the Baur-Strassen theorem (Baur and Strassen, 1983; Morgenstern, 1985). In the following we shall describe the practical applications of results (a) and (b).

3 Sensitivity Analysis

Let us consider $J(v)$ a function of the output parameters $\mathbf{v}$. We assume J to be "simple" in that an analytic expression of the gradient J with respect to $\mathbf{v}$, $\nabla_{\mathbf{v}}J$ is available (note that this is not a restriction, it is always possible to consider J as a simple function of something, e.g., $J = J$).

Let us introduce an inner product $\langle,\rangle_{\mathbf{v}}$ in the space of the output parameters $\mathbf{v}$.

By definition of $\nabla_{\mathbf{v}} J$, one has for any perturbation $\delta\mathbf{v}$ (and to first order)

$$\Delta J = J(v + \delta v) - J(v) = \langle \nabla_{\mathbf{v}} J,\ \delta \mathbf{v} \rangle_{\mathbf{v}}.$$

$\delta\mathbf{v}$ may be related to $\delta\mathbf{u}$ using (2) which leads to

$$\Delta J = \langle \nabla_{\mathbf{v}} J,\ \mathbf{G}'_{\mathbf{u}} \delta \mathbf{u} \rangle_{\mathbf{v}}$$

denoting $\langle , \rangle_{\mathbf{u}}$ the inner product in the space of the input parameters $\mathbf{u}$, by definition of the adjoint operator

$$\Delta J = \langle \mathbf{G}'^{*}_{\mathbf{u}} \nabla_{\mathbf{v}} J,\ \delta \mathbf{u} \rangle_{\mathbf{u}}$$

which implies that

$$\nabla_{\mathbf{u}} J = \mathbf{G}'^{*}_{\mathbf{u}} \nabla_{\mathbf{v}} J. \tag{4}$$

As a consequence of result (b), it is then possible to compute the sensitivity of J with respect to the input parameters $\mathbf{u}$: the gradient of J with respect to $\mathbf{u}$ is obtained by applying the adjoint of the tangent-linear operator $\mathbf{G}'^{*}_{\mathbf{u}}$ to the gradient of J with respect to $\mathbf{v}$.

a *Sensitivity to initial conditions*

Here we consider $\mathbf{u}$ as the initial conditions of a numerical weather prediction model and J as one output. Courtier (1987), using a shallow-water model, identified a tidal wave problem in nonlinear normal mode initialization. Errico and Vukicevic (1992) studied a case of lee cyclogenesis and concluded that a more intense synoptic wave (stronger ridge over the Atlantic and deeper over France) would lead to a more intense lee cyclogenesis.

Rabier et al. (1992) studied the sensitivity of the baroclinic instability of Simmons and Hoskins (1978) to the initial conditions 24 hours before, during the 24 hours of most intense cyclogenesis. They showed that it is easy to eliminate the gravity wave signal present in the sensitivity pattern using the adjoint of nonlinear normal mode initialization as can be seen comparing Fig. 1 (their Fig. 3) and Fig. 2 (their Fig. 5). This gravity wave signal was not visible at the 36-hour range in the Errico and Vukicevic work reported in the previous paragraph since it had already been dissipated at the boundaries (however, it was visible at very short range).

The vertical structure of the sensitivity showing a maximum in the low troposphere is depicted in Fig. 3 (their Fig. 10) which is consistent with other baroclinic instability studies.

Rabier et al. (1993b) performed a feasibility study of the use of the adjoint equations for routinely monitoring the sensitivity of the short-range (48-hour) forecast errors with respect to the initial conditions. The inner product used in the definition of the cost function corresponds to the quadratic invariant of the linearized primitive equations in the vicinity of a state of rest and the cost function is the square

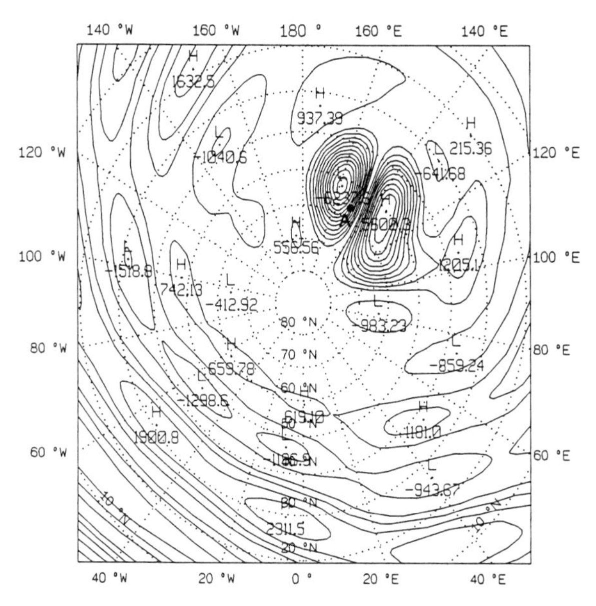

Fig. 1 Sensitivity of the logarithm of the surface pressure at point A (middle of the surface pressure low) to the 500 hPa meridional wind 24 hours earlier (in $m^{-1}s$ (1.0 E+9), isolines every 500). (Reproduced with permission from Rabier et al., 1992). The inner product is the mean square of the wind field.

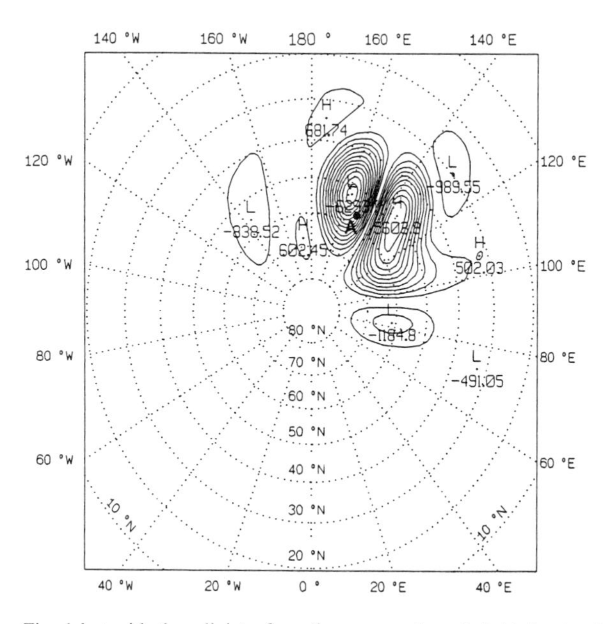

Fig. 2 Same as Fig. 1 but with the adjoint of nonlinear normal mode initialization included. (Reproduced with permission from Rabier et al., 1992).

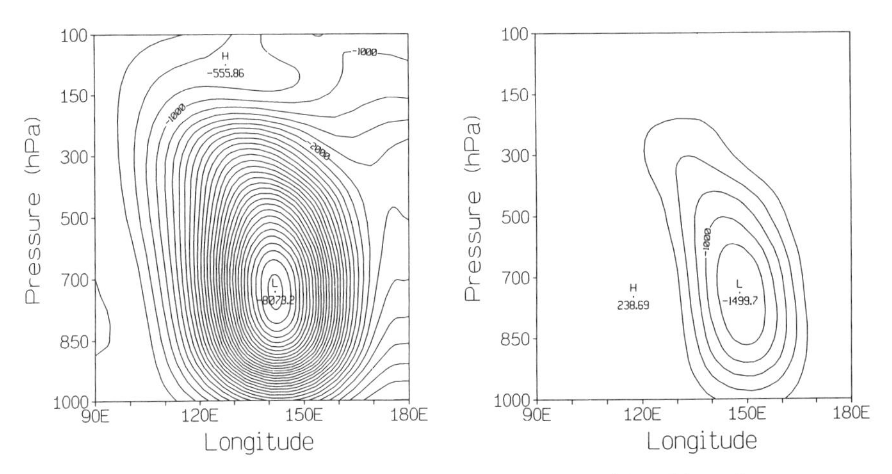

Fig. 3 Cross-section at 52°N of the gradient of the average of the logarithm of the surface pressure over the low with respect to the initial conditions 24 hours earlier. Left: vorticity (in s (1.0E+3), isolines every 250), right: temperature (in K^{-1} (1.0E+8), isolines every 250). (Reproduced with permission from Rabier et al., 1992). The inner product is the mean square of the vorticity (respectively temperature) field.

norm of the difference **e** between the operational 48-h forecast and the verifying analysis.

$$J = \frac{1}{2}\langle \mathbf{e_{48}},\ \mathbf{e_{48}} \rangle$$

$\mathbf{e_{48}}$ is assumed to be related to the initial condition errors $\mathbf{e_0}$ through the integration of the tangent linear version of a T63L31 adiabatic model (with horizontal and vertical diffusion and a surface drag) denoted $\mathbf{M}'$, linearized in the vicinity of an adiabatic trajectory originated from the ECMWF analysis valid 48 hours before.

$$\mathbf{e_{48}} = \mathbf{M}'(48,\ 0)\mathbf{e_0}$$

The developments of Section 2 may be used to obtain the gradient of J with respect to the initial conditions $\nabla_0 J$.

$$\nabla_0 J = \mathbf{M}'^*(48,\ 0)\nabla_{48} J = \mathbf{M}'^*(48,\ 0)\mathbf{e_{48}}$$

At the end of the adjoint integration, the adjoint of nonlinear normal mode initialization is performed. The inner product with respect to which the gradients are displayed is defined by $\langle , \rangle_G = \sum_i \sum_l w_i \mathbf{x}_i^l \mathbf{y}_i^l$ where the summation extends over all the horizontal levels l and all grid points i with w_i representing the Gaussian weight at grid point i, i.e. the percentage of the globe surface occupied by this grid point.

Such gradient computations were performed for every day of the month of Jan-

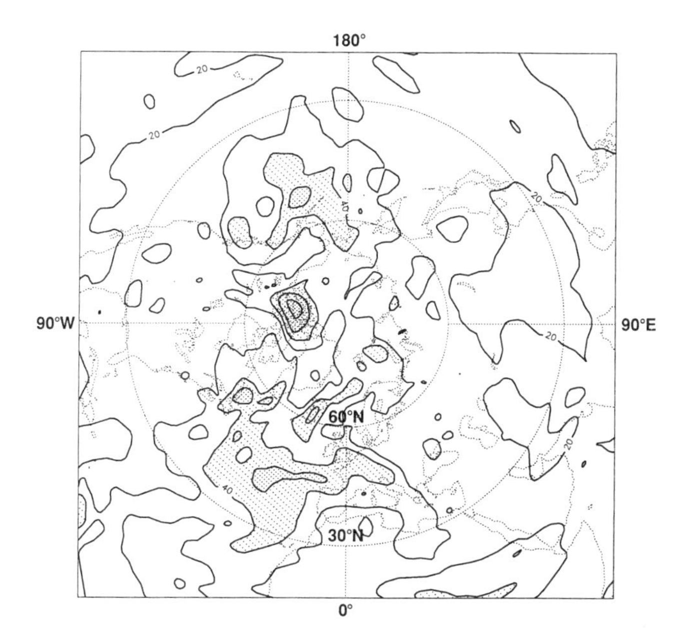

Fig. 4 Rms computed over the month of January of the gradient of the 48-h forecast errors with respect to the vorticity at model level 11 (−250 hPa) of the initial conditions. See Section 3a for the mathematical definition. The units are kg m^3 s^{-3}. Isolines are every 20. (Reproduced with permission from Rabier et al., 1993b).

uary 1993 and Figs 4, 5 and 6 (extracted from Rabier et al. 1993b) present the rms over the month of January 1993 of the vorticity fields at model level 11 (~250 hPa), level 18 (~500 hPa) and level 26 (~850 hPa) of $\nabla_0 J$. In agreement with Rabier et al. (1992), the gradient is stronger with respect to mid troposphere vorticity. It is stronger over the oceans than over the continents which is consistent with better data coverage over the continents.

Figure 7 shows the average over the month of January 1993 of the analyzed 500-hPa geopotential height. The maxima in the Atlantic of Fig. 5 are located in the cyclone track going from Newfoundland to the British Isles, with little sensitivity in the different areas over Scandinavia and central Europe. The maximum located north of Hudson Bay corresponds to the descending branch of the Arctic jet.

Rabier et al. (1993b) looked at the temporal variation of the cost function (forecast errors square norm), of the square norm of the gradient and of the rms of the gradient with respect to the vorticity at level 26 (not shown). There is little day-to-day variability in the forecast error norms. However, the sensitivity to the initial conditions can vary a lot and by as much as a factor of 4. There is no apparent correlation with the day-to-day variability of the medium range scores.

The rms of the gradient with respect to vorticity for each model level and for each day was computed. The sensitivity is maximum at around 850 hPa, remaining

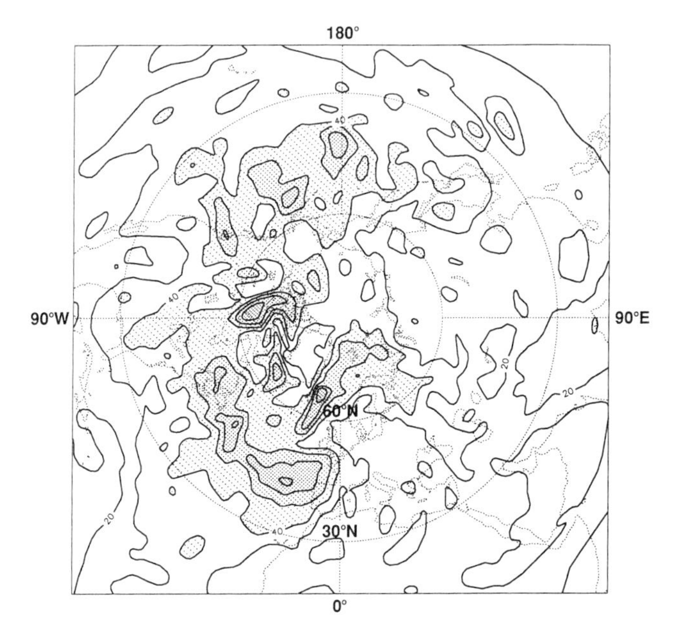

Fig. 5 Same as Fig. 4 but for model level 18 (–500 hPa). (Reproduced with permission from Rabier et al., 1993b).

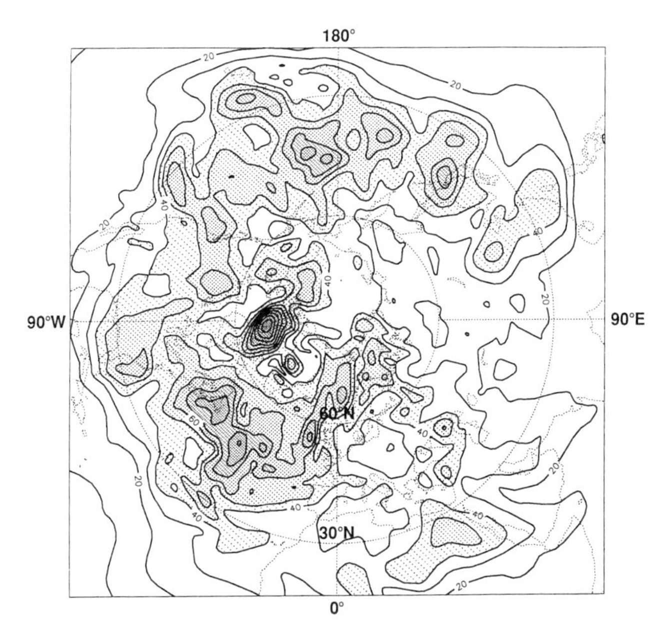

Fig. 6 Same as Fig. 4 but for model level 26 (–850 hPa). (Reproduced with permission from Rabier et al., 1993b).

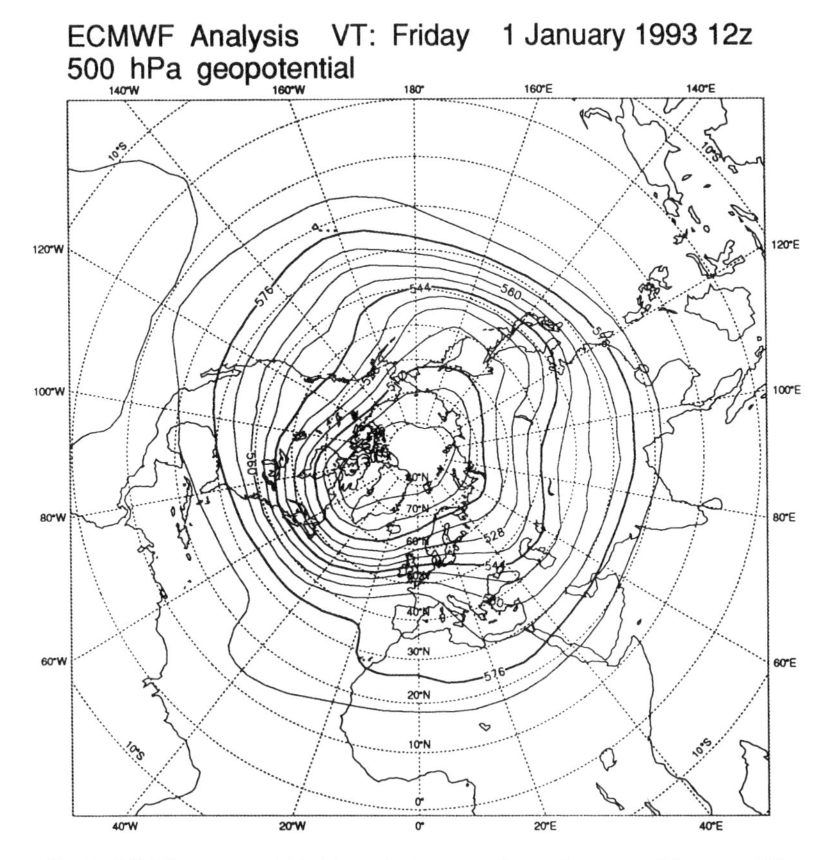

Fig. 7 500-hPa geopotential height analysis averaged over the month of January 1993.

almost constant up to the jet level. The sensitivity is low at the surface and in the middle of the stratosphere. The large sensitivity at 850 hPa contrasts with the vertical variation of the forecast errors which show a continuous increase from the surface to the jet level. This is consistent with the results obtained by Rabier et al. (1992) on the maximum sensitivity at the steering level in baroclinically unstable conditions. Furthermore, the day-to-day variability is significant (±25%) with, in addition, a few exceptionally large sensitivities located at either 850 hPa or at the jet level. This feasibility study led to an operational implementation of the sensitivity computations in February 1994 at ECMWF. The results of this routine monitoring and case studies are described in Rabier et al. (1996).

b *Sensitivity to parameters*

The input parameters **u** are not necessarily the initial conditions of a numerical weather prediction model but may rather be some parameters like the orography for an atmospheric model or the atmospheric forcing for an ocean circulation model. Hall et al. (1982) applied the approach to compute the sensitivity of a radiative convective model to 312 parameters (including basic physical constants). Courtier (1987) computed the sensitivity of short-range forecast errors to orography in a shallow-water model.

4 Estimation

The adjoint method allows computation of the gradient of a cost function with respect to some parameters in an efficient way. It is then possible to minimize the cost function using descent algorithms like conjugate gradient or quasi-Newton (Le Dimet and Talagrand, 1986). This may be applied to several problems such as:

a *Variational assimilation*

Pailleux (1990) proposed a 3D variational analysis scheme (3D-Var) as an alternative to the Optimal Interpolation (OI) scheme. A cost function measuring the departure between the state to be estimated and the various sources of information (background, observations, slow-manifold) is minimized. The adjoint of the operators involved in the distance to the background, to the observations or to the slow manifold allows computation of the gradient of the cost function. 3D-Var is under pre-operational evaluation at ECMWF. It is described in detail in Courtier et al. (1993a). Similar ideas have been implemented operationally at NMC (National Meteorological Center) (Parrish and Derber, 1992).

The main a priori advantages of 3D-Var as compared to the OI are, first, more flexibility for the description of the spatial structure of the short-range forecast error: both the NMC and ECMWF implementations use spatial correlations which are a non-separable (between the vertical and horizontal direction) function of the spatial distance. The second advantage lies in the easy use of weakly nonlinear observation operators, which is the case for most of the satellite data.

Another strength of the variational formulation is that it readily extends to the time dimension: 4D-Var. One then seeks a model trajectory which best fits the available information (minimize a cost function measuring the distance of the model trajectory to the available information). The gradient of the cost function is evaluated integrating the adjoint of the forecast model. Thépaut et al. (1993) and Rabier et al. (1993a) demonstrated the ability of 4D-Var to generate flow dependent structure functions. Being iterative, 4D-Var is expensive since it requires several integrations of the model and its adjoint. Ideas to reduce the cost by introducing simplifications in the 4D-Var formulation are discussed in Courtier et al. (1994): the incremental approach approximates the full-size minimization problem by a quadratic problem involving a lower resolution model. The cost reduction is sufficient to foresee operational implementation within a few years.

b *Parameter estimation - inverse problems*

Variational assimilation can be seen as a particular inverse problem (Tarantola, 1987) where one estimates some geophysical parameters from observations.

Thépaut and Moll (1990) developed the adjoint of a fast radiative transfer model for inverting Tiros Operational Vertical Sounder (TOVS) radiances. Smedstad and O'Brien (1991), together with data assimilation, retrieved parameters of their ocean model. Marais and Musson-Genon (1992) estimated soil parameters fitting a vertical column model to screen level observations.

c *Nonlinear equilibration*

Vautard and Legras (1988) were looking for "weather regimes" defined as the large-scale patterns of the flow which are on average stationary (a statistical equilibration occurs between self interaction and feedback from the small scales). They solve a nonlinear optimization problem in which the cost function is the statistical average of the large scale tendencies. With the statistical averaging defined using ergodicity hypothesis through long time interval averaging, the authors faced a difficulty: while the time interval was becoming longer and longer, the cost function was converging to an asymptotic value, but not its gradient.

A trivial (and non meteorological) example is the following:
Let $f(t, x) = x \cos tx$ where x is the phase space variable and t time. We have

$$\frac{1}{T}\int_0^T f(t, x)dt = \frac{1}{T} \sin Tx.$$

Since $\sin Tx$ is bounded, we have

$$\lim_{T\to\infty} \frac{1}{T} \sin Tx = 0$$

whereas its derivative with respect to x: $\cos Tx$ has no limit when T tends to infinity.

Vautard and Legras (1988) solved the problem using an ensemble mean. This points to a difficulty in using the adjoint technique for climatic application: the adjoint is still useful for computing the gradient of a cost function with respect to its input parameters. However, care has to be taken in the definition of the cost function for its gradient to maintain a physical meaning.

5 Kalman filtering

Assuming that the forecast error evolution is exactly governed by the tangent-linear model, the evolution from time t to time $t+T$ of the covariances of forecast errors $\mathbf{B}$ reads

$$\mathbf{B}(t+T) = \mathbf{M}'(t+T, t)\mathbf{B}(t)\mathbf{M}'^T(t+T, t) \tag{5}$$

where $\mathbf{M}'(t+T, t)$ is the resolvent of the tangent-linear model between times t and $t+T$ (see, for example, Jaszwinksi, 1970). Here we have assumed the model to be perfect, no source term is present (the reader will easily generalize to an imperfect model in the following discussion).

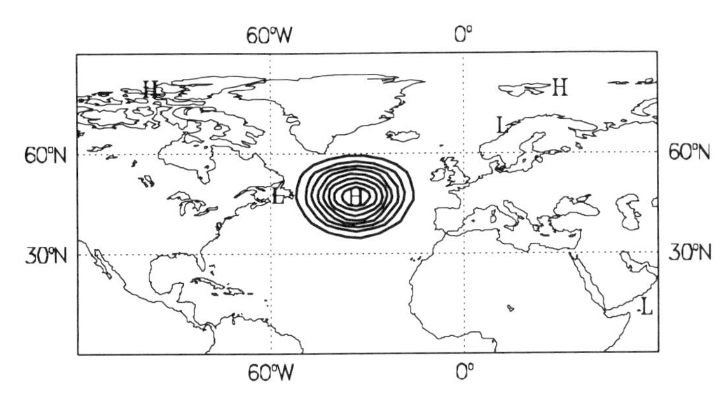

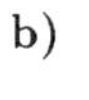

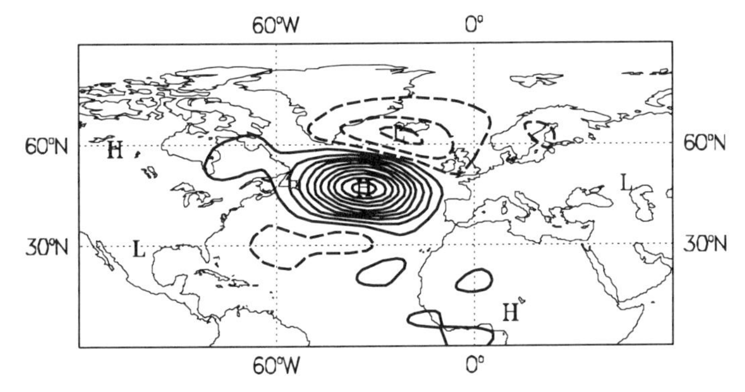

c)

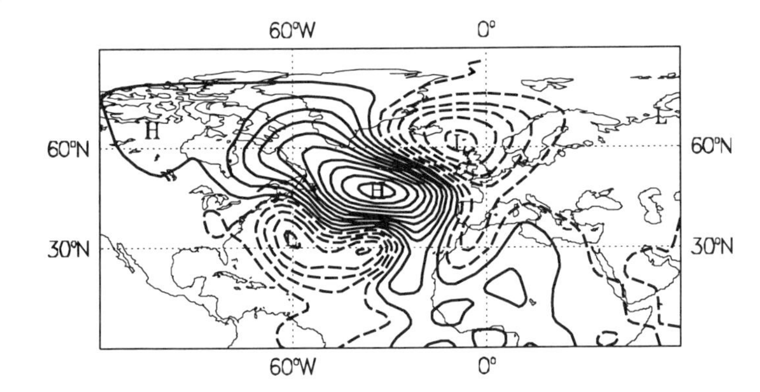

Fig. 8 Evolution of the autocorrelation field relative to point (45°N, 35°W) in a barotropic vorticity equation model (reproduced with permission from Bouttier, 1993). a) initial state; b) 6-hour forecast; c) 24-hour forecast. The interval is 0.1 with the highest isoline at 0.95.

This equation is often solved as

$$\mathbf{B}(t+T) = \mathbf{M}'(t+T,\ t)(\mathbf{M}'(t+T,\ t)\mathbf{B}(t))^T$$

by doing product of matrices. This generally requires less CPU time than the original formulation since $\mathbf{M}'$ has to be evaluated only once. However, it requires the storage of $\mathbf{M}'(t+T,t)\mathbf{B}(t)$.

Let us now consider the vector $\mathbf{e_i}$ which has zeroes everywhere except at the ith position where it is a 1. Then $\mathbf{B}(t+T)\mathbf{e_i}$ provides the covariances of forecast errors between $\mathbf{e_i}$ and all the other model variables: if $\mathbf{e_i}$ represents, for example, the surface pressure at a given location, then $\mathbf{B}(t+T)\mathbf{e_i}$ provides the covariances of the forecast errors of the surface pressure at this particular location with the surface pressure and all other variables at all grid points of the model. Thus we also obtain the variance of error of the surface pressure at this point. $\mathbf{M}'(t+T,t)$, $\mathbf{B}(t)$ and $\mathbf{M}'(t+T,t)^T$ being available as operators, $\mathbf{B}(t+T)\mathbf{e_i}$ is computed efficiently. This relies on the use of the adjoint model $\mathbf{M}'(t+T,t)^T$.

Bouttier (1993) applied this approach to a global vorticity equation model. Figure 8 (his Fig. 4) presents the autocorrelation function of a point with its neighbour at the initial time and for 6- and 24-hour forecasts. One should notice the significant deformation induced by the flow. In order to go from covariances to correlations, he had to compute the diagonal of the covariance matrix (and then the full matrix).

He also computed the 24-hour prediction of the variances of errors induced by idealized flows and by a real situation. Figure 9 (his Fig. 20) shows a case of difluence. He relates the maximum of error to the position of the maximum of the jet through barotropic instability which is then advected (and not to difluence itself). Figure 10 (his Fig. 21) presents the forecast of variances of error in the case of a large-scale wave. The maximum variance is located in the eastern part of the trough. It is consistent with a result obtained by Barkmeijer (1992). As in Veyre (1991), the approach has been applied to a real situation. The maximum amplification of error is located in the areas of barotropic instability.

The Kalman filter also contains an analysis equation which involves the observation operators. Gauthier et al. (1993) used the adjoint of the observation operator to efficiently compute its matrix.

6 Singular vectors

Let $\mathbf{M}'(t+T,t)$ be the tangent-linear model integrated from time t to time $t+T$. Let $\langle , \rangle$ be a norm which measures the forecast errors, for example the quadratic invariant of the primitive equations linearized in the vicinity of a state of rest. Let $\delta\mathbf{u}(t)$ be the initial errors and $\delta\mathbf{u}(t+T)$ the errors at time $t+T$. We have:

$$\begin{aligned}\langle \delta\mathbf{u}(t+T),\ \delta\mathbf{u}(t+T)\rangle &= \langle \mathbf{M}'(t+T,\ t)\delta\mathbf{u}(t),\ \mathbf{M}'(t+T,\ t)\delta\mathbf{u}(t)\rangle \\ &= \langle \mathbf{M}'^{*}(t+T,\ t)\ \mathbf{M}'(t+T,\ t)\delta\mathbf{u}(t),\ \delta\mathbf{u}(t)\rangle\end{aligned}$$

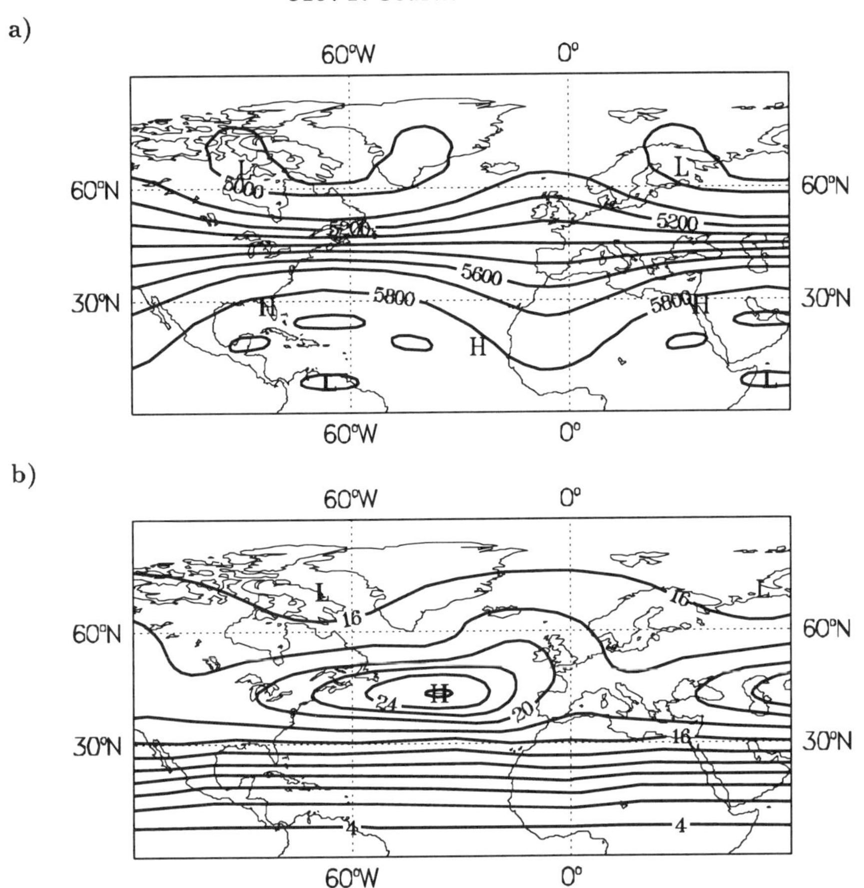

Fig. 9 Forecast of height error standard deviation (panel b) in the vicinity of the idealized meteorological situation (panel a) (reproduced with permission from Bouttier, 1993). Isoline is 100 m in panel a, 2 m in panel b.

which is extremal for the considered norm when $\mathbf{M}'^{*}(t+T,t)\ \mathbf{M}'(t+T,t)\boldsymbol{\delta}\mathbf{u}(t)$ is proportional to $\boldsymbol{\delta}\mathbf{u}(t)$.

The initial perturbations $\boldsymbol{\delta}\mathbf{u}(t)$ which will lead to the largest error according to the chosen norm are then given by the dominant eigenvectors of $\mathbf{M}'^{*}(t+T,t)$ $\mathbf{M}'(t+T,t)$. The eigenvectors/eigenvalues of $\mathbf{M}'^{*}(t+T,t)\ \mathbf{M}'(t+T,t)$ are called the singular vectors/values of $\mathbf{M}'(t+T,t)$ in linear algebra. Finding the most unstable perturbations is thus reduced to an eigenvalue problem which may be solved using a Lanczos iterative algorithm. This requires only the ability to compute $\mathbf{M}'^{*}(t+T,t)$ $\mathbf{M}'(t+T,t)$ applied to some vectors which makes the algorithm tractable, even for a large-scale problem.

The important point to note is that the eigenvalues of $\mathbf{M}'^{*}(t+T,t)\mathbf{M}'(t+T,t)$ can be very different from the square of those of $\mathbf{M}'(t+T,t)$. Let us consider the simple

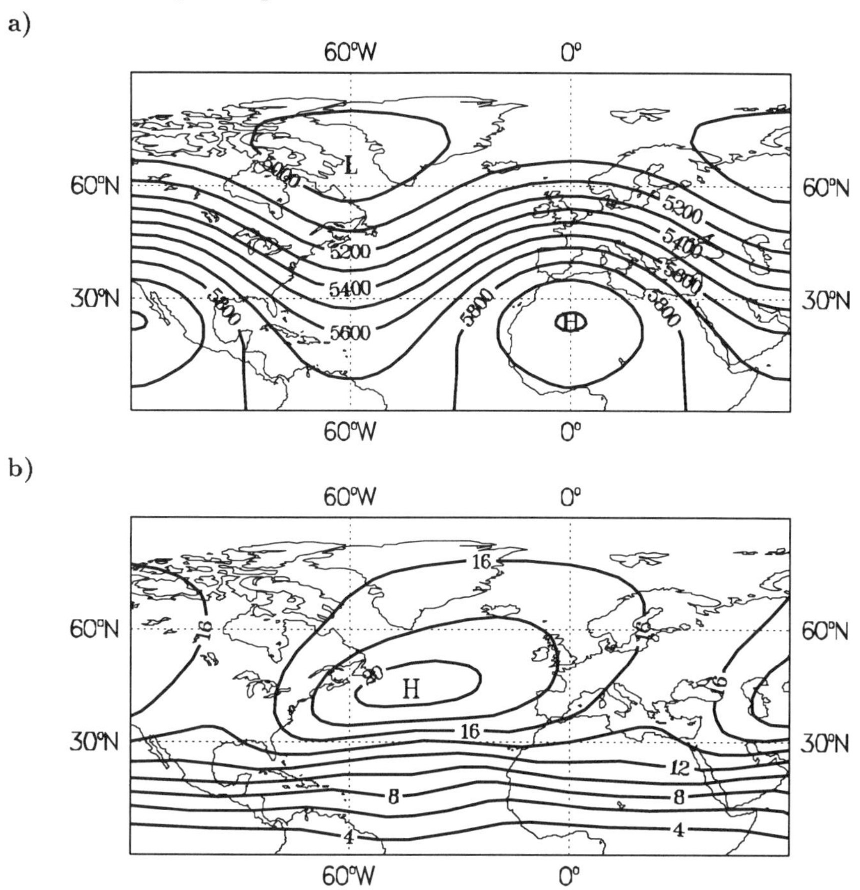

Fig. 10 (a) Large-scale height field. (b) Standard deviation of the height field error. (Reproduced with permission from Bouttier, 1993).

and purely dissipative example illustrated in Fig. 11. The trajectory starts from a circle of equi-energy. As there is strong dissipation along the horizontal axis, the trajectory reaches the axis of weak dissipation without moving significantly along that direction. Then it converges toward the state of rest. This illustrates that, even in a purely dissipative system, it is possible to have growth of energy for a finite time. This happens when two eigenvectors of $\mathbf{M}'$ are close to parallel but associated to significantly different eigenvalues.

This is a common feature of meteorology and Farrel (1989) used this concept for understanding baroclinic instability. Molteni and Palmer (1993) computed the singular vector of the tangent-linear version of a 3-level quasi-geostrophic model. They found that the singular vectors computed depend significantly on the time

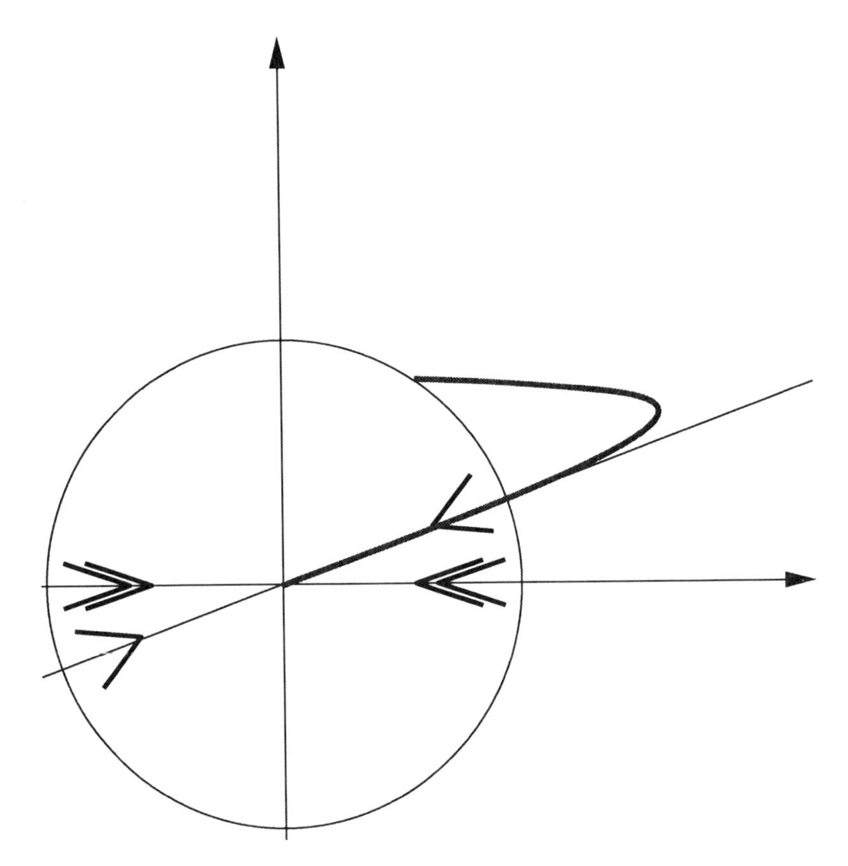

Fig. 11 Schematic phase space diagram of a purely dissipative 2D dynamical system. The » « denotes a strong dissipation in this direction whereas the ⟩ ⟨ denotes a weak dissipation. The trajectory starting from a circle of equi-energy starts with an energy increase and then converges toward the state of rest.

interval chosen: 12 hours, two days and eight days; their subsequent evolution is also different, with a strong impact of being optimal for a given range.

Buizza (1993) computed the singular vectors of a T21L19 primitive equation model. He found it necessary to include a simple vertical diffusion and surface drag in order to prevent spurious structures at low level. Figure 12 (his Fig. 5) presents the spectrum of the eigenvalues detained for a different time interval. When the time interval is increased, the separation between the first few singular vector increases. The meteorological structures are stable from a one-day time interval up to three days. 36 hours is a good compromise between the cost of the method and the meteorological significance of the singular vectors.

He presents the 6 dominant eigenvectors for a 36-hour time interval at time t_0 and at time t_0 + 36 hours. They all grow very fast and propagate eastward. Their

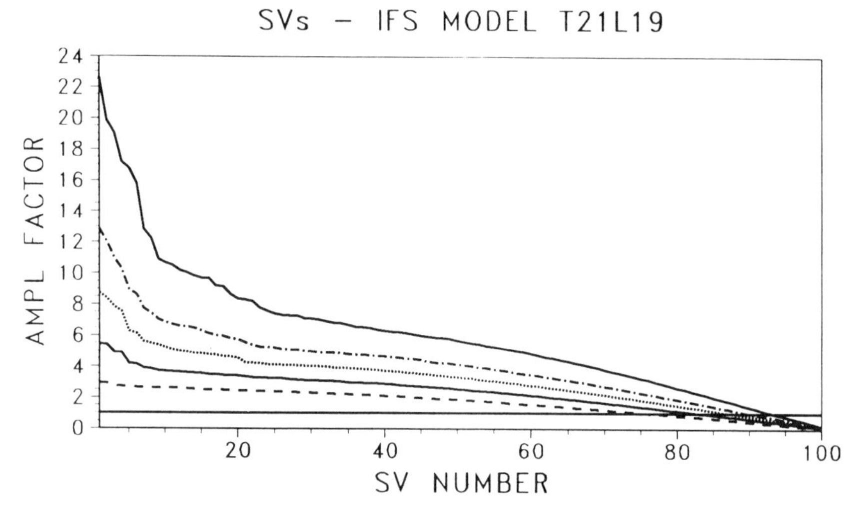

Fig. 12 Spectrum of the singular values for the situation 17.01.1989 and for different time intervals; dashed 12 h, middle solid 24 h, dotted solid 36 h, chain dashed 48 h, upper solid 72 h. (Reproduced with permission from Buizza, 1993).

vertical structure depicted in Fig. 13 (his Fig. 7) shows a maximum at around 600 hPa at time t_0 which is then located at 300 hPa at time t_0 + 36 hours. This is consistent with baroclinic instability theories and with the sensitivity experiments reported in Section 3.

7 Conclusion

The concepts presented in this paper were introduced in the early days of numerical weather prediction. Sasaki (1958) may be seen as a precursor to 3D-Var, Thomson (1969) introduces the ideas behind 4D-Var, Epstein (1969) pioneers error propagation and Jones (1965) introduces Kalman filtering while Lorenz (1965) uses the singular vectors. The adjoint models bring the feasibility of the practical implementation for large-scale problems. Other applications may emerge in the future if applying the transpose of a matrix to a vector is required for an efficient algorithm. At this stage, we should mention the second order adjoint (Wang et al., 1992) which allows the computation of the Hessian of a cost function applied to a vector in a minimization problem.

The use of the adjoint model has a major limitation: it requires the tangent-linear model to be meteorologically realistic for a finite amplitude perturbation. 4D-Var results and, quite convincingly, the recent sensitivity study of Rabier et al. (1996) indicate that this is indeed the case for the adiabatic evolution of perturbations of an order of magnitude comparable to the analysis error, over a 2-day time interval. The stiffness of the adiabatic primitive equations does not seem to be a critical problem.

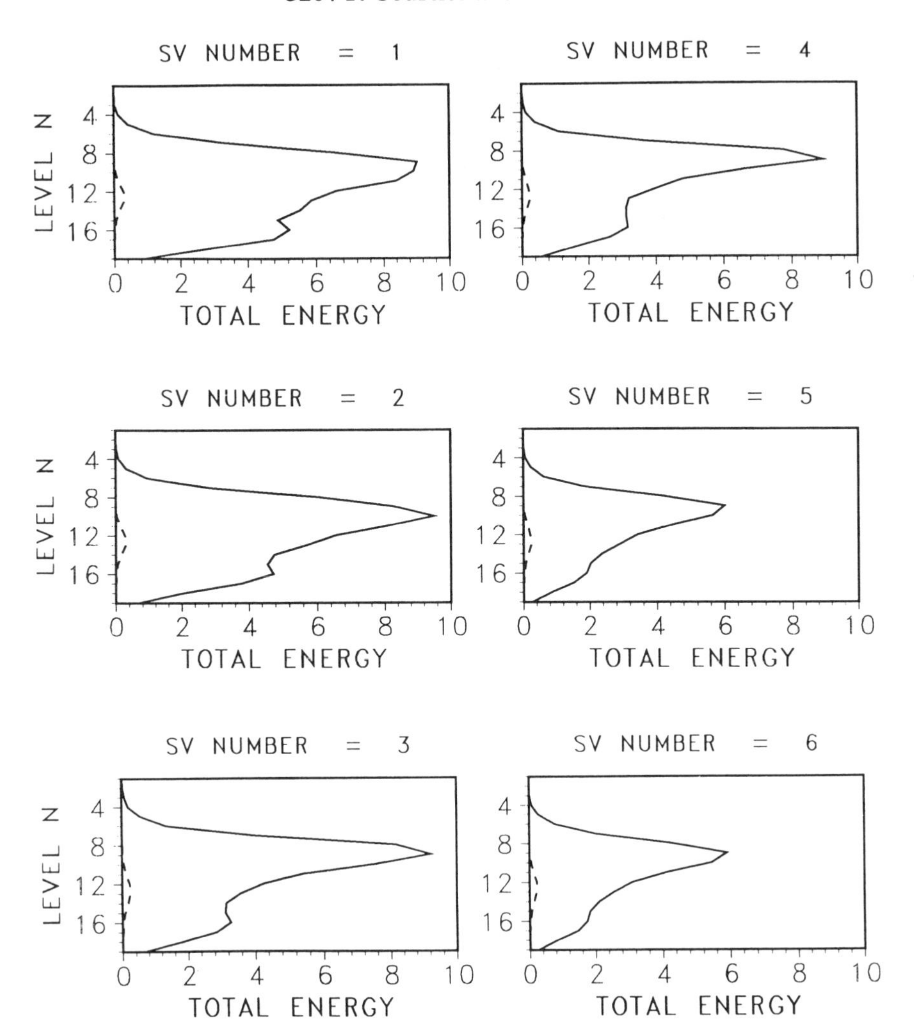

Fig. 13 Vertical structure of the dominant singular vectors. (Reproduced with permission from Buizza, 1993). The dashed line is the initial value of the total energy, the solid line is the total energy after 36 h of evolution.

However, the introduction of physical parametrizations seems to be more critical. Significant progress is being made by Vukicevic and Errico (1993), Zou et al. (1993) and Zupanski (1993), but it is probably fair to say that much work remains to be done in this area. This might even impact on the physical parametrization design with regularization widely introduced in the formulation.

Acknowledgements

All the results presented here have been obtained with the ARPEGE/IFS numerical weather prediction system developed jointly by ECMWF and Météo-France. The ARPEGE and IFS teams are warmly acknowledged. Carole Edis carefully typed the manuscript.

References

BARKMEIJER, J. 1992. Local error growth in a barotropic model. *Tellus*, **44A**: 314–323.

BAUR, W and V. STRASSEN. 1983. The complexity of partial derivatives. *Theoretical Comp. Sci.* **22**: 317–330.

BOUTTIER, F. 1993. The dynamics of error covariances in a barotropic model. *Tellus*, **45A**: 408–423.

BUIZZA, R. 1993. Impact of a simple vertical diffusion scheme and of the optimization time interval on optimal unstable structures. ECMWF Tech. Memo No. 192 (available from ECMWF), 25 pp.

COURTIER, P. 1987. Application du contrôle optimal à la prévision numérique en météorologie. Thèse de Doctorat de l'Université Pierre et Marie Curie, Paris, France, 327 pp.

———; E. ANDERSSON, W. HECKLEY, G. KELLY, J. PAILLEUX, F. RABIER, J-N. THÉPAUT, P. UNDÉN, D. VASILJEVIC, C. CARDINALI, J. EYRE, M. HAMRUD, J. HASELER, A. HOLLINGSWORTH, A. MCNALLY and A. STOFFELEN. 1993a. Variational assimilation at ECMWF. ECMWF Tech. Memo 194 (available from ECMWF), 84 pp.

———; J. DERBER, R. ERRICO, J-F. LOUIS and T. VUKICEVIC. 1993b. Important literature on the use of adjoint, variational methods and the Kalman filter in meteorology. *Tellus*, **45A**: 342–357.

———; J-N. THÉPAUT and A. HOLLINGSWORTH. 1994. A strategy for operational implementation of 4D-Var, using an incremental approach. *Q. J. R. Meteorol. Soc.* **120**: 1367–1387.

EPSTEIN, G.S. 1969. Stochastic dynamic prediction. *Tellus*, **21**: 739–759.

ERRICO, R. and T. VUKICEVIC. 1992. Sensitivity analysis using an adjoint of the PSU-NCAR mesoscale model. *Mon. Weather Rev.* **120**: 1644–1660.

FARREL, B.F. 1989. Optimal excitation of baroclinic waves. *J. Atmos. Sci.* **46**: 1193–1206.

GAUTHIER, P.; P. COURTIER and P. MOLL. 1993. Assimilation of simulated wind lidar data with a Kalman filter. *Mon. Weather Rev.* **121**: 1803–1820.

HALL, M.C.G.; D.G. CACUCI and M.E. SCHLESINGER. 1982. Sensitivity analysis of a radiative-convective model by the adjoint method. *J. Atmos. Sci.* **39**: 2038–2050.

JASZWINKSI, A.H. 1970. *Stochastic processes and filtering theory.* Academic Press, New York, 376 pp.

JONES, R.H. 1965. Optimal estimation of initial conditions for numerical prediction. *J. Atmos. Sci.* **22**: 658–663.

LE DIMET, F-X. and O. TALAGRAND. 1986. Variational algorithms for analysis and assimilation of meteorological observations and theoretical aspects. *Tellus*, **38A**: 97–110.

LORENZ, E.N. 1965. A study of the predictability of a 28-variable atmosphere model. *Tellus*, **17**: 321–333.

MARAIS, L. and L. MUSSON-GENON. 1992. Forecasting the surface weather elements with a local dynamical adaptation method using a variational technique. *Mon. Weather Rev.* **120**: 1035–1049.

MARCHUK, G.I. 1974 (Russian version 1967). *The numerical solution of problems of atmospheric and oceanic dynamics*, Gidrometeoizdat, Leningrad, USSR, 387 pp. (English translation, Rainbow systems, Alexandria, va.).

MOLTENI, F. and T.N. PALMER. 1993. Predictability and finite-time instability of the Northern Hemisphere circulation. *Q. J. R. Meteorol. Soc.* **119**: 269–298.

MORGENSTERN, J. 1973. Algorithmes linéaires tangents et complexité. Compte rendu de l'Academie des Sciences t.277:367.

———. 1985. How to compute a function and all its derivatives fast, a variation on the theorem of Baur-Strassen. *Sigact. News,* **16**: 60–62.

PAILLEUX, J. 1990. A global variational assimilation scheme and its application for using TOVS radiances. Proceedings World Meteorological Organisation, International Symposium on Assimilation of Observations in Meteorology and Oceanography, Clermont-Ferrand, France, 9-13 July 1990, pp. 325–328.

PARRISH, D.F. and J.C. DERBER. 1992. The National Meteorological Center's Spectral Statistical Interpolation analysis system. *Mon. Weather Rev.* **110**: 1747–1766.

RABIER, F.; P. COURTIER and O. TALAGRAND. 1992. An application of adjoint models to sensitivity analysis. *Beitr. Phys. Atmos.* **65**: 177–192.

———; ———, J. PAILLEUX, O. TALAGRAND and D. VASILJEVIC. 1993a. A comparison between four-dimensional variational assimilation and simplified sequential assimilation relying on three-dimensional variational analysis. *Q. J. R. Meteorol. Soc.* **119**: 845–880.

———; ———, M. HERVEOU, B. STRAUSS and A. PERSSON. 1993b. Sensitivity of forecast error to initial conditions using the adjoint model. ECMWF Tech. Memo 197, available from ECMWF, 29 pp.

———; E. KLINKER, P. COURTIER and A. HOLLINGSWORTH. 1996. Sensitivity of forecast errors to initial conditions. *Q. J. R. Meteorol. Soc.* **122**: 121–150.

SASAKI, Y. 1958. An objective analysis based on the variational method. *J. Meteorol. Soc. Jpn.* **36**: 77–88.

SIMMONS, A.J. and B. HOSKINS. 1978. The life cycles of some nonlinear baroclinic waves. *J. Atmos. Sci.* **35**: 414–432.

SMEDSTAD, O.M. and J.J. O'BRIEN. 1991. Variational data assimilation and parameter estimation in the equatorial Pacific Ocean. *Prog. Oceanog.* **26**: 179–241.

TARANTOLA, A. 1987. *Inverse problem theory, method for data fitting and model parameter estimation.* Elsevier, Amsterdam, 613 pp.

THÉPAUT, J-N. and P. MOLL. 1990. Variational inversion of simulated TOVS radiances using the adjoint technique. *Q. J. R. Meteorol. Soc.* **116**: 1425–1448.

———; R. HOFFMAN and P. COURTIER. 1993. Interaction of dynamics and observations in a four-dimensional variational assimilation. *Mon. Weather Rev.* **121**: 3393–3414.

THOMSON, P.N. 1969. Reduction of analysis error through constraints of dynamical consistency. *J. Appl. Meteorol.* **8**: 738–742.

VAUTARD, R. and B. LEGRAS. 1988. On the source of midlatitude low-frequency variability. Part II: nonlinear equilibration of weather regimes. *J. Atmos. Sci.* **45**: 2845–2867.

VEYRE, P. 1991. Direct prediction of error variances by the tangent-linear model: a way to forecast uncertainty in the short range. ECMWF workshop on new developments in predictability, 13-15 November 1991, ECMWF, Reading, U.K., pp 65–86.

VUKICEVIC, T. and R.M. ERRICO. 1993. Linearization and adjoint of parametrized moist diabatic processes. *Tellus*, **45A**: 493–510.

WANG, Z.; I-M. NAVON, F-X. LE DIMET and X. ZOU. 1992. The second order adjoint analysis: theory and application. *Meteorol. Atmos. Phys.* **50**: 3–20.

ZOU, X.; I.M. NAVON and J.G. SELA. 1993. Variational data assimilation with moist threshold processes using the NMC spectral model. *Tellus*, **45A**: 370–387.

ZUPANSKI, D. 1993. The effects of discontinuities in the Betts-Miller cumulus convective scheme on four-dimensional variational data assimilation. *Tellus*, **45A**: 511–524.

Error Propagation and Observability for the Constituent Transport Equation in Steady, Non-Divergent, Two-Dimensional Flow

Roger Daley*
Atmospheric Environment Service
4905 Dufferin Street
Downsview, Ontario M3H 5T4, Canada

[Original manuscript received 15 October 1994; in revised form 4 August 1995]

ABSTRACT *There is increasing interest in the application of data assimilation procedures to the analysis of atmospheric trace gas observations from ground- and space-based instruments. Although early constituent assimilation experiments will use conventional statistical interpolation, three-dimensional variational or successive corrections algorithms, it can be expected that, ultimately, more powerful techniques such as the four-dimensional variational or Kalman filter algorithms will be used. One of the benefits of Kalman filter algorithms is that they generate complete second moment error statistics, which makes them ideal for investigating basic data assimilation problems. In this study, Kalman filters (based on discretizations of the constituent transport equation) are used to examine constituent data assimilation in simple two-dimensional, non-divergent flow fields – parallel shear flow, axisymmetric flow and confluent/diffluent flow. The evolution (i.e. propagation) of second moment error statistics using discrete transport algorithms was compared with the corresponding continuous form discussed by Cohn (1993). Observability, i.e. the ability to determine a unique state from a finite sequence of observations using a perfect model, was first established with a Kalman filter using stationary observation networks and the results were explained by analysis of the eigenstructures of the discrete transport algorithms. It was found that, in general, very few observation stations were required for observability of constituents. However, in some pathological cases, observability was found to be strongly dependent on the geometry of the flow with respect to the model grid, as well as on the numerics of the transport algorithm.*

RÉSUMÉ *L'application des procédures d'assimilation des données à l'analyse des gaz atmosphériques en trace, observés à partir dl instruments au sol et de télédétection, retient de plus en plus l'intérêt. Bien que les premières expériences d'assimilation de constituants utiliseront une interpolation statistique conventionnelle avec des algorithmes de correction à trois dimensions, variables ou successifs, on peut prévoir qu'un jour des techniques plus*

---

*Present Address: Naval Research Laboratory, 7 Grace Hopper Avenue, Monterey, California 93943-5502, USA.

puissantes, telles que des algorithmes variationnels à quatre dimensions ou à filtre de Kalman seront utilisés. Un des avantages des algorithmes à filtre de Kalman est la génération de statistiques d'erreurs du second ordre, ce qui les rend idéaux pour l'étude des problèmes d'assimilation des données de base. Dans cette étude, les filtres de Kalman (basés sur la discrétisation de l'équation de transport des constituants) sont utilisés pour examiner l'assimilation des données des constituants dans des champs simples d'écoulement non divergeant en deux dimensions: cisaillement parallèle, axisymétrique et confluent/diffluent. On compare l'évolution (c.-à-d., la propagation) des statistiques d'erreurs du second ordre lorsqu'on utilise des algorithmes de transport discrets, aux formes continues correspondantes de Cohn (1993). On a d'abord établi l'observabilité (habilité de déterminer, à l'aide d'un modèle parfait, un état unique à partir d'une séquence d'observations) avec un filtre de Kalman utilisant des réseaux d'observation stationnaires, et ensuite expliqué les résultats par l'analyse des structures propres des algorithmes de transport discrets. En général, on a constaté que très peu de stations d'observation étaient nécessaires pour l'observabilité des constituants sauf dans certains cas pathologiques où cette dernière était fortement dépendante de la géométrie de l'écoulement par rapport à la grille du modèle ainsi que de la discrétisation de l'algorithme de transport.

---

1 Introduction

There is increasing interest in monitoring atmospheric trace gases because of their direct or indirect effects on economic, social or other human activities. In addition to ongoing ground-based monitoring programs, there are a number of new space-borne missions being planned or implemented. These chemical constituent measurements are very different from the traditional meteorological observations, and advanced procedures will have to be devised in order to analyze them.

Data assimilation has proven to be a very useful tool for obtaining an accurate four-dimensional representation of the atmospheric state, at least with respect to the traditional meteorological variables (wind, temperature and moisture). Data assimilation attempts to optimally combine all current observations with model generated estimates based on all available past information. The time-dependent assimilating model acts as a constraint on the time sequences of atmospheric analysis, providing temporal continuity, coupling between different variables, internal consistency and filling in data voids. Assimilating models have to be comprehensive, including all atmospheric processes thought to be of consequence, and the more accurate the model, the more successful the data assimilation.

Most present atmospheric data assimilation procedures are intermittent procedures in which there is an attempt to minimize the expected analysis error variance given the observations, forecasts from previous observations and some knowledge of the observation and forecast error statistics. These essentially static procedures include optimum or statistical interpolation, successive correction methods and the three-dimensional variational algorithm. However, a new class of more powerful assimilation algorithms (four-dimensional variational – 4DVAR and the Kalman filter) has recently been developed and could in principle be exploited for the problem of analyzing atmospheric chemical constituents.

Work has already begun on the analysis of stratospheric constituents (Salby, 1987; Schoeberl and Lait, 1992), and data assimilation techniques have recently been applied to this problem by Austin (1992). Daley (1995, Sections 1–4) used one-dimensional transport models and extended Kalman filters to investigate aspects of the constituent data assimilation problem.

However, one-dimensional studies, though revealing, are both unrealistic and restrictive. Many constituent transport studies impose the condition of non-divergence on the wind; which, in one dimension, implies windfields which are constant in time and space. Two-dimensional, non-divergent windfields, on the other hand, can include rotation, shear, confluence/diffluence etc.; and when applied to constituent transport models can generate much more complicated, interesting and realistic constituent distributions.

It is the purpose of this study to examine constituent data assimilation in a two-dimensional context using three different types of specified non-divergent windfields – parallel shear flow, axisymmetric flow and confluent/diffluent flow. The emphasis is on the analysis of the constituent field, given constituent observations only – there is no attempt here, to determine the wind field as well (as in Daley, 1995, Sections 5–6). Two simple types of constituent transport models will be described and tested. The analysis algorithm will be the standard Kalman filter. Since the problem is strictly linear, the standard Kalman filter is formally equivalent to the 4DVAR algorithm. The experiments will be based entirely on the examination of second moment error statistics – which, under the present conditions, completely reveal the properties of the system.

The constituent data assimilation system is described in Section 2. In Section 3, we discuss the temporal evolution of the second moment error statistics. In Sections 4 and 5, we discuss the observation systems required to successfully analyze two-dimensional constituent distributions.

2 Constituent transport models and Kalman filter equations

This section will introduce the two-dimensional constituent transport equation and two numerical discretizations of this equation. Three basic steady, non-divergent, two-dimensional windfields are introduced – parallel shear flow, confluent/diffluent flow and axisymmetric flow. Finally, the Kalman filter equations appropriate for data assimilation with the constituent transport equations are introduced.

a *The constituent transport equation*

Abiabatic transport of constituents occurs only along isentropic surfaces which are quasi-horizontal. Adiabatic flow may be a reasonable approximation for transport problems on timescales of a few days to a week, depending on the timescales of chemical and diabatic processes. We consider here, the two-dimensional problem (implying flow along a quasi-horizontal isentropic surface). The constituent density (mass of constituent per unit volume) is denoted as $C(x, y, t)$, where t indicates time. We define two-dimensional, non-divergent windfields with x and y components u

and v. Then, in the absence of constituent sources or sinks, the constituent density is assumed to satisfy the constituent transport equation,

$$\frac{\partial C}{\partial t} + \frac{\partial (Cu)}{\partial x} + \frac{\partial (Cv)}{\partial y} = 0. \tag{1}$$

Thus, we are simulating a chemical constituent which is transport dominated. We further assume that the model (1) is perfect. Since the flow is non-divergent, $\partial u/\partial x + \partial v/\partial y = 0$ and (1) can also be written,

$$\frac{\partial C}{\partial t} + u\frac{\partial C}{\partial x} + v\frac{\partial C}{\partial y} = 0. \tag{2}$$

It might be noted, that for non-divergent flow and in the absence of sources or sinks, equation (2) is also valid for other constituent variables such as the number density.

b *Specified windfields*

The winds (u, v) must be specified in (1) and (2). Such winds could be specified from objective wind analyses or by integrating a model such as the filtered barotropic equation. However, we are primarily interested in examining the problem at a fundamental level using very simple windfields. Consequently, we specify the winds analytically. Furthermore, we specify time-invariant winds $u = u(x, y)$ and $v = v(x, y)$. This last assumption is not very realistic, but it is necessary for the observability analysis in Sections 4 and 5.

Three basic windfields on two different domains were tested. The first domain is $-\pi \leqslant x \leqslant \pi$ and $0 \leqslant y \leqslant \pi$, with no normal flow $(v = 0)$ at $y = 0$ and $y = \pi$, and periodic in $x(C(-\pi, y, t) = C(\pi, y, t)$ and similarly for u and v).

The first windfield is shear flow parallel to the x axis ($v(x, y) = 0$ and $u(x, y)$ independent of x). The y profile of u for this flow is given by $u(y) = A_1 \sin^2(y) - A_2$ and is illustrated in Fig. 1 for two choices of the specified constants A_1 and A_2. In curve "a", $A_1 = 1$ and $A_2 = 0$, corresponding to a jet profile which has the same sign everywhere. Curve "b" shows a case in which there are sign reversals (and lines of constant y with zero flow).

Figure 2 illustrates the streamfunction for a confluent/diffluent flow. As in Fig. 1, the domain has no normal flow at $y = 0, \pi$ and is periodic in x. The direction of the wind is indicated by the arrow at the point of maximum wind ($x = 0, y = \pi/2$). The streamfunction in Fig. 2 has the form,

$$\psi(x, y) = -B_1(y - \pi/2) + B_2 \sin(2y)\exp(-x^2/2l_w^2), \tag{3}$$

where $l_w = \pi/4$, $B_1 = 1/2$ and $B_2 = 1/4$. $u(x, y) = -\partial\psi/\partial y$ and $v(x, y) = \partial\psi/\partial x$.

The final windfield is (essentially) axisymmetric. The domain is square with no normal flow across the boundaries. The domain is $0 \leqslant x \leqslant \pi$ and $0 \leqslant y \leqslant \pi$ with

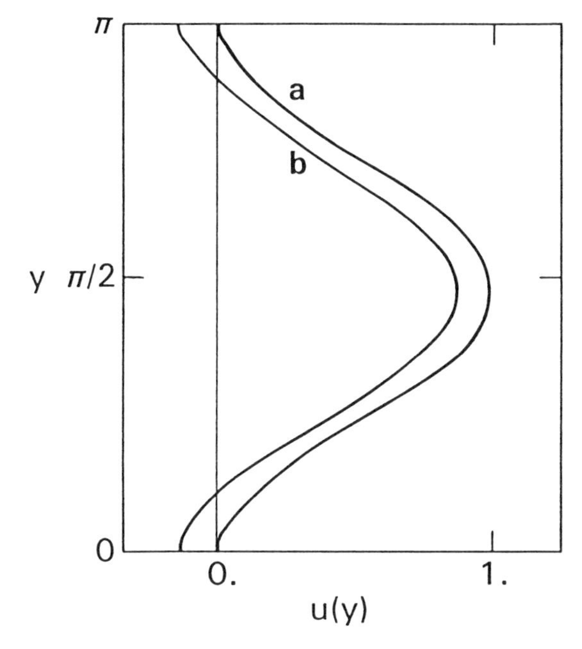

Fig. 1 Wind profiles $u(y)$ for parallel shear flow.

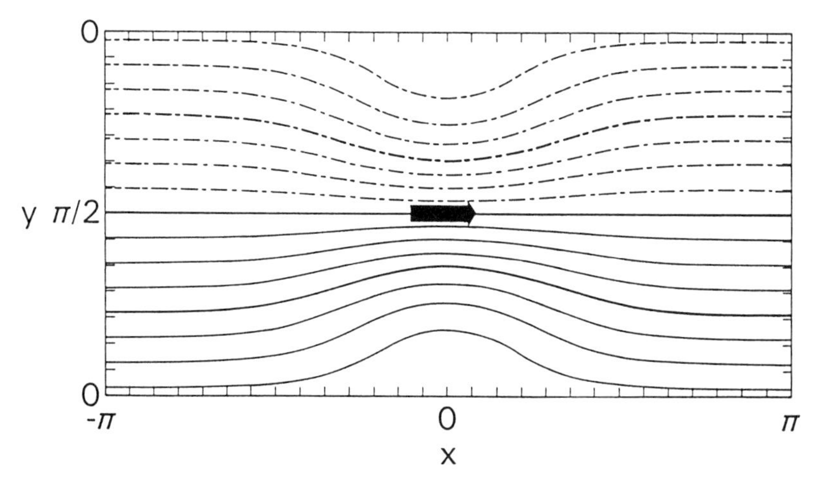

Fig. 2 Streamfunction for confluent/diffluent wind field.

$u(0, y) = u(\pi, y) = 0$ and $v(x, 0) = v(x, \pi) = 0$. The streamfunction has the form,

$$\psi(x, y) = D_1 \sin(x) \sin(y) \exp\{-[(x - \pi/2)^2 + (y - \pi/2)^2]/2l_w^2\}. \tag{4}$$

Figure 3a shows the case with $D_1 = -0.3$ and $l_w = \pi/8$. The arrows show the direction of the flow and are located at the radius of maximum wind. This flow becomes vanishingly small near the boundaries and is purely axisymmetric. Figure 3b shows the case where $D_1 = -0.6$ and $l_w = \pi/2$. Again the arrows indicate the direction of flow at the radius of maximum wind. In this case there is flow right out to the boundaries of the domain and the flow is not purely axisymmetric. In both Figs 3a and 3b, the boundary is a streamline.

c *Two discrete models of the transport equation*

Equation (1) was discretized using two very different discretization algorithms. These algorithms were chosen because of their simplicity, because they employed two time levels (advantageous for constructing a Kalman filter) and because they represented completely opposite design philosophies. Discretization algorithms for integration of the constituent transport equation have been reviewed by Rood (1987), and neither of the algorithms tested here should be seriously considered for implementation in a real problem. However, the subject of this paper is *not* the construction of better transport algorithms, and the deficiencies of the two transport discretizations are *not* particularly detrimental to the actual goals of this study.

The first model is a semi-Lagrangian discretization of equation (2). It is basically the same algorithm used, in a one-dimensional context, by Daley (1992) and Ménard (1994). It is a two-step scheme which is forward in time. Consider the domain with boundaries which are periodic in x and have no normal flow in y. Define the following grid,

$$\{x_i, y_j\} = \{-\pi + \Delta x/2 + (i - 1)\Delta x,\ \Delta y/2 + (j - 1)\Delta y\},$$

$$1 \leqslant i \leqslant I,\ 1 \leqslant j \leqslant J, \tag{5}$$

where $\Delta x = 2\pi/I$ and $\Delta y = \pi/J$. Then, a field $\mathbf{C}(x_i, y_j)$ can be expanded in a double Fourier series on this grid (in complex form),

$$\mathbf{C}(x_i, y_j) = \sum_{l=-(I-1)/2}^{(I-1)/2} \sum_{m=0}^{J-1} \hat{\mathbf{C}}_{\mathbf{m}}^{l} \exp(ilx_i) \cos(my_j), \tag{6}$$

where $\hat{\mathbf{C}}_{\mathbf{m}}^{l}$ is complex. We can write (6) in real vector form as,

$$\mathbf{C} = \mathbf{F}\hat{\mathbf{C}}, \tag{7}$$

where $\mathbf{C}$ is a column vector of length IJ of elements $\mathbf{C}(x_i, y_j)$ and $\hat{\mathbf{C}}$ is a column

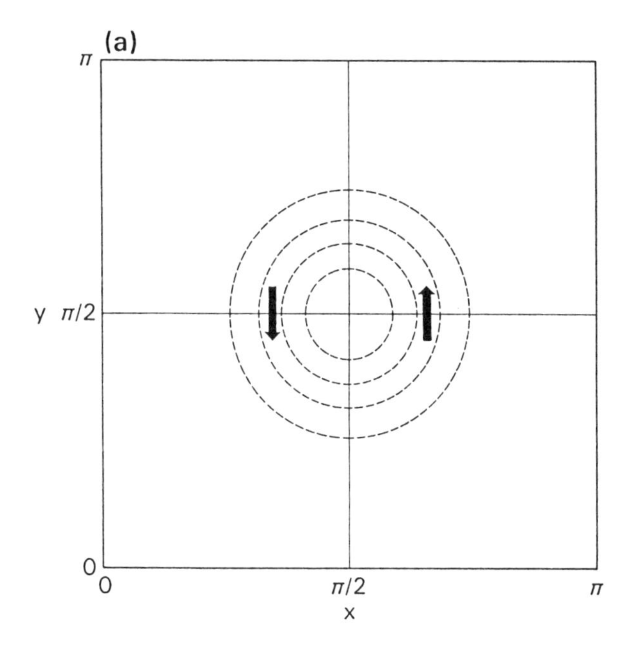

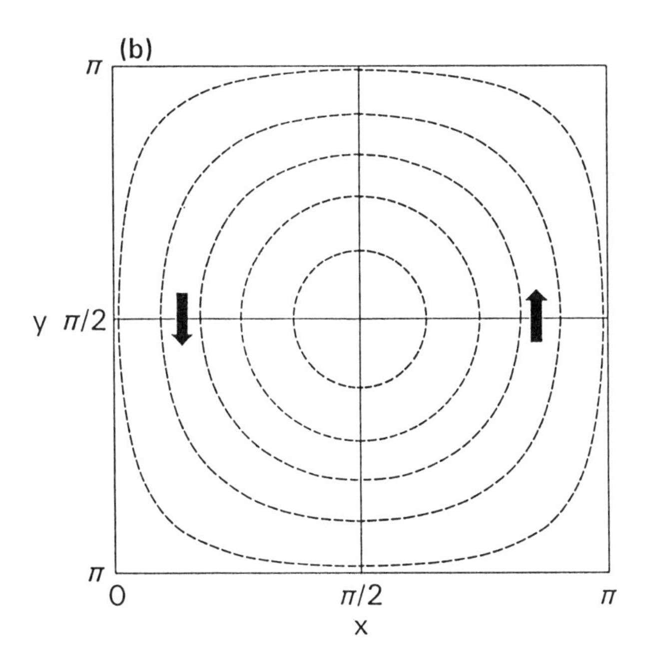

Fig. 3 Streamfunctions for axisymmetric flow. Panel (a) $l_w = \pi/8$ and panel (b) $l_w = \pi/2$.

vector of length IJ of the real elements corresponding to the $\hat{\mathbf{C}}_{\mathbf{m}}^{l}$ of (6). The matrix $\mathbf{F}$ is an $IJ \times IJ$ square matrix with elements of the form $\cos(lx_i)\cos(my_j)$ and $\sin(lx_i)\cos(my_j)$. On the grid $\{x_i, y_j\}$, $\mathbf{F}$ is orthogonal, and by appropriate normalization, we can write,

$$\hat{\mathbf{C}} = \mathbf{F}^{\mathbf{T}}\mathbf{C} \text{ and } \mathbf{F}\mathbf{F}^{\mathbf{T}} = \mathbf{I} \text{ (identity matrix).} \tag{8}$$

Define a time step Δt, and for every grid point $\{x_i, y_j\}$, define a departure point,

$$\{x_i^d, y_j^d\} = \{x_i - u(x_i, y_j)\Delta t,\ y_j - v(x_i, y_j)\Delta t\}. \tag{9}$$

Then, we discretize (2) as,

$$\mathbf{C_{n+1}} = \mathbf{M_s}\mathbf{C_n}, \text{ with } \mathbf{M_s} = \mathbf{F_d}\mathbf{F}^{\mathbf{T}}, \tag{10}$$

and $\mathbf{F_d}$ is the Fourier transform of the form in (6) and (7), but defined at the departure points $\{x_i^d, y_j^d\}$. The subscripts "n" and "$n+1$" refer to times t_n and $t_{n+1} = t_n + \Delta t$. The $IJ \times IJ$ square matrix $\mathbf{M_s}$ (subscript "s" stands for semi-Lagrangian) depends on the grid definition and the (time-invariant) specified windfields. This formulation is periodic in x, but not in y, and is appropriate for the windfields of Figs 1 and 2. For the windfield of Fig. 3, there is no normal flow at the boundaries and the formulation is slightly modified by replacing the form $\exp(ilx_i)$ in (6) by $\cos(lx_i)$ and the corresponding sum over l by a sum $0 \leqslant l \leqslant I$. In practice, $\Delta x = \Delta y$. The windfields illustrated in Figs 1–3 are, in fact, Courant numbers, $u(x, y)\Delta t/\Delta x$ and $v(x, y)\Delta t/\Delta y$.

This simplest of all semi-Lagrangian algorithms is related to that of Bates and McDonald (1982); with interpolation of the constituent field to the departure point done by appropriate (depending on the boundary condition) Fourier interpolation. A standard test for transport algorithms is one-dimensional, non-divergent (i.e. constant in time and space) flow (Allen et al., 1991, Section 2). This algorithm has no phase or amplitude errors for this simple one-dimensional test (although there may be such errors for the windfields of Figs 2 and 3). This semi-Lagrangian algorithm is not Courant-Friedrich-Levy (CFL) limited, which has some advantages when the transport algorithm is used to recover windfields from constituent observations (Daley, 1995, Section 6).

However, this semi-Lagrangian algorithm does have several disadvantages. The algorithm is not monotonic (i.e. it may create spurious maxima and minima) and there is no guarantee that the constituent density will remain positive. The total constituent is not strictly conserved (in the absence of constituent sources or sinks). As we shall show subsequently, this particular semi-Lagrangian algorithm (with its simple boundary treatment) may have a very weak instability for the windfields of Figs 2 or 3. It is possible to develop more sophisticated semi-Lagrangian algorithms which do not have these drawbacks.

The second model that we will consider is simple upstream-differencing of equa-

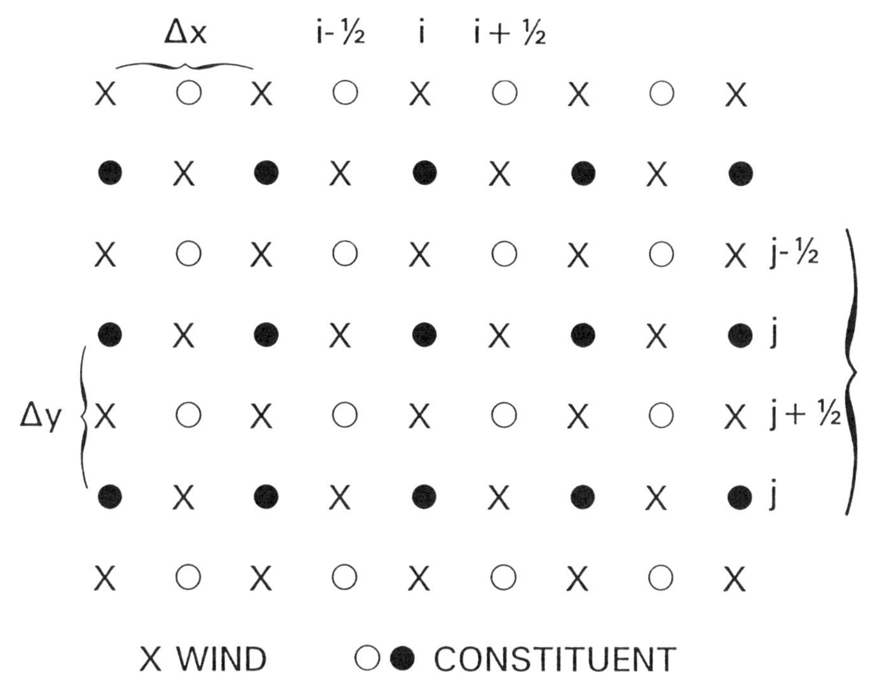

Fig. 4 Grid for upstream transport model. Solid and open circles mark constituent grid points and X's indicate wind grid points.

tion (1). The layout of the grid is shown in Fig. 4. The winds (u and v) are located at points marked with an "X", while there are two sets of constituent points – those marked with solid circles and those with open circles. The gridlengths Δx and Δy are the same in both directions and are indicated on the diagram. Also shown are the x and y indices (i and j).

Then, the values of the constituent density at the (solid circle) point $\{x_i, y_j\}$ and the (open circle) point $\{x_{i+1/2}, y_{j+1/2}\}$ at time t_{n+1} are given by,

$$C_j^i(t_{n+1}) = C_j^i(t_n) - \frac{\Delta t}{\Delta x}[F_{i+1/2}^u - F_{i+1/2}^u] - \frac{\Delta t}{\Delta y}[F_{j+1/2}^v - F_{j-1/2}^v] \tag{11}$$

$$C_{j+1/2}^{i+1/2}(t_{n+1}) = C_{j+1/2}^{i+1/2}(t_n) - \frac{\Delta t}{\Delta y}[F_{i+1}^u - F_i^u] - \frac{\Delta t}{\Delta x}[F_{j+1}^v - F_j^v].$$

Here, F_i^u, F_j^v are constituent fluxes evaluated at time t_n, given by,

$$F_i^u = u_{j+1/2}^i C_{j+1/2}^{i-1/2},\ u_{j+1/2}^i \geqslant 0 \text{ and } F_i^u = u_{j+1/2}^i C_{j+1/2}^{i+1/2},\ u_{j+1/2}^i < 0, \tag{12}$$

$$F_j^v = v_j^{i+1/2} C_{j-1/2}^{i+1/2},\ v_j^{i+1/2} \geqslant 0 \text{ and } F_j^v = v_j^{i+1/2} C_{j+1/2}^{i+1/2},\ v_j^{i+1/2} < 0.$$

In (11) and (12), $C_j^i = C(x_i, y_j)$ and $u_{j+1/2}^i = u(x_i, y_{j+1/2})$ etc. At the boundaries, relations, (11) and (12), are suitably modified depending on whether the boundary condition is periodic or no normal flow. As with the semi-Lagrangian model, the winds used are actually Courant numbers $-u(x,y)\Delta t/\Delta x$ and $v(x,y)\Delta t/\Delta y$.

Equations (11) and (12) can be written in matrix form in a manner analagous to (10),

$$\mathbf{C_{n+1}} = \mathbf{M_u C_n}, \tag{13}$$

where $\mathbf{C_n}$ is a column vector of constituent density values $\mathbf{C}(x_i, y_j)$ and $\mathbf{C}(x_{i+1/2}, y_{j+1/2})$ evaluated at time t_n and $\mathbf{M_u}$ is the model matrix for the upstream-differencing model.

The upstream-differencing algorithm is monotonic and never creates negative values of the constituent density. It also conserves the total constituent. On the other hand, this algorithm is well-known to be very diffusive. Thus, when applied to the one-dimensional constituent transport problem with a constant wind, there are very significant amplitude and phase errors for the small scales. This algorithm also has a strict CFL limit, and no winds can be allowed to exceed a Courant number of 1.

It might be noted that the continuous form of the windfields shown in Figures 1 to 3 is non-divergent. This may not be strictly true for the gridpoint operators $\mathbf{M_u}$ and $\mathbf{M_s}$.

The continuous equations (1) and (2) are completely non-diffusive, while $\mathbf{M_s}$ is slightly diffusive and $\mathbf{M_u}$ is strongly diffusive. This means that the discrete equations are advective/diffusive approximations to a continuous, inviscid, advective equation. Thus, discretization changes the fundamental character of the equation. This can't be helped. It would seem that most discretizations of the constituent transport equation have some spurious dissipation. However, this difference between the continuous and discrete cases, has important implications for the observability properties of the constituent transport equations, as we shall discuss in Sections 4 and 5.

d *The Kalman filter*

For data assimilation purposes, we will be using a standard Kalman filter, because the windfields are specified and so equations (1) and (2) and their discrete approximations, (10) and (13), are linear. The Kalman filter equations are derived in many places in the literature, but our notation will follow that of Daley (1992). We make the perfect model approximation for both models $\mathbf{M_s}$ and $\mathbf{M_u}$. (We have ignored the philosophical contradiction that two different models cannot both yield the "truth".) Then, the second moment error statistics for the Kalman filter are given by,

$$\mathbf{P^f_{n+1}} = \mathbf{M_s P^a_n M^T_s} \text{ or } \mathbf{P^f_{n+1}} = \mathbf{M_u P^a_u M^T_u}, \text{ (error covariance propagation)}$$

$$\mathbf{P^a_n} = [\mathbf{I} - \mathbf{K_n H_n}]\mathbf{P^f_n}, \text{ (analysis error covariance equation)}, \tag{14}$$

$$\mathbf{K_n} = \mathbf{P^f_n H^T_n}[\mathbf{H_n P^f_n H^T_n} + \mathbf{R_n}]^{-1}, \text{ (Kalman gain)}.$$

Here $\mathbf{P_n^f}$ and $\mathbf{P_n^a}$ are the forecast and analysis error covariances at time t_n, $\mathbf{H_n}$ is the forward interpolation matrix at t_n for interpolation from model grid to observation location(s), $\mathbf{R_n}$ is the observation error covariance, and $\mathbf{K_n}$ is the matrix of analysis weights which minimizes the expected analysis error variance. We have assumed here that the observation errors are unbiased and are not serially correlated. The only observations in this system will be constituent observations, and thus all errors are constituent errors. $\mathbf{P_n^f}$ and $\mathbf{P_n^a}$ are square, positive-definite covariance matrices whose order is the number of grid points. $\mathbf{R_n}$ is a square positive-definite matrix, whose order is the number of observations at time t_n. Equations (14) form a closed system which yields a complete description of the time evolution of the constituent error covariances. It is not actually necessary to perform data assimilation per se; and, thus, we have not included the first moment equations for the forecast and analyzed constituent fields.

This completes the description of the constituent transport models, the specified windfields and the Kalman filter. We are now ready to discuss the first experiments on propagation of constituent error.

3 Constituent error propagation with two discrete transport models

Before considering data assimilation we consider the evolution of the error covariances in the absence of observations. Thus, in this section, we are concerned only with the first equation of (14),

$$\mathbf{P_{n+1}} = \mathbf{M_s P_n M_s^T} \quad \text{or} \quad \mathbf{P_{n+1}} = \mathbf{M_u P_n M_u^T}, \tag{15}$$

where we have dropped the superscripts "a" and "f", because there are no observations and the analyses and forecast are identical. We write the constituent error covariance matrix $\mathbf{P_n}$ in the form,

$$\mathbf{P_n} = \mathbf{d_n p_n d_n}, \tag{16}$$

where $\mathbf{p_n}$ is the constituent error correlation matrix and $\mathbf{d_n}$ is the diagonal matrix, whose elements are equal to the square roots of the diagonal elements (the variances) of $\mathbf{P_n}$.

At initial time ($t = 0$), we assume that the constituent error covariances are homogeneous and isotropic and are given by a simple second order autoregressive function. Then, the initial error covariance between two points $\{x, y\}$ and $\{x', y'\}$ would be given by,

$$P_0^C(x, y, x', y') = d^2[1 + r/l_c]\exp[-r/l_c], \tag{17}$$

with $r = [(x - x')^2 + (y - y')^2]^{1/2}$. Here d^2 is the initial constituent error variance which is, of course, spatially invariant. The one-dimensional spectral form of (17) was used in Daley (1995), but in the present two-dimensional case, the physical-space form (17) was used. Covariance, (17), is positive-definite for an infinite, two-dimensional plane (Daley, 1991, Section 3.3); but positive-definiteness cannot be guaranteed for the finite domains introduced in Section 2.

As noted in Section 2, there are fundamental differences between the continuous and discrete cases. Thus, before discussing the experimental results with the discrete models, we first consider what should happen for a continuous constituent transport model (Equation (2)). To do this we apply the theory of Cohn (1993). In particular, we consider the two-dimensonal variance and correlation evolution equation given by Cohn (1993, Equations 3.48 and 3.54). Define the constituent error variance at the point $\{x, y\}$ at time t as $d^2(x, y, t)$. Now define the constituent error correlation between the points $\{x, y\}$ and $\{x', y'\}$ at time t as $p(x, y, x', y', t)$. Then, the error variance equation consistent with (2) is,

$$\frac{\partial d^2}{\partial t} + u(x, y)\frac{\partial d^2}{\partial x} + v(x, y)\frac{\partial d^2}{\partial y} = 0, \tag{18}$$

and the corresponding constituent error correlation equation is,

$$\frac{\partial p}{\partial t} + u(x, y)\frac{\partial p}{\partial x} + u(x', y')\frac{\partial p}{\partial x'} + v(x, y)\frac{\partial p}{\partial y} + v(x', y')\frac{\partial p}{\partial y'} = 0. \tag{19}$$

If the initial constituent error variance is homogeneous (as in equation (17)), with initial error variance d^2 then from (18), it will remain homogenous and equal to d^2 everywhere for all time.

The correlation equation (19) resembles a four (spatial) dimensional version of (2); although, it must be remembered that $p(x, y, x', y')$ is symmetric in $x - x'$ and $y - y'$ and is always equal to 1 when $x = x'$ and $y = y'$. However, the initial correlation between any two points is maintained in a Lagrangian sense. For example, consider two points $\{x, y\}$ and $\{x', y'\}$ initially very close to each other and with a correlation just less than 1. Then, the constituent parcel which is initially at $\{x', y'\}$ will maintain its high (error) correlation with the parcel initially at $\{x, y\}$, even though the two parcels may later become widely separated. This means that only correlations which are present at initial time, will be present at later time. In particular, if the initial error correlation is everywhere positive, as in (17), then it will remain everywhere positive. Whether or not these properties hold with the discrete models (10) and (13) remains to be seen.

In this section we will only consider the evolution of the constituent error *correlations*. The evolution of the *forecast error variances* can be more easily understood using the eigenstructures of the models $\mathbf{M_s}$ and $\mathbf{M_u}$, which are discussed in the next two sections.

We will give examples of constituent error evolution using the flows of Figs 2 and 3. We consider first, the application of the semi-Lagrangian model $\mathbf{M_s}$ to the axisymmetric flow of Fig. 3a. In this case, the model grid is $21 \times 21 = 441$ constituent grid points and the maximum Courant number (which occurs at the radius marked by the arrows in Fig. 3a) is approximately 0.5. The initial constituent error covariance is homogeneous, isotropic and given by (17), with $l_c = \pi/20$.

The constituent correlation at a given time has (essentially) four spatial dimen-

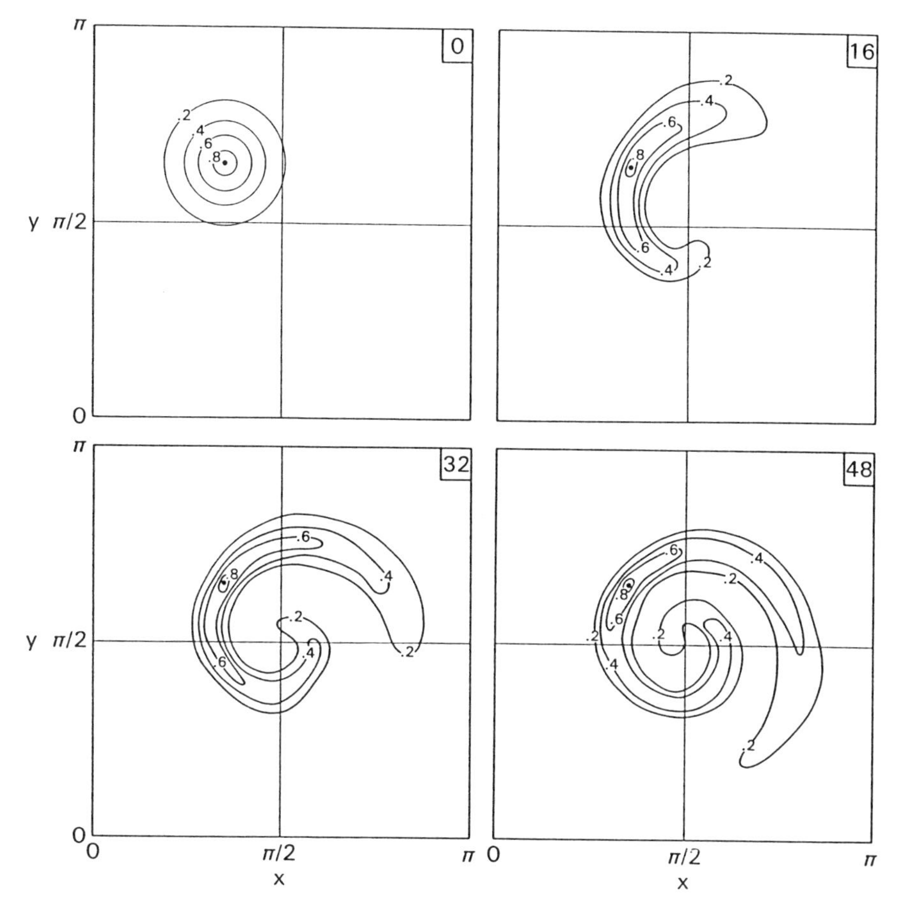

Fig. 5 Evolution of a single row of the constituent error correlation matrix for the semi-Lagrangian model and the windfield of Fig. 3a. The correlation is shown at time steps 0, 16, 32 and 48 and the contours are 1.0, 0.8, 0.6, 0.4 and 0.2.

sions, so it cannot be easily represented by two-dimensional plots. We will attempt to show this correlation as a function of time in both an Eulerian and Lagrangian framework. In the Eulerian description, we plot the error correlation with *a fixed point in space* as a function of time. Thus, a single row of the constituent error correlation matrix at 0, 16, 32 and 48 time steps is shown in the four panels of Fig. 5. In all panels, the correlations are with respect to the grid point marked with a solid circle – the correlation at that point is always equal to 1. It can be seen that the initial isotropic correlation at $t = 0$, becomes rapidly anisotropic as neighbouring constituent parcels gradually drift apart because of the windshear. After a time, a spiral structure is formed, and there are points which are a considerable distance from the solid circle, but, which are fairly highly correlated with it. Any other row of the correlation matrix, would yield a different correlation evolution.

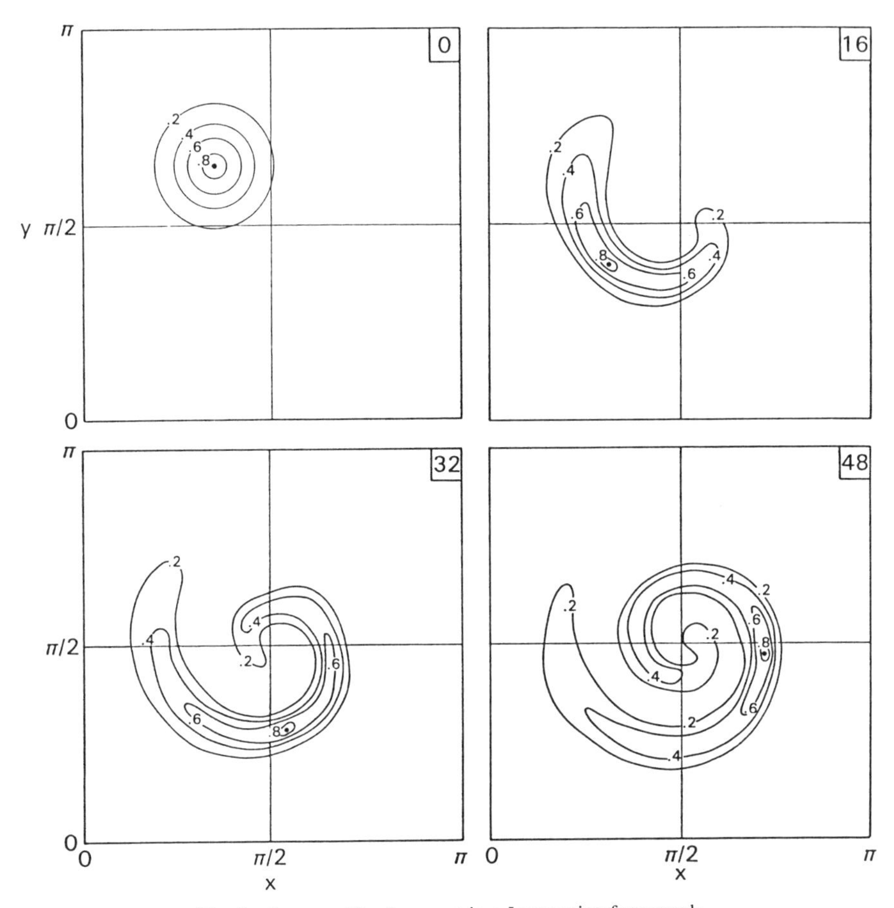

Fig. 6 Same as Fig. 5, except in a Lagrangian framework.

It might be noted that these diagrams are hand drawn and do not show correlations less than 0.2. Negative correlations do occur with the semi-Lagrangian model, particularly in later time, when the model has insufficient resolution to resolve the correlation evolution. While negative correlations are not non-physical, they are inconsistent with an initial constituent covariance which was everywhere positive.

In a Lagrangian description of the error correlation evolution, the correlation is *with a given constituent parcel as it moves in the flow*. In the present case, because the flow is axisymmetric and the initial error covariances are homogeneous and isotropic, it is straightforward to obtain the Lagrangian correlation from the Eulerian correlation by simply rotating the correlations in Fig. 5. The constituent parcel at the point marked with a solid dot in Fig. 5a takes 80 time steps to complete one rotation about the centre of the vortex. Figure 6 (in the same format as Fig. 5) is produced by simply rotating the 16 time step panel of Fig. 5 by 72 degrees

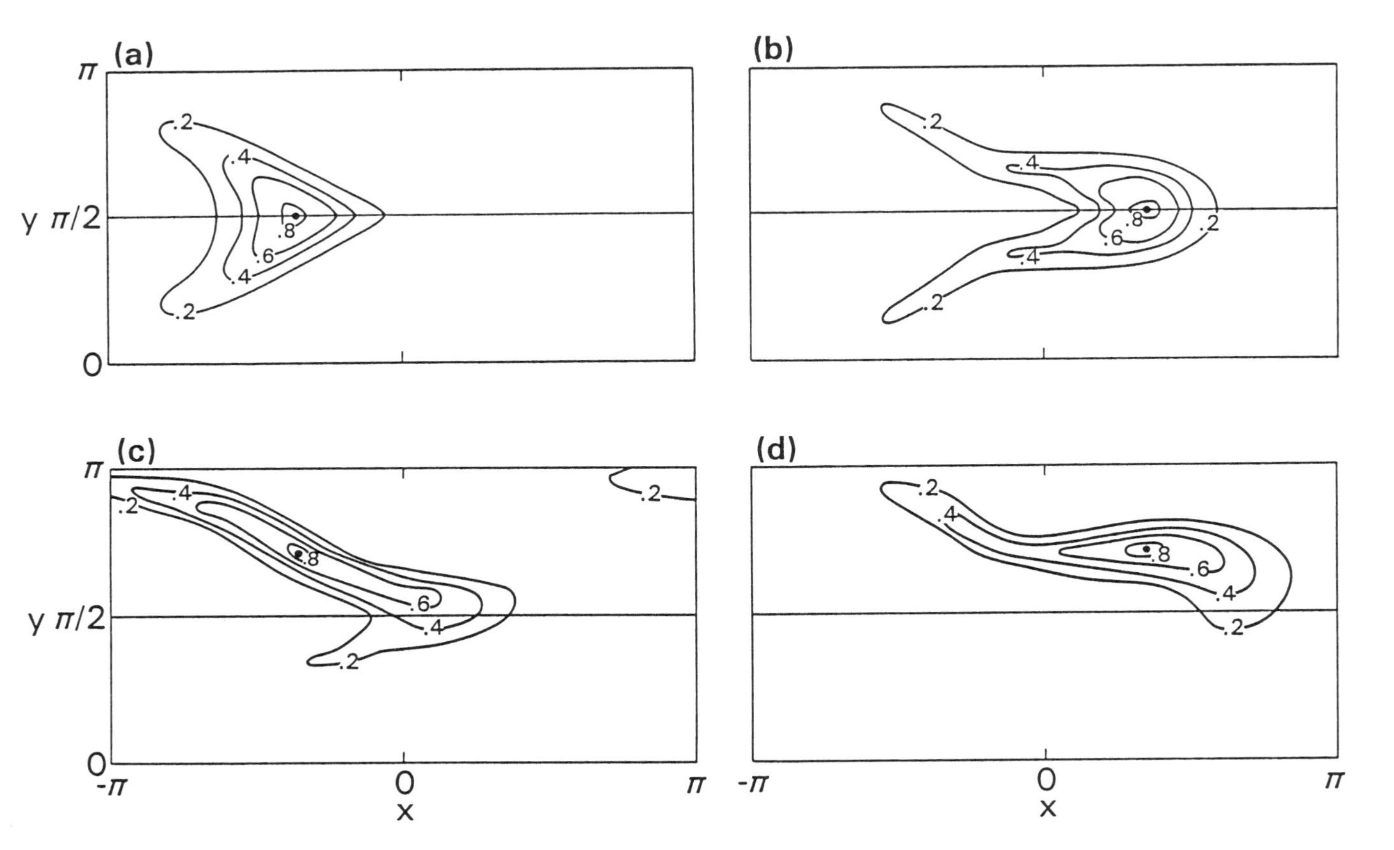

Fig. 7 Four rows of the constituent error correlation matrix at time step 96 for the semi-Lagrangian model and the windfield of Fig. 2. Contour intervals as in Fig. 5.

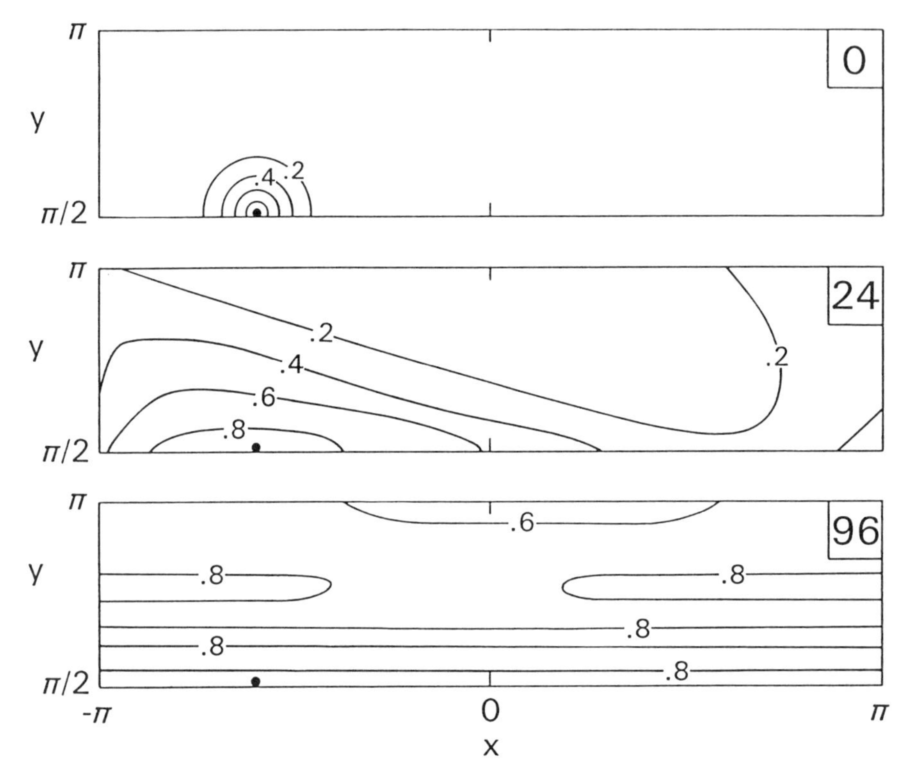

Fig. 8 Evolution of a single row of the constituent error correlation matrix for the upstream-difference model and the windfield of Fig. 2. Contour intervals as in Fig. 5.

counterclockwise, with 144 degree and 216 degree counterclockwise rotations at 32 and 48 time steps respectively.

Figure 7 shows 4 different rows of the constituent error correlation matrix at time step 96, for the confluent/diffluent windfield of Fig. 2 and the semi-Lagrangian model. The model has 29 grid points in the x direction and 15 in the y direction (435 grid points). The maximum Courant number (indicated by the arrow in Fig. 2) is 1.0. The initial constituent error covariance is given by (17), $l_c = \pi/12$. In each panel, the correlation is with the grid point marked with a solid circle and the correlation contours are at 0.8, 0.6, 0.4 and 0.2. In panels (a and b) the initial point is on the line $y = \pi/2$ and the correlations are symmetric about that line.

Figure 8 shows the evolution at time steps 0, 24 and 96 of a single row of the constituent error correlation matrix for the confluent/diffluent windfield of Fig. 2 and the upstream-difference model $\mathbf{M_u}$. Since the correlation is with a point on the line $y = \pi/2$, it is symmetric about that line and only the section $\pi/2 \leqslant y \leqslant \pi$ is shown. There are 225 grid points in this integration. The initial constituent error correlation is given by (17), with $l_c = \pi/20$. The evolving correlation structure demonstrates the highly diffusive nature of this algorithm. After 96 time steps, the correlation is greater than 0.6 almost everywhere. Figure 8 (panel at 96 hours) can be compared with Fig. 7a (semi-Lagrangian model) because the correlation point is at approximately the same location. The error correlation is always positive in this case, as it should be. The alternating bands of correlations less than/greater than 0.8 at 96 time steps are due to the fact that solid circle points in Fig. 4 are more highly correlated with other solid circle points than they are with open circle points (and vice versa). We will return to this point in Section 5.

We are now ready to consider the introduction of constituent observations into the system and the application of the full Kalman filter equations (14). In the following two sections, we will separately consider the semi-Lagrangian model and the upstream-difference model.

4 Observability for the semi-Lagrangian model

An important question in data assimilation is the adequacy or inadequacy of an actual or proposed observing system to determine the state of the atmosphere. Under certain circumstances, this rather loose notion can be given a very precise mathematical formulation, and has come to be known as *observability theory*. The original ideas for observability (and the related concepts of constructability, controllability and reachability) first appeared in the pioneering work of Kalman and Bucy (1961).

Strictly speaking, observability is the ability of a data assimilation system with perfect model and perfect observations to determine a unique initial state from a finite time observation sequence. However, the ideas can easily be extended to include imperfect models and imperfect observations. The observability and controllability of a data assimilation system are necessary for the asymptotic stability of the Kalman filter. For a perfect (stable or damped) linear model, and a spatially-fixed observing network with imperfect observations; the analysis error variance for the Kalman filter will converge to zero (at least geometrically), *if the system is observable*. Under the same conditions, except for an imperfect or unstable model, the analysis error variance will converge to non-zero but finite values.

The concept of observability was introduced to the meteorological community by Cohn and Dee (1988), and has also been employed by Ménard (1994) and Daley (1995). Cohn and Dee (1988) considered time-independent linear models and spatially-fixed observation networks. They showed that observability depended only on the eigenstructure of the model or transition matrix ($\mathbf{M_s}$ or $\mathbf{M_u}$ in the present case). Cohn and Dee's (1988) examination of a number of discretizations of the one-dimensional linear advection equation, yielded two results which we will apply in the present and following sections. Consider the eigenstructure of a model matrix $\mathbf{M}$. Then,

(1) the observability of a fixed-network data assimilation system requires that the number of observing stations be equal to the maximum multiplicity of the eigenvalues of $\mathbf{M}$,

(2) if **e** is any eigenvector of **M**, then this eigenvector is not observable if $\mathbf{He} = 0$, where **H** is the forward interpolation matrix introduced in (14).

In the experiments of Sections 4 and 5, we consider observing systems in which spatially-fixed constituent observation stations coincide with constituent grid points of models $\mathbf{M_s}$ or $\mathbf{M_u}$. If there are K observations and IJ grid points, then **H** is a $K \times IJ$ time-independent rectangular matrix whose elements are all equal to "1" or "0". It is a simple extension to the more general case, where the observation stations do not coincide with the model grid points; but, this extension is not considered here.

We consider first, the axisymmetric windfields of Fig. 3. We determine the eigenstructure of $\mathbf{M_s}$ for the windfield of Fig. 3b, with $21 \times 21 = 441$ grid points. The maximum Courant number of the flow (indicated by the arrows in Fig. 3b) is approximately 0.5. In this case, we find that there are both real eigenvalues and complex pairs. All eigenvalues are distinct and the modulus of the great majority of eigenvalues is very close to 1, as one would expect in a neutral system. There are four eigenvalues whose modulus exceeds 1. The eigenvalue with largest modulus has a modulus of 1.00041, indicating that the semi-Lagrangian model is very weakly unstable with the windfield of Fig. 3a. This weak instability, which is unusual for semi-Lagrangian algorithms, appears to be due to the particularly simple implementation of the wall boundary conditions. This weak instability of the semi-Lagrangian model, while somewhat unsatisfying, doesn't really affect the conclusions of this study.

Before considering the observability problem, we re-visit the problem of *error variance propagation*, which was introduced, but not completed, in Section 3. Since the semi-Lagrangian model has a weak instability with this windfield, it is obvious that the initial homogeneous error variance d^2 will *not* be maintained by the model. We show, in Fig. 9, the modulus of the most unstable eigenvector (as a function of x and y). The values have been normalized, so that the largest value is equal to 1. The largest values occur around the boundaries, and this eigenvector is identically zero at the centre of the vortex $\{\pi/2, \pi/2\}$. At large t, the constituent error variance would be dominated by this eigenvector. Clearly, the semi-Lagrangian model will *not* maintain the homogeneous error variance, as does the continuous transport equation (2).

The spatial structures of the eigenvectors in this case can be divided into two classes. In the first class, there is a non-zero element at every grid point. In the second class, there is a non-zero element at every grid point except at $\{\pi/2, \pi/2\}$ – the centre of the vortex. The eigenvector, whose modulus is plotted in Fig. 9, is an example of this second class of eigenvector. Applying the two criteria above, leads to the conclusion that the axisymmetric windfield with the semi-Lagrangian model will be completely observable if there is a single observation, provided this information is not at the centre of the vortex. The point $\{\pi/2, \pi/2\}$ is a special point, where there is no flow and constituent information cannot be propagated from this point out into the rest of the flow field.

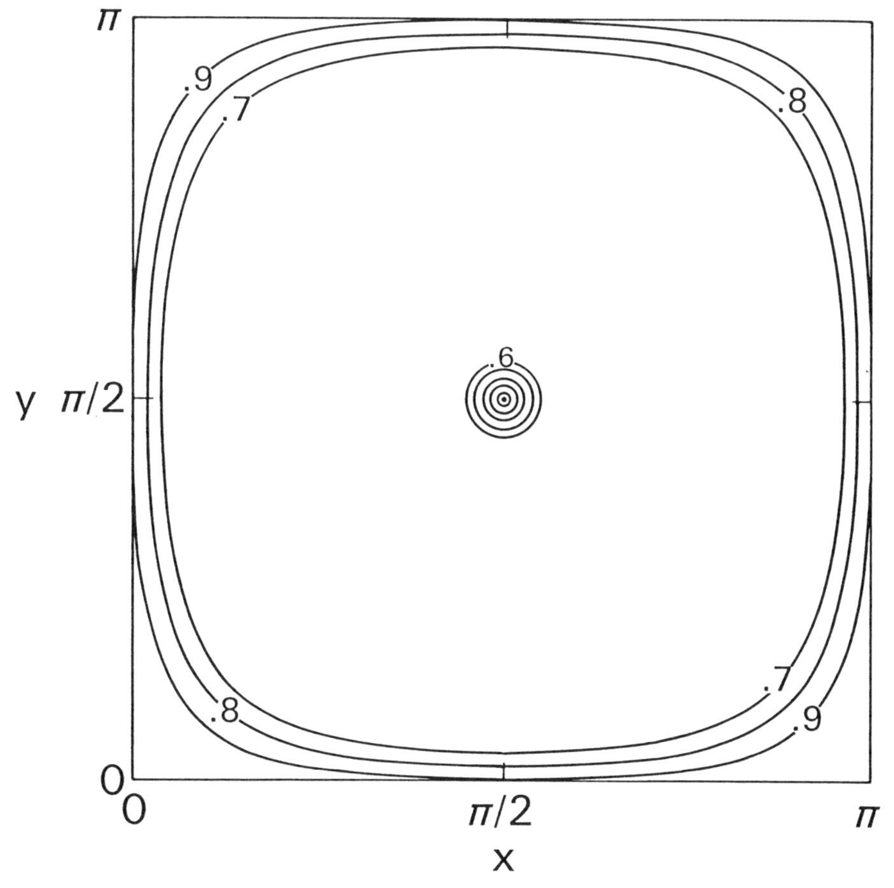

Fig. 9 Modulus of the eigenvector (as a function of x and y) for the eigenvalue with largest modulus, for the windfield of Fig. 3b and the semi-Lagrangian model with $21 \times 21 = 441$ grid points.

These ideas were tested by running a Kalman filter, using Equations (14). The initial constituent error covariance was homogeneous and isotropic (Equation (17), with $l_c = \pi/20$ and $d^2 = 1.0$). The windfield is shown in Fig. 3b and the grid was 21×21. There was a single observation station with an expected constituent observation error variance equal to 0.01. (This value of the observation error variance was used in all subsequent experiments.) Figure 10 shows the time evolution of the analysis error variance at 0, 32, 64 and 192 time steps as a function of x and y. The position of the observation station is marked with a solid circle in each panel and is obviously not located at $\{\pi/2, \pi/2\}$. Initially the analysis error is equal to 1 everywhere, except in the immediate vicinity of the observation station. After a period of time, the analysis error gradually decreases over the domain, as the information from the single observing station propagates throughout the domain.

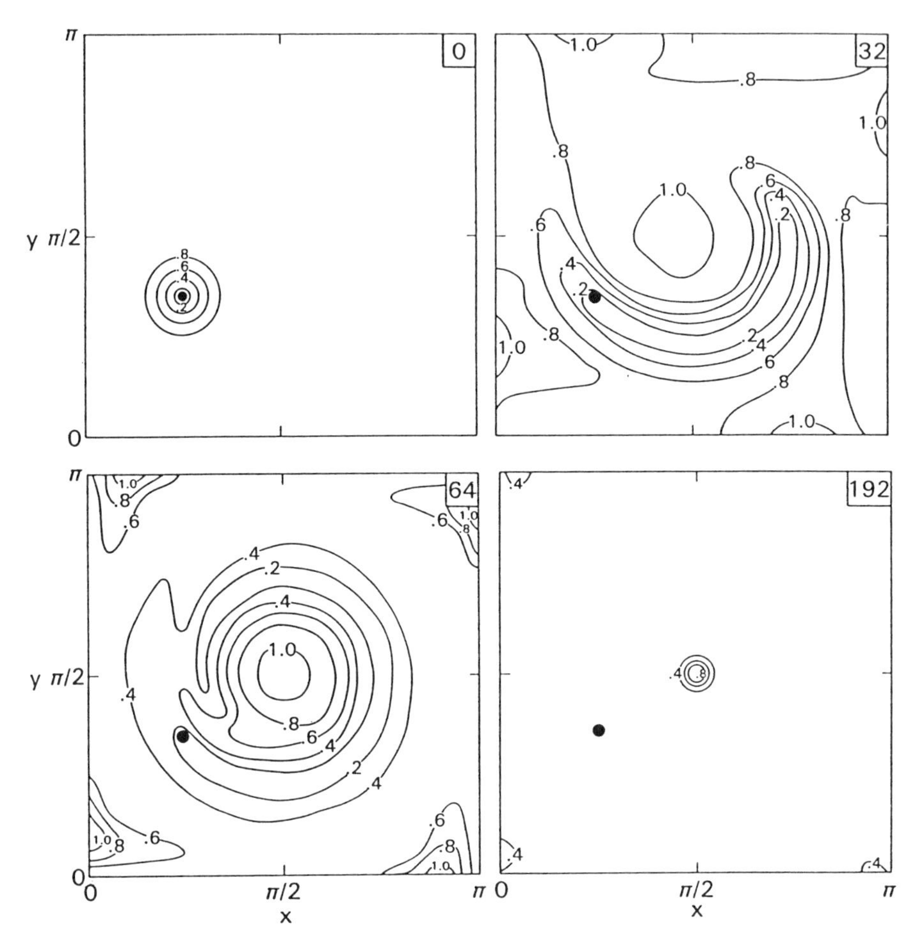

Fig. 10 Evolution of the analysis error variance at 0, 32, 64 and 192 time steps for a single observation station marked with the solid circle. Contour intervals at 1.0, 0.8, 0.6, 0.4 and 0.2.

The error at the centre of the vortex decreases very slowly; but ultimately, it too, becomes vanishingly small. A similar experiment with a single observation station at the centre of the vortex ultimately becomes unstable, because this observation station has no projection on the most unstable mode of Fig. 9. These results are in accord with the eigensystem analysis above.

The eigensystem of the semi-Lagrangian model with the confluent/diffluent windfield of Fig. 2 was also examined. It was found that the eigenvalues were distinct, but there were a small number of eigenvalues with modulus greater than 1. The eigenvectors divided into two classes – those which had non-zero elements everywhere, and those which had non-zero elements everywhere except along the line

$y = \pi/2$. Following the two criteria above, we interpreted the eigenanalysis to imply that the system was observable with a single observation, provided this observation was not located along the jet axis at $y = \pi/2$. This result was confirmed by running the Kalman filter.

We also considered the two windfields of Fig. 1. In this case the y component of the wind is equal to zero and the x component is independent of x. The model $\mathbf{M}_s$ can be written as a block diagonal matrix, with J $I \times I$ blocks along the main diagonal. In effect, the model degenerates into a series of one-dimensional models in x, at each y_j, $1 \leqslant j \leqslant J$. Consequently, the eigensystem can be analyzed as J separate eigensystems for J one-dimensional constituent transport models. For a given gridline y_j, if $u(y_j)$ is not identically equal to zero, only one observation on that gridline will be required to make that gridline observable. If $u(y_j) = 0$ on the gridline y_j, then it will require I observations along that gridline to render it observable (Cohn and Dee, 1988). Thus, if J_z is the number of gridlines where $u(y) = 0$, then the total number of observation stations required is $J + J_z(I - 1)$. It should be noted that, although the model decouples in this case, the various gridlines y_j are still coupled through the analysis equation.

This is illustrated in Fig. 11. In the left three panels, we show the evolution of the analysis error variance at time steps 0, 20 and the asymptotic limit, for the windfield of Fig. 1 (curve a) which is positive everywhere. The initial error covariance is given by equation (17) (with $l_c = \pi/12$ and $d^2 = 1.0$). The observation station locations are indicated with solid circles and the analysis error variance contours are at 0.8, 0.6, 0.4 and 0.2. The model grid is 29×15. From the eigensystem analysis above, one would expect that the asymptotic analysis error would not vanish along those y gridlines where there were no observations. This is, in fact what happens. It is interesting to note, that where the asymptotic analysis error does not vanish, it becomes independent of x.

In the three right panels, we consider a more complete observation system, with the windfield of Fig. 1 (curve b). In this case, the winds are identically zero at y_2 and y_{J-1}. We show analysis error variance at time steps 0 and 30 and the asymptotic limit. As would be expected, from the above eigenanalysis, this system is also not observable and the asymptotic analysis error does not vanish. If the zero wind lines are located in between gridlines y_j, then only J observations (1 along each line y_j) are required for observability.

It might be wondered why observability required only one constituent observation station for the, essentially, axisymmetric flow (Fig. 10), while many stations were required for the shear flow (Fig. 11). After all, the shear flow (Fig. 1) is one-dimensional and the axisymmetric flow (Fig. 3) in polar coordinates also has reduced dimensionality. The explanation of this puzzle is found by considering the continuous equation (2).

Consider the flow of Fig. 3 for the continous (inviscid) transport equation and a single observation station which observes continuously in time. What is observable is the trajectory of a constituent parcel which follows the streamline passing through

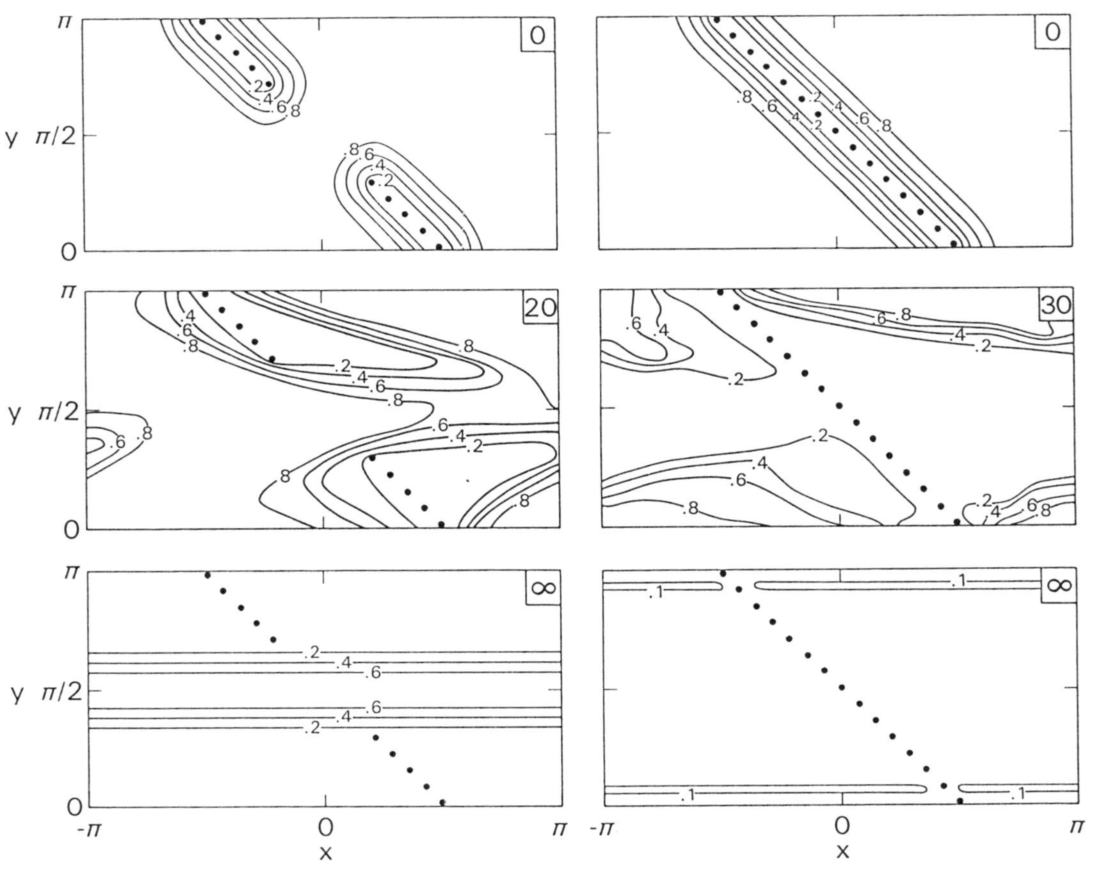

Fig. 11 Evolution of the analysis error variance for the semi-Lagrangian model and the windfields of Fig. 1 (curve a) – left panels and Fig. 1 (curve b) – right panels.

the observation station. This trajectory is not dense in the domain, but follows a simple closed curve. This means that the constituent field which does not lie along this streamline *is not observable by this single observation.*

What makes the whole field observable for the discrete model $\mathbf{M}_s$ with a single observation station, is the, spurious, diffusion introduced by the (semi-Lagrangian) discretization. For the shear flow (Fig. 1), the model $\mathbf{M}_s$ has no dissipation in the direction normal to the streamlines, and hence no way to spread the influence of an observation across streamlines. Thus, observability is limited to the streamline which passes through the observation station. For the case of axisymmetric flow, if the model $\mathbf{M}_s$ had been formulated in polar coordinates (centred on the vortex centre), then there would have been no dissipation in the radial direction (i.e. normal to the streamline). Thus, again observability would have been limited to the streamline which passed through the observation station.

The greatly enhanced observability for the model $\mathbf{M}_s$ noted in Fig. 10 (for a Cartesian grid), occurs because (a) of the inherent dissipation of the model $\mathbf{M}_s$ and (b) the Cartesian grid lines are not streamlines. Some observation patterns do not ensure observability (Fig. 11), because with these patterns, the discretization becomes non-diffusive in some directions. However, changing the grid could lead back to observability (as in Fig. 10).

The simple observability analysis of this section can only be performed for steady windfields. For time-dependent winds, trajectories and streamlines do not coincide and the trajectories may cover more of the domain, possibly leading to enhanced observability.

In conclusion, discretization of the constituent transport equation with steady windfields introduces (spurious) dissipation and this greatly enhances observability vis-à-vis the continuous equation. For the semi-Lagrangian model, only one observation station is generally required for complete observability. There is reduced observability for observation stations at the centre of a vortex, or where the flow is strictly parallel to one of the grid axes, or where the wind vanishes.

We will now apply this knowledge to an analysis of the observability of the upstream-difference model.

5 Observability of the upstream-difference model

Observability for the upstream-difference model is more complicated than for the semi-Lagrangian model. We first note a property of the upstream-difference model, (11) to (13), that will be important in the ensuing discussion. In Fig. 4, there are two sets of constituent grid points, those marked with open circles and those marked with solid circles. From equations (11) to (12), it is easy to see that the constituent time tendency at an open circle point does not depend on the constituent values at any solid circle points, and vice versa. In other words, the solid circle and open circle grids are completely independent of one another. This property of the upstream-difference grid is the reason for the alternating bands of high and low correlation in Fig. 8.

We first consider observability for the model $\mathbf{M_u}$ with respect to the axisymmetric windfield of Fig. 3b. We consider the case where the centre of the vortex is at a solid circle point (point $\{i,j\}$ in Fig. 4). The eigenvalues of $\mathbf{M_u}$ generally occur in complex pairs, but there are a few real eigenvalues. There are two real eigenvalues equal to 1.0; the moduli of all the remaining eigenvalues are less than 1.0, which is to be expected in such a highly damped algorithm. The eigenvectors can be broken into three classes – eigenvectors involving (i.e. having non-zero values only at) open circle points, eigenvectors involving only the centre point and eigenvectors involving all solid circle points except the centre point. One of the two eigenvectors whose eigenvalue is equal to 1.0, is an eigenvector which involves only the centre point, and the other involves all solid circle points except the centre point. All eigenvectors which involve only open circle points are damped to greater or lesser degree.

The two observability criteria introduced in Section 4 suggest that this system will not be observable unless there are at least two observation stations – one at the centre point and another at any other solid circle point. No observation stations are required at the open circle points.

This result was confirmed by running a Kalman filter. In Fig. 12, we show the total constituent analysis error variance summed over all solid circle points (except the centre point) in panel (a), summed over all open circle points in panel (b) and at the centre point in panel (c) as a function of time step (abscissa). These plots summarize three different experiments. The solid curves indicate a single observation station at a solid circle point (not the centre point). The dashed curves indicate a single observation station at an open circle point and the dash-dot curves indicate a single observation station at the centre point. None of these experiments are observable, as would be expected from the eigenanalysis. The error on the open circle points always decays, regardless of whether they are observed or not. If there is an observation station at the centre point, then the other solid circle points are not observable and vice versa.

Before going on to the next windfield, it might be noted that the fact that the moduli of most of the eigenvalues of $\mathbf{M_u}$ are less than 1.0, means that for the pure error propagation experiments of Section 3, the upstream-difference model would lose constituent error variance rapidly, which is very different than for the continuous transport model (1).

The observability properties of the upstream-difference model with respect to the winfields of Fig. 1, is the same as for the semi-Lagrangian model. While the detailed time evolution and asymptotic distributions of the constituent analysis error variances differ between the two models, the requisite number and distribution of observation stations is the same. This result is to be expected because there is no cross-streamline dissipation in this case.

The final case considered is the confluent/diffluent windfield of Fig. 2, and the application of the upstream-difference model. Referring to Fig. 4, suppose that the line marked "j" corresponds to the symmetry line $y = \pi/2$ in Fig. 2. Thus,

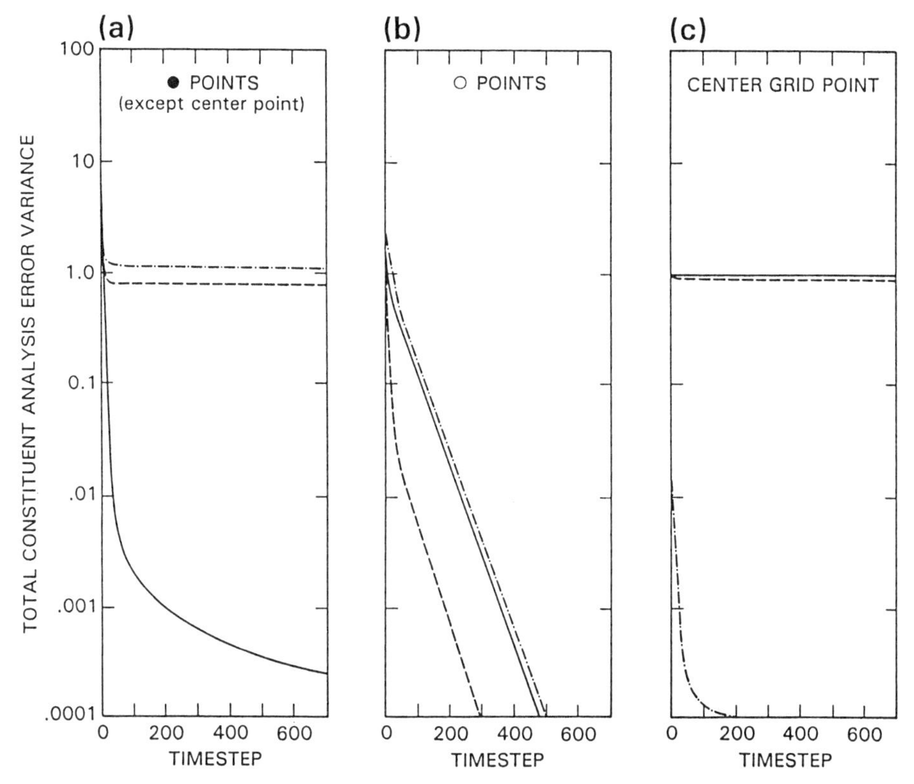

Fig. 12 Total constituent analysis error variance (ordinate) versus time step (abscissa) for the upstream-difference model applied to the axisymmetric windfield (Fig. 3b). Panel (a) is for all solid circle points (except the centre point), panel (b) is for all open circle points and panel (c) is for the centre point.

along this line $v = 0$ and constituent points are solid circle points. Examination of Equations (11) and (12) demonstrates that not only are solid and open circle points decoupled from each other, but open circle points with $y > \pi/2$ are disconnected from open circle points with $y < \pi/2$.

Examination of the eigensystem of the upstream-difference model for the confluent/diffluent windfield yields the following results. All eigenvalues have moduli less than 1, except for three eigenvalues which are equal to 1. The eigenvectors fall into three classes – those involving only solid circle points, those involving open circle points for which $y < \pi/2$ and those involving open circle points with $y > \pi/2$. Application of the two criteria of Section 4 to this case indicates that a minimum of three observation stations are necessary for observability of the system. One of the observation stations must be a solid circle point and the other two must be open circle points which lie on either side of the line $y = \pi/2$.

This result was verified by applying the Kalman filter to the upstream-difference

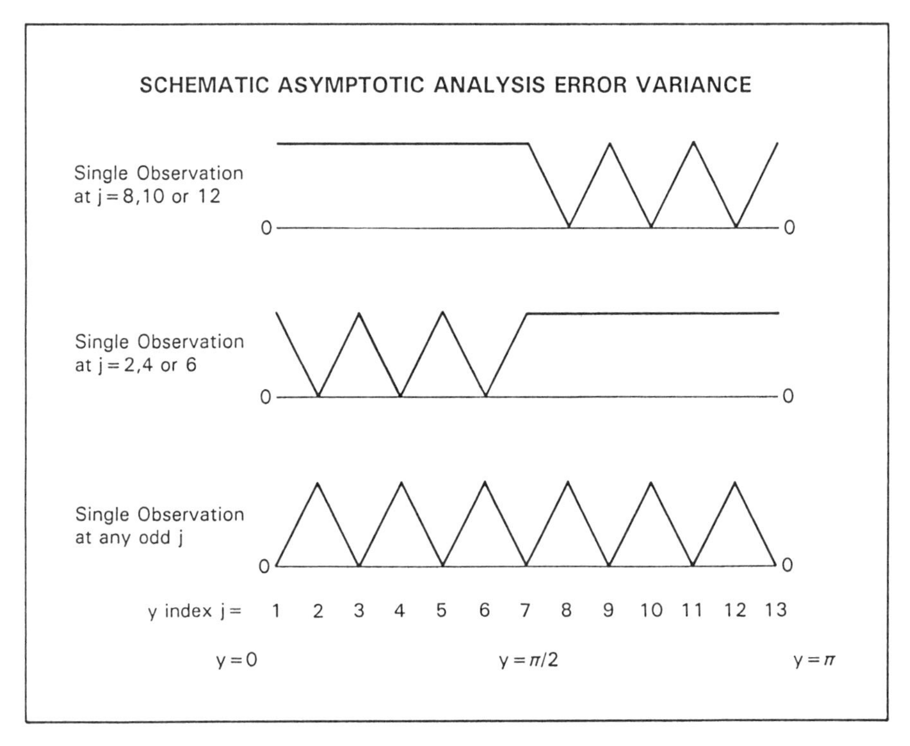

Fig. 13 Asymptotic constituent analysis error variance (schematic) as a function of the y index j for the upstream-difference model and the confluent/diffluent windfield. Each of the three cases corresponds to a single observation station.

model. $J = 13$ and thus there were open circle points at $j = 2, 4, 6, 8, 10, 12$ and solid circle points at $j = 1, 3, 5, 7, 9, 11$ and 13. Three experiments were performed with single observation stations – one with an observation station at a grid point along the lines $j = 8$, 10 or 12, another with an observation station at a grid point along the lines $j = 2$, 4 or 6 and another with an observation station along the lines $j = 1, 3, 5, 7, 9, 11$ or 13. The asymptotic constituent analysis error variance as a function of j is indicated schematically in Fig. 13, and it can be seen that the results are consistent with the eigenanalysis.

Thus, for the upstream difference model, complete observability generally requires at least two observation stations because there are essentially two decoupled grids. (It should be noted, however, that all observation stations in these experiments were located exactly at grid points. Observation stations that were located in between grid points, might project on both of the decoupled grids, thus effectively coupling them and increasing the observability.) There is reduced observability for observation stations at the centre of a vortex, or where the flow is strictly parallel to one of the grid axes, or where the windspeed vanishes.

6 Summary and conclusions

Chemical constituent data assimilation is likely to become an increasingly important activity over the next few decades, as a number of constituent instruments are developed and deployed. At this time, constituent assimilation has received little attention; and yet, as shown in Daley (1995), the subject is both complex and interesting.

Two aspects of the constituent assimilation problem were examined in this paper. The effect of several different two-dimensional steady windfields on error propagation by constituent transport algorithms was considered first. Then, observability theory was applied to the constituent assimilation problem, to determine how a constituent observation system should be deployed to measure transport-dominated constituent distributions.

The windfields for these experiments were all specified, and were assumed to be known; there was no attempt to determine the windfield as in Daley (1995, Sections 5–6). The winds were two-dimensional, non-divergent and steady and were of three types, namely, parallel shear flow, axisymmetric flow and confluent/diffluent flow. Two models of the two-dimensional constituent transport equations were constructed. The first model was a simple semi-Lagrangian model, which used Fourier interpolation at the departure points. The second model was a simple upstream-difference model. The upstream-difference model was monotonic and did not generate negative constituent values. Both models were diffusive/advective discretizations of an inviscid advective model, although the upstream-difference model was considerably more diffusive. Data assimulation was performed (or at least the second moment error statistics were calculated) using a standard Kalman filter.

In the first experiments, there were no observations, and the error covariance propagation equation of the Kalman filter was integrated to determine constituent error variances and correlations as a function of time. These results could be compared with the known properties of the continuous form of the error variance and correlation equations (appropriate for constituent transport) using the theory of Cohn (1993). For both the upstream-difference and the semi-Lagrangian algorithms, the error propagation was quite different than what would be expected from the continuous transport equation. Not surprisingly, the upstream-difference algorithm lost variance rapidly and quickly generated very broad-scale error correlations. The semi-Lagrangian algorithm produced the increasingly complex and filament-like error correlations that would be expected from the continuous transport equation, but did not preserve the total error variance.

Observability theory was applied to the constituent assimilation problem to determine the number and positioning of observation stations required to analyze the constituent distribution. This was done by examination of the eigenstructures of the transition (model) matrices corresponding to the semi-Lagrangian and upstream-difference algorithms. In particular, two properties from the theory of Cohn and Dee (1988) were exploited: the maximum multiplicity of the eigenvalues and the projection of the forward interpolation matrix onto the eigenvectors. The predictions of the observability theory were then verified by running the Kalman filter until an asymptotic state was reached.

The results showed that the constituent fields could be observed with either transport algorithm and for many of the imposed windfields with between 1 and 3 fixed observation stations. Generally speaking, the upstream-difference model required more observation stations than the semi-Lagrangian model because it suffered from grid-decoupling. There was reduced observability at the centre of vortices, in regions of low windspeed, or where the flow was parallel to one of the grid axes. It might be noted that these results would be the same for the 4DVAR algorithm; because it is equivalent to the standard Kalman filter in these experiments. For these simple windfields, it was clear that the continuous (inviscid) transport model would be much less observable than either of the discrete models. The enhanced observability with the discrete approximations could be entirely explained by their (spurious) dissipation.

One implication of these experiments was that for constituent observations with steady, axisymmetric windfields, the observability was greater with a Cartesian discretization than for a polar discretization. This might have some applicability to constituent assimilation in the vicinity of the polar vortex. If the polar vortex is coincident with the Earth's pole, then a Cartesian (i.e. polar stereographic) discretization is likely to lead to greater observability than a latitude-longitude discretization. This result has since been confirmed with a global Kalman filter constituent assimilation system running on an isentropic surface (Ménard et al., 1995).

Now, these conclusions are undoubtedly over-optimistic because they depend on the optimality of the Kalman filter. The results of this paper would not be strictly valid if the model were not perfect (unless the model error statistics were known precisely), or if there were important chemical interactions, or if the windfield were time dependent and/or imprecisely known, or if the observation network were time dependent. Unfortunately, observability theory does not cover all of these more realistic cases; leaving observation system simulation experiments (OSSE) or experiments with a full Kalman filter under more realistic physical and observation conditions as the only ways to obtain this information. One could anticipate that such experiments would demonstrate that more constituent observations would be required to determine the constituent distribution than is indicated by the present work. That being said, however, it remains encouraging how few observations seem to be necessary to determine the distribution of transport-dominated constituents.

A companion study using these same models and windfields has considered the recovery of non-divergent windfields from constituent observations (Daley, 1996).

Acknowledgements

The author would like to thank Richard Ménard and a referee (Philippe Courtier) for their many helpful comments on the manuscript. The diagrams were drafted by Tom Chivers of the Atmospheric Environment Service.

References

ALLEN, D.; A. DOUGLASS, R. ROOD and P. GUTHERIE. 1991. Application of a monotonic up-stream based transport scheme to three-dimensional constituent transport calculations. *Mon. Weather Rev.* **119**: 2456–2464.

AUSTIN, J. 1992. Toward the four dimensional assimilation of stratospheric chemical constituents. *J. Geophys. Res.* **97 D18**: 2569–2588.

BATES, J. and A. MCDONALD. 1982. Multiple-upstream, semi-Lagrangian advection schemes with minimized disipation and dispersion errors. *Mon. Weather Rev.* **112**: 1831–1842.

COHN, S.E. 1993. Dynamics of short-term univariate forecast error covariances. *Mon. Weather Rev.* **121**: 3123–3149.

——— and D. DEE. 1988. Observability of discretized partial differential equations. *SIAM J. Numer. Anal.* **25**: 586–617.

DALEY, R. 1991. *Atmospheric data analysis*. Cambridge University Press, 457pp.

———. 1992. The lagged innovation covariance: a performance diagnostic for atmospheric data assimilation. *Mon. Weather Rev.* **120**: 179–196.

———. 1995. Estimating the windfield from chemical constituent observations: experiments with a one-dimensional extended Kalman filter. *Mon. Weather Rev.* **123**: 181–198.

———. 1996. Recovery of one and two dimensional windfields from chemical constituent observations using the constituent transport equation and an extended Kalman filter. *Meteorol. Atmos. Phys.* **60**: 119–136.

KALMAN, R. and R. BUCY. 1961. New results in linear filtering and prediction theory. ASME Trans. Part D **83**: *J. Basic. Eng.* 95–108.

MÉNARD, R. 1994. Kalman filtering of Burgers' equation and its application to atmospheric data assimilation. Ph.D. Thesis, McGill University, Montréal, Québec, Canada, 211pp.

———; P. LYSTER, L.P. CHANG and S.E. COHN. 1995. Middle atmosphere assimilation of UARS constituent data using Kalman filtering: Preliminary results. World Meteorological Organization Second International Symposium on Assimilation of Observations in Meteorology and Oceanography. 13–17 March 1995, Tokyo, Japan, WMO/TD – No. 651. pp. 235–239.

ROOD, R. 1987. Numerical transport algorithms and their role in atmospheric transport and chemistry models. *Rev. Geophys.* **25**: 71–100.

SALBY, M. 1987. Irregular and diurnal variability in asynoptic measurements of stratospheric trace species. *J. Geophys. Res.* **92 D12**: 14781–14805.

SCHOEBERL, M. and L. LAIT. 1992. Conservative-coordinate transformations for atmospheric measurements. *In*: The use of EOS for studies of Atmospheric Physics. Proc. of the International School of Physics «Enrico Fermi», J. Gille and G. Visconti (Eds), North Holland Press, pp. 419–431.

Improvement of Spin-up of Precipitation Calculation with Use of Observed Rainfall in the Initialization Scheme

Takayuki Matsumura[1], Isao Takano[2], Kazumasa Aonashi[2] and Takashi Nitta[3]†
[1]*Numerical Prediction Division, Japan Meteorological Agency (JMA)*
[2]*Forecast Research Laboratory, Meteorological Research Institute, JMA*
[3]*Faculty of General Culture, Tokai University*

[Original manuscript received 10 October 1994; in revised form 4 August 1995]

ABSTRACT *A new approach is proposed for the initialization of the primitive equations model (hereafter, NWP model), taking account of the effect of non-adiabatic heating due to release of the latent heat into humidity and divergent wind fields. In the present paper, an operational model of the Japan Meteorological Agency (JMA), the Japanese Spectral Model (JSM) is taken up as the NWP model.*

For the purpose of the present study, rainfall amount is estimated, with a resolution of about 5-km mesh, from radar-AMeDAS composite precipitation data (hereafter, R-A data), where AMeDAS stands for the Automated Meteorological Data Acquisition System which is a network of automated surface observation stations including raingauge measurements. The observed rainfall is considered to represent the mesoscale precipitation field.

The total heating rate calculated from the R-A data is assumed to be partitioned parabolically in the vertical between the lifting condensation level (LCL) and the cloud top (TBB CT) estimated by the GMS (Japanese geostationary meteorological satellite), where TBB means the black body temperature. The diabatic initialization method proposed here consists of two steps. The first step is a physical initialization of the water vapour field, where the relative humidity field is changed to be more moist over the area of rainfall (>1 mm hr^{-1}). The moistening is performed in such a way that, if the air is not moist enough over the area of rainfall, the lapse rate at each model layer from the LCL to the TBB CT becomes a critical value presumably required by the moist convective adjustment scheme used in the NWP model concerned. On the other hand, over the area where no rainfall is observed, the relative humidity is left untouched. The second step is a nonlinear normal mode initialization (NNMI) which includes the non-adiabatic heating rate obtained from the R-A data among other factors. The first four vertical modes with the period less than 6 hours are adopted. It should be remarked that the divergent component of wind field which is necessary to continuously maintain the condensation due to mesoscale rainfall is also initialized at the stage of NNMI with the large-scale condensation and the moist convective adjustment.

Improvement in the operational performance of the NWP model with the use of the

†Corresponding author.

proposed diabatic initialization method is demonstrated for a case study and statistical verification. The forecasts with and without the present diabatic initialization are compared. Particularly, amelioration of the spin-up of mesoscale precipitation calculation for the first 6 hours after the start is clearly seen. The initialization method shows stable execution and seems to be useful for the prediction of the mesoscale phenomena, particularly in the tropics where meteoroligical data are relatively sparse.

RÉSUMÉ *On propose une nouvelle approche d'initialisation des modèles de prévision du temps aux équations primitives tenant compte de l'effet du réchauffement diabatique produit par un apport de chaleur latente dans les champs d'humidité et de vents divergents. Le modèle utilisé est le modèle spectral en exploitation au Japon, de la Japan Meteorological Agency (JMA).*

Dans le cadre de cette étude, on estime la quantité de pluie, pour une résolution de maille d'environ 5 km, à l'aide d'un composite de données de précipitation radar-AMeDAS, données R-A. AMEDAS (Automated Meteorological Data Acquisition System) est un réseau de stations d'observation en surface incluant des mesures pluviométriques. La quantité de pluie mesurée représente le champ de précipitation à la mésoéchelle.

On présume que le taux de réchauffement total obtenu des données R-A est réparti paraboliquement à la verticale entre le niveau de condensation par ascendance (LCL) et le sommet du nuage (TBB CT) estimé par le GMS (satellite météo géostationnaire japonnais); TBB étant la température du corps noir.

La méthode d'initialisation diabatique proposée comprend deux étapes. La première est une initialisation physique du champ de vapeur d'eau, où le champ d'humidité relative est rendu plus humide sur la zone de précipitation (>1 mm h^{-1}). On procède à cette humidification de telle façon que, si l'air n'est pas assez humide sur la zone de précipitation, le gradient adiabatique à chaque couche du modèle, du LCL au TBB CT, devient une valeur critique pouvant être nécessaire au schéma d'ajustement convectif humide utilisé dans le modèle pertinent; par contre, sur la zone sans précipitation, l'humidité relative n'est pas touchée. La deuxième étape est une initialisation par modes normaux non linéaires (IMNN) incluant le taux de réchauffement diabatique obtenu, entre autres, des données R-A; les quatres premiers modes verticaux avec une période de moins de 6 heures sont affectés. On doit noter que la composante divergente du champ de vent, nécessaire pour maintenir continuellement la condensation due à la pluviosité de mésoéchelle, est aussi initialisée au stade de l'IMNN avec la condensation à grande échelle et l'ajustement convectif humide.

On démontre l'amélioration de la performance du modèle en exploitation utilisant la méthode d'initialisation proposée par une étude de cas et une vérification statistique. on compare les prévisions avec et sans l'initialisation diabatique. On remarque particulièrement l'amélioration de la durée de relaxation du calcul de la précipitation à la méscéchelle pour les premières 6 heures de prévision. La méthode d'initialisation se montre d'exécution stable et semble utile pour la prévision des phénomènes de mésoéchelle, surtout dans les tropiques, là où les données météorologiques sont relativement rares.

---

1 Introduction

In recent years, attention has been particularly focused on the prediction of mesoscale disturbances, among other things, in connection with very short-term forecasting. For this purpose, the resolution of a numerical weather prediction (NWP) model should be increased, physical and dynamical processes of the model should be refined and the initial data should be improved to detect and accurately represent the mesoscale disturbances at the initial time.

However, one of the obstacles to preparing better initial data is the sparsity of conventional observation data with adequate resolution for the scale of the target disturbances. Therefore, one has to utilize various types of available observations, including those obtained by remote-sensing techniques such as weather radar and meteorological satellite.

Several authors have so far attempted (e.g., Krishnamurti et al., 1984; Ueno et al., 1987; Takano and Segami, 1993; Aonashi, 1993) to incorporate observed precipitation measured by remote-sensing techniques in order to depict mesoscale disturbances accurately at the initial time. From the practical point of view, the spin-up problem of mesoscale precipitation calculation in NWP is particularly important when observed rainfall data are incorporated.

Recently, significant progress has been made in ameliorating the spin-up of precipitation calculation against the impact of observed rainfall data (Takano and Segami, 1993; Aonashi, 1993; Kasahara et al., 1994).

The present study is an attempt to combine the assimilation and initialization methods proposed by Takano and Segami (1993) and Aonashi (1993) into a single procedure of the diabatic initialization in order to examine its effectiveness on the mesoscale prediction, particulary from the point of view of improvement of the spin-up of the precipitation forecast. The procedure is designed to make use of precipitation observations from a composite radar-automated raingauge network over Japan by the JMA. The network is called AMeDAS and has an average 17×17 km resolution (Fig. 1). (Hereafter, the composite radar-AMeDAS data is abbreviated as the R-A data.) The R-A data have a horizontal resolution of 5×5 km and the radar echo is calibrated by the AMeDAS data and converted into rainfall amount so that one can obtain a composite precipitation distribution with higher resolution (Takase et al., 1988). As is clearly explained by Kasahara et al. (1994), the following three steps have to be taken into account to obtain the smooth and realistic spin-up of precipitation calculation, (1) the divergent adjustment, (2) the adjustment of moisture and temperature and (3) the diabatic NNMI.

The diabatic initialization method proposed and tested in the present paper consists, in principle, of two steps, i.e., a physical initialization of the water vapour field using the R-A data after the objective analysis and a diabatic NNMI. The first step corresponds to step (2) mentioned above, while the second step corresponds to steps (1) and (3).

In Section 2, the diabatic initialization method to unify the above three steps for the inclusion of the R-A data is proposed. Sections 3 and 4 present the results of the forecast experiments using an operational NWP model of JMA with the proposed method for a case study and statistical verification. Comparison is made between the forecasts with and without the proposed diabatic initialization. (The latter forecast uses only an adiabatic NNMI and the R-A proxy data in the objective analysis

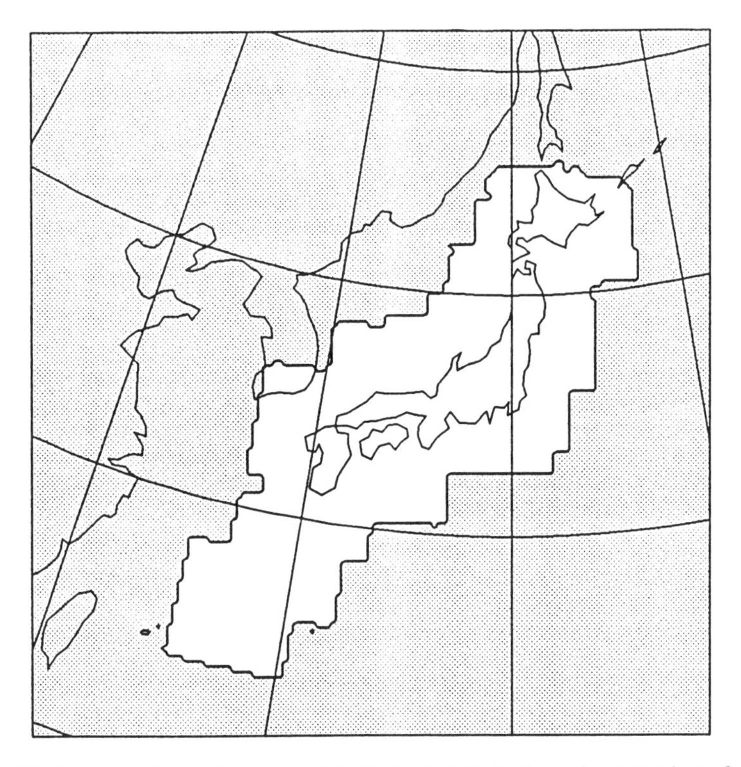

Fig. 1 JSM forecast region and coverage of the Radar-AMeDAS observation network as of 1993.

of the moisture field.) Conclusions and considerations for further development are given in Section 5.

2 A diabatic initialization method for incorporation of the R-A data

a *Description of the NWP model used for the experiment*

In this study, an operational regional spectral model of JMA, the JSM is used. The model has 23 vertical sigma levels and a horizontal grid interval equivalent to 30 km at 60°N. The grid system consists of 129 × 129 grid points with a maximum wavenumber of 83 × 83. The horizontal domain of the model with the grid system and the coverage of the R-A data are shown in Fig. 1. The model adopts the regional spectral method developed by Tatsumi (1986) for the dynamic part and has regular physical processes including the moist convective adjustment scheme proposed by Gadd and Keers (1970). The initial state is prepared by the optimum interpolation method in the objective analysis, using the 12-hour forecast of the JSM as the first guess field.

Historically speaking, the present JSM was upgraded on 10 March 1992. Major points changed at that time were:

(1) Intermittent data assimilation method with an interval of 12 hours (hereafter called the forecast-analysis cycle) is introduced,
(2) The horizontal and vertical resolutions are enhanced from 40 to 30 km for the horizontal grid interval and from 19 to 23 sigma levels in the vertical,
(3) Full radiation scheme is incorporated,
(4) Shallow convection scheme is adopted,
(5) Envelope orography is used for the model terrain,
(6) The R-A data are utilized as proxy data for the objective analysis of the moisture field.

For details of the other points of the model refer to Segami et al. (1989).

The forecast-analysis cycle and the inclusion of the R-A data for moisture analysis had a great positive impact on the improvement of the precipitation forecast, particularly during the beginning stage of the forecast. However, the adiabatic NNMI attenuated the intensity of predicted precipitation in such a way that the divergence field associated with the convective system could not be maintained and turned out to be weakened. This problem is the motivation for the present study.

b *Vertical profile of the heating assumed to be due to the observed precipitation*
The heating rate (δT) is obtained using the following equation:

$$\delta T(x, y, \sigma) = w(x, y, \sigma)\frac{L}{C_p}\frac{g}{P_s}\rho_w R(x, y) \tag{1}$$

where R is the precipitation rate, ρ_w the density of liquid water, g the acceleration due to gravity, P_s the pressure of the model's surface, L the latent heat and C_p the specific heat at constant pressure. The precipitation rate R is horizontally interpolated onto the model's grid point and suitably averaged.

The total heating rate calculated from the R-A data is assumed to be partitioned parabolically in the vertical between the LCL and TBB CT (the cloud top estimated by the GMS using the black body temperature) (Fig.2). If the observed cloud top level is higher than the tropopause level (σ_{top}) of the model atmosphere, σ_{top} is used as σ_{TBB}. The vertical weight coefficient $w(x, y, \sigma)$ is set to zero if the precipitation rate is smaller than 1 mm hr^{-1} or the cloud amount observed by the GMS is smaller than 80%.

c *Physical initialization of water vapour*
As mentioned earlier, the diabatic initialization method proposed here consists of two steps. The first step is a physical initialization of the water vapour field, where the relative humidity field is changed to be more moist over the area of rainfall (>1 mm hr^{-1}) observed by the R-A following the scheme proposed by Aonashi (1993). Namely, the moistening is performed in such a way that, if the air is not moist enough over the area of rainfall, the lapse rate at each model layer from the LCL (σ_{LCL}) to the TBB CT (σ_{TBB}) becomes a critical value presumably required by

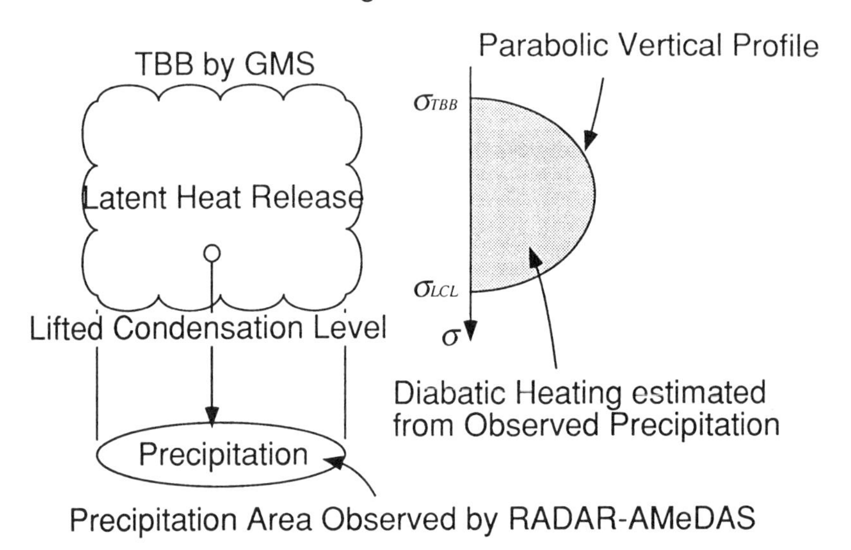

Fig. 2 The vertical profile of the heating function and the precipitation area estimated from the R-A data, where the following two conditions are met: (1) precipitation rate $\geqslant$ 1 mm/hr, (2) cloud amount observed by GMS $\geqslant$ 80%.

the moist convective adjustment (MCA) scheme used in the NWP model concerned. On the other hand, over the area where no rainfall is observed, the relative humidity is left untouched. The physical initialization mentioned above may be formulated in the following way. The relative humidity B required by the MCA is described as follows:

$$B = \begin{cases} \dfrac{1}{\Gamma_d - \Gamma_m}\{B_c(\Gamma_d - \Gamma) + B_D(\Gamma - \Gamma_m)\} & \Gamma_m \leqslant \Gamma \leqslant \Gamma_d \qquad \text{(2a)} \\ B_c & \Gamma < \Gamma_m \qquad \text{(2b)} \end{cases}$$

where Γ_d is the dry adiabatic lapse rate, Γ_m the moist adiabatic lapse rate and Γ the lapse rate of the model atmosphere. The constants B_C and B_D are set to be 100% and 50%, respectively (Fig. 3). The relative humidity provided by the objective analysis without the R-A proxy data is set to be B at all grid points where $w(x, y, \sigma)$ has a non-zero value, except for the grid point where the air is wet enough.

d *NNMI with diabatic heating*

The second step is the diabatic NNMI using two kinds of heating rate, i.e., the non-adiabatic heating rate (δT) obtained from the R-A data and the model-calculated one by the large-scale condensation and the moist convective adjustment. Details

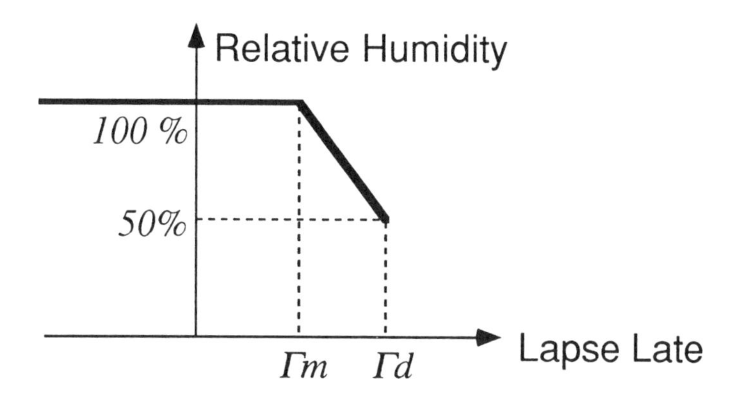

Fig. 3 Physical initialization of relative humidity with the R-A data (after Aonashi, 1993).

of the solving procedures of the NNMI are described in the paper by Takano et al. (1990) for the adiabatic case. The first four vertical modes whose periods are shorter than 6 hours are initialized. This limitation is required in order to satisfy the convergence condition of the iteration of the scheme by Machenhauer (1977). Therefore, a convective system whose top level is so high as to reach the tropopause is integrated into the initial state. The iteration is performed four times. Only the first iteration step uses the observed heating while the remaining steps use the model precipitation scheme. The divergence field corresponding to the R-A data is, to a sufficient extent, created at the first iteration step. It should be noted that the divergent component of the wind field is necessary to continuously maintain condensation for the mesoscale rain. Then, the three iteration steps that follow are executed so as to relieve the difference between the observed heating and the predicted diabatic heating of the model at the early stage of forecasting. This procedure also mitigates disparity between the R-A data affected area and the no R-A data area. It restrains the occurrence of gravity noises caused by the gap.

3 Forecast experiment: A case study

A disturbance formed on the Baiu front (i.e. the stationary front located near Japan during the rainy season) is used as an example (Fig. 4). The disturbance was located near Kyushu district, in the western part of Japan, on 13 July 1992. Precipitation over 20 mm hr^{-1} was observed around the Goto Islands at 0000 UTC 13 July (see, Fig. 8). The horizontal scale of the cloud cluster exceeded 300 km and its top was found at the tropopause level.

Results of the prediction made by the numerical model with the adiabatic initialization are shown in Fig. 5(a), where the predicted precipitation accumulated during the first 3 hours and the 3-hour forecasts of sea level pressure and surface wind are illustrated. Comparison with the verifying R-A data in Fig. 5(b) shows

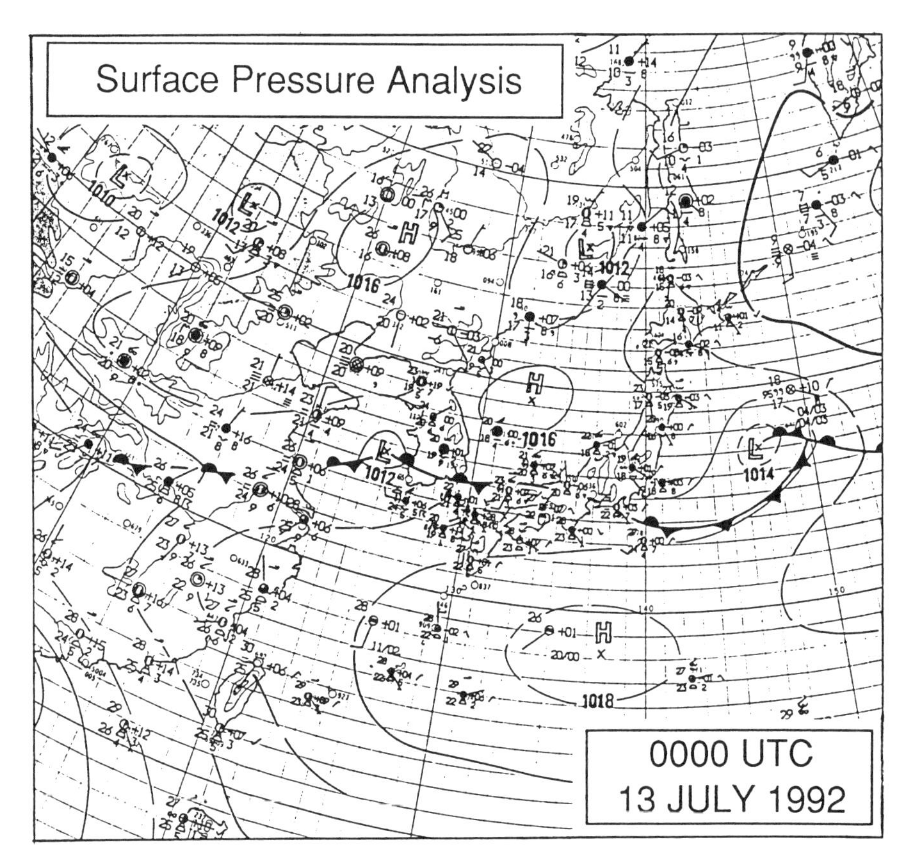

Fig. 4 The surface weather map at 0000 UTC 13 July 1992. The initial surface pressure distribution for the case study.

that very little rainfall was predicted and the majority of the rainfall area was inaccurately located. Figure 6 is the same as Fig. 5(a), but for the 15-hour forecasts from 1200 UTC 12 July. It should be noted that both forecasts have the same valid time of 0300 UTC 13 July 1992. Although the model required nearly 6 hours for the spin-up, a reasonable amount of precipitation is achieved in the 15-hour forecast.

Next, let us examine the divergence field in terms of the vertical motion. Figure 7 shows the vertical p-velocity at the 700-hPa level and the wind field at the 850-hPa level, respectively. Figure 7(a) shows those at the initial time, i.e., 0000 UTC 13 July and Fig. 7(b) for the 12-hour forecast starting from 1200 UTC 12 July. It should be noted that the fields shown in Fig. 7(b) were used as the first guess field of the objective analysis for Fig. 7(a). Vertical p-velocity of about -80 hPa hr^{-1} is predicted at the centre of the precipitation area in the first guess. However, it was reduced to about half in the initial state as a result of the use of the adiabatic NNMI.

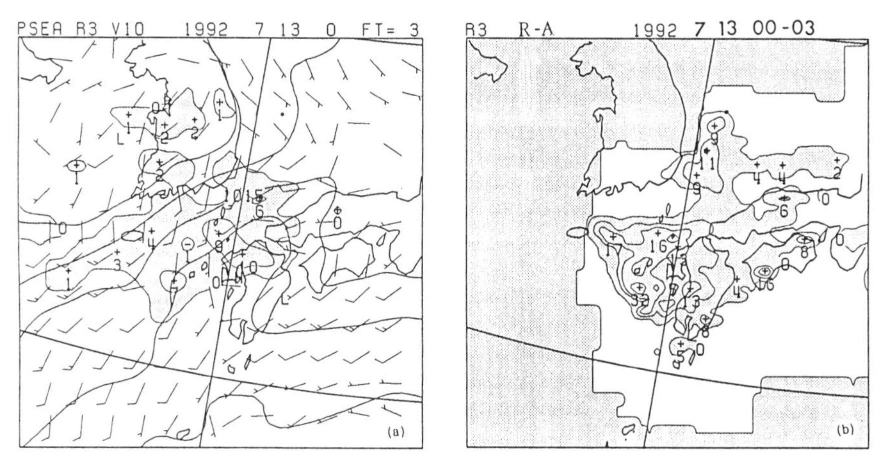

Fig. 5 (a) The precipitation during the first three hours of the forecast and the sea level pressure and surface winds at the 3-hour forecast using the adiabatic initialization from 0000 UTC 13 July 1992. (Valid time: 0300 UTC 13 July 1992.) (b) Verifying Radar-AMeDAS observation corresponding to the precipitation of (a).

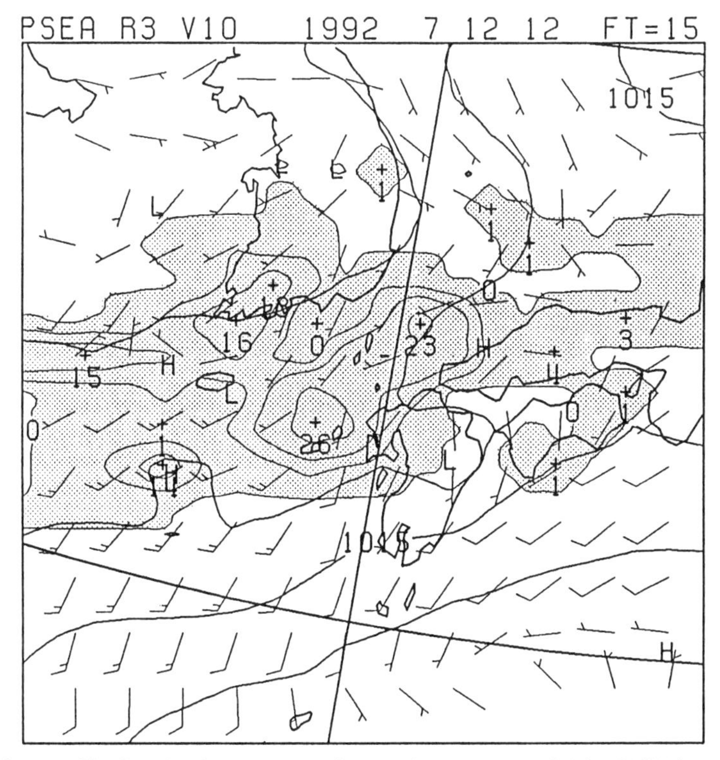

Fig. 6 Same as Fig. 5(a), but for the 15-hour forecast from 1200 UTC 12 July. (Valid time: 0300 UTC 13 July 1992).

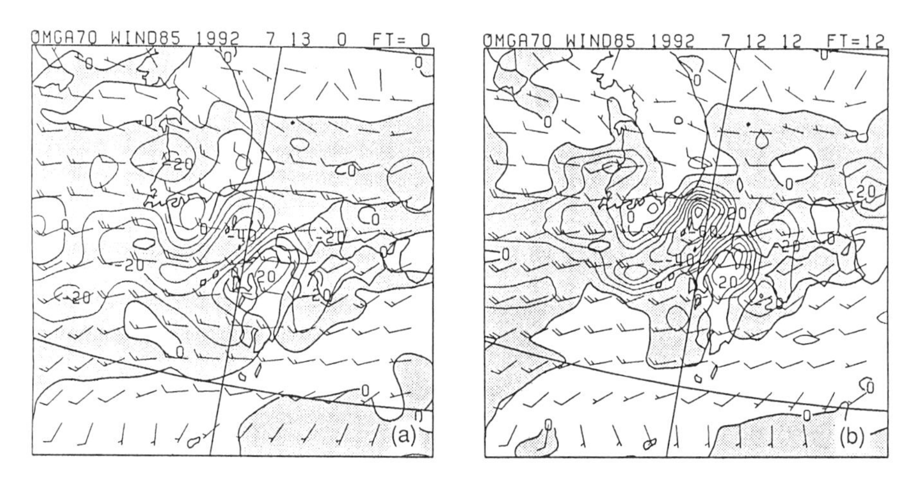

Fig. 7 (a) The vertical p-velocity at 700 hPa and the wind at 850 hPa for the initial state of 0000 UTC 13 July. (b) Same as (a), but for the 12-hour forecast using the adiabatic initialization from 1200 UTC 12 July.

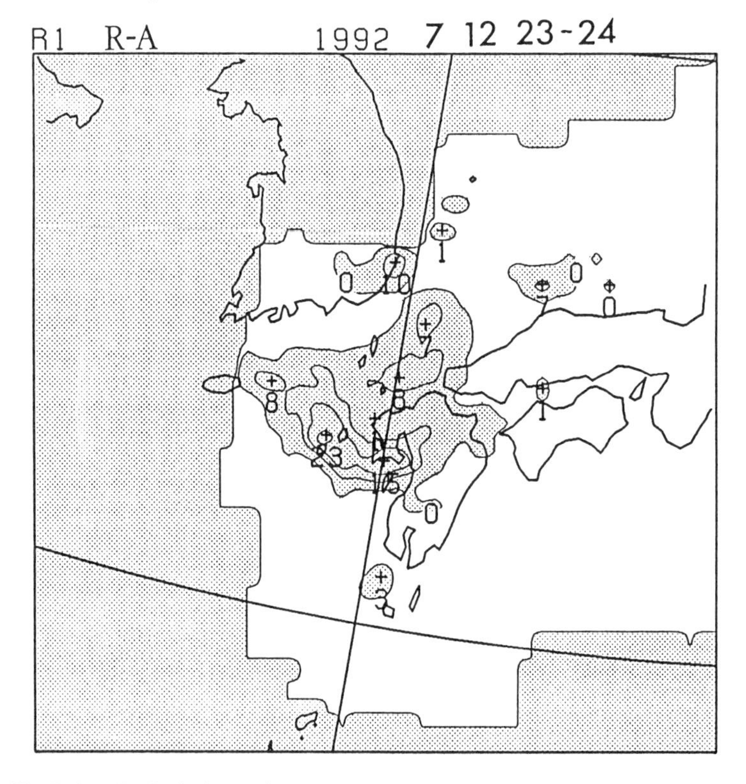

Fig. 8 The Radar-AMeDAS observation at 0000 UTC 13 July 1992 used for estimating the heating rate for NNMI.

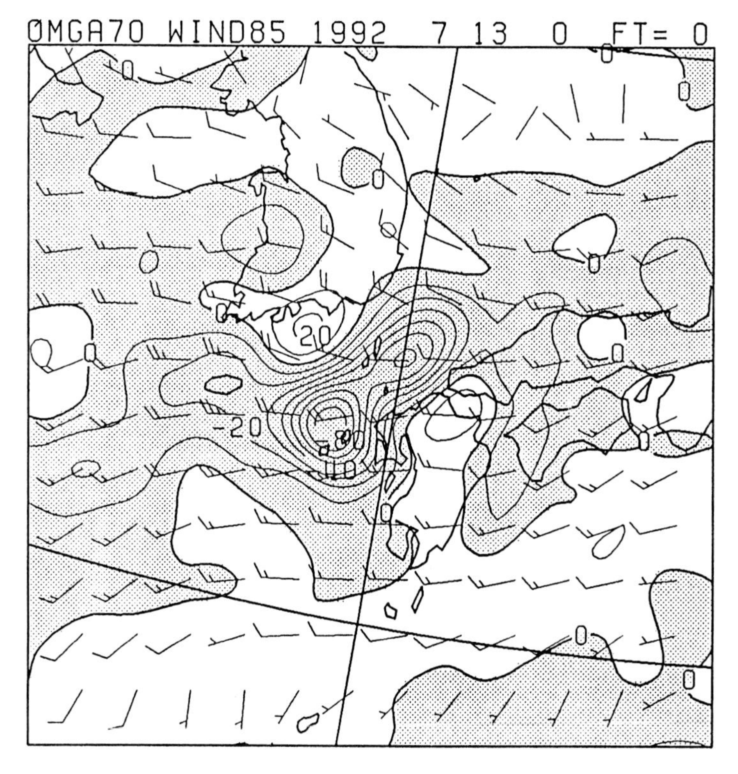

Fig. 9 Same as Fig. 7(a), but for the initial state of 0000 UTC 13 July prepared by the present diabatic initialization.

Figure 8 shows the R-A data used for the estimation of the heating rate for NNMI. Figure 9 illustrates the same field as Fig. 7, but for the initial state at 0000 UTC 13 July with the use of the proposed initialization method. The vertical p-velocity is maintained as large as that of the first guess and shows good correspondence to the observed precipitation area. Figure 10 is the same as Fig. 5(a), but for the forecasts with the diabatic initialization which shows remarkable skill. A large amount of precipitation is predicted in this case in contrast to the prediction with the adiabatic initialization, and the location of the rainfall area is also better forecast.

4 Forecast experiment: Statistical verification

In order to confirm the superiority of the diabatic initialization proposed in the present paper to the adiabatic initialization, the 6-hour precipitation forecasts in 32 cases during the rainy (Baiu) season from 20 June to 5 July 1991 were verified against the observed rainfall amount by AMeDAS averaged over an area of 80 × 80 km. Figures 11 and 12 illustrate the results of verification for the 6-hour

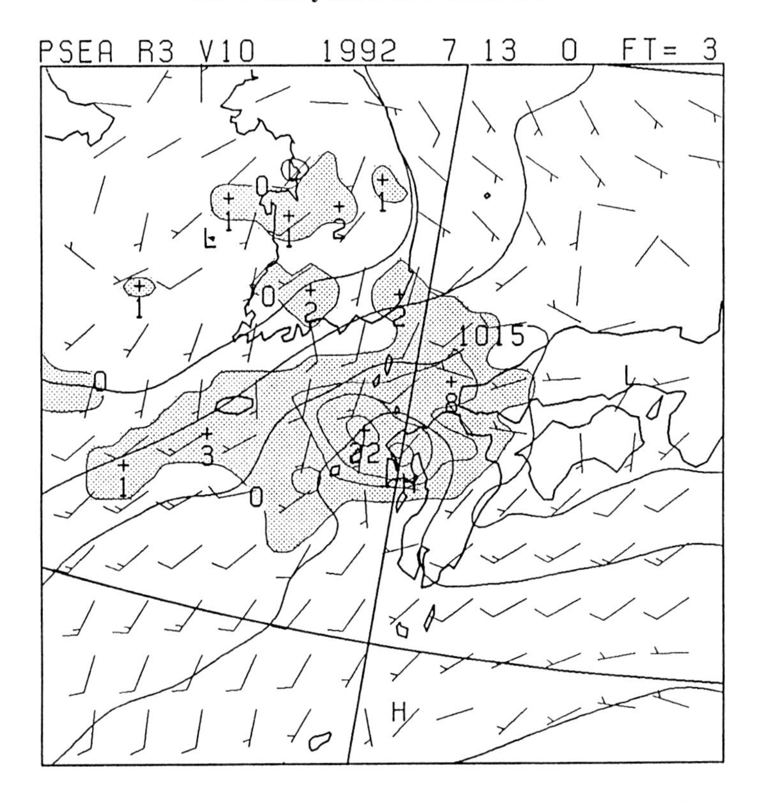

Fig. 10 Same as Fig. 5(a), but for the 3-hour forecast using the initial state prepared by the present diabatic initialization.

precipitation forecast up to 24 hours in terms of the threat score and the bias score. The thick solid line denotes the forecast with the present diabatic initialization, the thin solid line that with the adiabatic initialization and the dashed line that with the adiabatic initialization excluding the forecast-analysis cycle at the stage of objective analysis. The scores are classified for two threshold values, i.e., 1 mm/6 hrs and 5 mm/6 hrs.

It is readily pointed out in the threat score that the present diabatic initialization results in remarkable amelioration in the spin-up of 6-hour precipitation forecasts. However, the bias score shows little improvement for the first 6 hours. It is presumed that a combination of better threat score and unimproved bias score means better position forecast but narrow (or weak) rainfall area.

5 Conclusions and considerations for further development

The performance of the fine-mesh, limited area model is improved by the combined initialization method consisting of the physical initialization of water vapour and the

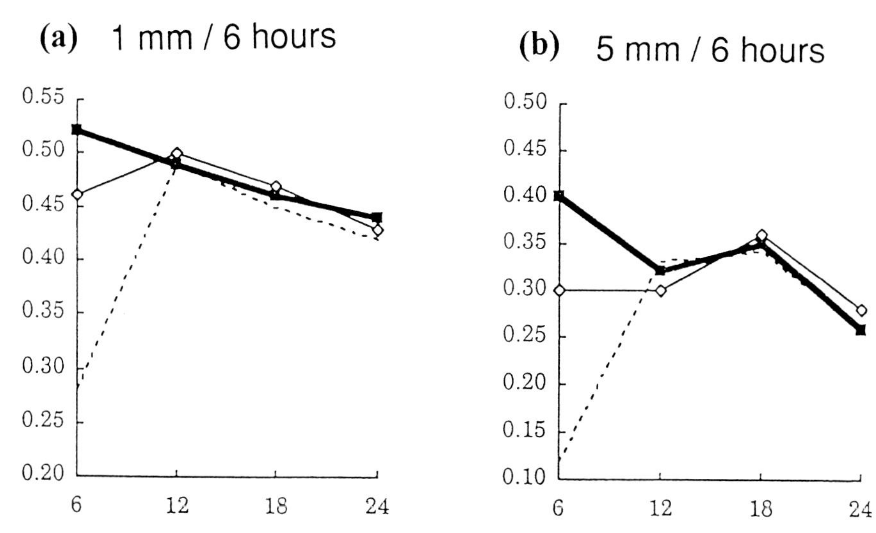

Fig. 11 The threat score of the 6-hour precipitation forecasts up to 24 hours for 32 cases between 20 June and 5 July 1991. The thick solid line denotes the forecast with the present diabatic initialization, the thin solid line that with the adiabatic initialization and the dashed line that with the adiabatic initialization excluding the forecast-analysis cycle at the stage of objective analysis, respectively. (a) Threshold value = 1 mm/6 hrs, (b) threshold value = 5 mm/6 hrs.

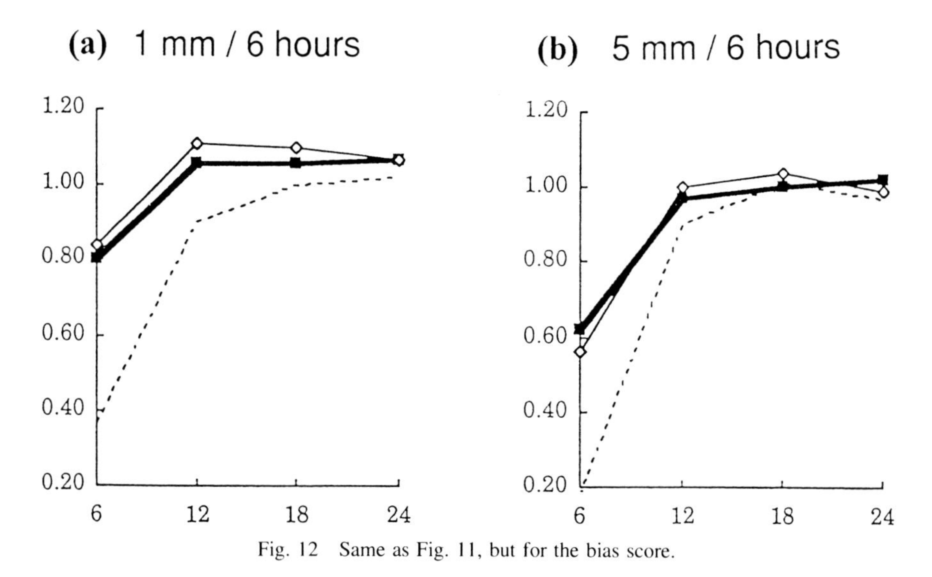

Fig. 12 Same as Fig. 11, but for the bias score.

NNMI with use of the observed heating. Particularly, the spin-up of precipitation forecast is much ameliorated compared with that using the adiabatic initialization.

In order to complement the conventional observation data the R-A data, which to a large extent represent the mesoscale convective rainfall, are introduced at the initial time. The generation of an adequate divergence field around the observed rainfall area is a result of NNMI with the diabatic heating. The physical initialization of water vapour contributes to maintaining the precipitation at the early stage of the forecast together with the divergent component of the wind field mentioned above. The initialization scheme presented in this paper was incorporated into the operational JSM at the JMA on 11 May 1993.

Next, one may say that although the diabatic initialization provides us with a positive contribution to ameliorate the spin-up of precipitation forecast, there are several points to be further improved in the initialization method in future. Namely, (1) introduction of a more realistic vertical profile of non-adiabatic heating due to condensation estimated from observation, (2) inclusion of higher vertical modes in the NNMI to represent a more accurate vertical profile of the vertical p-velocity, (3) extension of the data source of the precipitation observations to other remote-sensing observations such as those by meteorological satellites, (4) use of a more refined cumulus parametrization scheme, and (5) replacement of an intermittent data assimilation scheme (i.e., the forecast-analysis cycle) to a continuous data assimilation scheme.

It seems to us doubtful whether the parabola is appropriate for representing the vertical profile of the condensation heating. One of the practical strategies to finding a solution to this problem may be the direct use of PR (precipitation radar) observations for the purpose of making a vertical profile of the heating rate due to condensation more realistic. It is planned to load the radar on the TRMM (Tropical Rainfall Measuring Mission) satellite scheduled for launch in summer 1997. Since the PR/TRMM data are expected to yield the vertical distribution of rainfall intensity with very fine resolution, we may retrieve the vertical profile of the heating rate from the observed data of rainfall intensity under several reasonable assumptions.

Development of an iteration scheme that can handle higher vertical modes is also needed. Since the structure which is described only by the first four modes is quite simple, the vertical mode to be initialized should be extended to higher modes in order to further utilize the observed vertical profile of the heating and also to represent a more accurate vertical profile of the vertical p-velocity. A scheme is now being developed at the JMA on the basis of Kudoh's idea (1984).

It should be emphasized that extension of the data source of the precipitation observations is another task to be realized. The R-A data are available only in the limited area covered by the weather radar (see Fig. 1). Therefore, one should consider utilizing satellite data such as TRMM data and GMS cloud images. As is well known, there have been many attempts to retrieve precipitation information from cloud images obtained by meteorological satellites (e.g., regarding precipitation estimation: Adler and Negri, 1988; Puri and Miller, 1990; Turpeinen, 1990; Janowiak and Arkin, 1991; Mathur et al., 1992; the humidity analysis: Baba, 1987; Puri and Davidson, 1992; Kuma, 1993). However, the difficulty in converting the cloud images into rainfall amount remains unsolved.

Refinement of the cumulus parametrization scheme may be a kind of permanent theme for meteorologists. Therefore, in principle, one should always try to adopt a more sophisticated and realistic scheme in expectation of ameliorating the spin-up of forecast calculation among other merits. In this case, one needs to adopt a more complicated physical initialization method corresponding to the upgraded cumulus parametrization scheme. In many parametrization schemes, it is considered to be difficult to obtain meteorological fields as a unique solution inversely from the rainfall rate. However, it may presumably be possible to yield proper dynamic, thermodynamic and moisture fields under the condition that the calculated rainfall rate is equal to that observed.

Acknowledgements

One of the co-authors, Takashi Nitta, cordially thanks Environment Canada for inviting him to the André J. Robert Symposium on Numerical Methods in Atmospheric and Oceanic Sciences, together with the financial support. We are indebted to the staff members of the Numerical Prediction Division, JMA for their warm encouragement given to us during the course of the present study, particularly to Dr. T. Kitade, former Director of the Division, for his kind advice.

Finally, in order to take account of asynoptic data more smoothly, it is desirable to use a continuous 4-dimensional data assimilation scheme such as the nudging method or the adjoint method at the stage of objective analysis instead of NNMI.

References

ALDER, R.F. and A.J. NEGRI. 1988. A satellite infrared technique to estimate tropical convective and stratiform rainfall. *J. Appl. Meteorol.* **27**: 30–51.

AONASHI, K. 1993. An initialization method to incorporate precipitation data into a mesoscale numerical weather prediction model. *J. Meteorol. Soc. Jpn.* **71**: 393–406.

BABA, A. 1987. Improvement of the estimation method of moisture data from satellite cloud sounding. JMA/NPD Technical Report No. 16, 54 pp.

GADD, A.J. and J.F. KEERS. 1970. Surface exchanges of sensible and latent heat in a 10-level model atmosphere. *Q. J. R. Meteorol. Soc.* **96**: 297–308.

JANOWIAK, J.E. and P.A. ARKIN. 1991. Rainfall variations in the tropics during 1986–1989, as estimated from observations of cloud-top temperature. *J. Geophys. Res.* **96**: (Supplement) 3359–3373.

KASAHARA, A.; A.P. MIZZI and L.J. DONNER. 1994. Diabatic initialization for improvement in the tropical analysis of divergence and moisture using satellite radiometric imagery data. *Tellus*, **46A:** 242–264.

KRISHNAMURTI, T.N.; K. INGLES, S. COOKE, T. KITADE and R. PASCH. 1984. Details of low-latitude, medium-range numerical weather prediction using a global spectral model. Part 2: Effects of orography and physical initialization. *J. Meteorol. Soc. Jpn.* **62**: 613–648.

KUDOH, T. 1984. Normal model initialization for a spectral model. Technical Memorandum, Numerical Prediction Division, JMA (Suutiyohouka-houkoku), No. 30, pp. 20–22 (In Japanese) (Available from Numerical Prediction Division, JMA).

KUMA, K. 1993. The impact of satellite moisture data upon the numerical prediction of Australian Monsoon onset. *J. Meteorol. Soc. Jpn.* **71**: 545–551.

MACHENHAUER, B. 1977. On the dynamics of gravity oscillations in a shallow water model with application to normal mode initialization. *Beit. Phys. Atmos.* **50**: 253–271.

MATHUR, M.B.; H.S. BEDI, T.N. KRISHNAMURTI, M. KANAMITSU and J.S. WOOLLEN. 1992. Use of satellite derived rainfall for improving tropical forecasts. *Mon. Weather Rev.* **120**: 2540–2560.

PURI, K. and M.J. MILLER. 1990. The use of satellite data in the specification of convective heating for diabatic initialization and moisture adjustment in numerical weather prediction models. *Mon. Weather Rev.* **118**: 67–93.

——— and N.E. DAVIDSON. 1992. The use of infrared satellite cloud imagery data as proxy data for moisture and diabatic heating in data assimilation. *Mon. Weather Rev.* **120**: 2329–2341.

SEGAMI, A.; K. KURIHARA, H. NAKAMURA, M. UENO, I. TAKANO and Y. TATSUMI. 1989. Operational mesoscale weather prediction with Japan Spectral Model. *J. Meteorol. Soc. Jpn.* **67**: 907–924.

TAKANO, I.; H. NAKAMURA and Y. TATSUMI. 1990. Non-linear normal mode initialization for a spectral limited-area model. *J. Meteorol. Soc. Jpn.* **68**: 265–280.

——— and A. SEGAMI. 1993. Assimilation and initialization of a mesoscale model for improved spinup of precipitation. *J. Meteorol. Soc. Jpn.* **71**: 377–391.

TAKASE, K.; Y. TAKEMURA, K. AONASHI, N. KITABATAKE, Y. MAKIHARA and Y. NYOMURA. 1988. Operational precipitation observation system in Japan Meteorological Agency. In: *Tropical Rainfall Measurements*, J.S. Theon and N. Fugono (Eds) A. Deepack Publishing, pp. 407–413.

TATSUMI, Y. 1986. A spectral limited-area model with time-dependent lateral boundary conditions and its application to a multi-level primitive equations model. *J. Meteorol. Soc. Jpn.* **64**: 637–663.

TURPEINEN, O.M. 1990. Diabatic initialization of the Canadian Regional Finite-Element (RFE) Model using satellite data. Part II: Sensitivity to humidity enhancement, latent-heating profile and rain rates. *Mon. Weather Rev.* **118**: 1396–1407.

UENO, M.; R. TAIRA and K. KUDO. 1987. A dynamic assimilation method for a mesoscale model using observed rainfall rates. In: *Short- and Medium-Range Numerical Weather Prediction*. T. Matsuno (Ed.), Meteorological Society of Japan, pp. 573–584.

Physical Initialization

T.N. Krishnamurti, H.S. Bedi, G.D. Rohaly, D.K. Oosterhof,
R.C. Torres and E. Williford
Department of Meteorology
Florida State University
Tallahassee, FL 32306-3034

and

N. Surgi
National Hurricane Center
1320 S. Dixie Highway
Coral Gables, FL 33146

[Original manuscript received 27 October 1994; in revised form 12 July 1995]

ABSTRACT *In this overview we present a summary of results on the impact of physical initialization on various aspects of numerical weather prediction. These include the improvements in the nowcasting and short range forecasting skill, the organization of mesoconvective precipitating elements as typhoons form calibration of surface fluxes using the TOGA-COARE datasets during the intensive observation period from its intensive flux array, the improvements in the hydrological budgets during flooding events, the spread reduction for an ensemble of hurricane track forecasts and an overall improvement of cloud forecasts. The paper concludes with an emphasis on the need for a direct incorporation of the process within four-dimensional variational analysis procedures.*

RÉSUMÉ *Nous présentons un résumé des résultats de l'influence de l'initialisation physique sur divers aspects de la prévision météorologique numérique. Parmi ceux-ci, les améliorations apportées aux capacités de visualisation du temps et à la prévision à court terme; l'organisation des éléments précipitants à l'échelle convective lors de la formation des typhons; la calibration des flux de surface en utilisant les ensembles de données provenant du réseau concentré de mesure de flux de la période intensive d'observation des expériences TOGA-COARE; les améliorations des bilans hydrologiques durant les épisodes d'inondation; la réduction de l'étalement d'un ensemble de prévisions de trajets d'ouragans; et l'amélioration générale de la prévision des nuages. On termine en soulignant le besoin d'une incorporation directe du processus à l'intérieur des procédures d'analyse variationnelle à quatre dimensions.*

1 Introduction

The physical initialization procedure developed by Krishnamurti et al. (1991) assimilates 'observed' measures of rain rates into an atmospheric forecast model.

TABLE 1. List of Acronyms and Symbols

TOGA-COARE	TOGA-Coupled Ocean Atmospheric Response Experiment
$\hat{Q}_1$	Vertically integrated apparent heat source
$\hat{Q}_2$	Vertically integrated apparent moisture sink
$\hat{Q}_R$	Radiative heat source
OLR	Outgoing Longwave Radiation
SSM/I	Special Sensor Microwave Imager
FSU	Florida State University
IOP	Intensive Observation Period
ISS	Integrated Sounding System
$\hat{E}$	Evaporation rate
$\hat{P}$	Precipitation rate
SST	Sea Surface Temperature
T106, T170, T213	Triangular Truncation at wavenumber 106, 170, 213
UTC	Coordinated Universal Time
ITCZ	Intertropical Convergence Zone
NMC	National Meterological Center
NASA-GODDARD	National Aeronautics and Space Administration, Goddard
ECMWF	European Centre for Medium-Range Weather Forecasts
GFDL	Geophysical Fluid Dynamics Laboratory
TOGA	Tropical Ocean and Global Atmospheric Program
FGGE	First GARP Global Experiment
GARP	Global Atmospheric Research Program
IFA	Intensive Flux Array
GDAS	Global Data Assimilation System

During this process the surface moisture flux, the vertical distribution of the humidity variable, the vertical distribution of mass divergence, the vertical distribution of convective heating, the apparent moisture sink (following Yanai et al., 1973) and the surface pressure experience a spin-up consistent with the model physics and the imposed (observed) rain rates.

This is accomplished through a number of reverse physical algorithms within the assimilation mode, these include: a reverse similarity algorithm, a reverse cumulus parametrization algorithm and an algorithm that restructures the vertical distribution of the humidity variable to provide a match between the model-calculated outgoing longwave radiation and its satellite-based observations.

The reverse similarity algorithm is structured from the vertically integrated equations for the apparent moisture sink ($\hat{Q}_2$) and the apparent heat source ($\hat{Q}_1$), (a list of acronyms and symbols appear in Table 1). Using the observed rain rates, the surface evaporative fluxes can be obtained from the sum of the apparent moisture sink ($\hat{Q}_2$) and the observed rain rate (P). The surface sensible heat fluxes can also be obtained from a knowledge of the apparent heat source ($\hat{Q}_1$) and the net radiative heating ($\hat{Q}_R$). The details are presented in Krishnamurti et al. (1991). During the assimilation $\hat{Q}_1$, $\hat{Q}_2$, $\hat{Q}_R$ continually evolve from the insertion of the observed rain rates, and the resulting surface fluxes tend to exhibit a consistency with the observed rain rates, which is an important component of the reverse similarity theory (Krishnamurti et al., 1991, 1993, 1994). These fluxes are then used

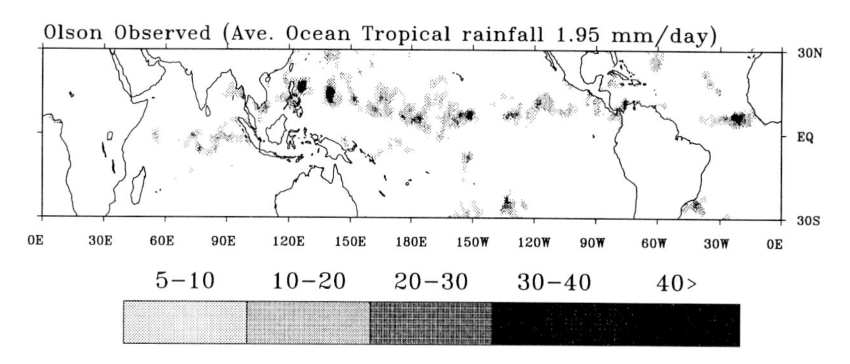

Fig. 1 An example of rainfall for a 24 hour period (mm/day) based on the mixed (SSM/I, OLR, rain-gauge) algorithm.

within the similarity theory, and one solves for the potential temperature and the humidity variable (assumed to be unknowns) at the top of the constant flux layer. The assimilation of this database provides a consistency among the observed rain rates and the surface fluxes; a robust coupling of the ocean and the atmosphere was also seen to result from this approach (Krishnamurti et al., 1993).

The starting point of this analysis is a rain rate input. Figure 1 shows an observed rainfall field based on a mixed algorithm (Gairola and Krishnamurti, 1992). This scheme utilizes a mix of data from polar orbiting satellites, the SSM/I and OLR, and the rain-gauge observations. The rain rate algorithm utilizes a sharpened OLR-based rainfall as a first guess and incorporates the radiometer data of SSM/I into the analysis over the oceanic regions and the surface rain-gauge data over the land areas. The SSM/I algorithm, we use is based on Olson et al. (unpublished manuscript).

Figure 2, from Gairola and Krishnamurti (1992), describes a flow chart of the components of this algorithm. A first phase of this algorithm addresses a statistical procedure for the sharpening of the OLR rain to the levels of the SSMI/I rain. This is done by regression based on a large volume of diverse rain rate datasets over the 10°S and 10°N latitudes. The cirrus anvils of OLR generally produce too much of a broadening of the rainfall signatures when the OLR algorithms are used. The SSM/I based data comes from a number of microwave radiometric channels (85 GHz, 37 GHz, 22 GHz and 19 GHz). These do not exhibit such a cirrus broadening.

The footprint of the SSM/I varies roughly between 35 and 55 km^2. Thus, a much higher resolution rainfall signature is possible from the SSM/I. We have examined three different rain rate algorithms based on the SSM/I datasets; these were developed by Olson et al. (unpublished manuscript), Kummerow and Giglio (1994) and Spencer (1993). These are essentially based on validation of the use of multi-channel SSM/I data against ground truth provided by surface-based measurements from radar and rain-gauges. The global tropical precipitation intensity amounts

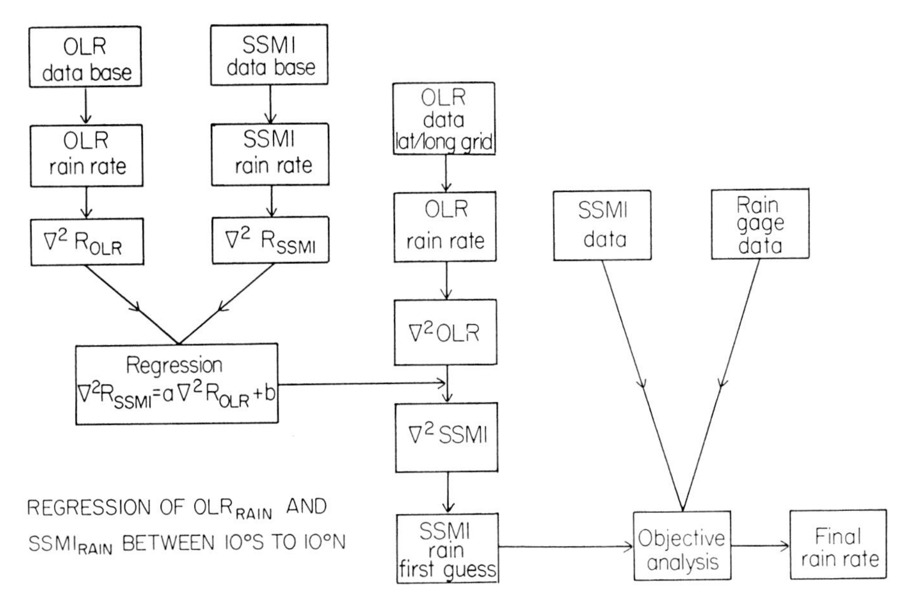

Fig. 2 Flow chart for the mixed rain rate algorithm; OLR, SSM/I and rain-gauge.

were quite different in these approaches. These are illustrated in Figs 3a, b and c, where we show an example of the rainfall distribution from three algorithms. It is interesting to note the current uncertainties in the rainfall measurement. The mutual tropical correlations of these fields are of the order of 0.6. We noted that when we carried out the physical initialization using these three fields to define the initial distributions, we experienced a somewhat different response from the three methods in medium range forecasts. It is apparent that further work is needed to improve the satellite-based rain rate algorithms.

The model rainfall is a function of horizontal resolution; in that context we found that some rain rate algorithms were better suited than others for high resolution global model integrations. There are important hydrological constraints on the globally averaged measures of surface evaporation and rainfall. Global average rainfall values such as 2.75 mm/d are generally accepted from global hydrological considerations. In that context we found that some algorithms were better suited than others for describing the global budgets. Based on these considerations, we have largely been using the algorithms of Olson et al. (unpublished manuscript) over the oceanic portions. The land area rainfall description has been largely based on rain-gauge observations and the sharpened OLR rainfall estimates. The sharpened OLR provides a first guess of the global tropical rainfall for our analysis.

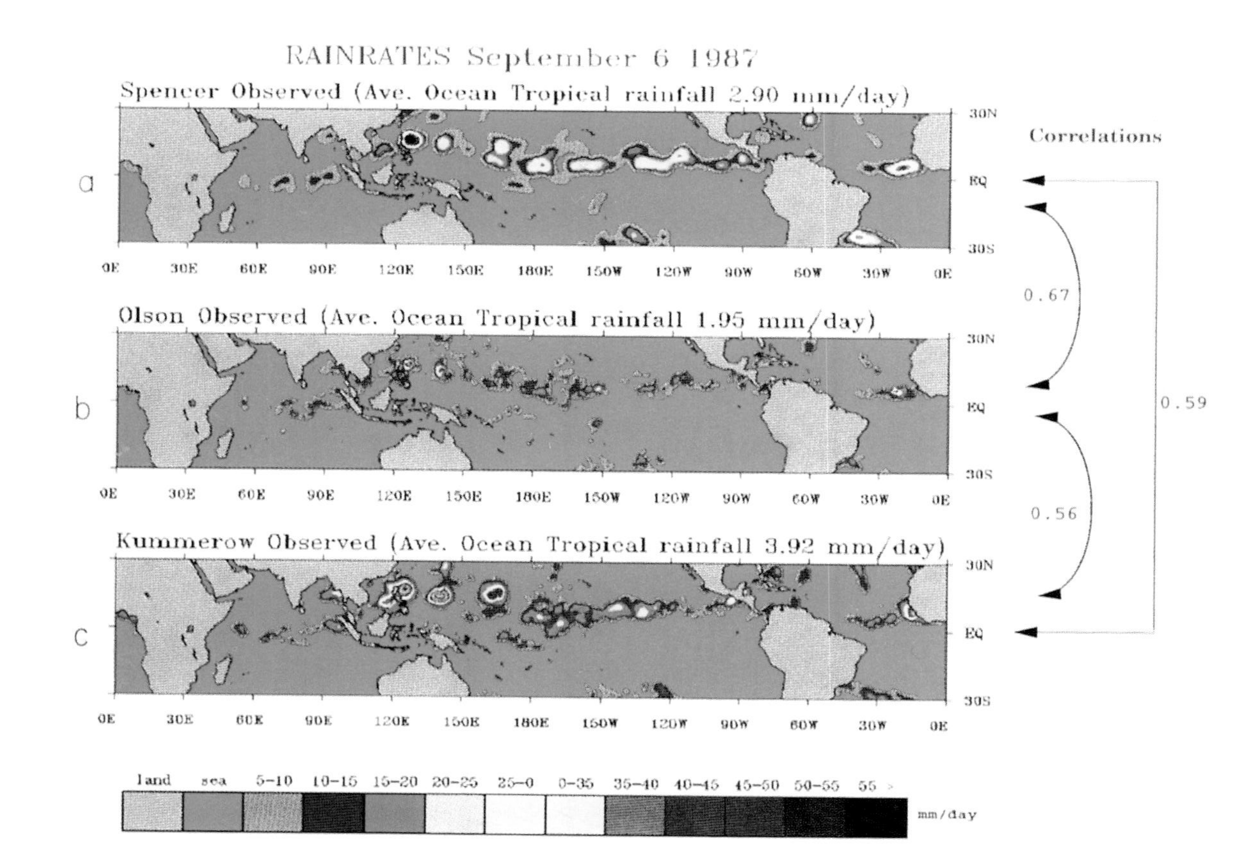

Fig. 3 24 hour rainfall total based on three rain rate algorithms, units: mm/day a) Spencer (1993), b) Olson et al. (1990) and c) Kummerow and Giglio (1994).

A regression of the Laplacian of the OLR algorithm rain rates and the Laplacian of the SSM/I based (and collocated) datasets between 10°S and 10°N were used to define a sharpening function. Thus, given the OLR and the SSM/I based rainfall distributions, a regression among these two provides a sharpening of the Laplacian of the OLR rain and the inverse problem, i.e. the solutions of the Poisson equation, in turn, provides the sharpened rain. That process is defined in the left-hand side of the flow chart in Fig. 2. Thus given the datasets of the OLR, SSM/I and the rain-gauge for a particular period, rainfall analyses were carried out following Gairola and Krishnamurti (1992). This sharpened OLR rainfall covers the entire tropical belt between 30°S and 30°N. Next the SSMI/I based rainfall on satellite swaths over the oceanic regions plus the rain-gauge estimates from the surface reports over land areas were objectively analyzed to obtain the final rain rate descriptions. Space-time bins, 6 hours in time and a spectral transform grid square in space, define the resolution of the 'observed' rainfall which is assimilated via the physical initialization procedure.

2 Rainfall Prediction Skill

If we follow a sequence of OLR matching, reverse similarity and a reverse cumulus parametrization within an assimilation, we find a major improvement in the nowcasting and one day forecasting skill of rainfall (Krishnamurti et al., 1994). We performed the physical initialization for the entire month of October 1991 at a resolution T106 for our global model. This is outlined in the Appendix of this paper.

The month long skill of rainfall initialization is shown in Fig. 4a from Krishnamurti et al. (1994). A much higher skill (of the order of 90%) compared to those of real time operations can be seen here. In Fig. 4b we show the one day forecast skill. We find that the use of physical initialization leads to much higher one day forecast correlations, i.e. of the order of 60% for this ensemble of 30 forecasts. This compares with a skill of roughly 35% for the one day forecasts from the various operational centres and those for the persistence and for a control experiment conducted with the FSU model, where no physical initialization was invoked.

There appears to be only a marginal improvement in useful skill for rainfall forecasts from the physical initialization on forecasts beyond two days.

3 TOGA-COARE Fluxes

Calibration of model-based surface fluxes to some rich database, such as those provided by TOGA-COARE, is desirable for the physical initialization.

During the field phase of the TOGA-COARE the intensive observation period (IOP) included November 1992 through February 1993. Within this period an intensive flux array (IFA) that provided oceanic surface fluxes over the warm pool of the tropical Pacific ocean was set up for higher resolution observations. This special dataset was useful for the calibration of model-based estimates of surface fluxes. The IFA is located within the warm pool close to the equator (2°S, 156°E).

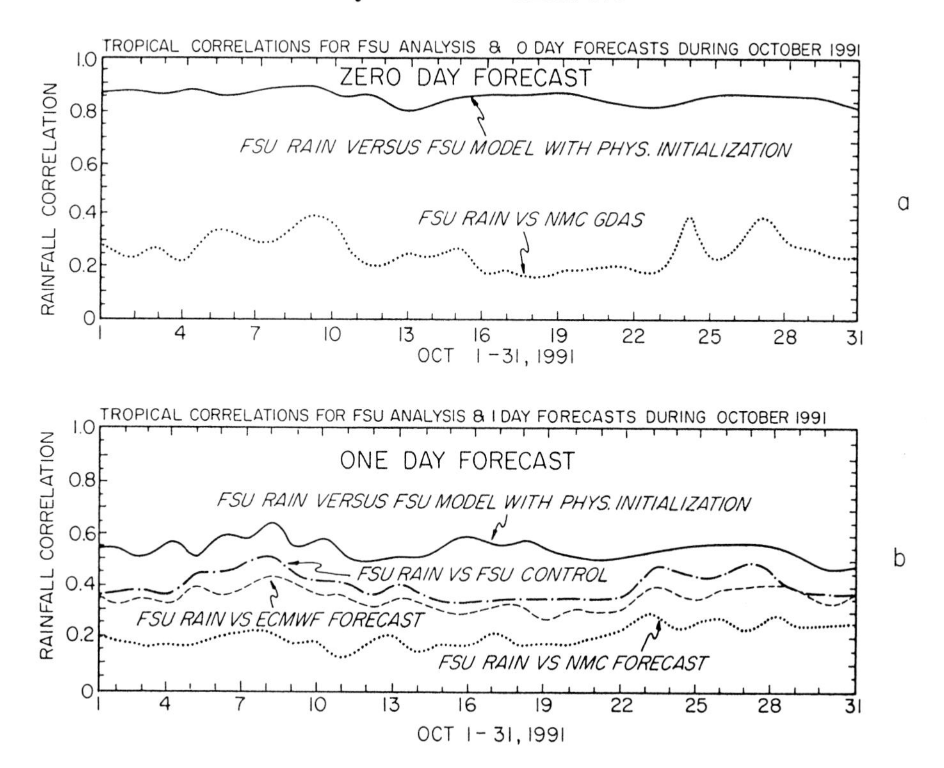

Fig. 4 Month long skill i.e. correlation of predicted and observed rain over the tropical belt, 30°S to 30°N: a) for the FSU model (based on physical initialization) for day 0, also shown is the skill of the NMC/GDAS rain compared to observed rain. b) for the FSU (physically initialized and control), NMC and ECMWF models for the one day forecast.

A polygonal array of observational sites, weather stations at Kapingawa range and Kavieng and ships located at (2°S, 158°E), (4°S, 155°E) and at the centre (2°S, 156°E) constitute the IFA. An integrated sounding systems (ISS) is part of the flux array; it includes a profiler, a radio acoustic sounding system, an omega based Navaid sounding system and an enhanced surface station (Parsons et al., 1994). A total of six ISS were implemented during the TOGA-COARE IOP phase.

Physical initialization includes a reverse similarity theory within its precipitation assimilation. Over most of the transform grid points of the global model the vertical integral of the apparent moisture sinks following Yanai et al. (1973), provides a measure of the surface flux of moisture over the rain areas, i.e.

$$\hat{Q}_2 = \hat{E} - \hat{P}$$

Given an estimate of precipitation from the satellites and the rain-gauge, the assimilation constantly improves the estimates of $\hat{Q}_2$ and $\hat{E}$.

In order to provide consistency among precipitation and evaporation we use

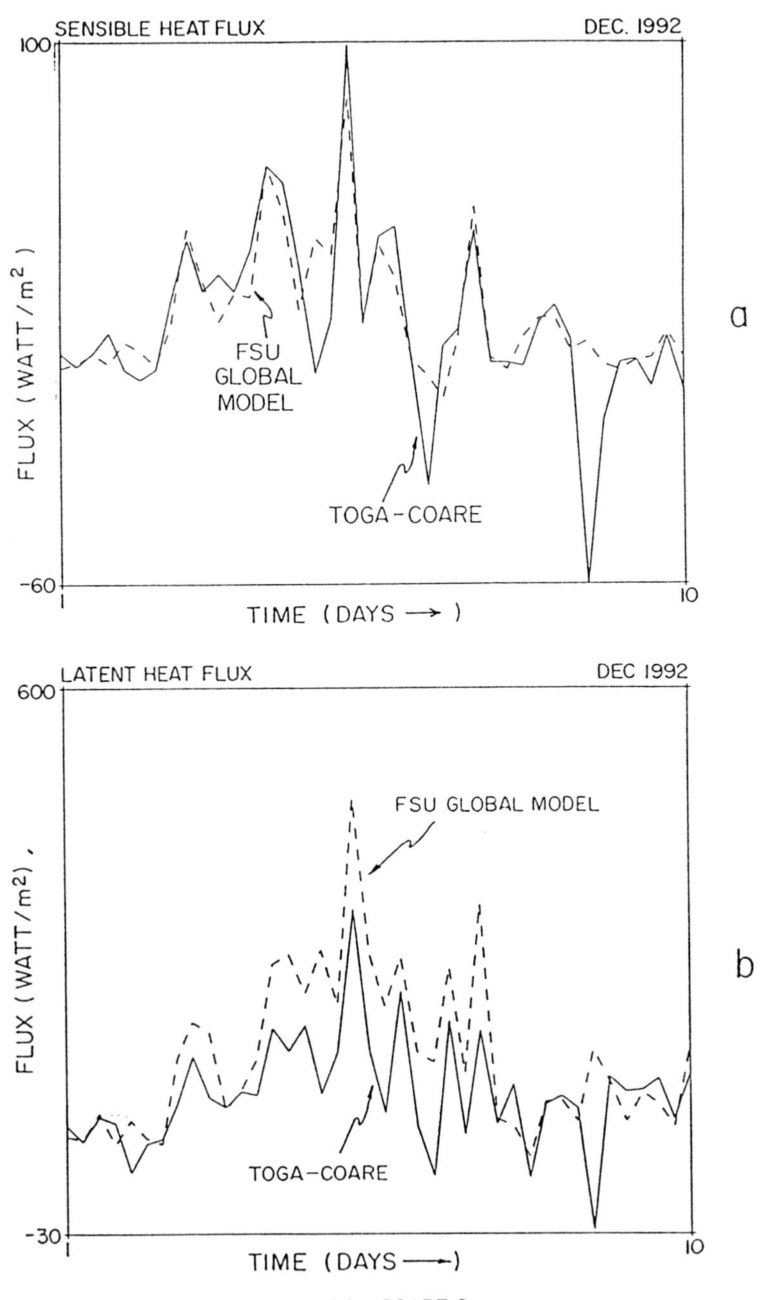

Fig. 5 TOGA-COARE fluxes

a reverse similarity theory (Krishnamurti et al., 1991) to calculate the data on top of the constant flux layer that makes use of the aforementioned Yanai fluxes. Assimilation of this data brings about this consistency. However, several modelling approximations implicit in the surface similarity theory show parameter sensitivity to the modelled fluxes. The depth of the constant flux layer is one such parameter. It is possible to calibrate an optimal value of this depth with respect to the data from the intensive flux array. Figure 5 illustrates a comparison of TOGA-COARE fluxes with those from our model.

4 The Impact of Physical Initialization on Implicit Cloud Forecasts

In many numerical weather prediction models the cloud-radiative interactions are carried out from an implicit definition of cloud fractions. Clouds are defined to exist if the relative humidity at a model level exceeds a threshold value (usually called a critical relative humidity). These threshold values are usually determined by examining the satellite measures of outgoing longwave radiation for clear line-of-sight, low, middle or high cloud targets. Therefore, with collocated temperature/humidity soundings, one can find the prevailing humidities during occurrences of low, middle or high clouds. The Nimbus 7 satellite data and the FGGE soundings have been used to define these threshold values globally (Lee and Krishnamurti, 1995).

The humidity analysis is one of the weakest components of data assimilation at operational analysis centres. The first guess field is based on the model forecast and the available radiosonde stations provide the real time data, which in fact are very sparsely distributed over the tropics. This leads to large uncertainties in the coverage of low, middle and high clouds. The model-based definition of the net OLR, as a consequence, is subject to large errors from the misrepresentation of clouds. Within the physical initialization, a matching of the model-calculated and the satellite-observed OLR is accomplished. The vertical structure of the humidity variable in the upper trophosphere, where the humidity observations are sparse and unreliable, are defined by a single parameter structure function. This parameter is determined from a minimization of the difference between the satellite and the model OLR values using a bisection method. Lacking a more sophisticated approach this simple method has been found to be fairly useful in improving the humidity analysis over the upper troposphere.

Figure 6 (a, b and c) shows an example illustrating the improvement in the definition of the OLR field. Panel a) shows satellite observations; Panel b) shows a control run which did not include the physical initialization; and Panel c) shows the results from the physical initialization experiment. These results (Krishnamurti et al., 1991), demonstrate that this method produces a close match of the observed and the model OLR, see Figs 6a and c. The correlation coefficients of the OLR (satellite) and the OLR (control) is 47%; whereas the correlation coefficient of the OLR (satellite) and the OLR (physical initialization) is 97%.

Physical initialization, as a consequence, appears to provide a strong positive impact on the prediction of clouds. This is apparent from the root mean square

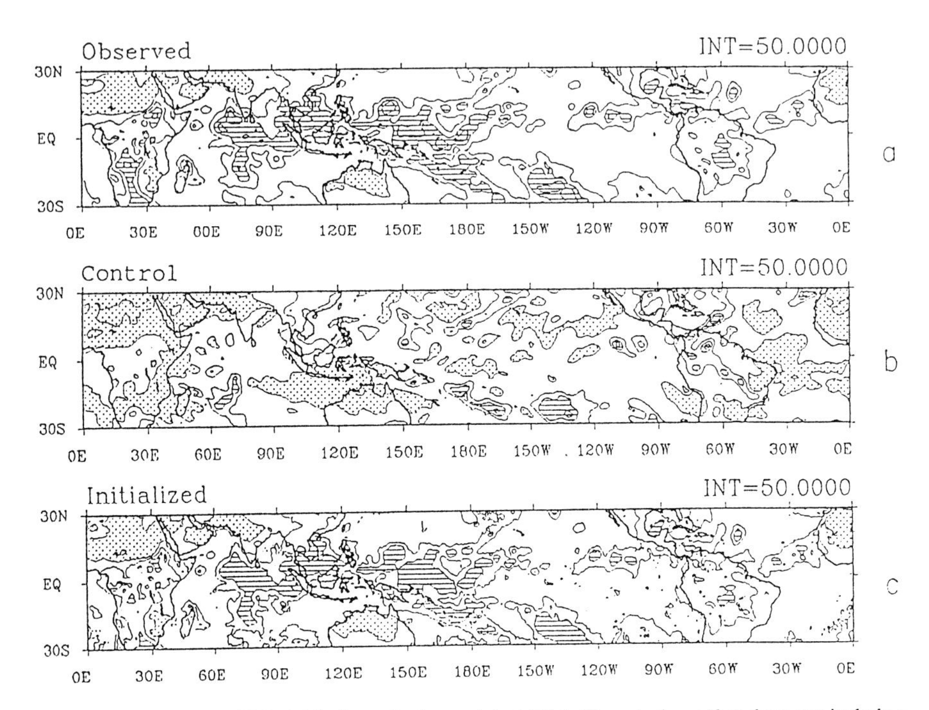

Fig. 6 Typical Initial OLR fields from a) observed (satellite), b) control run that does not include a physical initialization and c) physical initialization. Units: watts m^{-2}.

errors of the predicted low, middle and high clouds. The question arises as to how we should define an observed cloud fraction. Since physical initialization produces a much superior initial OLR field, it seems reasonable that the observed cloud fraction should be calculated from physically initialized fields for the respective forecast verification periods. Figure 7 shows the spin-up of clouds for the control and physical initialization experiments. A near constancy of the forecast skill of the middle and high clouds throughout the forecast period was noted for the physically initialized experiment. The low cloud fractions continue to exhibit a spin-up even after the physical initialization was involved. This is understandable because these low clouds are largely non-precipitating and are not handled by our procedure. Further work is needed to improve the initialization of low cloud fractions. The treatment of shallow convection parametrization is based on a simple formulation of diffusion for heat and moisture (Tiedke, 1984). The control experiment, however, did exhibit a spin-up for these cloud fractions. The predicted skill of total cloud cover for the control and the physical initialization (normalized with respect to persistence error) are shown in Fig. 8. A value greater than one denotes a skill less than persistence. The control experiment appears to have a useful skill of roughly 1.5 days whereas the physical initialization appears to improve this situation to roughly 3.5 days. Prediction of cloud cover is apparently one of the most difficult areas

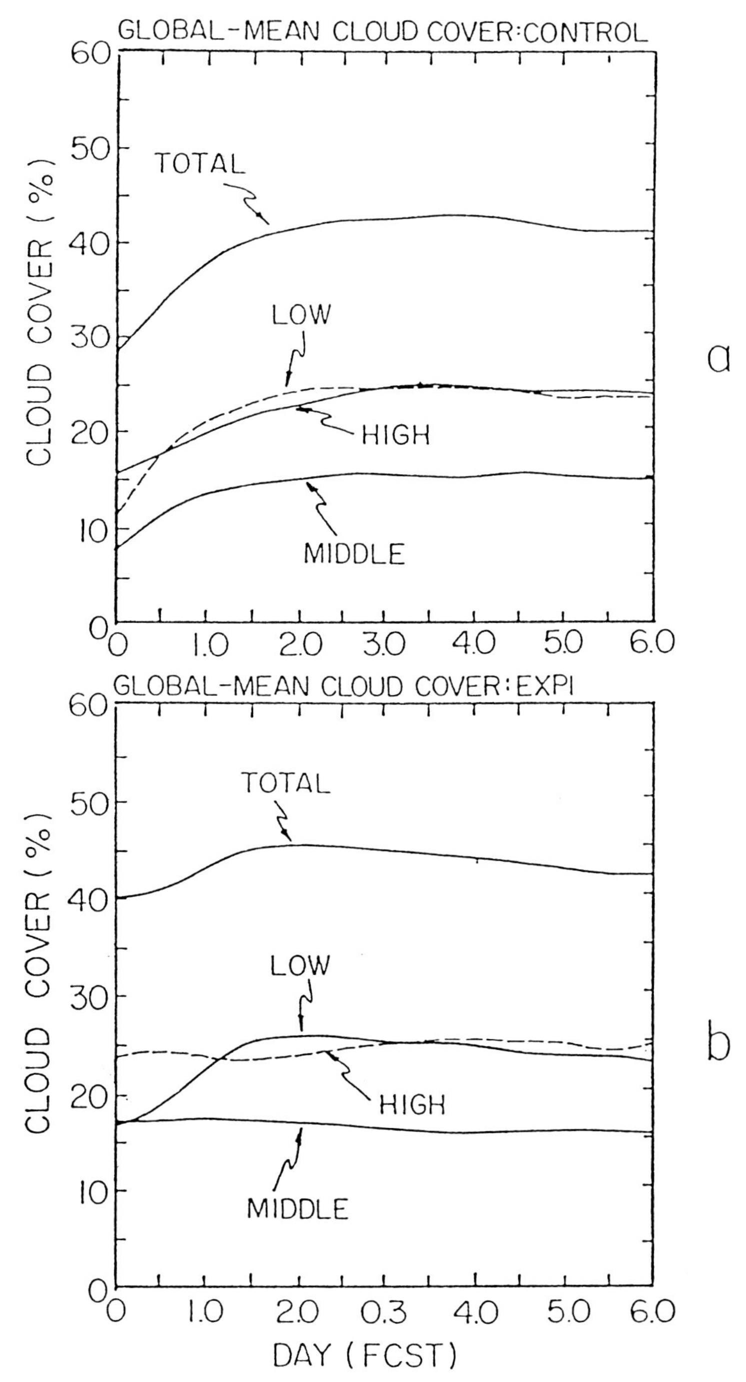

Fig. 7 Spin-up of the fractional cloud cover for a) the control and b) the physical initialization experiments. Low, middle and high cloud fractions are illustrated.

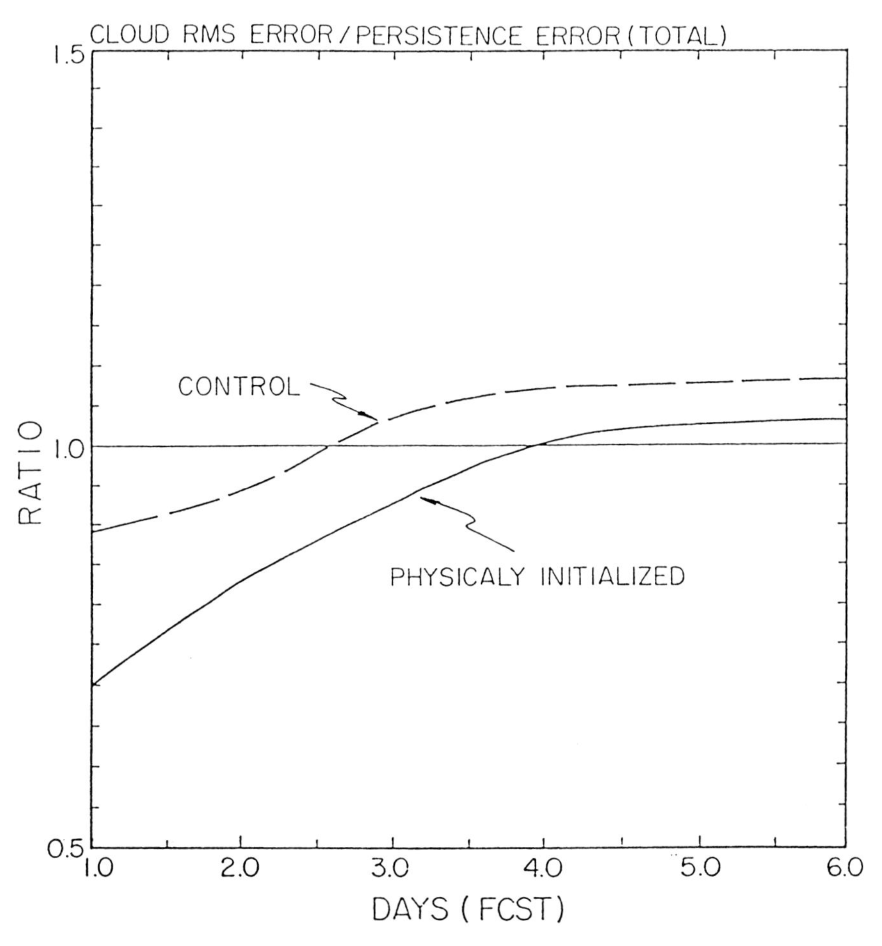

Fig. 8 Predicted skill, normalized to that of persistence, for the total cloud cover is shown against the forecast days. A value greater than one denotes a skill less than that of persistence. Results for the control and the physical initialization are shown here.

of medium-range prediction. The move towards the explicit handling of clouds, rather than their definition via threshold relative humidity, appears to be the new approach. This approach relies on the definition of the formative and dissipative processes for convective and non-convective clouds. These are areas where further work will be required prior to their assessment within physical initialization.

5 Signatures of Meso-Convective Systems

For a long time observations from satellite, aircraft and radar have revealed the amplification of tropical disturbances from the organization of tropical convection. This organization of convection, in fact, appears to occur on meso-convective scales. If the placement of meso-convective systems organizes into a quasi-circular geometry, then an intensification of the zonally symmetric scale (wavenumber zero) about a local cylindrical frame of reference is possible. In the tropical latitudes (roughly between 5 and 15 degrees off the equator), given a low level cyclonic vorticity, warm SST's ($\geqslant 27°C$) and weak vertical shear of the horizontal wind, such an organization of deep convection generally contributes to the generation of available potential energy and its conversion to kinetic energy with an amplification of axially symmetric motions. When we perform the physical initialization at a very high resolution, such as T213, it is possible to retain the mesoscale signatures of rainfall as seen from the SSM/I. The transform grid separation at this resolution (T213) is comparable to the footprint of the radiometers of the SSM/I instrument.

Figure 9 (a, b, and c) illustrates a typical match between the 'observed', the 'physically initialized' and the 'control experiment' based rainfall results at the resolution of T213. The model is able to initialize almost all of the meso-convective precipitating elements. There were roughly 47 elements around the global tropics on this date (22 June 1992). The tropical correlation of the initialized rain, with respect to the observed rain, over all of the transform grid squares is of the order of 0.9. On the other hand the correlations for control experiments that do not invoke physical initialization is around 0.35.

Each of these meso-convective elements are initialized by the physical initialization algorithm. Each of these systems appears to acquire a robust vertical structure from the physical initialization. Figures 10, 11, 12 and 13, following Krishnamurti et al. (1995), show some examples of mesoscale structures of the various precipitating elements of Fig. 9. Figure 10 illustrates results for an element over the northern end of New Guinea. This element is located close to the equator around 140°E. The control run describes a straight flow, Fig. 10a at the 850 mb surface. This region is located to the west of the south Pacific convergence zone. In panel b, the flow field at 850 mb based on physical initialization illustrates a near equatorial vortical circulation. Also superimposed on this panel are the initialized rain rates (maximum rainfall is of the order of 43 mm/d). The winds respond to the heating and a vortical circulation results. A marked change in the initial circulation results from the physical initialization. Figure 10 also illustrates the vertical profiles of vertical motion and relative vorticity over the mesoscale precipitating area for the control and for the physically initialized experiments. It is apparent that these precipitating clusters in the physical initialization experiment imply a more robust lower tropospheric convergence, a strong upper tropospheric divergence, stronger upward motions, slightly enhanced positive relative vorticity over the lower troposphere and a somewhat enhanced negative relative vorticity in the upper troposphere. Figure 11 illustrates another similar sequence for a mesoscale precipitating cluster over India for 22 August 1992, 1200 UTC. Panels a, b and c show a robust initial response for the mesoscale divergence, vorticity and vertical motion from the physical initialization (solid lines); the dashed lines show the initial results from the control experiment where the mesoscale precipitation was absent.

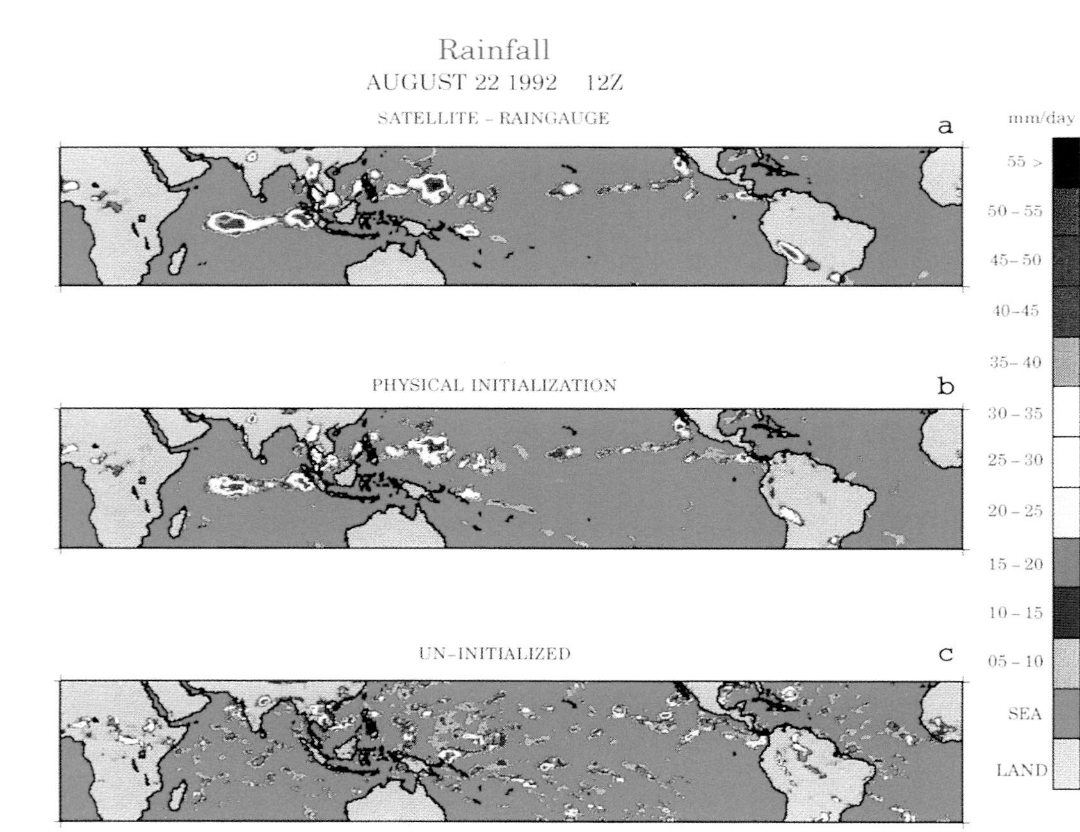

Fig. 9 a) Observed (based on SSM/I, OLR and rain-gauge) rainfall for a 24 hour period. b) Physically initialized rain for the same 24 hour period. c) Control experiment based rain for the same period. Units: mm/day

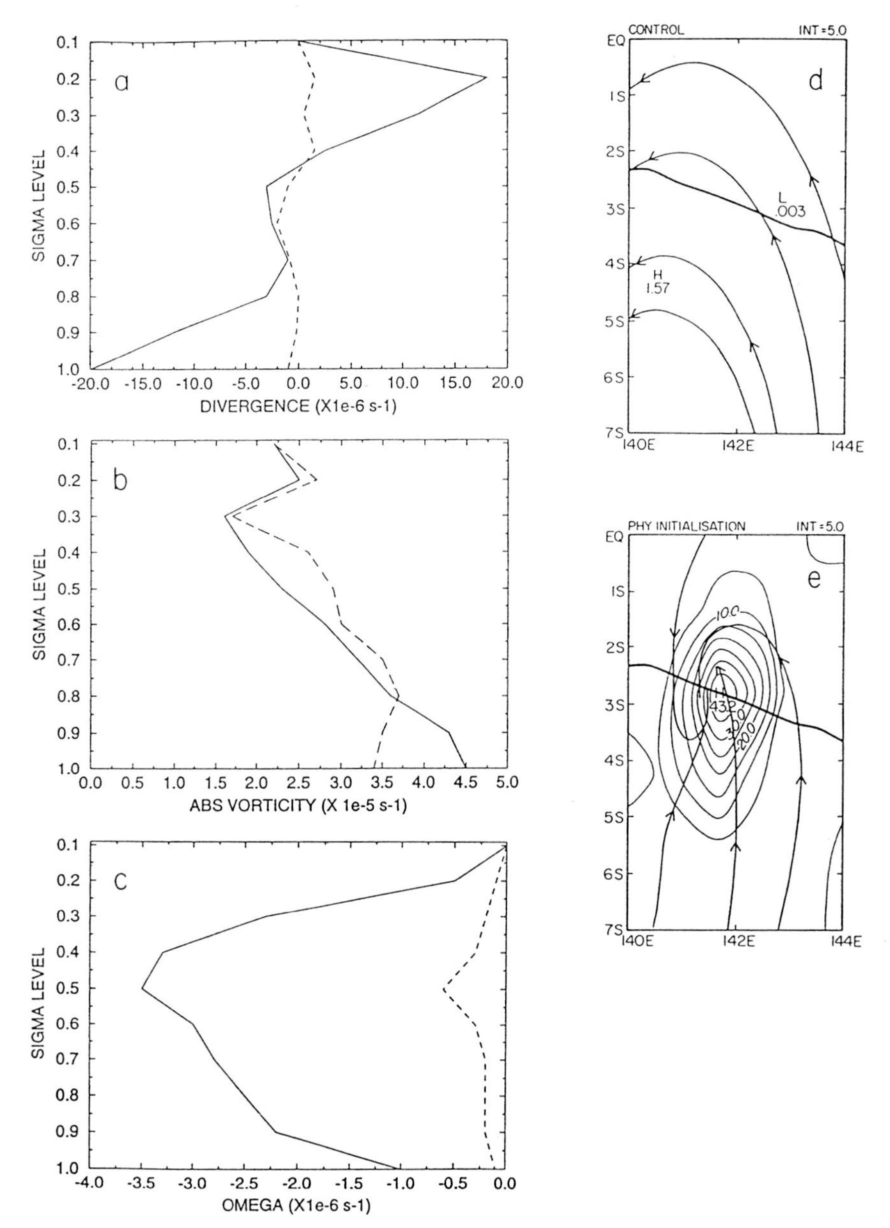

Fig. 10 Structure of a meso-convective element near New Guinea, 22 August 12 UTC, 1992. a) Vertical structure of divergence with (solid line) and without (dashed line) physical initialization. b) Vertical structure of absolute vorticity with (solid line) and without (dashed line) physical initialization. c) Vertical structure of vertical velocity omega with (solid line) and without (dashed line) physical initialization. d) Initial 850 mb streamlines for the control experiment without physical initialization. Weak initial rainfall amounts mm/3 h are highlighted for the control experiment. e) Initial streamlines (with arrows) for the experiment with physical initialization. Also shown is the physically initialized rainfall (mm/3 h).

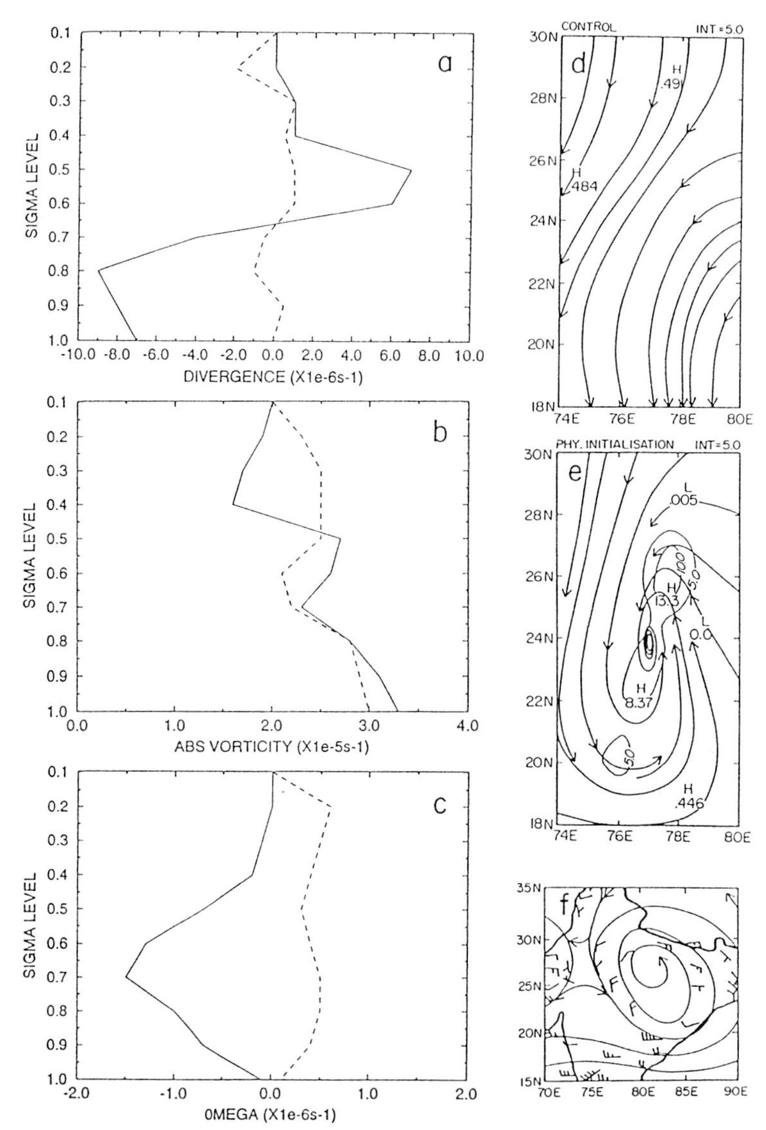

Fig. 11 The structure of a meso-convective element over north-central India at around 24°N and 77°E on 22 August 12 UTC, 1992. a) Vertical profile of divergence with (solid line) and without (dashed line) physical initialization. b) Vertical profile of absolute vorticity with (solid line) and without (dashed line) physical initialization. c) Vertical profile of vertical velocity with (solid line) and without (dashed line) physical initialization. d) Flow field at 850 mb derived from operational analysis at ECMWF. The initial rainfall amounts 0.484 and 0.491 mm/3 h show the maximum rainfall over northern India for the control experiment. e) Flow field at 850 mb obtained after physical initialization. A monsoon depression is clearly evident in this analysis. The initialization rain (mm/3 h) is superimposed. f) 850 mb weather map obtained from India Meterological Department for 22 August 1992 at 1200 UTC. The depression is located near 80°E and 27°N. The observed winds are plotted at station locations.

The flow field at 850 mb (panel d) for the control run shows a basic northerly flow. The initial rainfall amounts implied by the control run are of the order of 0.5 mm/3h. When the precipitation input is introduced via physical initialization, the initial flow field exhibits a cyclonic circulation (show in panel e). The superimposed rainfall rates are of the order of 13 mm/3h, which developed a strong divergent wind response defining what looks like a mesoscale monsoon depression. Panel f shows an analysis, based on delayed data from the India Meteorological Department which was valid for 22 August 1992 at 1200 UTC. Overall we note that the physical initialization can improve data deficiencies of the operational datasets. In a recent paper on these mesoscale signatures of physical initialization (Krishnamurti et al., 1995), we have illustrated several such examples around the global tropical belt. Basically the entire tropical analysis is modified by the physical initialization from the introduction of meso-convective precipitating elements.

The occasional organization of meso-convective elements is another important aspect of very high resolution global modelling. As we proceed westward along the undisturbed trades we see a gradual build-up of towering cumulus and the occurrence of meso-convective elements. The low-level circulation tends to advect (sweep) and organize these elements. Developing tropical systems tend to advect and thus organize these elements into synoptic scale systems. Inner convection tends to draw the outer air from great distances and mesoscale intensification occurs from the angular momentum principle i.e. spin-up of the rotational flow at the inner radii. Although angular momentum is not conserved, due to losses from pressure and frictional torques, hurricane development does seem to follow along these lines. We can, in fact, illustrate much of this line of thinking from the detailed output of very high resolution global models. Next we shall illustrate some results of global prediction experiments carried out at the resolution T213.

Figure 12 (a to g) illustrates the motion (sweeping), coalescence, decay and new formation of meso-convective precipitating elements. This sequence illustrates rainfall forecasts (with superimposed 850 mb predicted streamlines) at intervals of 3 hours between hours 30 and 48 of forecast. Typhoon Omar formed over the tropical Pacific ocean from a tropical depression during this period of the forecast. The predicted amplitude of the 850 mb wind increased from 15 m s^{-1} to 23 m s^{-1} during this period. The precipitating mesoscale elements clearly exhibit an organization as the storm develops. Next we shall study the longer time history of a single element. A cluster, denoted as number 1, was followed during the period it moved from the trade wind circulation into the storm circulation. The vertical distribution of divergence, vertical motion and vorticity between hours 30 and 48 for the meso-convective precipitating element 1 are shown at 6 hourly intervals in Fig. 13. We note that the vertical motion increases considerably as the cluster moves into the storm's circulation. The lower tropospheric convergence and the upper tropospheric divergence amplify during this transition. The mesoscale vorticity does not seem to exhibit an analogous increase (Krishnamurti et al., 1995). This was an important result. In a series of recent articles, based on the shallow

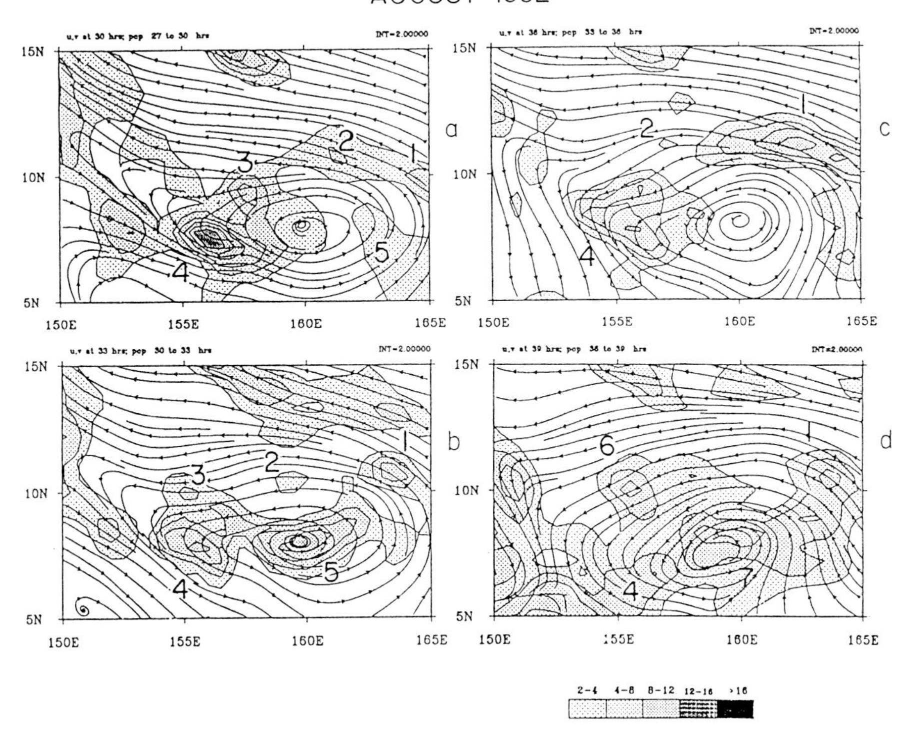

Fig. 12 (a though g) Flow field at 850 mb at intervals of 3 hours between hours 30 and 48 of forecast. Superimposed are the precipitation (mm/3 h) contours, interval of analysis 2 mm/3 h. The meso-convective elements are identified by numerical labels. Shading scales for rainfall identified below figure.

water equations, it was demonstrated that mesoscale vorticity could be swept by the circulations of a larger scale vortex, leading to the intensification of the latter (Holland and Dietachmayer, 1993; Lander and Holland, 1993; Ritchie and Holland, 1993). Our global model studies did not exhibit such a behaviour for the vorticity of individual mesoscale elements (Krishnamurti et al., 1995). The storm (Omar) did intensify from the sweeping of the divergence, vertical motion and heating of these mesocale precipitating elements. The organization of these elements into a circular geometry enhances the lower tropospheric convergence for the axially symmetric mode (wavenumber 0 for a local cylindrical frame of reference), the covariance of convergence and vorticity on this scale leads to the intensification of the storm vorticity. In the present global model, at the resolution T213, we note more of a direct thermodynamical role for these meso-convective precipitating elements. This of course needs to be looked at further, with high resolution meso-

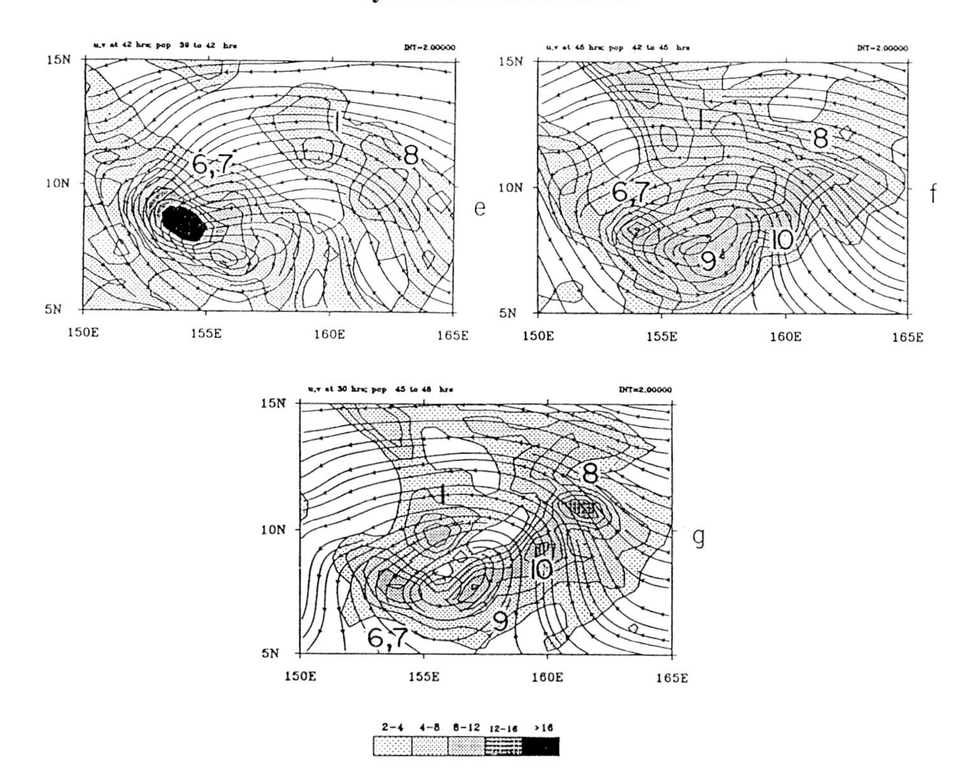

Fig. 12 *Concluded.*

convective non-hydrostatic models with detailed microphysics. We also noted, after examining many of these meso-convective precipitating clusters, that the magnitude of divergence was in general larger than the magnitude of the relative vorticity on these scales.

We also noted that a resolution such as T213 was necessary to resolve the meso-convective precipitating elements, we did not see such individual meso-convective precipitating clusters at the resolution T106, and they were poorly defined at the resolution T170. Figure 14 (a, b and c) illustrates the physically initialized rainfall at the resolution T106, T170 and T213. We note a gradual improvement in the definition of meso-convective elements as we initialize at the higher resolutions. Initially the rainfall over the oceans on this scale, as defined from the SSM/I radiometric data, was closely initialized because of the match between the footprint of the SSM/I data and the transform grid separation of the global model. Figure 15 shows the variation of physically initialized rainfall along a quasi-zonal line of maximum precipitation at different resolutions. As one follows the tropical axis of heaviest rainfall from west to east, the differences among the three different resolutions become more apparent. The graphs show large rainfall rates at the

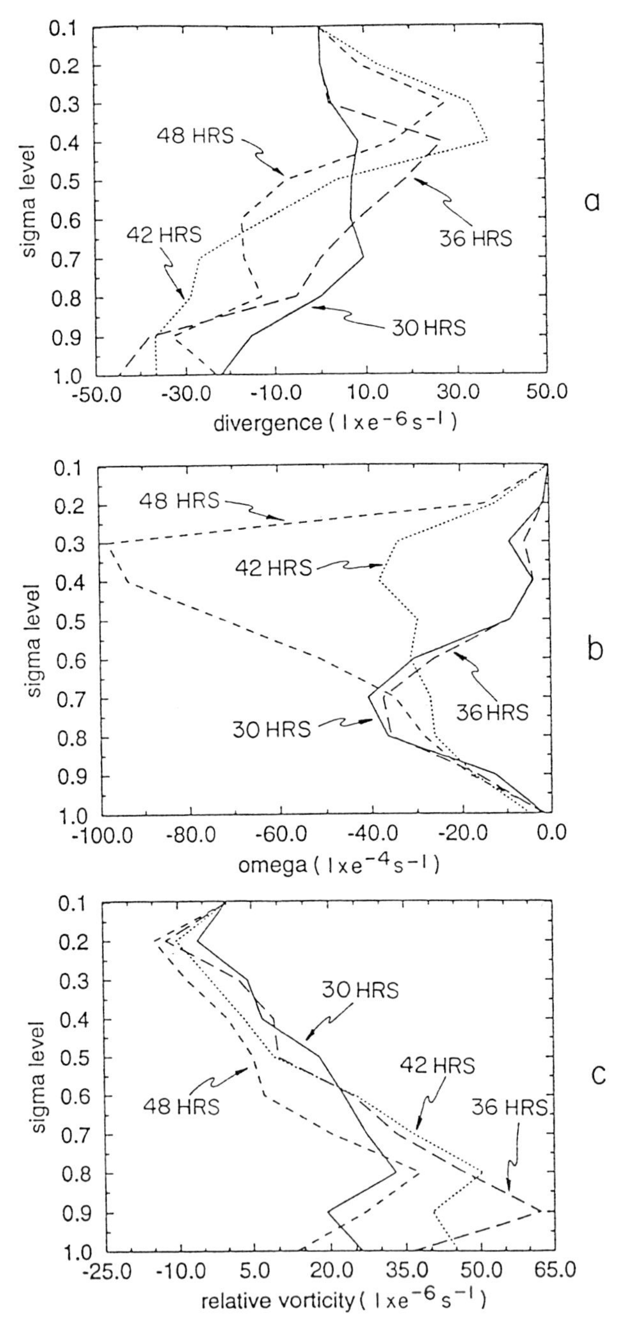

Fig. 13 Time history following an individual meso-convective element between hours 30 and 48. a) Vertical profile of divergence, b) Vertical profile of vertical velocity, c) Vertical profile of relative vorticity

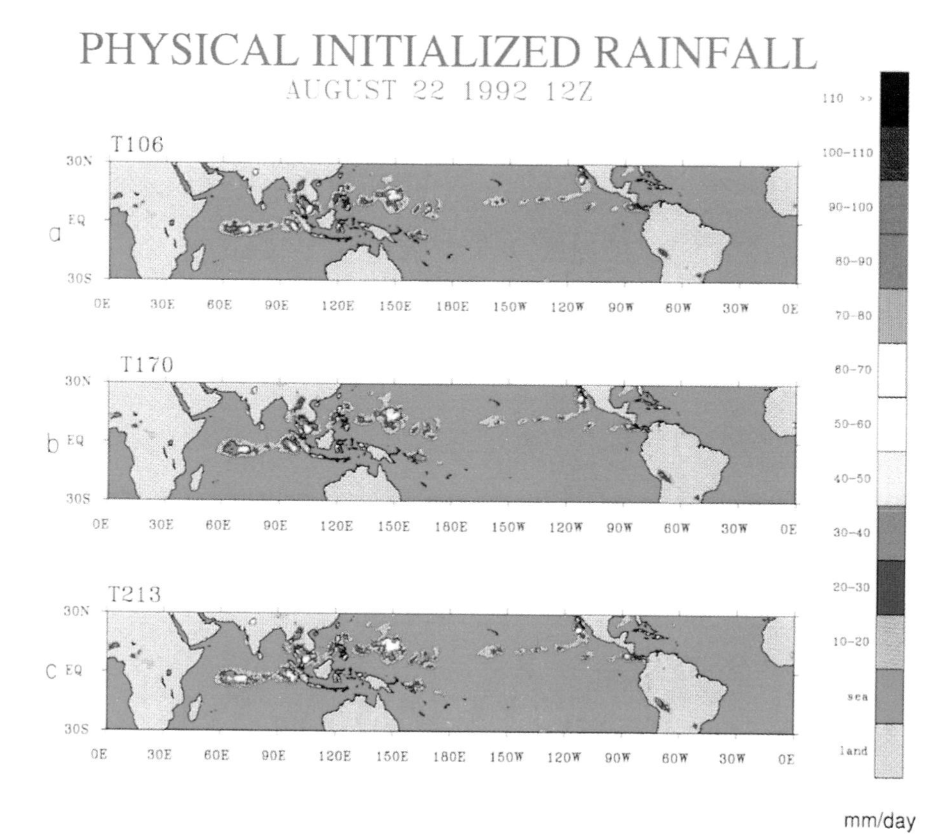

Fig. 14 24 hour physically initialized rainfall total (units: mm/day) at different spectral resolutions. a) T106; b) T170; and c) T213.

resolution T213 compared to the other resolutions. The spikes of larger rainfall at the resolution T213 denote the mesoscale systems (scales of the order of a few hundred kilometres) that can be followed in time within the high resolution model integrations.

The model at the high resolution exhibits the demise of the initial generation of meso-convective precipitating elements and the birth of a new generation of elements. The life-span of these meso-convective elements is of the order of one day. The model's one day forecast skill (Fig. 4) is quite high (a correlation of 55% for the observed versus predicted rain) and demonstrates that meso-convective elements are, in fact, being advected reasonably in the high resolution global model. The high skill of the rainfall covers some 40 to 50 meso-convective precipitating clusters.

An important question is raised in reference to the future of modelling. There is a large thrust to expand the scope of non-hydrostatic, very high resolution models

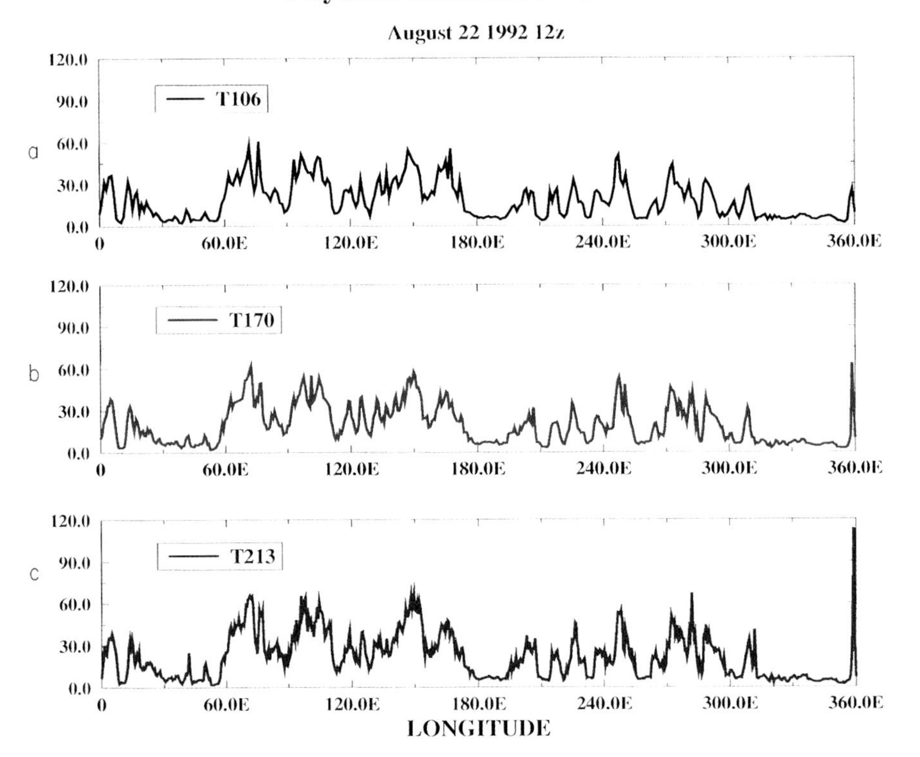

Fig. 15 West-east variation of physically initialized rainfall along a quasi-zonal line of maximum precipitation at different resolutions, (units: mm/day). a) T106, b) T107, and c) T213.

including microphysical processes. To address meso-convective elements whose scales are of the order of few hundred kilometres, a non-hydrostatic model with explicit cloud physics, with a grid resolution of the order of a kilometre may be minimally necessary. Prediction of such elements would require the definition of the initial state at these resolutions. Lacking adequate 3-dimensional definition of the initial state, forecast error will no doubt be present. The question arises as to whether such models could have nowcasting and one-day forecasting skills above those seen in Fig. 4. To achieve such skills over the entire tropics the computing and initial data requirements would be exceptionally demanding for the non-hydrostatic models. This appears to be the future direction of mesoscale modelling. For the present, the implicit methods for the handling of convective processes may be competitive with the explicit non-hydrostatic microphysical models for the handling of the entire tropics.

6 The Improvements in the Hydrological Budgets

The Bangladesh region encounters some of the most severe floods on an annual basis. Although floods are more common in the monsoon months, June through

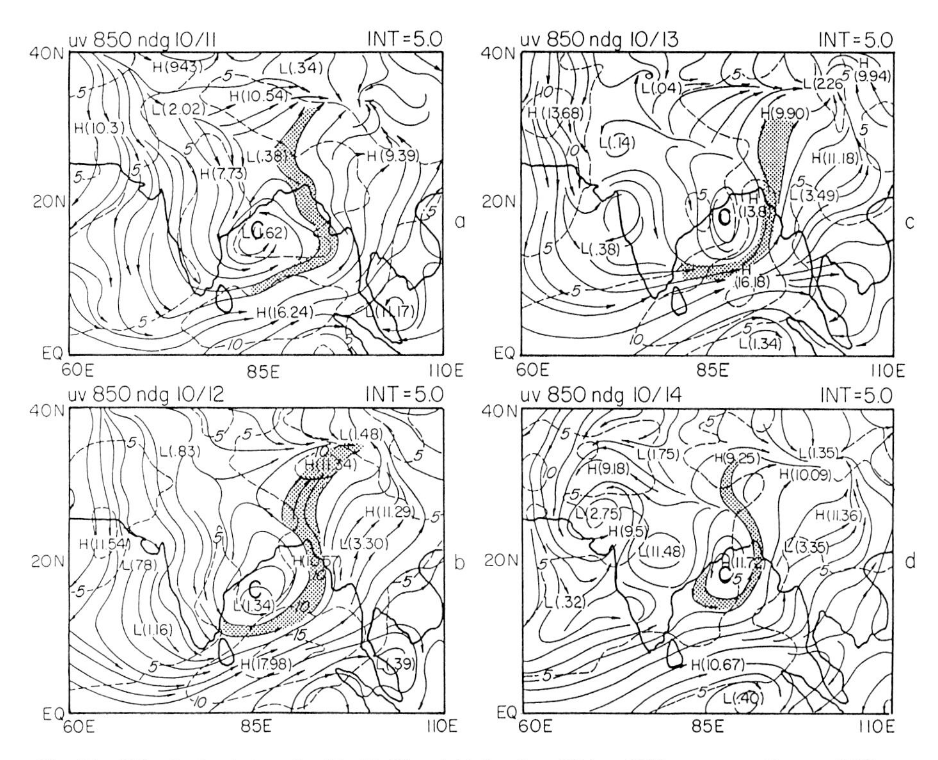

Fig. 16 850 mb circulations for 11, 12, 13 and 14 October 1991 at 1200 UTC streamlines: solid lines; isotachs: dashed lines (m s^{-1}).

early September, flooding related to the passage of tropical depressions and tropical storms have been historically more catastrophic. Estimates of hydrological budgets during such episodes can be accomplished with an improved consistency among the estimates of moisture convergence, evaporation and the observed rain rates with the physical initialization. An example of the episodic nature of these flooding events will be illustrated. If a short duration event such as the passage of a tropical depression causes heavy flooding, then useful interpretation of the hydrological budget can only be obtained by focusing on the duration of that event. Use of month-long datasets can be somewhat misleading.

We carried out physical initialization for the entire month of October, 1991, during this month a post-monsoon season flooding event occurred which in fact lasted for roughly 4 days from 11 October to 14 October. This was related to the passage of a monsoon depression that formed over the Bay of Bengal along the ITCZ at roughly 15°N and moved northwards. Figure 16 shows the 850 mb circulation illustrating its northward passage. On the eastern flank of this storm the southerly flow brought in a large amount of moist air from the Bay of Bengal. The convergence of flux of this moist air, during this period, resulted in substantial rainfall and flooding.

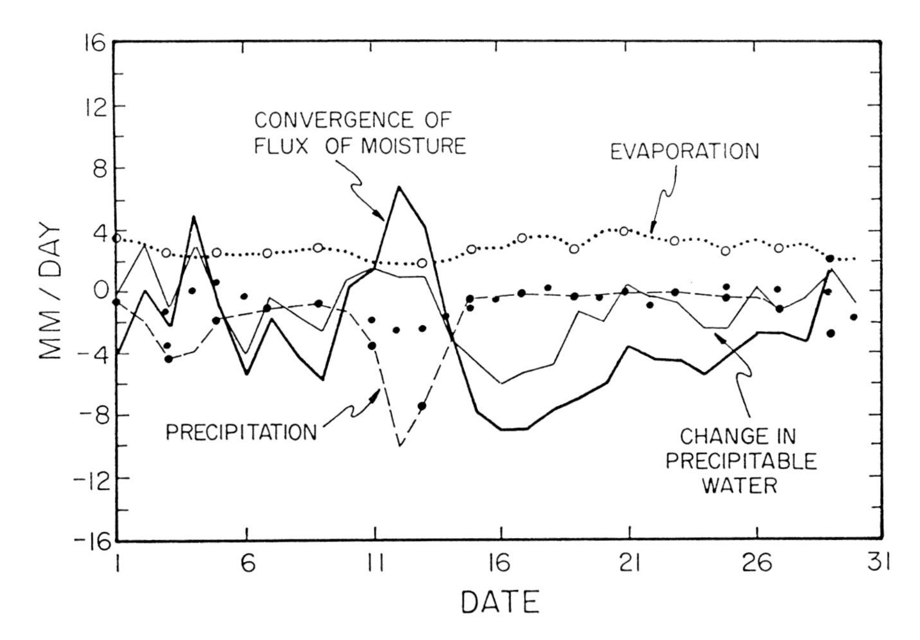

Fig. 17 Daily time history of the hydrological budget components during October 1991 over Bangladesh. The different components are labeled in the diagram. The unconnected dots denote the diffusion terms.

The daily history of the evaporation, precipitation, rain rates and convergence of the flux of moisture for the Bangladesh region are shown in Fig. 17. This includes the results for the entire month of October, 1991. Krishnamurti et al. (1994) examined this period in some detail and demonstrated that physical initialization with a high resolution global model provides a consistency among the elements of the hydrological budget. The estimates of evaporation presented are based on surface similarity theory which invokes stability-dependent exchange coefficients. Also included is a low wind speed correction, based on Beljaars and Miller (1990) for the exchange coefficients. This involves vertical eddy turbulent kinetic energy in its estimates of the exchange coefficients for wind speeds less than 5 m s^{-1}. The rain rate estimates come from the previously mentioned mixed algorithm (Gairola and Krishnamurti, 1992).

If one considers the hydrological budget for the entire month of October 1991, a completely different perspective is apparent compared to that found during the period of the flooding, 11 through 14 October. October is usually a dry, post-monsoon month over the Bangladesh region. The hydrological budget based on month long data for this region is shown in Fig. 18a. A divergence of the moisture flux is noted along with a decrease in the total precipitable water (for the entire month of October). This does not convey the occurrence of heavy flooding which

WATER BUDGET OCTOBER 1991

BANGLADESH

$$\frac{\delta \bar{q}}{\delta t} = -\overline{\nabla \cdot Vq} + E - P + (Res.)$$

$$-1.25 = -2.69 + 2.95 - 1.75 + (0.24)\ \mathrm{mm/day}$$

$$-5.33 = -11.48 + 12.58 - 7.48 + (1.05)\ 10^8\ \mathrm{mton/day}$$

a

MOISTURE CONVERGENCE BUDGET

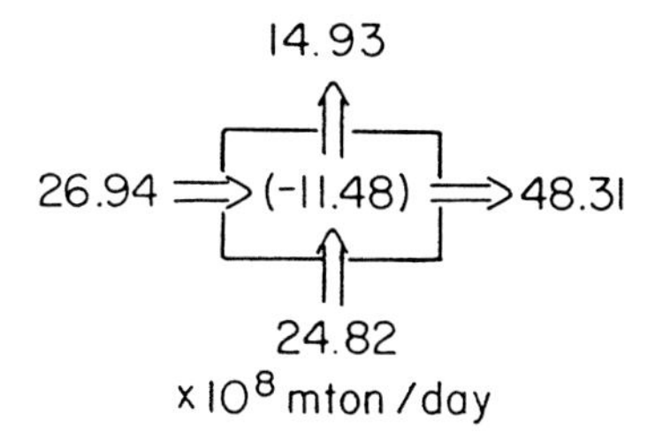

WATER BUDGET 11-14 OCTOBER 1991

$$\frac{\delta \bar{q}}{\delta t} = -\overline{\nabla \cdot Vq} + E - P + (Res.)$$

$$0.15 = 2.51 + 2.45 - 6.00 + (1.19)\ \mathrm{mm/day}$$

$$0.63 = 10.72 + 10.45 - 25.62 + (5.08) \times 10^8\ \mathrm{mton/day}$$

b

MOISTURE CONVERGENCE BUDGET

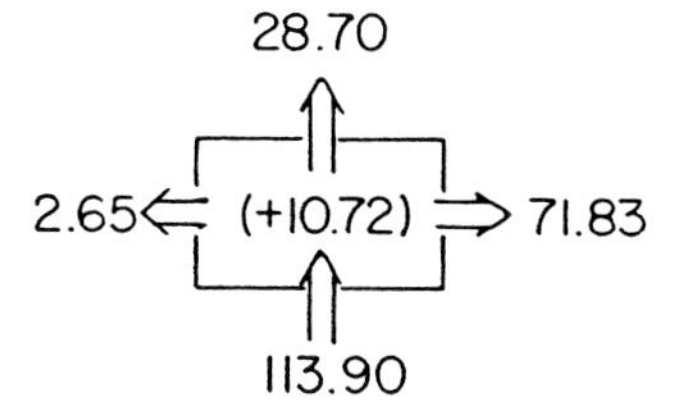

Fig. 18 Hydrological budget over the Bangladesh region. a) month long datasets, October 1991. b) For the period of heavy rains, 11 to 14 October 1991.

was of short duration. If one looks at the hydrological budget of the episode, averaged over the 4 days (11 October through 14) we find a net convergence of moisture flux, a moistening within the column enhancement of total precipitable water and precipitation exceeding the net evaporation, Figure 18b. This appears to reflect the flooding situation better than using the entire month of data.

7 Ensemble Forecasts of Tropical Cyclone Tracks

The use of observed rain rates within the physical initialization modifies the definition of divergence (the divergent wind), the vertical distribution of latent heating and moisture and the pressure tendency throughout the assimilation period. If several different analyses from diverse data assimilation groups were subjected to physical initialization, using the same observed rain rate, one might expect to see a reduction in the spread of ensemble forecasts especially for tropical storm tracks. To examine this hypothesis 16 medium-range forecast experiments were completed. These include 8 control experiments and 8 experiments which include the physical initialization. Among these we have 4 analyses, from NASA-GODDARD, NMC, ECMWF and GFDL at 0000 UTC and 4 analyses from each of these same groups at 1200 UTC. Rain rates were prepared from a mix of OLR and rain-gauge data similar to Gairola and Krishnamurti (1992) for a 24 hour period preceding the start times of the respective forecasts.

Figure 19 (a and b) illustrates the spread of predicted tracks for Hurricane Frederic of 1979 with and without physical initialization.

Figure 19a shows the forecasts for the control experiments and Fig. 19b shows the results from forecasts including physical initialization. It is clearly apparent from these forecasts that the ensemble spread for the medium range forecasts is considerably reduced by the inclusion of physical initialization. This improvement has come from the incorporation of similar initial information on divergence, convective heating, moisture and surface pressure tendency for the entire ensemble of 8 experiments.

The forecast skill of these experiments was evaluated. The position errors for the storms are calculated by averaging the forecast positions of the ensemble members and calculating the difference between ensemble-averaged positions (forecast tracks from outliers were excluded in the averaging process) and the best track positions provided by the National Hurricane Center and Joint Typhoon Warning Center. We noted that the forecasts which included the physical initialization reduced the position errors when compared to the control forecasts.

8 Concluding Remarks

Physical initialization appears to be a powerful procedure for the nowcasting and short-range forecasting of tropical precipitation. This procedure incorporates additional observations over the data sparse tropics. These data are the observed precipitation measurements obtained from surface rain-gauges and the satellite-based indirect estimates of rain rates which come from OLR and the microwave radiometer. The physical initialization enhances the definition of mesoscale divergence, vorticity, vertical motion, convective heating and the surface pressure tendencies.

Physical initialization can only be as good as the rain rates we provide to its reverse algorithms and to the formulations of physical parametrizations. In spite of that major limitation, within a prescribed set of rainfall input and prescribed

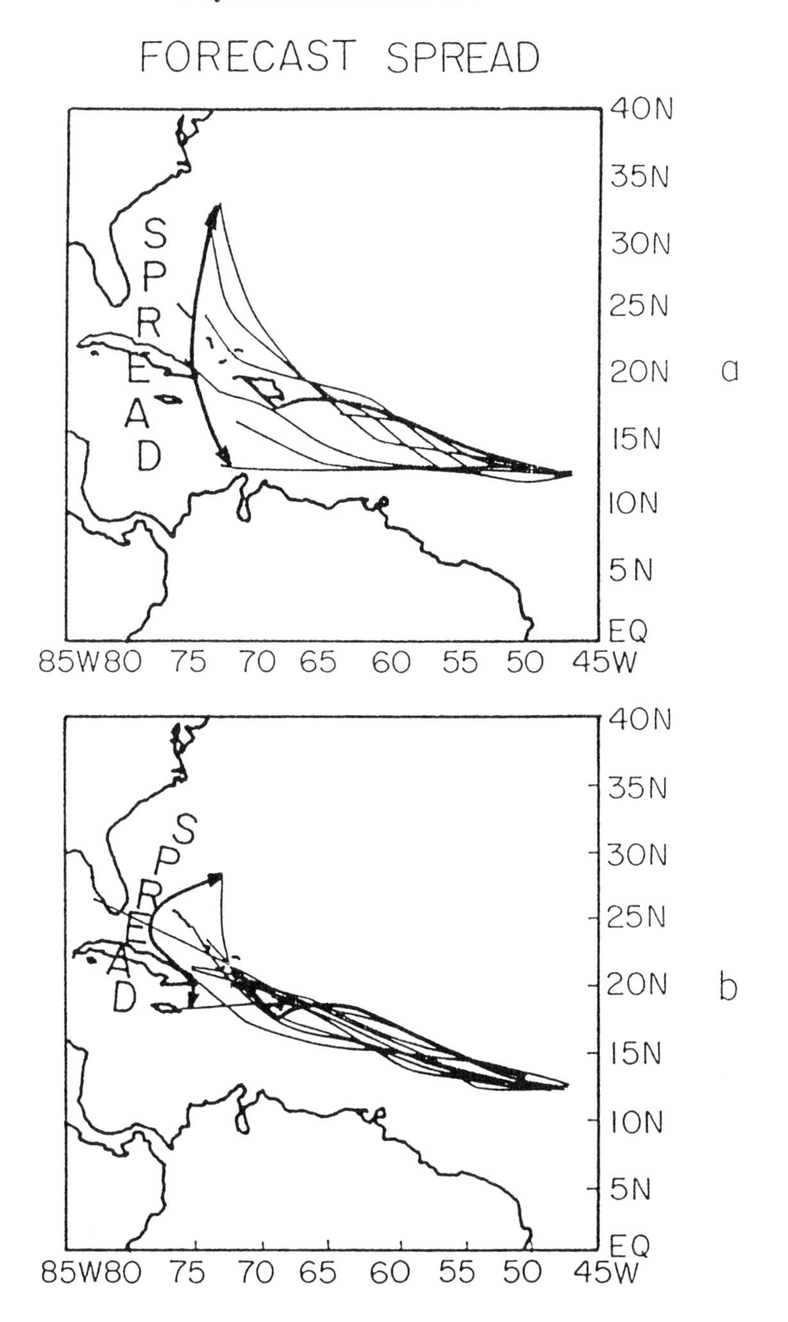

Fig. 19 Ensemble forecasts for Hurricane Frederic of 1979 over the Atlantic ocean. Start date is 1 September 1979. Predicted tracks are based on different initial analyses. The spread of forecasts on day 6 are highlighted.

parametrizations of physical processes, the method appears to enhance the quality of tropical initialization and short range prediction markedly. Physical initialization appears to provide skills above those of the operational centres for nowcasting and short-range prediction over the tropics.

We also note that for a high resolution global model, such as T213, the separation of the transform grid roughly matches the footprint of the SSM/I. Thus the physical initialization appears to capture the initial and the subsequent motions of meso-convective precipitating elements quite well. Since most of the rainfall occurs from the passage of meso-convective elements, at this resolution we see a higher skill for the one day rainfall forecasts from the passage of these meso-convective elements. In this context, we illustrated the development of Typhoon Omar of 1992 which developed from a tropical depression. We showed an interesting organization of convection into a near circular geometry. This contributed to the initial development of Typhoon Omar. It had been difficult to see such mesoscale contributions to storm developments from coarse resolution models.

Physical initialization appears to provide a consistency among the precipitation, evaporation, convergence of the flux of moisture and the local changes in the precipitable water. The improvements in the convergence of the moisture flux arise from the improvements in the definition of the divergent component of the wind and the vertical variations of the humidity variable. As a consequence, it has been possible to improve the computations of hydrological budgets. This was illustrated for a short duration Bangladesh flood event. Such budgets executed from datasets that are not initialized for the observed rain rate fail to define the correct sign of the moisture convergence – largely due to data inadequacies.

We have seen a benefit of physical initialization to ensemble forecasts of hurricane tracks. The spread of an ensemble of track forecasts is reduced. That reduction is related to the fact that all members of the ensemble are subjected to the rain rate input in the physical initialization. The enrichment of the similar additional data (divergence, convective heating and moisture) for each member acts to limit the spread somewhat.

The future of physical initialization must lie in a proper implementation of this entire process within a four-dimensional variational assimilation where the adjoints for the precipitation processes need to be incorporated requiring that the revere algorithms incorporate the observed rain rates.

Acknowledgments

The work reported here was supported by the following grants: NSF ATM 9312537, ONR 00014-93-10243, NASA 37 WA 0361, NOAA NA 37WA0361. Computations reported here were carried out on the CRAY YMP's at NCAR and FSU. We dedicate this study to the memory of Dr. Andre Robert who was our inspiration during the last three decades. He was a friend, a teacher and a great scholar. We shall miss him.

Appendix: An Outline of the FSU Global Spectral Model

The global model used in this study is identical in all respects to that used in Krishnamurti et al. (1991). The following is an outline of the global model:

(a) Independent variables (x, y, σ, t).
(b) Dependent variables: vorticity, divergence, surface pressure, vertical velocity, temperature and humidity.
(c) Horizontal resolution: Triangular 213 waves.
(d) Vertical resolution: 15 layers between roughly 50 and 1000 mb.
(e) Semi-implicit time differencing scheme.
(f) Envelope orography (Wallace et al., 1983).
(g) Centred differences in the vertical for all variables except humidity which is handled by an upstream differencing scheme.
(h) Fourth order horizontal diffusion (Kanamitsu et al., 1983).
(i) Kuo-type cumulus parametrization (Krishnamurti et al., 1983).
(j) Shallow convection (Tiedke, 1984).
(k) Dry convective adjustment.
(l) Large-scale condensation (Kanamitsu, 1975).
(m) Surface fluxes via similarity theory (Businger et al., 1971).
(n) Vertical distribution of fluxes utilizing diffusive formulation where the exchange coefficients are functions of the Richardson number (Louis, 1979).
(o) Long- and short-wave radiative fluxes based on a band model (Harshvardan and Corsetti, 1984; Lacis and Hansen, 1974).
(p) Diurnal cycle.
(q) Parametrization of low, middle and high clouds based on threshold relative humidity for radiative transfer calculations.
(r) Surface energy balance coupled to the similarity theory (Krishnamurti et al., 1991).
(s) Nonlinear normal mode initialization – 5 vertical modes (Kitade, 1983).
(t) Physical initialization (Krishnamurti et al., 1991).

References

BELJAARS, A. and M. MILLER. 1990. A note concerning the evaporation from the tropical oceans: sensitivity of the ECMWF model to the transfer coefficient of moisture at low wind speed. ECMWF Research Department Technical Memo 170, ECMWF, Reading, 19 pp.

BUSINGER J.A.; J.C. WYNGAARD, Y. IZUMI and E.F. BRADLEY. 1971. Flux profile relationship in the atmospheric surface layer. *J. Atmos. Sci.* **28**: 181–189.

GAIROLA, R.K. and T.N. KRISHNAMURTI. 1992. Rain rates based on SSM/I, OLR and raingauge data sets. *J. Meteorol. Atmos. Phys.* **50**: 165–174.

HARSHVARDAN and T.G. CORSETTI. 1984. Longwave parameterization for the UCLA/GLAS GCM NASA Tech. Mem. 86072, Goddard Space Flight Center, Greenbelt, MD.

HOLLAND, G.J. and G.S. DIETACHMAYER. 1993. On the interaction of tropical-cyclone-scale vortices. III: Continuous barotropic vortices. *Q. J. R. Meteorol. Soc.* **119**: 1381–1391.

KANAMITSU M. 1975. On numerical prediction over

a global tropical belt. Report No. 75-1, pp. 1-282. (Available from the Dept. of Meteorology, Florida State University, Tallahassee, FL-32306.)

———; K. TADA, T. KUDO, N. SATO and S. ISA. 1983. Description of the JMA operational spectral model. *J. Meteorol. Soc. Jpn.* **61**: 812–828.

KITADE, T. 1983. Nonlinear normal mode initialization with physics. *Mon. Weather Rev.* **111**: 2194–2213.

KRISHNAMURTI, T.N.; S.K.R. BHOWMIK, D. OOSTERHOF, G. ROHALY and N. SURGI. 1995. Meso-scale structure implied by physical initialization. *Mon. Weather Rev.* **123**: 2771–2790.

———; G. ROHALY and H.S.BEDI. 1994. On the improvement of precipitation forecast skill from physical initialization. *Tellus*, **46A**: 598–614.

——— H.S. BEDI and K. INGLES. 1993. Physical initialization using SSM/I rain rates. *Tellus*, **45A**: 247–269.

——— J. XUE, H.S. BEDI, K. INGLES and D. OOSTERHOF. 1991. Physical initialization for numerical weather prediction over the tropics. *Tellus*, **43AB**: 53–81.

KUMMEROW, C. and L. GIGLIO. 1994. A passive microwave technique for estimating rainfall and vertical structure information from space. Part 1: Algorithm description. *J. Appl. Meteorol.* **33**: 3–18.

LACIS, A.A. and J.E. HANSEN. 1974. A parameterization for the absorption of solar radiation in the earth's atmosphere. *J. Atmos. Sci.* **31**: 118–133.

LANDER, M. and G.J. HOLLAND. 1993. On the interaction of tropical-cyclone-scale vortices. I: Observations. *Q. J. R. Meteorol. Soc.* **119**: 1347–1361.

LEE, H.S. and T.N. KRISHNAMURTI. 1995. Impact of physical initialization on cloud forecasts. *J. Meteorol. Atmos. Phys.* **56**: 261–273.

LOUIS, J.F. 1979. A parametric model of vertical eddy fluxes in the atmosphere. *Boundary-Layer Meteorol.* **17**: 187–202.

PARSONS, D.; W. DABBERDT, H. COLE, T. HOCK, C. MARIN, A. BARRETT, E. MILLER, M. SPOWART, M. HOWARD, W. ECKLUND, D. CARTER, K. GAGE and J. WILSON, 1994. The Integrated Sounding System: Description and Preliminary Observations from TOGA COARE. *Bull Amer. Meteorol. Soc.* **75**: 553–568.

RITCHIE, E.A. and G.J. HOLLAND. 1993. On the interacting of tropical-cyclone-scale vortices. II: Discrete vortex. *Q. J. R. Meteorol. Soc.* **119**: 1363–1379.

SPENCER, R.W. 1993. Global oceanic precipitation from MSU during 1979–91 and comparison to other climatologies. *J. Clim.* **6**: 1301–1326.

TIEDKE, M. 1984. The sensitivity of the time-mean large-scale flow to cummulus convection in the ECMWF model. *In*: Workshop on Convection in Large-Scale Numerical Models. ECMWF, 28 Nov.–1 Dec. 1983, pp. 297–316.

WALLACE, J.M.; S. TIBALDI and A.J. SIMMONS. 1983. Reduction of systematic forecast errors in the ECMWF model through the introduction of envelope orography. *Q.J.R. Meteorol. Soc.* **109**: 683–718.

YANAI, M.; S. ESBENSEN and J.H. CHU, 1973. Determination of bulk properties of tropical cloud clusters from large-scale heat and moisture budgets. *J. Atmos. Sci.* **30**: 611–627.

Impact of Resolution and of the Eta Coordinate on Skill of the Eta Model Precipitation Forecasts

Fedor Mesinger and Thomas L. Black
National Meteorological Center, 5200 Auth Road
Camp Springs, Maryland 20746-4304, U.S.A.

and

Michael E. Baldwin
General Sciences Corporation, Laurel, Maryland

[Original manuscript received 13 December 1994; in revised form 1 August 1995]

ABSTRACT *During the last somewhat more than a year of the operational running of the Eta Model at the U.S. National Meteorological Center (NMC) a considerable body of statistics has been accumulated demonstrating the model's significantly increased accuracy in forecasting precipitation compared to that of NMC's Nested Grid Model (NGM). The model has shown a smaller but just as consistent advantage in skill against that of NMC's global spectral model. An obvious question is whether there are design features of the Eta Model which could be identified as responsible for this generally improved performance.*

One feature on which we have data is spatial resolution. In an experiment we performed, a sample of 148 forecasts was formed from each of four models: three versions of the Eta Model differing in resolution only, and the NGM. The Eta Model of that time had achieved a substantial advantage over the NGM with no increase in resolution at all of our eight precipitation categories except at the lightest "rain/no rain" category. Thus, reasons other than resolution are responsible for most of the advantage of the Eta Model over the NGM. An increase in vertical resolution from 17 to 38 layers resulted in an additional modest improvement of skill at the medium and at most of the higher intensity categories. An increase in horizontal resolution from 80 to 40 km brought greater improvements, this time at the lightest and at the medium intensity categories.

Another feature we have investigated is the use of the eta coordinate. As a sequel to our earlier tests, we have recently done 16 forecasts running the Eta Model as an eta and also as a sigma system model. In contrast to the earlier experiments, in the sigma mode no "eta-like" discretization of mountains was performed. At all categories the "Eta" has achieved higher scores than the "Eta/sigma", its advantage being particularly large for more intense precipitation.

A fitting question on this occasion is, "What are our views on possible further progress in weather prediction through developments in numerical methods?" From the vantage point of the Eta Model, which we take as a state-of-the-art standard, we identify five areas which are offering or may offer promise of additional increase in skill. These are the two areas referred to above: resolution and the choice of the vertical coordinate, and those of the choice of the vertical grid, of the numerics of the propagation of gravity-inertia waves, and of the box-average vs. point-sample treatment of predicted variables. We comment on prospects for benefits to be achieved via efforts in the direction of each of the five areas listed.

RÉSUMÉ *Depuis un peu plus d'un an d'intégration en temps réel on a accumulé un important ensemble de statistiques démontrant que la précision de la prévision des précipitations du modèle Eta en exploitation au National Meteorological Center (NMC) des États-Unis est grandement améliorée par rapport au modèle du NMC à grilles emboîtées (NGM). Le modèle a montré un avantage d'habileté plus petit mais tout aussi consistant par rapport au modèle spectral mondial du NMC. On peut se demander si il n'y a pas des éléments du modèle Eta, responsables de cette performance améliorée, qui pourraient être identifiés.*

La résolution spatiale est un des éléments sur lequel nous possédons des données. Dans un essai, on a exécuté un ensemble de 148 prévisions à partir de quatre modèles : trois versions du modèle Eta, dont seule la résolution était différente, et le NGM. Le modèle Eta courant a démontré des avantages notables sur le NGM, sans aucune augmentation de la résolution dans nos huit catégories de précipitation sauf la catégorie la plus faible "pluie/pas de pluie". On peut donc dire que la résolution n'est qu'une petite partie de l'avantage du modèle Eta. Une augmentation de la résolution verticale de 17 à 38 couches a produit une faible amélioration supplémentaire de l'habileté pour les catégories d'intensité moyenne et la plupart de celles de plus haute intensité. Une augmentation de la résolution spatiale de 80 à 40 km a produit de plus grandes améliorations, cette fois aux catégories d'intensité la plus faible et moyenne.

On a aussi examiné l'utilisation de la coordonnée Eta. Suite aux essais précédents, on a récemment produit 16 prévisions avec le modèle Eta tournant dans le mode eta et aussi dans le mode signa. Contrairement aux autres essais, dans le mode sigma on n'a pas utilisé une discrétisation du genre eta pour les montagnes. Pour toutes les catégories, le Eta a produit de meilleurs résultats que le Eta/sigma, surtout remarquables pour les précipitations plus intenses.

Quels autres progrès concernant la prévision météorologique, associés au développement de méthodes numériques, peut-on envisager en cette occasion? Sous l'angle du modèle Eta, qui pour nous est le plus avancé, on peut identifier cinq domaines offrant ou laissant entrevoir des amélioration additionnelles d'habileté. Il y a les deux domaines ci-dessus, la résolution et le choix des coordonnées verticales, en plus du choix de la grille verticale, de la numérisation de la propagation des ondes de gravité-inertie, et du traitement des variables prévues par moyenne de la boîte versus échantillonnage ponctuel. On souligne ce que pourraient être les bénéfices possibles en rapport avec les efforts nécessaires quant à chacun des cinq domaines suggérés.

---

1 Introduction: Comparison of precipitation scores

Comparison of predictions from different models is a powerful diagnostic tool available at operational centres which run more than one forecasting model starting with the same initial condition. At the U.S. National Meteorological Center (NMC), at the time of this writing three operational models are run each day on 0000 and 1200 UTC data, and – in the case of the Eta Model – on at least three occasions comparison of results from different models and known model features has led to an identification of a model problem. In all of these cases, the problems were removed via a model change resulting in improved performance.

The three operational models, their aliases and a few of their most basic characteristics are as follow.

- The Nested Grid Model, NGM, or RAFS (Regional Analysis and Forecasting System). This is a grid point sigma system model, with about an 80-km inner grid nested inside its own coarser outer grid. Both grids have 16 layers in the vertical. It employs fourth-order accuracy schemes, along with a periodic application of a fourth-order Shapiro filter. The model as well as its analysis system have been "frozen" as of August 1991 (DiMego et al., 1992), making the model an attractive tool in calibrating progress of developing models;
- The Medium Range Forecasting Model, MRF, or Aviation ("Avn") model is a global spectral sigma system model. During the time we are concerned with here it was run with the triangular 126 truncation (T126), and 28 layers (e.g., Kanamitsu et al., 1991; Pan and Wu, 1994). The two names, MRF and Avn, refer to the same model but to different data cutoff times: twice daily, at 0000 and 1200 UTC, the model is run 72 h ahead with an early data cutoff under the name Aviation model; at 0000 UTC the Avn run is followed by a later data cutoff, "MRF model" run.
- The Eta Model is a grid point, step-mountain vertical coordinate model. Its dynamical part is designed following the approach of Arakawa of maintaining a variety of chosen integral constraints of the continuous atmosphere (e.g., Janjić, 1984; Mesinger et al., 1988). Differencing schemes are of the second- or approximately second-order accuracy. As of 9 June 1993, the model is run operationally with an 80-km resolution and 38 layers in the vertical, as the so-called "NMC Early Run". The term "early" refers to an early data cutoff, aimed at providing guidance as quickly as possible.

All three models have comprehensive physics packages, with substantial differences among them. An exception to this general rule is the common radiation package used by the MRF and the Eta Model developed by the Geophysical Fluid Dynamics Laboratory (GFDL, e.g., Black, 1994).

Comparison of the results of the operational as well as experimental models takes of course a variety of forms, including verification statistics of various kinds for individual as well as for samples of forecasts. Regarding the latter, by far the most emphasis has been placed upon verification of precipitation forecasts in the case of the Eta Model. There are a number of obviously attractive features that precipitation verification exhibits in comparison with some of the more usual verification quantities such as the anomaly correlation and root mean square errors. For instance, precipitation analyses are not "contaminated" (Ramage, 1993) by model information via model-based assimilation schemes. It also appears very unlikely that one can significantly improve precipitation scores by artificially forcing the prediction model to produce smoother circulation fields, with amplitudes reduced compared to the analyzed amplitudes, or by smoothing the model output prior to verification. One is tempted to add to this list the feature of precipitation patterns being of a much smaller spatial scale than various circulation fields, perhaps resulting in skill above random being already "practically useful" skill as opposed to an arbitrarily chosen quantity such as the ubiquitous 60% limit used for the anomaly correlation coefficient. Note that when it comes to precipitation scores, modifying definitions of various skill scores such as the threat score to have skill equal to zero for a random forecast is a straightforward exercise (e.g., Schaefer, 1990; Gandin and Murphy, 1992).

The precipitation analysis system of the NMC's Development Division is based on data provided by the National Weather Services's River Forecast Centers (RFCs). The data consist of reports of accumulated precipitation for each 24-h period ending at 1200 UTC. The analysis covers the area of the contiguous United States with reports from about ten thousand RFCs rain gauge stations. In areas of poor coverage, RFCs data are augmented by radar precipitation estimates if rain gauge data are available in the vicinity to calibrate the radar data. Radar estimates derived in this way are subsequently treated as equivalent to station data. Data are analyzed to grid-boxes of the model to be verified by simple grid-box averaging if the grid box contains at least one reporting station or radar estimate. For the three models considered here about 90% of the area of the contiguous United States is covered in this way by grid-box average precipitation values, with a total number of model grid-box values ranging from about 860 in the case of the global model to about 1160 in the case of the highest horizontal resolution model, the NGM. Finally, note that the 80-km Eta and NGM produce 48-h forecasts twice daily; thus, with no archiving or other problems each 24-h verification period is covered by three forecasts by each of the three operational models.

At the time of writing, archiving the results of all three operational models from three short-range verification periods (00-24, 12-36 and 24-48 h forecasts) has been in place for somewhat more than a year. A summary of the results for the first 12 months of this period, September 1993 – August 1994, is shown in Fig. 1. For extra information, results for the 12-month sample are shown split into two 6-month subsamples, September–February, upper panel, and March–August, lower panel. The values shown are those of the equitable threat score (Schaefer, 1990; Mesinger and Black, 1992),

$$T_e = [n_{fo} - E(n_{fo})]/[n_{fno} + n_{nfo} + n_{fo} - E(n_{fo})],$$

where n_{fo} is the number of grid points at which a considered precipitation event is forecast and also observed ("hits"), n_{fno} is the number of points at which the event is forecast but not observed ("false alarms"), n_{nfo} is the number of points at which the event is not forecast but has been observed ("misses"), and $E(n_{fo})$ is the expected number of hits for a random forecast. This number one can readily evaluate as

$$E(n_{fo}) = n_f n_o / N,$$

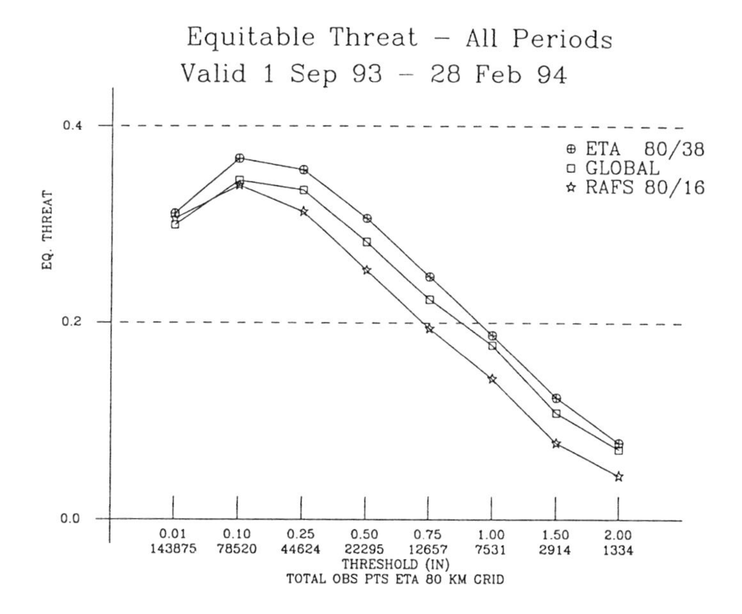

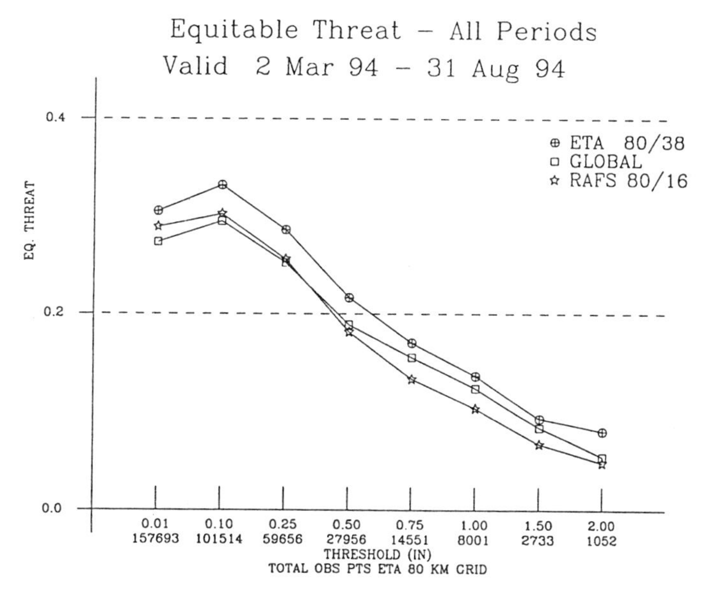

Fig. 1 Equitable precipitation threat scores for three National Meteorological Center's operational models for September 1993 – August 1994, and for various precipitation thresholds. Scores for the "winter" period of the sample, September–February, are shown in the upper panel, and those for the "summer" period, March–August, in the lower panel. ETA stands for the operational Eta Model (80 km/38 layers); GLOBAL for NMC's global spectral (T126), or Avn/MRF model and RAFS for Regional Analysis and Forecast System, of which the NGM is the forecast model.

n_f and n_o being the total number of "forecast" and "observed" points, respectively, and N the total number of points in the verification domain. Note that the equitable threat score is zero for a random forecast of any areal coverage, and one for a perfect forecast. A precipitation event here is defined as precipitation greater than a given threshold. In Fig. 1 eight thresholds are entered, shown along the abscissa (top number), in inches per 24 h. To assess significance, the total number of observed grid points on the 80-km Eta grid is also shown (bottom number). Results for a sample of 906 verifications by each of the models are shown; 474 in the "winter" half of the 12-month period, September–February, and 432 in the "summer" half, March–August. The missing verifications were rendered unusable for a three-way comparison by archiving or postprocessing failures with at least one of the models. The "global" model archiving and results displayed are of the MRF model runs when available, for 0000 UTC data; and of the Aviation runs for 1200 UTC data.

The size of the sample – reflected in the smoothness of the lines plotted – seems to warrant considerable confidence in the apparent advantage of the Eta Model over both the NGM and the global model for perhaps all of the eight precipitation categories. Indeed, even if instead of the two 6-month half-periods our 12-month sample is stratified into the individual forecast periods, 00–24, 12–36 and 24–48 h, the Eta Model still wins all of the eight categories in each of the three sub-samples which are formed in this way, just as it has in the two sub-samples shown in Fig. 1.

As the models are different, an obvious question is whether there are design features of the Eta Model which could be identified as responsible for its generally improved performance. A number of aspects, however, stand in the way. Firstly, models change. In this respect the NGM as a frozen model serves as an excellent control. Furthermore, there are aspects other than the model design which affect the verification scores. Three such aspects come to mind: differences in the sizes of verification grids, in the data cut-off time, and in the model analysis/initialization schemes. Note that verification on a higher resolution grid tends to adversely affect the precipitation scores because the impact of the increased noisiness of the analysis using the grid-box average approach outweighs a possible increase in accuracy due to the increased horizontal resolution of the model. Thus, in all three of the aspects just referred to, the Eta is at somewhat of a disadvantage compared to the Avn/MRF in terms of the scores one would expect it to achieve. Comparing the Eta to the NGM the situation is more mixed, with the Eta benefiting from its larger verification grid size, and the NGM from its later data cut-off and perhaps from its more comprehensive analysis/initialization effort. Model differences are thus a promising option in searching for primary causes of the Eta's edge in precipitation scores shown in the figure.

2 The model resolution

We have obtained evidence for the contributions which a number of Eta Model design features make in achieving this improved skill by setting up "parallel" runs

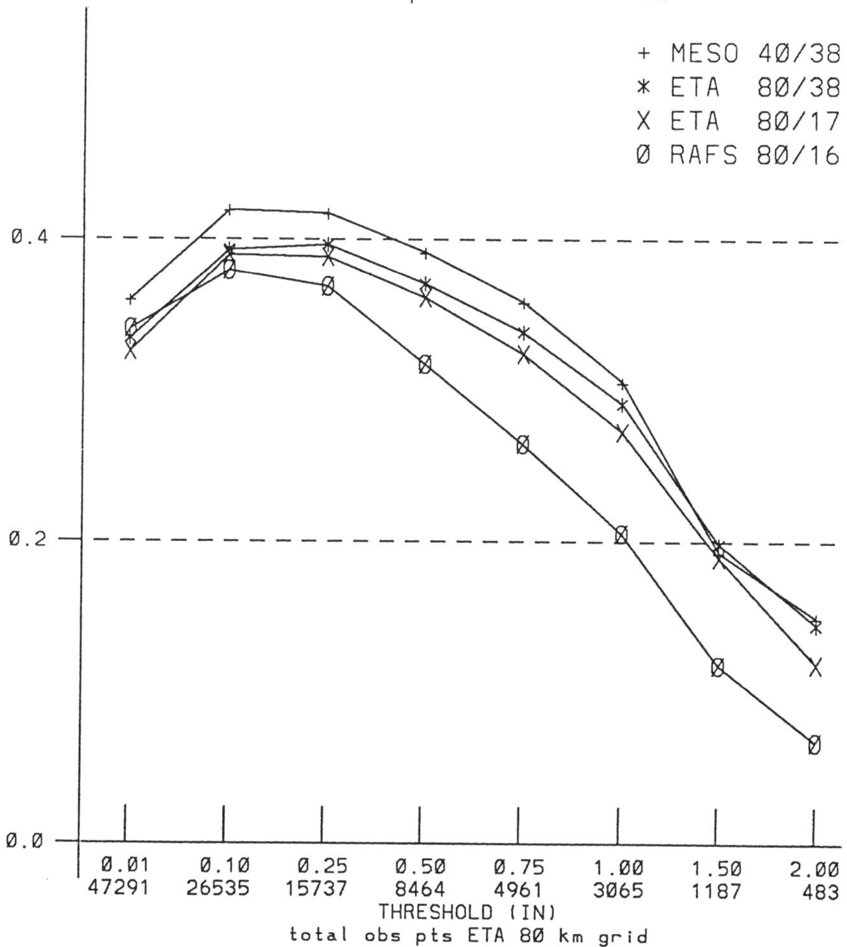

Fig. 2 Equitable precipitation threat scores for three versions of the Eta Model ("MESO", 40 km/38 layers; ETA 80 km/38 layers, and ETA 80 km/17 layers) and for the NGM, for 148 forecasts during September–December 1992. Forecasts started at 1200 UTC and those started at 0000 UTC are verified for 00–24 h and for 12–36 h accumulated precipitation, respectively.

in which models are verified differing in one feature only. One such feature is model resolution. In Fig. 2 precipitation threat scores are shown for a sample of 148 forecasts during September–December 1992 by each of four models: three versions of the Eta Model differing in resolution only, and the NGM. In this plot all three Eta Models have been verified on the 80-km eta grid, to avoid having the 40-km model be at the disadvantage it would be facing were it to be verified on its own grid due to noisiness created in the precipitation analysis by a fourfold reduction in the number of reports per grid box. The remapping of the 40-km forecasts to the 80-km grid is done in a way which maintains the total precipitation to a desired degree of accuracy.

It is seen that the Eta Model of late 1992 with no increase in resolution had achieved a substantial advantage over the NGM at all eight precipitation categories except at the lightest "rain/no rain" category. An increase in vertical resolution from 17 to 38 layers resulted in moderate improvement of skill at the medium and at most of the higher intensity categories. An increase in horizontal resolution from 80 to 40 km brought greater improvements, this time at the lightest and at the medium intensity categories.

A long-standing problem of the Eta Model, visible in this resolution experiment, was that it generally did not outperform the NGM at the lightest precipitation category unless a higher horizontal resolution was used. This problem appears to have been solved in April 1993 by the introduction of a modest degree of "horizontal diffusion" which, inadvertently, was practically absent in the model until that time (Mesinger et al., 1993). The motivation for our introduction of some "second-order" horizontal diffusion was not one of filtering numerically generated noise, but that of attempting to parametrize the effect of unresolved advective processes on scales greater than those of turbulence (Mellor, 1985).

3 The eta coordinate: Precipitation threat scores

A factor on which we have already published some results is the use of the eta vs. sigma coordinate (Mesinger and Black, 1992). These results were obtained by taking advantage of an attractive feature of the eta coordinate (e.g., Mesinger et al., 1988) whereby a simple redefinition of its surface values suffices to change it into a terrain-following ("sigma") coordinate so that the same code can be run as both the eta and a sigma system model. We shall present here additional material on this topic by reporting on statistics for a more recent sample of forecasts, and by discussing in some detail two of the cases on which we have made experiments.

The sample of forecasts in which the Eta Model was run as an eta and also as a sigma system model by Mesinger and Black (1992) was limited. Its sigma mode mountains were identical to the eta mountains of the time, thus having the same elevation for groups of four neighbouring points, discretized to the nearest reference elevation of the eta layers. This discretization, coupled with the modest vertical resolution of the time of only 16 layers, may have detrimentally affected the skill of the sigma mode forecasts.

For an experiment largely free of these deficiencies, a sample of 16 forecasts was created in which the 80 km/38 layer Eta Model was run as a sigma system model. In its sigma mode, the Eta Model was run for this experiment with the current "single-point" eta mountains but without the Eta Model's discretization of mountains in the vertical for which there is no need when using the sigma system. Precipitation threat scores of the "Eta" and the "Eta/sigma" models, along with

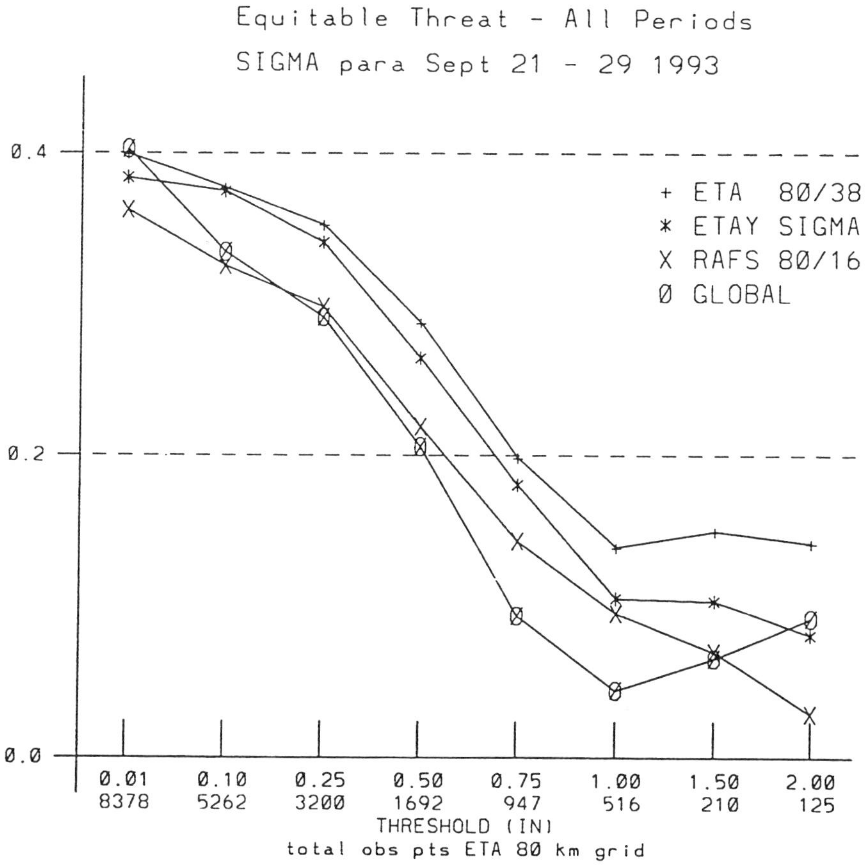

Fig. 3 Equitable precipitation threat scores for two versions of the Eta Model: Eta 80 km/38 layers ("ETA"), and the same version of the Eta Model but run using sigma coordinate ("ETAY"), and for the NGM (RAFS), and the Avn/MRF ("global") Model; for a sample of 16 forecasts verifying 1200 UTC 21 September through 1200 UTC 29 September 1993. Eight forecasts are each verified once, for 12–36 h, and the remaining eight each twice, for 00–24 and for the 24–48 h accumulated precipitation.

those of the NGM and the Avn/MRF model for the same period, are shown in Fig 3. At all categories the Eta has achieved higher scores than the Eta/sigma, its advantage being particularly large for more intense precipitation.

It may be worth noting at this point that the Eta when run in its sigma mode, has outscored both sigma system models of Fig. 3 in most of the categories by a margin greater than the benefit that the Eta in Fig. 2 displays from its higher vertical resolution. To the extent that results from our 16-forecast sample of Fig. 3 are significant, this suggests that features other than the model numerics contribute to the Eta's advantage in precipitation scores. One might hope to glean additional information in this sense from the relative differences between models in scores for the "winter" and for the "summer" half of the year, shown in the two panels of Fig. 1. The Eta's overall advantage over the two sigma models appears to be greater in the summer because the number of Eta wins that are only marginal is clearly greater in the winter. This appears indicative of the importance of model "physics" in particular given the results obtained at the European Centre for Medium-Range Weather Forecasts (Jarraud et al., 1987) where a distinct benefit from the "envelope" orography, which presumably removes some of the sigma system problems, was found in winter but not in summer. Since convection should be playing a much greater role in summer than in winter, we find that the plots of Fig. 1 reinforce our feeling that the Eta Model's convection scheme (Janjić, 1994) is making a significant contribution to the model's advantage in precipitation scores.

Scatter plots of individual Eta and Eta/sigma verifications for four of the eight categories, >0.01, >0.25, >1.0 and >2.0″/24 h, are shown in Fig. 4. The two upper panels contain 24 verifications each; recall that this is the total number of verifications for the sample. The lower left panel contains 22 verifications (precipitation >1.0″/24 h not being analyzed at the last verification period). Finally, the lower right panel contains 16 verifications (precipitation >2.0″/24 h not being analyzed at three out of the nine verification periods of the sample).

The overall advantage of the eta over the sigma mode is also readily visible in the scatter-plot representation of this figure, with many more points found above than below the diagonal line, and, in particular, with points above the line showing a strong tendency to be more frequently further away from it than those below the line. In addition, the plots once more illustrate the tendency for the benefit from the eta coordinate to be more pronounced for more intense precipitation. In this sense the 2.0″ plot may be especially convincing, in that all three verifications in which at least one of the modes had skill registered a substantial advantage of the eta over the sigma mode.

Regarding this last plot, one may be disappointed by the fact that of the 16 verifications as many as 13 are clustered in a super-point of zero and slightly below zero skill for both of the modes. One can however take some comfort in the information that the three forecasts in which the eta mode had skill account for more than half of the total number of the analyzed 2″ points, all three verifying at 1200 UTC 25 September. This of course illustrates the well-known point that forecasting an event at some of the grid-boxes at which it is also analyzed becomes increasingly less likely as the size of the area at which the event is analyzed decreases. At 1200 UTC 25 September, when all three eta mode forecasts had skill, accumulated precipitation of 2″ and greater was analyzed over an area of as many as 22 eta 80-km points, which covered a major part of Missouri and extended into Kansas, Oklahoma and Arkansas, and was associated with substantial flooding in the region. We shall analyze in some detail one of these forecasts in the following section.

4 A frontal wave development case

Events which to a substantial degree are caused by mountains are of course prime

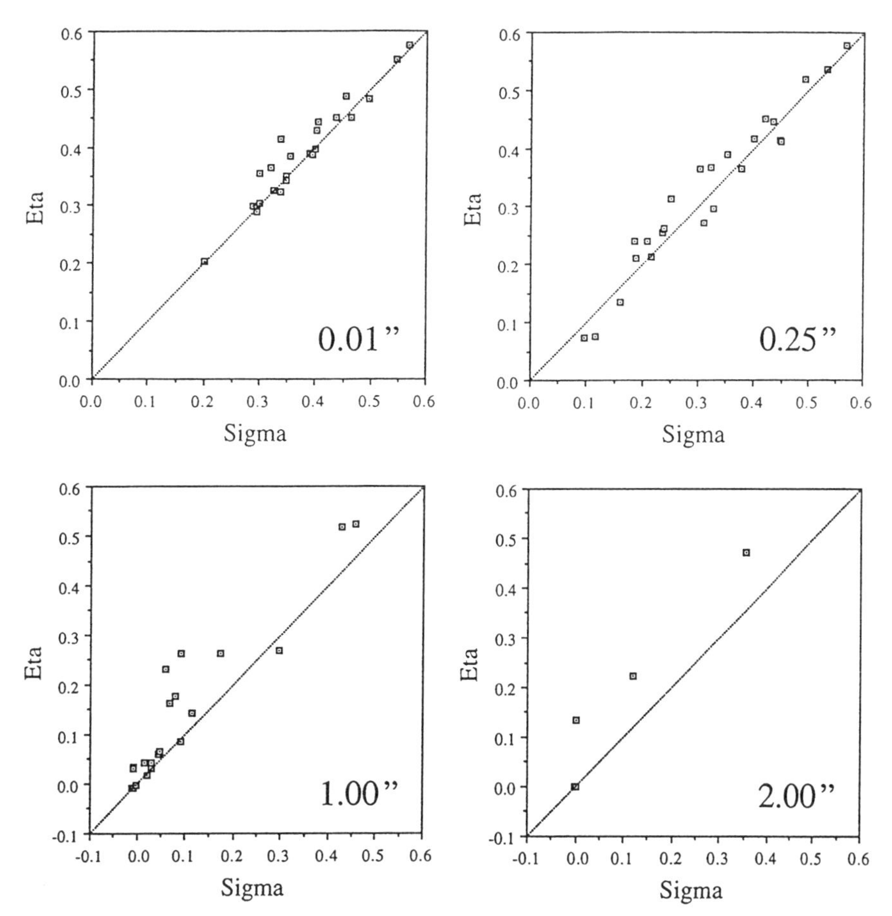

Fig. 4 Equitable precipitation threat scores of individual "Eta" and "Eta/sigma" verifications for four of the categories plotted in Fig. 3, >0.01, >0.25, >1.0 and >2.0″/24 h. There are 24 verifications in the upper two panels, 22 in the lower left and 16 in the lower right panel; note that in the lower two panels some of the verifications have clustered at or close to the (0,0) points.

candidates for a possible impact from the eta coordinate. Indeed, in monitoring the performance of the Eta and other NMC models and making occasional experiments with the Eta Model's sigma feature we have noted at least three mountain-related events in which (a) errors of the NMC's sigma system models were absent or were much smaller in Eta, while subsequently (b) these errors were largely reproduced by the Eta when it was run in its sigma mode. One of these events is the low in the lee of the Rockies discussed by Mesinger and Black (1992; "the April 1991 case"). In this case, NMC's sigma system models were placing the centre of a low in the lee of the Rockies at the Oklahoma-Texas border or even further north inside Oklahoma while the Eta Model was placing it in east-central Texas, approximately as analyzed.

The second of our three events is that of a frontal wave development in the lee

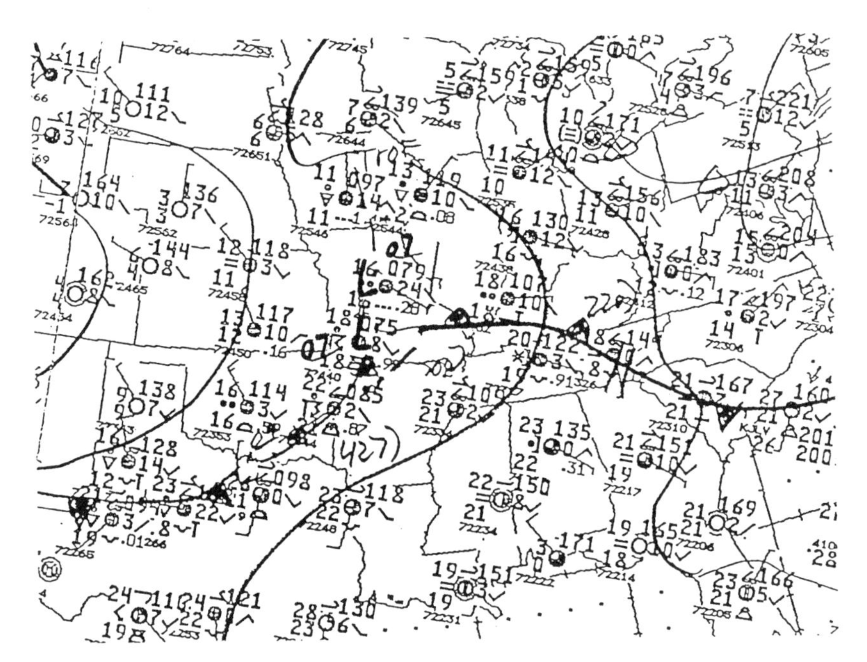

Fig. 5 A section of the NMC's "Northern Hemisphere" surfaces analysis, valid at 1200 UTC 25 September 1993. Isobars are drawn at intervals of 4 mb.

of the Rockies, resulting in extremely heavy rains over Missouri that we referred to above verifying at 1200 UTC 25 September 1993. This case seems to typify the benefit from the eta coordinate in the sense that relatively modest but consistent improvement in the position of the responsible synoptic-scale system resulted in large benefits in the objective precipitation scores in all three of the Eta forecasts verifying at this time. As stated, these are the three forecasts with positive eta-mode skill in the 2″ plot of Fig. 4; at the same time two of these forecasts, 24- and 36-h, have resulted in the scores displayed next to the upper right corner of the 1″ plot of the figure.

A section of the NMC's "Northern Hemisphere" analysis valid at that time is shown in Fig. 5. The analyst has placed two low centres over Missouri, one at its southwestern corner and the other to the north-northeast of the first one. They are both analyzed at 1007 mb (07 in the figure). Note the position of the associated cold front, cutting across the middle of the Red River part of the Texas-Oklahoma border. Figure 6 shows the 24-h accumulated precipitation analysis verifying at the same time. Very heavy precipitation is seen over northeastern Oklahoma, southeastern Kansas and southern Missouri, with the maximum of 6″ (150 mm) /24 h over the southernmost part of the Kansas-Missouri border, and the area of >2″/24 h covering perhaps half of Missouri, and extending into Kansas, Oklahoma and Arkansas.

We shall show here the 36-h forecasts by the three models as well as by the

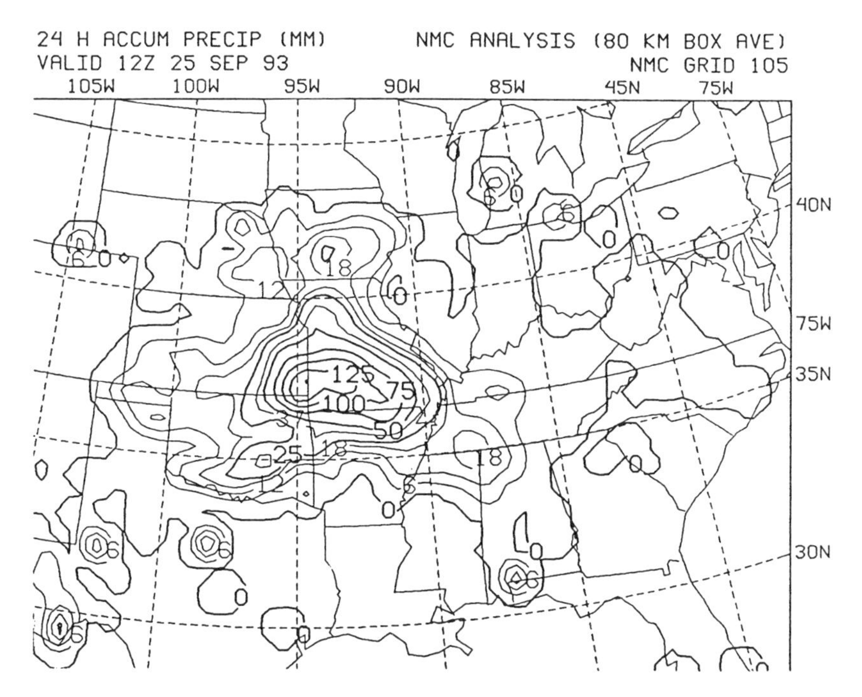

Fig. 6 A section of the NMC's 24-h accumulated precipitation analysis, valid at 1200 UTC 25 September 1993. Contours are drawn at intervals of 6.25 mm up to 25 mm, at 12.5 mm between 25 and 50 mm, and at 25 mm for amounts greater than 50 mm/24 h.

Eta/sigma; the Eta and the Eta/sigma forecasts were, at 2″, the least skillful of the three forecasts with positive skill for at least one of the models in the lower right panel of Fig. 4. Sections of the sea level pressure forecasts by each of the four models are shown in Fig. 7. Following the appearance of an inverted trough with perhaps an incipient frontal development centred over northeastern New Mexico 12 h earlier, all models are seen to have forecast the development of an organized low. They tend, however, to place the main centre too far to the east-northeast, the error being particularly conspicuous in the NGM forecast, upper left panel.

The feature that may have had the most impact on the associated precipitation forecasts however is the position of the front as it cuts across the Oklahoma-Texas region. Note that the front is much too slow in both the NGM and the Eta/sigma forecasts, upper left and lower left panel, respectively. The Avn model, upper right panel, had moved the front faster. The placement of the front by the Eta, lower right panel, is still more accurate; demonstrating that most of the Eta/sigma error in the position of the front was due to its use of the sigma coordinate.

Sections of the associated precipitation forecasts by each of the four models are shown in Fig. 8. The Eta/sigma model had forecast maximum precipitation less than 2″/24 h, and had its maximum placed to the east of the analyzed location.

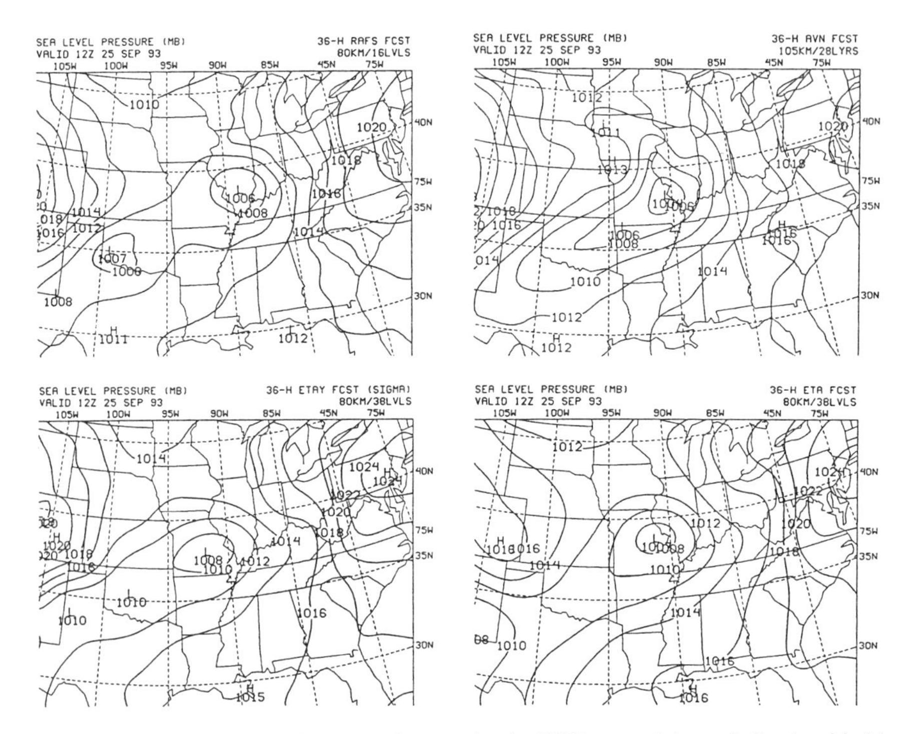

Fig. 7 Sections of the sea level pressure forecasts by the NGM, upper left panel, the Avn Model, upper right panel, the "Eta/sigma" model, lower left panel, and the Eta Model, lower right panel, all verifying at 1200 UTC 25 September 1993. Isobars are plotted at intervals of 2 mb. Eta/sigma is using a "standard" downward reduction of pressure to sea level, similar to those of the NGM and the Avn models; as opposed to the "horizontal" reduction of the Eta.

Using the eta coordinate, the model produced precipitation greater than 2″/24 h, with the maximum shifted westward and the 2″ contour encompassing the analyzed maximum. We anticipate that these changes have resulted from a more accurate position of the front, allowing for an intensified flow of moist air from the Gulf region, feeding into the developing cyclone.

An issue which might be raised looking at Fig. 8 is that of spurious precipitation at mountain tops in the sigma mode, given the precipitation maximum of the Eta/sigma forecast, lower left panel, over the highest Appalachians in southwestern North Carolina. Indeed, the bias scores of the Eta/sigma forecasts were typically a few percent higher than those of the Eta, which may primarily have resulted from horizontal diffusion along sloping sigma surfaces. For an assessment of the possible impact of this error one may look at the category with a maximum bias difference of the two modes, of 0.75″/24 h. At that category the bias difference was roughly 13%, with 1241 forecast points in the sigma mode vs. only 1102 points in the eta mode. In spite of this substantial difference in the number of forecast points, the

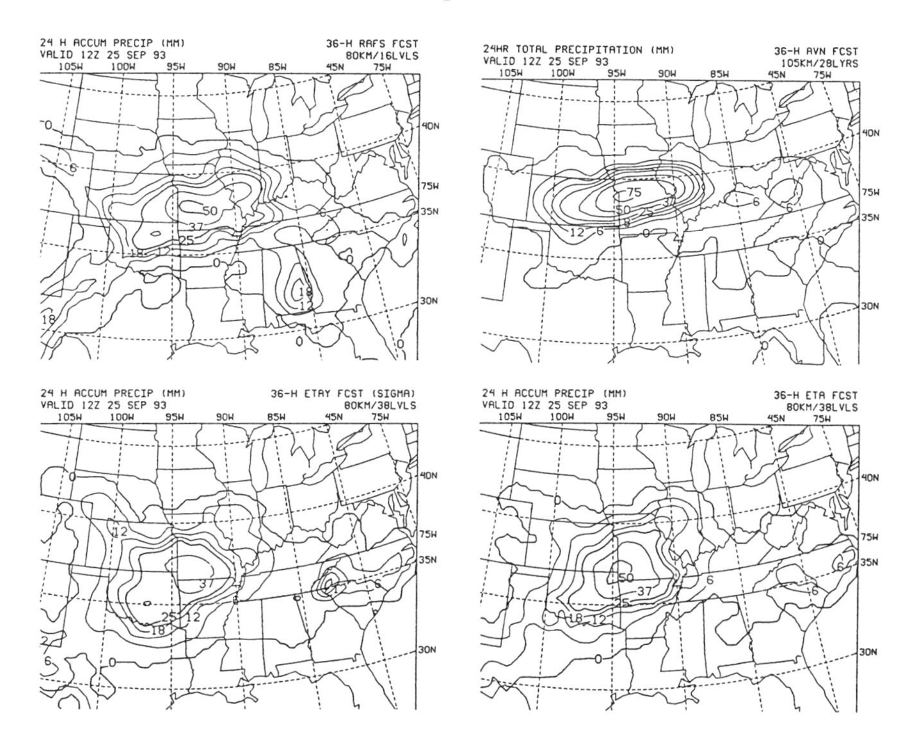

Fig. 8 Sections of the 12–36 h accumulated precipitation forecasts by the NGM, upper left panel, the Avn Model, upper right panel, the "Eta/sigma" model, lower left panel, and the Eta Model, lower right panel, all verifying at 1200 UTC 25 September 1993. Contours are drawn at the same intervals as used for the analysis in Fig. 6.

number of correctly forecast points above chance was greater in the eta mode, 324, compared to 319 in the sigma mode. Thus, we feel that the horizontal diffusion error did not have a major impact on the results shown in Figs 3 and 4.

For a more detailed view of the area of maximum precipitation of Fig. 8, magnified sections of the two Eta Model forecasts are shown in Fig. 9. The analyzed contour of 1″/24 h is also shown (heavy dashed lines), as well as areas where precipitation >1″/24 h was forecast but not observed (dotted), and observed but not forecast (hatched). The sigma mode forecast of precipitation >1″/24 h is seen to cover an area generally displaced southwestward with respect to the analyzed area. This problem is removed in the eta mode forecast by way of an overall northeastward displacement of the forecast area to a position in which the existence of a specific placement error of the forecast area is virtually invisible. The area covered by the forecast precipitation >1″/24 h has thereby remained the same, covering 35 grid boxes, or points, in both cases. The number of hits has however increased, from 23 to 26 points. This has resulted in a significant improvement in the equitable threat score, from 0.426 in the sigma mode, to 0.516 in the eta mode. Note that for

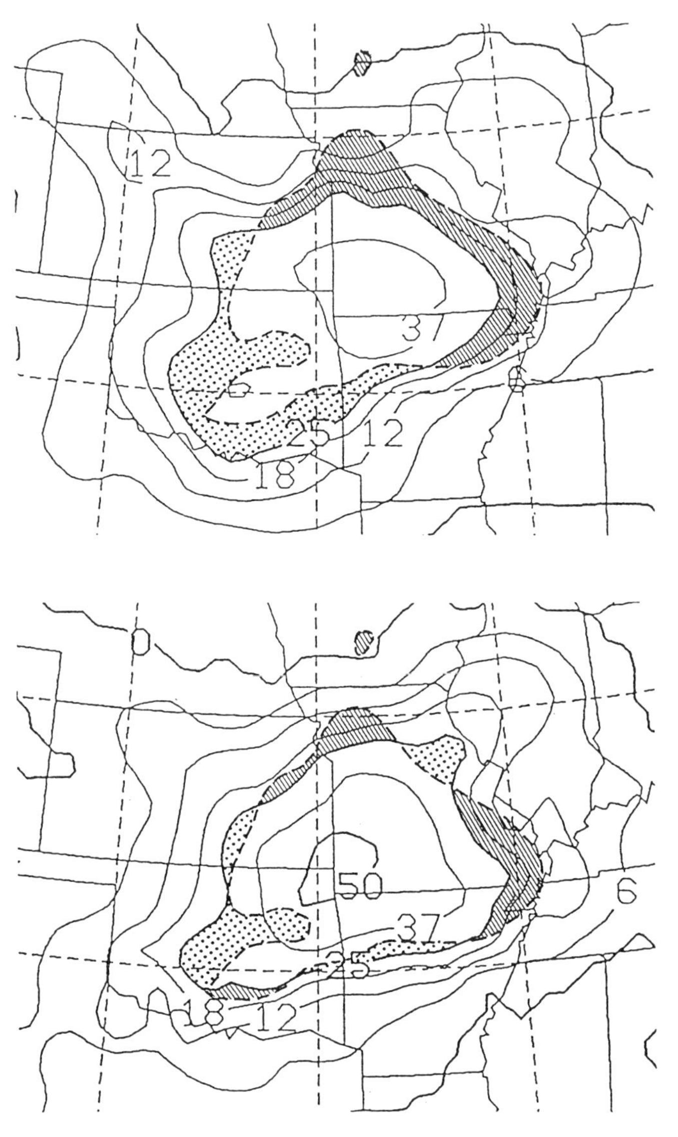

Fig. 9 Magnified view of sections of the Eta Model forecasts shown in Fig. 8, covering the area of maximum analyzed precipitation. The sigma mode forecast is shown in the upper panel, and the eta mode in the lower panel. The analyzed contour of 1″ (25 mm)/24 h, heavy dashed line, is also shown. Dots denote areas where precipitation >1″/24 h was forecast but not observed, while hatching denotes areas where the opposite occurred, that precipitation >1″/24 h was observed but not forecast. Area inside the shaded areas is one where precipitation >1″/24 h was forecast and also observed.

each mode both are the second best scores of the 1″ category, shown in the lower left panel of Fig. 4.

It is perhaps worth noting that the change from sigma to the eta mode has resulted in a displacement of the area of more intense precipitation, that of >1.5″ (37 mm)/24 h and the maximum itself, westward, in a direction which is nearly opposite to the northeastward displacement of the >1″ area. The placement of each of these features was more accurate in the eta mode forecast. A particularly large improvement of the score resulted from the expansion of the >1.5″ (37 mm) area westward, northward and eastward, with an increase in hits from 7 to 17 and in the equitable threat score from 0.226 to 0.578, the second best score of this 16-forecast experiment.

5 A midtropospheric cutoff case

One more mountain-related case which caught our attention was that of the midtropospheric cutoff over the southern Rockies on 12 March 1994, associated with very rapid ridging to the north of it off the U.S. west coast and northwestern states. The situation had the NMC's operational forecasters face a need to choose which to favour among different 48-h forecasts from three operational models verifying at 1200 UTC 12 March: NGM was the slowest in intensifying the ridge and had the deepest cutoff, centred at the Arizona-New Mexico border (Fig. 10, upper left panel); the Aviation model, at the same verification time, placed the cutoff further east over southern New Mexico (middle left panel), while the Eta placed it farthest to the west, over southern central Arizona (lower left panel). The 36-h forecasts from the three models verifying at the same time (Fig. 10, right panels) were virtually identical to the 48-h forecasts, bringing no change in the features mentioned. This occurred even with the 24-h forecasts (Fig. 11, left panels), each model in a way insisting on its forecast for three consecutive initial conditions. Finally, at 12 h prior to the 1200 UTC verification time, the models produced forecasts little different from each other, all three placing the cutoff over southern Arizona. This, of course, is what actually happened, as seen in the NMC's 500-mb geopotential height analysis shown in Fig. 12.

We have rerun the 48-h Eta Model forecast using the sigma-coordinate option of the code; the relevant section of the 500-mb geopotential height forecast is shown in Fig. 13. The forecast has largely reproduced the errors of the NGM (Fig. 10, upper left panel), centring the cutoff roughly at the Arizona-New Mexico border, and failing to intensify the ridge to the north of it fast enough. The cutoff is also too deep, even though not quite as deep as that of the NGM forecast.

6 Concluding remarks and outlook for further progress

It may be appropriate on this occasion to briefly present our views on outlook for further progress regarding the Eta Model and in weather prediction in general through developments in numerical methods. To this end we shall first make a few

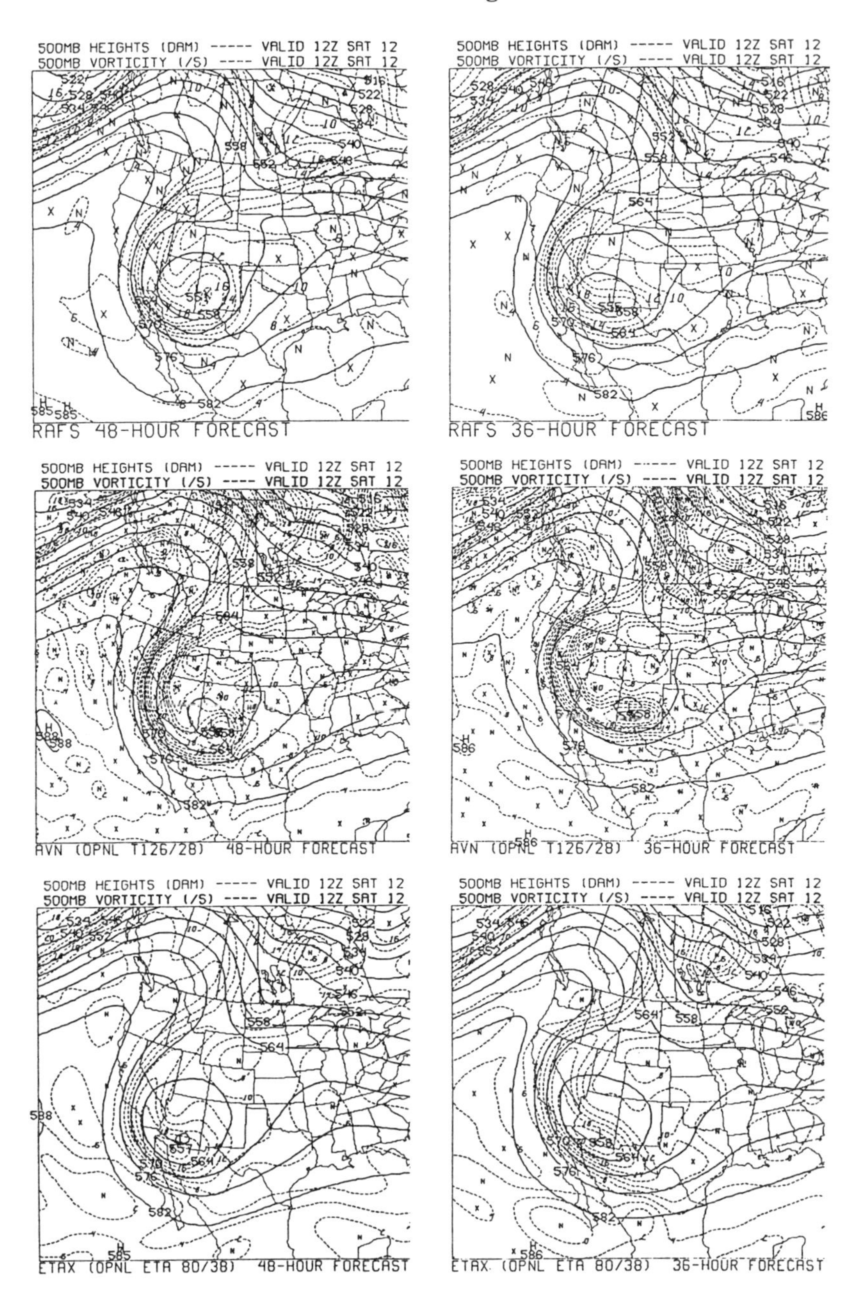

Fig. 10 Sections of the 500-m geopotential height (dam) and vorticity (s^{-1}) forecasts by the NGM, upper panels, the Avn Model, middle panels, and the Eta Model, lower panels, all verifying at 1200 UTC 12 March 1994. Left panels show 48-h and the right panels 36-h forecasts.

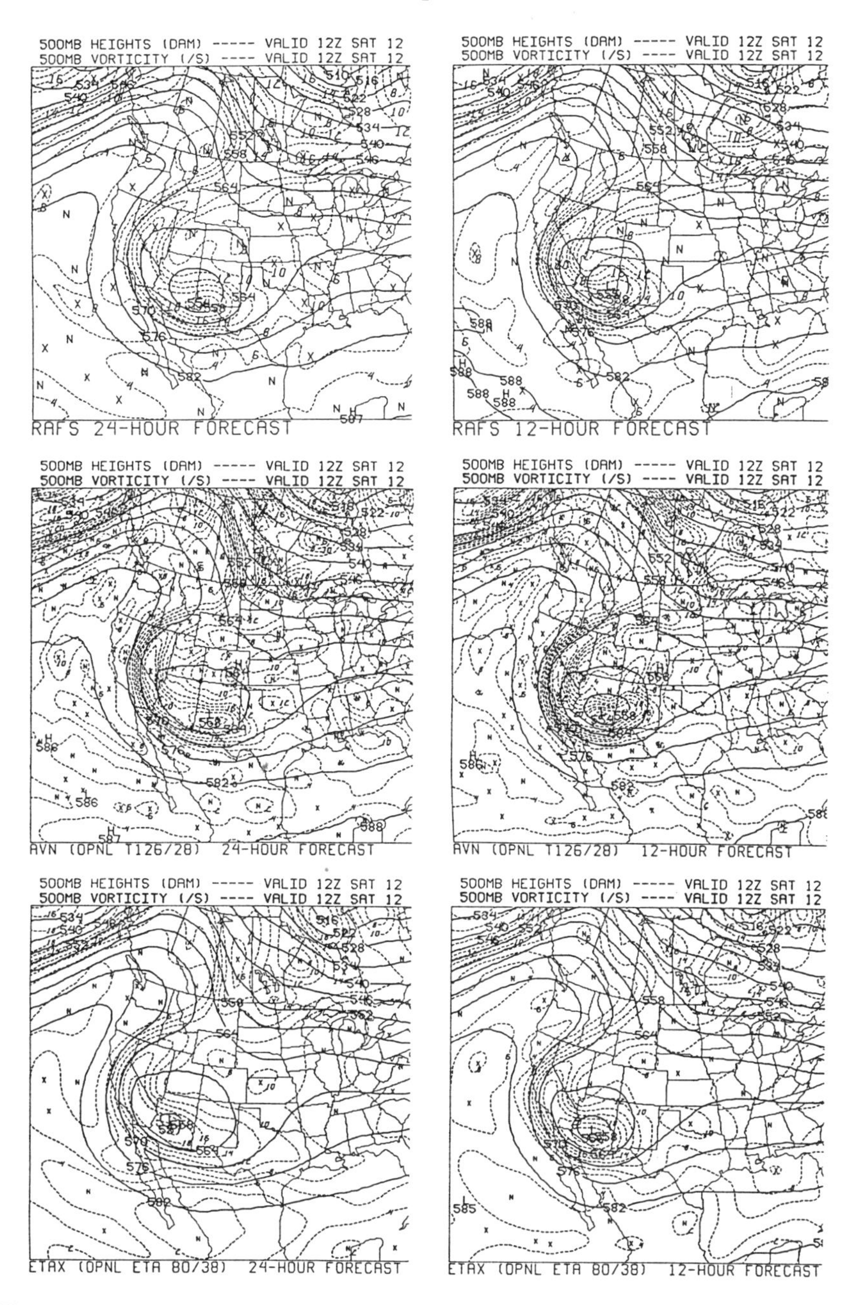

Fig. 11 Same as Fig. 10, except for the 24-h forecasts (left panels) and 12-h forecasts (right panels).

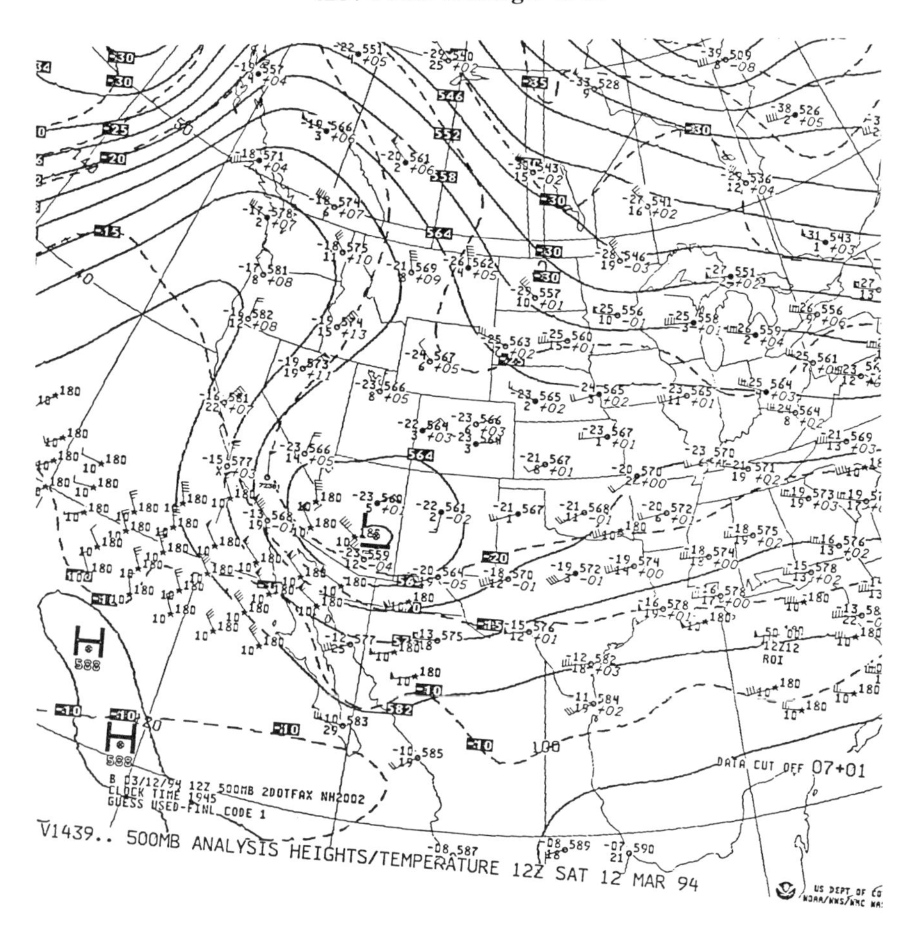

Fig. 12 A section of the NMC's 500-mb geopotential height and temperature analysis, valid at 1200 UTC 12 March 1994.

concluding remarks on two numerical methods factors from the preceding sections, namely, resolution and the choice of the vertical coordinate. We shall then comment on three additional areas which we believe are offering or may offer promise of additional increase in skill. These are the choice of the vertical grid, numerics of the propagation of gravity-inertia waves, and the issue of the box-average vs. point-sample treatment of predicted variables.

• One might consider it obvious that an increase in resolution – horizontal and in case of the eta coordinate also the vertical – has to result in an increase in skill because of the mere fact of achieving a more realistic representation of mountains. This, however, cannot be taken for granted as convergence problems might appear associated with an increase in resolution. For example, it seems to be general experience that models are becoming noisier as the resolution increases

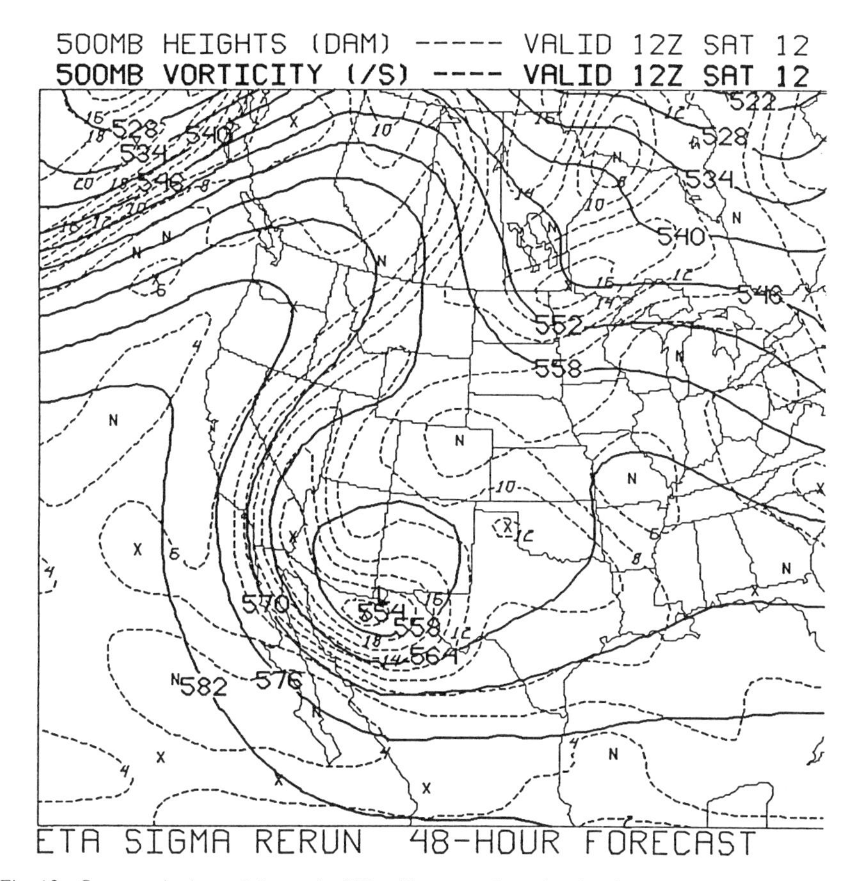

Fig. 13 Same as the lower left panel of Fig. 10, except for using the sigma-coordinate option of the Eta Model code.

(e.g., Haugen, 1994), requiring the use of a greater diffusion coefficient – in contrast to expectations of Mellor (1985). Therefore, we see the results of our resolution experiment shown in Fig. 2 as encouraging in the sense that convergence problems were not encountered. We anticipate additional benefits from increased resolution to be achieved by the 29 km/50 layers version of the Eta Model, which at the time of this writing is about to be implemented at NMC on a semi-operational basis. It will be run on a 70 × 50 deg longitude × rotated latitude domain, and will have six three-dimensional prognostic variables (Black, 1994; Zhao and Black, 1994).

- While we believe we have presented convincing evidence of substantial benefits derived from the eta vertical coordinate, at the same time more than one group has been advocating an isentropic/sigma approach (e.g., Bleck and Benjamin, 1993; Arakawa and Konor, 1994). An isentropic/eta approach is an obvious path to explore with the idea of capturing the best of both worlds to the extent that this can be done. We expect that the eta system would have to be retained in the lower troposphere, and that a predominantly isentropic coordinate could be beneficial as one approaches the tropopause and of course further on in the stratosphere.

Work on the use of the eta coordinate in a GCM is in progress at GFDL (B. Wyman, personal communication). An eta global model with no Fourier filtering – say via a revival of Sadourny's expanded-cube approach – could be a very promising next milestone for which to strive (Mesinger and Janjić, 1990). Highly encouraging results have in fact recently been obtained by Rančić et al. (1996) in shallow-water integrations using two versions of horizontal grid on the expanded cube, and performing the customary Rossby-Haurwitz tests.

- Increasing evidence seems to be appearing of the overall advantages of the Charney-Phillips as opposed to the Lorenz vertical staggering (Arakawa and Konor, 1994, 1996; Cullen et al., this volume). Obviously, this is a direction which can be explored in the case of the Eta Model.
- In the Eta Model a lattice-coupling scheme is used preventing separation of gravity wave solutions on two C-subgrids of the Arakawa E-grid (Mesinger, 1973; Janjić 1979). The scheme has damping properties. While the damping seems negligible for the geostrophic mode and becomes appreciable only for the smallest-scale gravity waves (e.g., Janjić and Mesinger, 1984), removal of this damping might nevertheless be beneficial. An elementary dynamic meteorology relationship states that in a hydrostatic atmosphere the ratio of the internal to gravitational potential energy is constant at approximately 71 to 29, except for the effect of a boundary term which is zero for a horizontal lower boundary. Consider now in a thought experiment a grid-column of a hydrostatic model receiving an amount of heat ΔQ as a result of a "physics" call: 29% of this heat will instantaneously be converted into the gravitational potential energy of the column. As a result the column will have an excess of potential energy relative to surrounding columns, generating outward-directed pressure gradient forces which will initiate a geostrophic adjustment process. While we of course do not know what fraction of the energy increment ΔQ will be retained by the column when the geostrophic adjustment is completed, one perhaps should not be surprised if it were not radically different from the original increment of the internal energy only. In other words, a substantial fraction of the heating resulting from model physics may go into energy which is radiated outward from various model grid columns by way of gravity-inertia waves so that the issue of these waves being damped to a considerable degree due to properties of the numerical schemes used could be a matter of importance.

Note that this subject, from the point of view of what happens in the continuous atmosphere as well as in various models, has recently started to receive increased attention (e.g., Nicholls and Pielke, 1994a, 1994b; Shutts, 1994). One ought to keep in mind that compared to the actual atmosphere in numerical models this issue is exacerbated by the fact that forcing by model physics is done at individual grid boxes while at the same time various numerical schemes are radically different in the way they treat such a forcing (e.g., Janjić and Mesinger, 1989).

• While difference schemes for the most part treat grid-point values as point samples of functions which can be expanded into Taylor series, model physics is done with predictive variables considered to represent averages over model grid boxes. Note that implied discontinuities at grid-box boundaries make the validity of Taylor expansions questionable. For example, the issue could be raised as to the reasons for the fourth-order accuracy of the NGM (Juang and Hoke, 1992) clearly not making an overwhelming contribution to the performance of the model in comparison with the second-order accuracy Eta. Characteristics based schemes can allow for discontinuities between grid boxes and thus do not suffer from this inconsistency (e.g., Carpenter et al., 1990) but they have problems of their own in the form of a difficulty in the two dimensional case. While it is not now obvious what specific harm might be due to this difference in treatment or what options it would be useful to pursue in order to achieve a higher degree of consistency between the two, the mere awareness of a potential problem could encourage efforts towards a better understanding of the issues involved.

One might hope that problems of this kind, if present, would diminish as model resolution is increased. As this is by no means obvious, we believe that tests against model-independent data such as those of our Figs 1–4 will be essential in efforts to establish preference among various directions such as those on which we have commented here. Verification on grid-box average precipitation data, consistent with the formulation of model physics and resistant to deceptive improvement via artificial smoothing of model fields, we find exceptionally useful in searching for further progress.

Acknowledgments

This research would not have been possible without recourse to the extraordinary research environment of the Development Division and of other unique facilities at the U. S. National Meteorological Center, Camp Springs, Maryland. We want to express our gratitude to Drs. Eugenia Kalnay and Ron McPherson for their ongoing support of this effort.

References

ARAKAWA, A. and C.S. KONOR. 1994. A generalized vertical coordinate and the choice of vertical grid for atmospheric models. In: *The Life Cycles of Extratropical Cyclones*, Vol. III, Bergen, Norway, Geophys. Inst., Univ. Bergen, pp. 259–264.

——— and ———. 1996. Vertical differencing of the primitive equations based on the Charney-Phillips grid in hybrid σ-p vertical coordinate. *Mon. Weather Rev.* **124**: 511–528.

BLACK, T.L. 1994. The new NMC mesoscale Eta Model: Description and Forecast Examples. *Weather Forecast.* **9**: 265–278.

BLECK, R. and S.G. BENJAMIN. 1993. Regional weather prediction with a model combining terrain-following and isentropic coordinates. Part I: Model Description. *Mon. Weather Rev.* **121**: 1770–1785.

CARPENTER, R.L. JR.; K.K. DROEGEMEIER, P.W. WOODWARD and C.E. HANE. 1990. Application of the piecewise parabolic method (PPM) to meteorological modeling. *Mon. Weather Rev.* **118**: 586–612.

DIMEGO, G.J.; K.E MITCHELL, R.A. PETERSEN, J.E. HOKE, J.P. GERRITY, J.J. TUCCILLO, R.L. WOBUS and H.-M.H. JUANG. 1992. Changes to NMC's regional analysis and forecast system. *Weather Forecast.* **7**: 185–198.

GANDIN, L.S. and A.H. MURPHY. 1992. Equitable skill scores for categorical forecasts. *Mon. Weather Rev.* **120**: 361–370.

HAUGEN, J.E. 1994. Experience with horizontal diffusion in DNMI LAM models. Results from parallel runs of the Eulerian and semi-Lagrangian HIRLAM models. HIRLAM 3 Workshop on Numerical Integration Techniques, Oslo, Norway, 26–27 January Swedish Meteorological and Hydrological Institute, Norrköping, Sweden, pp. 32–33.

JANJIĆ, Z.I. 1979. Forward-backward scheme modified to prevent two-grid-interval noise and its application in sigma coordinate models. *Contrib. Atmos. Phys.* **52**: 69–84.

———. 1984. Non-linear advection schemes and energy cascade on semi-staggered grids. *Mon. Weather Rev.* **112**: 1234–1245.

———. 1994. The step-mountain eta coordinate model: Further developments of the convection, viscous sublayer, and turbulence closure schemes. *Mon. Weather Rev.* **122**: 927–945.

——— and F. MESINGER. 1984. Finite difference methods for the shallow water equations on various horizontal grids. *In:* Numerical Methods for Weather Prediction, Vol. 1. Seminar 1983, ECMWF, Shinfield Park, Reading, U.K., pp. 29–101.

——— and ———. 1989. Response to small-scale forcing on two staggered grids used in finite-difference models of the atmosphere. *Q.J.R. Meteorol. Soc.* **115**: 1167–1176.

JARRAUD, M.; A.J. SIMMONS and M. KANAMITSU. 1987. The concept, implementation and impact of an envelope orography. Observation, Theory and Modelling of Orographic Effects, Vol. 2, Seminar/workshop 1986, ECMWF, Shinfield Park, Reading, U.K., pp. 81–127.

JUANG, H.-M.H. and J.E. HOKE. 1992. Application of fourth-order finite differencing to the NMC Nested Grid Model. *Mon. Weather Rev.* **120**: 1767–1782.

KANAMITSU, M.; J.C. ALPERT, K.A. CAMPANA, P.M. CAPLAN, D.G. DEAVEN, M. IREDELL, B. KATZ, H.-L. PAN, J. SELA and G.H. WHITE. 1991. Recent changes implemented into the global forecast system at NMC. *Weather Forecast.* **6**: 425–435.

MELLOR, G.L. 1985. Ensemble average, turbulence closure. *Advances in Geophysics, Issues in Atmos. and Ocean Modeling. Part B: Weather Dynamics*. S. Manabe (Ed.), Academic Press, pp. 345–357.

MESINGER, F. 1973. A method for construction of second-order accuracy difference schemes permitting no false two-grid-interval wave in the height field. *Tellus*, **25**: 444–458.

———; Z.I. JANJIĆ, S. NICKOVIC, D. GAVRILOV and D.G. DEAVEN. 1988. The step-mountain coordinate: model description and performance for cases of Alpine lee cyclogenesis and for a case of Appalachian redevelopment. *Mon. Weather Rev.* **116**: 1493–1518.

——— and ———. 1990. Numerical methods for the primitive equations (space). *In:* 10 Years of Medium-Range Weather Forecasting, Vol. I, Seminar 1989, ECMWF, Shinfield Park, Reading, U.K., pp. 205–251.

——— and T.L. BLACK. 1992. On the impact on forecast accuracy of the step-mountain (eta) vs. sigma coordinate. *Meteorol. Atmos. Phys.* **50**: 47–60.

———; ——— and M. BALDWIN. 1993. Simulations of a December 1992 severe East Coast storm ("Nor'easter 1") with the Eta Model. Part I: Sensitivity to model formulations. Thirteenth Conf. on Weather Analysis and Forecasting, Vienna, VA, 2-6 August 1993, Am. Meteorol. Soc., pp. 164–165.

NICHOLLS, M.E. and R.A. PIELKE. 1994a. Thermal compression waves. I: Total-energy transfer. *Q.J.R. Meteorol. Soc.* **120**: 305–332.

——— and ———. 1994b. Thermal compression waves. II: Mass adjustment and vertical transfer of total energy. *Q.J.R. Meteorol. Soc.* **120**: 333–359.

PAN, H.-L. and W.-S. WU. 1994. Implementing a mass-flux convective parameterization package for the NMC Medium-Range Forecast Model. Tenth Conf. on Numerical Weather Prediction, Portland, OR, 18-22 July 1994, Am. Meteorol. Soc., pp. 96–98.

RAMAGE, C.S. 1993. Forecasting in meteorology. *Bull. Am. Meteorol. Soc.* **74**: 1863–1871.

RANČIĆ, M.; R.J. PURSER and F. MESINGER. 1996. A

global shallow-water model using an expanded spherical cube: Gnomonic versus conformal coordinates. *Q.J.R. Meteorol. Soc.* **122**: 959–982.

SCHAEFER, J.T. 1990. The critical success index as an indicator of warning skill. *Weather Forecast.* **5**: 570–575.

SHUTTS, G. 1994. The adjustment of a rotating, stratified fluid subject to localized sources of mass. *Q.J.R. Meteorol. Soc.* **120**: 361–386.

ZHAO, Q. and T.L. BLACK 1994. Implementation of the cloud scheme in the Eta Model at NMC. Tenth Conf. on Numerical Weather Prediction, Portland, OR, 8-22 July 1994, Am. Meteorol. Soc., pp. 331–332.

An Overview of Numerical Methods for the Next Generation U.K. NWP and Climate Model

M.J.P. Cullen, T. Davies, M.H. Mawson, J.A. James and S.C. Coulter
Meteorological Office
London Road, Bracknell, Berks., U.K.

and

A. Malcolm
Department of Mathematics,
University of Reading,
Whiteknights, Reading, U.K.

[Original manuscript received 1 November 1994; in revised form 31 July 1995]

ABSTRACT *The U.K. Meteorological Office now uses a single model for atmospheric simulation and forecasting from all scales from mesoscale to climate. The constraints which numerical methods for such a model have to satisfy are described. A new version of the model is being developed with the aims of improving its accuracy by better treatment of the 'balanced' part of the flow, and increasing its applicability by including non-hydrostatic effects. Unusual features of this version are the use of the Charney-Phillips grid in the vertical, to improve the geostrophic adjustment properties, and the method of constructing the semi-implicit algorithm for solving the fully compressible equations. Idealized tests of these two aspects of the scheme are presented, showing that the Charney-Phillips grid reduces spurious gravity wave generation without compromising the treatment of the atmospheric boundary layer, and that the semi-implicit integration scheme can give stable solutions without the need for added temporal diffusion.*

RÉSUMÉ *Le Bureau météorologique du Royaume-Uni utilise maintenant un seul modèle pour la simulation et la prévision atmosphérique, de la mésoéchelle à l'échelle climatique. On décrit les contraintes que les méthodes numériques employées dans un tel modèle doivent satisfaire. On est à élaborer une nouvelle version du modèle qui devrait améliorer sa précision par un meilleur traitement de la partie « équilibrée » de l'écoulement et par une augmentation de son applicabilité, en incluant les effets non hydrostatiques. Cette version possède des caractéristiques inhabituelles, telles que la grille Charney-Phillips dans la verticale pour améliorer les propriétés d'ajustement géostrophique, et la méthode de contruire l'algorithme semi-implicite pour solutionner les équations entièrement compressibles. On présente des essais théoriques des deux aspects du schéma qui montrent que la grille Charney-Phillips réduit la génération d'ondes de gravité non essentielles sans compromettre le traitement de la couche limite atmosphérique et que le schéma d'intégration semi-implicite peut donner des solutions stables sans la nécessité d'ajouter une diffusion temporelle.*

1 Introduction

The U.K. Meteorological Office has used the same atmospheric model for all forecasting and climate simulation applications since 1992 (Cullen, 1993). This is both a result of the need to provide a large range of forecast and advisory services efficiently, and also because of the belief that the same scientific methods of simulating atmospheric behaviour will be appropriate regardless of the application. The need for such 'universal methods' is widely recognized and is even more essential when considering models which have large resolution variation within a single run, such as the stretched grid Action de Recherche Petite Echelle Grande Echelle (ARPEGE) system used operationally by Météo-France and the European Centre for Medium-Range Forecasts (Courtier and Geleyn, 1988).

In this paper we discuss the numerical methods used in this 'unified' model. We initially summarize the methods used in the first version of the model, and the reasons for the choices. Though the model was introduced operationally in 1991–2, the choices of formulation to be used had to be largely finalized in 1989. Since then, there has been a great deal of development of improved numerical algorithms. An example is the greatly increased acceptance of the semi-Lagrangian method for treatment of advection. In this paper we therefore discuss the numerical techniques proposed for the next major upgrade of the 'unified' model. The two main themes are, seeking improved model performance by more accurate treatment of the balanced part of the flow, and increasing the applicability of the model to small scales by including non-hydrostatic effects. The resulting design is described in Sections 3 and 4, and idealized tests of some aspects of it are illustrated in Section 5.

2 The current unified model integration scheme

Any integration scheme used for a forecast model has to be very efficient, because timeliness is a key factor, and the benefits of high horizontal and vertical resolution have been frequently demonstrated. In climate modelling, accuracy at low resolution is very important. Conservation properties are also desirable, both to ensure satisfactory long-term integration behaviour, and to allow proper studies of the thermodynamic and energy budgets from control and perturbation runs in climate change experiments. The integration scheme used in the initial version of the unified model was chosen to be as close to the existing methods used in the U.K. Meteorological Office as possible, while meeting the above requirements. Thus the model used finite difference methods, with a latitude-longitude grid. The algorithm was based on the very efficient split-explicit scheme of Gadd (1978) which was already used for global and limited area forecasting. This was adapted to meet climate model requirements by making it conservative. The key steps were to compute the gravity wave terms, including vertical advection of a basic state potential temperature profile, in short time steps, and using the average mass-weighted velocity from the short time steps in calculating the advection terms. Time-smoothing is applied to the fields within the sequence of short time steps. The advection terms are approximated by a two step second or fourth order Heun scheme. The method is described in detail by Cullen and Davies (1991). Fourier filtering is used to keep the model stable at high latitudes. This is done conservatively by filtering mass-weighted increments to the thermodynamic variables and mass-weighted velocity fields. A conservative diffusion term is used to remove small scale noise. The use of a deformation dependent nonlinear diffusion scheme of the form $\frac{1}{\mu}\nabla\cdot\mu K(u)\nabla$, where μ is a mass-weighting term, was found to be insufficiently scale selective for use at low resolution, and a scale selective form $(\frac{1}{\mu}\nabla\cdot\mu\nabla)^n$, where n is usually chosen to be 2 or 3, was used instead. In order to increase the accuracy, particularly of global climate integrations, the more accurate form of the primitive equations discussed by White and Bromley (1995) was used.

The performance of the combined advection, diffusion and filtering scheme is illustrated on one of the test problems introduced by Williamson et al. (1992). The fourth order approximation to advection is used. The advection scheme used on its own generates large oscillations if the advecting velocity is not parallel to a line of latitude, but it is the combination of the schemes that is actually approximating the transport in the full numerical model. Figure 1 compares the performance of the scheme with the alternative of using a 'monotone' advection scheme (Morton and Sweby, 1987), and a semi-Lagrangian advection scheme (Bates et al., 1990) where the advecting velocity is at an angle of 15°N to the lines of latitude. A 96 × 73 grid has been used. Both the unified model scheme (Fig. 1a) and the semi-Lagrangian scheme (Fig. 1c) produce overshoots, as indicated by the oscillations in the 1000 contour. Those produced by the unified model scheme are more coherent as expected from the consistent phase error for short wavelengths. The monotone scheme (Fig. 1b) produces no overshoots. The unified model scheme performs better than the others in retaining peak amplitude, but is the worst in the distortion of the shape. Figure 2 shows the variation of the r.m.s. error with diffusion coefficient. Note that the error for the optimum range of diffusion coefficient is lower than that given by the monotone scheme. The unified model scheme requires a time step much lower than that required for linear stability. However, when the scheme is used in the complete model, dispersion of noise by gravity waves allows the full expected time step to be used and there is no evidence that the performance is significantly improved by reducing the time step below the value needed for stability. The use of the fourth order Heun scheme is essential to obtain results of this quality in the test problem. However, the sensitivity of the complete model to the choice between second and fourth order schemes at forecast resolutions (grid lengths less than 100 km) has been slight.

The performance of the unified model is found to be remarkably insensitive to horizontal resolution in many respects. Figure 3 illustrates the simulation of the southern hemisphere circumpolar jet from 10 year integrations using 96 × 73 and 288 × 217 grids as compared with a climatology derived from U.K. operational analyses. Many other large scale aspects of the model performance, such as the zonal mean temperature cross-sections, are similarly insensitive. This suggests that

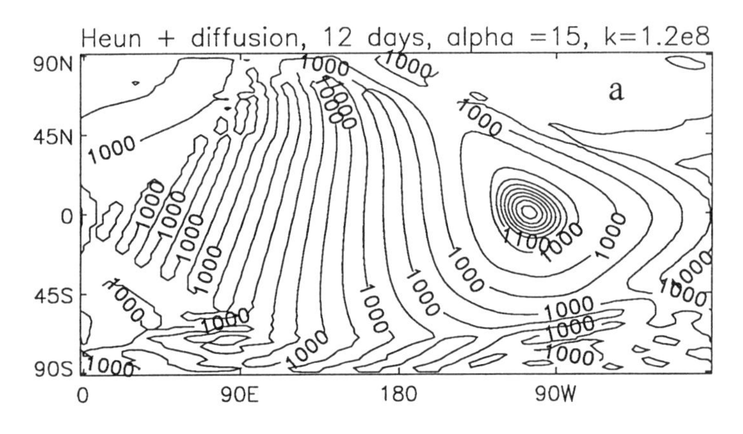

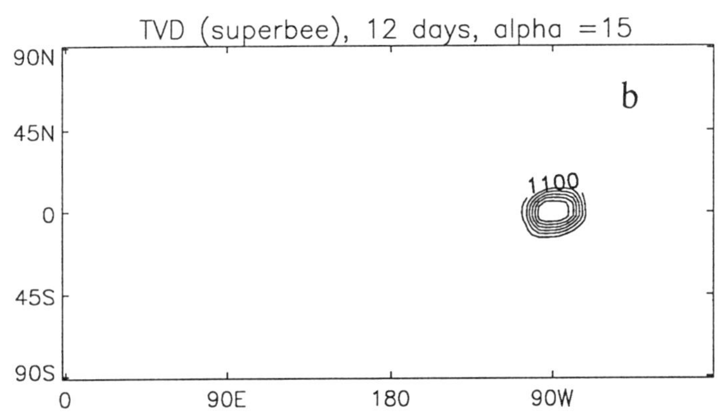

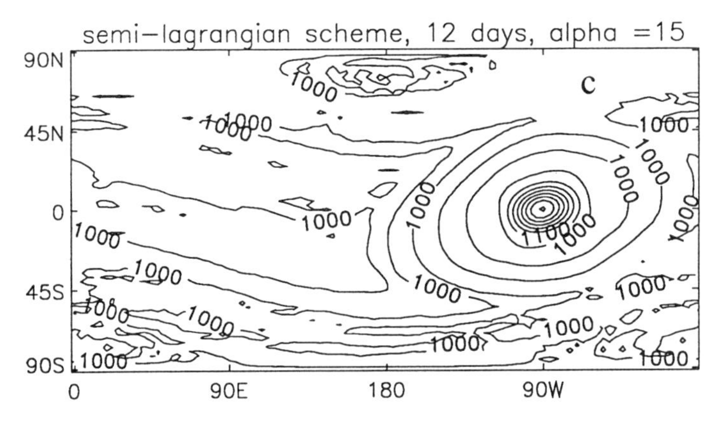

Fig. 1 Solutions for advecting a cosine bell once round a sphere at an angle of 15° to the equator. The initial maximum value is 2000 units. (a) Fourth order Heun scheme with Fourier filtering and fourth order Laplacian diffusion, coefficient 1.2×10^8. (b) Total Variation Diminishing (TVD) scheme with Superbee limiter. (c) Semi-Lagrangian scheme.

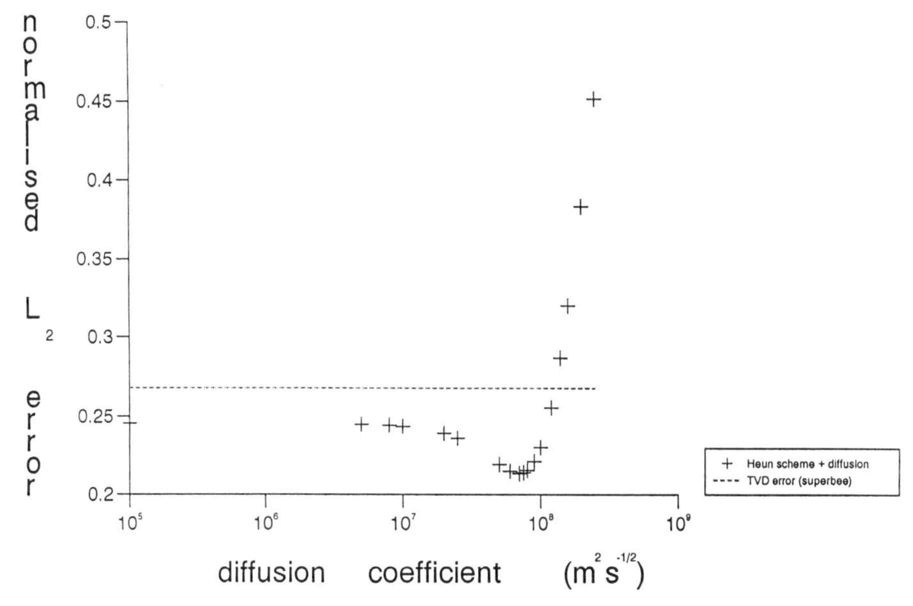

Fig. 2 Root mean square error after one revolution using different diffusion coefficients in the problem of Fig. 1. The dotted line indicates the error of the Superbee scheme.

in seeking further improvements to the model performance it is necessary to review the numerical methods and the physical parametrizations, as well as increasing the resolution further to allow more detail to be simulated.

3 Issues for the proposed new integration scheme

a *Overall requirements*

The purpose of the new scheme is to improve the performance of the model and to increase its scope by including non-hydrostatic effects. Noting the limited sensitivity of the high resolution versions of the model to the simulation of advection, a major attempt is made to improve the simulation of the balanced part of the flow, in particular of the geostrophic adjustment process. As well as improving the model's performance in forecast mode, it is hoped that the performance of the data assimilation will be improved, since much forecast error results from inaccurate analyses. Both requirements lead to the use of semi-implicit integration schemes. It is then natural to consider the use of semi-Lagrangian advection. This allows the maximum time step to be used commensurate with accuracy, and as illustrated in Section 2, reduces distortion of the advection when the flow is not aligned with the grid. There have been doubts about the accuracy of semi-Lagrangian methods when applied in low resolution models. However, the detailed investigation reported by

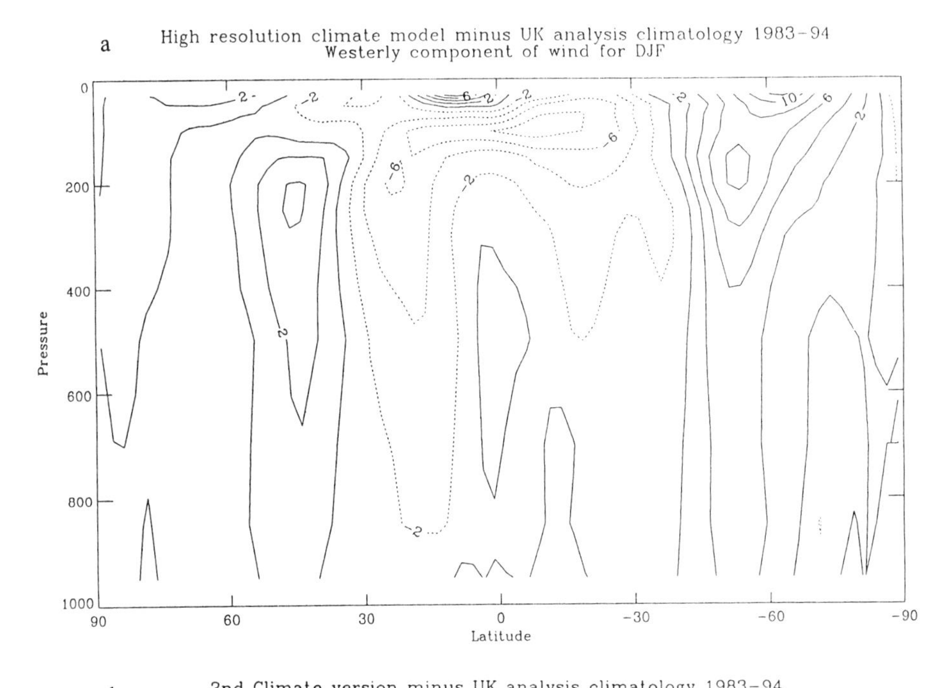

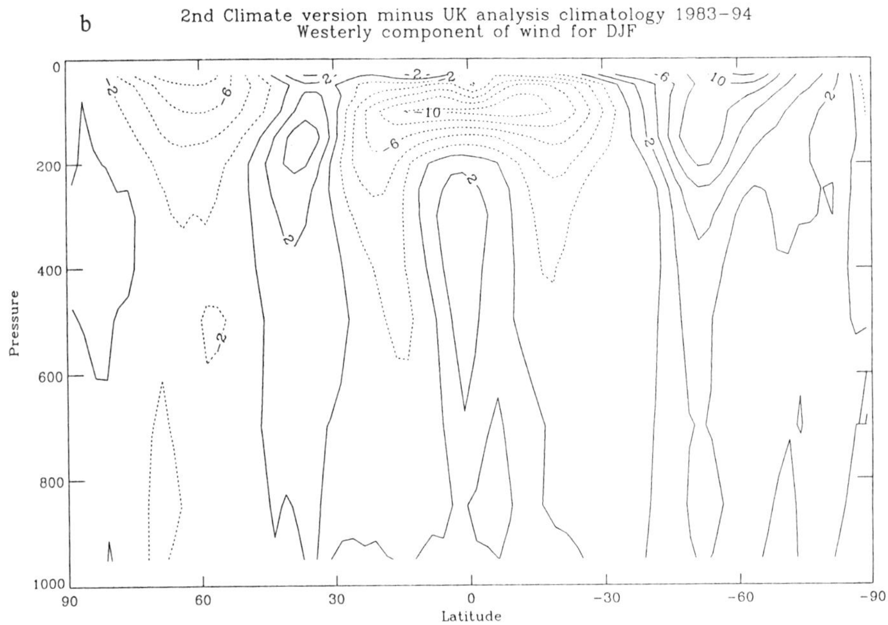

Fig. 3 Cross-section of the difference in zonal wind between multiyear climate model integrations and climatology obtained from U.K. Meteorological Office analyses. (a) High resolution, 288 × 217 grid. (b) Standard resolution 96 × 73 grid.

Williamson and Olson (1994) shows that these losses in accuracy have resulted from using insufficiently accurate interpolation procedures within the method, usually in order to save computer time. In addition, it has been shown, for instance by Priestley (1993), that it is possible to modify the methods so that they satisfy the conservation properties important for climate modelling.

b *Treatment of balanced flow*

There has been considerable study of the finite difference treatment of the geostrophic adjustment process using the shallow water equations, for instance Arakawa and Lamb (1977). This shows that, for non-time-staggered schemes and provided the grid length is less than the Rossby radius of deformation, the 'C' grid arrangement of variables is best, followed by the 'B' grid. Schemes based on vorticity and divergence are at least as good in this respect. There has been much less study of the appropriate treatment of variables in the vertical, because of the tendency to think of a three-dimensional model as a set of shallow water models generated by a decomposition into vertical eigenmodes. However, recent studies such as Arakawa and Moorthi (1987) and Leslie and Purser (1992) have shown that the geostrophic adjustment properties of the 'Charney-Phillips' vertical arrangement of variables are superior to those of the 'Lorenz' arrangement, Fig. 4. This is illustrated by the problem, important in data assimilation, of calculating height increments to balance wind increments. Figure 5 illustrates the error made if we convert wind increments to geostrophically consistent height increments, using a best fitting algorithm on the Lorenz grid, and then recalculating the wind increments geostrophically from the height increments (P. Andrews, private communication). Errors of up to 50% result. On the Charney-Phillips grid this process can be carried out without error.

Though the use of the 'C' grid is well established in finite difference models, the Charney-Phillips grid has been unfashionable recently. One exception is the new Canadian regional model (Tanguay et al., 1990). The reasons are the difficulty of ensuring energy conservation, important in climate modelling, and the more awkward interface to the physics calculations because different variables are held at different places in the vertical column. In particular, the calculations within the boundary layer parametrization may require extra averaging. A method of solving the energy conservation problem is described below and idealized tests of the boundary layer representation are illustrated in Section 5.

Cullen (1989) described and validated numerical methods for the semi-geostrophic equations in a vertical cross-section. This work should be a guide to other aspects of numerical methods important in treating balanced flow accurately. The use of the Charney-Phillips vertical grid was found essential to obtain stable results. Because the equations were implicit in some variables, a semi-implicit method had to be used. Within the method an elliptic equation for a pressure correction was derived and solved, and the results substituted back to complete the update of the other variables. It was necessary to use flow dependent coefficients in the

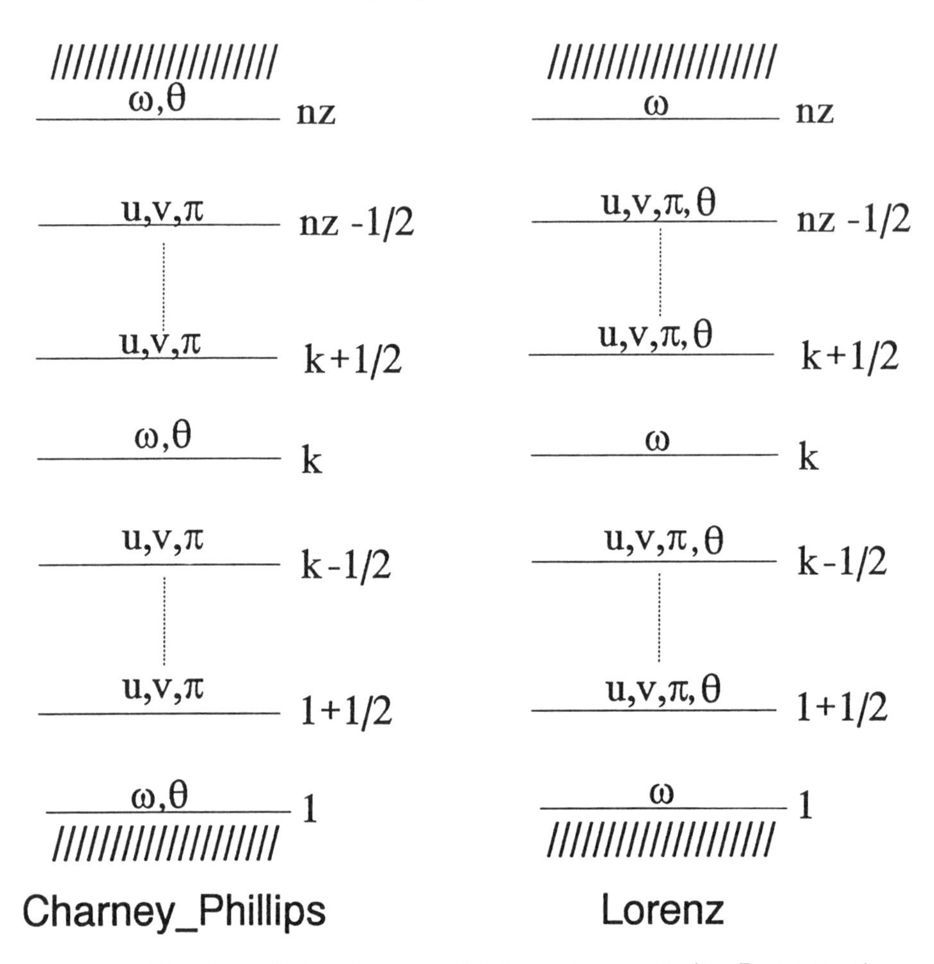

Fig. 4 Position of the variables of the Charney-Phillips and Lorenz grids. (π = Exner pressure).

terms treated implicitly, rather than extracting a constant coefficient problem in the usual way. The coefficients had to be selected to maximize the ellipticity of the pressure correction equation. In addition, accurate treatment of all the components of the pressure gradient term within the implicit step was found necessary to obtain satisfactory solutions over orography. In the two-dimensional problems solved in that paper it was sufficient to solve the variable coefficient elliptic equation by iterating a constant coefficient solver. However, this might not be adequate in a three-dimensional problem.

c *Non-hydrostatic integration schemes*

A number of atmospheric models have recently been extended to include non-hydrostatic effects. Techniques where a pressure-based coordinate is retained have

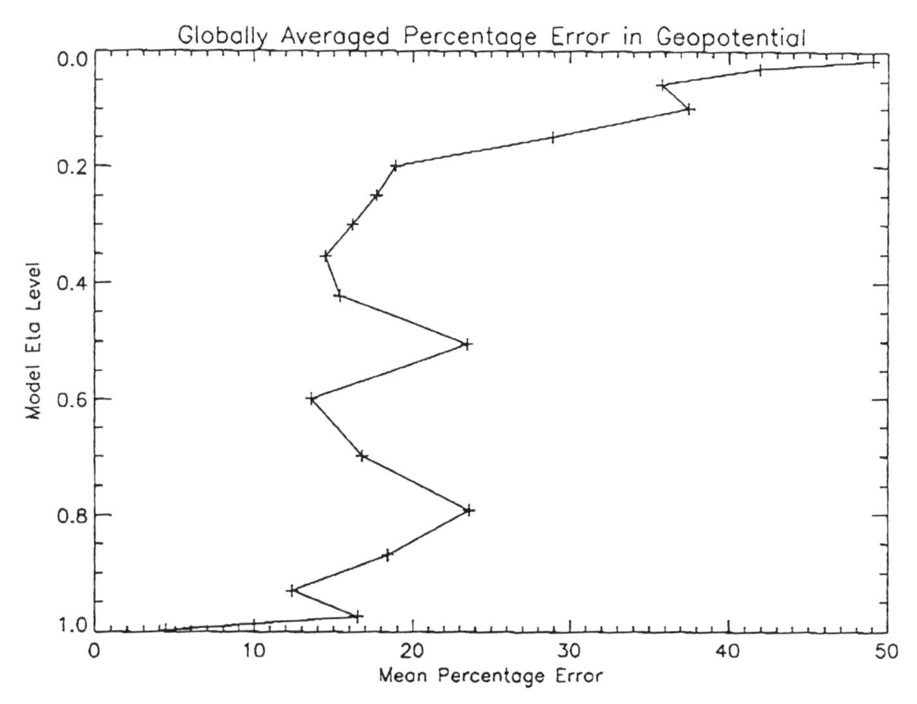

Fig. 5 Comparison of retransformed geopotential increment field with wind derived geopotential increment field. The original geopotential field is derived from non-surface wind observations. The percentage error is determined by dividing the rms difference between the two fields by the rms value of the original field, and multiplying by 100%. Note that a model eta level of 1 corresponds to the surface, and of zero to outer space.

been popular, because the conversion job is easier (Dudhia, 1993; Laprise, 1992). The 'unified' model currently uses a hybrid pressure-based coordinate from which the pressure at each level has to be recalculated every time step. Since it is more natural to use a height based vertical coordinate, especially if the full compressible Navier-Stokes equations are used and since pressure and height are then both available at all points, the interface to the physics routine does not need radical change to accommodate this. An area of difficulty, however, is the use of the two time-level schemes. In the unified model, a two time-level scheme is used with great advantage, as different time steps can easily be employed for different processes, and coupling the atmosphere to other models is also simpler because a single time-level of data provides a well-defined interface. Skamarock and Klemp (1992) demonstrated and analyzed instabilities in many two time level split schemes for the non-hydrostatic equations. Golding (1992) used such a scheme successfully, but found it necessary to use a basic state temperature profile in the semi-implicit method very close to the actual state. This would not be practical in a global model.

It is therefore proposed that a basic state profile is not used when selecting those parts of the equations to be treated implicitly. This requires the use of a variable coefficient solver for the implicit equations.

d *Implementation aspects*

Both requirements that the new scheme seeks to satisfy lead to the need to solve a variable coefficient elliptic equation. Efficient solution methods for these are therefore necessary. One of the most robust and efficient methods for use in computational fluid dynamics is considered to be the multigrid method. Application of this to meteorological problems is discussed by Fulton (1986). However, there are still doubts about its robustness for problems such as flow over orography, and other methods may yet prove superior.

The use of this type of semi-implicit method provides a way of satisfying the energy conservation requirement on the Charney-Phillips grid. Proof of energy conservation requires that the updates to the variables comprising the equation of state are calculated consistently. On the Charney-Phillips grid, the vertical staggering of the potential temperature from the pressure and density causes the difficulty. The solution is to generate the pressure correction equation by using an estimate of the vertically averaged potential temperature at the new time level calculated as

$$\overline{\theta^{n+1}}^{z} = \overline{\theta^{n}}^{z} + \underline{u} \cdot \overline{\nabla\theta}^{z} \tag{1}$$

rather than

$$\overline{\theta^{n+1}}^{z} = \overline{\theta^{n}}^{z} + \overline{\bar{\underline{u}}^{z} \cdot \nabla\theta}^{z} \tag{2}$$

The update of θ using the corrected pressure and consequently corrected winds is made using the normal (semi-Lagrangian) advection.

4 Description of the proposed scheme

The fully compressible equations of motion are used, with a basic vertical coordinate r defining distance from the centre of the Earth, and a terrain following coordinate η derived from it taking values $\eta = 1$ at the upper boundary of the model and $\eta = 0$ at the Earth's surface. The equations are written in terms of a density ρ scaled by r^2. The equations take the following time-discretized forms

$$u^{n+1} - u_d^n + \Delta t\left[- 2\Omega(\bar{v}' \sin\phi - \bar{w}' \cos\phi) + \frac{C_p\theta_v^n}{r\cos\phi}\left(\frac{\partial\bar{\Pi}'}{\partial\lambda} - \frac{\partial\bar{\Pi}'}{\partial r}\frac{\partial r}{\partial\lambda}\right)\right] = \frac{1}{\rho}\frac{\partial\bar{\tau}_x'}{\partial r} \tag{3}$$

$$v^{n+1} - v_d^n + \Delta t\left[2\Omega\bar{u}' \sin\phi + \frac{C_p\theta_v^n}{r^n}\left(\frac{\partial\bar{\Pi}'}{\partial\phi} - \frac{\partial\bar{\Pi}'}{\partial r}\frac{\partial r}{\partial\phi}\right)\right] = \frac{1}{\rho}\frac{\partial\bar{\tau}_y'}{\partial r} \tag{4}$$

$$w^{n+1} - w_d^n + \Delta t\left[-2\Omega u^n \cos\phi + \left(g\frac{(1+q_T)}{(1+q)}\right)^n + C_p\overline{\theta_v}' \frac{\partial\bar{\Pi}'}{\partial r}\right] = 0 \tag{5}$$

$$\rho_y^{n+1} - \rho_y^n + \frac{\Delta t}{\partial r/\partial\eta}\left[\frac{1}{\cos\phi}\frac{\partial}{\partial\lambda}\left(\frac{\rho_y}{r}\bar{u}'\frac{\partial r}{\partial\eta}\right)\right.$$
$$\left. + \frac{1}{\cos\phi}\frac{\partial}{\partial\phi}\left(\frac{\rho_y}{r}\bar{v}'\cos\phi\frac{\partial r}{\partial\eta}\right) + \frac{\partial}{\partial\eta}\left(\rho_y\bar{\dot{\eta}}'\frac{\partial r}{\partial\eta}\right)\right] = 0 \tag{6}$$

$$\theta_L^{n+1} - \theta_{Ld}^n = \Delta t\left[\left(\frac{\theta}{TC_p}\right)Q + \frac{\kappa\theta\omega}{pTC_p}(L_c q_{CL} + (L_c + L_f)q_{CF})\right] \tag{7}$$

$$q_T^{n+1} - q_{Td}^n = S\Delta t \tag{8}$$

$$\Pi^{\frac{\kappa-1}{\kappa}}\theta_v\rho = \frac{p_0 r^2}{\kappa C_p} \tag{9}$$

Suffix d denotes the departure point for the semi-Lagrangian scheme and superscripts n and $n + 1$ refer to time levels. The overbar indicates time averaging. The metric terms which would appear in the Eulerian form of the horizontal momentum equations are absorbed into the departure point calculation (Bates et al., 1990). Equation (9) is the equation of state. The notation is set out in Table 1. The implicit treatment of the advecting velocity and certain of the Coriolis terms is required to ensure conservation.

Solution of (3) and (9) as they stand would require coupled solution for all variables as three dimensional variables at the new time-level. This would be extremely expensive, and without better understanding of the structure of the equations, possibly ill-conditioned, resulting in failure of the integration scheme. The proposed method is to solve the implicit equations approximately using a predictor-corrector method, which can be thought of as using the first iteration only of an iterative method. The predictor step estimates corrections u', v' w', ρ', θ', q', p' using a subset of the terms treated implicitly in (3) to (9) chosen so that they can be reduced to a single three-dimensional equation which is strongly elliptic and therefore can be reliably solved. This also allows use of the form (1) instead of (2) for the estimate θ'. The terms used are the pressure gradient terms, including implicit treatment of the coefficient θ_v multiplying the pressure gradient term in the vertical velocity equation, the velocities in the continuity equation, a diagonal approximation to the boundary layer friction terms, the Coriolis terms, and the terms representing the

TABLE 1. Symbols used in equations

Symbol	Meaning
(r, λ, ϕ)	Spherical polar coordinates relative to centre of Earth
η	Terrain following scaled vertical coordinate
u, v, w	Velocity components
$\dot{\eta}$	$\frac{D\eta}{Dt}$
Ω	Earth's rotation rate
p, p_0	Pressure and reference value of pressure
Π	Exner pressure $\left(\frac{p}{p_0}\right)^{\kappa}$
C_p	Specific heat of dry air at constant pressure
κ	Gas constant over C_p
ω	$\frac{Dp}{Dt}$
ρ, ρ_y	Density of moist and dry air scaled by r^2
T	Temperature
θ	Potential temperature
θ_v	Virtual potential temperature
θ_L	Liquid water potential temperature
q	Specific humidity
q_T	Total water content
τ_x, τ_y	Horizontal components of turbulent stress

advection of the vertical component of vorticity. Further details are given in internal reports available from the authors.

The differences from the current unified model equations are

i) Use of the fully compressible non-hydrostatic equations, but, as with the existing unified model scheme, the vertical Coriolis terms are retained and the shallow atmosphere approximation is not used.

ii) Use of height as a vertical coordinate.

iii) No artificial horizontal diffusion, as a result of using monotone semi-Lagrangian advection.

5 Tests of the proposed scheme

The scheme is being tested on the full suite of shallow water test problems described by Williamson et al. (1992) and various published test problems which address the performance of the scheme in the vertical. We first describe results to test the proposed use of the Charney-Phillips vertical grid staggering. In order to take advantage of published results, these tests were carried out within the non-hydrostatic model of Golding (1992). This uses most of the features of our proposed integration scheme, except that it uses a fixed reference temperature profile in the semi-implicit integration scheme.

a *Eady wave test*

This is a simulation of the Eady-wave model of cyclogenesis in which a growing wave forms from a finite perturbation to a baroclinically unstable atmosphere. The experiment is similar to that of Nakamura and Held (1989). Their results were obtained using a hydrostatic primitive equation model, and were chiefly concerned with the process of equilibriation which occurs after the magnitude of the wave peaks at around day 7. The process of equilibration is complex and in a recent paper (Nakamura, 1994), it is suggested that the details are dependent on the form of horizontal diffusion. Since a semi-Lagrangian model has no added diffusion (and different intrinsic diffusion associated with the interpolation scheme), we compare results only for the first seven days of the simulation.

The non-hydrostatic equations are solved in a vertical (x, z) cross section on an f-plane at 45°N. All the variables are periodic in x with the domain length equal to the wavelength of the initial disturbance. The basic state is the same as that in Williams (1967) and consists of vertically sheared zonal flow in thermal wind balance with potential temperature. The pressure field is in hydrostatic balance with the temperature. All fields are assumed independent of y (the north-south coordinate) except for the basic state potential temperature and pressure.

The domain size was 4000 km in length and 10 km deep. The grid lengths used in the simulation were 31.25 km in x and 240 m in z. The basic state satisfies $\frac{\partial\theta}{\partial y} = -10^{-5}$, $\frac{\partial\theta}{\partial z} = 3.9 \times 10^{-3}$, $f = 10^{-4}$. The perturbation to the basic state coincides with the fastest growing eigenmode, as in Williams (1967). A short time step of 100 s was required at the end of the evolution when the gradients and velocities associated with the wave were very large.

The solutions on the two vertical grids were very similar for the first five days, when the fields are quite smooth. Figure 6 compares the results after 6 and 6.25 days, which are illustrative of the differences during days 5 to 7. There is considerably more noise in the Lorenz grid solution, especially near the upper boundary. The vertical velocity results (not shown) show stronger gravity wave activity on the Lorenz grid just below the upper boundary. The increased gravity wave activity is typical of the Lorenz grid solutions throughout the latter stages of the evolution. By day 7 the solution on the Charney-Phillips grid is also beginning to suffer from noise. The results support those of Arakawa and Moorthi (1987), also showing that their conclusions apply to non-hydrostatic models.

b *Boundary layer treatment*

A possible disadvantage of the Charney-Phillips grid is the implementation of the boundary layer scheme. Since the velocities are held at different levels from the thermodynamic variables, extra interpolations are required in implementing most standard parametrization schemes. Tests have therefore been carried out to assess whether these interpolations degrade the simulations.

The boundary layer scheme tested is based on that used by Golding (1993). The scheme calculates a turbulent kinetic energy (TKE) with the shear $\partial u/\partial z$ and vertical stability $\partial\theta/\partial z$ acting as the main generating factors. The tests used the version of the scheme with prognostic and diagnostic TKE, only the latter is illustrated. On the Charney-Phillips grid, the vertical shear and vertical stability are

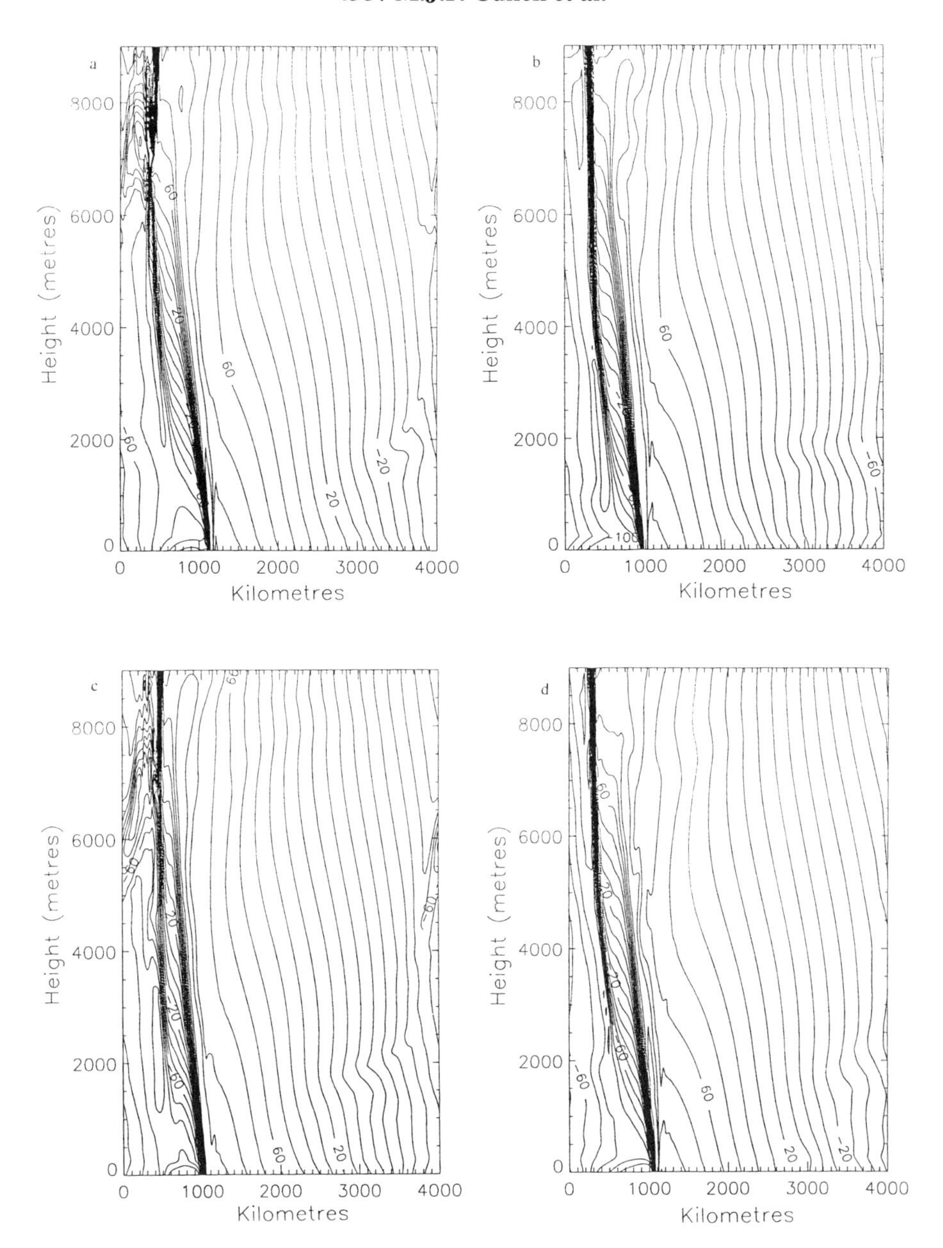

Fig. 6 y velocity components in Eady wave integration. a) Lorenz grid at 6 days, b) Charney-Phillips grid at 6 days, c) Lorenz grid at $6\,^1\!/_4$ days, d) Charney-Phillips grid at $6\,^1\!/_4$ days.

naturally calculated at different levels. There is a choice at which of these levels to hold the TKE, determining whether velocity or temperature variables have to be interpolated in calculating the generation terms. Both choices were tested.

The first experiment illustrated is a one-dimensional simulation of the evolution of the boundary layer at Wangara following data taken on 16 August 1967. This is a standard test bed for boundary layer parametrizations, e.g. Yamada and Mellor (1975), Golding (1993). Simulations using the Lorenz grid are compared with the two methods of implementation on the Charney-Phillips grid.

The integration is initialized at 0600 LST. Potential temperature, TKE, and u velocity were inspected at regular intervals up to 2400 LST, illustrated in Fig. 7. The experiment shows the response of the boundary layer as the solar heating of the land surface produces TKE and a subsequently well-mixed boundary layer. The land surface then cools and the TKE dies away, allowing a shallow inversion approximately 10 metres deep to form. At this point there is a little TKE above the immediate surface generated by wind shear.

All the integrations give similar results up to 1800 LST. After this time the Charney-Phillips integration which interpolates the velocity variables retains more TKE giving too much mixing and failure to form an inversion. The integration interpolating temperature variables gives similar results to the Lorenz grid.

A second experiment illustrates the interaction of the boundary layer scheme with the dynamics in a two-dimensional simulation. The simulation is of the development of fog at Perth (Western Australia) on 27 April 1990. The domain represents a cross section normal to the coast, with an idealized representation of the orography. The coast is 60 km from the western boundary, with 30 km of flat plain at 1 m above sea level to its east. This is terminated by a 10 km wide scarp rising linearly to a plateau at 300 m. The horizontal resolution used was 5 km. Experiments were performed using the vertical resolution of Golding (1993, Appendix 1). North-south derivatives are ignored except for a fixed pressure gradient term in the y-momentum equation. The synoptic situation and surface roughness and moisture availability are as described by Golding. The radiative forcing was, however, simplified to use constant prescribed day- and night-time heating rates.

The simulations are initiated at 1900 LST. Nine hours of simulated nocturnal cooling is represented by a fixed rate of downward radiative flux of 314 W m^{-2}. The downward radiative flux is then increased linearly over one hour to 1000 W m^{-2} at which it is held for a further two hours as a representation of dawn and daytime heating. Golding describes the nocturnal evolution as follows: 'Winds coming off the sea are approximately westerly with speed 8 m s^{-1}. The rough land surface rapidly decelerates the near-surface air allowing the surface temperature to drop. A highly turbulent boundary layer has developed on the scarp together with a weak easterly drainage flow that locally raises wind speed and temperatures and reduces humidity where it flows out onto the plain. These effects are diluted by surface cooling as it spreads onto the plain, but the enhanced shear generates turbulence. Between westerly winds from the sea and the easterly drainage flow, a stagnation

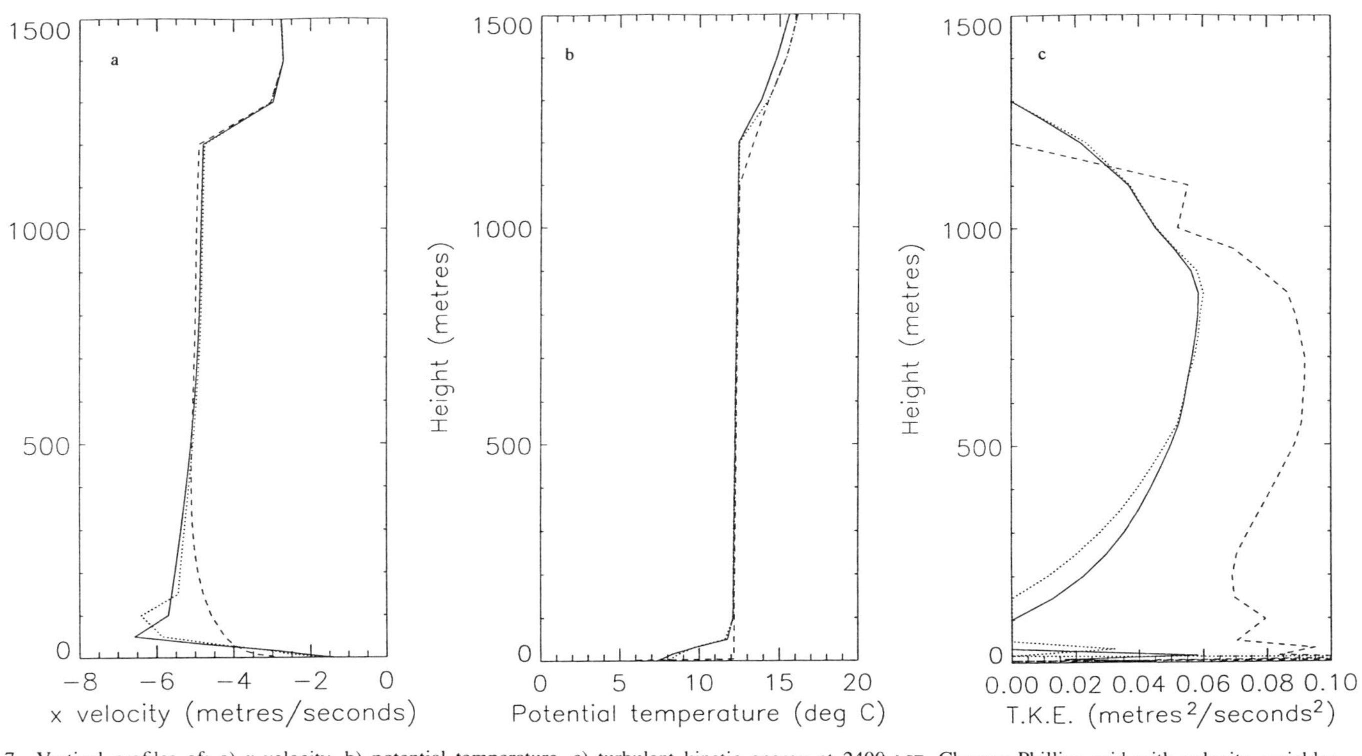

Fig. 7 Vertical profiles of: a) x-velocity, b) potential temperature, c) turbulent kinetic energy at 2400 LST. Charney-Phillips grid with velocity variables interpolated (dashed line), the Charney-Phillips grid with temperature variables interpolated (dotted line) and Lorenz grid (continuous line).

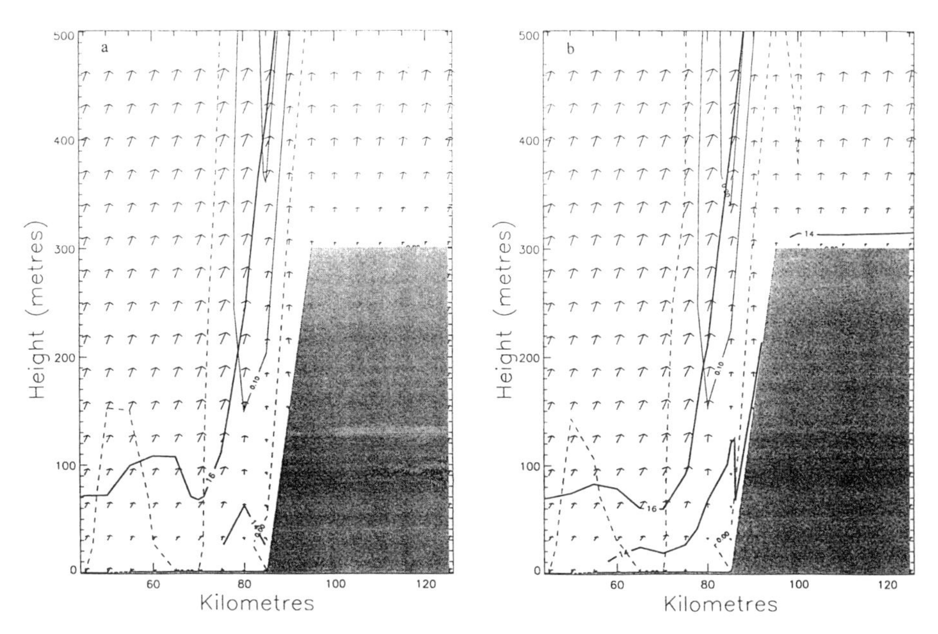

Fig. 8 Results from the Perth fog simulation experiment at 400 LST; a) Charney-Phillips grid, b) Lorenz grid. Horizontal wind velocity and direction (arrows), vertical velocity (light contours and zero contour dashed), potential temperature (heavy contours).

point has formed with associated weak uplift connected to the main scarp-driven ascent. This localized reduction in horizontal winds, and the associated drop in turbulent mixing, allow saturation to occur in the lowest model layers. At the same time, the upward motion associated with the convergent wind flow assists in deepening the saturated layer.'

Figure 8 shows the wind and potential temperature cross-sections using the Charney-Phillips grid with temperature variables interpolated and the Lorenz grid at 0400 LST. At this time the drainage flow is reaching a maximum. The differences are small. The area of descent is slightly greater if the Charney-Phillips grid is used. It is not possible to state which solution is preferable. A similar conclusion applies at other times. If the Charney-Phillips grid is used with momentum variables interpolated (not shown), there are considerable differences, including large vertical oscillations in the TKE. This is consistent with the results from the one-dimensional tests.

It is felt that these results show that the boundary layer simulation is not degraded by using the Charney-Phillips grid.

c *One dimensional behaviour of compressible model*

The proposed implementation of the integration scheme recommended in Section 3 uses a different way of constructing the semi-implicit scheme from that of Golding

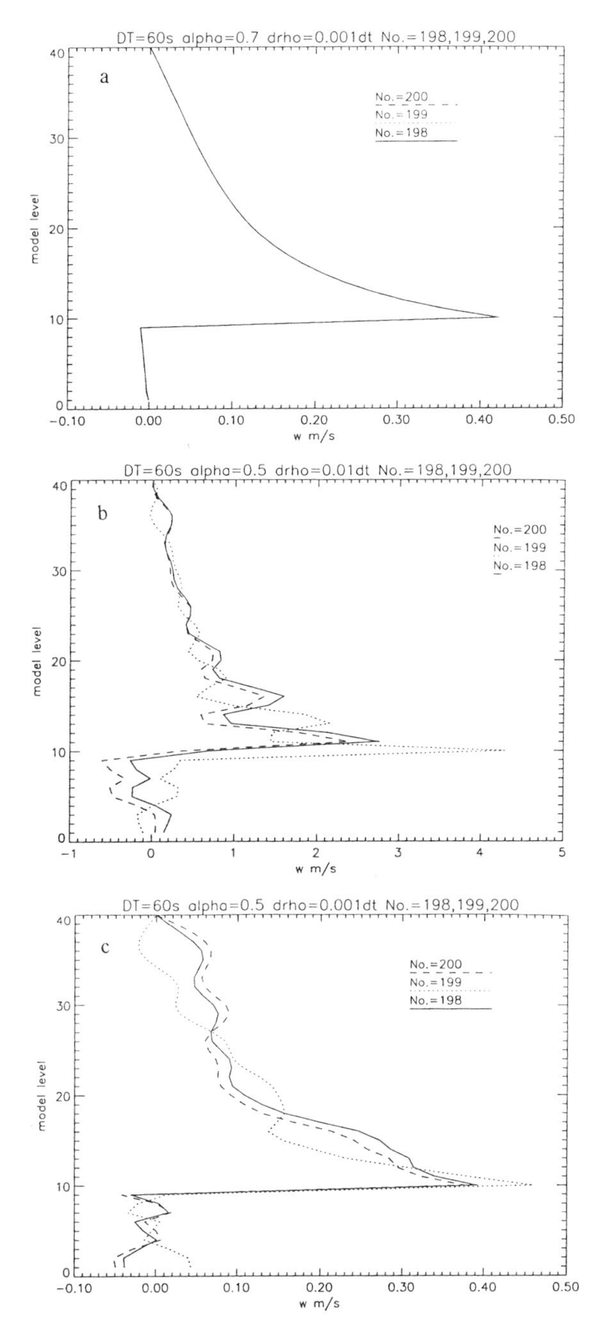

Fig. 9 Vertical velocity profiles in perturbed columns; (a) with backward implicit weighting, (b) no damping, (c) with initial perturbation 10 times larger. Superimposed curves are for successive time steps.

(1992), in particular working from a residual in the equation of state. The test described by Golding is thus repeated, in which perturbations to a vertical column of air 16 km high are simulated with a grid of 40 points and rigid upper and lower boundary conditions. The column is initially at rest in hydrostatic balance and is perturbed with a steady fractional mass source of 0.001 s^{-1} at the 10th point above the bottom. This source corresponds to major diabatic forcing. A 60 s time step was used for the integrations. The steady state response is an almost universally increasing pressure, with a slight gradient required to support the vertical motion required to redistribute the mass. On this are superposed the sound waves from the initial start-up. In the atmosphere the sound wave transients would not be of significant amplitude, and the open upper boundary condition would prevent resonance which is possible in this model.

Figures 9a) and b) show results comparable to Fig. 2 (right hand pair of profiles) of Golding (1992). The vertical velocity field is shown at three consecutive time steps. With centred implicit time differencing oscillations similar to those of Golding occur, but there is no tendency for them to amplify. With backward weighted time differencing ($\alpha = 0.7$) the transients disappear. Figure 9c) shows the results with centred time differencing using a mass source 10 times larger. There is again no unstable behaviour.

These results suggest that the basic structure of the time differencing scheme should be satisfactory.

6 Summary

We have presented a non-hydrostatic integration scheme which should be suitable for all applications of the unified model. Tests of some of the less usual aspects have been presented, giving satisfactory results. Further standard idealized tests are in progress and will be reported in due course.

References

ARAKAWA, A. and V.R. LAMB. 1977. Computational design of the basic dynamical processes of the UCLA general circulation model. In: *Methods in Comp. Phys.* Vol. **17**, Academic Press, pp. 174–265.

——— and S. MOORTHI. 1987. Baroclinic instability in vertically discrete systems. *J. Atmos. Sci.* **45**: 1688–1707.

BATES, J.R.; F.H.M. SEMAZZI, R.W. HIGGINS and S.R.M. BARROS. 1990. Integration of the shallow water equations on a sphere using a vector semi-Lagrangian scheme with a multigrid solver. *Mon. Weather Rev.* **118**: 1615–1627.

COURTIER, P. and J.-F. GELEYN. 1988. A global numerical weather prediction model with variable resolution: Application to shallow water equations. *Q.J.R. Meteorol. Soc.* **114**: 1321–1346.

CULLEN, M.J.P. 1989. Implicit finite difference methods for modelling discontinuous atmospheric flows. *J. Comp. Phys.* **81**: 319–348.

———. 1993. The unified forecast/climate model. *Meteorol. Mag.* **122**: 81–93.

——— and T. DAVIES. 1991. A conservative split-explicit scheme with fourth order horizontal advection. *Q.J.R. Meteorol. Soc.* **117**: 993–1002.

DUDHIA, J. 1993. A non-hydrostatic version of the Penn State-NCAR mesoscale model: validation tests and simulation of an Atlantic cyclone and cold front. *Mon. Weather Rev.* **121**: 1493–1513.

FULTON, S.R. 1986. Multigrid methods for elliptic

problems: a review. *Mon. Weather Rev.* **114**: 943–959.

GADD, A.J. 1978. A split-explicit scheme for numerical weather prediction. *Q.J.R. Meteorol. Soc.* **104**: 569–582.

GOLDING, B.W. 1992. An efficient non-hydrostatic forecast model. *Meteorol. Atmos. Phys.* **50**: 89–103.

———. 1993. A study of the influence of terrain on fog development. *Mon. Weather Rev.* **121**: 2529–2541.

LAPRISE, R. 1992. The Euler equations of motion with hydrostatic pressure as an independent variable. *Mon. Weather Rev.* **120**: 197–207.

LESLIE, L.M. and R.J. PURSER. 1992. A comparative study of the performance of various vertical discretisation schemes. *Meteorol. Atmos. Phys.* **50**: 61–73.

MORTON, K.W. and P.K. SWEBY. 1987. A comparison of flux limited difference methods and characteristic Galerkin methods for shock modelling. *J. Comp. Phys.* **73**: 203–230.

NAKAMURA, N. 1994. Nonlinear equilibriation of two-dimensional Eady waves: simulations with two-dimensional geostrophic momentum equations. *J. Atmos. Sci.* **51**: 1023–1035.

——— and I.M. HELD. 1989. Nonlinear equilibriation of two-dimensional Eady waves. *J. Atmos. Sci.* **46**: 3055–3064.

PRIESTLEY, A. 1993. A quasi conservative version of the semi-Lagrangian advection scheme. *Mon. Weather Rev.* **121**: 621–629.

SKAMAROCK, W.C. and J.B. KLEMP. 1992. The stability of time-split numerical methods for the hydrostatic and the non-hydrostatic elastic equations. *Mon. Weather Rev.* **120**: 2109–2127.

TANGUAY, M.; A. ROBERT and R. LAPRISE. 1990. A semi-implicit semi-Lagrangian fully compressible regional forecast model. *Mon. Weather Rev.* **118**: 1970–1980.

WHITE, A.A. and R.A. BROMLEY. 1995. Dynamically consistent, quasi-hydrostatic equations for global models with a complete representation of the Coriolis force. *Q.J.R. Meteorol. Soc.* **121**: 399–418.

WILLIAMS, R.T. 1967. Atmospheric frontogenesis: a numerical experiment. *J. Atmos. Sci.* **29**: 3–10.

WILLIAMSON, D.L.; J.B. DRAKE, J.J. HACK, R. JAKOB, and P.N. SWARTZRAUBER. 1992. A standard test set for numerical approximations to the shallow water equations in spherical geometry. *J. Comp. Phys.* **102**: 211–224.

——— and J.G. OLSON. 1994. Climate simulations with a semi-Lagrangian version of the NCAR community climate model. *Mon. Weather Rev.* **122**: 1594–1610.

YAMADA, T. and G.L. MELLOR. 1975. A simulation of the Wangara atmospheric boundary layer data. *J. Atmos. Sci.* **32**: 2309–2329.

Application of the Semi-Lagrangian Method to Global Spectral Forecast Models

Harold Ritchie
Recherche en prévision numérique
Service de l'environnement atmosphérique
Dorval, Québec, Canada H9P 1J3

[Original manuscript submitted 22 February 1995; in revised form 27 September 1995]

ABSTRACT *Since the original demonstration of the efficiency advantage of the semi-Lagrangian semi-implicit method over a decade ago by André Robert, this numerical integration scheme is being used in an increasing range of atmospheric models. Most of the applications have been in grid point models, where it has been shown that this method permits the use of time steps that are much larger than those permitted by the Courant-Friedrich-Levy (CFL) criterion for the corresponding Eulerian models. In this paper we concentrate on its application in spectral models. A review of the steps towards its operational implementation in global spectral forecast models is presented. Linear stability and geometric aspects are considered for the problem of simple advection on the Gaussian grid that is used in spectral models. Nonlinear stability, accuracy and efficiency of the approach are illustrated by its application to a spectral model of the shallow water equations. Application in multilevel spectral primitive equations models is demonstrated with the Canadian Global Spectral Forecast Model and a high resolution version of the ECMWF Forecast Model.*

RÉSUMÉ *Depuis qu'André Robert a démontré le gain en efficacité associé à la méthode semi-implicite semi-lagrangienne il y a plus d'une décennie, ce schéma d'intégration numérique est en train d'être utilisé dans une grande variété de modèles atmosphériques. La plupart des applications ont eu lieu dans des modèles à points de grille, pour lesquels on a démontré que cette méthode permet d'utiliser des pas temporels qui sont beaucoup plus grands que ceux qui respectent le critère Courant-Friedrich-Levy (CFL) pour les modèles eulériens correspondants. Dans cet article nous examinons son application dans les modèles spectraux. Un rappel des étapes menant à son exploitation dans des modèles spectraux mondiaux de prévision numérique du temps est présentée. La stabilité linéaire et des aspects géométriques sont considérés pour le problème d'advection simple sur la grille gaussienne qui est employée dans les modèles spectraux. La stabilité nonlinéaire, la précision, et l'efficacité de l'approche sont illustrées à l'aide d'un modèle spectral des équations St-Venant. Son application dans des modèles spectraux baroclines des équations primitives est démontrée en utilisant les modèles spectraux mondiaux de prévision utilisés au Canada et au Centre européen pour les prévisions météorologiques à moyen terme.*

1 Introduction

This paper reviews the application of the semi-implicit semi-Lagrangian method to global spectral forecast models. One reason for including this overview in the André Robert Memorial Symposium volume is that André supervised the early phases of this work, and the operational implementation of this method marks the completion of an area of research and development which he followed with great interest.

The efficiency advantage of the semi-Lagrangian semi-implicit scheme was demonstrated over a decade ago by André Robert (1981, 1982) in the context of gridpoint models of the shallow water equations. Since that time considerable work has been done in applying the method in baroclinic models, and several centres are now using it in operational weather forecasting models. The first step for spectral models was to examine the semi-Lagrangian treatment of advection on the Gaussian calculation grid used in such models (Ritchie, 1987), which addressed the geometric and linear stability questions. The next step was to combine the semi-Lagrangian approach with the semi-implicit scheme to get a stable treatment of both Rossby and gravity waves in a spectral model of the shallow water equations (Ritchie, 1988). The extension to baroclinic models was then examined through an application to a multilevel spectral primitive equations model, where several issues related to vertical discretizations were addressed (Ritchie, 1991). Following further optimization (Tanguay et al., 1992; Ritchie and Beaudoin, 1994), semi-Lagrangian semi-implicit spectral models have now been implemented at the Canadian Meteorological Centre and ECMWF (European Centre for Medium-range Weather Forecasts) (Ritchie et al., 1995), where they have enhanced the technical and meteorological performance. The following sections trace the evolution of these developments.

2 Semi-Lagrangian advection on a Gaussian grid

Here we review the characteristics of two types of semi-Lagrangian advection that were applied to this problem of solid body rotation (Ritchie, 1987). A schematic diagram of the experiment configuration is shown in Fig. 1. We consider rotation with constant angular velocity about an axis passing through the centre of the earth and a point P′ on the earth's surface. To facilitate the interpretation of the results we introduce a stereographic plane which is tangent to the earth's surface at P′. Since the motion corresponds to solid body rotation about P′, when viewed in this plane the advected field should simply rotate around P′ without any distortion. NP is the projection in the stereographic plane of the north pole of the Gaussian grid, EQ is the projection of the equatorial point E, and the X-axis of the stereographic (X,Y) plane is also indicated. The analytic solution is centred over EQ initially and its centre passes over NP after half a revolution. The initial field is chosen to be a "Gaussian hill" function which has a peak value of 100 units at its centre and decays towards zero with a Gaussian profile with a length scale of L, which is approximately the diameter of the contour whose value is 10 percent of the

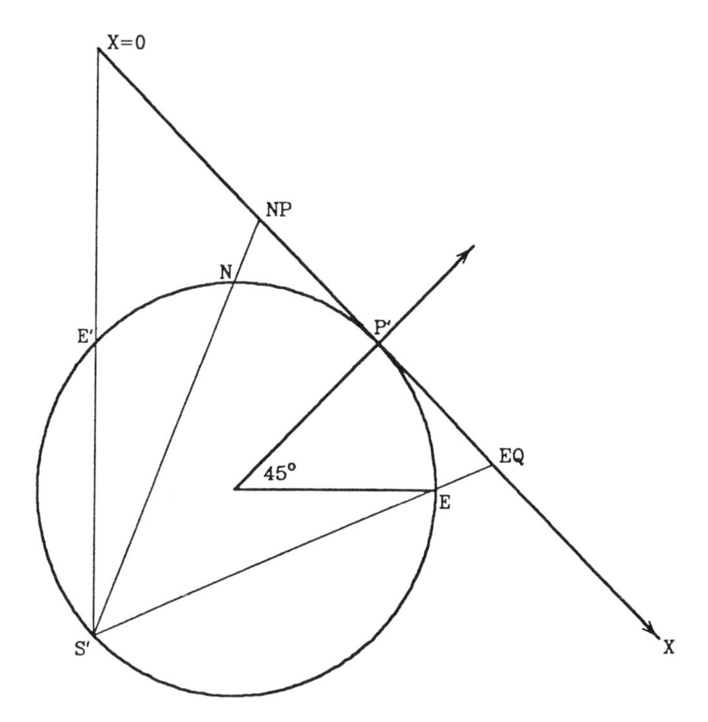

Fig. 1 Schematic diagram showing the experiment configuration. Labelled points are described in the text. (Reproduced with permission from Ritchie, 1987.)

peak value. The angular velocity is chosen to correspond to one rotation in 20 days, which gives a typical synoptic scale advecting velocity of 16.4 m/s through points E and N shown in Fig. 1. A stereographic map of an analytic solution with L = 2500 km after one rotation is shown in Fig. 2a.

An Eulerian spectral model was used to solve the solid body rotation problem. The spectral representation used triangular truncation with a maximum wavenumber N = 42, requiring 128 grid points in the longitudinal direction and 64 grid points in the latitudinal direction to avoid quadratic aliasing. The Courant-Friedrichs-Levy (CFL) stability limit for this configuration restricts the time step to 1.82 h. For convenience, the time step for the Eulerian spectral model integrations was chosen to be 1.5 h. The Eulerian model numerical solution corresponding to Fig. 2a is shown in Fig. 2b and the error (analytic-numerical) field is shown in Fig. 3a. We note that the dispersion associated with Eulerian model time truncation error has caused a noticeable distortion in the numerical solution, resulting in a downstream wake in the error pattern. Here the maximum absolute error value is in excess of 25 units.

Semi-Lagrangian schemes using high-order interpolation have much smaller dispersion errors than Eulerian schemes. This can be illustrated by an interpolating

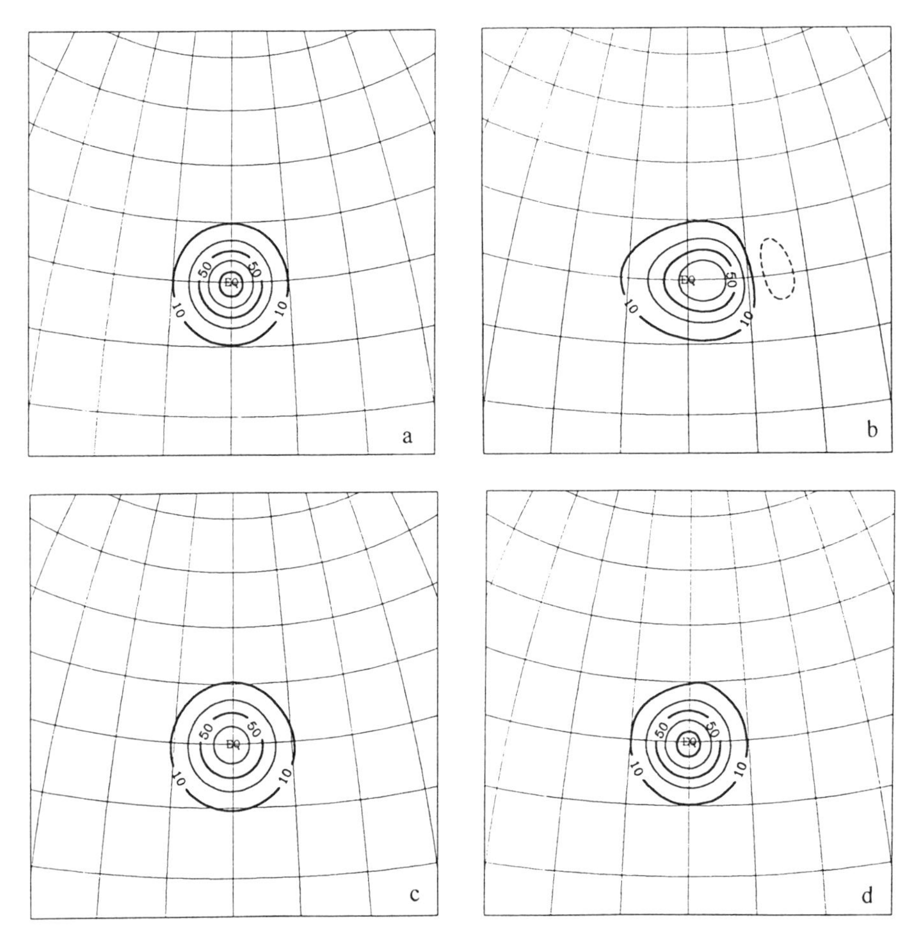

Fig. 2 Solutions for solid body rotation problem at 480 hours: a) Analytic solution; b) Eulerian spectral model solution; c) Interpolating semi-Lagrangian model solution; d) Noninterpolating semi-Lagrangian model solution. Latitude and longitude lines are drawn every 10 degrees. Contour interval 20 units. (Adapted with permission from Ritchie, 1987.)

semi-Lagrangian model solution of the solid body rotation problem. The model grid and parameters were the same as those used for the Eulerian spectral model except that the time step was six hours, which is four times that used by the Eulerian spectral model and violates the CFL stability limit by a factor of roughly 3.3. The numerical solution is shown in Fig. 2c, and the error field is presented in Fig. 3b. As indicated by the concentric shape of the error pattern, the semi-Lagrangian scheme gives a much more accurate treatment of the phase speed of the solution. Also, the maximum value of the error is about 16 units, which is much less than for the corresponding Eulerian field (Fig. 3a). However, the error pattern also shows

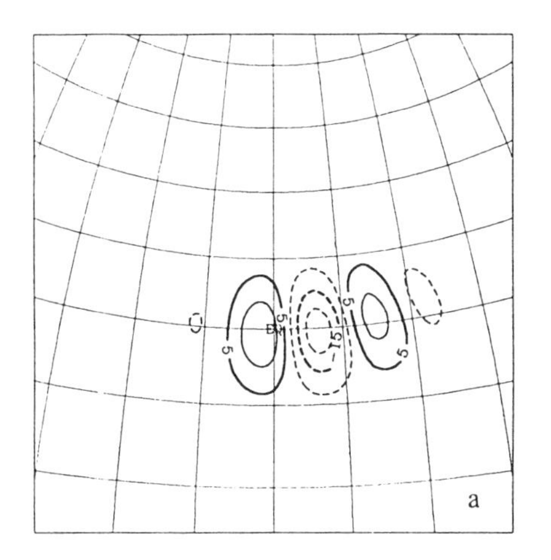

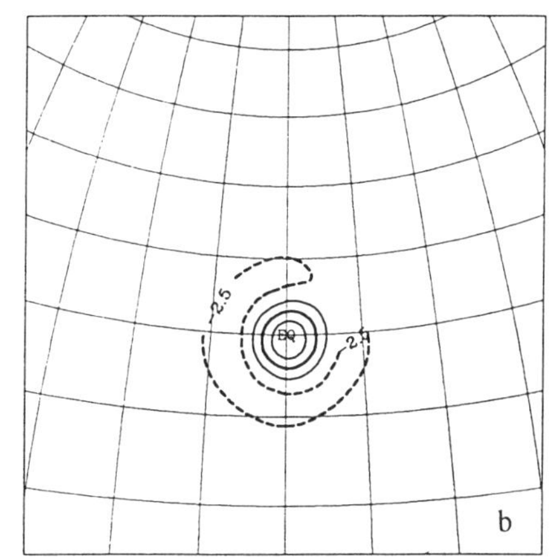

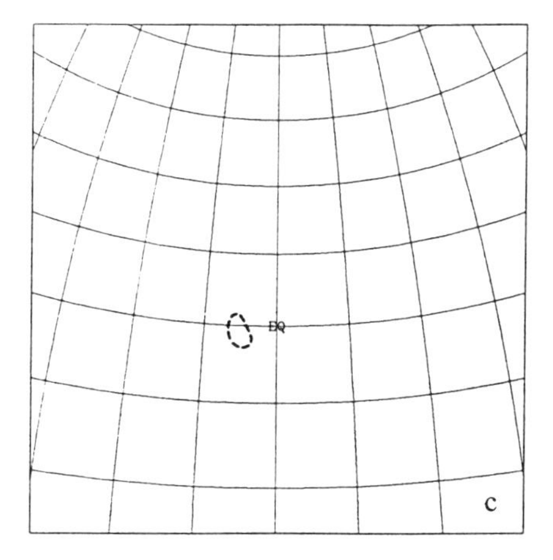

Fig. 3 Errors in model solutions at 480 hours: a) Eulerian spectral model, contour interval 10 units; b) Interpolating semi-Lagrangian model, contour interval 5 units; c) Noninterpolating semi-Lagrangian model, contour interval 5 units. Dashed contours indicate negative values. (Adapted with permission from Ritchie, 1987.)

clear evidence of damping due to the spreading out of the field as a result of interpolation errors.

There is another version of the semi-Lagrangian scheme which does not require interpolation in order to pick off the upstream values of the field being advected. This scheme, which will be referred to as "noninterpolating", does not suffer from significant damping errors, but retains the attractive stability property. The numerical solution produced using a model with this scheme is shown in Fig. 2d and the error field in Fig. 3c. The model grid and truncation were the same as those used for the Eulerian spectral model, and the time step was 6 hours as in the

TABLE 1. Error results for the model intercomparison as explained in the text. EUL denotes the Eulerian spectral model (time step 1.5 h), ILS denotes the interpolating semi-Lagrangian model (time step 6 h), and NISL denotes the noninterpolating semi-Langrangian model (time step 6 h). The errors are expressed in percent.

L (km)	Model	Prognosis (h) 120	240	360	480
10000	EUL	0.22	0.34	0.55	0.68
10000	ISL	0.18	0.35	0.52	0.70
10000	NISL	0.31	0.55	0.77	1.01
5000	EUL	1.55	2.73	4.24	5.48
5000	ISL	0.57	0.96	1.36	1.90
5000	NISL	0.62	1.05	1.31	1.67
2500	EUL	15.2	26.4	38.0	48.2
2500	ISL	6.31	8.02	10.6	15.2
2500	NISL	4.71	6.44	8.08	9.72

(Reproduced with permission from Ritchie, 1987).

interpolating semi-Lagrangian model. The maximum absolute error value in Fig. 3c is just over 2.5 units, which is much less than the corresponding error fields (Figs 3a and 3b) for the other two models. Although the 10 unit contour in Fig. 2d indicates some slight distortion due to the phase speed error, comparing to Fig. 2b and Fig. 3a shows that the phase error for this length scale is much less than with the Eulerian model. Also, in Fig. 3c there is no evidence of the concentric error pattern associated with damping as in the corresponding interpolating semi-Lagrangian integration (Fig. 3b).

In order to provide a more systematic and quantitative intercomparison, all three models were run with initial fields having length scales L = 10000 km, 5000 km and 2500 km. The grid and model parameters were the same as already outlined. The errors at various times were calculated by comparing with the analytic solution and, as a measure of the accuracy in each case, the integral of the absolute value of these errors was expressed as a percent of the integral of the analytic solution. Table 1 contains the percentage errors for each case at 120, 240, 360 and 480 h, which corresponds to one full rotation of the analytic solution. It is seen that all three models perform very well for L = 10000 km, with the noninterpolating semi-Lagrangian version being slightly poorer than the other two. For this length scale, the field is well resolved by all models, so there is very little dispersion or damping error. As the length scale is reduced, there is a marked deterioration in the performance of the Eulerian model due to the increased dispersion as discussed earlier. The performance of the interpolating semi-Lagrangian model deteriorates less, although the damping becomes noticeable for the short scales. At short scales, the noninterpolating semi-Lagrangian model gives the best overall performance because it has less dispersion than the Eulerian model and less damping than the interpolating semi-Lagrangian model.

This demonstrates that the semi-Lagrangian scheme can be stably and accurately applied to treat advection on the Gaussian grid with time steps that far exceed the CFL limit for the Eulerian spectral model. The interpolating and the noninterpolating versions both compare favourably with the Eulerian one in terms of accuracy, and the noninterpolating scheme seems to have some advantage for the treatment of short scales.

3 Application of the semi-Lagrangian method to a spectral model of the shallow water equations

In a study of the barotropic vorticity equation, Sawyer (1963) showed that the semi-Lagrangian treatment is unconditionally stable for Rossby waves and gives an accuracy comparable to that of Eulerian methods. Robert et al. (1972) demonstrated stability and accuracy of the semi-implicit method for gravity waves in an Eulerian model. By virtue of these properties of the semi-implicit scheme for the treatment of gravity waves, Robert (1981, 1982) succeeded in combining the semi-implicit scheme with the semi-Lagrangian scheme to produce stable and accurate integrations of gridpoint models of the shallow water equations using large time steps. Having examined the properties of semi-Lagrangian advection on a Gaussian grid, the next step for its application in spectral models was to combine the semi-Lagrangian approach with the semi-implicit scheme to get a stable treatment of both Rossby and gravity waves in a spectral model of the shallow water equations (Ritchie, 1988).

An intercomparison test was performed in which several models were run to produce five-day forecasts. The models were global, used no filters, and started from a First GARP Global Experiment (FGGE) analysis valid at 0000 UTC 21 December 1978. Eulerian, interpolating semi-Lagrangian, and noninterpolating semi-Lagrangian formulations were run using a triangular 126-wave (T126) truncation. In order to avoid an instability associated with an explicit treatment of the metric term in semi-Lagrangian models in spherical polar coordinates, the vector form of the horizontal momentum equation was treated with a semi-implicit semi-Lagrangian discretization. The Eulerian model was run with a time step of 10 min, which respects the CFL limit, while the semi-Lagrangian models used a time step of one hour. An integration of the Eulerian model with a T213 truncation and a time step of 6 min was used as a control run for the experiment. The global area-weighted root-mean-square (rms) differences among the forecasts of the 500 mb geopotential height were calculated after each 24-hour interval. In Fig. 4 each T126 model is compared to the control run, with the results for the Eulerian model being given by the solid curve, while the long dashed curve presents the interpolating semi-Lagrangian model results, and the short dashed curve gives the noninterpolating semi-Lagrangian ones. These results are expressed in metres and show that the

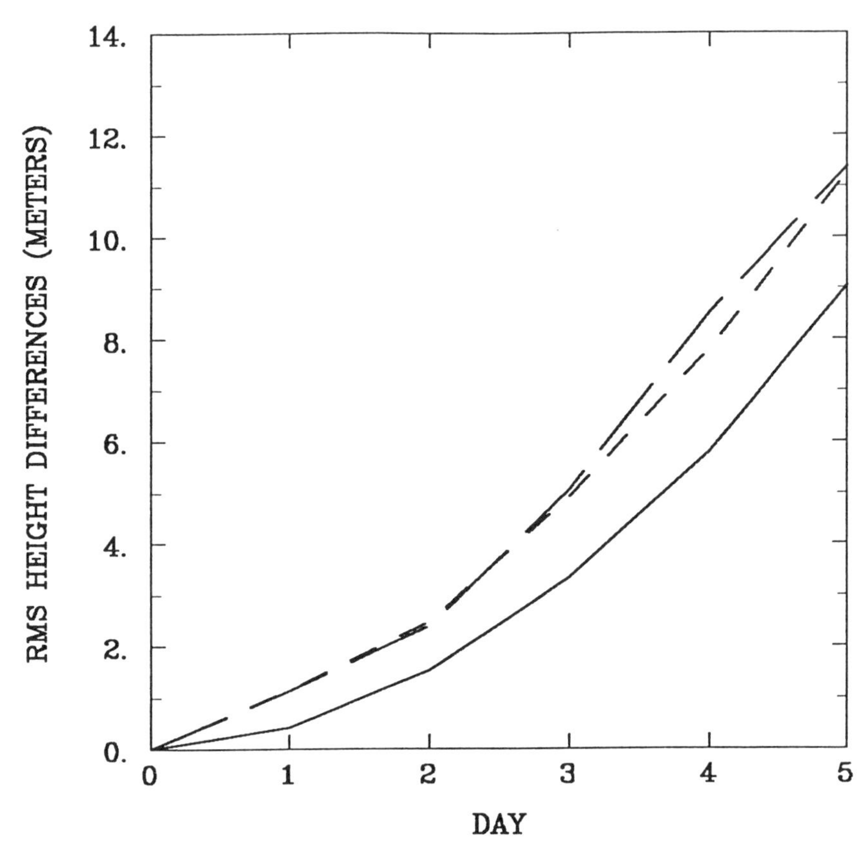

Fig. 4 Global rms differences between T126 and control run forecasts of 500 mb geopotential height in metres. Curves are plotted for Eulerian model with 10 min time step (solid), interpolating semi-Lagrangian model with 60 min time step (long dashed), and noninterpolating semi-Lagrangian model with 60 min time step (short dashed) as discussed in the text. (Reproduced with permission from Ritchie, 1988.)

semi-Lagrangian models give an accuracy that is quite acceptable in comparison with the Eulerian model at the same resolution. The rms difference between any one of these models and the control run is small in comparison with the 500 mb geopotential height errors for typical model forecasts.

An example of the energy conservation properties of the semi-implicit semi-Lagrangian schemes is found in Fig. 5. In order to examine the issue of the energy conservation properties of these schemes for longer integrations, the interpolating and noninterpolating semi-Lagrangian models were run out to 20 days with a 60 min time step at resolutions of T63 and T126. The interpolating model had no filters, while the noninterpolating one had a Robert time filter with a weak coefficient of 0.04 at T63 and 0.02 at T126 (using the definition as given in the analysis by Asselin (1972)). The evolutions of the potential energy per unit mass (P), the

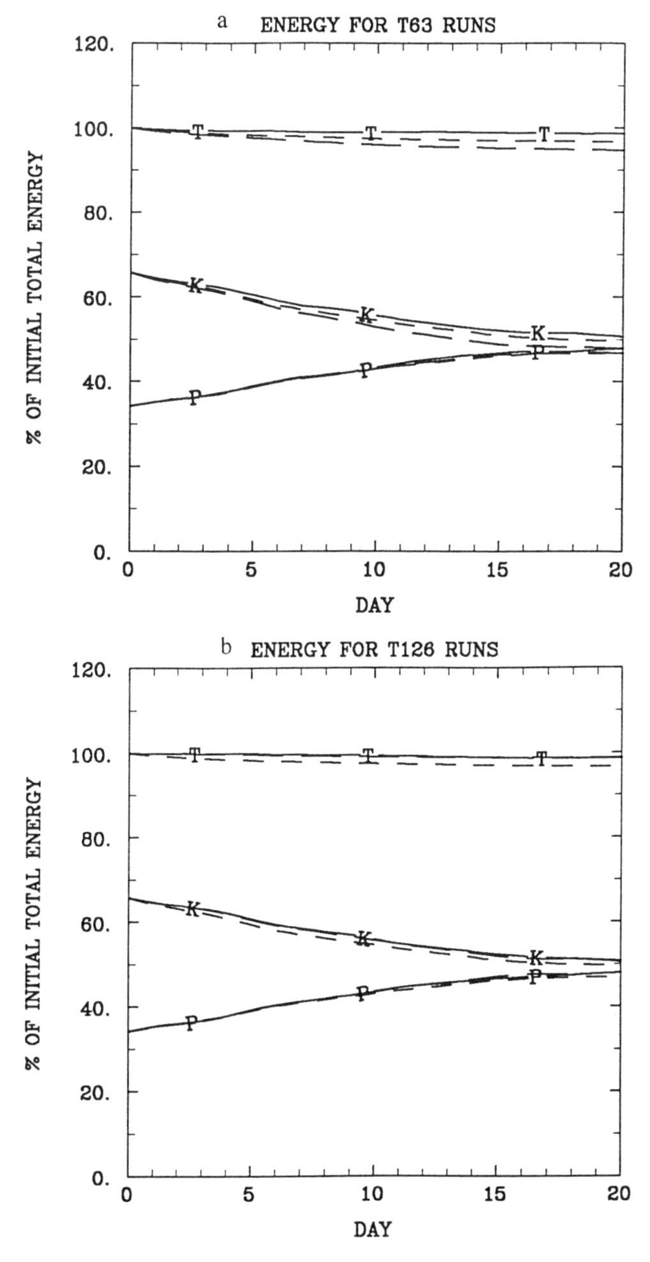

Fig. 5 Evolution of potential (P), kinetic (K), and total(T) energy in 20-day integrations as discussed in text. Solid curves are for the noninterpolating semi-Lagrangian model, and long dashed curves are for the interpolating one. Short dashed curves give the evolution for the T106 Eulerian model with diffusion and are included to indicate the behaviour in typical medium-range forecasts. Energies are expressed as a percentage of the initial total energy for the run: a) For T63 integrations; b) For T126 integrations; (Adapted with permission from Ritchie, 1988.)

kinetic energy per unit mass (K), and the total energy (T = P + K) are presented in Fig. 5a for the T63 runs and Fig. 5b for the T126 runs. The energies for each run are expressed as a percentage of the initial total energy for the run. Here the solid curves are for the noninterpolating model and the long dashed curves are for the interpolating one. As an indication of the energy conservation behaviour that is considered to be acceptable in typical medium range forecasts, the Eulerian model was run at a resolution of T106 with a time step of 12 min, and a fourth order spectral diffusion with a coefficient of 10**15 m**4/s was applied to the vorticity, divergence and perturbation geopotential. The evolutions of P, K and T for this model are given by the short dashed curves in Fig. 5. At all resolutions, all models have very similar evolutions of potential energy and the models appear to be evolving towards a state of equipartition of kinetic and potential energy. For the T106 Eulerian run, the loss in total energy after 20 days is about 3%. At the lower resolution (T63, Fig. 5a) the decay of kinetic energy for the interpolating model is a little excessive, resulting in a loss of about 5% in total energy, while the noninterpolating model conserves total energy to within about 1%. At the higher resolution (T126, Fig. 5b), the effect of damping in the interpolating scheme is much smaller and both semi-Lagrangian models conserve total energy to within about 1%. Note that the two semi-Lagrangian models are almost indistinguishable (the solid and long dashed curves are almost superimposed) in this figure, and both give better energy conservation than the T106 Eulerian model (short dashed curve). All these runs have good mass conservation characteristics, conserving mean perturbation geopotential to within half a percent.

4 Application of the semi-Lagrangian method to a multilevel spectral primitive equations model

The interpolating and noninterpolating semi-Lagrangian advection alternatives proved to be relevant in the next step: applying the semi-implicit semi-Lagrangian method to a multilevel spectral primitive equations model (Ritchie, 1991). This can be illustrated by intercomparison tests that were performed during the course of this work.

The integrations started from a FGGE analysis valid at 1200 UTC 12 February 1979. Nonlinear normal mode initialization was used, and in subsequent intercomparisons involving semi-Lagrangian models with large time steps, it was found that initialization was important for good agreement between models, particularly during the first two days. Hemispheric integrations were performed with a triangular 79-wave truncation and 20 equally spaced sigma levels ranging from 0.05 at the top to 1.0 at the surface. As in Daley et al. (1976), the physical processes included Cressman's (1960) drag formulation in the lowest layer, a moist convective adjustment based on the principles of Manabe et al. (1965), and orographic effects. Vertical momentum diffusion was included in the top layer. A Robert time filter (with a coefficient of 0.1 using the definition given in the analysis by Asselin (1972)) and a sixth order horizontal diffusion with a coefficient of 10**26 m**6/s were applied.

An Eulerian formulation (UVEUL) was converted to one that uses the interpolating semi-Lagrangian approach in the horizontal, while retaining the Eulerian treatment of the vertical advection (2DISL). The UVEUL version was integrated using a time step of 12 min which respects the CFL limit. In order to test the stability and accuracy of this 2DISL model, the time step was increased to 20 min and a 120-hour forecast was performed. The evolution of the rms differences compared with the UVEUL model showed acceptable values in the 10–15 m range in the mid-atmosphere at 120 hours.

In addition to the CFL limit associated with the horizontal advection, there is also one associated with vertical advection. An attempt was made to increase the time step to 30 min with the 2DISL model, but the integration failed, apparently because of a violation of the vertical CFL criterion. This illustrates the fact that, in order to avoid the vertical CFL limit in models that have high vertical resolution to support a sophisticated boundary layer physical parametrization, it is necessary to use fully 3-dimensional semi-Lagrangian treatments (3DISL). The 120-hour forecast was repeated with the 3DISL scheme using a 20 min time step, and the evolution of the rms differences with respect to the UVEUL model showed large differences that develop in the upper atmosphere early in the integration and then spread downwards, contaminating the whole domain. A closer examination of the model outputs suggested that this is a consequence of excessive vertical smoothing of the fields in the vicinity of the tropopause, where all of the model fields change abruptly in the vertical. In order to clearly identify the problem as originating from the vertical interpolation of the tropopause, a test was performed with a 2-1/2-dimensional semi-Lagrangian (2-1/2DISL) version that uses the 3DISL scheme around the tropopause (i.e. for sigma = 0.20, 0.25, 0.30, 0.35) and the 2DISL scheme elsewhere. The 120-hour forecast was repeated with this 2-1/2DISL model using a 20 min time step, and the evolution of the rms differences with respect to the 2DISL model with a 20 min time step is shown in Fig. 6a. The results are in metres and are presented for 10 equally spaced pressure levels, varying from 100 mb at the top to 1000 mb at the bottom of the figure, and at 12-hour intervals, from 0 hours at the left to 5 days at the right of the figure. The results show that, at least for the Lagrange cubic interpolator and the resolution used here, vertical interpolation of the fields around the tropopause introduces a large difference that spreads upwards and downwards during the integration.

As we saw earlier, one way of avoiding the excessive damping arising from interpolating rapidly varying fields is to modify the semi-Lagrangian scheme to use the noninterpolating approach. It was decided to apply this approach to the vertical advection, while retaining the 2DISL treatment in the horizontal. The resulting method is referred to as noninterpolating semi-Lagrangian in the vertical (NISLV). In order to test the stability and accuracy of this NISLV model, the time step was increased to 40 min and the 120-hour forecast was performed. The evolution of the rms differences with respect to the UVEUL run in Fig. 6b shows that at each level the difference grows roughly linearly in time, and reaches mid-atmospheric levels

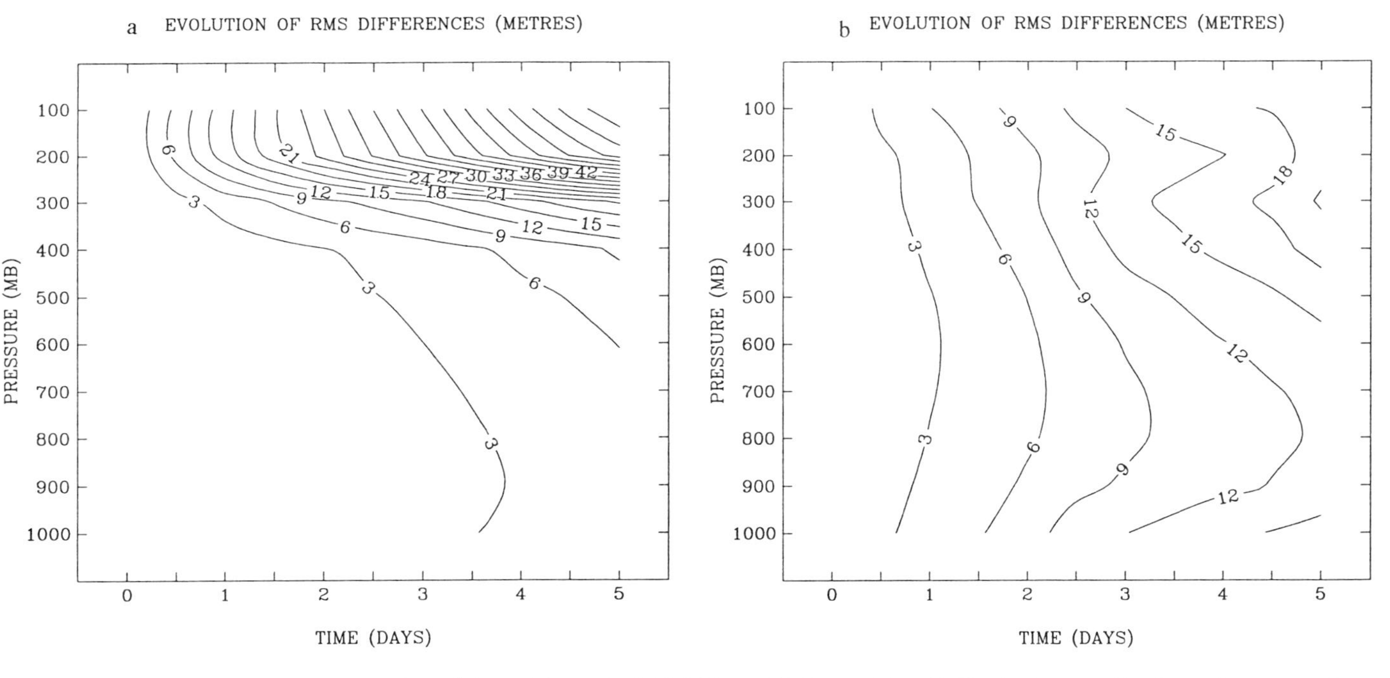

Fig. 6 Evolution of rms differences between baroclinic models: a) For 2-1/2 DISL and 2DISL models with 20 min time step. Results are in metres and are presented for 10 equally-spaced pressure levels, varying from 100 mb at the top to 1000 mb at the bottom of the figure, and at 12-hour intervals, from 0 at the left to 5 days at the right of the figure. Contour interval 3 m; b) For NISLV model with 40 min time step versus UVEUL model with 12 min time step. (Adapted with permission from Ritchie, 1991.)

of about 15 m after 5 days, confirming that the NISLV model produces stable and accurate results with a time step that far exceeds the CFL stability limit for the corresponding Eulerian model. The time step was further increased to 60 min. The model remained stable, but the rms differences indicated a marginal accuracy at this time step, with mid-atmospheric differences of about 20 m after 5 days.

5 Approximations and sensitivity experiments for operational implementation

These experiments showed that the semi-implicit semi-Lagrangian approach can be applied accurately and stably in a baroclinic spectral model using time steps that far exceed the CFL limit for the corresponding Eulerian model. However, this was demonstrated by means of intercomparison experiments for a single case using a hemispheric model with simple physical parametrizations. This model has been further developed and implemented as the global data assimilation and medium-range forecast model at the Canadian Meteorological Centre (CMC). This section summarizes the main optimization and model configuration extensions (Tanguay et al., 1992; Ritchie and Beaudoin, 1994) that were made in preparing the model for operational use.

Due to the overhead of the extra calculations required by the semi-Lagrangian technique for each time step, the net gain in efficiency can be significantly less than the factor by which the time step is increased. In particular, the study in the previous section used cubic three-dimensional interpolations throughout and also used very accurate algorithms for the trigonometric calculations in finding the upstream positions, and in the transformations required to treat the vector form of the equation of motion. It has been possible to use some approximations that significantly increase the efficiency of the semi-Lagrangian calculations without degrading the quality of the forecasts.

By combining the original semi-implicit semi-Lagrangian formulation and the spatial averaging of all the nonlinear terms that are treated explicitly in such three-time-level schemes, it was possible to reduce the distortion of topographically forced waves that are produced when the CFL number exceeds unity, as well as to increase model efficiency by significantly reducing the number of interpolations required by the algorithm (Tanguay et al., 1992). With this formulation in the NISLV version of the model, the only remaining three-dimensional interpolations are those used in the calculation of the upstream locations. The meteorological impact of reducing the accuracy of the vertical interpolation from cubic to linear, while retaining bicubic in the horizontal, was found to be negligible. For the horizontal interpolations the accuracy was reduced from cubic to linear on a first iteration and to quadratic on the second iteration, as justified by McDonald (1987), with negligible meteorological impact. Further optimizations were obtained by using spherical trigonometry and accurate approximations for the previously mentioned great circle trajectory calculations (Ritchie and Beaudoin, 1994). These changes have no significant meteorological consequences, but in a typical semi-Lagrangian model step they reduce the time spent in the semi-Lagrangian calculations to about 30%. Optimizing and multitasking this semi-Lagrangian spectral model enabled it to respect the operational deadlines at CMC and to be implemented as the global data assimilation and medium-range forecast model on 12 March 1991.

Model configuration sensitivity tests were also performed using a global version of the model with more sophisticated physics parametrizations and on a wider range of meteorological situations than were used in the previous study. In the following, results are presented for the Northern Hemisphere averaged over four FGGE cases: 2 January, 21 January, 12 February and 18 May 1979. The control model has T79 truncation in the horizontal and 21 levels with variable spacing in the vertical, and uses the semi-implicit semi-Lagrangian time integration scheme with a 30 minute time step. A Robert time filter (with a coefficient of 0.1 using the definition in the analysis given by Asselin (1972)) and a del-squared (10**5 m**2/s coefficient) horizontal diffusion are applied. This dynamical model is coupled with a physical parametrization that includes a planetary boundary layer based on turbulent kinetic energy, a surface layer based on similarity theory, solar and infrared radiation, large-scale precipitation, Manabe-type moist convection and gravity wave drag (see Girard et al., 1991 for more detail).

Figure 7a shows the evolution of the Northern Hemisphere rms differences between the semi-Lagrangian version of the model and an Eulerian version of the model, both using a 10 minute time step. It is seen that, even in these global runs with full physical parametrizations, the Eulerian to semi-Lagrangian conversion satisfies the sensitivity criterion of less than three metres per day at mid-atmospheric levels.

Figure 7b shows the evolution of the Northern Hemisphere rms differences between semi-Lagrangian runs using a 10 minute and a 30 minute time step, averaged over the four cases. This result suggests that, on average and with full physics, 30 minutes is currently about the upper limit for the time step based on acceptable time truncation errors, whereas 40 minutes was acceptable with simpler parametrizations (Fig. 6b). The CFL limit for the corresponding Eulerian model is about 10 minutes, and the overhead of the semi-Lagrangian calculations is about 30%, so that there is a significant gain in efficiency.

The control model has T79 truncation in the horizontal, whereas the former operational spectral model at CMC was run at T59. The sensitivity to horizontal resolution was tested by comparing the control model integrations to others run at T59, T99 and T119. In performing these experiments, the horizontal diffusion coefficient was adjusted to give the same e-folding time for the shortest resolvable scale. The main impact was found to be in the geopotential height biases, for which the day 5 profiles are presented in Fig. 8, where the variably dashed curve is for the T79 control model, the long dashed curve is for T59 resolution, the short dashed curve is for T99 resolution, and the solid curve is for T119 resolution. It is seen that there was improvement in going from T59 to T79 and that there is still more

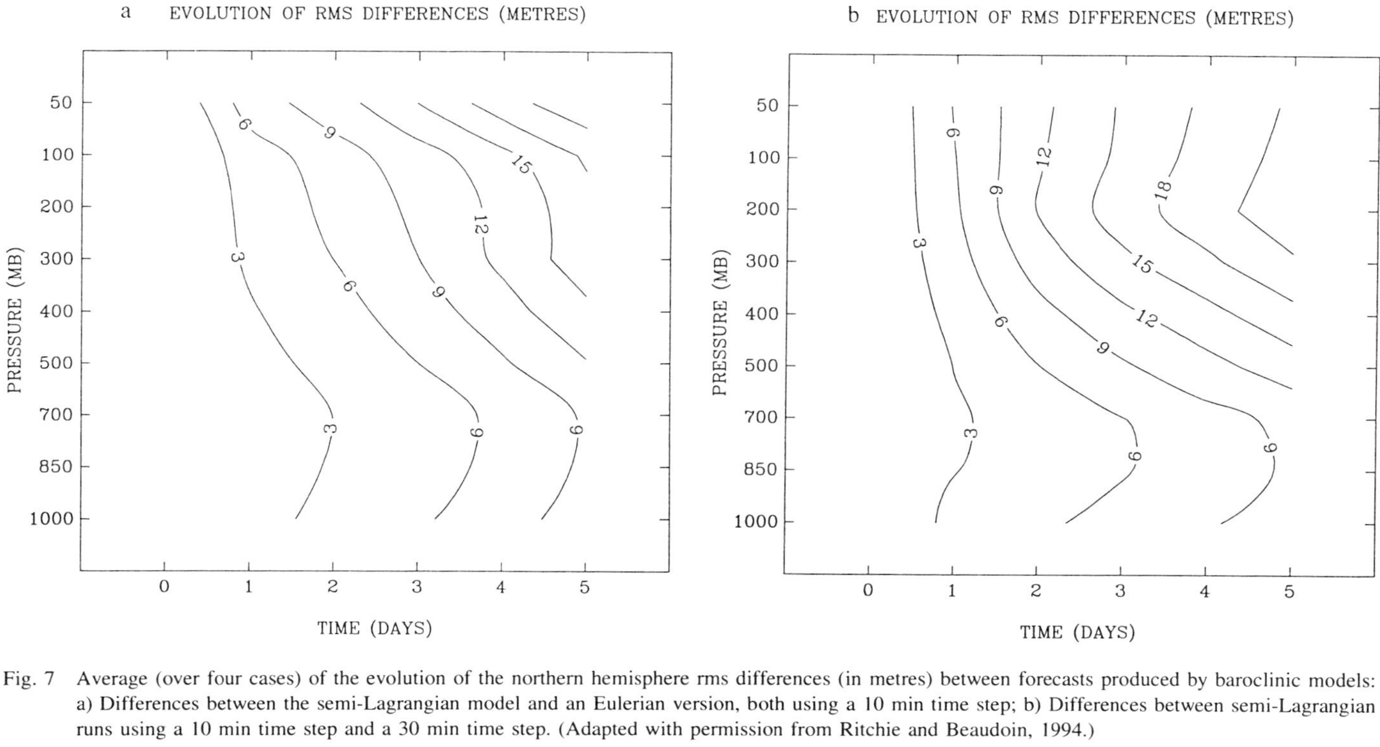

Fig. 7 Average (over four cases) of the evolution of the northern hemisphere rms differences (in metres) between forecasts produced by baroclinic models: a) Differences between the semi-Lagrangian model and an Eulerian version, both using a 10 min time step; b) Differences between semi-Lagrangian runs using a 10 min time step and a 30 min time step. (Adapted with permission from Ritchie and Beaudoin, 1994.)

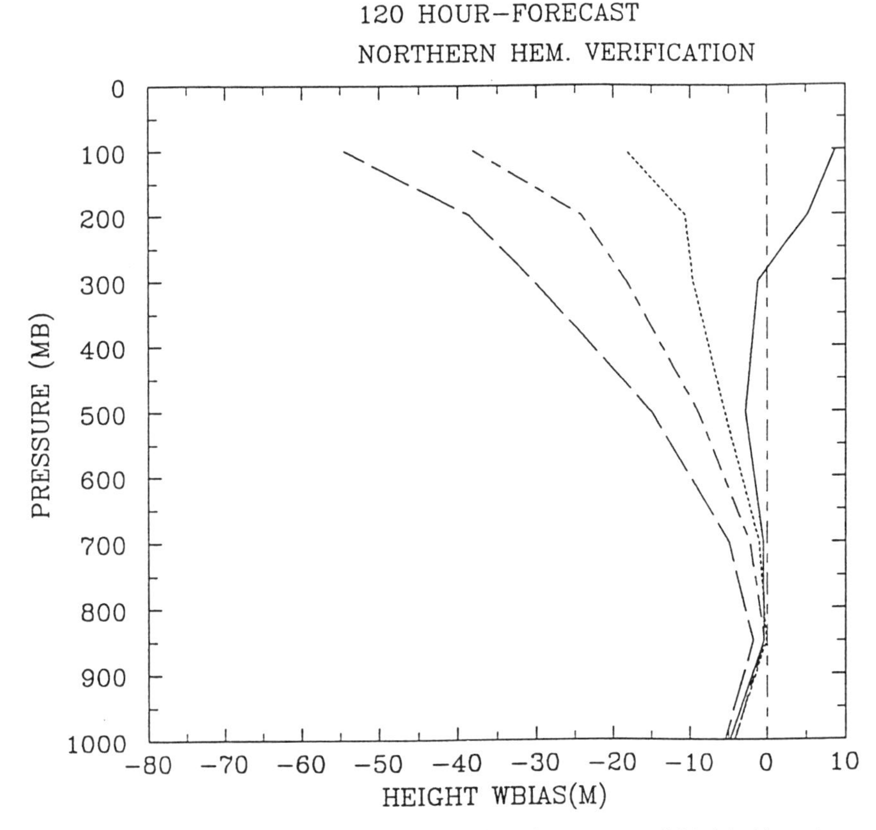

Fig. 8 Average (over four cases) of the vertical profiles of the geopotential height biases (metres, compared to analyses) of 5-day forecasts at various horizontal resolutions: T59 (long dashed), T79 (variably dashed), T99 (short dashed), and T119 (solid). (Adapted with permission from Ritchie and Beaudoin, 1994.)

improvement to come by further increasing the resolution to T119 and perhaps beyond.

6 Implementation in a high resolution version of the ECMWF forecast model

To date, the clearest test of the high resolution impact of the semi-Lagrangian scheme in spectral models has come with its implementation in the ECMWF forecast model (Ritchie et al., 1995). Novel aspects include the formulation for the ECMWF hybrid coordinate in the vertical, and its use in a baroclinic spectral model in conjunction with a reduced Gaussian grid in the horizontal as examined earlier in the context of simpler models (Hortal and Simmons, 1991). The former Eulerian vorticity-divergence formulation was first converted into a momentum-

TABLE 2. Analysis of CPU time (%)

	Eulerian	Fully interpolating semi-Lagrangian	Vertically non-interpolating semi-Lagrangian
Dynamics	21	15	17
Physics	53	42	45
FFT	6	3	4
Legendre transforms	20	13	14
Semi-Lagrangian	–	27	20

(Reproduced with permission from Ritchie et al., 1995).

equation formulation which is considerably more economical, thanks in part to the incorporation of Legendre transform efficiencies that were previously demonstrated for the shallow-water equations (Temperton, 1991).

The impact of the formulation changes has been assessed via numerical experiments on a set of 12 independent cases starting from operational analyses on the 15th of each month during the first year following the implementation of the semi-Lagrangian model on 17 September 1991. Unless otherwise stated, the experiments reported on here were performed with a high resolution version of the ECMWF forecast model having a T213 horizontal representation and 31 levels in the vertical (L31). The baseline semi-Lagrangian version is the vertically non-interpolating scheme (here denoted as VNISL) which has been used operationally at ECMWF since August 1992. The following figures compare the skill (mean anomaly correlations and rms height errors at 1000 and 500 hPa, averaged over the 12 cases) obtained with a control version and an experimental version of the model.

The main motivation for using a semi-Lagrangian formulation is to permit the use of time steps that far exceed the CFL stability criterion for the corresponding Eulerian model, thus enhancing the model efficiency, provided that the additional time truncation error does not significantly decrease the accuracy. Figure 9 shows the mean objective scores for the northern hemisphere comparing the Eulerian version with a 3 minute time step (solid) and the semi-Lagrangian version with a 15 minute time step (dashed). It is seen that the accuracies are almost equivalent, particularly for forecasts whose skill exceeds the 60% threshhold. Thus, even at this high resolution, the semi-Lagrangian scheme permits a fivefold increase in time step with no significant degradation in the quality of the forecasts. Table 2 shows the percentage breakdown of CPU time for the Eulerian, fully interpolating and vertically non-interpolating versions of the model at T213/L31. It is seen that the overhead of the semi-Lagrangian scheme in the baseline model is approximately 20%, so the semi-Lagrangian version gives an efficiency improvement of about a factor of four relative to the Eulerian. The results in Table 2 also suggest that the spectral method is still perfectly viable at this resolution, and that considerably higher resolutions can be achieved before the cost of the Legendre transforms becomes a matter for serious concern (e.g., Côté and Staniforth, 1990).

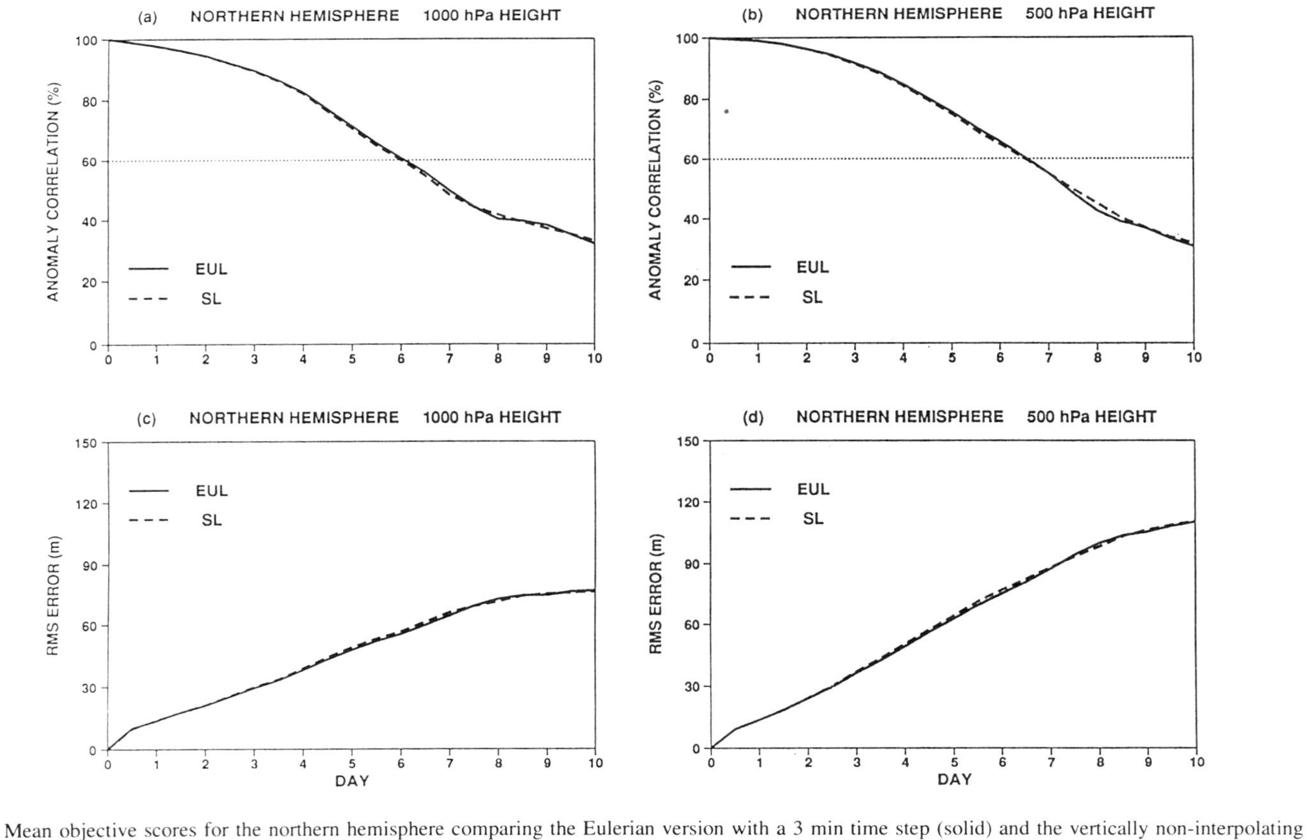

Fig. 9 Mean objective scores for the northern hemisphere comparing the Eulerian version with a 3 min time step (solid) and the vertically non-interpolating semi-Lagrangian version with a 15 min time step (dashed): (a) anomaly correlation of 1000 hPa height; (b) anomaly correlation of 500 hPa height; (c) rms error of 1000 hPa height (metres); (d) rms error of 500 hPa height (metres). (Reproduced with permission from Ritchie et al., 1995.)

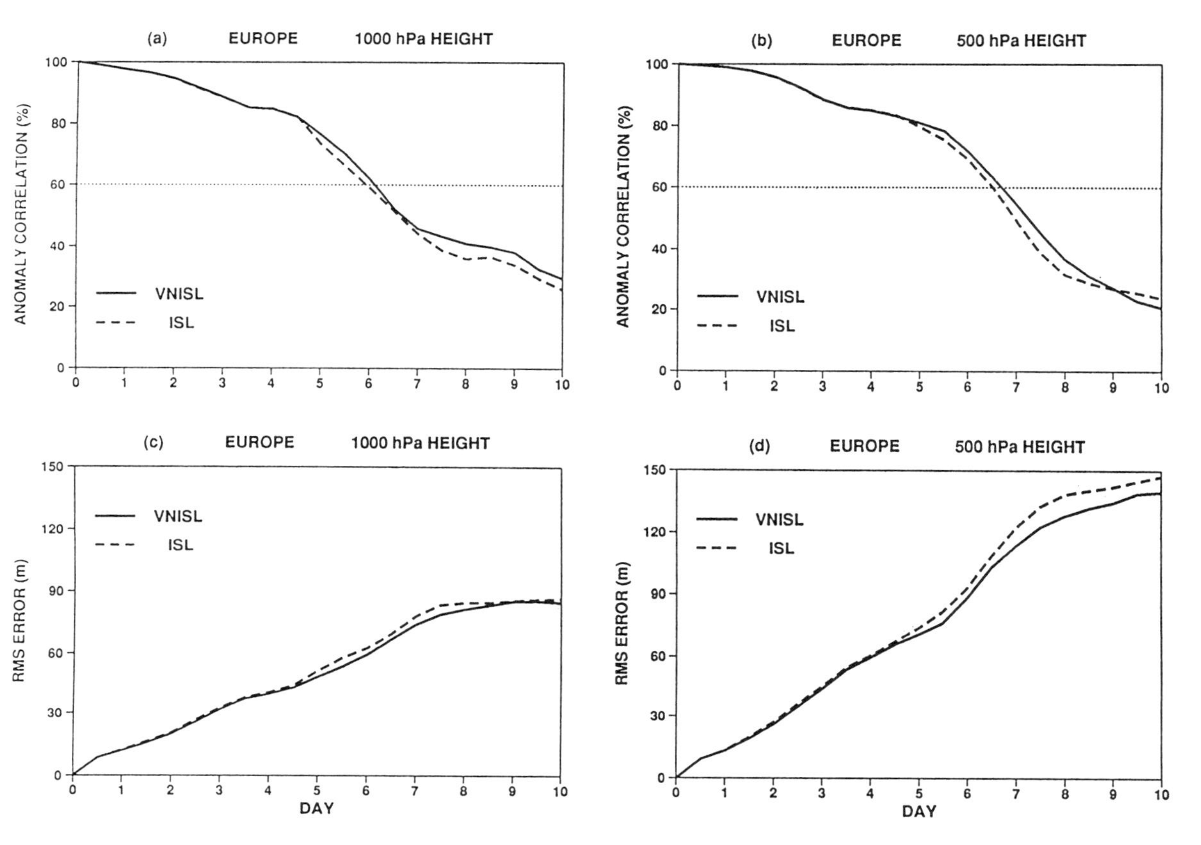

Fig. 10 As in Fig. 9 except for comparing vertically non-interpolating (solid) and fully interpolating (dashed) versions for the European region. (Reproduced with permission from Ritchie et al., 1995.)

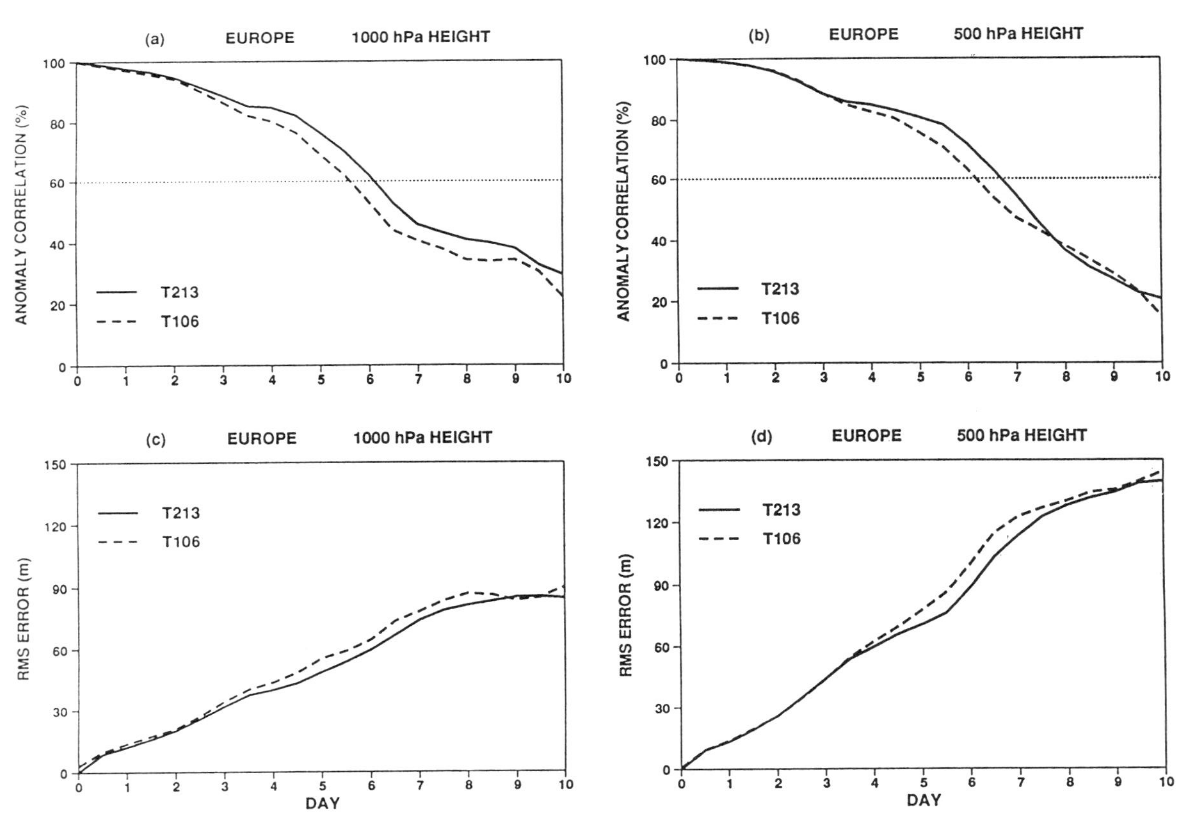

Fig. 11 As in Fig. 9 except for comparing T213 horizontal resolution (solid) and T106 horizontal resolution (dashed) for the European region. (Reproduced with permission from Ritchie et al., 1995.)

The measures of skill also indicate an improvement arising from the use of the vertically non-interpolating version as opposed to the fully interpolating form. This is evident from Fig. 10 which presents the comparison for the European region.

By virtue of its increased efficiency, incorporation of the semi-Lagrangian scheme was very important in enabling an increase in horizontal resolution from T106 to T213 in the operational ECMWF forecast model. Figure 11 documents the impact of this increase in horizontal resolution for the set of 12 cases studied here, for the European region. In these tests the same 31 level configuration was used for both horizontal resolutions. It is seen that this increase in horizontal resolution indeed had a very significant positive impact, which, in fact, is substantially greater than the impact of any of the other changes that were examined in this study.

7 Conclusion

This paper has reviewed the steps that have been followed in applying the semi-Lagrangian method to global spectral forecast models. After an initial examination of the geometric and linear stability questions in the treatment of advection on the Gaussian calculation grid used in spectral models, the semi-Lagrangian approach was combined with the semi-implicit scheme to get a stable treatment of both Rossby and gravity waves in a spectral model of the shallow water equations. These early phases of the work were done under André Robert's supervision and paralleled his previous demonstration of the efficiency advantage of the semi-implicit semi-Lagrangian method for gridpoint models. The extension to baroclinic models was then examined through an application to a multilevel spectral primitive equations model, where several issues related to vertical discretization were addressed. Following further optimization, semi-implicit semi-Lagrangian models have been implemented as operational forecast models at the Canadian Meteorological Centre and at the European Centre for Medium-Range Weather Forecasts. At ECMWF the semi-Lagrangian scheme described above has been operationally implemented in the Integrated Forecast System (IFS), developed in collaboration with Météo-France where it is known as ARPEGE (see Courtier et al. (1991) for an account of this project). In these applications the semi-Lagrangian method has realized its potential in spectral models by permitting the use of time steps that are much larger than those permitted by the CFL stability constraint for the corresponding Eulerian models. The increased efficiency has enabled the use of higher spatial resolutions, which have enhanced the meteorological performance of these models as evidenced by the improved scores of their medium-range forecasts. Other work currently in progress includes higher spectral resolution studies using aliased Gaussian grids permitted by replacing the strong nonlinear advection terms in conventional Eulerian spectral models by the semi-Lagrangian treatment of advection, following the initial demonstration by Côté and Staniforth (1988) for a shallow water spectral model.

The semi-Lagrangian method is also being investigated for use in climate simulations. In particular, Williamson and Olson (1994) have examined it for the NCAR

Community Climate Model (CCM2) model, and the impact of the aliased Gaussian grid has been studied in the context of climate simulations (Williamson, this volume). It is quite likely that the increased efficiency offered by the semi-Lagrangian method will benefit numerical models in a wide range of applications in atmospheric and oceanic sciences for years to come.

Acknowledgements

The author wishes to express his deep gratitude to André Robert for his supervision of the early phases of this work, and for his continuing interest, enthusiasm and encouragement until its completion. In a companion paper documenting an interview in 1987 (Ritchie and Robert, this volume), André expressed his long-term career objective as being several decades of high productivity, followed by the chance to pass his knowledge and experience on to the younger generation. André certainly succeeded in achieving this goal, with simultaneous productivity in many areas and teaching in many ways, and has left a very rich legacy for the whole numerical modelling community. Thanks are also expressed to other colleagues at RPN and ECMWF who have been co-authors on the research that has been summarized here: Christiane Beaudoin, Terry Davies, David Dent, Mats Hamrud, Mariano Hortal, Adrian Simmons, Monique Tanguay, Clive Temperton and Evhen Yakimiw. Monique Tanguay also thoughtfully reviewed this manuscript.

References

ASSELIN, R. 1972. Frequency filter for time integrations. *Mon. Weather Rev.* **100**: 487–490.

CÔTÉ, J. and A. STANIFORTH. 1988. A two-time-level semi-Lagrangian semi-implicit scheme for spectral models. *Mon. Weather Rev.* **116**: 2003–2012.

——— and ———. 1990. An accurate and efficient finite-element global model of the shallow-water primitive equations. *Mon. Weather Rev.* **118**: 2707–2717.

COURTIER, P.; C. FREYDER, J.-F. GELEYN, F. RABIER and M. ROCHAS. 1991. The ARPEGE project at Météo-France. *In*: Proc. ECMWF Seminar on Numerical Methods in Atmospheric Models, Reading, United Kingdom, 9–13 September 1991. ECMWF, pp. 192–231.

CRESSMAN, G.P. 1960. Improved terrain effects in barotropic forecasts. *Mon. Weather Rev.* **88**: 327–342.

DALEY, R.; C. GIRARD, J. HENDERSON and I. SIMMONDS. 1976. Short term forecasting with a multi-level spectral primitive equations model. ATMOSPHERE, **14**: 98–134.

GIRARD, C.; Y. DELAGE, J. MAILHOT, L. GARAND, B. BILODEAU, N. BRUNET and G. PELLERIN. 1991. Physical parameterization for the first Canadian global forecast model. *In*: Preprint Vol. Extended Abstracts: Ninth Conference on Numerical Weather Prediction, Denver, Colorado, 14–18 October 1991, pp. 502–505.

HORTAL, M. and A.J. SIMMONS. 1991. Use of reduced grids in spectral models. *Mon. Weather Rev.* **119**: 1057–1074.

MANABE, S.; J. SMAGORINSKY and R.F. STRICKLER. 1965. Simulated climatology of a general circulation model with a hydrologic cycle. *Mon. Weather Rev.* **93**: 769–798.

MCDONALD, A. 1987. Accuracy of multiply-upstream semi-Lagrangian advective schemes II. *Mon. Weather Rev.* **115**: 1446–1450.

RITCHIE, H. 1987. Semi-Lagrangian advection on a Gaussian grid. *Mon. Weather Rev.* **115**: 608–619.

———. 1988. Application of the semi-Lagrangian method to a spectral model of the shallow water equations. *Mon. Weather Rev.* **116**: 1587–1598.

———. 1991. Application of the semi-Lagrangian method to a multilevel spectral primitive-equations model. *Q. J. R. Meteorol. Soc.* **117**: 91–106.

——— and C. BEAUDOIN. 1994. Approximations and sensitivity experiments with a baroclinic semi-Lagrangian spectral model. *Mon. Weather Rev.* **122**: 2391–2399.

———; C. TEMPERTON, A. SIMMONS, M. HORTAL, T. DAVIES, D. DENT and M. HAMRUD. 1995. Implementation of the semi-Lagrangian method in a high resolution version of the ECMWF forecast model. *Mon. Weather Rev.* **123**: 489–514.

ROBERT, A. 1981. A stable numerical integration scheme for the primitive meteorological equations. ATMOSPHERE-OCEAN, **19**: 35–46.

———. 1982. A semi-Lagrangian and semi-implicit numerical integration scheme for the primitive meteorological equations. *J. Meteorol. Soc. Jpn.* **60**: 319–325.

———; J. HENDERSON and C. TURNBULL. 1972. An implicit time integration scheme for baroclinic models of the atmosphere. *Mon. Weather Rev.* **100**: 329–335.

SAWYER, J.S. 1963. A semi-Lagrangian method of solving the vorticity advection equation. *Tellus*, **15**: 336–342.

TANGUAY, M.; E. YAKIMIW, H. RITCHIE and A. ROBERT. 1992. Advantages of spatial averaging in semi-Lagrangian schemes. *Mon. Weather Rev.* **120**: 113–123.

TEMPERTON, C. 1991. On scalar and vector transform methods for global spectral models. *Mon. Weather Rev.* **119**: 1303–1307.

WILLIAMSON, D.L. and J.G. OLSON. 1994. Climate simulations with a semi-Lagrangian version of the NCAR CCM2. *Mon. Weather Rev.* **122**: 1594–1610.

Preliminary Study on SST Forecast Skill Associated with the 1982/83 El Niño Process, Using Coupled Model Data Assimilation

K. Miyakoda, J. Ploshay and A. Rosati
GFDL/NOAA Princeton University,
Princeton, NJ 08542, U.S.A.

[Original manuscript received 1 May 1995; in revised form 22 August 1995]

ABSTRACT *A previous study by Rosati et al. (1997) has concluded that the specification of an adequate thermocline structure along the equatorial Pacific ocean is most crucial for El Niño forecasts. In that paper, the oceanic initial condition was generated by a data assimilation (DA) system (Derber and Rosati, 1989). However, the initial condition for the atmospheric part was taken from the National Meteorological Center's (NMC) operational analysis, which was simply attached to the oceanic part for the coupled model forecasts.*

In the present paper, both the atmospheric and oceanic initial conditions are generated by a coupled DA system applied to a coupled air-sea general circulation model (GCM). The assimilation for the ocean is performed by the same system as mentioned above, in which the SST (sea surface temperature) and the subsurface temperatures are injected into a 15 vertical level oceanic GCM. The upper boundary condition, such as surface wind stress, is specified by the atmospheric DA. The assimilation for the atmosphere is performed by the continuous injection method of Stern and Ploshay (1992), using an 18 vertical level atmospheric GCM. The lower boundary condition, such as SST, is specified by the oceanic DA. The coupled model assimilations are carried out by switching the DA processes alternately every 6 hours between the ocean and the atmosphere.

The emphases of this study are: firstly, the effect of coupled air-sea model DA on the performance of subsequent forecasts; secondly, the impact of the coupled assimilation on improvement of the "spin-up" behaviour of forecasts, i.e. to see whether a smooth start to the forecast is achieved by the coupled model DA process; and thirdly, investigation of the effect that the "spring barrier" has on predictability in the coupled GCM system. Preliminary results indicate that, in order to answer these questions, ensemble forecasts are necessary. Besides, the coupled assimilation could be important in improving the overall behaviour of El Niño and La Niña forecasts.

RÉSUMÉ *Une étude antérieure de Rosati et al. (1955) a opiné que la prévision du El Niño dépend grandement de la spécification d'une structure thermocline adéquate sur l'océan Pacifique équatorial. Dans cette étude, la condition initiale océanique utilisait le système d'assimilation des données (DA) de Derber et Rosati (1989). Toutefois, la condition initiale de la partie atmosphérique était constituée de l'analyse régulière du NMC (National Meteorological center) qui était simplement rattachée à la partie océanique pour les prévisions du modèle couplé.*

Ici les conditions initiales, tant atmosphériques qu'océaniques, sont générées par un système DA couplé appliqué à un modèle de circulation générale (GCM) couplé air-mer. L'assimilation pour l'océan est effectuée par le même système que ci-dessus, dans lequel la température superficielle de la mer (SST) et les températures sous la surface sont injectées dans un GCM océanique à 15 niveaux verticaux. La condition à la couche limite supérieure, p. ex. la force d'entraînement du vent de surface, est déterminée par le système DA atmosphérique. L'assimilation pour l'atmosphère est accomplie par la méthode d'injection continue de Stern et Ploshay (1992), utilisant un GCM atmosphérique de 18 niveaux verticaux. Les conditions à la limite inférieure, telle que la SST, sont déterminées par le système DA océanique. Les assimilations du modèle couplé sont effectuées en enclenchant alternativement les processus d'assimilation de l'océan et de l'atmosphère à toutes les six heures.

L'étude porte surtout sur : l'effet du modèle avec système DA couplé air-mer sur la performance des prévisions subséquentes; l'impact de l'assimilation couplée sur l'amélioration du comportement de départ des prévisions (p. ex., un départ doux de la prévision résulte-t'il du système DA du modèle couplé) ; l'investigation de l'effet que la «barrière printannière» a sur la prévision dans le système de GCM couplé. Les premiers résultats indiquent que des prévisions d'ensembles sont nécessaires afin de répondre à ces questions. En outre, l'assimilation couplée pourrait être importante pour améliorer le comportement général de la prévision du El Niño et de LA Niña.

---

1 Introduction

Tropical oceanic-atmospheric forecasts have been considerably improved, using simple dynamical models or statistical methods (see Barnett et al., 1988). In this paper, however, the approach will be exclusively based on the coupled atmosphere-ocean general circulation models (GCMs). The issue is the seasonal forecasts, which treat the global oceanic and atmospheric states on the time ranges of about one year.

Concerning the activities related to the low frequency variation of the atmosphere, it is known (see for example, Brankovic et al., 1994) that the El Niño/Southern Oscillation signal is most dominant. This suggests that the forecast of ENSO should be of greatest concern in order to unravel the possibility of seasonal forecasting, and thereafter, further search for predictable elements should be pursued over other areas of the globe.

The simple model approach for El Niño prediction, such as Cane and Zebiak (1985) and Cane et al. (1986), treats the anomaly components of variables. On the other hand, the GCM approach has to treat the total components. This aspect presents one of the difficult and yet challenging problems in forecasts. An example is obtaining a good seasonal cycle of SST in the eastern equatorial Pacific, i.e., NINO-3 (150°W–90°W, 5°N–5°S) region, which is a formidable task. According to GCM experience (see for example, Neelin et al., 1992; Miyakoda et al., 1993), the coupling process is extraordinarily sensitive to the character of the atmospheric part of the physics and orographic or coast-line specifications.

Rosati et al. (1997) have shown successful forecasts, using a coupled oceanic-atmospheric GCM. The results are very encouraging. However, there are several issues that have to be made clear. They are: the limit of predictability, the spring barrier and the reduction of forecast errors associated with the initial spin-up problem. In order to investigate these issues, ensemble forecasts are useful or even required. If so, an adequate scheme for constructing the initial condition is essential. The main objective of this paper is to describe a DA system with a coupled oceanic-atmospheric GCM.

2 Background

Rosati et al. (1996) reported the results of 13-month forecasts, using a coupled atmospheric-oceanic GCM with fine equatorial resolution.

a *Model*

The atmospheric part has spectral triangular truncation at zonal wavenumber 30, corresponding to 4.0° longitude × 4.0° latitude (see Laprise, 1993) and 18 vertical levels. On the other hand, the oceanic part has a 1° × 1° longitudinal and meridional resolution outside of 10°N–10°S and 1/3° inside of the equatorial zone (Philander and Seigel, 1985). A set of adequate subgrid-scale physics for the atmosphere and ocean model is included. The prediction domain is the entire global atmosphere and the world ocean between 65°N and the Antarctic. All forecasts in this paper *do not* include flux corrections (Sausen et al., 1988).

The atmosphere and ocean models are coupled by exchanging the fluxes of momentum, heat, radiation and the SST specification with each other.

b *Initial conditions*

The inclusion of adequate information in the initial conditions is of considerable importance for weather forecasts; it may also be true for seasonal forecasts.

As the first step toward coupled model prediction, a scheme of oceanographic DA was developed by Derber and Rosati (1989). This was based on the variational method, in which various constraints are included to specify dynamical formulae and observational accuracies. Through these constraints, the observed SST and subsurface data, XBT (expendable bathythermograph) and others, are injected into the ocean model through the optimum interpolation scheme (OI). The upper boundary conditions of the ocean GCM are specified by surface wind stress, atmospheric heat flux, moisture flux and incoming shortwave and net longwave radiation. Most of these data are taken from the NMC (National Meteorological Center) operational analysis, with the exception that the radiation is given by the seasonally varying climatologies. The ocean model is integrated forward in time for many years (1979–1988), while these data are continuously injected and assimilated into the model (see Rosati et al., 1995 for more detail).

c *Cases*

The forecast cases consist of 7 episodes starting from January or July every year for the 7 years of 1982–1988. During this 7-year period, there are two distinct El

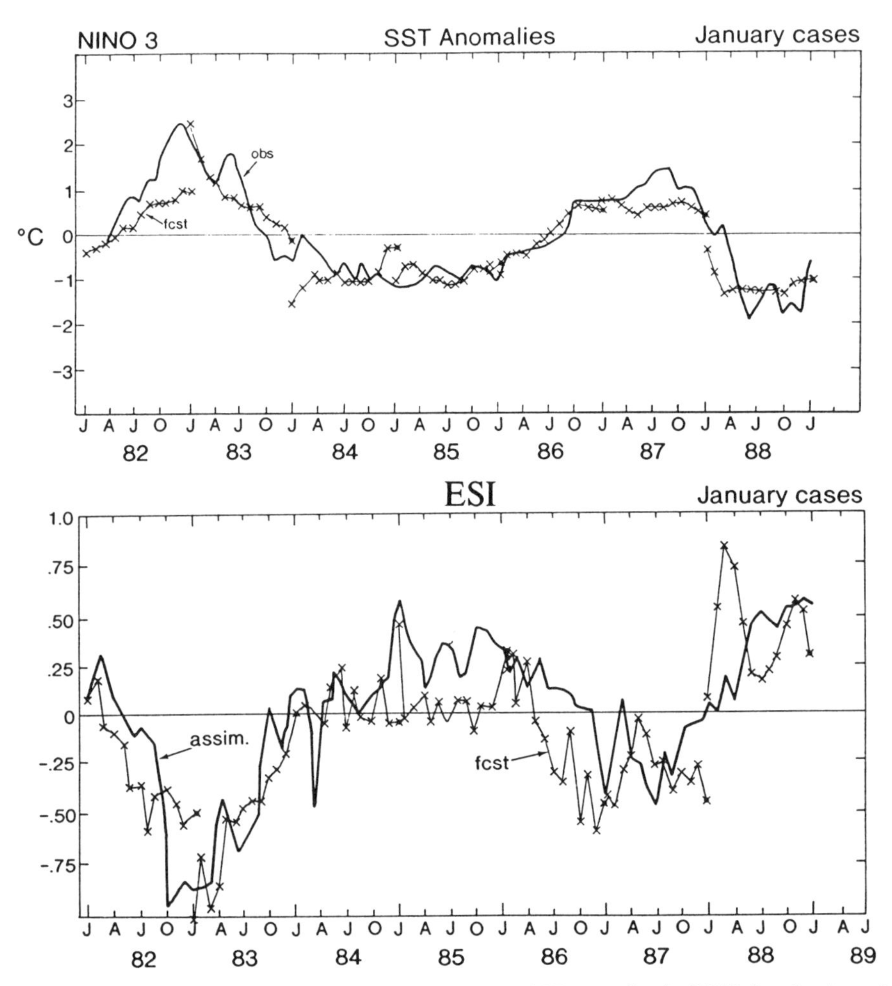

Fig. 1 Performance of the 13-month forecasts is shown by SST anomalies for NINO-3 region (upper) and ENSO indices, i.e. *ESI* (lower). Thick curves are the observations, and the curves connected with crosses are the forecasts (after Rosati et al., 1997).

Niño events and two events of La Niña. The initial condition time is 0000 GMT on the first day of January or July. The time range of the forecasts is 13 months.

d *Prediction skill*

The results of the 7 January forecasts are displayed in Fig. 1. For each month of the forecasts, the monthly mean SST anomaly was computed over the NINO-3 region and is shown in the upper panels. The lower panel is the zonal mean heat content anomaly index or ENSO index, i.e. ESI, which will be explained below.

For the atmosphere alone, the SOI (Southern Oscillation Index) is a traditional indicator of the low-frequency atmospheric oscillation, which operates over the semi-global domain (Troup, 1965; Trenberth, 1976). On the other hand, for the coupled system, other indices appear to be more appropriate. The first measure is of the west-east tilt of the 20° isotherm along the equatorial Pacific, which is defined by

$$GRD = (D_w - D_e)/110 - 1 \tag{1}$$

where D is the depth of the 20° isotherm, and D_w and D_e are its depths at 160°E and 90°W respectively. The mean value of $D_w - D_e$ is 110 metres, i.e. $160 - 50 = 110$. When *GRD* is negative, the situation corresponds to the warm phase in the eastern tropical Pacific, i.e. El Niño, while when the *GRD* is positive, the situation corresponds to the cold phase, i.e. La Niña.

In order to eliminate the seasonality of the DA, the anomaly of *GRD* is calculated, in a similar way as for the SOI. This anomaly is referred to as the ENSO index, or simply *ESI*, i.e.

$$ESI = GRD - \overline{GRD} \tag{2}$$

where $\overline{GRD}$ is the average over multiple years for the respective month, and it is, therefore, a function of Julian day (or month).

The continuous thick line curves in Fig. 1 are observed SST anomalies for the NINO-3 region (upper panel), after Reynolds (1988), and the *ESI* from the DA (lower panel). The thin line curves with crosses are the SST and *ESI* anomalies from the 7 forecast cases, each run for 13 months. The terminal points of the 13-month forecasts are indicated by black dots with crosses. Although the forecasts started from the DA, the first month does not agree exactly, because it is a monthly mean of the forecast.

The SST forecasts, shown in the upper panel, agree quite well with the observations, capturing the temperature increases in 1982/83 and 1986/87 and also the temperature decreases in 1983/84 and 1987/88. During 1984/85 when the anomalies had little change the forecasts again did well. The lower panel is the *ESI* and it should be noted how well it correlates with the NINO-3 SST anomaly. The forecasted *ESI* seems to capture the anomalous behaviour of the heat content associated with the rise and fall of the thermocline and hence yields a good SST prediction.

An additional measure is "the zonal mean of thermocline anomalies" (Li, personal communication), as opposed to the gradient of the thermocline anomalies, i.e. *ESI*. The second measure is of the zonal mean of the 20°C isotherm along the equatorial Pacific, which is defined by

$$MEAN = (D_w + D_E)/210 \tag{3}$$

where $210 = 160 + 50$ see (1). In order to eliminate the seasonality, the anomaly of *MEAN* is calculated. This anomaly is referred to here as the *memory index.*

$$memory\ index = MEAN - \overline{MEAN} \tag{4}$$

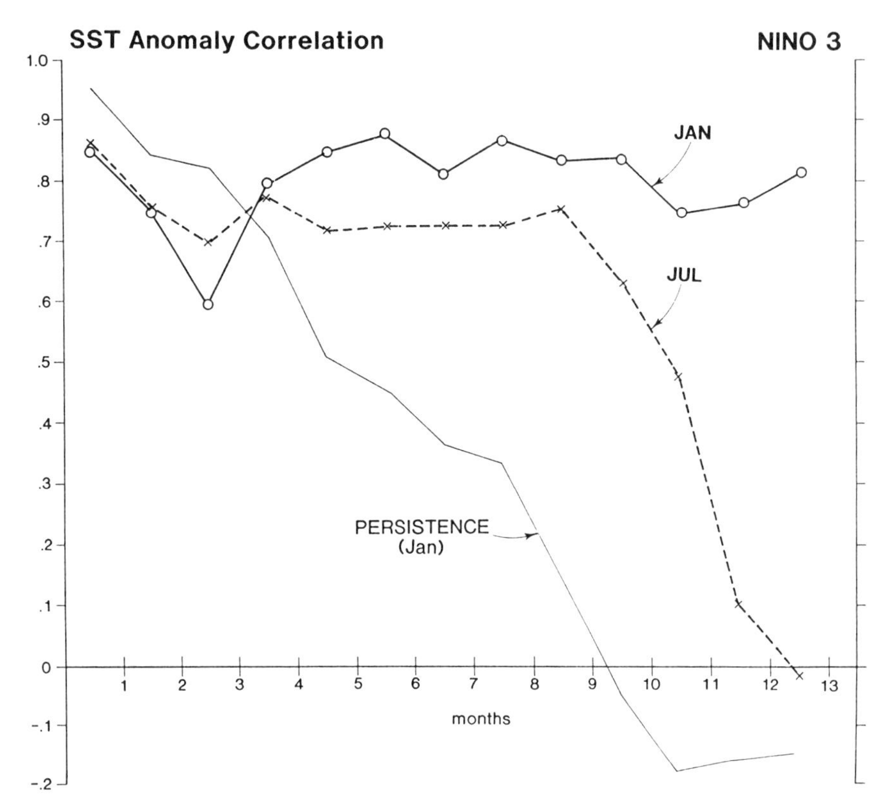

Fig. 2 NINO-3 SST anomaly correlations over the 13-month forecast period. Persistence is also shown for comparison (after Rosati et al., 1997).

According to Schneider et al. (1995), the variables, i.e. the zonal mean heat content anomalies, (4), are not in equilibrium with the wind stress anomalies, implying that they indicate the time tendency of heat content, whereas the heat content anomalies with the zonal mean removed, i.e. *ESI*, are approximately in equilibrium with the wind stress anomalies. This measure will be used later for showing the predictive skill of forecasts.

Figure 2 is a summary of the upper panel in Fig. 1, presenting the NINO-3 skill score curves against prediction time for 7 January and 7 July cases. This diagram confirms the already mentioned facts about the longevity of ENSO forecasts; the scores are significantly above the persistence except for the first 3 months. On the other hand, the far-field forecast scores, i.e. outside of the equatorial zone (not shown here), are limited up to six months, and besides the skill is very low; the scores are below the persistence.

The July score in Fig. 2 is almost comparable to the January score up to the 9th month, and thereafter, it decays rapidly, whereas the persistence curves are almost the same for both the January and July cases. It may be worth noting that the scores are worse than persistence in the first 3 months, and that this feature is very similar both in the January and July cases.

3 Issues on forecast skill

There are a number of issues related to the ENSO forecasts. The subjects we are interested in here are: Forecast error growth, Spring barrier, and Initial dip of forecast skill.

a *Forecast error growth*

According to some model-twin experiments (Gent and Tribbia, 1993), if initial perturbations are given to the vertical temperature distribution of the ocean, the SST error increases rapidly, for example, from 10^{-4}°C to 10^{-1}°C in about two weeks, and as a result, the correlation coefficients (15°N–15°S Pacific basin) drop to 0.5 after 4.5 months. This degree of error growth is normally inevitable, and is known as *predictability decay*. On the other hand, the correlation curves in Fig. 2 decay more slowly than those of Gent and Tribbia. The reason for the high correlation in Fig. 2 may be the fact that El Niño and La Niña forecasts are included, and that the successful forecasts of these extreme events have contributed to raising the forecast skill curves higher than that of Gent and Tribbia.

b *Spring barrier*

It has been argued, based on the Cane-Zebiak model (Cane and Zebiak, 1985), that there is a "prediction barrier" in the spring season and, as a consequence, that the forecasts starting from January are worse than those from July (see Zebiak and Cane, 1987; Blumenthal, 1991; Latif and Graham, 1991; Goswami and Shukla, 1993), though the view has recently been revised (Dr. Busalacchi, personal communication). Webster and Yang (1992) also show, from the lag-lead correlations of the SOI, that May and April emerge as the discontinuity season. As Balmaseda et al. (1994) mentioned, based on the analysis of their simple model forecasts, the correlation skill shows a pronounced drop during spring, often followed by a recovery, due to its memory of the ocean heat content. In other words, the information from the heat content and SST is not lost simultaneously. Comparing the forecasts between the January and July cases in Fig. 2, it appears that the skill score is higher in the January than in the July cases, and that the July score drops in April. However, the sample number in Fig. 2 is too limited to confirm the barrier issue.

c *Initial dip of forecast skill*

In Fig. 2, an inferior performance is evident in the first 3 months. Two possible causes can be considered. One is the "climate drift", because of the model's bias compared with the truth. Another is the improper adjustment of the initial condition to the model's climatology.

The first issue is outside of the scope of the present paper. Concerning the second issue, it appears that a substantial improvement of the initial data and its initialization is desired, perhaps including the precursor of the westerly bursts appropriately (Luther et al., 1983), and refining the model's interactive clouds properly (Gent and Tribbia, 1993). This inferior performance is a definite drawback in the current scheme of Rosati et al. (1997). Investigation should be made as to what extent the coupled model DA can improve this spin-up behaviour.

4 Data assimilation for the coupled system

In order to improve the spin-up problem, and facilitate the ensemble forecasts, a system of data assimilation is developed for the coupled atmospheric-oceanic model.

a *Schemes of the coupled model DA*

The oceanic DA was described in Section 2. The level II oceanic data are: COADS (Comprehensive Ocean Atmospheric Data Set), MOODS (Master Ocean Observation Data Set), NODC (the National Oceanic Data Center), and TOGA (Tropical Ocean and Global Atmosphere Project), which provide the data of surface ocean temperature and the vertical temperature profiles through the XBT.

The atmospheric DA is also based on a continuous injection method but in a different way (Stern and Ploshay, 1992). Level II meteorological data, such as radiosonde measurements, satellite soundings, aircraft reports, etc., are assimilated into the atmospheric GCM, where these data are injected continuously through the OI scheme into the atmospheric model with the help of linear normal mode initialization (Daley and Puri, 1980). In this process, the observed data are treated by taking only the incremental part beyond the first guess, i.e. forecast, and applying the increments to the linear balance in the multivariate framework. This atmospheric system and the resulting analyses have proven to be of comparable accuracy to those of operational centres in 1985 (Ploshay et al., 1992).

In particular, it is a salient feature of this continuous scheme that the initial condition does not produce any spin-up or spin-down effect. On the other hand, the intermittent scheme produces this transient character through the non-linear normal mode initialization, though there has been an effort to reduce this deficiency by introducing non-adiabatic effects (Wergen, 1987).

Using these continuous DA schemes of atmosphere and ocean, two systems of coupled model DA are developed.

1 THE FIRST COUPLED DA

The assimilation is performed through switching the DA for the ocean and the atmosphere alternately at a 6-hour interval. Namely the atmospheric DA is run for 6 hours by inserting the level II meteorological data. The 6-hour averages of wind stress, heat fluxes, and long and short wave incoming radiation are saved to force the ocean model. The ocean model is then run for 6 hours by inserting the level II oceanographic data, and returns the 6-hour averages of SST to be used during the next 6 hours by the atmospheric model.

This coupled DA method requires a 20% increase in computer memory above coupled forecasts. The technical difficulty was to set up the switching process between the atmosphere and the ocean DA. In any event, this technique has been completed and tested successfully. The computer time for the coupled DA is about three times that for the coupled simulation.

2 THE SECOND COUPLED DA

In order to quickly see the effect of coupled DA, a different method was also tested. Using the same DA systems of the atmosphere and the ocean, described under the first coupled DA, the oceanic and the atmospheric DA were processed separately for 10 days, and this separate process was iterated twice. The computer time is therefore twice that for the coupled simulation.

b *Results of the DAs*

The coupled model DAs were applied to construct the 1982 initial conditions. The DA process for the coupled system was started at 10 days before 0000 GMT 1 January 1982. However, the oceanic DA alone had already been run from 1979.

Figures 3a, 3b and 4 show the comparison of resulting atmospheric analysis between the NMC's DA and the coupled model DAs (the first version). The variables are temperature (Fig. 3a) and the zonal wind at the 1000-hPa level (Fig. 3b). There is a tendency for the zonal wind in the coupled DA to be more intense than those of NMC. Figure 4 shows the vertical sections of ocean temperature along the equator, which delineate the thermocline structure in the Pacific. In January 1982, there is already a tendency for the eastern part of the thermocline to be depressed. The top panel is based on the observed atmospheric forcing of the NMC analysis, which is the original DA (Rosati et al., 1995). The middle panel is the result of the first coupled model DA, and the bottom panel is that of the second coupled DA. The difference between the first and the second method is very subtle. The thermoclines (20°C isotherm, for example) are shallower at around 120°W in the coupled DA than in the original DA (top). Furthermore, the second version (bottom) is even shallower than the first version, implying that the structure is more adjusted to the thermocline climatology of this particular coupled model.

It is customarily considered that one of the best ways to evaluate the DA products is to utilize them for forecasts. Using these analyses, 13-month forecasts were carried out. Figures 5a and 5b are Hovmöller (longitude-time) diagrams representing the time evolution of the SST (top), and the depth of 20°C isotherms (bottom). Among them, Fig. 5a does not show forecasts, but the oceanic DA (left), and the ocean simulation (right). The SST fields (left top) are based on Reynolds (1988), and the 20°C isotherm depth (left bottom) is based on Rosati et al. (1995). On the other hand, the top and bottom of the right hand side figures show the results of ocean simulation. Both the ocean DA and simulation are performed by forcing the

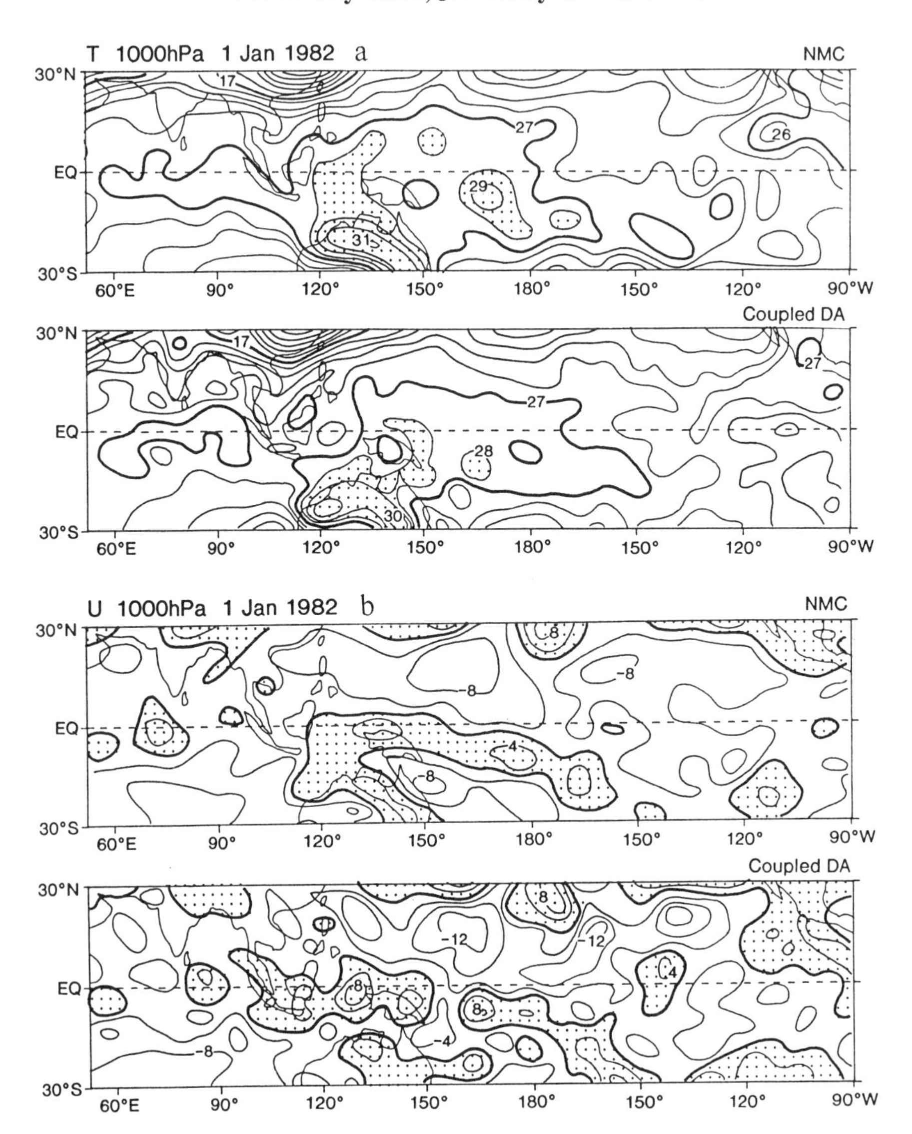

Fig. 3 (a) Comparison of air temperature at 1000-hPa level between the NMC analysis (left) and the coupled DA (right). Units are °C. (b) The same as Fig. 3a but for the zonal wind stress at 1000-hPa level. Units are m s^{-1}.

same observed atmospheric DA at the ocean surface. The only difference between the oceanic DA and the simulation is that the XBT data are used in the former but not in the latter. It may be surprising to note that the patterns of the 20°C isotherm are quite different from each other; it is simply the consequence of the XBT data injection in the former but not in the latter.

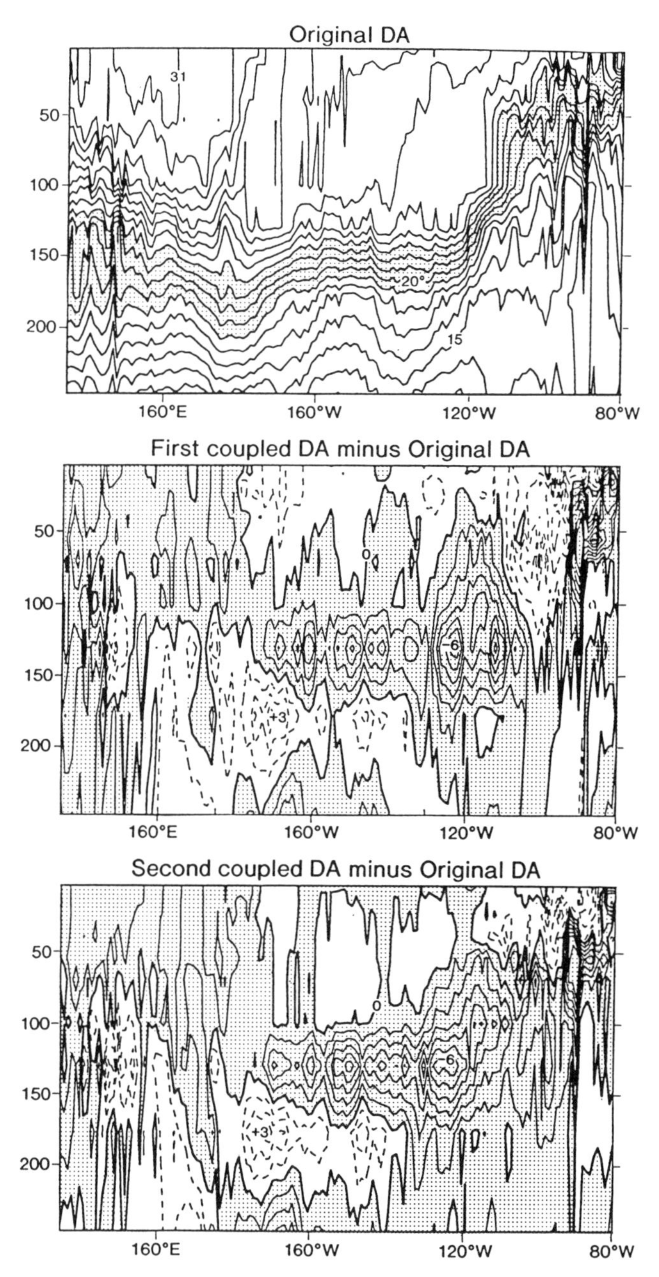

Fig. 4 Longitude-depth section of ocean temperature for January 1982. The oceanic DA (top), the difference of the first coupled DA from the top (middle), and the same for the second coupled DA (bottom). Contour interval is 1°C. Stippling is 19°–23°C (top), and negative values (middle and bottom).

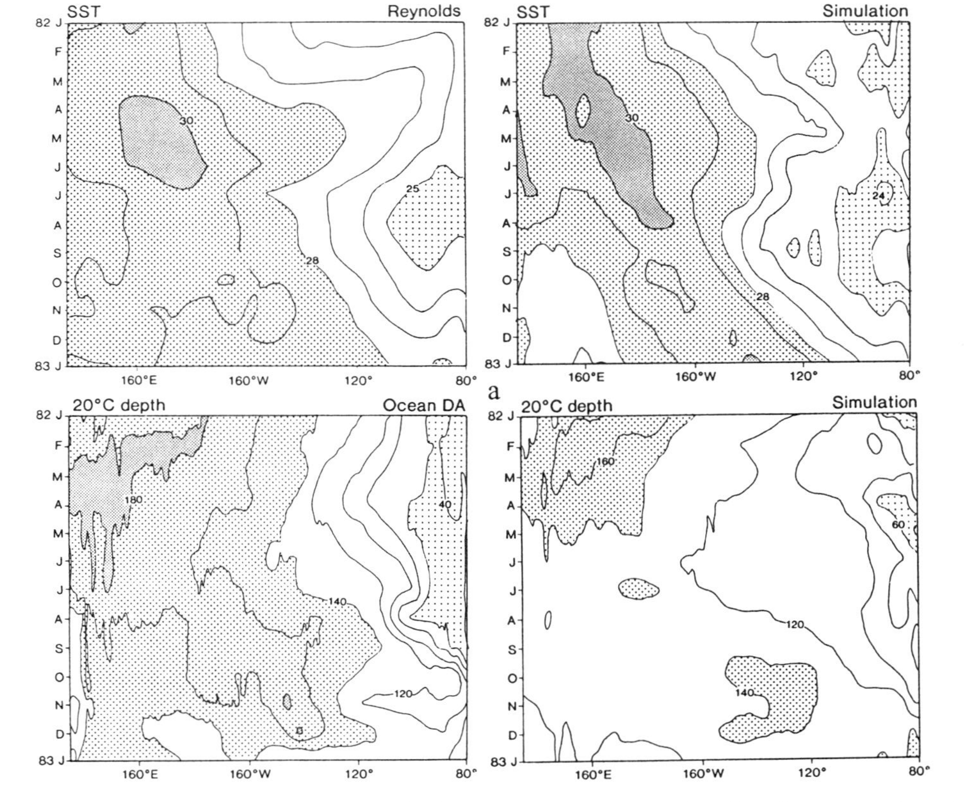

Fig. 5 (a) Longitude-time (Hovmöller) diagrams of SST (top) and 20°C isotherm depth (bottom). The SST of Reynolds analysis (left top), the 20°C depth based on the oceanic DA (left bottom), the SST in simulation (top right), and the 20°C depth (bottom right). Contour intervals are 1°C for SST and 10 m for 20°C depth. (b) The same as Fig. 5a, but all are the results of forecasts. The original (left) is the forecast from the initial conditions of oceanic DA and NMC atmospheric DA; and the forecasts (middle) and (right) are based on the first coupled DA and the second coupled DA.

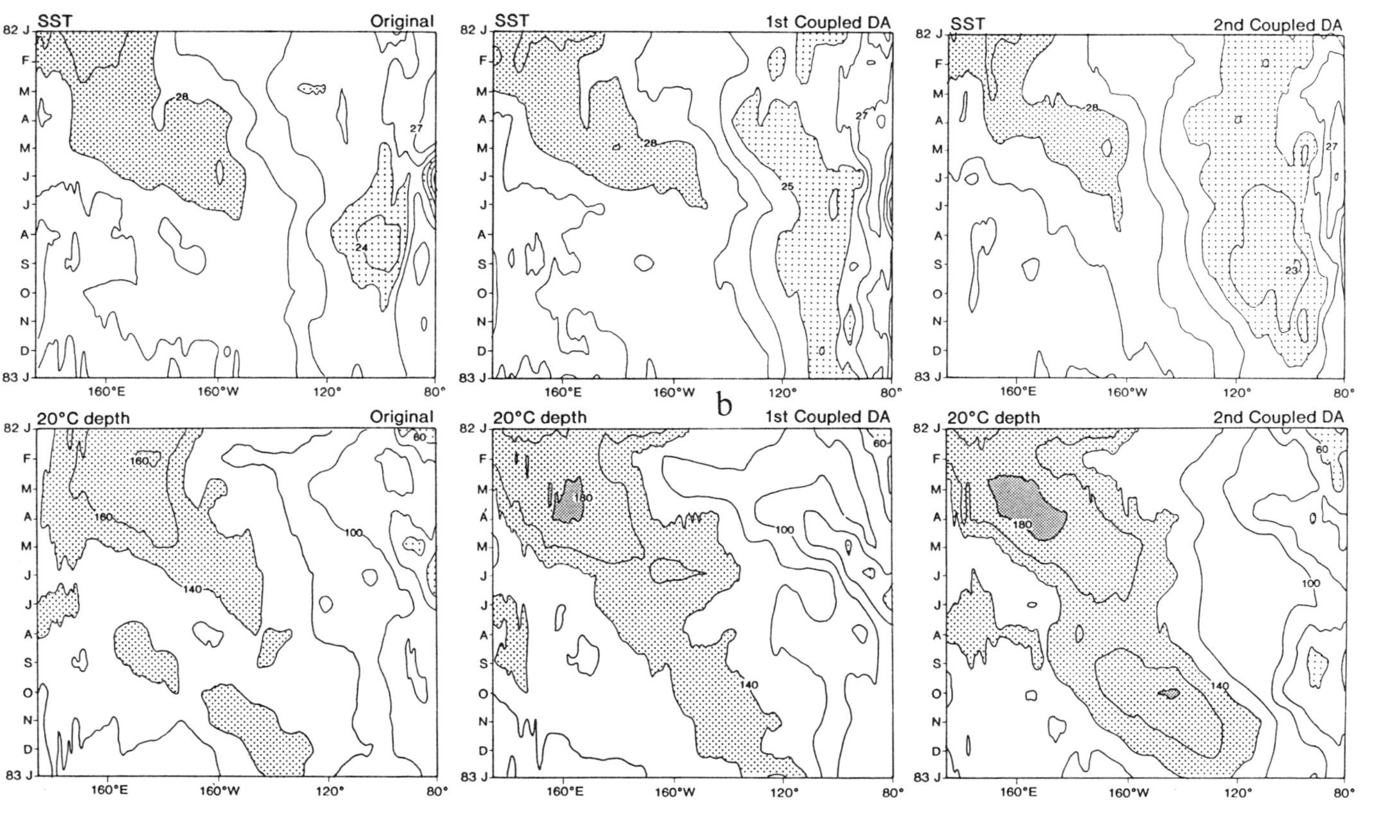

Fig. 5 *(concluded).*

Figure 5b shows the three forecasts. These results are compared with those of Fig. 5a. The forecast at the left is the same as that of Rosati et al. (1996), which is based on the NMC and the oceanic DA analyses for the initial condition. The forecasts in the middle and at the right are the results from the two initial conditions, i.e. the first and the second coupled DAs. It is a common characteristic in this particular coupled model that the SST in forecasts are considerably lower than the observation (see Fig. 1). This is due to the weaker atmospheric forcing in forecasts by this model (systematic error). Particularly the second coupled DA (far right) gives a different solution from others (far left and middle). The heat content, represented by the 20°C isotherms, propagates eastward more strongly in the second version than in the first version. The last feature is favourable, but from the standpoint of the SST, the forecast (right) deviates considerably from the observation.

To summarize, (a) with respect to the eastward propagation of warm SST, the best forecast is the original one (left), the second is the one based on the initial condition of the first coupled DA (middle), and the third is the one of the second coupled DA (right); (b) on the other hand, the order is just opposite with respect to the development of 20°C isothermal depth. In other words, the second coupled DA gives more model-adjusted and less data-oriented analysis than the first coupled DA or the original DA. This reasoning is consistent with the thermocline structure in Fig. 4. In this respect, the spin-up aspect is best in the second version; a problem is that the SST deviates most from the observation. The only way to improve the situation from both standpoints is through improved physics and resolution in the coupled model.

5 Ensemble forecasts

Using the analyses based on the first version of the coupled DA, forecasts are performed from the initial conditions of 0 day, −2 day, −4 day, and +2 day, for the subsequent 13 months. This decision is not because this version gives the superior forecast, but because we feel that a coupled DA system is the desirable method of the future at GFDL.

a *The case of 1982/83*

Figure 6 displays five curves of SST anomalies in the NINO-3 region; Fig. 7 displays four curves of the ENSO index, *ESI*; and Fig. 8 shows four curves of the memory index. The model's climatology is borrowed from that of Rosati et al. (1996). The five curves (Fig. 6) start to spread in April, and stay separated from each other, the spread being almost the same until January 1983. The ensemble of SST anomalies is considerably lower in amplitude than the observation, the reason being described earlier. It is noted that the forecasts based on the second version of the coupled DA (not shown here) are substantially and unacceptably lower than the group in this figure. The spread (standard deviation) of the ensemble can be calculated, showing that in April, the spread is slightly greater but not substantially large.

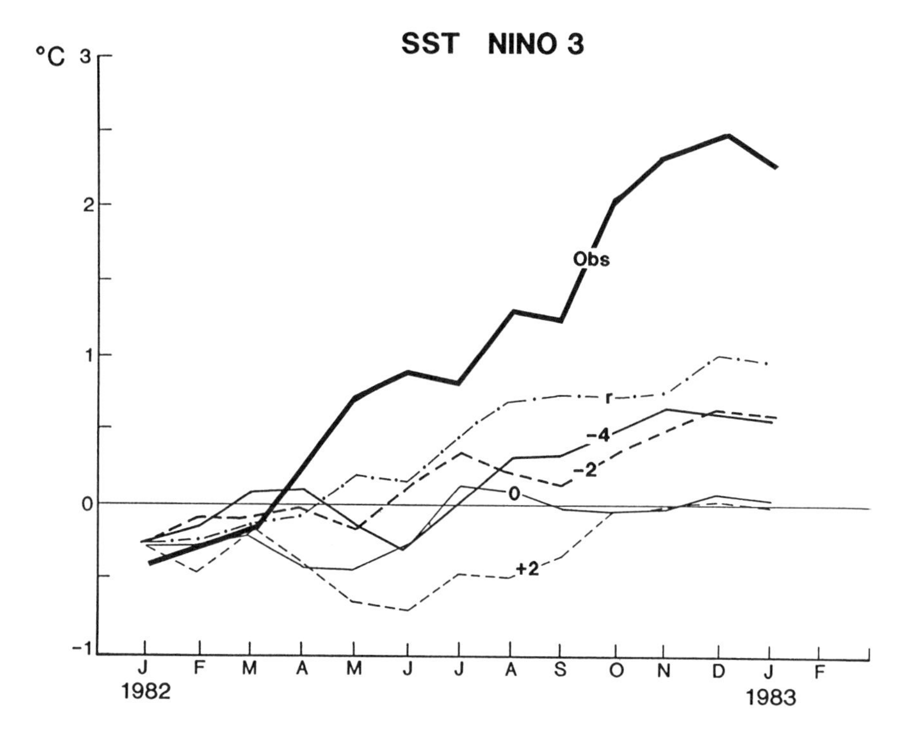

Fig. 6 Performances of 13-month forecasts are shown by SST anomalies for NINO-3 region. The observation (thick curve), and four forecasts from day 0, day −2, day −4, and day +2, based on the initial conditions of the first coupled DA, and a forecast denoted by r, based on the original initial condition, i.e. the oceanic DA and NMC atmospheric DA (Rosati et al., 1997).

Figure 7 shows the ensemble of *ESI*; each member of the four curves behaves similarly; the spread is not large. Here the undulation of three waves may be noticeable; presumably this is arbitrarily caused by a different model climatology that has been borrowed from other forecasts. In any event, it is clear that the *ESI* of these predictions is not sufficiently low, compared with the DA (see Fig. 1). Figure 8 includes the ensemble of the memory index.

b *Preliminary results*

Based on the four member forecasts, two issues are overviewed as a preliminary investigation. Concerning the spread of multiple solutions, there is no serious divergence, meaning that the spring barrier is not recognized, though this is a single case. It is also difficult to conclude that there is an improvement in the spin-up problem. Multiple cases are needed.

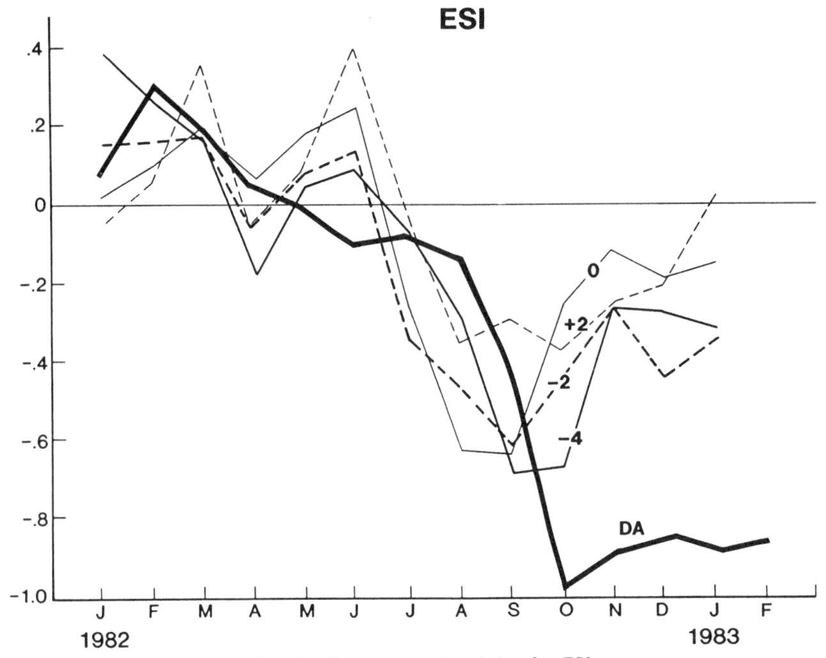

Fig. 7 The same as Fig. 6, but for *ESI*.

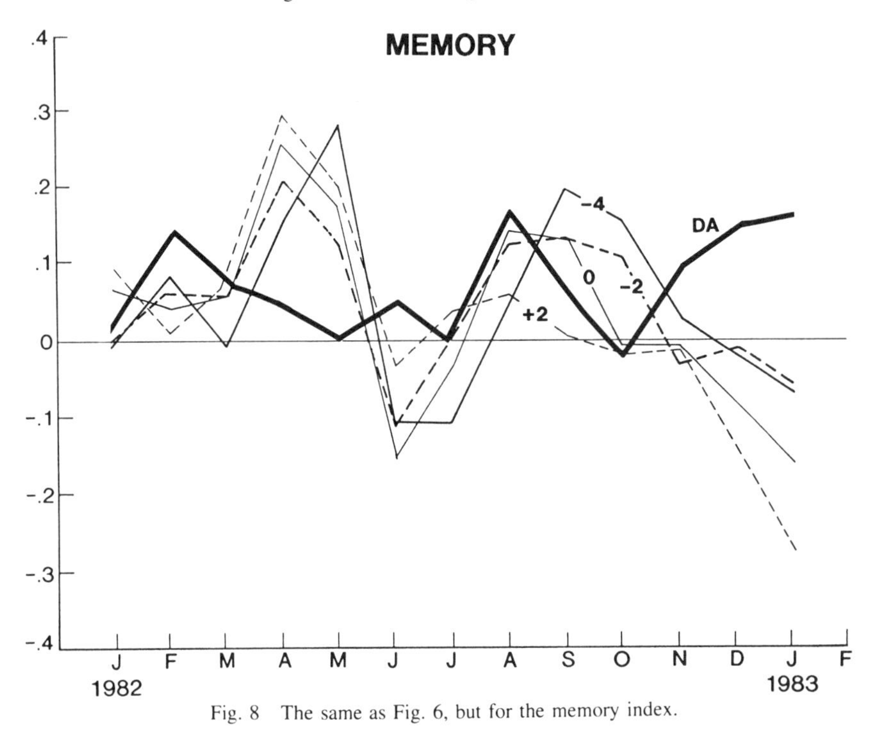

Fig. 8 The same as Fig. 6, but for the memory index.

6 Conclusions and comments

In order to investigate and improve the SST forecast skill associated with El Niño/La Niña events, a coupled atmospheric-oceanic model DA has been developed. The scheme is of the continuous data injection type, in which the GCMs are used for the first guess and the incremental components of observed data beyond the first guess are inserted into the models through the OI. In the atmospheric DA, the dynamic balancing is forced continuously for the incremental parts, while in the oceanic DA, any balancing is not forced (though desirable). These atmospheric and oceanic DAs are applied to the observed data alternately every 6 hours for 10 days, and the initial conditions for the coupled system are obtained (the first coupled DA).

This set of initial conditions is used for the ensemble forecasts as a sample case during the 1982/83 El Niño period. There are two objectives. One is to improve the forecast in the first 3 months, and the other is to investigate whether the "spring barrier" for the equatorial forecasts really exists.

It is tentatively concluded, based on the 4 member forecasts, that the spread of ensemble forecasts for SST is not small for the 13-month forecast range; the spread becomes noticeable after 6 months; however, the spread for the *ESI* or the memory index is not large; the spring barrier is not recognized clearly in this study (see the comment below); and, the spin-up issue is not conclusive due to the limited number of cases. In particular, it is necessary to establish the climatology for this prediction system. In order to discuss the spin-up issue, more forecast cases are needed.

With respect to the spring barrier, an hypothesis is postulated. As is known, the Southern Oscillation has biennial character (see for example, Barnett, 1991), and the separation between the warm and the cold phases is considered. If a 13-month forecast range is within the same phase, the "spring barrier" is not a problem, such as January 1982–January 1983. On the other hand, if the 13-month range crosses from one phase to the other, for example, from January 1983–January 1984, the "spring barrier" may have a disturbing effect. This is an assumption. It will be interesting to apply the ensemble forecasts to other cases.

As has been mentioned with respect to the first coupled DA, the current drawback is due to the large systematic bias of the particular coupled GCM; the SST is substantially lower than the observation, and the surface wind is biased in a certain way. In other words, this GCM is not of a great help for DA. The DA scheme should be modified with respect to the model bias utilizing climatological information.

Acknowledgment

The authors wish to thank William Stern and Richard Gudgel for useful advice. Significant suggestions for improved presentation of the results were offered by Tim Li, and useful advice was provided by Yoshio Kurihara and Jerry Mahlman. Special thanks go to Cathy Raphael, Jeff Varanyak and Wendy Marshall who produced the illustration figures and the manuscript.

References

BALMASEDA, M.A.; M.K. DAVEY and D.L.T. ANDERSON. 1994. Seasonal dependence of ENSO prediction skill. Climate Research Tech. Note, No. 51, Hadley Centre, England, 21 pp.

BARNETT, T.P. 1991. The interaction of multiple time scales in the tropical climate system. *J. Clim.* **4**: 269–285.

———; N. GRAHAM, M. CANE, S. ZEBIAK, S. DOLAN, J. O'BRIEN and D. LEGLER. 1988. On prediction of the El Niño of 1986–87. *Science*, **241**: 109–196.

BLUMENTHAL, M.B. 1991. Predictability of a coupled ocean-atmosphere model. *J. Clim.* **4**: 766–784.

BRANKOVIC, C.; T.N. PALMER and L. FERRANTI. 1994. Predictability of seasonal atmospheric variations. *J. Clim.* **7**: 217–237.

CANE, M.A. and S.E. ZEBIAK. 1985. A theory for El Niño and the Southern Oscillation. *Science*, **228**: 1084–1087.

———; ——— and S.C. DOLAN. 1986. Experimental forecasts of El Niño. *Science*, **321**: 827–832.

DALEY, R. and K. PURI. 1980. Four-dimensional data assimilation and the slow manifold. *Mon. Weather Rev.* **108**: 85–99.

DERBER, J. and A. ROSATI. 1989. A global oceanic data assimilation system. *J. Phys. Oceanogr.* **19**: 1333–1347.

GENT, P.R. and J.J. TRIBBIA. 1993. Simulation and predictability in a coupled TOGA model. *J. Clim.* **6**: 1843–1858.

GOSWAMI, B.N. and J. SHUKLA. 1993. Predictability of a coupled ocean-atmospheric flow. *J. Atmos. Sci.* **4**: 3–22.

LAPRISE, R. 1993. The resolution of global spectral models. *Bull. Am. Meteorol. Soc.* **73**: 1453–1454.

LATIF, M. and N.E. GRAHAM. 1991. How much predictive skill is contained in the thermal structure of an OGCM? *TOGA Notes*, **2**: 6–8.

LUTHER, D.S.; D.E. HARRISON and R.A. KNOX. 1983. Zonal winds in the central equatorial Pacific and El Niño. *Science*, **222**: 327–330.

MIYAKODA, K.; A. ROSATI and R. GUDGEL. 1993. Toward the GCM El Niño simulation. In: *Prediction of Interannual Climate Variations*, J. Shukla (Ed.), NATO ASI Series, 16, Springer-Verlag, Berlin and Heidelberg, pp. 125–151.

NEELIN, J.D.; M. LATIF, M.A.F. ALLAART, M.A. CANE, U. CUBASCH, W.L. GATES, P.R. GENT, M. GHIL, C. GORDON, N.C. LAU, C.R. MECHOSO, G.A. MEEHL, J.M. OBERTUBER, S.G.H. PHILANDER, P.S. SCHOPF, K.R. SPERBER, A. STERL, T. TOKIOKA, J. TRIBBIA and S.E. ZEBIAK. 1992. Tropical air-sea interaction in general circulation models. *Clim. Dyn.* **7**: 75–104.

PHILANDER, S.G.H. and A.D. SEIGEL. 1985. Simulation of El Niño of 1982-83. In: *Coupled Ocean-Atmosphere Models*, J.G.J. Nihoul (Ed.), Elsevier Oceanography Series, No. 40, pp. 517–541.

PLOSHAY, J.J.; W.F. STERN and K. MIYAKODA. 1992. FGGE reanalysis at GFDL. *Mon. Weather Rev.* **120**: 2083–2108.

REYNOLDS, R.W. 1988. A real time global sea surface temperature analysis. *J. Clim.* **1**: 75–86.

ROSATI, A.; R.G. GUDGEL and K. MIYAKODA. 1995. Decadal analysis produced from an ocean data assimilation system. *Mon. Weather Rev.* **123**: 2206–2228.

———; K. MIYAKODA and R.G. GUDGEL. 1997. The impact of oceanic initial conditions on ENSO forecasting with a coupled model. *Mon. Weather Rev.* **125**: 255–273.

SAUSEN, R.; K. BARTHELS and K. HASSELMANN. 1988. Coupled ocean-atmospheric models with flux correction. *Clim. Dyn.* **2**: 154–163.

SCHNEIDER, E.K.; B. HUANG and J. SHUKLA. 1995. Ocean wave dynamics and El Niño. *J. Clim.* **8**: 2415–2439.

STERN, W.F. and J.J. PLOSHAY. 1992. A scheme for continuous data assimilation. *Mon. Weather Rev.* **120**: 1417–1432.

TROUP, A.J. 1965. The Southern oscillation. *Q. J. R. Meteorol. Soc.* **91**: 490–506.

TRENBERTH, K.E. 1976. Spatial and temporal variations of the Southern Oscillation. *Q. J. R. Meteorol. Soc.* **102**: 639–653.

WEBSTER, P.J. and S. YANG. 1992. Monsoon and ENSO. Selectively interactive systems. *Q. J. R. Meteorol. Soc.* **118**: 877–926.

WERGEN, W. 1987. Diabatic non-linear normal mode initialization for a spectral model with a hybrid vertical coordinate. ECMWF Tech. Rept., 59; available from ECMWF, Reading, UK, 83 pp.

ZEBIAK, S.E. and M.A. CANE. 1987. A model El Niño-Southern Oscillation. *Mon. Weather Rev.* **115**: 2262–2278.

A C-grid Ocean General Circulation Model: Model Formulation and Frictional Parametrizations

Weimin Xu[1], Charles Lin[1,2] and André Robert[3]
[1]*Department of Atmospheric and Oceanic Sciences and Centre for Climate and Global Change Studies McGill University, Montreal, Quebec, Canada H3A 2K6*
[2]*Centre de recherche en calcul appliqué Montréal, Québec, Canada H3X 2H9*
[3]*Département de physique, Université de Montréal à Québec Montréal, Québec, Canada H3C 3P8*

[Original manuscript received 6 February 1995; in revised form 30 July 1995]

ABSTRACT *We describe the formation of an ocean circulation model using the Arakawa C-grid. The C-grid has been shown to give more accurate results than the B-grid in the treatment of geostrophic adjustment and linear convection. The model is formulated with β-plane geometry, and temperature is the only state variable. We include in the model a semi-implicit formulation of the Coriolis terms and a small scale dissipative term which depends on the horizontal divergence. The latter reduces the noise in the vertical motion field at coarse horizontal resolution. The results are compared to those obtained with other forms of frictional parametrizations.*

RÉSUMÉ *On décrit la mise au point d'un modèle de circulation océanique utilisant la grille C d'Arakawa. Il a été démontré que cette grille produit des résultats plus précis que la grille B dans le traitement de l'ajustement géostrophique et de la convection linéaire. Le modèle est conçu en fonction d'une géométrie de plan β et la température est la seule variable d'état. Un traitement semi-implicite des termes de Coriolis et un terme dissipatif dépendant de la convergence horizontale sont inclus. Ce dernier réduit le bruit dans le champ de mouvement vertical lorsque la résolution horizontale est faible. On compare les résultats à ceux obtenus par d'autres formes de paramétrage du frottement.*

1 Introduction

The world's oceans are an important component of the climate system. They cover more than 70% of the earth's surface and have a much larger heat capacity than the atmosphere. Ocean currents and their associated fields of pressure, temperature and density vary energetically in both time and space throughout the ocean. Such variability (eddies) in fact contains more energy than any other form of motions in the sea. These eddies have a characteristic spatial scale of a few hundred kilometres and a time-scale of weeks to months. Their importance for mixing, heat and mass transport as well as their potential role on climate are subjects of current research.

Eddy resolving general circulation modelling studies have mainly examined the North Atlantic (Semtner and Mintz, 1977; Cox, 1985; Bryan and Holland, 1989; Böning and Budich, 1992). Recently, Semtner and Chervin (1988, 1992) have formulated a global eddy resolving model which uses advanced supercomputing methods. Hurlburt et al. (1992) studied the eddy-resolving wind-driven circulation of the North Pacific. Most of these studies use the Arakawa B-grid for horizontal discretization. However, for horizontal grid sizes less than Rossby radius of deformation, the C-grid is more accurate (Arakawa and Lamb, 1977; Batteen and Han, 1981; Bryan, 1989; Xu and Lin, 1993).

There are existing C-grid ocean circulation models. Blumberg and Mellor (1987) developed a C-grid model for coastal oceans, which was recently extended to the North Atlantic (Ezer and Mellor, 1994). Andrich et al. (1988) developed a primitive equation model to study the dynamics of the tropical Atlantic and the western Mediterranean Sea. Haidvogel et al. (1991) formulated a semi-spectral model using orthogonal curvilinear horizontal coordinates, which has been used for various applications (e.g., Beckmann and Haidvogel, 1993; Fukumori et al., 1993). Dietrich et al. (1987) developed a gridpoint model and used it to examine the circulation in the Gulf of Mexico (Dietrich and Lin, 1994). Brugge et al. (1991) developed a non-hydrostatic model to study oceanic convection. Hurlburt et al. (1992) formulated a new layer formulation C-grid model based on Hurlburt and Thompson's (1980) model. Bleck and Smith (1990) developed an isopycnal model using the C-grid (see also Bleck et al., 1992). As noted in Semtner (1996a), the use of the C-grid generates more noisy fields than the B-grid at coarse resolution. This is especially true when the model domain includes the high latitudes where the Rossby radius of deformation is small. In this case, a high spatial resolution is needed to eliminate the noise, as the latter may eventually contaminate other model variables.

In this study, we present a new C-grid ocean circulation model. This work is motivated by two factors. We include a new form of viscosity in the momentum equations to reduce noise in the vertical velocity field; this viscosity formulation is particularly useful in the coarse resolution regime. The second factor is the use of a semi-implicit scheme to treat the Coriolis terms in order to relax the time step limitation imposed by external Rossby waves and inertial waves. This limitation applies to coarse resolution models; a similar semi-implicit method has been used for the B-grid (Bryan, 1969; Cox, 1984; Semtner, 1986b). Results from coarse resolution simulations can be used as initial conditions for eddy-resolving integrations, as it is still not feasible to integrate eddy-resolving models sufficiently long to obtain a statistical equilibrium with currently available computing resources. In addition, results from coarse resolution models can be compared with those of eddy-resolving models to identify the effects of eddies on various physical processes. An eventual goal is to use the model for eddy resolving simulations (Xu et al., 1995).

The governing equations and boundary conditions are given in Section 2. The method of solution and the numerical schemes are described in Sections 3 and 4 respectively. Section 5 discusses the use of different parametrizations of lateral viscosities. The conclusions are given in Section 6.

2 Model equations and boundary conditions

The growing equations are the primitive equations with the Boussinesq and hydrostatic approximations. They consist of the horizontal momentum equations, the temperature equation, the hydrostatic equation, the incompressibility condition and the equation of state. They are shown below as Equations (1)–(6) respectively. The notation used is standard, and a description of the symbols used is given in Appendix A.

$$\frac{\partial u}{\partial t} + L(u) = -\frac{1}{\rho_0}\frac{\partial p}{\partial x} + fv + \frac{\partial}{\partial z}\left(A_{MV}\frac{\partial u}{\partial z}\right) + F_x + \epsilon\lambda\frac{\partial D}{\partial x} \tag{1}$$

$$\frac{\partial v}{\partial t} + L(v) = -\frac{1}{\rho_0}\frac{\partial p}{\partial y} - fu + \frac{\partial}{\partial z}\left(A_{MV}\frac{\partial v}{\partial z}\right) + F_y + \epsilon\lambda\frac{\partial D}{\partial y} \tag{2}$$

$$\frac{\partial T}{\partial t} + L(T) = \frac{\partial}{\partial z}\left(A_{HV}\frac{\partial T}{\partial z}\right) + F_T \tag{3}$$

$$p_z = -\rho g \tag{4}$$

$$\nabla \cdot \mathbf{V} = 0 \tag{5}$$

For simplicity, we assume temperature is the only state variable in the equation of state,

$$\rho = \rho_0(1 + \alpha(T_0 - T)). \tag{6}$$

Here, $L(\ )$ is the advection operator, while F_x, F_y are the horizontal viscosity terms in the x and y directions respectively; F_T represents the horizontal diffusion terms in the temperature equation; T_0 is a reference temperature, and α is the expansion coefficient of water. The expressions for the operators $L(\ )$, F_x, F_y, F_T are

$$L(\sigma) = \frac{\partial(u\sigma)}{\partial x} + \frac{\partial(v\sigma)}{\partial y} + \frac{\partial(w\sigma)}{\partial z}$$

$$F_x = \frac{\partial}{\partial x}\left(A_{MH}\frac{\partial u}{\partial x}\right) + \frac{\partial}{\partial y}\left(A_{MH}\frac{\partial u}{\partial y}\right)$$

$$F_y = \frac{\partial}{\partial x}\left(A_{MH}\frac{\partial v}{\partial x}\right) + \frac{\partial}{\partial y}\left(A_{MH}\frac{\partial v}{\partial y}\right)$$

$$F_T = \frac{\partial}{\partial x}\left(A_{HH}\frac{\partial T}{\partial x}\right) + \frac{\partial}{\partial y}\left(A_{HH}\frac{\partial T}{\partial y}\right).$$

An additional friction term is added in the horizontal momentum equations. It is multiplied by the multiplier ϵ, which takes on a value of either 0 or 1; λ is an eddy diffusivity which operates on the horizontal divergence (D) (Sadourny, 1975). This term is added to reduce the noise which may be present in the vertical velocity field due to gravity waves (Batteen and Han, 1981). As we will show later, the inclusion of the divergence term effectively eliminates such noise.

To investigate whether other forms of lateral viscosity parametrizations can be used to eliminate the noise in the vertical velocity, we also examine the effects of biharmonic diffusion and the Smagorinsky (1963) form of eddy diffusivity. Both the biharmonic and Smagorinsky diffusive formulations are more scale selective than Laplacian friction with constant diffusivity.

The horizontal boundaries are insulating with no normal flow. At the western and eastern boundaries, we thus have

$$(u,\ v_n,\ T_n) = 0$$

and at the southern and northern boundaries

$$(u_n,\ v,\ T_n) = 0.$$

Here, $(\ )_n$ indicates a local derivative with respect to the coordinate normal to the boundary of the walls. An additional boundary condition is required for biharmonic diffusion, which is the same as above, but with u, v and T replaced by their corresponding Laplacian derivatives.

The top boundary at $z = 0$ is rigid, with a specified idealized wind stress distribution. The surface boundary condition for temperature can be of either restoring (Haney, 1971) or flux form. These conditions at $z = 0$ can be expressed as follows:

$$A_{MV}(u_z,\ v_z) = (\tau_x,\ \tau_y)$$

$$A_{HV}T_z = Q^T$$

$$w = 0.$$

The bottom boundary at $z = H$ is flat and insulating with a slip condition.

$$A_{MV}(u_z,\ v_z) = (\tau_x^B,\ \tau_y^B)$$

$$A_{HV}T_z = 0$$

$$w = 0$$

where τ_x^B and τ_y^B are the bottom friction in the x and y directions respectively, parametrized as linear Rayleigh friction.

Our model does not include salinity, the latitudinal variation of the metric factors and the spherical curvature terms. We use a domain which is similar in size to the North Atlantic and an idealized climatology of surface thermal forcing and wind stress, to examine the effects of different frictional parametrizations on the large scale circulation.

3 Method of solution

The use of a top rigid boundary condition eliminates external gravity waves. For models with a rigid top boundary, the motion can be divided into barotropic and baroclinic components. We denote the sea surface height and atmospheric pressure by ξ_1 and p_a respectively. The pressure in the ocean is then

$$p(x, y, z) = p_a(x, y) + \rho g \xi_1 + \int_z^0 \rho g d\xi = p_s + p_h \tag{7}$$

where p_h is the hydrostatic component of pressure at depth z, and is given by the integrated mass field between that depth and the surface; $p_s = p_a + \rho g \xi_1$ is the surface pressure, which includes the contribution from the atmospheric pressure and sea surface topography. p_s is usually not known explicitly in this type of model. A standard way to proceed is to introduce the mass transport streamfunction, and then solve the vertically integrated vorticity equation (Bryan, 1969; Cox, 1984; Haidvogel et al., 1991). An alternate method is to solve for the surface pressure directly by using the diagnostic divergence equation (Dietrich et al., 1987, 1990; Dukowicz et al., 1993); the primary variables in this case are the pressure and velocity. We have chosen the latter approach due to its relative ease of implementation (Harlow and Welch, 1965; Dukowicz et al., 1993). In addition, the only reliable observational data for a basin scale ocean is satellite altimetry; it is thus promising to develop a model which can easily accommodate these data. The surface pressure method is also easier to use in a domain with islands, as solving for the barotropic streamfunction would require the determination of the streamfunction along the boundary of each island. We conducted some experiments with both methods and found identical results.

The use of a rigid top boundary requires a method to ensure the vertically integrated flow be non-divergent.

$$\hat{D} = \frac{\partial \hat{u}}{\partial x} + \frac{\partial \hat{v}}{\partial y} = 0 \tag{8}$$

where $(\hat{\ })$ denotes the vertical averaging operator,

$$(\hat{\ }) = \frac{1}{H} \int_{-H}^{0} (\ ) dz.$$

By taking the vertical average of Equations (1) and (2), we obtain

$$\frac{\partial \hat{u}}{\partial t} = -\frac{1}{\rho_0}\frac{\partial p_s}{\partial x} + \hat{G}_x \tag{9}$$

$$\frac{\partial \hat{v}}{\partial t} = -\frac{1}{\rho_0}\frac{\partial p_s}{\partial y} + \hat{G}_y \tag{10}$$

where

$$G_x = -\frac{1}{\rho_0}\frac{\partial p_h}{\partial x} - L(u) + fv + \frac{\partial}{\partial z}\left(A_{MV}\frac{\partial u}{\partial z}\right) + F_x + \epsilon\lambda\frac{\partial D}{\partial x} \tag{11}$$

$$G_y = -\frac{1}{\rho_0}\frac{\partial p_h}{\partial y} - L(v) - fu + \frac{\partial}{\partial z}\left(A_{MV}\frac{\partial v}{\partial z}\right) + F_y + \epsilon\lambda\frac{\partial D}{\partial y} \tag{12}$$

From Equations (9) and (10), we obtain a prediction equation for the vertically averaged horizontal divergence.

$$\frac{\partial \hat{D}}{\partial t} = -\frac{1}{\rho_0}\nabla^2 p_s + \frac{\partial \hat{G}_x}{\partial x} + \frac{\partial \hat{G}_y}{\partial y} \tag{13}$$

To eliminate external gravity waves, we set the left hand side of Equation (13) to zero. This results in an elliptic equation for p_s. Its solution forms one step in the time integration procedure. After obtaining the surface pressure, the horizontal velocity fields can be determined using the momentum equations, and the vertical velocity from the continuity equation. The temperature is then obtained from the thermodynamic equation, and the complete pressure distribution from the equation of state and the hydrostatic equation. Further details of the time and space differencing schemes are discussed below.

4 Finite difference schemes

The spatial finite differencing is based on the Arakawa C-grid. The distribution of the variables is shown Fig. 1. On a horizontal plane, the u and v points are staggered, while p, T are carried at the centre of the grid cell. These variables all reside at the mid-point of a vertical layer. The vertical velocity w is defined at the centre of grid cells in each layer.

We use the label (i,j,k) to denote the 3-dimensional index of the centre of a grid cell. Given the values of a model variable η at the adjacent points $(i+{}^1\!/_2, j, k)$ and $(i-{}^1\!/_2, j, k)$, we define the difference and average operators in the x-direction as follows:

$$(\delta_x \eta)_{i,j,k} \equiv \frac{\eta_{i+1/2,j,k} - \eta_{i-1/2,j,k}}{\delta x} \tag{14}$$

$$\overline{\eta_{i,j,k}}^{x} \equiv \frac{1}{2}(\eta_{i+1/2,j,k} + \eta_{i-1/2,j,k}). \tag{15}$$

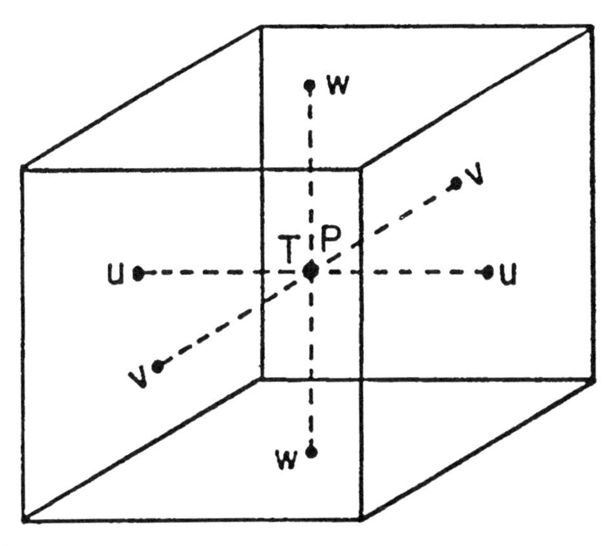

Fig. 1 The grid structure of the C-grid, showing the position of the velocity (u, v, w), pressure (p) and temperature (T) points.

Similar operators are defined in the y- and z-directions. For time differencing, we have

$$(\delta_t \eta)_{i,j,k} \equiv \frac{\eta(t+\delta t) - \eta(t-\delta t)}{2\delta t} \tag{16}$$

We also define

$$\overline{\eta_{i,j,k}}^{xy} \equiv \overline{\overline{\eta_{i,j,k}}^{x}}^{y} \tag{17}$$

We now describe the time integration procedure, starting with the temperature equation. Assuming that the variables at times steps n and $n-1$ are known, we use them to predict their new values at time step $n+1$. Superscripts are used to denote the time level. The finite difference form of this equation can be written as,

$$(\delta_t T)^n = -L(T)^n + F_T^{n-1} + (A_{HV}\delta_z T)_z^{n+1} \tag{18}$$

The model uses a leapfrog time step for advection and a forward time step for horizontal diffusion. An implicit scheme (Richtmyer and Morton, 1967) is used for vertical diffusion. In case of unstable stratification, convective adjustment is done by locally increasing the vertical diffusivity from the standard value of 1 $cm^2\ s^{-1}$ to $10^4\ cm^4\ s^{-1}$ (Cox, 1984). In this way, T^{n+1} can be obtained from known fields at time steps n and $n-1$. The density at time step $n+1$ can be obtained from the temperature using the equation of state, and the pressure can then be computed using the hydrostatic equation.

$$p_{k+1} = p_k + g\rho_{k+1/2}\delta z_{k+1/2} \tag{19}$$

where $\delta z_{k+1/2} = (\delta z_k + \delta z_{k+1})/2$, and $\rho_{k+1/2} = (\rho_k + \rho_{k+1})/2$. Note that p_s, being independent of height, is included in Equation (19).

In the momentum equations, the model uses a leapfrog time step for advection and pressure gradient terms and a forward time step for both horizontal and vertical diffusion. A semi-implicit scheme is used for the Coriolis terms. This use of the semi-implicit method is more complicated in the C-grid than in the B-grid (Bryan, 1969) formulation, thus we give a detailed description here.

The Coriolis term in the x-momentum Equation (1) is represented by

$$f\,\overline{[av^{n+1} + (1-a)v^{n-1}]}^{xy}$$

while that in the y-momentum Equation (2) is represented by

$$-\overline{f\,\overline{[au^{n+1} + (1-a)u^{n-1}]}^{x}}^{y}$$

where $0 \leqslant a \leqslant 1$. The spatial averages of the velocities in the Coriolis terms are required to conserve kinetic energy (Arakawa and Lamb, 1977). The finite difference form of the horizontal momentum equations can then be written as

$$\frac{u^{n+1} - u^{n-1}}{2\Delta t} - af\,\overline{(v^{n+1} - v^{n-1})}^{xy} = Q_1 \tag{20}$$

$$\frac{v^{n+1} - v^{n-1}}{2\Delta t} + \overline{af\,\overline{(u^{n+1} - u^{n-1})}^{x}}^{y} = Q_2 \tag{21}$$

where

$$Q_1 = \left[-\frac{1}{\rho_0}\frac{\partial p^n}{\partial x} - L(u)^n + f\,\overline{v^{n-1}}^{xy} + \frac{\partial}{\partial z}\left(A_{MV}\frac{\partial u^{n-1}}{\partial z}\right) + F_x^{n-1} + \epsilon\lambda\frac{\partial D^{n-1}}{\partial x}\right]$$

$$Q_2 = \left[-\frac{1}{\rho_0}\frac{\partial p^n}{\partial y} - L(v)^n - \overline{f\,\overline{u^{n-1}}^{x}}^{y} + \frac{\partial}{\partial z}\left(A_{MV}\frac{\partial v^{n-1}}{\partial z}\right) + F_y^{n-1} + \epsilon\lambda\frac{\partial D^{n-1}}{\partial y}\right]$$

Taking spatial averages $\overline{f\,\overline{(\;)}^{x}}^{y}$ and $\overline{(\;)}^{xy}$ of Equations (20) and (21) and substituting them back into the equations, we obtain

$$u^{n+1} = u^{n-1} + 2\Delta t\,\frac{Q_1 + 2a\Delta t f\,\overline{Q_2}^{xy}}{1 + (2a\Delta t f)^2} \tag{22}$$

$$v^{n+1} = v^{n-1} + 2\Delta t\,\frac{Q_2 - 2a\Delta t\,\overline{f\,\overline{Q_1}^{x}}^{y}}{1 + (2a\Delta t)^2\overline{f^2}^{y}} \tag{23}$$

This set of equations is similar to that of Bryan (1969) for his B-grid model. However, we have assumed here that the spatial average of u^{n+1} and v^{n+1} at a grid point can be represented initially by its local value. An iteration procedure is thus needed to get the solution. The Newton-Raphson method is used to solve for the horizontal velocities; this is usually a very efficient method because the initial guess is quite close to the solution. By choosing $a \geqslant 1/2$, it is possible to exceed time step limitations imposed by external Rossby waves and inertial waves (Bryan, 1969; Cox, 1984; Takano, 1974).

The two-dimensional Poisson equation for the surface pressure is solved using the direct method of Dietrich et al. (1987, 1990).

The schemes used here conserve mass, momentum, energy, temperature and temperature variance (Arakawa and Lamb, 1977). A Robert/Asselin time filter (Robert, 1966; Asselin, 1972) is used to remove the computational mode associated with the leapfrog scheme.

5 Parametrization of lateral friction

The zonal extent, meridional extent and depth of the model are $L_x = 4400$ km, $L_y = 5500$ km and $H = 4000$ m respectively. The mid-point of the grid in the meridional direction of the model domain is taken to be 45°N. Figure 2 shows the equivalent atmospheric temperature and wind stress forcing distributions at the upper boundary. The former is the temperature distribution to which the top model layer temperature is restored with a time-scale of $\tau_R = 30$ days (Haney, 1971). The surface forcing is an idealized form of the climatological forcing which was used in an earlier study (Gough and Lin, 1992).

Different time steps are used for the momentum and temperature equations: 30 minutes for the former and 1 day for the latter (Bryan, 1984). This is done to speed up the integration of the model; due to the large thermal capacity of the ocean compared to the atmosphere, ocean models require a much longer spin-up time. Another difference is that the baroclinic Rossby radius of deformation in the ocean is only about 50 km or less, compared to about 1000 km for the atmosphere. Here, we conduct experiments with coarse horizontal resolution, with the horizontal grid length larger than the Rossby radius.

Xu (1994) has compared the results of many experiments of the model to published results, primarily those of Bryan-Cox B-grid model. The details will not be reported here. For example, Fig. 3 shows the mean meridional circulation, or the thermohaline circulation. The main cell reaches a maximum amplitude of 14 Sv, with sinking concentrated in a narrow band in the high latitudes due to the surface cooling, and rising motion throughout the rest of the domain. Near the surface, there are weak cells which are due to the Ekman transport produced by the surface wind stress. Many more experiments have been performed to test the sensitivity of the model to different parameters. The behaviour is quite similar to that reported by Bryan (1987).

The use of a divergence friction term is a new feature of our model. Although it has been used with success in atmospheric modelling (Robert, 1981), it has not been used in ocean models. At coarse horizontal resolution not resolving the radius of deformation, a large viscosity coefficient must be used in order to avoid numer-

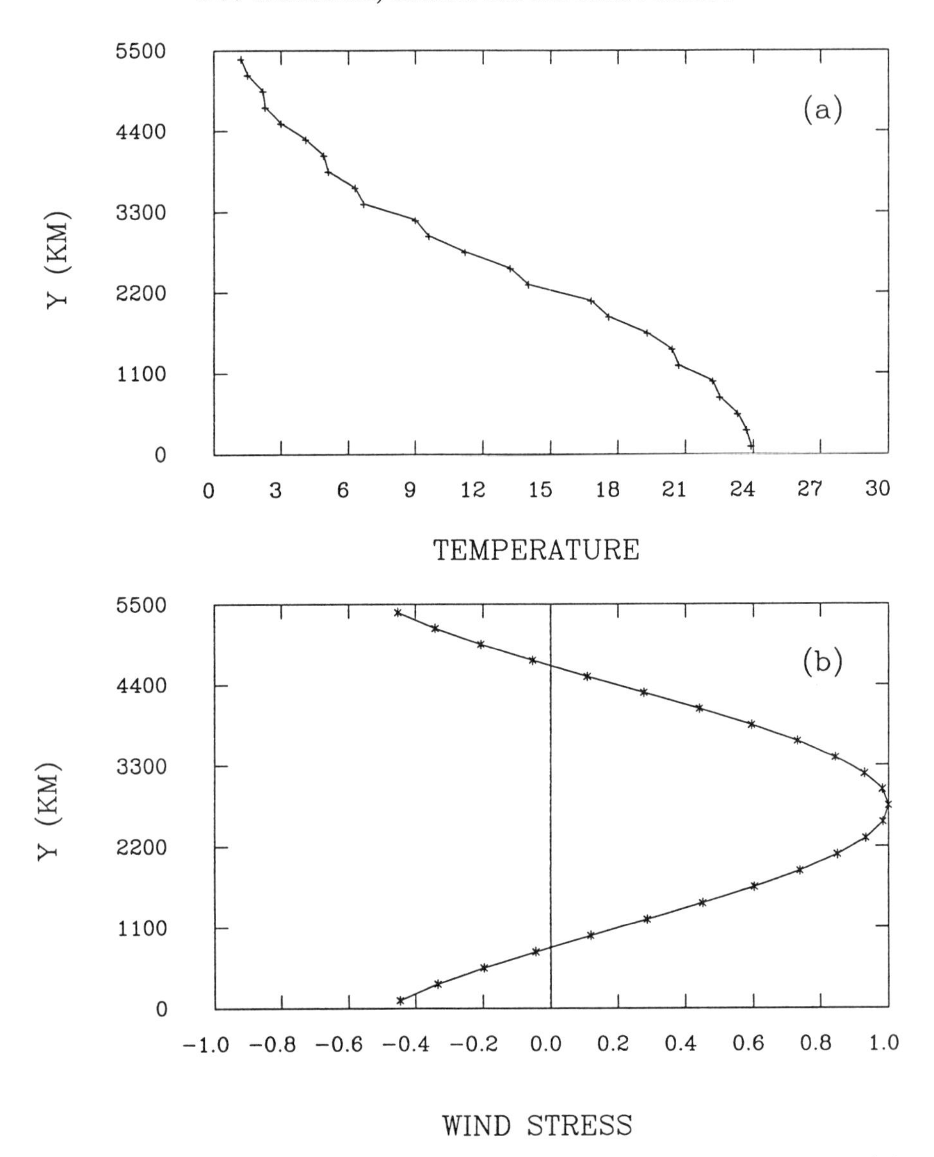

Fig. 2 The latitude distribution of the (a) apparent atmospheric temperature (°C), and (b) zonal wind stress (dyne cm^{-2}), used as surface forcings. There is no variation with longitude.

ical instability. The resolved motion is thus quasi-geostrophic, with the horizontal divergence being an order smaller than the vorticity. This provides a physical basis for adding this friction term.

We now compare the effects of two other lateral frictional parametrizations, biharmonic and Smagorinsky friction, to the divergence dissipation. The biharmonic

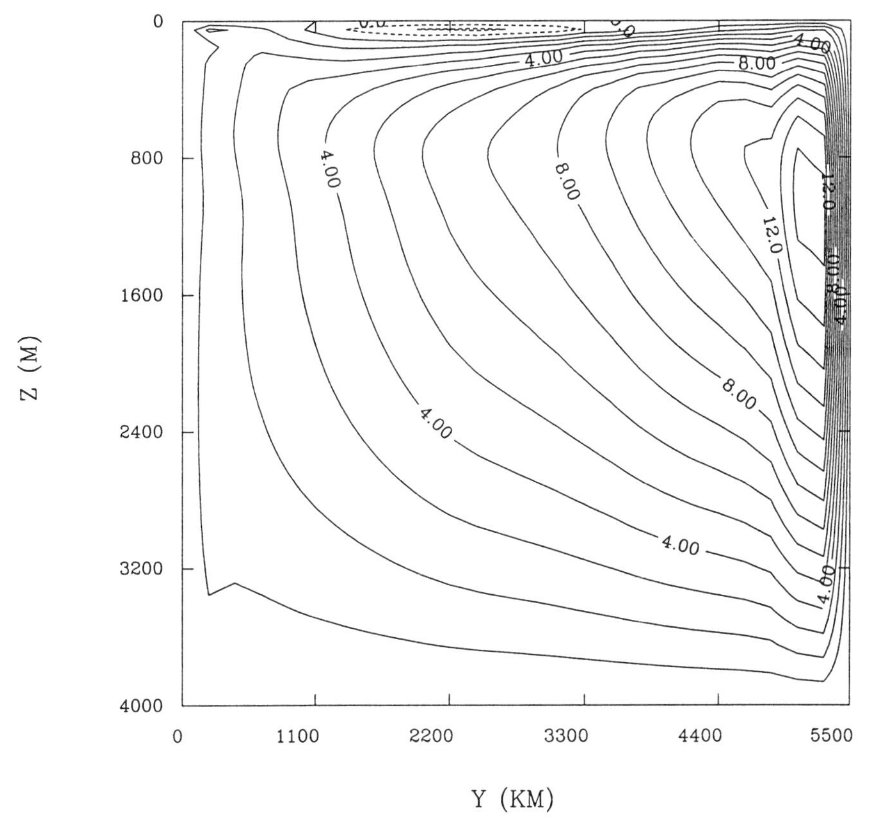

Fig. 3 The mean meridional circulation, i.e., the thermohaline circulation (Sv), with dashed lines indicating anticyclonic circulation, and solid lines indicating cyclonic circulation.

and Smagorinsky parametrizations are both more scale selective than Laplacian dissipation.

Biharmonic friction is much more effective at damping short, high frequency waves than Laplacian friction, but is less effective for the large scales. It has usually only been used in eddy-resolving models (Holland, 1978; Semtner and Mintz, 1977; Cox, 1985; Bryan and Holland, 1989). To compare with Laplacian friction, we relate the coefficients of the two forms of friction as (Semtner and Mintz, 1977):

$$A_{MHB} = -\frac{1}{4} A_{MH} L^2 \tag{24}$$

where A_{MH} is as before, and A_{MHB} is the viscosity coefficient for biharmonic friction. This scaling relation implies that the effects of dissipation are comparable at the scale L for both biharmonic and Laplacian frictions. A similar relation

TABLE 1. The vertical coordinate of the model and the depth of each level.

level	level depth (m)	thickness (m)
1	25	50
2	75	50
3	150	100
4	250	100
5	400	200
6	600	200
7	850	300
8	1200	400
9	1600	400
10	2000	400
11	2400	400
12	2800	400
13	3250	500
14	3750	500

applies to the diffusivities as well. For the shortest resolve scale ($L = 2\Delta x$), a biharmonic friction coefficient of $A_{MHB} = \Delta x^2 A_{MH}$ would have the same effect as Laplacian friction at the scale of $2\Delta x$. For scales increasingly large compared to $2\Delta x$, the dissipative effect of biharmonic friction becomes less effective compared to Laplacian friction.

The Smagorinsky frictional parametrization depends on the fluid deformation and was first used in an atmospheric primitive equation model (Smagorinsky, 1963). It has recently been used in coastal ocean modelling (Mellor, 1992). We reformulate the Smagorinksy viscosity and diffusivity as:

$$A_{MH} = A_{MH0} + C_M \Delta x \Delta y \sqrt{\left(\frac{\partial u}{\partial x}\right)^2 + \frac{1}{2}\left(\frac{\partial v}{\partial x} + \frac{\partial u}{\partial y}\right)^2 + \left(\frac{\partial v}{\partial y}\right)^2} \tag{25}$$

$$A_{HH} = A_{HH0} + C_H \Delta x \Delta y \sqrt{\left(\frac{\partial u}{\partial x}\right)^2 + \frac{1}{2}\left(\frac{\partial v}{\partial x} + \frac{\partial u}{\partial y}\right)^2 + \left(\frac{\partial v}{\partial y}\right)^2} \tag{26}$$

where Δx and Δy are the grid sizes, and C_M and C_H are dimensionless constants. We used the velocity field from a preliminary experiment to compute A_{MH} with $C_M = 10$ and $A_{MH0} = 0$; the resulting viscosity varies by several orders of magnitude depending on the location, so base values A_{MH0}, A_{HH0} were included in the above formulae. Note that in coarse resolution models, A_{MH} must be sufficiently large to resolve the Munk layer, while A_{HH} can be much smaller.

The horizontal resolution used here is $\Delta x = 100$ km, $\Delta y = 115$ km. Table 1 shows the vertical structure, and Table 2 shows a summary of the horizontal/vertical, viscosity/diffusivity (A_{MH}, A_{HH}, A_{MV}, A_{HV}) and the divergence dissipation parameter (λ) for the different cases. We have used the same symbols A_{MH}, A_{HH} to denote

TABLE 2. A summary of the dissipative parameters used in the numerical experiments. All parameters have units of $cm^2 s^{-1}$, except for A_{MH}, A_{HH} in Case 3, which have units of $cm^4 s^{-1}$.

Case	A_{MH}	A_{HH}	A_{MV}	A_{HV}	λ
1	8×10^8	10^7	20	0.5	0
2	8×10^8	10^7	20	0.5	2×10^9
3	-10^{23}	-10^{21}	20	0.5	0
4	8×10^8 ($C_M = 10$)	10^7 ($C_H = 0.1$)	20	0.5	0

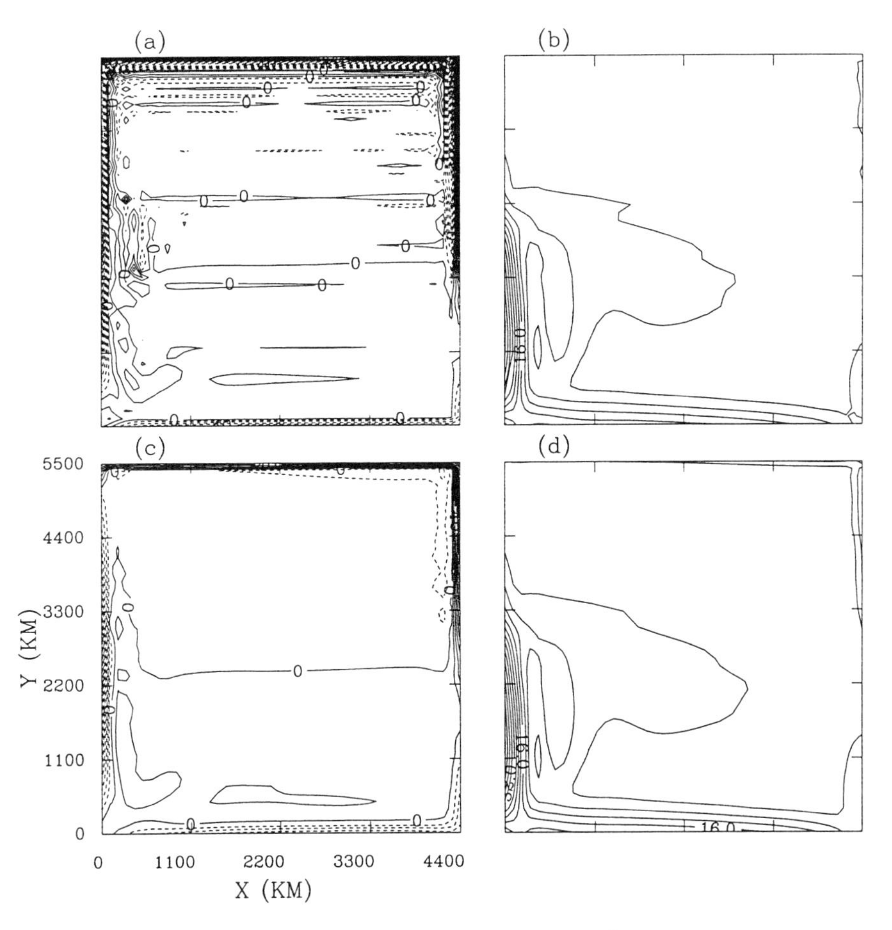

Fig. 4 The horizontal distribution of (a) the vertical velocity (cm s^{-1}, C.I.=4) at the bottom of the first layer, and (b) the root-mean-square horizontal velocity (cm s^{-1}, C.I.=4) of the first layer for Case 1. (c), (d) are similar to (a), (b) respectively, but for Case 2.

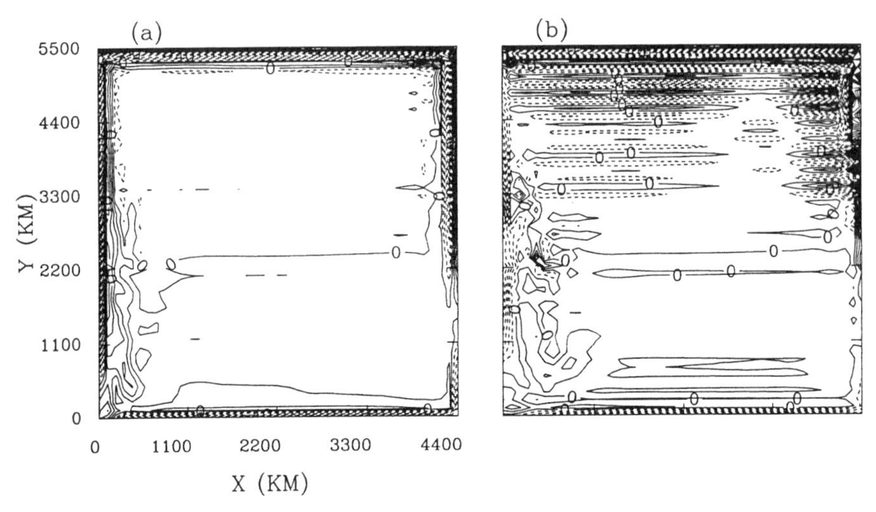

Fig. 5 The horizontal distribution of the vertical velocity (cm s^{-1}, C.I=10^{-4}) at the bottom of the first layer for (a) Case 3; (b) Case 4.

the coefficients for Laplacian and biharmonic friction, with units of $cm^2\ s^{-1}$ and $cm^4\ s^{-1}$ respectively. In the case of Smagorinsky friction, the symbols denote the base values of Equations (25) and (26), with units of $cm^2\ s^{-1}$.

Case 1 has Laplacian friction without the divergence dissipation, while the latter is included in Case 2. Figure 4 shows the vertical velocity field and the root-mean-square horizontal velocity for the two cases. We see that the divergence dissipation is very effective at removing noise in the vertical velocity field, but leaves the horizontal motion field undisturbed. Other fields (not shown) are also relatively unchanged. Thus small scale waves are effectively removed by the divergence dissipation parametrization.

Cases 3 and 4 correspond to biharmonic and Smagorinsky friction respectively. The viscosity coefficients are chose so that both cases are equivalent to Laplacian friction with no divergence dissipation (Case 1). For biharmonic friction, the coefficients are chosen according to Equation (24) so that the frictional effects are comparable to Laplacian friction at the smallest resolved scale of $2\Delta x$. For Smagorinsky friction, the background base coefficients are identical to the viscosities and diffusivities of Case 1. We see from Fig. 5 that considerable noise still remains in the vertical velocity field with these two parametrizations. We can also choose to determine the biharmonic frictional coefficient by matching the dissipative effect at a scale which is larger than the smallest resolved scale, say at $6\Delta x$ or $8\Delta x$ instead of $2\Delta x$. This results in a larger biharmonic coefficient. However, this leads to some damping of the horizontal velocity. This is an undesirable effect as coarse resolution models already underestimate the magnitude of horizontal currents.

In all experiments described in this section, the standard stability criteria are satisfied by our choice of viscosities. In each case, the Munk layer is resolved to avoid chessboard noise (Takano, 1975),

$$A_{MH} > \beta \left(\frac{2\sqrt{3}\Delta x}{3\pi} \right)^3$$

due to the use of centred differences in the diffusive terms; the grid Peclet number is small enough to supress 2-grid point waves,

$$A_{MH} > \frac{U \Delta x}{2}$$

and the use of a forward time integration scheme with diffusion requires

$$A_{HH} < \frac{\Delta x^2}{2\Delta t}.$$

6 Summary

In this study, we have presented the formulation of a C-grid primitive equation ocean circulation model. The motivation for using the C-grid, rather than the B-grid, is that the former performs better at high resolution (Arakawa and Lamb, 1977; Batteen and Han, 1981; Bryan, 1989; Xu and Lin, 1993). A semi-implicit treatment of the Coriolis term is introduced, in order to allow for time steps larger than that allowed by external Rossby and gravity waves. A divergence dissipation (Sadourny, 1975) is also introduced, which effectively removes noise in the vertical velocity field and leaves other fields relatively undisturbed. This is important as noisy vertical motion fields have been a difficulty with the C-grid. An ultimate goal is to use the model for eddy resolving experiments (Xu et al., 1995). Note that the divergence dissipation parametrization also works to dissipate noise in eddy resolving basin scale models which include the high latitudes (Xu, 1994). The radius of deformation becomes very small at these locations due to the weak stratification and the large Coriolis parameter.

We compared the effects of two other lateral frictional parametrizations, biharmonic and Smagorinsky friction, to the divergence dissipation used in our study. It was shown that the latter is the most effective at removing noise in the vertical velocity fields. Biharmonic friction can also be effective at eliminating noise due to 2-grid length waves, but is less effective at longer wavelengths, due to its highly scale selective feature. The Smagorinsky frictional parametrization is not as effective as the other two formulations at removing such noise.

Acknowledgements

This work is supported by grants to CAL from the Natural Sciences and Engineering Research Council (NSERC), and the Atmospheric Environment Service (AES) of

Canada. Computing time on the NEC SX3 at AES was provided by Dr. Michel Béland. The use of the computing facilities of CERCA (Centre de recherche en calcul appliqué) is also gratefully acknowledged.

Appendix A

Listed below are the principal symbols used in this paper.

x, y, z:	eastward, northward, upward Cartesian coordinates
u:	x component of velocity (eastward)
v:	y component of velocity (northward)
w:	z component of velocity (upward)
p:	pressure
t:	time
f:	Coriolis parameter
ρ:	density
ρ_0:	reference density
T:	temperature
α:	thermal expansion coefficient, 2.5×10^{-4} $°C^{-1}$
D:	horizontal divergence
A_{MH}, A_{HH}:	horizontal viscosity and diffusivity
A_{MV}, A_{HV}:	vertical viscosity and diffusivity
λ:	eddy viscosity operating on the horizontal divergence
τ_x, τ_y:	surface wind stresses in x, y directions
T_A:	equivalent atmospheric temperature
T_1:	surface ocean temperature
Δz_1:	the top layer model depth
τ_R:	restoring time constant for surface thermal forcing
Q^T:	surface heat flux, $\Delta z_1 (T_A - T_1)/\tau_R$.

References

ANDRICH P.; P. DELECLUSE, C. LEVY and G. MADEC. 1988 A Multi-tasked General Circulation Model of the Ocean, Report of Laboratoire d'Oceanographie Dynamique et de Climatologie, University Paris 6, 75002 Paris, France.

ARAKAWA, A. and V.R. LAMB. 1977 Computational design of the basic dynamical processes of the UCLA general circulation model. In: *Methods of Computational Physics*, vol. **17**, Academic Press, New York, pp. 173–265.

ASSELIN, R.A. 1972. Frequency filter for time integration. *Mon. Weather Rev.* **100**: 487–490.

BATTEEN, M.L. and Y.-J. HAN. On the computational noise of finite difference schemes used in ocean models. *Tellus*, **33**: 387–396.

BECKMANN, A. and D.B. HAIDVOGEL 1993. Numerical simulation of flow around a tall, isolated seamount. Part I: Problem formulation and model accuracy. *J. Phys. Oceanogr.* **23**: 1736–1753.

BLECK, R. and L.T. SMITH. 1990 A wind-driven isopycnal coordinate model of the North and equatorial Atlantic ocean. 1. Model development and supporting experiments. *J. Geophys. Res.* **95**: 3273–3285.

———; C. ROOTH, D. HU and L.T. SMITH. 1992. Salinity-driven thermocline transients in a wind- and thermohaline-forced isopycnal coordinate model of the North Atlantic. *J. Phys. Oceanogr.* **22**: 1486–1505.

BLUMBERG, A.F. and G.L. MELLOR. 1987. A description of a three-dimensional coastal circulation model. In: *Three-Dimensional Coastal Ocean Models*, vol 4, N. Heaps (Ed.), American Geophysical Union, Washington, D.C., pp. 208.

BONING, C.W. and R.G. BUDICH. 1992. Eddy dynamics in a primitive equation model: Sensivity to horizontal resolution and friction. *J. Phys. Oceanogr.* **22**: 361–381.

BRUGGE, R.; H.L. JONES and J.C. MARSHALL. 1991. Non-hydrostatic ocean modeling for studies of open-ocean deep convection. *In*: Proc. Workshop Deep Convection and Deep Water Formation in the Oceans, Monterey, CA, Elsevier Science, pp. 325–340.

BRYAN, F.O. 1987. Parameter sensitivity of primitive equation ocean general circulation models. *J. Phys. Oceanogr.* **17**: 970–985.

——— and W.R. HOLLAND. 1989. A high resolution simulation of the wind- and thermohaline-driven circulation in the North Atlantic Ocean. *In*: "Parameterisation of Small-Scale Processes", Proc. of 'Aha Huliko'a Winter Workshop 1989, P. Muller and D. Henderson (Eds). Hawaii Institute of Geophysics, University of Hawaii, Hawaii, 1989, pp. 99–116.

BRYAN, K. 1969. A numerical method for the study of the circulation of the world ocean. *J. Comput. Phys.* **4**: 347–376.

———. 1984. Accelerating the convergence to equilibrium of ocean-climate models. *J. Phys. Oceanogr.* **14**: 666–673.

———. 1989. The design of numerical models of the ocean circulation. In: "*Oceanic Circulation Models: Combining Data and Dynamics*", Anderson, D.L.T. and J. Willebrand (Eds), Kluwer Academic Publishers, The Netherlands, pp. 465–500.

COX, M.D. 1984. A primitive equation, three dimensional model of the ocean. GFDL Ocean Tech. Report No. 1, Princeton, NJ, 141 pp.

———. 1985. An eddy-resolving numerical model of the ventilated thermocline. *J. Phys. Oceanogr.* **15**: 1312–1324.

DIETRICH, D.E.; M.G. MARIETTA and P.J. ROACHE. 1987. An ocean modeling system with turbulent boundary layers and topography: numerical description. *Int. J. Numer. Meth. Fluids*, **7**: 833–855.

———; P.J. ROACHE and M.G. MARIETTA. 1990. Convergence studies with the Sandia Ocean Modeling System. *Int. J. Numer. Meth. Fluids*, **11**: 127–150.

——— and C.A. LIN. 1994. Eddy shedding in Gulf of Mexico. *J. Geophys. Res.* **99**: 7599–7615.

DUKOWICZ, J.K.; R.D. SMITH and R.C. MALONE. 1993. A reformulation and implementation of the Bryan-Cox-Semtner ocean model on the Connection Machine. *J. Atmos. Oceanic Tech.* **10**: 195–208.

EZER, T. and G.L. MELLOR. 1994. Diagnostic and prognostic calculations of the North Atlantic circulation and sea level using a sigma coordinate ocean model. *J. Geophys. Res.* **99**: 14,159–14,171.

FUKUMORI, I.; J. BENVENISTE, C. WUNSCH and D.B. HAIDVOGEL. 1993. Assimilation of sea surface topography into an ocean circulation model using a steady-state smoother. *J. Phys. Oceanogr.* **23**: 1831–1855.

GOUGH, W. and C.A. LIN. 1992. The response of an ocean general circulation model to long time scale surface anomalies. ATMOSPHERE-OCEAN, **30**: 653–674.

HAIDVOGEL, D.B.; J. WILKIN and J. YOUNG 1991. A Semi-spectral Primitive Equation Ocean Circulation Model Using Vertical Sigma and Orthogonal Curvilinear Horizontal Coordinates. *J. Comput. Phys.* **94**: 151–185.

HANEY, R.L. 1971. Surface thermal boundary condition for ocean circulation model. *J. Phys. Oceanogr.* **1**: 241–248.

HARLOW, F.H. and J.E. WELCH. 1965. Numerical calculation of the time-dependent viscous incompressible flow of fluid with free surface, *Phys. Fluids*, **8**: No. 12, 2182–2189.

HOLLAND, W.R. 1978. The role of mesoscale eddies in the general circulation of the ocean – numerical experiments using a wind-driven quasi-geostrophic model. *J. Phys. Oceanogr.* **8**: 363–392.

HURLBURT, H.E. and J.D. THOMPSON. 1980. A numerical study of Loop Current intrusions and eddy shedding. *J. Phys. Oceanogr.* **10**: 1611–1651.

———; A.J. WALLCRAFT, Z. SIRKES and E.J. METZGER. 1992. Modelling of the global and Pacific oceans: On the path to eddy-resolving ocean prediction. *Oceanogr.* **5**: 9–18.

MELLOR, G. 1992. USER'S Guide for A Three-Dimensional, Primitive Equation, Numerical Ocean Circulation Model. Atmospheric and Oceanic Sciences Program, Princeton University, Princeton, NJ 08540.

RICHTMYER, R.D. and K.W. MORTON. 1967. *Difference Methods for Initial Value Problems*, Interscience Publisher, New York, 405 pp.

ROBERT, A. 1966. The integration of a low order spectral form of the primitive meteorological equations. *J. Meteorol. Soc. Jpn.* **44**: No. 5, 237–245.

———. 1981. A stable numerical integration scheme for the primitive meteorological equations. ATMOSPHERE OCEAN, **19**: 35–46.

SADOURNY, R. 1975. The dynamics of finite-difference models to the shallow water equations. *J. Atmos. Sci.* **32**: 680–689.

SEMTNER, A.J. 1986a. History and methodology of modelling the circulation of the world ocean. In: *Advanced Physical Oceanographic Numerical Modelling*, James J. O'Brien (Ed.), D. Reidel Publishing Company, 1986, pp. 23.

———. 1986b. Finite-difference formulation of a world ocean model. In: *Advanced Physical Oceanographic Numerical Modelling*, James J. O'Brien (Ed.), D. Reidel Publishing Company, 1986), pp. 187.

——— and Y. MINTZ. 1977. Numerical simulation of the Gulf Stream and mid-ocean eddies. *J. Phys. Oceanogr.* **7**: 208–230.

——— and R.M. CHERVIN. 1988. A simulation of the global ocean circulation with resolved eddies. *J. Geophys. Res.* **93**: 15,502–15,522.

——— and ———. 1992. Ocean general circulation from a global eddy-resolved model. *J. Geophys. Res.* **97**: 5,495–5550.

SMAGORINSKY, J. 1963. General circulation experiments with the primitive equations: I. The basic experiments. *Mon. Weather Rev.* **91**: 99–164.

TAKANO, K. 1974. A general circulation model for the world ocean. Technical Report No. 8, Dept. of Meteorology, Univ. of California, Los Angeles.

———. 1975. Relationship between the grid size and coefficient of lateral eddy viscosity in the finite difference computation of the linear vorticity equation in the ocean. *J. Ocean. Soc. Jpn.* **31**: 105–108.

XU, W. 1994. A C-grid ocean general circulation model and eddy simulation. Ph.D. Thesis, McGill University, 157 pp.

——— and C.A. LIN. 1993. A numerical solution of the linear Rayleigh-Bernard convection equations with B- and C-grid formulations. *Tellus*, **45A**: 193–200.

———; R.J. GREATBATCH and C.A. LIN. 1995. The sensitivity of an eddy resolving model to the surface thermal boundary conditions. *J. Geophys. Res.* **100(58)**: 15,899–15,914.

Global Modelling of the Ocean and Atmosphere Using the Spectral Element Method

Dale B. Haidvogel, Enrique Curchitser, Mohamed Iskandarani
and Rowan Hughes
Institute of Marine and Coastal Sciences
Rutgers University
P.O. Box 231
New Brunswick, NJ 08903-0231, U.S.A.

and

Mark Taylor
Department of Meteorology
University of Maryland
College Park, MD 20742, U.S.A.

[Original manuscript received 17 October 1994; in revised form 23 August 1995]

ABSTRACT *The use of spectral methods now has a long history in global atmospheric modelling wherein the attractive properties of Fourier series on spheres, including higher-order convergence rates and efficient implementation via the transform method, have proven advantageous. Partially offsetting these advantages, however, are several competing disadvantages. Two of these, the appearance of Gibbs oscillations for localized processes (e.g., orographic interactions) and the difficulty of mapping spectral techniques onto parallel computer architectures, are inherent to the global nature of these techniques. A third drawback, the restriction of these methods to regular geometries, has severely limited their application to the modelling of the large-scale ocean circulation.*

We describe a global circulation model that has, in principle, none of these limitations. The model utilizes the spectral element method that combines the geometrical flexibility of traditional finite element methods with the rapid convergence rates of spectral approximation techniques. Simple test problems drawn from both oceanic and atmospheric modelling are used to demonstrate that the resulting model is exponentially convergent, yet allows effective representation of irregular geometry and efficient grid refinement in regions of dynamical interest. Lastly, performance characteristics on the nCUBE/2 and Cray T3D architectures confirm that the element model is ideally suited to the parallel computing environment.

RÉSUMÉ *L'utilisation de méthodes spectrales a déjà une longue histoire dans la modélisation mondiale de l'atmosphère. Dans ces méthodes, les propriétés attrayantes de la série de Fourier sur les sphères, des taux de convergence d'ordre élevé et une application efficace grâce à la méthode de transformation, se sont montrées avantageuses. Toutefois, plusieurs désavantages concurrents perturbent quelque peu la situation. Deux de ceux-ci, l'apparition d'oscillations de Gibbs pour les processus localisés (p. ex., interactions orographiques), et la difficulté du Nappage des techniques spectrales dans les multiprocesseurs, sont inhérents à la nature sphérique de ces techniques. Aussi, ces méthodes n'emploient que des géométries régulières ce qui limite grandement leur application dans la modélisation de la circulation océanique à grande échelle.*

On décrit un modèle mondial de circulation océanique qui, en principe, n'a aucune de ces limites. Le modèle utilise la méthode de l'élément spectral, combinant la flexibilité géométrique des méthodes traditionnelles des éléments finis avec les taux rapides de convergence des techniques d'approximation spectrales. On utilise des problèmes d'essais simples, issus de la modélisation océanique et atmosphérique, pour démontrer que le modèle obtenu est exponentiellement convergent mais qu'il permet quand même une représentation effective de la géométrie irrégulière et un raffinement de la grille dans les régions d'intérêt pour la dynamique. Enfin, les caractéristiques des performances sur multiprocesseurs nCUBE/2 et Cray T3D confirment que le modèle élément est idéal pour un environnement de calcul parallèle.

1 Introduction

Modellers of the global atmosphere and ocean face unique dynamical and computational challenges. As a consequence, although some variety of algorithmic approaches exists within both modelling communities, rather different numerical approaches have become traditional in each. These choices reflect differing compromises among such issues as algorithmic complexity, convergence rates, computational performance on vector and parallel machines and geometrical flexibility.

A key distinction between global atmospheric and oceanic modelling, which has largely dictated differing choices of horizontal discretization scheme, is the geometry within which each fluid is contained. The natural geometry for the atmosphere, the sphere, has no boundary and for boundaryless or periodic domains spectral methods are known to have high accuracy. For the sphere, the natural basis functions are spherical harmonics; and indeed, the majority of all major climate models in use today are spherical harmonic-based spectral methods (Boer et al., 1992; Gates, 1992).

In addition to high accuracy, spherical harmonic methods also provide a completely isotropic representation of a scalar function on the sphere. Any finite difference model cannot have this property, since the only isotropic grids would be those with grid points chosen to be the vertices of a platonic solid, and the largest such grid comes from the dodecahedron which has only 20 vertices. The common latitude-longitude grid is extremely nonisotropic, with points clustered near the poles resulting in severe time-stepping constraints. This "pole problem" is naturally avoided by the use of spherical harmonics. Another advantage of spherical harmonics is that they are eigenfunctions of the Laplace operator on the sphere. This property makes it very easy to invert the linear elliptic operators required for implicit time-stepping schemes.

Spherical harmonics do have some disadvantages, the most notable being their high cost. Unlike the Fast Fourier Transform (FFT), the discrete spherical harmonic transform (DST) is computed by direct methods since it is currently unknown if a fast transform exists. Consequently, at high resolutions the transforms dominate the computations. In addition, the DST is also more difficult to implement on distributed memory machines because of the global communications involved. Lastly, it is difficult, if not impossible, to incorporate local mesh refinement in global spectral models, as might be desirable to resolve, for example, local orographic processes.

In contrast to the atmosphere, the geometry of ocean basins features both irregular continental boundaries and tall topographic features which often extend through a large fraction of the fluid column. The existence of complex bathymetry and sidewalls introduces fundamental dynamical constraints, absent in the atmosphere, which lead, for example, to the existence of narrow, intense current systems on the western margins of the ocean basins, as well as to a wide range of coastal circulation phenomena. As a consequence, modellers of the global ocean circulation have had to contend not only with irregular sidewalls, but also with a higher degree of lateral inhomogeneity than in the atmosphere.

The irregularity of the basin geometry clearly rules out the use of global spectral methods such as those based on spherical harmonics. The alternative traditionally adopted by modellers of the global ocean has been that of lower-order finite difference and (to a lesser extent) finite element methods – for example, the well known Geophysical Fluid Dynamics Laboratory (GFDL) ocean circulation model originated by Bryan (1969) and elaborated on by many investigators since. Interestingly, this choice makes exactly the opposite compromises from those accepted traditionally by atmospheric modellers. First, whereas spherical harmonics are an exponentially convergent approximation with no trouble near the pole, finite difference methods on latitude/longitude grids are only algebraically convergent and do suffer a Courant-Friedrich-Levy (CFL) pole problem. Second, while global spectral methods are difficult to adapt to complex geometry and to local mesh refinement, low-order finite difference methods are readily configured for geometrically complex and/or spatially inhomogeneous flows. Lastly, the high communications overhead incurred by spherical harmonic techniques on parallel computers is absent in the lower-order methods typically used in today's ocean circulation models.

Here, we describe a third approach to spatial representation of the atmosphere and ocean. This technique, the spectral element method, accommodates, in principle, all of the desirable properties of both the global spectral and the low-order finite difference schemes – that is, exponential convergence to the true solution (and hence high accuracy), approximately isotropic resolution on the sphere (and hence no pole problem), flexible representation of complex geometries, straightforward regional mesh refinement and nearly perfect scalability on highly parallel computing platforms. It therefore represents an attractive alternative to the modelling of the coupled atmosphere-ocean system.

The plan of the paper is as follows: Section 2 introduces the concept of the spectral element technique and discusses the derivation of the discrete form of the shallow water equations. The accuracy and convergence properties of the spectral element approximation in the atmospheric situation of spherical geometry is assessed in Section 3 by conducting the first three shallow-water test problems of Williamson et al. (1992). Application of the new model to the simulation of the wind-driven circulation in the Pacific Ocean on a global, non-uniform grid is discussed in Section 4. The parallel performance of the spectral element model is explored in Section 5. Lastly, a concluding section offers a summary and discussion.

2 The spectral element technique

The spectral finite element (SFE) method has been developed during the past decade to combine the accuracy and convergence rates of spectral methods and the geometrical flexibility of traditional finite element methods. The method can be classified as an *h–p* type finite element method, where convergence to the solution can be achieved either by increasing the number of elements, keeping the order of the interpolation fixed, or by keeping the elemental partition fixed and increasing the order of the interpolation polynomial. The SFE method has been successfully applied to the solution of fluid flows in engineering, as well as to several simplified problems in ocean circulation modelling (Ma, 1993; Iskandarani et al., 1995). Our current model solves the shallow-water equations; it is meant to be the prototype for an eventual three-dimensional primitive equation model.

a *Overview*

In the spectral element method, the computational domain is broken up into an unstructured set of quadrilateral regions called elements. The quadrilateral elements are actually logical rectangles, which may be isoparametrically mapped to fit an irregular geometry. Within each of these elements, dependent variables are approximated by an expansion in polynomials or test functions. The discrete equations are derived using an integral form of the governing equations, in conjunction with a suitable set of test functions and quadrature formulae (Maday and Patera, 1988).

The procedure used here serves as an example. First, an integral form of the equations is obtained by multiplying both sides of the governing equations by a test function and then integrating over the entire computational domain (for example, the surface of the sphere). The sphere is then tiled with rectangular elements and the integrals are written as the sum of integrals over each element. The continuous integrals are then approximated by making particular choices for the quadrature rules and the test functions, i.e., Gauss-Lobatto quadrature within each element and the associated Gauss-Lobatto cardinal functions, respectively. The result is a standard Legendre spectral transform method within each element. Derivatives are calculated with Legendre transforms on the Gauss-Labatto grid. Because of the choice of grid and the associated cardinal functions, the only communication between elements occurs at the element boundaries, where neighbouring elements share common points. At these points, the terms appearing in the equations are obtained by weighted averaging.

The contrast with more traditional discretization methods is noteworthy. The spectral element formalism replaces the global spectral representation of conventional spherical harmonic models with a series of local polynomial transforms at the elemental level. Rapid convergence to the continuous solution is therefore maintained. However, unlike global spectral methods, the underlying quadrilateral grid allows straightforward mesh refinement and use within irregular geometries, much like a lower-order finite difference or finite element method. Lastly, since the sharing of information is sparse across element boundaries, the spectral element method maps naturally and efficiently onto high performance parallel computer architectures, as described below.

In summary, at the sub-element level, the SFE method is algorithmically analogous to the spherical harmonic approach used in the majority of global atmospheric models. At the level of individual elements, information exchange is between adjacent neighbours only, as is typical for the lower-order schemes used in traditional ocean circulation models.

b *Derivation of the Shallow Water Equations*

Let Ω be the two-dimensional region occupied by the fluid and let Γ denote its boundary. The shallow water equations in Ω are given by

$$\frac{\partial \mathbf{u}}{\partial t} + \mathbf{u} \cdot \nabla \mathbf{u} + \mathbf{f} \times \mathbf{u} + g \nabla \zeta + \gamma \mathbf{u} - \frac{\nabla \cdot [\nu (h + \zeta) \nabla \mathbf{u}]}{(h + \zeta)} = \frac{\boldsymbol{\tau}}{\rho (h + \zeta)} \tag{1}$$

$$\frac{\partial \zeta}{\partial t} + \nabla \cdot [(h + \zeta) \mathbf{u}] = 0 \tag{2}$$

where $\mathbf{u}$ is the horizontal velocity vector; h, the resting depth of the fluid; ζ, the free surface elevation; $\mathbf{f}$, the vertical component of the Coriolis force; g, the gravitational acceleration; γ, the bottom drag coefficient; ν, the lateral viscosity coefficient; ρ, the density of the fluid; $\boldsymbol{\tau}$ the wind stress acting on the surface of the fluid; and, ∇, the two-dimensional gradient operator. The boundary conditions are Dirichlet conditions on $\mathbf{u}$ and/or ζ:

$$\mathbf{u} = \mathbf{u}^b \text{ on } \Gamma_D, \quad \zeta = \zeta^b \text{ on } \Gamma_D^\zeta \tag{3}$$

and Neumann conditions on $\mathbf{u}$:

$$\nu \nabla \mathbf{u} \cdot \mathbf{n} = \mathbf{q} \text{ on } \Gamma_N. \tag{4}$$

Here, $\mathbf{n}$ is the unit normal to the boundary; $\mathbf{q}$ is the applied stress (if free-slip boundary conditions are applied on a rectangular domain, for example, $\mathbf{q} = 0$); and $\mathbf{u}^b$ and ζ^b are the values of $\mathbf{u}$ and ζ imposed on specified segments of the boundary. If $\nu = 0$, only the normal component of the velocity at the boundary can be specified: $\mathbf{u}\cdot\mathbf{n} = \mathbf{u}^b \cdot \mathbf{n}$. Γ_D and Γ_D^ζ are those portions of the boundary where Dirichlet conditions on $\mathbf{u}$ and ζ are applied. Γ_N denotes the boundary segments where Neumann conditions on $\mathbf{u}$ are applied; we must have $\Gamma_D \cap \Gamma_N = \emptyset$. We refer the reader to Bernardi and Pironneau (1991) for a discussion of the appropriate boundary conditions to apply for different forms of the shallow water equations.

The starting point of the spectral element model is the Galerkin formulation of the shallow water equations:

$$\int_A \frac{\partial \mathbf{u}}{\partial t} w dA = \int_A \left\{ \frac{\tau}{\rho(h+\zeta)} - \mathbf{u}\cdot\nabla\mathbf{u} - \mathbf{f}\times\mathbf{u} - g\nabla\zeta - \gamma\mathbf{u} \right.$$
$$\left. - \frac{\nu\nabla(h+\zeta)\cdot\nabla\mathbf{u}}{(h+\zeta)} \right\} wdA$$
$$- \int_A \nu\nabla\mathbf{u}\cdot\nabla w dA + \int_{\Gamma_N} \mathbf{q} w dS, \qquad \forall w \in V \tag{5}$$

$$\int_A \frac{\partial \zeta}{\partial t} w^p \, dA = - \int_A \nabla \cdot [(h+\zeta)\mathbf{u}] w^p \, dA \qquad \forall w^p \in Z \tag{6}$$

where V and Z are the Sobolev spaces defined by

$$V(\Omega) = \{v \in H^1(\Omega),\ v(\Gamma_D) = 0\}$$

$$Z(\Omega) = \{v \in H^1(\Omega),\ v(\Gamma_D^\zeta) = 0\}.$$

H^1 is the set of square integrable functions whose first derivatives are also square integrable; w and w^p are the weight functions associated with the velocity and the surface elevation, respectively.

When the surface displacement ζ is much smaller than the resting depth h, the flow is nearly incompressible. In this limit, the pressure interpolation must comply with the so-called Babuska-Brezzi or div-stability condition (Maday and Patera, 1988), otherwise spurious pressure modes appear and can destroy the solution. Several alternatives are available to circumvent the spurious modes. The one adopted here is to use a staggered mesh and interpolate the pressure with a polynomial of degree two less than that used for the velocity (Iskandarani et al., 1995). This staggering preserves the continuity of the pressure at the inter-element boundaries. If the flow is sufficiently divergent, this staggering is unnecessary (e.g., Ma, 1993).

The spatial discretization proceeds by subdividing the domain into a set of conforming quadrilateral isoparametric elements. Each element is mapped into the unit square in the computational domain (ξ, η), and the variables $\mathbf{u}$ and ζ are interpolated as:

$$\left\{ \begin{array}{l} \mathbf{u}(\xi,\eta) = \sum_{i=1}^{N^v} \sum_{j=1}^{N^v} \mathbf{u}_{i,j}(t) h_i^v(\xi) h_j^v(\eta) \\ \zeta(\xi,\eta) = \sum_{i=1}^{N^p} \sum_{j=1}^{N^p} \zeta_{i,j}(t) h_i^p(\xi) h_j^p(\eta) \end{array} \right. \tag{7}$$

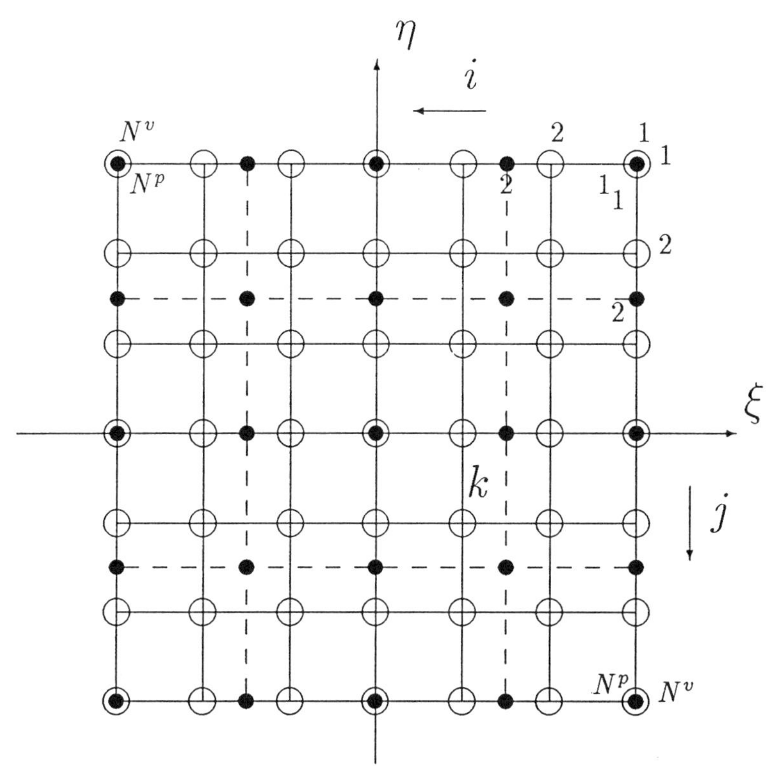

Fig. 1 Isoparametric element in computational plane. The Gauss-Lobatto points of the velocity nodes are shown by open circles and the pressure nodes by filled circles. The local node number of the velocity nodes are shown outside the element while those of the pressure nodes are shown inside, $N^p = N^v - 2$.

where $\mathbf{u}_{ij}$ is the velocity vector at the velocity collocation nodes $(\xi_{ij}^v, \eta_{ij}^v)$, $(i,j) = 1, \cdots, N^v$, and ζ_{ij} is the surface elevation at the pressure collocation nodes $(\xi_{ij}^p, \eta_{ij}^p)$, $(i,j) = 1, \cdots, N^p$. N^v and N^p are the number of nodes per element in the ξ and η directions for the velocity and pressure interpolation, respectively; therefore, $N^p = N^v - 2$.

The interpolation functions h_i^v are the Legendre Cardinal functions (Boyd, 1989):

$$h_i^v(\xi) = \frac{-(1-\xi^2)L'_{N^v-1}(\xi)}{N^v(N^v-1)L_{N^v-1}(\xi_i^v)(\xi-\xi_i^v)}, \quad i = 1, 2, \ldots, N^v. \tag{8}$$

L_{N^v-1} denotes the Legendre polynomial of degree $(N^v - 1)$ and L'_{N^v-1} denotes its derivative. The ξ_i^v are the N^v Gauss-Lobatto roots of L_{N^v-1}. The pressure interpolation functions h_i^p are defined similarly but with the superscript v replaced by p. Figure 1 shows a typical element and the layout of the velocity and pressure Gauss-Lobatto nodes. The Gauss-Lobatto nodes are not evenly spaced as sketched in the figure but are actually more crowded near the boundaries.

A system of ordinary differential equations (for $\mathbf{u}$ and ζ) is obtained after inserting (7) in (5) and (6) and substituting $h_i^v h_j^v$ for w and $h_i^p h_j^p$ for w^p:

$$M^v \frac{d\mathbf{u}}{dt} = \mathbf{a} \tag{9}$$

$$M^p \frac{d\zeta}{dt} = c. \tag{10}$$

The matrices M^v and M^p are the mass matrices associated with the velocity and pressure interpolation functions, respectively; they are defined as:

$$M^v_{ij,kl} = \int_A h_i^v(\xi) h_j^v(\eta) h_k^v(\xi) h_l^v(\eta)\, dA \tag{11}$$

$$M^p_{ij,kl} = \int_A h_i^p(\xi) h_j^p(\eta) h_k^p(\xi) h_l^p(\eta)\, dA. \tag{12}$$

The right hand side vectors are:

$$\begin{aligned} \mathbf{a}_{ij} = \int_A \Bigg\{ & \frac{\boldsymbol{\tau}}{\rho(h+\zeta)} - \mathbf{u}\cdot\nabla\mathbf{u} - \mathbf{f}\times\mathbf{u} - g\nabla\zeta - \gamma\mathbf{u} \\ & - \frac{\nu\nabla(h+\zeta)\cdot\nabla\mathbf{u}}{(h+\zeta)} \Bigg\} h_i^v(\xi) h_j^v(\eta)\, dA \\ & - \int_A \nu\nabla\mathbf{u}\cdot\nabla[h_i^v(\xi) h_j^v(\eta)]\, dA + \int_{\Gamma_N} h_i^v(\xi) h_j^v(\eta)\mathbf{q}\, dS \end{aligned} \tag{13}$$

$$c_{ij} = -\int_A \nabla\cdot[(\zeta+h)\mathbf{u}] h_i^p(\xi) h_j^p(\eta)\, dA. \tag{14}$$

Equations (9) and (10) hold at the elemental level. The assembly procedure adds the contribution of the different elements to the system of equations.

Note that even an explicit time-integration scheme requires the inversion of the matrices M^v and M^p. Fortunately, the mass matrices can be made diagonal by evaluating the integrals with Gauss-Lobatto quadrature. The diagonal form of the mass matrices leads to tremendous savings in computations and storage with negligible loss of accuracy.

The explicit time integration of equations (9) and (10) is performed with a third-order Adams-Bashforth (AB3) scheme which we outline briefly. Each of the equations in (9) and (10) can be written in the generic form $Md\mathbf{u}/dt = \mathbf{r}$ where $\mathbf{u}$ and $\mathbf{r}$ are the vector of unknowns and the vector of right hand sides, respectively, and M is one of the mass matrices. The AB3 scheme takes the form (Gear, 1971):

$$\mathbf{u}^{n+1} = \mathbf{u}^n + \Delta t M^{-1}\left[\frac{23}{12}\mathbf{r}^n - \frac{16}{12}\mathbf{r}^{n-1} + \frac{5}{12}\mathbf{r}^{n-2}\right]. \tag{15}$$

The calculations require information at two previous time levels and thus a start-up method is needed at the initial time step; we choose a fourth-order Runge-Kutta scheme. All computations are performed at the elemental level and only the vector **r** needs to be assembled.

The stability of the explicit scheme is determined by the smallest of the stability limits of the advection operator, Δt_a, and diffusion operator, Δt_d. Ma (1993) gives these stability limits as $\Delta t_a \leqslant \alpha_1 L/(UN^2K)$ and $\Delta t_d \leqslant \alpha_2 L^2/(\nu N^4 K^2)$. Here, U is the speed of the fastest moving wave in the problem (in oceanic flows this is the free surface gravity wave $U = \sqrt{gh}$); L, the length scale; ν, the kinematic viscosity; N, the number of polynomials in the interpolation; K, the number of elements; and α_a and α_d are proportionality constants. The ratio $(\Delta t_d/\Delta t_a)$ is proportional to $R_e/(N^2K)$ where $R_e = UL/\nu$ is the Reynolds number. Hence, as long as the Reynolds number is larger than N^2K, as is the case in most large-scale geophysical flows, the advection operator determines the largest allowable time step. Typical values for the ocean of $U = 200$ m/s, $L = 1000$ km, $\nu = 2000$ m^2/s, $N = 12$ and $K = 30$ yield a ratio of $10^5/4000 = 20$. The time step in a medium to high resolution grid varies from 2 to 6 minutes when the model is run in reduced gravity mode (surface wave speed of about 2–5m/s; Ma, 1993; Iskandarani et al., 1995) and from 5 to 15 seconds for a full gravity model with surface wave speeds between 100 and 200 m/s.

3 Tests problems on a sphere

To evaluate the suitability of spectral elements for atmospheric modelling, we implement the method in spherical geometry and measure its performance relative to other methods. To do so, we adopt the now standard suite of shallow water test cases proposed by Williamson et al. (1992). The shallow water equations inherit many of the difficulties associated with three-dimensional global atmospheric modelling, and as such, any potential global model should perform well on these tests.

a *Spectral Elements on the Sphere*

The main difficulty in extending the SFE method to full spherical geometry is the lack of a good, global coordinate system for the sphere. In spherical coordinates, the velocity vector is discontinuous at the poles, and thus poorly approximated by a spectral representation. This artificial singularity is due entirely to the coordinate system, and is easily avoided by using local Cartesian coordinates in each element.

The first step in applying the spectral element method to spherical geometry therefore is to tile the sphere with squares, or regions that can be easily mapped to squares. A simple, completely isotropic way to do this is to project a cube from the centre of the sphere onto the surface of the sphere, creating 6 large elements. For now, we are using only these 6 elements and a relatively high spectral degree within each element. Ideally one would further divide each of these elements into smaller elements, striking a balance between accuracy and computation cost by

Fig. 2 The cube projected onto the sphere.

adjusting the number of elements and the spectral degree used in each element. The projection onto the sphere of such a configuration is shown in Fig. 2. Each of the 6 faces of the cube has been shown divided into an 8×8 array of nodal points.

We use the natural Cartesian coordinates in each face of the cube. We begin by rewriting the momentum equations in the following equivalent form:

$$\frac{\partial \mathbf{u}}{\partial t} = -(\omega + f)\hat{k} \times \mathbf{u} - \nabla\left(\tfrac{1}{2}\,\mathbf{u}\cdot\mathbf{u} + g\zeta\right), \tag{16}$$

where $\omega = \hat{k}\cdot\nabla \times \mathbf{u}$ is the relative vorticity. Let x and y be the Cartesian coordinates in a particular face of the cube. On the sphere, we denote longitude by λ and latitude by θ. We also define the unit vectors associated with each coordinate by $\hat{x}$, $\hat{y}$, $\hat{\lambda}$ and $\hat{\theta}$. To solve the shallow water equations for the velocity, we must pick a set of basis vectors and solve for the coefficients of the velocity when expressed in terms of these vectors. The choice of the spherical coordinate system $\mathbf{u} = u_1\hat{\lambda}+u_2\hat{\theta}$ is particularly bad since the coordinates are discontinuous at the poles. The most natural choice is that associated with the coordinate system we are using for the independent variables x and y, $\mathbf{u} = v_1\hat{x} + v_2\hat{y}$. (There are other possibilities, such as using spherical coordinates in the elements which are near the equator and a rotated spherical coordinate system in those elements near the poles.)

In this work, we have chosen to solve the shallow water equations for ζ, v_1 and v_2. This coordinate system is not orthogonal, which makes the equations unmanageable unless they are written in a matrix-vector form. To do this, we start by defining the matrix $\mathbf{D}$, which represents the derivative of the mapping from $(x, y) \longrightarrow (\lambda, \theta)$,

$$\mathbf{D} = \begin{pmatrix} \cos(\theta)\lambda_x & \cos(\theta)\lambda_y \\ \theta_x & \theta_y \end{pmatrix}. \tag{17}$$

A subscript of either x or y represents differentiation with respect to x or y. The components of the velocity vector $\mathbf{u}$ in the (λ, θ) and (x, y) coordinate systems are related by

$$\begin{pmatrix} u_1 \\ u_2 \end{pmatrix} = \mathbf{D} \begin{pmatrix} v_1 \\ v_2 \end{pmatrix}.$$

We now take each vector and differential operator appearing in the shallow water equations and write them in the (x, y) coordinate system:

$$\begin{aligned}
\mathbf{u}\cdot\mathbf{u} &= \begin{pmatrix} v_1 \\ v_2 \end{pmatrix}^T \mathbf{D}^T\mathbf{D} \begin{pmatrix} v_1 \\ v_2 \end{pmatrix}, \\
\hat{k} \times \mathbf{u} &= \mathbf{D}^{-1} \begin{pmatrix} 0 & -1 \\ 1 & 0 \end{pmatrix} \mathbf{D} \begin{pmatrix} v_1 \\ v_2 \end{pmatrix}, \\
\nabla\zeta &= \mathbf{D}^{-1}\mathbf{D}^{-T} \begin{pmatrix} \zeta_x \\ \zeta_y \end{pmatrix}, \\
\nabla\cdot\mathbf{u} &= \begin{pmatrix} \dfrac{\partial}{\partial x} \\ \dfrac{\partial}{\partial y} \end{pmatrix}^T \begin{pmatrix} v_1 \\ v_2 \end{pmatrix} + \mathbf{p}^T \begin{pmatrix} v_1 \\ v_2 \end{pmatrix}, \\
\omega &= \left[\mathbf{D}^T \begin{pmatrix} 0 & -1 \\ 1 & 0 \end{pmatrix} \mathbf{D}^{-T} \begin{pmatrix} \dfrac{\partial}{\partial x} \\ \dfrac{\partial}{\partial y} \end{pmatrix}\right]^T \begin{pmatrix} v_1 \\ v_2 \end{pmatrix} + \mathbf{q}^T \begin{pmatrix} v_1 \\ v_2 \end{pmatrix},
\end{aligned} \tag{18}$$

where $\mathbf{p}$ and $\mathbf{q}$ are 2-vectors containing additional metric terms. For efficiency, $\mathbf{p}$, $\mathbf{q}$ and all the 2×2 matrices which appear in the above formulae are analytically derived and their values at each grid point are precomputed and stored. The final equations have the form

$$
\begin{aligned}
\begin{pmatrix} v_1 \\ v_2 \end{pmatrix}_t &= -(\omega + f)\mathbf{D}^{-1}\begin{pmatrix} 0 & -1 \\ 1 & 0 \end{pmatrix}\mathbf{D}\begin{pmatrix} v_1 \\ v_2 \end{pmatrix} - \mathbf{D}^{-1}\mathbf{D}^{-T}\begin{pmatrix} \phi_x \\ \phi_y \end{pmatrix}, \\
\zeta_t &= -\begin{pmatrix} \dfrac{\partial}{\partial x} \\ \dfrac{\partial}{\partial y} \end{pmatrix}^T \begin{pmatrix} (h+\zeta)v_1 \\ (h+\zeta)v_2 \end{pmatrix} + \mathbf{p}^T \begin{pmatrix} (h+\zeta)v_1 \\ (h+\zeta)v_2 \end{pmatrix},
\end{aligned}
\tag{19}
$$

where $\phi = \frac{1}{2}\,\mathbf{u}\cdot\mathbf{u} + g\zeta$.

Besides the modifications to the form of the equations outlined above, only one other addition is required. Each of the six elements uses a different coordinate system, so the communication between elements must be modified slightly. All vector quantities passed to an element from its neighbours must be mapped to the correct coordinate system. These mappings require 2×2 matrix multiplies where the matrices are again represented in terms of $\mathbf{D}$.

b *Test Case Results*

As a first test of the suitability of the spectral element model for spherical geometry, we use a standard set of shallow water test cases (Williamson et al., 1992). The existence of these test cases allows us easily to compare the spectral element model with several other methods without having to obtain and run the other codes. The test cases prescribe several forms of error and how they are to be computed so that any method can be objectively compared to several other methods by simply consulting the literature.

Here, we describe the first three test cases, and present a summary of results. We will briefly describe each test case, but rely on Williamson et al. (1992) for the detailed mathematical description of the cases and performance measures. The first three test cases primarily determine how well the method handles spherical geometry. They have analytic solutions making the numerical errors easy to calculate, but they have no small-scale structure. More realistic cases with complicated flows are represented in the latter test cases; however, these cases lack analytic solutions and the results must be compared with high resolution runs of other models. A full discussion of these tests is presented elsewhere (Taylor et al., 1997).

Test case 1 is the advection of a cosine bell with compact support around the sphere. This is the only case which does not involve the full set of shallow water equations. The velocity is fixed (rigid rotation about an axis at an angle α from the Earth's axis) and we solve only the advection equation for the height field for 12 days (one rotation). Several orientations of the velocity are specified by the parameter α: around the equator, over the poles and minor shifts from these two orientations. We also added an orientation which sends the cosine bell through the corners of the cube.

TABLE 1. Results from the spectral element model, the NCAR spectral model (Jakob-Chien et al. (1995)) and two finite difference models, Ross Heikes' twisted icosahedral grid and the Arakawa-Lamb model (Heikes (1993)).

Method	Resolution	l^{∞} Error		
		Case 1	Case 2	Case 3
Spec. Elem. $6 \times 16 \times 16$	1536	0.3	2×10^{-4}	3×10^{-3}
Spec. Elem. $6 \times 24 \times 24$	3456	0.05	6×10^{-7}	6×10^{-4}
Spec. Elem. $6 \times 32 \times 32$	6144	0.01	1×10^{-9}	8×10^{-5}
NCAR T42	1849	0.01	5×10^{-9}	1×10^{-9}
Twisted Icosahedral Grid	2562	0.5	2×10^{-3}	2×10^{-3}
Arakawa-Lamb 77×44	3168	0.5	1×10^{-3}	2×10^{-3}

Test case 2 is a steady-state solution of the full nonlinear shallow water equations. It again consists of solid body rotation for the velocity, but this time we use the corresponding geostrophic height field. Again several orientations are specified by the parameter α. The height field consists of lows over each pole. Test case 3 is similar to case 2 except that the wind field has compact support. It is a mid-latitude jet and the associated height field has a single low over the region encircled by the jet.

Our results along with results published in Jakob-Chien et al. (1995) and Heikes (1993) are summarized in Table 1. The errors listed in the table are normalized l_{∞} errors between the computed and analytic height field. The number listed for resolution represents the total number of grid points for gridpoint models, or the number of spectral coefficients retained for spectral models. For test cases 1 and 2, the spectral element model was relatively insensitive to α, and all results are given for the worst case α. Test case 3 was run for $\alpha = 0$.

Table 1 shows that the spectral element method is significantly more accurate than the finite difference models. As expected, the spectral element model with comparable resolution does not do as well as the spherical harmonic model, but one can see from the table that it is possible to achieve competitive accuracy with a modest increase in resolution. We should note that test cases 2 and 3 happen to be particularly trivial for global spectral models since the exact solutions are expressible with only a few spherical harmonics, yielding errors close to the round-off limit.

To verify the exponential convergence of the spectral element model, the data from the table is plotted in Fig. 3. The lines drawn are the least-squares fit to the data points which are plotted as circles. The lines show the almost perfect exponential decay of the error as a function of the number of grid points on a side per element.

Contour plots of the height field and error from test case 1 are shown in Fig. 4. The analytic and computed height fields are shown after one rotation of the cosine bell around the sphere. As specified in the test case, these are plotted on an ortho-

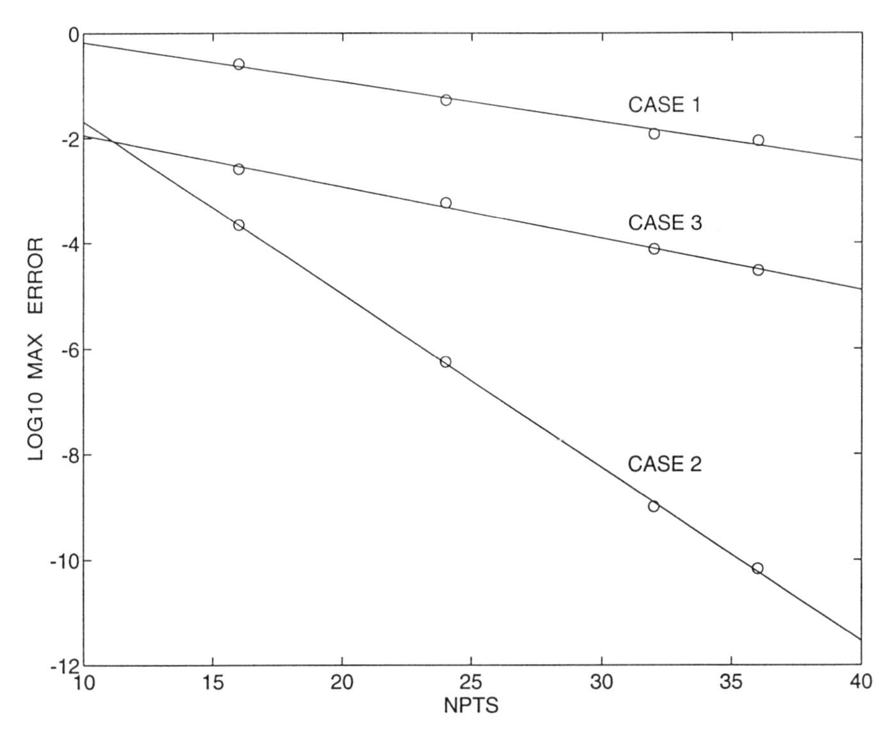

Fig. 3 Exponential convergence under mesh refinement for the spectral element method. Each element used a grid of size NPTS × NPTS.

graphic projection centred over the cosine bell, and only linear interpolation from the computation grid was used to generate the contours. The difference between these two is contoured in Fig. 4, where we observe the global, low-level noise typical of spectral methods.

4 Ocean modelling on unstructured grids

As previously discussed, ocean circulation models require flexibility and ease in the treatment of complex geometry, and convenient mesh refinement in regions of small-scale dynamical variability. Given their basis in h-type finite element methods, spectral elements naturally accommodate these requirements. A related property, of some importance to modellers of the ocean circulation on sub-global domains, is the prospect of producing high-resolution regional simulations on global, non-uniformly resolved grids. As illustrated by Fig. 5 and associated discussion, the need to invoke (poorly known) open boundary conditions can thereby be avoided at only modest additional computational expense.

Here, we describe an implementation of the shallow water, spectral element model to a reduced gravity simulation of the Pacific Ocean. The equatorial and

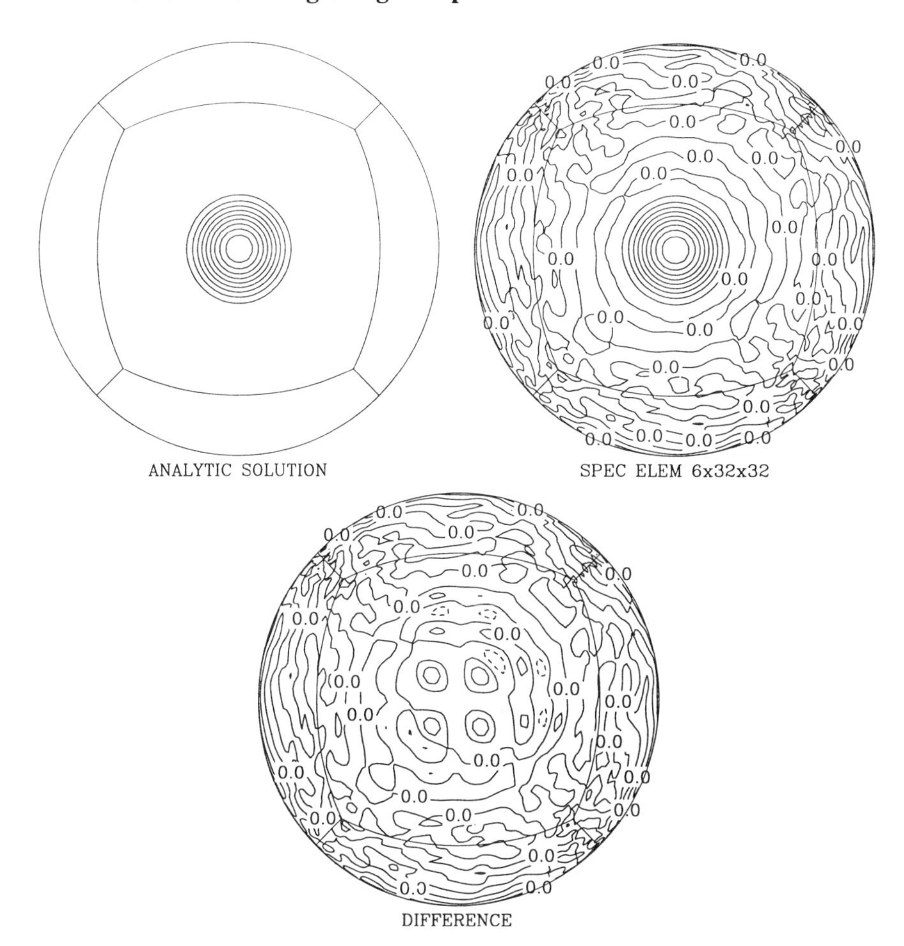

Fig. 4 Surface height field for shallow water test case 1 after 12 days for $\alpha = \pi/4$. Shown are the analytic solution, the computed result using the spectral element model and the difference between the two. The contour intervals are 100 metres for the analytic and SFE solutions, and 5 metres for the difference field.

midlatitude regions of the Pacific are the principal focus. To bypass the difficulties of setting variables at open boundaries, the global ocean is included, but with reduced resolution; elements outside of the Pacific account for no more than one-quarter of the computational effort. Figure 5 shows the velocity collocation grid on a Mercator projection. The average node spacing is 1.09° in the Pacific, 2.0° elsewhere, giving a global average of 1.32°. There are 337 elements in the Pacific and 111 elsewhere, for a total of 448. Each of the elements contains 7×7 velocity collocation nodes and 5×5 pressure nodes. Cyclic continuity around the Southern Ocean is attained merely by the logical connection between the sides of elements; it is not defined at particular elements and does not require additional code. Longitude

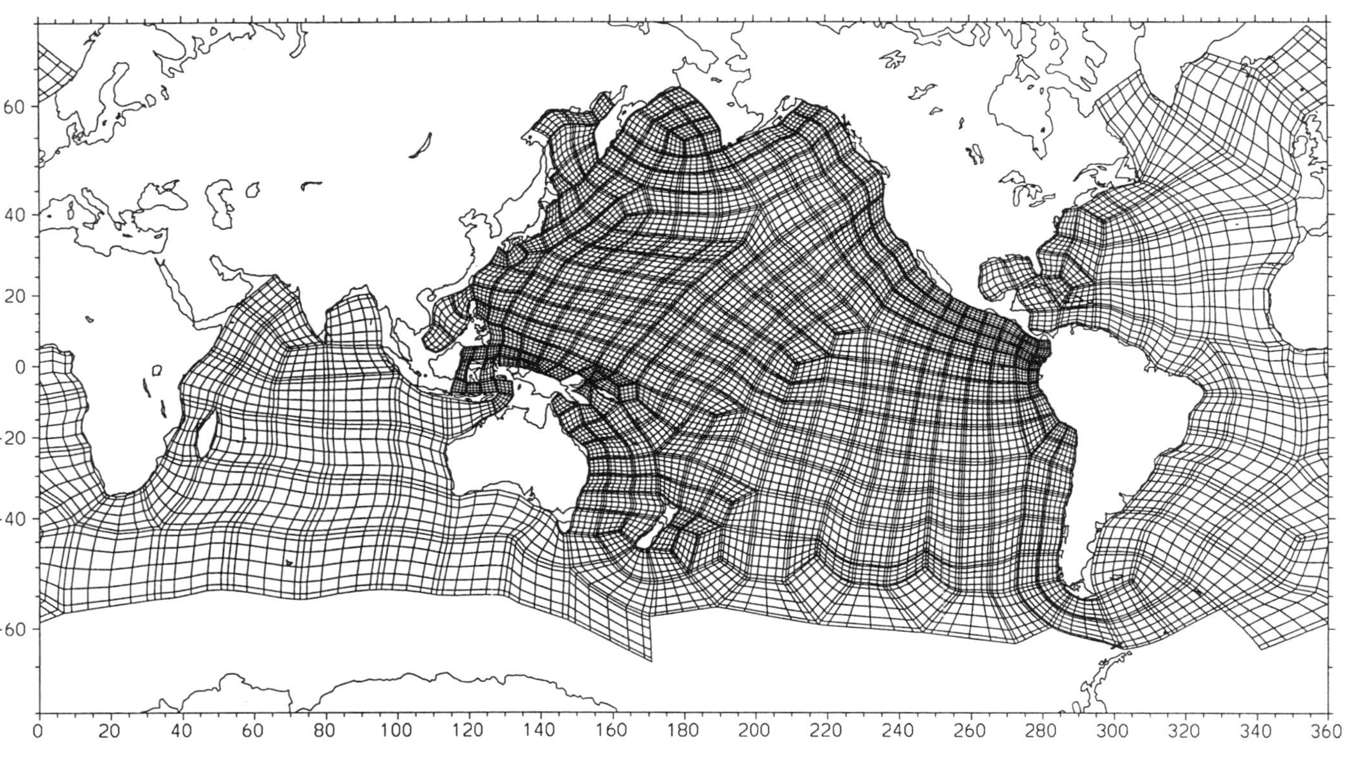

Fig. 5 Velocity collocation grid. Each of the 448 elements is comprised of a 7 × 7 mesh.

coordinates within a single element, however, must not include a branch point (i.e., crossing the Date Line) but this is trivially remedied by adding $\pm\pi$ to selected nodal longitudes.

The Indonesian passages are open, spanned by a line of elements which are too wide to be realistic, but are considered to be essential inclusions. Neither the Bering Strait nor the Arctic Ocean are included, since these are not considered essential to the large-scale circulation of the Pacific. Two islands are included in the Pacific: New Zealand and the Solomon Islands chain. An original grid extended to the Antarctic coastline, but elements south of the Antarctic Circumpolar Current (ACC) were removed to reduce artificial entrainment (see later). This has left an irregular southern boundary (Fig. 5), but is of no consequence. Areas less than 300 m in depth along the Asia–Japan coast have been excluded. The change from fine to coarse resolution is evident along a line joining Tasmania to Cape Horn, and around the Indonesian passages. Here, the utility of the elements is well demonstrated, with arbitrary connectivity and rapid changes in resolution, in notable contrast to finite difference grids.

A difficulty that often arises when using a reduced gravity model over large scales is that of layer outcropping. Variations of the layer thickness will be hundreds of metres, principally at high latitudes, while it is intended that the resting thickness be no more than this. Here, the first internal gravity mode celerity is required to be about 3 m s^{-1}, typical for the equator, giving a resting layer thickness of $h = 300$ m ($c = \sqrt{g'h}$, $g' = 0.03$ m s^{-2}). As time integration proceeds the layer will become progressively thinner in certain areas. This, of course, occurs in reality but is rather difficult to deal with in a simple one-layer model. We adopt the entrainment method of McCreary and Kundu (1988) to keep the layer thickness larger than a certain minimum. An entrainment velocity, w_e, is added to the right hand side of Equation 2:

$$w_e = \begin{cases} (h + \zeta - H_{\text{min}})^2/(\tau_e H_{\text{min}}) & h + \zeta \leqslant H_{\text{min}} \\ 0 & h + \zeta > H_{\text{min}}. \end{cases} \tag{20}$$

H_{min} is the desired minimum depth, 40 m in this case, and τ_e is a time constant, set here to 10800 s. During integration the minimum layer thickness is never more than a few metres less than H_{min}. It is readily seen that this method does not conserve volume when w_e is activated; however, conservative methods were incapable of preventing outcropping. To prevent an excessive increase in fluid volume, detrainment is also included in certain areas. This takes the same form as w_e but the numerator is replaced by a small negative constant.

Outcropping occurs in all areas south of the ACC flow path and in the centres of the northern subpolar gyres (North Pacific and North Atlantic). Entrainment is activated in these areas with the largest volume influx occurring south of the ACC. As mentioned above, many elements in this area were removed since they caused excessive entrainment and the layer thickness was uniformly thin. Bottom friction was also increased in these areas to help reduce flow velocities, which, via geostrophy, cause layer shallowing. Simple Rayleigh damping is used for bottom friction and has a time constant of eighty days. The time constant is smoothly decreased to 2.7 days south of 49°S. Similarly, time constants in the North Atlantic (north of 48°N) and North Pacific (north of 42°N) are decreased to two days. Detrainment is activated in the latitudinal band 22–40°S, but only across the Indian and Atlantic oceans. All bands of high friction and detrainment are smoothly merged with the remainder of the oceans.

Since there is a large variation in grid resolution, it is necessary to vary the viscosity accordingly. Simply setting the viscosity to a large value everywhere reduces the effectiveness of fine-resolution areas. To allow lower values of viscosity in regions of fine resolution, and larger values elsewhere, a nominal viscosity, ν, is computed for each element based on the square of the maximum length of the four sides. After the set of ν values are computed, bicubic splines are used to construct a smooth field of viscosity values for the velocity grid. The present ν varies from 1.2×10^3 m^2 s^{-1} to 6×10^4 m^2 s^{-1}.

The only external forcing is the wind stress. Wind stress data products from Hellerman and Rosenstein (1983) and the European Centre for Medium-Range Weather Forecasts (ECMWF) (NCAR dataset number ds110.1) are currently being used. The present model results have have been obtained by a hybrid wind field; the mean is that of Hellerman and Rosenstein while monthly variability is from the 1980–1987 ECMWF set. ECMWF data from this period are compiled into a set of monthly means, the eight-year mean is subtracted, and the long-term mean of Hellerman and Rosenstein is added. The resulting sets of monthly winds are then interpolated to the model's velocity grid.

Integration commences from a state of no motion and no wind stress, with a resting layer thickness of 300 m. The time step is 1800 s. The wind stress is increased linearly over a period of one hundred days. Variable wind forcing is used at all stages, and updates are performed at each time step. Winds are linearly interpolated from the two respective monthly means. Integration is performed for thirty-eight years and all model fields are archived at the start and midpoints of each calendar month. This period is sufficiently long for a Sverdrup balance to be established only in the equatorial, midlatitude and northern Pacific. Very slow adjustments are still occurring in the Southern Ocean. Since c is approximately 3 m s^{-1}, the period to attain a Sverdrup balance at 50°S is about 100 years.

Figure 6 shows the global distribution of ζ, i.e., the variation of the layer thickness about the 300 m mean. This is an instantaneous field from the last month of integration. The large-scale gyres and boundary currents are roughly correct. The North Pacific subpolar gyre, and the corresponding gyre in the North Atlantic, are weak and diffused. This is due to the large bottom friction, which is unavoidable unless a much larger resting depth is used. The Agulhas retroflection is weak, and results in much of the Agulhas transport traversing the South Atlantic. This result is similar to that found in a linear Sverdrup model (e.g., Godfrey, 1989), and is due to the coarse resolution of the local elements.

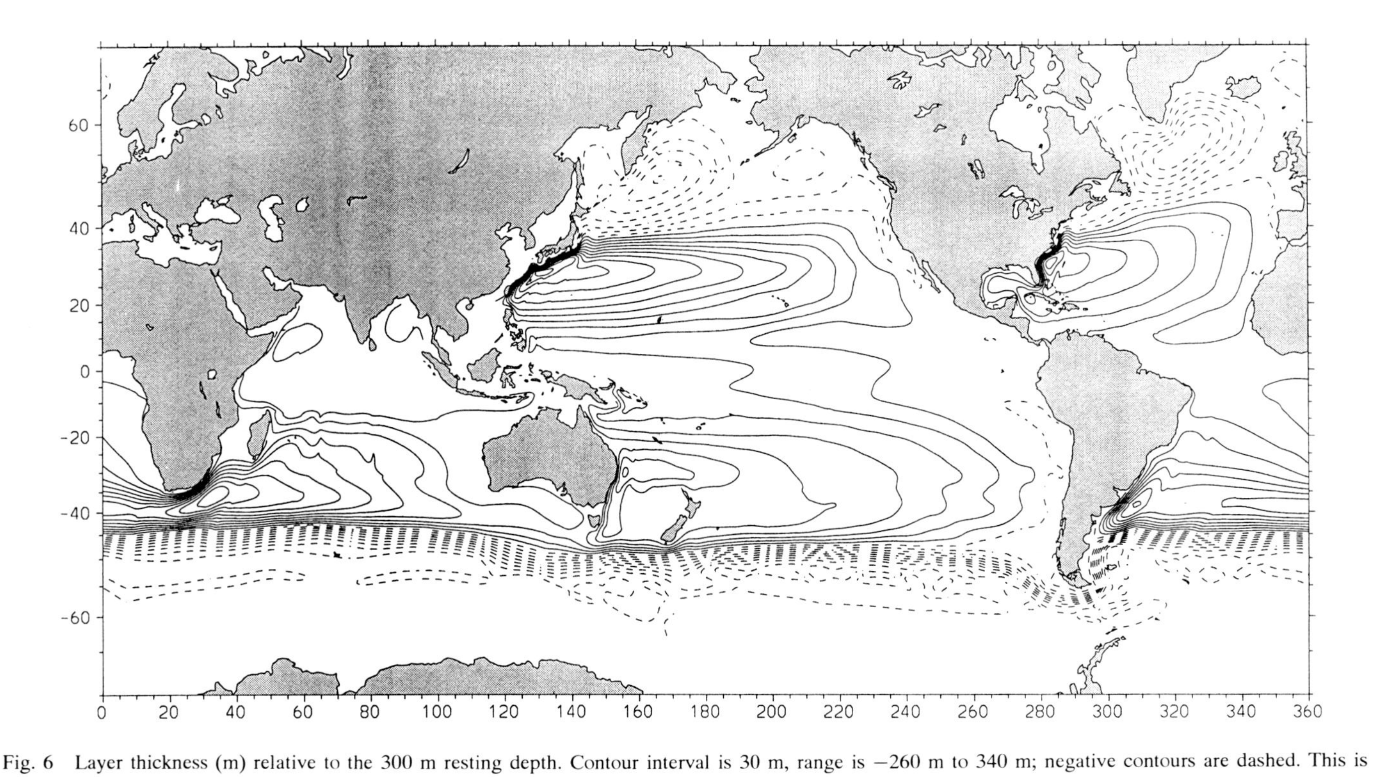

Fig. 6 Layer thickness (m) relative to the 300 m resting depth. Contour interval is 30 m, range is −260 m to 340 m; negative contours are dashed. This is an instantaneous field from July of year thirty-eight.

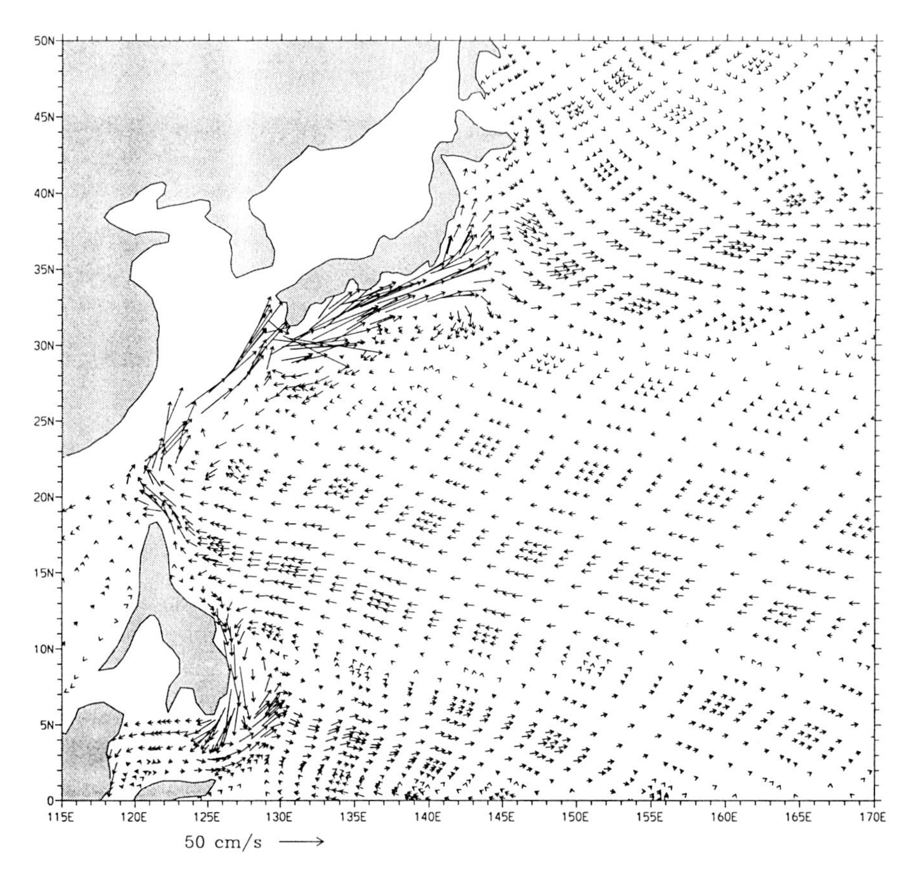

Fig. 7 Instantaneous velocity field in the western North Pacific, during January of year thirty-eight. The maximum vector length is 1.05 m/s. The NEC bifurcation occurs near 15°N and the Kuroshio departure latitude is 35–40°N.

The variation of layer thickness across the Pacific equator is nearly 90 m, being lower on the eastern side. In reality, this can be interpreted as the variation in thickness of the main thermocline. Tsuchiya's (1981) Figs 13 and 14 show that this variation is actually 150–200 m. The $200\,clt^{-1}$ isanostere may be approximated as the bottom of the main thermocline. An alternative interpretation of the model's layer thickness, multiplied by g'/g, is the sea surface steric height. The equatorial variation is then 27 cm, while in reality it is 60–70 cm. Thus the model's equatorial pressure gradient is too small by a factor of two. Given the complex vertical structure of actual equatorial flows, higher baroclinic modes are required to improve these model results.

Figure 7 shows the velocity field in the Philippines/Japan area. These are instantaneous velocities recorded during January of year thirty-eight. The Kuroshio and Mindanao boundary currents are clearly visible. The bifurcation point of the North

Equatorial Current occurs near 15°N on the Philippines coast, and is realistic. The departure latitude of the Kuroshio is spread between 35 and 40°N. In reality, this latitude is less than 35°N, and is more defined. Meanders and variations in the departure latitude are not observed in the model, nor are they expected with the given resolution and in the absence of bottom topography. These results are entirely in line with those of finite difference models. Western boundary current flows in the present model, however, do not suffer the effects of a stepped land boundary, as is the case in finite difference models.

5 Parallel implementation and performance

Because of the complexity of the computational kernel within each element, as well as the sparsity of inter-element communication, the spectral finite element technique is particularly suitable for implementation on multiprocessor computers (Fischer, 1989). As an example, we have developed a parallel version of the shallow water model described above. The parallel code uses an architecture-independent data parallel scheme. It is written in message passing form to make use of the variety of both homogeneous and heterogeneous computing networks available today. The issues pertinent to the parallelization of the code that are addressed here are time and space discretization, decomposition of the domain, load balancing, and efficiency of the communications scheme.

The space discretization of the spectral finite element scheme has as its most basic unit the element. Though each element has an expansion that makes it the equivalent of a small grid (7×7 for the velocity grid in our test case) it is advantageous not to distribute elements across processors. This helps with load balancing issues and serves to simplify the logic of the decomposition. The drawback of not reducing the granularity beyond one element is insignificant since most processors on today's parallel computers are equipped with $O(1\text{–}10^2)$ megabytes of RAM capable of handling dozens of elements each. To ensure load balancing, we need to create a grid that has a number of elements that is a multiple of the numbers of processors available. On the two computers our code was tested on, the nCUBE/2 and the Cray T3D, only power of two processors subcubes are allowed. Optimal grids will therefore have a number of elements equal to some multiple of a power of two, making each processor carry an equal number of elements (a load-balanced distribution).

As described in previous sections, the time discretization is done with an explicit time-marching scheme. Unlike implicit schemes, where there is a global data dependency, the explicit time-marching scheme is local and inter-processor communication is needed only when updating the collocation points on the edges of the elements that lie on the boundary of a processor's domain, i.e., those that are shared with other processors. Most domain boundary points are shared by two processors but element corner points can be shared by as many as six processors, depending on the connectivity of the element.

The communication scheme has two logical stages. The first, called the *Pre-*

Shuffle, is executed once only during the problem set-up stage. The second is the *Shuffle* which is executed inside the time-march loop. The pre-shuffle, executed by each processor, looks at the collocation points in its domain and first determines with which processors it will have to exchange data. Then it creates an array of sorted global indices which will be used to match the contributions coming in from adjacent physical domains. The shuffle routine is called once during each time iteration. Once every processor adds its contributions to its local elements, a global right hand side vector is assembled for each processor for the unknowns u, v and ζ. By assembling the global vectors, each field will need only one exchange with each adjacent domain, rather than multiple exchanges. The shuffle routine assembles all the points that need to be sent to a given processor into one array which is then sent. The packed array is read by the proper processor and unpacked according to the sorting evaluated in the pre-shuffle routine. The data received is then added to the local contributions of each field.

This scheme achieves two objectives. First, it sets up a communication pattern that is completely independent of the architecture of the machine. Each processor evaluates which other processors it needs to communicate with in a logical processor array; thus the actual physical location of the processor is irrelevant. Second, it efficiently packs all the interchangeable data so that any two processors that share collocation points do not need to communicate more than once during each iteration. A side effect of this scheme is that when the number of elements increases in a processor (and so also the length of the vectors that are communicated), the total communication time does not necessarily increase as would be expected. There are two reasons for this behaviour. One is that the number of data points that are exchanged between processors is so small that most of the communication cost is the start-up time, or overhead. Furthermore, decreasing the number of elements in a processor will not reduce the actual number of communications performed, and in some cases it will increase it, as some points will now be shared by more processors, e.g., at element corner points. With communication overhead as the largest component of the total communication cost, reducing the vector length does not necessarily result in total communications time savings.

Though the actual communication scheme is architecture-independent, if physically adjacent domains are distributed to physically distant processors the efficiency of the code will suffer. The number of collocation points that are shared between processors will be proportional to the perimeter length of the domain in each processor. To address this problem a domain decomposition technique was devised, based on the Breadth First Search (BFS) tree spanning algorithm (Kruse, 1987). This scheme, given the criteria for load balancing, will decompose a domain, as seen in Fig. 8, so as to minimize the perimeter-to-area ratio of each domain and thereby increase the computational load relative to the number of data points that have to be shared with other processors. It is possible to map the domains obtained from the BFS algorithm to a particular architecture, though the communications costs are already reduced to the point where an optimal mapping will not have a

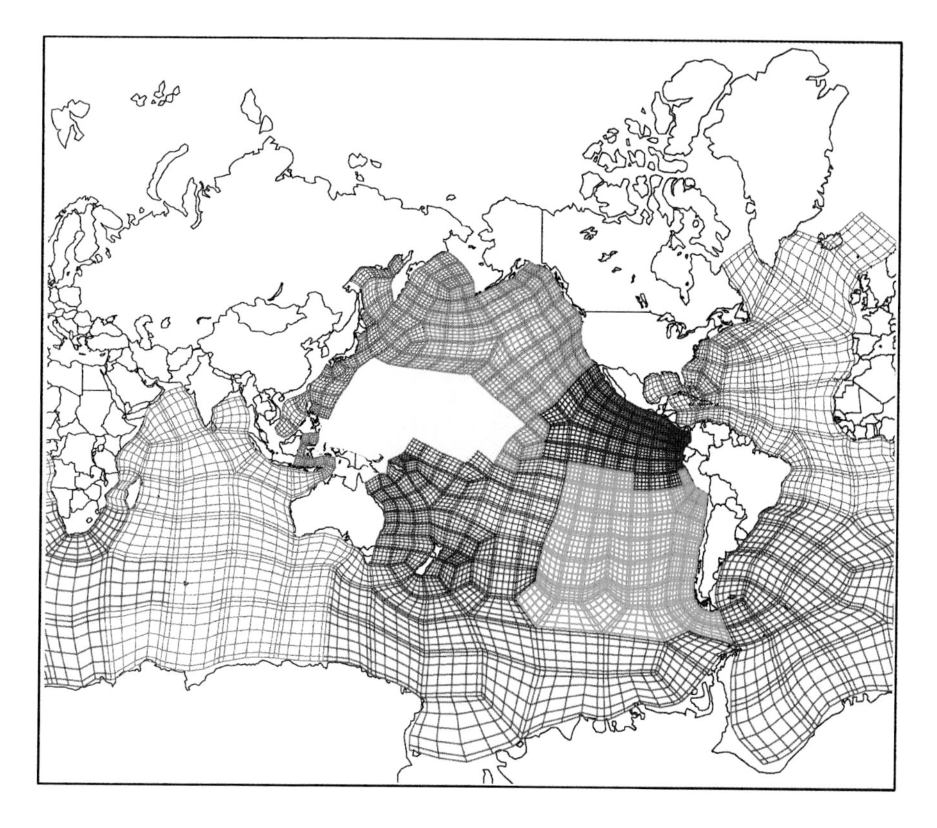

Fig. 8 Domain decomposition of global grid for eight processors.

significant impact on the total cost of the communications. This is also partly due to the high communication bandwidth of the currently available parallel computers.

Figure 9 shows the speedup obtained with the parallel algorithm. The code was run on an nCUBE/2 computer with 512 processors. (Comparable behaviour is observed on the Cray T3D.) We were able to test the range of 2 to 512 processors with a rectangular grid of 3072 elements, where each element has a 7×7 expansion for the velocity and a 5×5 for pressure. The solid line represents linear speedup. The circles are our actual execution times. As shown, the code deviates negligibly from linear speedup.

Parallel scalability of the SFE model (speedup divided by the number of processors) is not found to be sensitive to the exact number of elements assigned to each processor, so long as load balancing is maintained, nor to the specific partitioning algorithm used. Thus, scalabilities in excess of 90 percent are found on the nCUBE/2 for 6 or more elements per processor, while similar scalability on the T3D is attained for as few as 2 elements per processor. Note that any additional dynamical complexity – for example, the inclusion of prognostic equations

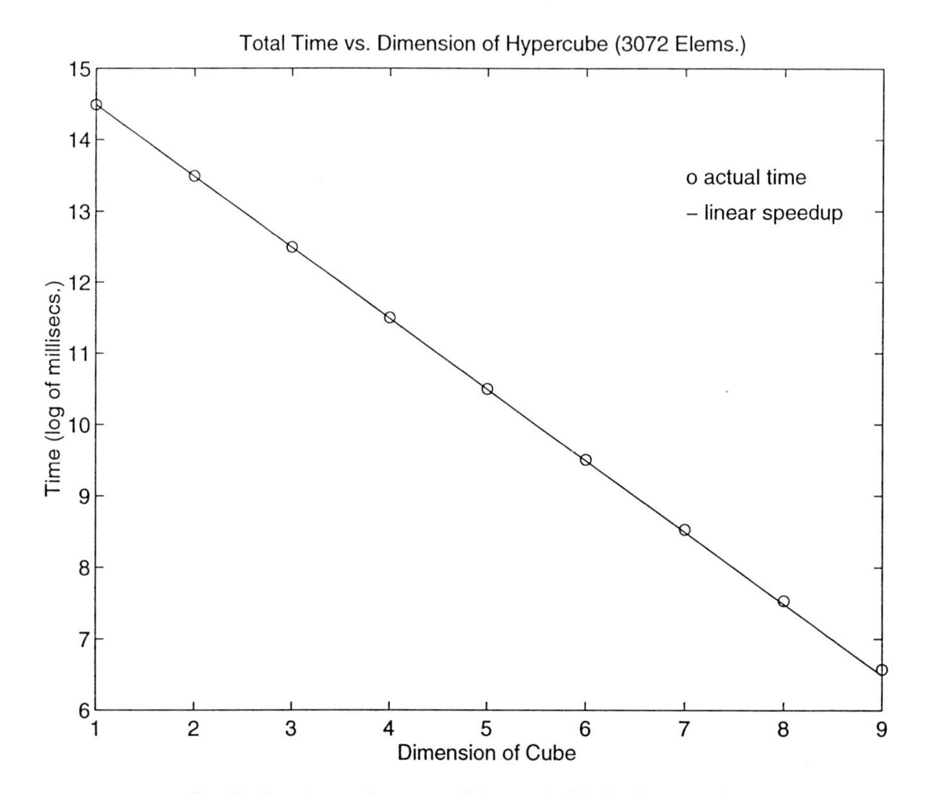

Fig. 9 Speedup performance of the spectral finite element scheme.

for temperature or salinity, or the incorporation of a third (vertical) dimension – will serve to *increase* these scalability figures. On the nCUBE/2, for instance, a thermodynamic tracer equation has been added to the SFE shallow water model, with no measurable impact on element-to-element communication costs.

Routine scalabilites of over 90 percent might raise suspicions of an inefficient algorithm, perhaps one with an artificially high operation count and therefore high but misleading scalability properties. Such is not the case for this implementation. Taking both additions and multiplications into account, the leading-order, per-gridpoint operation count for the SFE model is $24N^v$, where N^v is the number of velocity nodes per spectral element. This is to be compared with a per-gridpoint operation count of $4.2N^l$ for the NCAR spherical harmonic model, where N^l is the number of latitudinal gridpoints (Foster et al., 1992). Thus, the SFE model has a comparable operation count to the NCAR model for $N^v = N^l/5.7$. Routine spherical harmonic truncations lie between T42 and T83, for which a comparable operation count for the SFE model is achieved for an N^v of about 10 to 20, well within the rapidly convergent regime for the problems discussed here (see Fig. 3). Note that if even higher horizontal resolution is desired, the SFE model becomes progressively more competitive because elements may be added (holding N^v fixed) without affecting the per-gridpoint operation count. This clearly demonstrates that the SFE algorithm is cost-effective, relative to the standard of an optimized spherical harmonic model, and reinforces our intuition that accurate, yet affordable, grids for many problems can be reached at a moderate polynomial truncation (N^v of order 10).

6 Summary and discussion

We have described an alternative to traditional approaches for spatial discretization in numerical models of the global atmosphere and ocean. The spectral element technique, so-called because it merges the flexible geometric properties of low-order finite elements with the rapid convergence properties of global spectral methods, combines in principle all of the attractive features of both spherical harmonic expansions (accuracy, nearly isotropic resolution on a sphere) and lower-order finite difference schemes (ability to treat complex geometry, mesh refinement, low overhead on parallel computers).

Shallow water test problems on a sphere confirm that spectral elements have most of the advantages of spherical harmonics. They are of course not as accurate for a given horizontal resolution; however, they do converge exponentially fast under mesh refinement, and are superior to lower-order finite difference schemes. Spectral elements do not have any pole problems since the sphere can be tiled with squares of comparable size; thus, by using a local coordinate system within each element, any singularities in the coordinate system can be avoided. Lastly, since the number of elements and the degree of spectral interpolation within each element can be individually varied, the size of the spectral transforms can be kept small, thus keeping the $O(N^3)$ cost of the transforms under control. This latter point – involving the trade-offs between numbers of elements and the spectral truncation on the one hand, and accuracy and cost on the other – needs further exploration.

From the perspective of traditional ocean circulation modelling, the use of the spectral element method offers several immediate advantages. As our reduced gravity, wind-driven simulations of the Pacific Ocean indicate, the spectral element method allows basin-scale integrations to be efficiently carried out on global, spatially non-uniform grids. The resulting grids avoid the serious issues associated with open boundary conditions in regional domains, and represent a more straightforward alternative than the use of nested models. Lastly, the spectral element technique is more accurate than the approaches more traditionally adopted in global ocean modelling. This, together with its versatile mesh refinement capabilities, suggests that the spectral element technique will be superior to traditional approaches in dealing with the small-scale physical processes (convection, overflows, etc.) known to be poorly represented in today's global ocean models.

Spectral elements may well be ideally suited for implementation and application to parallel machines. With inter-element communication limited to exchanges of data between element boundary points only, and with dense (spectral) arithmetic necessary on the elemental level, communications overhead is found to be very low and scalability across many processors nearly linear over more than two orders of magnitude in the number of processors.

Lastly, we note that the use of spectral elements has the potential disadvantage that it may no longer be trivial to invert the linear differential operators required for implicit time stepping schemes. However, we believe that this will not represent a hard problem with modern iterative and multigrid methods. Indeed, temporally implicit versions of our shallow water model have now been developed, and have been shown to be competitive with the explicit, spectral element model described here. A fully three-dimensional version of the spectral element model has been implemented and is presently being evaluated.

Acknowledgments

Development of the spectral element model is supported by the Office of Naval Research (N00014-93-0197). Application of the spectral element model to the global atmosphere and ocean are sponsored by the Department of Energy (DOE/CHAMMP grant DE-FG05-91ER61219) and by the National Aeronautics and Space Administration (TOPEX/POSEIDON, NAS7-918). Parallel implementation of the spectral element model has been supported by the Institute of Marine and Coastal Sciences, and by a research grant from Cray Research, Inc. Computer time has been provided by the Arctic Region Supercomputer Center (Univ. of Alaska, Fairbanks) and by the Computer Aids in Industrial Productivity Center (Rutgers University).

References

BERNARDI, C. and O. PIRONNEAU. 1991. On the shallow water equations at low Reynolds number. *Commun. partial diff. equations*, **16**: 59–104.

BOER, G.J; K. ARPE, M. BLACKBURN, M. DE'QUE', W.L. GATES, T.L. HART, H. LE TREUT, E. ROECKNER, D.A. SHEIRIN, I. SIMMONDS, R.N.B. SMITH, T. TOKIOKA, R.T. WETHERALD and D. WILLAMSON. 1992. Some results from an intercomparison of the climates simulated by 14 atmospheric general circulation models. *J. Geophys. Res.* **97**: 12771–12786.

BOYD, J.P. 1989. Chebyshev and Fourier Spectral Methods. *Lecture Notes in Engineering*, C.A. Brebbia and S.A. Orszag (Eds), Springer-Verlag, New York.

BRYAN, K. 1969. A numerical method for the study of the circulation of the world oceans. *J. Comp. Phys.* **4**: 347–376.

FISCHER, P.F. 1989. Spectral Element Solution of the Navier-Stokes Equations on High Performance Distributed-Memory Parallel Processors. Ph.D. Thesis, MIT, Cambridge., Mass. 133 pp.

FOSTER, I.; W. GROPP and R. STEVENS. 1992. The parallel scalability of the spectral transform method. *Mon. Weather Rev.* **120**: 835–850.

GATES, W.L. 1992. AMIP: The atmospheric model intercomparison project. *Bull. Am. Meteorol. Soc.* **73**: 1962–1970.

GEAR, C.W. 1971. *Numerical Initial Value Problems in Ordinary Differential Equations*, Prentice-Hall, New Jersey, 253 pp.

GODFREY, J.S. 1989. A Sverdrup model of the depth-integrated flow for the world ocean allowing for island circulations. *Geophys. Astrophys. Fluid Dyn.* **45**: 89–112.

HEIKES, R. and D.A. RANDALL, 1995. Numerical Integration of the shallow water equations on a twisted icosahedral grid. Part I: Basic design and results of tests. *Mon. Weather Rev.* **123**: 1862–1880.

HELLERMAN, S. and M. ROSENSTEIN. 1983. Normal monthly wind stress over the world ocean with error estimates. *J. Phys. Oceanogr.* **13**: 1093–1104.

ISKANDARANI, M.; D.B. HAIDVOGEL and J.P. BOYD. 1995. A staggered spectral finite element model with application to the oceanic shallow water equations. *Int. J. Num. Meth. Fluids*, **20**: 393–414.

JAKOB-CHIEN, R.; J.J. HACK and D.L. WILLIAMSON. 1995. Spectral transform solutions to the shallow water test set. *J. Comp. Phys.* **119**: 164–187.

KRUSE, R. 1987. *Data Structures and Program Design*. Prentice Hall, New York, 586 pp.

MA, H. 1993. A spectral element basin model for the shallow water equations. *J. Comp. Phys.* **109**: 133–149.

MADAY, Y. and A.T. PATERA. 1988. Spectral Element Methods for the Incompressible Navier-Stokes Equations. In: *State of the Arts Surveys in Computational Mechanics*, A.K. Noor (Ed.), ASME, New York, 556 pp.

MCCREARY, J.P. and P.K. KUNDU. 1988. A numerical investigation of the Somali Current during the Southwest Monsoon. *J. Marine Res.* **46**: 25–58.

TAYLOR, M.; J. TRIBBIA and M. ISKANDARANI. 1997. The spectral element method for the shallow water equations on the sphere. *J. Comp. Phys.* **130**: 92–108.

TSUCHIYA, M. 1981. The origin of the Pacific equatorial 13°C water. *J. Phys. Oceanogr.* **11**: 794–812.

WILLIAMSON, D.L.; J.B. DRAKE, J.J. HACK, R. JAKOB and P.N. SWARZTRAUBER. 1992. A standard test set for the numerical approximations to the shallow water equations in spherical geometry. *J. Comp. Phys.* **102**: 211–224.

Steric Sea Level in the Los Alamos POP Code–Non-Boussinesq Effects

John K. Dukowicz
Theoretical Division, Group T-3
Los Alamos National Laboratory
University of California, Los Alamos, New Mexico 87454, U.S.A.

[Original manuscript received 5 December 1994; in revised form 29 August 1995]

ABSTRACT *The recent availability of accurate global ocean surface elevation data from satellites permits intercomparison with model results, and therefore raises the question of the accuracy of the computed sea level in ocean models. Most models, even those that incorporate a free surface, make use of the Boussinesq approximation – that is, they assume an incompressible fluid and conserve volume rather than mass. This results in an error in predicting steric sea level changes. The Los Alamos POP model, a free-surface Bryan-Cox type of ocean model developed for parallel computers, is in this category. We generalize the POP (Parallel Ocean Program) model to account for non-zero dilatation (non-zero velocity divergence – the primary non-Boussinesq effect) due to a variety of causes, such as expansion due to thermal and salinity mixing, as well as compressibility. We then compute both local and global changes of the sea level due to these effects and conclude that overall global dynamic effects are entirely negligible. There are some small local effects such as a very small increase in the Antarctic Circumpolar Current transport, for example. We also observe a global mean sea level decrease at a rate on the order of a half-metre per century. While in this case this is most likely due to the ocean relaxing toward equilibrium (the computation is not initiated from a state of equilibrium), this does indicate that mean ocean expansion/contraction may be significant from the point of view of climate and ought to be included in models.*

RÉSUMÉ *La récente disponibilité de données satellitaires, mondiales et précises, du niveau de la surface océanique a permis de les comparer aux résultats des modèles et a soulevé le problème de la précision des calculs du niveau de la mer des modèles. Presque tous les modèles, même ceux incorporant une surface libre, utilisent l'approximation de Boussinesq (ils considèrent le fluide incompressible et conservent le volume plutôt que la masse). Il en résulte une erreur dans la prévision des changements stériques du niveau de la mer. Le modèle POP (Parallel Ocean Program) de Los Alamos, modèle de l'océan à surface libre de type Bryan-Cox développé pour les multiprocesseurs, est l'un de ces modèles. On généralise POP pour tenir compte de la dilatation non nulle (vitesse de divergence non nulle – l'effet non Boussinnesq principal) résultant de plusieurs causes, dont l'expansion due aux mélanges thermique et salin ainsi que la compressibilité. On calcule alors les changements locaux et globaux du niveau de la mer entraînés par ces effets et détermine que les effets dynamiques globaux sont très négligeables. Il y a quelques effets locaux comme, par exemple, une très petite augmentation dans le transport du courant circumpolaire antarctique. On observe aussi une diminution globale du niveau moyen de la mer à un taux de l'ordre d'un demi-mètre par siècle. Bien que dans ce cas elle résulte probablement de l'océan retournant à l'équilibre (le calcul n'est pas initié à partir d'un état à l'équilibre), ceci indique que l'expansion/contraction moyenne de l'océan peut être significative du point de vue du climat et devrait être incluse dans les modèles.*

---

1 Introduction

The recent availability of global satellite sea level data with an accuracy of a few centimetres, aside from relatively large uncertainties in the geoid, makes possible detailed global-scale intercomparisons between models and data (Topex/Poseidon special issue, *J. Geophys. Res.* Vol. 99, No. C12, 1994, Vol. 100, No. C12, 1995). The question naturally arises as to whether current ocean models contain approximations that would lead to systematic errors in the predicted steric sea level, and therefore to systematic differences in such comparisons. One such approximation, which is common to most ocean circulation models, is the Boussinesq approximation. The Boussinesq approximation, among other things, assumes incompressibility or zero dilatation – that is, zero velocity divergence or zero rate of local volume expansion/contraction – and therefore conserves volume rather than mass. This is an excellent approximation locally – the local value of the dilatation in the ocean is indeed very small – however, when integrated over the depth or over the entire volume of the ocean, the cumulative effect on the sea level could be quite large. Indeed, simple estimates suggest that this effect might account for local sea level changes on the order of several centimetres.

Recently, Greatbatch (1994) discussed this effect theoretically. On the basis of the analogy with forcing by precipitation and evaporation in the barotropic continuity equation, he estimates that the dilatation forcing is an order of magnitude smaller than the E-P forcing, and therefore that the equilibrium response to dilatation forcing, as manifested by the production of Goldsbrough/Stommel gyres, is negligible. However, he notes that there is a net global expansion/contraction due to this effect which is important on climate time-scales.

Here we investigate the expansion effect directly (i.e., computationally) by incorporating non-zero dilatation into a Boussinesq ocean model – in effect, restoring mass conservation. This allows us to examine the dynamic as well as the equilibrium effects. We use the Los Alamos POP ocean model which is based on the Bryan-Cox z-level formulation but adapted for parallel computers and with greatly improved numerics (Dukowicz and Smith, 1994). The model includes a free-surface formulation which is convenient for studying surface elevation changes. We included all effects which contribute to the dilatation: compressibility, thermal and salinity mixing, and also the effects of convective mixing. This allowed us to see not only the effects which manifest themselves in the barotropic continuity equation – the surface elevation changes – but also baroclinic effects. To study these effects we performed a global computation for a period of two years, with and without dilatation, following a 33 year spin-up calculation.

In Section 2 of this paper we define the problem, describe the changes required to implement non-zero dilatation in POP and describe the simulations performed. In Section 3 we discuss the a priori estimates of the magnitude of the effect and the possibilities for diagnostic computation of the sea level changes. In Section 4 we discuss the results and in Section 5 we present a summary.

2 Relaxing the Boussinesq approximation

Most ocean models, including Bryan-Cox-type models (such as POP) make use of the Boussinesq approximation. From the point of view of an ocean model (Kundu, 1990) the Boussinesq approximation consists of two parts: (a) replacement of the density which divides the pressure gradient in the horizontal momentum equations by a constant representative density, and (b) the assumption of zero velocity divergence, i.e., $\mathbf{\nabla}_3 \cdot \mathbf{u}_3 = 0$, where this is the full three-dimensional divergence and $\mathbf{u}_3 = (\mathbf{u}, w)$ is the three-dimensional velocity, where $\mathbf{u}$ is the horizontal velocity and w is the vertical velocity. We retain the first part of the approximation because it has essentially no consequences with regard to the sea level; it only slightly modifies the dynamics. The second part of the approximation implies that

$$\frac{1}{v}\left(\frac{\partial}{\partial t} + \mathbf{u}_3 \cdot \mathbf{\nabla}_3\right) v = \frac{1}{v}\frac{Dv}{Dt} = \mathbf{\nabla}_3 \cdot \mathbf{u}_3 = 0, \tag{1}$$

where v is an infinitesimal Lagrangian volume, and globally, by Reynolds' theorem (Aris, 1962), that

$$\frac{D\mathcal{V}}{Dt} = \iiint \mathbf{\nabla}_3 \cdot \mathbf{u}_3 \, dV = 0, \tag{2}$$

where $\mathcal{V}$ is the total volume of the ocean. That is, the Boussinesq approximation implies both local and global volume conservation. The real ocean conserves mass, however, as given by

$$\mathbf{\nabla}_3 \cdot \mathbf{u}_3 = -\frac{1}{\rho}\frac{D\rho}{Dt} \equiv \varepsilon \; (\neq 0), \tag{3}$$

where ρ is the in-situ density, and where we have defined the right-hand-side to be the dilatation ε, (we have chosen to use the symbol ε to emphasize that the dilatation is very small compared to any other inverse time scales in the ocean). In this study we will retain a non-zero ε, i.e. the non-Boussinesq terms; these will modify the continuity equation (Eq. 1) to account for the effects of expansion/contraction. Note that an anelastic approximation retains some of these terms also (it only neglects the contribution of the time derivative of density on the rhs of Eq. 3).

Integrating (3) vertically, we obtain the barotropic continuity equation

$$\eta_t + \mathbf{\nabla} \cdot H\mathbf{U} = H(\tilde{\varepsilon} - \bar{\varepsilon}), \tag{4}$$

where η is the surface elevation above the mean height of the ocean, H is the nominal depth of the ocean, $\mathbf{U}$ is the two-dimensional barotropic velocity, $\tilde{\varepsilon}$ is the vertically-averaged dilatation, and $\bar{\varepsilon}$ is the mean dilatation, defined as follows,

$$\mathbf{U} = \frac{1}{H}\int_{-H}^{\eta} \mathbf{u}dz,\ \tilde{\varepsilon} = \frac{1}{H}\int_{-H}^{\eta} \varepsilon dz,\ \bar{\varepsilon} = \frac{1}{\mathcal{V}}\iiint \varepsilon dV,\ \text{and}\ \mathcal{V} = \iiint dV,$$

This equation includes the assumption that $\eta \ll H$ (Dukowicz and Smith, 1994). In addition,

$$h_t = \frac{\mathcal{V}}{A}\bar{\varepsilon}, \tag{5}$$

where h is the mean height of the ocean relative to a zero reference level, such that the total depth of the ocean is $H + h + \eta$, and A is the surface area. For the purpose of the present study we renormalize H, that is, we absorb h into H. This is done for convenience to minimize changes to the present code, and because h is indeed very small relative to the thickness of the top layer for the short integration time of the present study. For long duration calculations h may have to be continuously absorbed into the thickness of the top layer, thereby continuously redefining H. Equation (4) therefore replaces the present barotropic continuity equation. It retains the property that η is zero when averaged over the ocean surface. Equation (5) is solved diagnostically because it is uncoupled from the dynamics.

Taking into account the equation of state, $\rho = \rho(T, S, p)$, where T is the potential temperature, S is the salinity, and p is the pressure, the dilatation may be written in expanded form as

$$\varepsilon \equiv -\frac{1}{\rho}\frac{D\rho}{Dt} = \alpha\frac{DT}{Dt} - \beta\frac{DS}{Dt} - \frac{1}{\rho c_s^2}\frac{Dp}{Dt}, \tag{6}$$

where

$$\alpha \equiv -\frac{1}{\rho}\frac{\partial\rho}{\partial T}\bigg|_{S,p},\quad \beta \equiv \frac{1}{\rho}\frac{\partial\rho}{\partial S}\bigg|_{T,p},$$

are the thermal and salinity expansion coefficients, respectively, and c_s is the speed of sound. However, in the code we approximate the equation of state as follows: $\rho = \rho(T, S, z)$, where the hydrostatic approximation is used to approximate $p \sim -\rho_0 gz$, where ρ_0 is a representative density and g is the acceleration due to gravity. Assuming that the tracer transport equations are

$$\frac{DT}{Dt} = \mathcal{D}(T),\quad \frac{DS}{Dt} = \mathcal{D}(S),$$

where $\mathcal{D}(T)$ and $\mathcal{D}(S)$ are the mixing terms for potential temperature and salinity, respectively, this gives, finally,

$$\varepsilon = \alpha'\mathcal{D}(T) - \beta'\mathcal{D}(S) + \frac{wg}{c_s^2}, \tag{7}$$

where w is the vertical velocity and the expansion coefficients are approximated as

$$\alpha' \equiv -\frac{1}{\rho_0}\left.\frac{\partial\rho}{\partial T}\right|_{S,p}, \quad \beta' \equiv \frac{1}{\rho_0}\left.\frac{\partial\rho}{\partial S}\right|_{T,p}.$$

The first two terms account for the effects of thermal and salinity mixing, and the last term is the effect of compressibility. We assume a constant sound speed $c_s = 1.5 \times 10^5$ cm/s and $\rho_0 = 1000$ gm/cm^3. Note that in obtaining ε by using (7) (that is, from its sources) rather than directly from its definition in terms of the density (6), we are abandoning strict mass conservation. However, computing ε from density changes runs the risk of incurring large errors in ε due to the loss of precision which would inevitably happen when trying to obtain the very small dilatation from the near-cancellation of relatively large terms.

The terms $\mathcal{D}(T)$ and $\mathcal{D}(S)$ contain not only the vertical and horizontal mixing terms in the transport equations (Laplacian or biharmonic, as the case may be), but one must be careful to include all types of mixing, such as that involved in convective adjustment and surface forcing. Thus, taking the case of thermal mixing as an example, we write

$$\mathcal{D}(T) = A_H \nabla^2 T + A_V \frac{\partial^2 T}{\partial z^2} + \frac{1}{\Delta t}\Delta_{CVA} T + \frac{1}{\tau}(T_0 - T)\delta(\text{layer } 1), \tag{8}$$

where the first two terms on the right hand side represent the explicit horizontal and vertical mixing terms in the potential temperature equation (which may or may not be of exactly this form). The third term gives the contribution of the convective adjustment algorithm because it represents a mixing process, whereupon $\Delta_{CVA} T = T_{ACV} - T_{BCV}$, T_{ACV} is the temperature after convective adjustment, T_{BCV} is the temperature before convective adjustment, and Δt is the time step. The fourth term represents the effect of surface forcing, where T_0 is the surface temperature to which the model is being forced, and τ is the relaxation time-scale. The forcing, applied only in the top layer of the model (schematically indicated by the delta function – the model uses layers of constant thickness, except for the top layer whose thickness is modified by a variable free surface), represents the effect of a surface heat flux and as such must be viewed as part of the vertical mixing term (the second term on the right, above).

The discrete version of the above equations is simply (see Dukowicz and Smith, 1994, for the justification of the discretization in the absence of dilatation),

$$\text{Baroclinic: } \nabla \cdot \mathbf{u}^{n+1} = \varepsilon^n - \bar{\varepsilon}^n,$$

$$\text{Barotropic: } \frac{\eta^{n+1} - \eta^n}{\Delta t} + \nabla \cdot H\mathbf{u}^{n+1} = H(\tilde{\varepsilon}^n - \bar{\varepsilon}^n), \tag{9}$$

$$\text{Global: } \frac{h^{n+1} - h^n}{\Delta t} = \frac{\mathcal{V}}{A}\bar{\varepsilon}^n.$$

Note that the dilatation on the right hand side is treated explicitly and is not necessarily at the correct time level. This is again because of the smallness of the dilatation which makes the resulting time discretization error negligible.

The present study was performed at a nominal 1° resolution, using a global 384 × 144 × 20 grid extending from 78°S to 66°N, and a time step of 2 hours. The computations were forced at the surface using climatological monthly wind stress (Hellerman and Rosenstein, 1983) and monthly temperature and salinity fields (Levitus, 1982). The relaxation time τ for temperature and salinity was one month. The standard convective adjustment scheme (2 passes) and Richardson-number-based vertical mixing (Pacanowski and Philander, 1981) were used. Horizontal mixing used a Laplacian coefficient $A_H = 10^7$ cm^2/s. Computations with and without dilatation were carried out for a period of two years, beginning in the spring, and starting from a 33-year initializing calculation.

3 Estimating the effect of the dilatation

Gill (1982) shows that one can transform the barotropic equations so that all forcing terms that change sea level are combined on the right hand side of the continuity equation and hence can be compared to one another:

$$\eta'_t + \nabla \cdot H\mathbf{U}' = (\eta_a + \eta_{P/E} + \eta_E)_t, \tag{10}$$

where η' and $\mathbf{U}'$ are the transformed sea surface elevation and barotropic velocity, η_a is the surface elevation change due to an inverse barometer, i.e., due to atmospheric pressure fluctuations, $\eta_{P/E}$ is the displacement due to precipitation/evaporation, and η_E is the Ekman displacement due to wind forcing. Gill estimates that the displacement rate due to Ekman pumping is 30 m/yr, that due to precipitation/evaporation effects is 1 m/yr and that due to atmospheric pressure fluctuations is a few centimetres/yr. To the above we must add the effect of expansion, namely,

$$\eta'_t + \nabla \cdot H\mathbf{U}' = -(\eta_a + \eta_{P/E} + \eta_E)_t + H(\tilde{\varepsilon} - \bar{\varepsilon}). \tag{11}$$

Figure 1 shows a plot of $H(\tilde{\varepsilon} - \bar{\varepsilon})$, the depth-averaged zero-mean dilatation, at the end of the two-year integration, which is typical irrespective of season. The dilatation is large locally in the equatorial region, presumably because of active eddy mixing, and in the polar regions, at least partly because of convective adjustment. A region of intensive activity is located east of Drake Passage. This is a region of particularly high mesoscale eddy activity partly as a result of the confluence of the Falkland and Brazil currents. Other regions of eddy activity, such as the Gulf Stream and the Kuroshio, are visible but are much less extensive. The present calculation is only marginally eddy-resolving. Higher resolution calculations would yield much stronger eddy activity and therefore would show other regions of large local dilatation. The maximum magnitude in the active regions is plus or minus 10^{-4} cm/s, or 30 m/yr, and elsewhere it is of the order of 10^{-5} cm/s, or 3 m/yr. Thus, this suggests that expansion effects should be comparable to or larger than

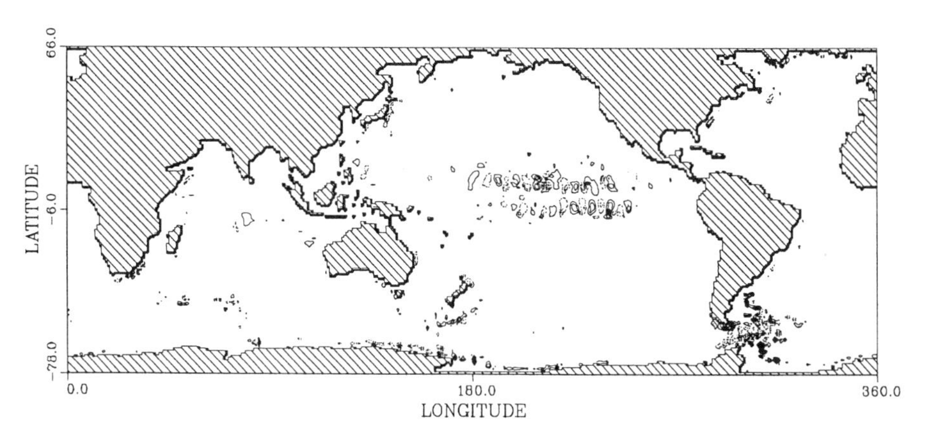

Fig. 1 The vertically-averaged zero-mean dilatation, multiplied by the depth, $H(\tilde{\varepsilon} - \bar{\varepsilon})$, at the end of the two year integration. The extreme contours are plus and minus 10^{-4} cm/s and the contour interval is 10^{-5} cm/s. Negative contours are in dotted lines. The zero contour has been eliminated to reduce clutter.

those due to precipitation/evaporation, at least locally. However, the maximum average dilatation is highly localized and occurs with relatively small spatial scales, as is apparent from Fig. 1. The small spatial scales of the dilatational fluctuations are presumably associated with the underlying eddy scales.

The smallness of the dilatation suggests that it might be possible to diagnose its effects. Assuming that the barotropic velocity is unchanged, we obtain

$$\eta_{D,t} + \nabla \cdot H\mathbf{U}_B = H(\tilde{\varepsilon} - \bar{\varepsilon}),$$

where $\mathbf{U}_B$ is the Boussinesq barotropic velocity, i.e., the barotropic velocity in the absence of dilatation, and η_D is the diagnosed elevation. But because $\nabla \cdot H\mathbf{U}_B = -\eta_{B,t}$, we have

$$\eta_D = \eta_B + H \int_0^t (\tilde{\varepsilon} - \bar{\varepsilon}) dt'. \tag{12}$$

That is, the diagnosed change in surface elevation amounts to integrating in time the vertically stacked dilatation. Figure 2 shows contours of the diagnosed elevation change, $\eta_D - \eta_B$, again at the end of the two year integration. The general pattern of the diagnosed elevation change is very nearly the same as that of the depth-averaged dilatation itself, shown in Fig. 1, indicating very little change in time. The maximum magnitude in the active regions is about ten centimetres, and elsewhere it is of the order of a few centimetres. This size of the effect, if it were realistic, would make it significant from the point of view of satellite observations. However, its actual size is far smaller, as we will now see.

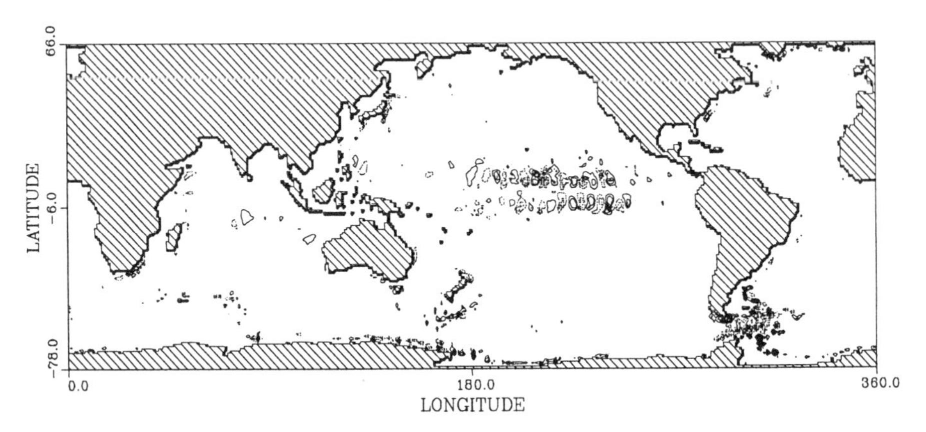

Fig. 2 The diagnosed elevation change, $\eta_D - \eta_B$, at the end of the two-year integration. The extreme contours are plus and minus 10 cm and the contour interval is 1 cm. Otherwise, same as Fig. 1.

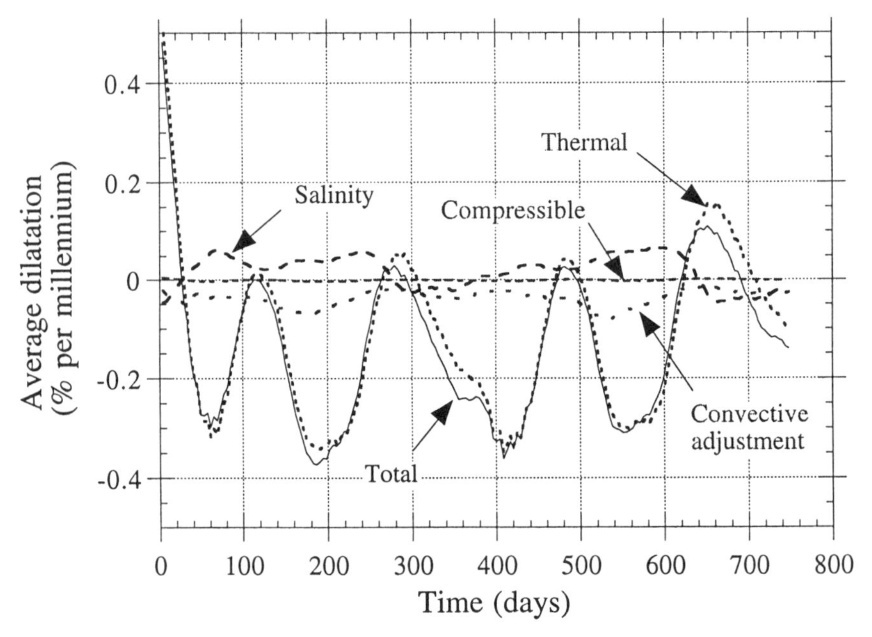

Fig. 3 The globally averaged dilatation, $\bar{\varepsilon}$, and its components, as a function of time during the integration.

4 Computational results

Figure 3 shows the globally averaged dilatation, $\bar{\varepsilon}$, and its various components, as a function of time during the integration. The scale on the vertical axis emphasizes the extremely small magnitude of the dilatation (% per millennium $\sim 3.2 \times 10^{-13}$ per second). It is obvious that the contribution of compressibility is entirely negligible, while the contributions due to convective adjustment and salinity mixing are small and comparable in size. The contribution due to thermal mixing is dominant and accounts for most of the dilatation. The surface forcing contributions are incorporated into the respective mixing contributions. Note that the globally averaged dilatation has a net negative value; we shall return to this later.

Figure 4 shows contours of the seasonally averaged surface elevation change, $\eta - \eta_B$, during the second year of the two-year integration. The magnitude of the change is of the order of a few millimetres; more than an order of magnitude smaller than the diagnostic prediction! As such, it is insignificant compared to other uncertainties in the satellite measurements. This is the principal observation as a result of this study. The general pattern of the elevation change is smoother than, and therefore quite different from, the pattern of the dilatation itself, Fig. 1, in contrast to the diagnostic prediction. This, and the smallness of the change compared to that predicted diagnostically, suggests that the dynamic response to the dilatation forcing is extremely effective. There is a small but noticeable northward gradient in the surface elevation in the vicinity of the Antarctic Circumpolar Current (ACC); we shall comment on this later.

Figure 5 compares the RMS surface elevation over the globe as a function of time for the calculation with expansion and for the reference calculation with no expansion. The difference caused by the presence of dilatation is minuscule. Similarly, a comparison of the global mean kinetic energy shows essentially no difference due to the presence of expansion. This corroborates the result found in Fig. 4, and further justifies the Boussinesq approximation.

The dramatic difference between the result of Fig. 2 and that of Fig. 4 comes about because of the absence of dynamic response in the former case, and its presence in the latter. Let us briefly investigate the perturbation of the barotropic mode by the presence of the dilatation. The linearized equations describing the response of the barotropic mode are

$$\eta'_t + \nabla \cdot H\mathbf{U}' = H\tilde{\varepsilon},$$

$$\mathbf{U}'_t + g\nabla\eta' = 0,$$

where the primed quantities are perturbations due to the presence of dilatation, and where we have neglected perturbations due to the advection and Coriolis terms. Eliminating $\mathbf{U}'$ gives us a forced external gravity wave equation, as follows,

$$\eta'_{tt} - \nabla \cdot gH\nabla\eta' = H\tilde{\varepsilon}_t.$$

Assuming constant depth H, the magnitude of the Fourier transform of the elevation perturbation $\hat{\eta}$ is found to be

$$|\hat{\eta}| = \frac{\omega H}{c^2\mathbf{k}_H^2 - \omega^2}\,|\hat{\varepsilon}|,$$

where ω is the frequency, $\mathbf{k}_H$ is the horizontal wavenumber vector, and $c^2 = gH$,

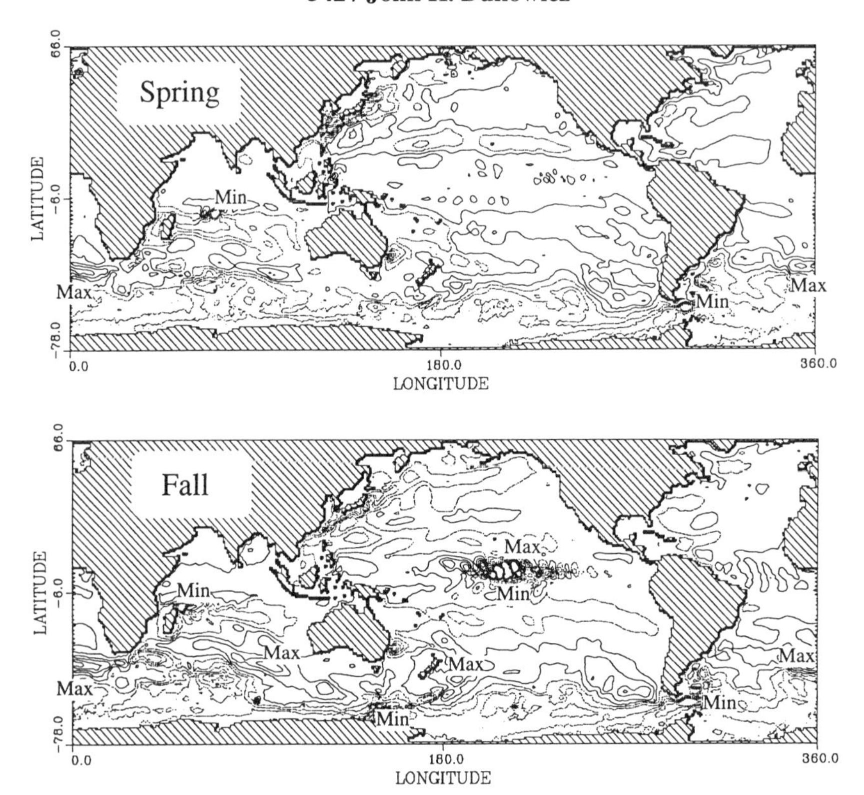

Fig. 4 Seasonally-averaged surface elevation change, $\eta - \eta_B$, in the second year of the integration. The extreme contours are plus and minus one millimetre, and the contour interval is 0.2 millimetre. The approximate location of the maximum and minimum contours are indicated by 'Max' and 'Min', respectively. Negative contours are in dotted lines.

the square of the external or barotropic gravity wave phase speed. Assuming $\omega^2 \ll c^2\mathbf{k}_H^2$, that is, sub-resonant conditions, this may be approximated as follows

$$|\hat{\eta}| \approx \left(\frac{\omega^2}{c^2\mathbf{k}_H^2}\right)\left(\frac{H|\hat{\varepsilon}|}{\omega}\right),$$

where the second factor on the right hand side is the response neglecting dynamics (i.e., that corresponding to Eq. 12), and the first factor gives the modification due to the dynamics, i.e., the dynamic response due to barotropic gravity waves. We have just observed computationally that this factor must be very small, which is consistent with and therefore justifies the above assumption. This means that c^2 is effectively much greater than $\omega^2/\mathbf{k}_H^2$, which is due both to the high barotropic gravity wave phase speed and also to the small value of $\omega^2/\mathbf{k}_H^2$. We have seen from Figs 1 and 2 that the magnitude of the Fourier transform $|\hat{\varepsilon}|$ must be strongly

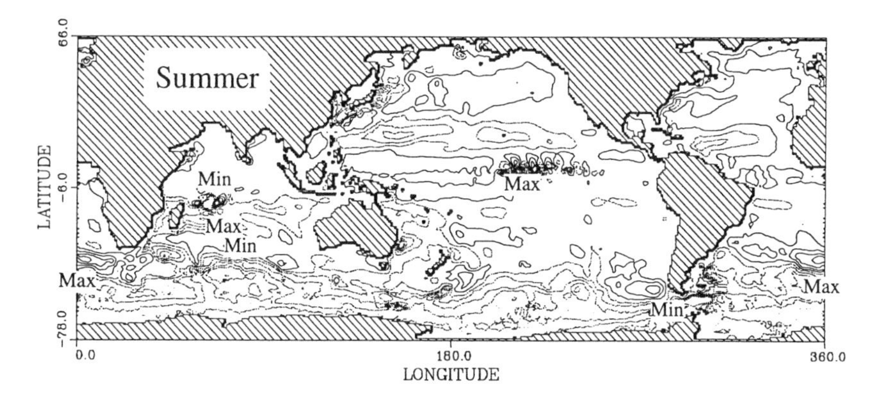

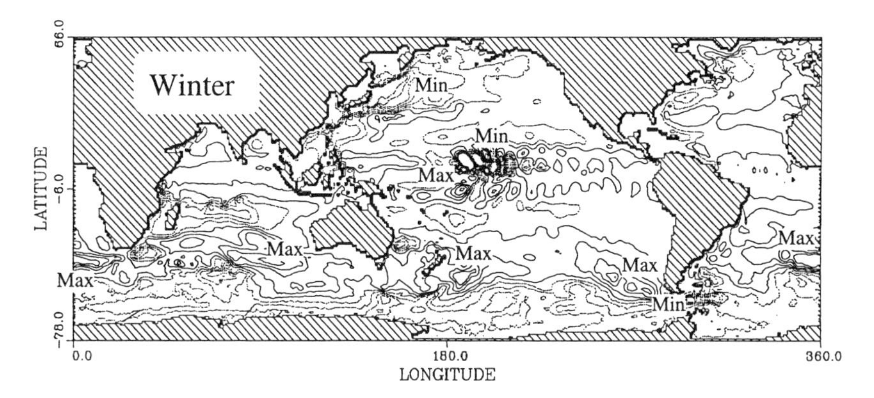

Fig. 4 *(Concluded).*

peaked at high wavenumbers and low frequencies, making the effective $\omega^2/\mathbf{k}_H^2$ small compared to c^2. This, therefore, implies that barotropic gravity waves are very effective in equilibrating the dilatational forcing.

Figure 6 shows the magnitude of the transport in the ACC as a function of time, with and without dilatation. There is a small but noticeable increase in the transport due to the expansion, ranging between about 0.1 Sv in the spring and 0.2 Sv in fall/winter. Geostrophically, this corresponds to the small northward gradient in the surface elevation change observable in the vicinity of the ACC in Fig. 4. Because the mechanism driving the ACC is largely baroclinic in nature, it is likely that this small increase in transport is due to the baroclinic effects of the expansion.

Perhaps the most significant effect of the expansion is illustrated in Fig. 7. We observed in Fig. 3 a net negative dilatation when averaged over the entire ocean. Therefore, integrating Eq. 5 in time we observe a decreasing mean surface elevation as shown in Fig. 7, resulting in a mean surface elevation about one centimetre lower at the end of two years. In this case this is a negligible effect, and it occurs presumably because the model is not fully equilibrated. One would

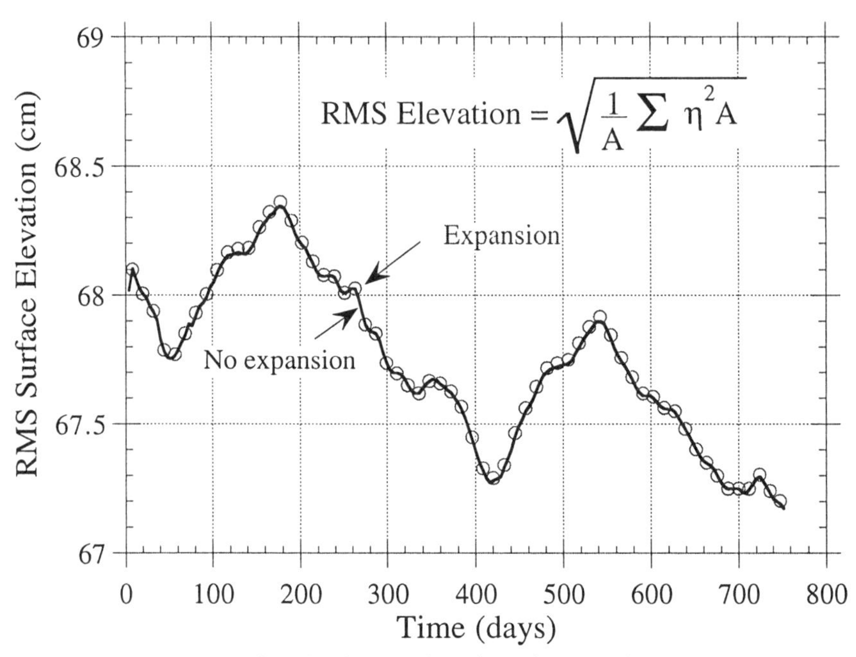

Fig. 5 Root-mean-square surface elevation over the entire surface area of the ocean, with and without dilatation. The elevation with dilatation is plotted using circles every 150 time steps.

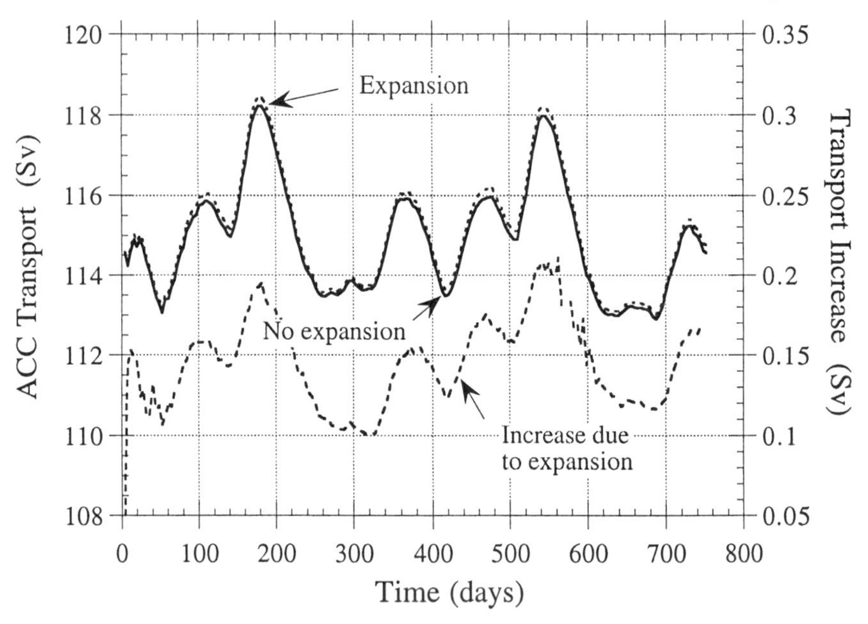

Fig. 6 The Antarctic Circumpolar Current transport, with and without dilatation, as a function of time, and the increase in ACC transport due to the expansion of the ocean (i.e., non-zero dilatation). Units are Sverdrups (10^6 m^3/s). The peaks occur between fall and winter.

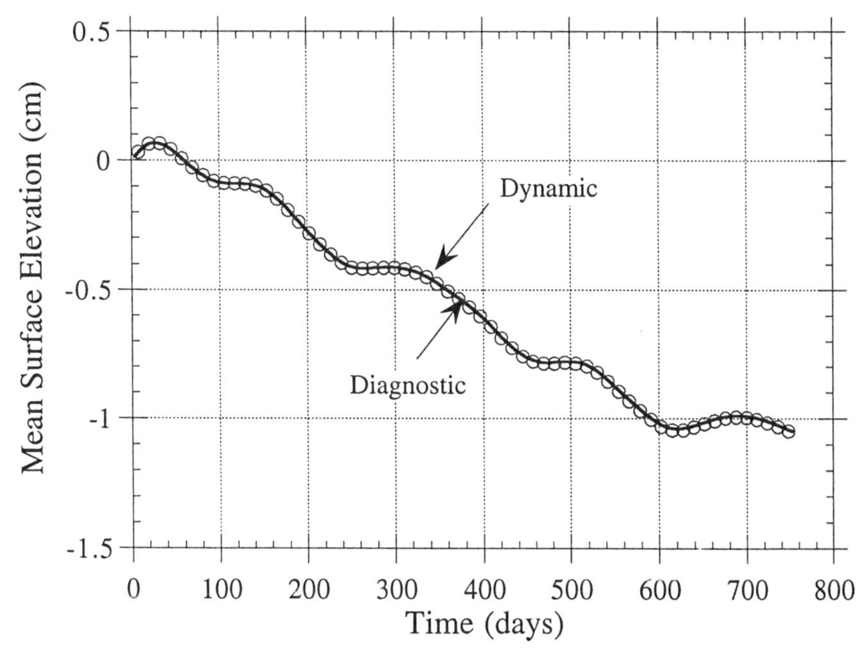

Fig. 7 The change in the mean surface elevation of the global ocean, h, during the course of the integration. The fully dynamic change is plotted using circles every 150 time steps. There is a steady decrease in ocean level, presumably because the ocean has not reached equilibrium, with small seasonal fluctuations.

expect that mean global surface elevation would not change when the model is fully equilibrated, providing the forcing does not change and the model conserves heat and salt. However, this illustrates the fact, pointed out by Greatbatch (1994), that rates of change of the global sea level of this magnitude are important if sustained on climate time-scales when the net heat or salinity flux changes. Sea level changes of this magnitude may be comparable to, and will augment those due to icemelt when caused by global warming (Church et al., 1991). These changes are missing in current Boussinesq ocean models, but they can be easily incorporated in such models. Figure 7 shows two curves: one showing the mean surface elevation derived from a fully-dynamic dilatation, and the other using a diagnostically-determined dilatation (that is, using Eq. 5, but with zero right hand side in Eq. 4). They are nearly indistinguishable from each other, as is to be expected from the negligible effect of the dilatation on the dynamics. This demonstrates that the global effect of expansion can be easily treated diagnostically.

5 Summary

The effect of conserving volume rather than mass as a result of the Boussinesq approximation in an ocean model is negligible from the point of view of changing surface elevation, a result which is somewhat surprising because diagnostic estimates suggest non-negligible effects. This is therefore evidence of the effectiveness of the barotropic adjustment process and it is at least partly due to the forcing being concentrated at high wavenumbers and low frequencies. There is a small, presumably baroclinic, increase in the ACC transport. This effect is too small to worry about in view of other uncertainties in the model, but it could be easily included if desired. These results justify the Boussinesq assumption, both locally and when integrated over the depth of the ocean.

However, the global change in sea level, the one important non-Boussinesq effect, at least from the climate point of view, is missing from most current models, but it has no dynamic consequences and it can be easily incorporated diagnostically.

Acknowledgments

The help of the author's colleagues, Rick Smith and Mat Maltrud, is gratefully acknowledged. This work was sponsored by the DOE CHAMMP program, and was performed in part using the resources located at the Advanced Computing Laboratory of Los Alamos National Laboratory.

References

ARIS, R. 1962. *Vectors, Tensors, and the Basic Equations of Fluid Mechanics.* Prentice-Hall, Englewood Cliffs, NJ, 286 pp.

CHURCH, J.A.; J.S. GODFREY, D.R. JACKETT and T.J. MCDOUGALL. 1991. A Model of Sea Level Rise Caused by Ocean Thermal Expansion, *J. Clim.* **4**: 438–456.

DUKOWICZ, J.K. and R.D. SMITH. 1994. Implicit Free-Surface Method for the Bryan-Cox-Semtner Ocean Model. *J. Geophys. Res.* **99**: 7991–8014.

GILL, A.E. 1982. *Atmosphere-Ocean Dynamics.* Academic Press, San Diego, pp. 340.

GREATBATCH, R.J. 1994. A Note on the Representation of Steric Sea Level in Models that Conserve Volume rather than Mass. *J. Geophys. Res.* **99**: 12767–12771.

HELLERMAN, S. and M. ROSENSTEIN. 1983. Normal Monthly Wind Stress over the World Ocean with Error Estimates. *J. Phys. Oceanogr.* **13**: 1093–1104.

KUNDU, P.K. 1990. *Fluid Mechanics.* Academic Press, San Diego, 638 pp.

LEVITUS, S. 1982. Climatological Atlas of the World Oceans, NOAA Prof. Paper 13, U.S. Govt. Print. Office, Washington, DC, 173 pp.

PACANOWSKI, R.C. and S.G.H. PHILANDER. 1981. Parameterization of Vertical Mixing in Numerical Models of Tropical Oceans. *J. Phys. Oceanogr.* **11**: 1443–1451.

Formulations of Lateral Diffusion in Geophysical Fluid Dynamics Models

Robert Sadourny and Karine Maynard
Laboratoire de Météorologie Dynamique du CNRS
École Normale Supérieure, 24, rue Lhomond, 75005 Paris

[Original manuscript received 1 March 1995; in revised form 24 August 1995]

ABSTRACT *We discuss several approaches to the problem of lateral diffusion in geophysical fluid dynamics models. It is shown that the diffusion model should include, in addition to the usual linear or nonlinear formulation involving deformations or iterated Laplacians, a gradient-of-divergence term representing sub-grid scale dynamic pressure effects related to interactions between vortical modes and inertia-gravity modes. General forms of the former terms, valid in particular for stretched-coordinate models, are also presented, as well as primitive equation formulations of the Anticipated Potential Vorticity Method, in entropy or other vertical coordinates.*

RÉSUMÉ *On examine plusieurs approches au problème de diffusion horizontale des modèles géophysiques de la dynamique des fluides. On montre que le modèle de diffusion devrait inclure, en plus des formes linéaire et non linéaire impliquant les déformations ou les laplaciens itérés, un terme de gradient de divergence représentant les effets de la pression dynamique inférieurs à la grille associés aux interactions entre les modes de rotation et les modes d'inertie-gravité. On présente aussi des formes générales de ces termes, valides, entre autres, pour les modèles aux coordonnées étirées, ainsi que des formes pour les équations primitives de la méthode du tourbillon potential anticipé, dans des coordonnées verticales de l'entropie ou autres.*

1 Introduction

The modelling of sub-grid scale diabatic and adiabatic processes is an essential part of geophysical fluid dynamics. The governing equations being in essence nonlinear, the sub-grid scale dynamics and diabatic forcing mechanisms ultimately affect the resolved scales through scale-to-scale interactions commonly summarized as cascade or inverse cascade processes. Here we shall consider adiabatic processes only, and concentrate on various ways of modelling the dynamical effect of sub-grid scale dynamics on the resolved scales. Even there we shall limit ourselves to what may be called "lateral" diffusion, by contrast to three-dimensional convective or turbulent diffusion. We may think of lateral diffusion as a quasi-two-dimensional process operating along well-behaved isentropic surfaces, approximately governed by quasi-geostrophic dynamics. The term "diffusion", therefore, has to be taken in a rather general sense to account for inverse cascade (negative diffusion) processes that are classical in this context (Kraichnan, 1967; Leith, 1968; Batchelor, 1969). Here we shall only discuss horizontal formulations, bearing in mind that all operators we consider are to be taken preferably on isentropic surfaces. We shall express these operators using a curvilinear coordinate formalism general enough to cover all spherical mappings, including the various stretched coordinate approaches used so far (Schmidt, 1977; Sharma et al., 1987; Courtier et al., 1991).

To put the following in adequate perspective, we first review, in Section 2, the general formulation and properties of molecular viscosity. Section 3 deals with the basic formulation of lateral diffusion as a substitute for the Reynolds tensor divergence, separating a generally overlooked isotropic term based on dynamic pressure, from the anisotropic contribution which stems from the Eliassen-Palm tensor. Section 4 discusses more scale-selective formulations, in the light of what has been intensively used in quasi-geostrophic turbulence models, consistently redesigned for the more general case of the primitive equations. Section 5 discusses a quite different approach, applying the so-called Anticipated Potential Vorticity Method (Sadourny and Basdevant, 1985), based on the formal conservation of potential enstrophy, to the primitive equations.

2 Velocity gradient, stress tensor and viscosity

a *The velocity gradient*

We consider orthogonal curvilinear coordinates x_1, x_2 arbitrarily defined on geopotential surfaces, with mapping factors h_1, h_2. A local, orthonormal reference frame in the tangent plane is then defined locally as $(\mathbf{e}_1, \mathbf{e}_2)$, $\mathbf{e}_i = \partial\mathbf{M}/h_i\partial x_i$, and the horizontal velocity field is expressed as $\mathbf{u} = u_1\mathbf{e}_1 + u_2\mathbf{e}_2$. The gradient tensor of the horizontal velocity field along a given reference surface (geopotential, isentropic or else) is then defined as

$$\nabla\mathbf{u} = \begin{bmatrix} \dfrac{1}{h_1}\left(\dfrac{\partial u_1}{\partial x_1} + \dfrac{u_2}{h_2}\dfrac{\partial h_1}{\partial x_2}\right) & \dfrac{1}{h_2}\left(\dfrac{\partial u_1}{\partial x_2} - \dfrac{u_2}{h_1}\dfrac{\partial h_2}{\partial x_1}\right) \\ \dfrac{1}{h_1}\left(\dfrac{\partial u_2}{\partial x_1} - \dfrac{u_1}{h_2}\dfrac{\partial h_1}{\partial x_2}\right) & \dfrac{1}{h_2}\left(\dfrac{\partial u_2}{\partial x_2} + \dfrac{u_1}{h_1}\dfrac{\partial h_2}{\partial x_1}\right) \end{bmatrix}, \tag{1}$$

where derivatives are taken along the reference surface. We recall that the trace of $\nabla\mathbf{u}$ (double of its scalar component) corresponds to the divergence or rate of surfacic dilation:

$$\delta = \nabla \cdot \mathbf{u} = \frac{1}{h_1 h_2}\left[\frac{\partial(h_2 u_1)}{\partial x_1} + \frac{\partial(h_1 u_2)}{\partial x_2}\right], \tag{2}$$

while the curl is defined as two times the skew-symmetric (pseudo-scalar) component

$$\omega = \nabla \wedge \mathbf{u} = \frac{1}{h_1 h_2}\left[\frac{\partial(h_2 u_2)}{\partial x_1} - \frac{\partial(h_1 u_1)}{\partial x_2}\right]. \tag{3}$$

The gradient of the velocity field along the reference surface is completely defined by δ, ω and the tensor of deformation rates $\mathbf{d}$ — a symmetric tensor with null trace — defined by

$$2\mathbf{d} = \mathbf{Def\ u} = \begin{bmatrix} d' & d'' \\ d'' & -d'' \end{bmatrix}, \quad \text{with} \quad \begin{cases} d' = \dfrac{h_2}{h_1}\dfrac{\partial}{\partial x_1}\dfrac{u_1}{h_2} - \dfrac{h_1}{h_2}\dfrac{\partial}{\partial x_2}\dfrac{u_2}{h_1} \\ d'' = \dfrac{h_2}{h_1}\dfrac{\partial}{\partial x_1}\dfrac{u_2}{h_2} + \dfrac{h_1}{h_2}\dfrac{\partial}{\partial x_2}\dfrac{u_1}{h_1} \end{cases}; \tag{4}$$

we must mention here that d' and d'' have no intrinsic meaning, the only intrinsic scalar quantity being the total deformation $d = \sqrt{d'^2 + d''^2}$. When rotating by an angle α the local reference frame, $\mathbf{d}$ transforms into

$$\mathbf{D} = \frac{1}{2}\begin{bmatrix} D' & D'' \\ D'' & -D' \end{bmatrix}, \quad \text{with} \quad \begin{cases} D' = d' \cos 2\alpha + d'' \sin 2\alpha \\ D'' = -d' \sin 2\alpha + d'' \cos 2\alpha \end{cases} \tag{5}$$

(a rotation by angle 2α). The angle α can be chosen either to get

$$D'' = 0: \quad \text{then} \quad \alpha = \frac{1}{2}\arctan\frac{d''}{d'}, \qquad D' = d, \tag{6}$$

the full deformation appearing then as pure longitudinal stretching; or to get

$$D' = 0: \quad \text{then} \quad \alpha = -\frac{1}{2}\arctan\frac{d'}{d''}, \qquad D'' = d, \tag{7}$$

the full deformation appearing then as pure lateral shear.

b *Stress tensor and viscosity*

The general form of the momentum equation in an inertial reference frame is

$$\rho\,\frac{D\mathbf{u}}{Dt} = \nabla \cdot \mathbf{s}, \tag{8}$$

where $\mathbf{s}$ is the stress tensor

$$\mathbf{s} = (-p + \lambda\delta)\mathbf{i} + 2\mu\mathbf{d}. \tag{9}$$

Here p is the thermodynamic pressure, λ and μ the molecular viscosity coefficients, $\mathbf{i}$ the unit tensor; λ, which expresses the stress exerted by dilation, is generally neglected. The right-hand side of (8) is expressed in orthogonal curvilinear coordinates as

$$\nabla \cdot (2\mu\mathbf{d}) = \frac{1}{h_1 h_2} \begin{bmatrix} \dfrac{1}{h_2}\dfrac{\partial(h_2^2\mu d')}{\partial x_1} + \dfrac{1}{h_1}\dfrac{\partial(h_1^2\mu d'')}{\partial x_2} \\ \dfrac{1}{h_2}\dfrac{\partial(h_2^2\mu d'')}{\partial x_1} - \dfrac{1}{h_1}\dfrac{\partial(h_1^2\mu d')}{\partial x_2} \end{bmatrix}, \tag{10}$$

a form corresponding to the divergence of a symmetric tensor with null trace. These definitions ensure that viscosity dissipates kinetic energy:

$$\frac{d}{dt}\int\int \frac{1}{2} u^2 \rho h_1 h_2 dx_1 dx_2 - \int\int p\delta h_1 h_2 dx_1 dx_2 = -\int\int \times [\lambda\delta^2 + \mu d^2] h_1 h_2 dx_1 dx_2. \tag{11}$$

On the other hand, the viscosity force being the divergence of the stress tensor has no effect on momentum or angular momentum except on the fluid boundaries; this property stems from the relation

$$\rho h_1 h_2 \frac{\partial}{\partial t}(h_1 u_1) = h_1 h_2 \frac{\partial(\lambda\delta)}{\partial x_1} + \frac{h_1}{h_2}\frac{\partial(h_2^2\mu d')}{\partial x_1} + \frac{\partial(h_1^2\mu d'')}{\partial x_2}, \tag{12}$$

taking x_1 as longitude and x_2 as a function of latitude (in that case h_1 and h_2 do not depend on x_1).

3 The general form of lateral diffusion as based on the Reynolds and Eliassen-Palm tensors

Separating the flow into its mean (i.e., resolved) and perturbation (i.e., sub-grid scale) components:

$$\mathbf{u} = \langle \mathbf{u} \rangle + \mathbf{u}' \tag{13}$$

and neglecting density fluctuations, we rewrite the mean momentum equation as

$$\frac{\partial}{\partial t}(\rho\langle\mathbf{u}\rangle) + \nabla\cdot(\rho\langle\mathbf{u}\rangle \otimes \langle\mathbf{u}\rangle) = \nabla(-\langle p\rangle + \lambda\langle\delta\rangle) + \nabla\cdot(2\mu\langle\mathbf{d}\rangle - \rho\langle\mathbf{u}'\otimes\mathbf{u}'\rangle). \tag{14}$$

The symmetric Reynolds tensor on the right-hand side can be split into its isotropic (scalar) and anisotropic parts:

$$\langle\mathbf{u}'\otimes\mathbf{u}'\rangle = \frac{1}{2}\langle u'^2\rangle\mathbf{i} + \mathbf{a}, \tag{15}$$

$$\mathbf{a} = \frac{1}{2}\begin{bmatrix} a' & a'' \\ a'' & -a' \end{bmatrix}, \quad \text{with} \quad \begin{cases} a' = \langle u_1'^2 - u_2'^2\rangle \\ a'' = 2u_1' u_2' \end{cases}; \tag{16}$$

The anisotropic part **a** can be refered to as the Eliassen-Palm tensor, a symmetric tensor with null trace. The mean momentum equation then reads

$$\frac{\partial}{\partial t}(\rho\langle\mathbf{u}\rangle) + \nabla\cdot(\rho\langle\mathbf{u}\rangle\otimes\langle\mathbf{u}\rangle) = -\nabla\left[\langle p\rangle + \frac{1}{2}\rho\langle u'^2\rangle - \lambda\langle\delta\rangle\right] + \nabla\cdot(2\mu\langle\mathbf{d}\rangle - \rho\mathbf{a}). \qquad (17)$$

When formulating lateral diffusion, we must relate the Reynold stress tensor to the mean velocity gradient. In doing so, we have to express its isotropic part (the dynamical pressure due to sub-grid scale motion) as a function of the mean divergence, and its anisotropic part as a function of the deformation tensor:

$$\frac{1}{2}\langle u'^2\rangle = F(\delta), \qquad \mathbf{a} = G(\mathbf{d}). \qquad (18)$$

In practice, we choose linear functions

$$\frac{1}{2}\langle u'^2\rangle = -2\lambda^*\delta, \qquad \mathbf{a} = -2\mu^*\mathbf{d}, \qquad (19)$$

with "turbulent" viscosity coefficients λ^* and μ^*. Forms (18) and (19) are indeed consistent in terms of tensor structure; a further sign of consistency is the fact that **a** transforms just like **d** when the local reference frame is rotated by an angle α:

$$\mathbf{A} = \frac{1}{2}\begin{bmatrix} A' & A'' \\ A'' & -A' \end{bmatrix}, \quad \text{with} \quad \begin{cases} A' = a'\cos 2\alpha + a''\sin 2\alpha \\ A'' = -a'\sin 2\alpha + a''\cos 2\alpha \end{cases}. \qquad (20)$$

Using (19) as a sub-grid scale model, we rewrite equation (17) as

$$\frac{\partial}{\partial t}(\rho\langle\mathbf{u}\rangle) + \nabla\cdot(\rho\langle\mathbf{u}\rangle\otimes\langle\mathbf{u}\rangle) = \nabla[-\langle p\rangle + (\lambda+\lambda^*)\langle\delta\rangle] + \nabla\cdot[2(\mu+\mu^*)\langle\mathbf{d}\rangle]. \qquad (21)$$

We may note that (21) is angular momentum conserving and kinetic energy dissipating. In the right-hand side of (21), isotropic stresses have an impact on the fluid motion only when it departs significantly from incompressibility. In fact, the "divergence damping" term in λ^* has intensively been used on heuristic grounds to favour geostrophic adjustment in geophysical fluid simulation or initialization. More precisely, it can be seen as a model of nonlinear interactions between inertia-gravity waves and rotational motion, preventing the spurious accumulation of inertia-gravity wave energy near the cut-off wavenumber illustrated for example by Farge and Sadourny (1989) or Warn (1986). The transfer of kinetic energy from the resolved scales to the subgrid scales modelled by equation (21) is expressed by (11) after replacing λ, μ by λ^*, μ^*. An expression for μ^* has been given by Smagorinsky (1963):

$$\mu^* = Kl^2 d, \qquad (22)$$

where l is the truncation scale and K a nondimensional constant. A similar expression can be given for λ^*:

$$\lambda^* = K' l^2 \delta. \tag{23}$$

Formulations (22, 23) are first order in the Laplacian and nonlinear; as such, their spectral formulations turn out to be relatively expensive.

4 Higher order diffusion

Higher order diffusion ("superviscosity") operators have been introduced first in quasi-geostrophic models (e.g., Basdevant et al., 1981; Basdevant and Sadourny, 1983). The motivation for reaching higher orders of derivation is the need to maximize the ratio of enstrophy to energy dissipation, as the theory of two-dimensional turbulence predicts the vanishing of energy dissipation rate at infinite Reynolds numbers. Placing ourselves from now in the infinite Reynolds number limit, we neglect molecular viscosity and formulate the superviscosity model as

$$\frac{\partial}{\partial t}(\rho\langle\mathbf{u}\rangle) + \nabla\cdot(\rho\langle\mathbf{u}\rangle\otimes\langle\mathbf{u}\rangle) = \nabla[-\langle p\rangle + \lambda^*\langle\delta\rangle] - (-\Delta)^m[\mu^*_{2m}(-\Delta)^m\langle\mathbf{u}\rangle] \tag{24}$$

for even-order ($2m$) diffusion, and

$$\frac{\partial}{\partial t}(\rho\langle\mathbf{u}\rangle) + \nabla\cdot(\rho\langle\mathbf{u}\rangle\otimes\langle\mathbf{u}\rangle) = \nabla[-\langle p\rangle + \lambda^*\langle\delta\rangle] + (-\Delta)^{m-1}\nabla\cdot[\mu^*_{2m-1}\mathbf{Def}(-\Delta)^{m-1}\langle\mathbf{u}\rangle] \tag{25}$$

for odd-order ($2m-1$) diffusion; here we have used $\nabla\cdot\mathbf{Def} = \Delta$, and $\mu^*_n = l^{*2n}/\tau^*$, where l^* is the cut-off scale and τ^* is the eddy-turnover time at l^*. Formulations (24) and (25) are valid for variable μ^*_n: this is the case when we use a nonlinear formulation like (22), or a stretched coordinate system. Note that we have not iterated the divergence term, nonlinear interactions between inertia-gravity waves and vortical structures being quite different from an enstrophy cascade process. It may be noted that, like (21), (24) and (25) do conserve total angular momentum except for lateral boundary effects. Also, the higher order diffusion terms in (24) and (25) can be rewritten using the identity $\nabla\cdot(\alpha\mathbf{Def}) = \nabla(\alpha\nabla\cdot) + \mathbf{n}\times\nabla(\alpha\nabla\times)$, where α is a scalar field. Energy is dissipated according to

$$\frac{d}{dt}\int\!\!\int\frac{1}{2}\langle u\rangle^2\rho h_1h_2dx_1dx_2 - \int\!\!\int\langle p\delta\rangle h_1h_2dx_1dx_2 = -\int\!\!\int \times\{\lambda^*\langle\delta\rangle^2 + \mu^*_{2m}[(-\Delta)^m\langle\mathbf{u}\rangle]^2\}h_1h_2dx_1dx_2 \tag{26}$$

for $2m$th-order diffusion, and

$$\frac{d}{dt}\int\!\!\int\frac{1}{2}\langle u\rangle^2\rho h_1h_2dx_1dx_2 - \int\!\!\int\langle p\delta\rangle h_1h_2dx_1dx_2 = -\int\!\!\int \times\{\lambda^*\langle\delta\rangle^2 + \mu^*_{2m-1}[\mathbf{Def}(-\Delta)^{m-1}\langle\mathbf{u}\rangle]^2\}h_1h_2dx_1dx_2 \tag{27}$$

for $(2m-1)$th-order diffusion. Expressions (26) and (27) are straightforward generalizations of (11), where energy dissipation is selectively trapped in shorter and shorter scales with increasing values of m.

5 Anticipated Potential Vorticity Model

The essence of the Antipated Potential Vorticity Model (APVM), originally introduced for quasi-geostrophic flows by Sadourny and Basdevant (1985) is to use an upstream value of Ertel's potential vorticity η in the equation of motion:

$$\frac{\partial_\theta \langle \mathbf{u} \rangle}{\partial t} = -\nabla_\theta \left(\langle c_p T + gz \rangle + \frac{1}{2} \langle u \rangle^2 - \lambda^* \langle \delta \rangle \right)$$
$$- \mathbf{n} \times \rho_\theta \langle \mathbf{u} \rangle [\langle \eta \rangle - \tau^* \Lambda_\theta (\langle \mathbf{u} \rangle \cdot \nabla_\theta \langle \eta \rangle)] \quad (28)$$

Equation (28) has been written using entropy or potential temperature θ as vertical coordinate; as it is valid only in the limit of very high Reynolds numbers, the molecular viscosity terms involving λ and μ have been neglected; $\rho_\theta = -1/g\partial p/\partial\theta$ is the density in θ coordinate, τ^* is a time scale and Λ_θ a (positive definite) space filter – for example, an adimensional iterated Laplacian: $\Lambda_\theta = (-l^{*2}\Delta_\theta)^n$ – operating along isentropic surfaces and modulating the upstream lag as a function of scale. Ertel's potential vorticity is defined as

$$\eta = \frac{\nabla_\theta \times \mathbf{u}}{\rho_\theta}. \quad (29)$$

The time lag is equal to τ^* at the cut-off scale $l = l^*$ and decreases rapidly as l increases. The vorticity equation implied by (28) reads as a scale-selective upstream numerical scheme for potential vorticity advection:

$$\frac{\partial_\theta}{\partial t}(\rho_\theta \langle \eta \rangle) = -\nabla_\theta \{ \rho_\theta \langle \mathbf{u} \rangle [\langle \eta \rangle - \tau^* \Lambda_\theta (\langle \mathbf{u} \rangle \cdot \nabla_\theta \langle \eta \rangle)] \}. \quad (30)$$

Formulations (28) or (30) ensure both potential enstrophy dissipation and energy conservation in the quasi-geostrophic limit ($\delta = 0$), a behaviour consistent with quasi-geostrophic theory at infinite Reynolds numbers. Energy conservation stems from the fact that the dissipation term is orthogonal to velocity; and potential enstrophy is dissipated according to

$$\frac{d}{dt} \int \int_{\theta=\theta_0} \frac{1}{2} \eta^2 \rho_\theta h_1 h_2 dx_1 dx_2 = - \int \int_{\theta=\theta_0}$$
$$\times (\langle \mathbf{u} \rangle \cdot \nabla_\theta \langle \eta \rangle) \tau^* \Lambda (\langle \mathbf{u} \rangle \cdot \nabla_\theta \langle \eta \rangle) \rho_\theta h_1 h_2 dx_1 dx_2. \quad (31)$$

The ability of the scheme to conserve energy while dissipating potential enstrophy means that it must be able to convert potential energy into kinetic energy and thus mimic, at least partly, the subgrid-scale baroclinic instability processes. It has been shown by Sadourny and Basdevant (1985) that this is indeed the case for quasi-geostrophic flow. Equation (28), however, may be difficult to use in practice, as most primitive equations models are not written using entropy as the vertical coordinate. We may rewrite it as

$$\rho \frac{D}{Dt}\langle \mathbf{u} \rangle + \nabla \langle p \rangle = \nabla_\theta(\lambda^* \nabla_\theta \cdot \langle \mathbf{u} \rangle) + \mathbf{n} \times \rho_\theta \langle \mathbf{u} \rangle \tau^* \Lambda_\theta(\langle \mathbf{u} \rangle \cdot \nabla_\theta \langle \eta \rangle), \tag{32}$$

where D/Dt refers to the Lagrangian derivative. The left-hand side here can be written using an arbitrary vertical coordinate ζ; then the differentiation operators in the diffusion terms on the right-hand side have to be defined as

$$\nabla_\theta = \nabla_\zeta - \nabla_\zeta \theta \frac{\partial}{\partial \theta}. \tag{33}$$

As suggested by Holloway and Benzi (quoted in Vallis and Hua, 1988), formulation (32) can be rewritten in an equivalent but perhaps slightly more convenient form by using two time levels t, $t + \Delta t$:

$$\rho \frac{D}{Dt}\langle \mathbf{u} \rangle + \nabla \langle p \rangle = \nabla_\theta(\lambda^* \nabla_\theta \cdot \langle \mathbf{u} \rangle)$$

$$- \mathbf{n} \times \rho_\theta \langle \mathbf{u} \rangle \frac{\tau^*}{\Delta t} \Lambda_\theta \left[\langle \eta \rangle_t - \langle \eta \rangle_{t-\Delta t} - \frac{\partial \eta}{\partial \theta}(\langle \theta \rangle_t - \langle \theta \rangle_{t-\Delta t}) \right]. \tag{34}$$

These forms can indeed be tested in general circulation models. The only complication over a standard diffusion scheme is the isentropic formulation, which is quite reasonable from dynamical arguments. A final comment is that the APVM is just one example of a class of numerical schemes of the primitive equations where lateral diffusion is treated by imposing a prescribed form of the transport equation for potential vorticity. Instead of using the scale-selective upstream scheme, one could use any well-behaved transport algorithm.

6 Diffusion of potential temperature

As both potential temperature and potential vorticity are Lagrangian invariants (in the absence of irreversible processes), they should in principle be treated with identical algorithms. For instance, if, according to equation (30), we use the APVM for potential vorticity, then potential temperature should be treated using the "mirror" equation:

$$\frac{\partial_\eta}{\partial t}(\rho_\eta \langle \theta \rangle) = -\nabla_\eta \cdot \{ \rho_\eta \langle \mathbf{u} \rangle [\langle \theta \rangle - \tau^* \Lambda_\eta(\langle \mathbf{u} \rangle \cdot \nabla_\eta \langle \theta \rangle)]\}, \tag{35}$$

or equivalently

$$\frac{\partial}{\partial t}(\rho_\eta \langle \theta \rangle) = -\nabla_\eta \cdot \left\{ \rho_\eta \langle \mathbf{u} \rangle \left[\langle \theta \rangle + \frac{\tau^*}{\Delta t} \Lambda_\eta(\langle \theta \rangle_t - \langle \theta \rangle_{t-\Delta t}) \right] \right\}. \tag{36}$$

Notations here are straightforward: derivations and filtering are performed along iso-potential vorticity surfaces, and $\rho_\eta = g^{-1}|\partial p/\partial\eta|$. The meaning is that potential temperature is advected along potential vorticity surfaces, and that diffusion should then occur along these surfaces. Just as above, the advective part of these equations can be written in any convenient vertical coordinate, only diffusion must be modelled explicitly on potential vorticity surfaces, using

$$\nabla_\eta = \nabla_\zeta - \nabla_\zeta \eta \, \frac{\partial}{\partial \eta}. \tag{37}$$

7 Conclusion

The present paper is not aimed at giving a really exhaustive review of lateral diffusion schemes for use in general circulation models of the atmosphere or ocean. Indeed, as we have noted incidently above, we have not touched upon a wide class of "implicit" diffusion schemes constructed on well-behaved transport schemes for Ertel's potential vorticity equations (and the entropy equation).

By inspecting the formal analogy between the Reynolds tensor and the velocity gradient tensor upon which molecular viscosity is based, we have shown that lateral diffusion should be modelled by separating an isotropic part depending on the large-scale divergence and related to vortical-inertia gravity modes interactions, from the usual term involving the divergence of the large-scale deformation tensor. This splitting is based on the decomposition of the Reynolds stresses into its isotropic part corresponding to the dynamical pressure term, and its anisotropic part corresponding to what we have called the Eliassen-Palm tensor. Based on this separation, we have given general formulations for scale selective "super-viscosity" schemes, valid for stretched coordinates in linear or nonlinear frames.

Finally, we have given appropriate forms of the Anticipated Potential Vorticity Model, which, up to now, has been used only for quasi-geostrophic models, but is currently adapted to general circulation models using entropy or other vertical coordinates – a sample of the wider class of schemes mentioned above. The symmetry between potential vorticity advection and potential temperature advection allows a parallel treatment of diffusion: potential vorticity should diffuse along isentropic surfaces, entropy should diffuse along potential vorticity surfaces, both with similar time lags.

Most of the formulations suggested above require elaborate algorithmic developments, constrained by nonlinear conservation or dissipation properties; also, they will be significantly more expensive to use than the usual simple diffusion operators. An analysis of the practical implementation of some of the above schemes and their computational interest will be the subject of forthcoming papers.

Acknowledgements

This research work was performed under contract EV5V-CT92-0125 of the European Communities Commission.

Appendix – Table of symbols

a, **A**	Eliassen-Palm tensor	Tensor
Def	Two-dimensional deformation operator	Vector to Tensor
d, **D**	Deformation of a two-dimensional vector field **u**	Tensor
d	Total deformation of the horizontal velocity field **u**	Scalar
g	Acceleration of gravity	Scalar
h_i	Mapping factor along coordinate x_i	Scalar
p	Pressure	Scalar
s	Stress tensor	Tensor
u	Horizontal velocity	Vector
x_i	Curvilinear horizontal coordinate	Scalar
l^*	Cut-off scale	Scalar
δ	Horizontal divergence	Scalar
η	Potential vorticity	Scalar
θ	Potential temperature	Scalar
λ	Viscosity coefficient associated with dilation effects	Scalar
μ	Viscosity coefficient associated with straining effects	Scalar
ρ, ρ_θ, ρ_η	Density in z, θ, η vertical coordinate	Scalar
ω	Vorticity of the two-dimensional vector field **u**	Pseudo-scalar

References

BASDEVANT, C.; B. LEGRAS, R. SADOURNY and M. BÉLAND. 1981. A study of barotropic model flows: Intermittency, waves and predictability. *J. Atmos. Sci.* **38**: 2305–2326.

——— and R. SADOURNY. 1983. Modélisation des échelles virtuelles dans la simulation numérique des écoulements turbulents bidimensionnels. *J. de Mécanique théorique et appliquée*. Numéro spécial, pp. 243–269.

BATCHELOR, G.K. 1969. High speed computing in fluid dynamics. *Phys. Fluids*, **12**: (Suppl. 2), 233–239.

COURTIER, P.; C. FREYDIER, J.-F. GELEYN, F. RABIER and M. ROCHAS. 1991. The ARPÈGE Project at Météo France. *In*: Numerical Methods in Atmospheric Models, II, European Centre for Medium Range Weather Forecasts Seminar Proceedings, 9–13 September 1991, pp. 193–231.

FARGE, M. and R SADOURNY. 1989. Wave-vortex dynamics in rotating shallow water. *J. Fluid Mech.* **206**: 433–462.

KRAICHNAN, R.H. 1967. Inertial ranges in two-dimensional turbulence. *Phys. Fluids* **10**: 1417–1423.

LEITH, C.E. 1968. Diffusion approximation for two-dimensional turbulence. *Phys. Fluids*, **11**: 671–673.

SADOURNY, R. and C. BASDEVANT. 1985. Parameterization of sub-grid scale barotropic and baroclinic eddies: Anticipated Potential Vorticity Method. *J. Atmos. Sci.* **42**: 1353–1363.

SHARMA, O.P.; H. UPADHYAYA, T. BRAINE-BONNAIRE and R. SADOURNY. 1987. Experiments on regional forecasting using a stretched-coordinate general circulation model. *J. Meteorol. Soc. Jpn.* Special Volume on Short- and Medium-Range Numerical Weather Prediction: 263–271.

SCHMIDT, F. 1977. Variable fine mesh in spectral global model. *Beitr. Phys. Atmos.* **50**: 211–217.

SMAGORINSKY, J. 1963. General circulation experiments with the primitive equations. I. The basic experiment. *Mon. Weather Rev.* **9**: 99–164.

VALLIS, G.K. and B.L. HUA. 1988. Eddy viscosity of the Anticipated Potential Vorticity Method. *J. Atmos. Sci.* **45**: 617–627.

WARN, T. 1986. Statistical mechanical equilibria of the shallow water equations. *Tellus*, **38 A**: 1–11.

Several Unresolved Issues in Numerical Modelling of Geophysical Flows

R.A. Pielke, M.E. Nicholls, R.L. Walko
Department of Atmospheric Science
Colorado State University,
Fort Collins, Colorado 80523

T.A. Nygaard
Institut for Energiteknikk
P.O. Box 40, 2007 Kjeller, Norway

and

X. Zeng
Department of Atmospheric Sciences
The University of Arizona, Tucson, AZ 85721

[Original manuscript received 7 November 1994; in revised form 4 October 1995]

ABSTRACT *This paper discusses several issues in the numerical modelling of atmospheric flow which have been left unresolved. These include the influence of partial model resolution of features that are represented in a model by only a few grid points and the implication of this inadequacy in terms of the parametrization of subgrid-scale processes. As a result of nonlinear interactions, this lack of resolution requires the introduction of new parametrization techniques such as "backscatter", as well as an assessment of the predictability and parametrizability of atmospheric flows as a result of the limited spatial resolution. Examples of these issues are presented in the paper. Among the conclusions is that for large eddy simulations a stochastic subgrid-scale parametrization is needed, while for well-defined surface forced mesoscale systems such as sea breezes, an ensemble-based subgrid-scale representation is adequate. Also introduced in this paper is an additional physical realism to atmospheric models (i.e., the compressibility effect) which provides an additional modelling degree of freedom to internal and externally imposed perturbations.*

RÉSUMÉ *On examine plusieurs problèmes de modélisation numérique de l'écoulement atmosphérique qui sont encore sans solution. Parmi ceux-ci, l'influence de la résolution partielle par le modèle d'entités représentées par seulement quelques points de grille et la répercussion de cette insuffisance en termes de paramétrage des processus d'échelle inférieure à la maille. À cause des interactions non linéaires, ce manque de résolution demande l'introduction de nouvelles techniques de paramétrage, la «rétrodiffusion» p. ex., ainsi qu'une évaluation de la possibilité de prévision et de paramétrage des écoulements atmosphériques tenant compte de la résolution spatiale limitée. Dans cet article on expose des exemples de ces problèmes. On trouve qu'un paramétrage stochastique de l'échelle inférieure à la maille est nécessaire pour les simulations de grands tourbillons alors que pour des systèmes bien définis de mésoéchelle forcés par la surface, tels que les brises de mer, une représentation de l'échelle inférieure à la maille basée sur un ensemble est adéquate. On introduit aussi un réalisme physique supplémentaire dans les modèles atmosphériques (p. ex., effet de compressibilité) qui fournit un degré de liberté supplémentaire de modélisation pour les perturbations internes et celles imposées de l'extérieur.*

---

1 Introduction

There is a range of issues in numerical modelling of geophysical flows which has received incomplete attention. While there has been considerable progress on the use of numerical models for the simulation of atmospheric flow on global to turbulence scales, there are substantial research issues that remain. These include the adequate understanding of a model's actual resolution with respect to its grid intervals, accurate procedures to represent partially and unresolved scale features, and the sensitivity of model results to these features as well as to initial and boundary conditions. In addition, the importance of the mathematical assumptions that are made in the model, such as the exclusion of soundwave effects, needs to be assessed.

Specific examples of each of these issues and suggested procedures to address these aspects of numerical modelling are presented in this paper. With respect to model resolution, a framework to define resolution in a model is presented. It is shown that there are spatial and temporal scales that are partially resolved, such that the dynamics and thermodynamics involved with these scales are explicitly resolved, but the remainder of this influence needs to be parametrized since these smaller scales cannot be represented using the selected grid increments. The introduction of a stochastic-based subgrid parametrization is shown to be a more physically-based formulation and is shown to be consistent with the concept in large eddy simulation modelling referred to as "backscatter". We also show that strong boundary forcing can substantially extend predictability from what would be expected based on the more traditional evaluation of sensitivity to initial conditions.

Finally, compressibility effects are shown to permit a more physically accurate representation of pressure adjustments in atmospheric flows. This concept also provides an effective procedure to diagnose diabatic heating remotely using the propagation of low frequency thermally-forced compression waves.

2 Resolution as related to grid increment

In the atmosphere, as discussed in the next several paragraphs by Pielke (1984), there are spatial scales in which kinetic energy is being produced and scales in which this kinetic energy is being dissipated into heat by molecular interactions. In the first case, scales of motion are on the order of a hundred metres to thousands of kilometres, whereas the size of motion significantly affected by molecular interactions is a centimetre or less. Somewhere in between exist scales of motion that are not directly influenced by either molecular dissipation or the forces generating the kinetic energy in the first place. Thus, it is expected that in this region the kinetic energy per unit wavenumber per unit mass as a function of wavenumber $E(k)$ is proportional to only the spatial scale of the motion (as specified by its wavenumber) and by the rate at which energy is being removed at the much smaller scales. Moreover, since kinetic energy does not accumulate once the larger-scale forcing is terminated, the energy must be transformed with time into smaller and smaller scales of motion until they are of a centimetre or less in size and can be removed by molecular interactions. From dimensional arguments, if $E(k)$ is dependent only on wavenumber and dissipation rate:

$$E(k) = a\epsilon^{2/3}k^{-5/3},$$

where a is a proportionality constant and ϵ has units of energy per unit time per unit mass. This region, where kinetic energy is independent of the original forcings of the motion and of its dissipation by molecular viscosity, is called the *inertial subrange*. This name arises because the advective terms (in this context these terms are also called the inertial terms) transfer kinetic energy among the components of velocity as well as generate smaller and smaller sizes of circulations. Lumley and Panofsky (1964) have provided a detailed discussion of the transfer of kinetic energy by turbulence. Gage (1979), Tennekes (1978), Lilly (1983), and Moran (1992, Chapter 2) discussed the observed occurrence of such a $k^{-5/3}$ relation in mesoscale and larger atmospheric features.

In a numerical model, however, this cascade of energy to smaller scales cannot occur because the smallest feature that can be resolved has a wavelength of two times the grid spacing. If, for example,

$$\phi_1 = \phi_0 \cos k_1\Delta x, \quad \text{and} \quad \phi_2 = \phi_0 \cos k_2\Delta x$$

represent two waves in a model with amplitudes ϕ_0, wavenumber k_1 and k_2, and grid increment of Δx, then a nonlinear interaction between the two can be represented by:

$$\phi_1\phi_2 = \phi_0^2 \cos k_1\Delta x \cos k_2\Delta x$$

or, using trigonometric identities,

$$\phi_1\phi_2 = \frac{1}{2}\phi_0^2[\cos(k_1 + k_2)\Delta x + \cos(k_1 - k_2)\Delta x].$$

Thus, two waves result from this interaction with wavenumbers $k_1 + k_2$ and $k_1 - k_2$.

TABLE 1. Result of nonlinear interaction of two waves.

Interactive wavelengths	Should produce	Will produce due to aliasing
$2\Delta x$ and $2\Delta x$	$1\Delta x$	Add a constant to the entire model
$2\Delta x$ and $4\Delta x$	$1.33\Delta x$	$4\Delta x$
$2\Delta x$ and $6\Delta x$	$1.5\Delta x$	$3\Delta x$
$2\Delta x$ and $8\Delta x$	$1.6\Delta x$	$8\Delta x$
$2\Delta x$ and $10\Delta x$	$1.67\Delta x$	$5\Delta x$

Suppose, as an example, a $2\Delta x$ and a $4\Delta x$ wave interact ($k_1 = 2\pi/2\Delta x, k_2 = 2\pi/4\Delta x$); the resultant waves are given by:

$$\phi_1\phi_2 = \frac{1}{2}\phi_0^2\left(\cos 2\pi\left(\frac{6}{8}\right)\Delta x + \cos 2\pi\left(\frac{1}{4}\right)\Delta x\right),$$

which corresponds to a $1.33\Delta x$ and a $4\Delta x$ wave. The latter size wave, of course, can be resolved but the $1.33\Delta x$ wave cannot. Instead it will be fictitiously seen as a $4\Delta x$ since that size is the first integer multiple of $\frac{4}{3}\Delta x$ equal to $n\Delta x$, where n is also an integer and $n \geq 2$. Waves that appear erroneously in this fashion are said to have aliased or folded to longer wavelengths. As seen from the above equation, to have aliasing, one of the waves must be less than $4\Delta x$ in length in order to generate physical solutions less than $2\Delta x$. Listed in Table 1 are examples of wave–wave interactions that will produce aliased waves.

Figure 1 illustrates how a $1.33\Delta x$ wave would be misinterpreted as a $4\Delta x$ wave on a computational grid. Even if no $2\Delta x$ waves are initially present they will be created since the interaction of two $4\Delta x$ waves will generate a $2\Delta x$ wave, whereas longer wave–wave interaction will produce $4\Delta x$ waves. Two interactive waves, both with $2\Delta x$ wavelengths will not produce a wave of $1\Delta x$ because identical values will result at each grid point. In this case the energy in the $1\Delta x$ wave will be seen fictitiously as the addition of a constant value of energy to the model.

Thus, when wave interactions occur in the real world, smaller and larger wavelengths result. Eventually, the smaller waves attain a size in which molecular dissipation can eliminate motion. In a numerical model, however, which has a discrete grid, waves smaller than $2\Delta x$ are erroneously seen as larger-scale waves. These erroneous larger scales interact and again transfer their energy to larger and smaller scales. Because the proper cascade of energy to smaller and smaller scales is interrupted, a fictitious energy buildup occurs as energy continues to be added to the model through the forcing terms, but with its dissipation improperly represented.

Therefore, even if a computational solution technique is linearly stable, the results can degrade into physically meaningless computational noise. Indeed, with many computational solution schemes, this erroneous accumulation of energy can cause the model dependent variables to increase in magnitude without bound – an error that is referred to as nonlinear instability.

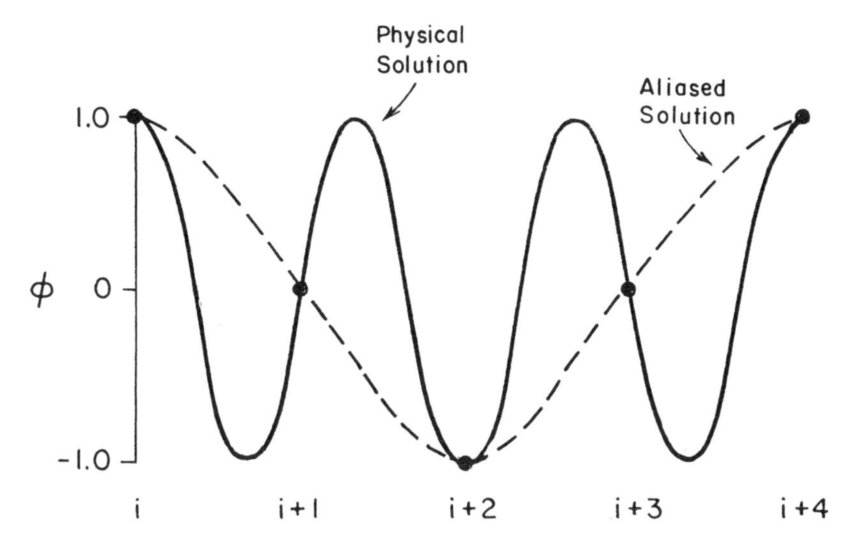

Fig. 1 Schematic illustration of how a physical solution with a wavelength of $1.33\Delta x$, caused by the nonlinear interaction of waves of $2\Delta x$ and $4\Delta x$ in length, is seen as a computational $4\Delta x$ wave in the computational grid (from Pielke, 1984).

Wavelengths less than $4\Delta x$ are required for aliasing to occur. As shown in Pielke (1984), Derickson (1992), and discussed in Pielke (1991) and Laprise (1992), such short waves are inadequately resolved on a computation grid and even in the linearized equations are poorly represented in terms of amplitude and/or phase. For these reasons, and because they are expected to cascade to even smaller scales anyway, it is desirable to remove these waves.

There are two methods to accomplish this task:

(1) by the proper parametrization of the subgrid-scale correlation terms (i.e., $\overline{u_i'' u_j''}$, $\overline{\theta'' u_j''}$, etc. where u_i'' and θ'' are subgrid-scale velocity and potential temperature respectively) so that energy is extracted from the averaged equations in a manner consistent with reality; or

(2) by the use of a spatial smoother (also referred to as a filter) which removes the shortest waves, but leaves the longer ones relatively unaffected.

The first method is the most attractive, of course, because it is based on fundamental physical concepts. Unfortunately, however, except for models with fine-scale grid increments in all three spatial directions (e.g., Large Eddy Simulation (LES) models) only the vertical subgrid-scale correlation terms are reasonably well known and can be parametrized accurately in terms of the dependent variables. Moreover, even for LES models, as discussed in Section 3, the subgrid-scale terms must include the influence of stochastic downscaling of kinetic energy. Horizontal subgrid mixing in larger-scale models can be estimated only crudely since the horizontal averaging scale in such models is typically much larger than the vertical scale (i.e.,

$\Delta x \simeq \Delta y \gg \Delta z$). Moveover, there has been little theoretical or observational studies on the structure of horizontal mixing over heterogeneous ground surfaces.

There is the additional complication that even for wavelengths equal to or greater than $2\Delta x$, depending on what data points are sampled at any given time step, only a fraction of the wave is often resolved, as shown in Fig. 2. The remaining fraction at that time period would need to be parametrized. This fraction will vary over time, and is a partial explanation as to why a "backscatter" parametrization has been proposed in LES models (see Section 3 of the paper). As seen in Fig. 2, for a $2\Delta x$ feature, up to 64% could be resolved at any time, but the fraction resolved can be zero. Even for a $6\Delta x$ feature, at least 4% of the influence of the wave would need to be parametrized.

One approach to explore is the use of spatial-averaged numerical solution techniques (e.g., such as the type proposed by Tanguay et al., 1992 and Derickson, 1992) in order to further minimize the influence of these smallest-scale resolved waves. Similarly, the intrinsic skill of semi-Lagrangian solution methods, such as that discussed by Robert (1982) and Staniforth and Côté (1991), in which numerical phase and amplitude accuracy are consistent in terms of skill with each other, minimizes the influence of these short spatial scales. (Although the influence of these shorter waves on a model result are lost.)

The main conclusion from this analysis is that, while it is generally recognized that spatial features less than a model grid increment must be parametrized, there has been insufficient study as to whether an ensemble-based or stochastic subgrid-scale parametrization should be used. Also, even for features larger than a grid increment a portion of a modelled feature must still be parametrized, and this portion will vary in time and space.

3 Ensemble versus realization-based subgrid-scale parametrizations in Large Eddy Simulations (LES) of microscale atmospheric flows

a *Definition of LES and its use in numerical modelling of geophysical flows*

In the previous section, it has been demonstrated that a fraction of a feature with a wavelength equal to or greater than $2\Delta x$ still needs to be parametrized. This technique may be fulfilled by a technique called the backscatter method. This method is discussed in this section on the LES of atmospheric flows.

Numerical simulation of atmospheric flow on the microscale is of practical interest in a number of applications. Examples are estimation of loads on buildings, bridges and sites of wind turbines in complex terrain, as well as the interaction between the atmosphere and the natural land surface including vegetation and the ocean. These applications require a detailed knowledge of the turbulent characteristiçs of the flow. In LES, a fine resolution allows an explicit description of the large eddies, from production scale and into the inertial subrange. In the interior of the flow, the resolved eddies account for most of the turbulent kinetic energy, and the results have traditionally been assumed to be insensitive to the turbulence model describing the subgrid-scale stresses (SGS). Simple eddy-viscosity closures,

$\theta = \sin\frac{\partial\pi x}{n\Delta x}$

n is a real number $\quad \overline{\theta} \int_{x-\Delta x/2}^{x+\Delta x/2} \theta dx/\Delta x = \frac{n}{\pi}\left[\sin\frac{\partial\pi x}{n\Delta x}\sin\frac{\pi}{n}\right]$

Δx

Maximum resolvable	n	Minimum to be Parameterized = $w(n)$	Example: $\Delta x = 10$ km
0	1	100%	10 km feature
64%	2	36%	20 km feature
83%	3	17%	30 km feature
90%	4	10%	40 km feature
94%	5	6%	50 km feature
96%	6	4%	60 km feature

$$\text{e.g.,}\quad \frac{\partial\overline{\theta}}{\partial t} = -\overline{u}_j\frac{\partial\overline{\theta}}{\partial x_j} - \frac{1}{\rho_0}\frac{\partial}{\partial x_j}\rho_0\overline{u_j''\theta''} + \overline{S}_\theta \Longrightarrow \overline{S}_\theta = w(n)Q_{\text{cumulus process function}}$$

- Parameterization of subgrid scale effects must also include all features of the variables which are excluded by the limited resolution.

Fig. 2 Grid volume averages.

describing the SGS stresses as deterministic functions of the resolved fields have therefore been chosen. Such a formulation is referred to as an ensemble-based parametrization because given the same value of the resolved dependent variable yields the same subgrid-scale parametrization results.

LES applications include representations of meteorological events such as downbursts, thunderstorms or tornadoes. This can be done with a mesoscale model with nested grids, as shown by Nicholls et al. (1993) and Costigan (1992). Within the restricted area of the fine grid, it is an open question whether eddies with the appropriate length scales will form as a feature propagates from a coarser to finer grid mesh. Another problem, perhaps not so important, is aliasing when the fine-grid mesh waves enter the course grid, where they are resolved only as very short wave features.

Most structures of engineering interest are located in the surface layer. Here, the basic assumptions for an LES are particularly hard to satisfy, since the size of the eddies become smaller as the surface is approached. Typically, the vertical grid spacing is reduced towards the surface in order to describe the quasi two-dimensional shear. Description of the three-dimensional eddies is still restricted by the horizontal grid spacing, however, which is usually chosen to describe the energy-containing eddies well above the ground. One may therefore expect problems in the surface layer. Less of the eddies are resolved and the SGS eddies are not necessarily in the inertial subrange. A simple eddy viscosity closure would be expected to have shortcomings in this region, except very close to the surface where production and dissipation are in balance.

Mason and Thomson (1992) have addressed this problem and have developed supplements to the traditional (Smagorinsky, 1963) parametrization of subgrid fluxes in the neutral turbulent boundary layer. This supplement involved the introduction of a stochastic contribution to the subgrid scale which is referred to as backscatter.

b *The Smagorinsky subgrid model*

In LES, a filter operation in space (here denoted by an overbar) is applied to the governing equations. For the neutral turbulent boundary layer, the filtered momentum equations can be approximated by:

$$\frac{\partial \overline{\mathbf{u}_i}}{\partial t} + \frac{\partial}{\partial x}\overline{\mathbf{u}_i}\,\overline{\mathbf{u}_j} = -\frac{1}{\rho o}\frac{\partial \bar{\mathbf{p}}}{\partial \overline{\mathbf{x}_i}} + +g\boldsymbol{\lambda}_{i3}\frac{\partial}{\partial \mathbf{x}_j}(-\mathbf{L}_{ij} + \boldsymbol{\tau}_{ij}) - f(\bar{\mathbf{G}}_j - \bar{\mathbf{u}}_i)\boldsymbol{\epsilon}_{ij3} \qquad (1)$$

$$\mathbf{L}_{ij} = \overline{\bar{\mathbf{u}}_i\bar{\mathbf{u}}_j} - \overline{\mathbf{u}_i}\,\overline{\mathbf{u}_j} \qquad (2)$$

$$\boldsymbol{\tau}_{ij} = -\overline{(\overline{\mathbf{u}_i}\mathbf{u}_j' + \mathbf{u}_i'\overline{\mathbf{u}_j} + \mathbf{u}_i'\mathbf{u}_j')} \qquad (3)$$

Deviations from the filtered values are denoted by primes; $\mathbf{u}_i$ is the vector velocity, p is pressure, ρ_o is the basic state density (function of height only), $\mathbf{G}_j$ is the geostrophic wind, g is the gravitational constant, $\mathbf{L_{ij}}$ is the Leonard stress (the difference between filtered and true grid-scale inertial terms), and $\boldsymbol{\tau_{ij}}$ is the true subgrid stress.

Leonard (1974) showed that the Leonard stress term is a significant part of the total drain of turbulent kinetic energy. This term can be estimated by application of a filter operation, but is usually either ignored or absorbed into the modelling of $\boldsymbol{\tau_{ij}}$.

The Smagorinsky model for the subgrid stress tensor is an ensemble-based eddy-viscosity closure:

$$\tau_{ij} = \kappa \left(\frac{\partial \overline{u_i}}{\partial x_j} + \frac{\partial \overline{u_j}}{\partial x_i} \right) \tag{4}$$

$$\kappa = l^2 S \tag{5}$$

$$S^2 = \frac{1}{2} \left(\frac{\partial \overline{u_i}}{\partial x_j} + \frac{\partial \overline{u_j}}{\partial x_i} \right)^2 \tag{6}$$

In this formulation, the same distribution of $\overline{u_i}$ results in an identical value of τ_{ij} as long as l is specified as described below. Lilly (1967) relates the subgrid mixing length scale, l to a filter scale l_f by:

$$l_f = l / C_s \tag{7}$$

which usually is taken as a typical grid spacing.

Theoretical estimates indicates that C_s should be around 0.2. In practice, C_s is chosen to control numerical noise, and to match observations of turbulence levels in experiments.

For simulations of the neutral turbulent boundary layer, Mason and Thomson (1992) have used the following matching towards the lower boundary with an aerodynamic roughness, z_o:

$$\frac{1}{l^n} = \frac{1}{l_o^n} + \frac{1}{(\kappa(z + z_o))^n} \tag{8}$$

l_o is the free-stream subgrid mixing length:

$$l_o = C_s (\Delta x \Delta y \Delta z)^{1/3} \tag{9}$$

for a typical grid spacing well above the ground.

This forces the subgrid mixing length towards Prandtl's suggested value as the lower boundary is approached, where most of the stress should be provided by the subgrid model. Diffusion of scalars are treated in a similar way.

c *Discussion of a stochastic subgrid-scale parametrization in LES*
Leslie and Quarini (1979) examined the usefulness of the classical subgrid turbulence closures for LES, by analysis in Fourier space of homogeneous isotropic flow. Different filters for turbulence spectra were used to estimate the eddy viscosity for different wavenumbers.

The main conclusions follow.

- The eddy viscosity is reasonably constant for different wavenumbers, but is in error for the long waves.
- The Leonard stress accounts for typically 14% of the total drain of turbulent kinetic energy.
- In cases with anisotropic grid-scale turbulence, the classical subgrid model should work well in the flow interior given that the subgrid turbulence is isotropic.
- The estimation of the dissipation from the grid-scale strain should be based on a statistical or volume average rather than the instantaneous values.
- Modelling the gross drain and the backscatter separately is recommended. Backscatter is the return of some of the turbulent kinetic energy from the subgrid scales to the grid scales. In wavenumber space, this is seen as interactions between subgrid-scale and grid-scale waves, as discussed in Section 1 of this paper. In physical space this can be seen as fluctuations in the subgrid stresses.

Mason and Thomson (1992) have shown that in LES of the neutral turbulent boundary layer with the Smagorinsky model, the solution lacks resolved stresses in a matching region, where subgrid and resolved stresses are comparable. They assume that this lack of resolved stress is caused by the deterministic (i.e., ensemble) evaluation of the subgrid stress (from the filtered velocities); a conclusion that is supported by the arguments of Leslie and Quarini (1979). The method of stochastic backscatter has been proposed to provide a more realistic parametrization of subgrid-scale effects. This technique permits a representation of the variable amount of a wave that is resolved at any time, as discussed in Section 2 of this paper. (In this context, while we will continue to refer to subgrid-scale effects; more accurately we are parametrizing "non-resolvable effects" which include subgrid spatial scales (up to $2\Delta x$) and the portion of a wave larger than $2\Delta x$ that is not resolved at any given time step.)

Supported by dimensional and statistical arguments, it is assumed that the turbulent kinetic energy backscatter rate is:

$$\frac{\partial E}{\partial t} = C_B \left(\frac{l}{l_o}\right)^5 \epsilon = (\sigma_{a_1}^2 + \sigma_{a_2}^2 + \sigma_{a_3}^2)\Delta t \tag{10}$$

where E is the resolved energy, C_B is the constant of proportionality, ϵ is the dissipation rate and σ_{a_i} is the standard deviation of the random acceleration a_i. Δt is the time step.

Omitting subgrid transport processes, this backscatter is equal to the difference between subgrid turbulent energy production and dissipation.

$$\frac{\partial E}{\partial t} = \kappa S^2 - \epsilon \tag{11}$$

The time scale of the fluctuations are of the order of the viscous stability limit on the time step. In the application of the backscatter technique the exact magnitude of this time scale is ignored, and accelerations are added for every second time step.

The procedure for inclusion of stochastic backscatter is:

(1) Generate three 3-D random velocity fields with zero mean.
(2) Scale amplitudes with position-dependent scale factors based on Eqs (10) and (11), also taking into account the loss of variance in the filter and curl operation.
(3) Filter the fields with a 1:2:1 filter in all three directions.
(4) Take curl to obtain non-divergent fields. These accelerations are added into the tendencies.

A similar procedure is used for the scalar diffusion, except that the scalar flux is obtained by taking the divergence of the random flux vector field.

Chasnov (1991) has estimated C_B to be 1.4. A series of simulations (Mason and Thomson, 1987) suggests that $n = 2$ in Eq. (8) is the optimum value.

Tests with the Regional Atmospheric Modelling System (RAMS; Pielke et al., 1992; Nicholls et al., 1995) show an improvement for LES modelling of the neutral static stability planetary boundary layer when the stochastic backscatter technique is utilized. The model has been run with a grid consisting of $50 \times 50 \times 50$ nodes, with horizontal grid spacing 20 m. The vertical grid spacing increases from 1 m at the ground to 40 m in the upper region. Using this limited domain in an attempt to resolve eddies just above the surface layer, some production-scale mechanisms are not, however, described.

The stochastic backscatter scheme has been implemented somewhat differently than reported by Mason and Thomson (1992). Using a 1:2:1 filter and taking the curl on a grid with different horizontal and vertical grid spacing can lead to anisotropy of the backscatter. This (sometimes desired) anisotropy could be controlled by applying a different vertical filter. In the first tests with RAMS, a triangular filter with the same shape and width as the horizontal filter is used before evaluating the curl. Figures 3 to 5 show the velocity variances without backscatter, normalized with respect to the friction velocity squared obtained in the same simulation. The u variance has a high peak near the surface. Mason and Thomson (1992) explain this as the model's lack of turbulent kinetic energy transfer from u to the other velocity components in the unresolved small eddies near the surface.

Figures 6 to 8 show velocity variances with backscatter. The geostrophic drag coefficient is 1.60×10^{-3}, which is lower than the value of 1.89×10^{-3} in the Mason and Thomson (1992) simulations. Without backscatter Mason and Thomson

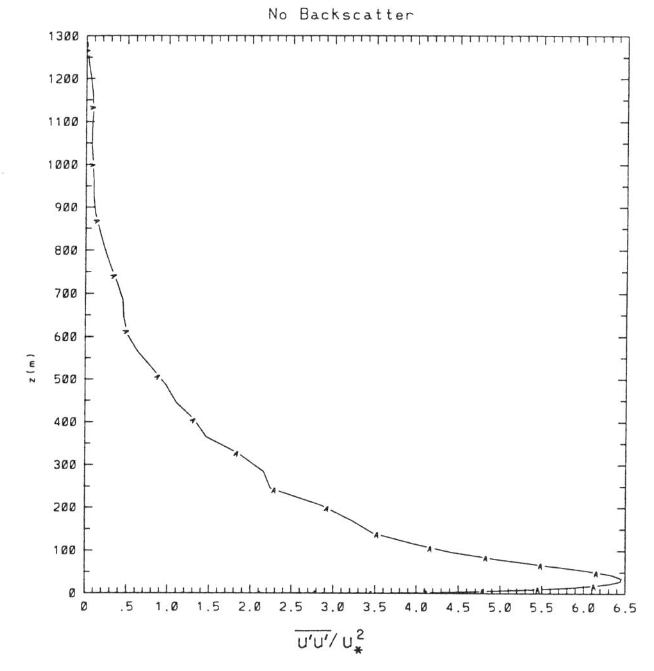

Fig. 3 *u* velocity variance without backscatter.

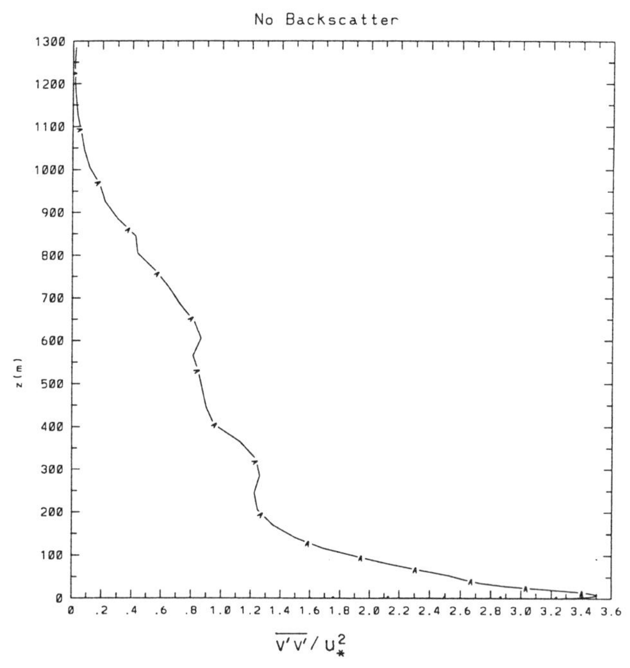

Fig. 4 *v* velocity variance without backscatter.

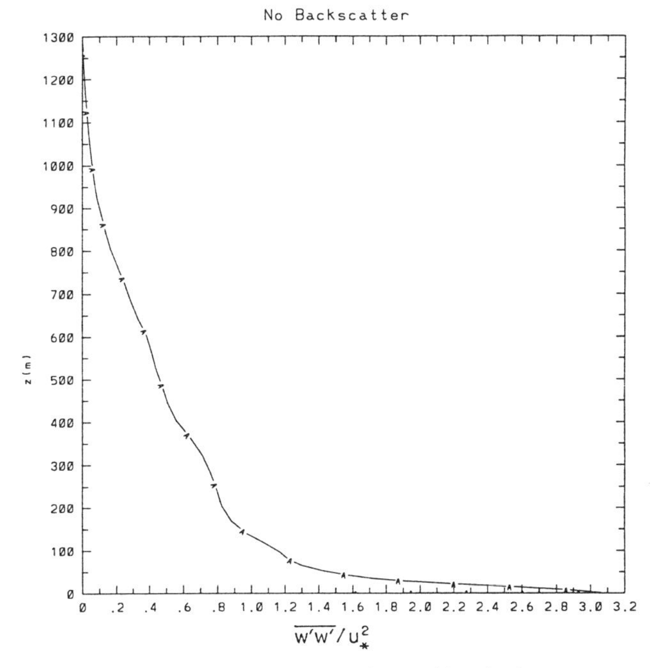

Fig. 5 Vertical velocity variance without backscatter.

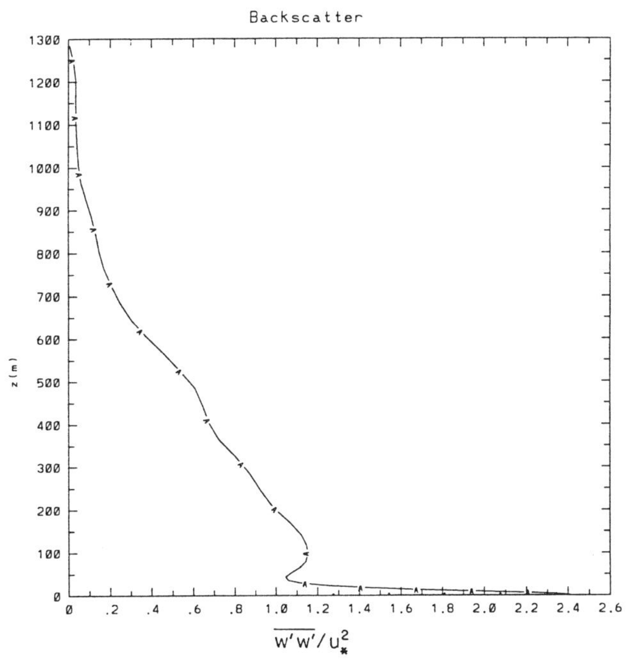

Fig. 6 Vertical velocity variance with backscatter.

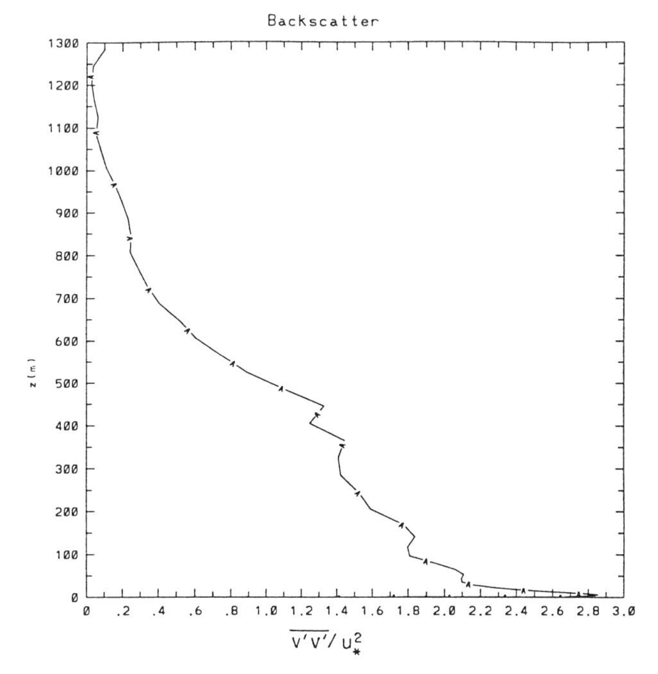

Fig. 7 v velocity variance with backscatter.

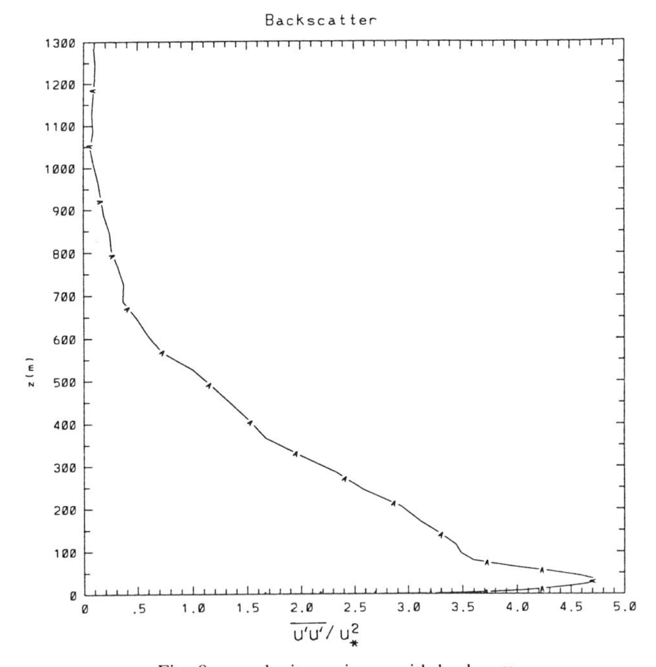

Fig. 8 u velocity variance with backscatter.

(1992) obtained a geostrophic drag coefficient of 1.41×10^{-3}. This lower value is probably caused by the modification of the vertical filter. These results are preliminary (half-hour average on a still evolving flow), so a detailed comparison with the Mason and Thomson results cannot yet be made. It is, however, encouraging to note that the peak on the u variance is more realistic and that the profiles above the surface layer look reasonable.

The results of this Section demonstrate that the details of the subgrid-scale parametrization critically influence LES results. An ensemble based subgrid-scale representation is not adequate; rather than stochastic-based parametrization must be used so that the variable percent of resolved features at any given time step can be represented in a statistical manner in terms of the subgrid-scale parametrization.

d *Future work*

The optimism about future applications of LES and the recent success of the stochastic backscatter model is expected to encourage further efforts in this area. Developments are expected both in the basic formulation of LES, the Smagorinsky model and supplements to the Smagorinsky model such as the stochastic backscatter method. Based on the recommendations given in the reviewed papers, the following topics seem to be of particular interest:

- Examination of the Leonard stress term. Should this term be evaluated explicitly by a filter operation, or can this term be successfully modelled by the Smagorinsky model?
- A filtering/averaging operation when evaluating total deformation versus instantaneous point values (Smagorinsky model).
- Correct time and space scales of the random accelerations in the stochastic backscatter method.
- Empirical corrections for deficiencies near the boundaries.
- Usefulness of nested grids in LES, and possible coupling of meso- and microscale simulations.

4 Predictability and parametrizability of geophysical flows

In the previous section, the backscatter method in LES modelling has been discussed. Another important issue in numerical modelling (including LES) of geophysical flows is their predictability. For the subgrid parametrization in numerical models, an important issue is the parametrizability of the subgrid processes, which is related to the predictability of these processes. The concept of chaos is also closely related to the predictability of a flow.

In numerical modelling of geophysical flows, appropriate boundary conditions and model parameters need to be determined. In general, the predictability of geophysical flows in these numerical simulations depends on not only the initial conditions but also boundary conditions and model parameters. Just as the initial conditions cannot be determined accurately for a complicated system such as the atmosphere and the oceans, the boundary conditions and model parameters cannot be determined accurately. However, uncertainties in boundary conditions and model parameters introduce uncertainties at every moment rather than just at the beginning of the integration of the governing equations of the system. Boundary conditions, of course, can also enhance predictability by continuously reinforcing the system.

In order to demonstrate that strong boundary forcing can substantially extend predictability from what would be expected based on the more traditional evaluation of sensitivity to initial conditions, we discuss the interaction between inhomogeneous surfaces and convective boundary layer turbulence. Using RAMS, systematic two-dimensional numerical simulations are performed (Zeng and Pielke, 1993, 1995a, b). The surface consists of alternating water and land strips with strip size L, and the large-scale wind U ranges from 0 to 9 m s^{-1}. Figure 9 shows the time-averaged RMSC(w)/RMSP(w) as a function of the large-scale wind U and the patch size L. RMSC(w) is the signal of the vertical velocity defined as the root-mean-square (RMS) difference of w between time t and the initial time in the control run, and RMSP(w) is the noise of the vertical velocity defined as the RMS of w in a perturbed run (i.e., with initial perturbations) from the control run (i.e., without initial perturbations). The ratio of RMSC(w)/RMSP(w), $\bar{r}(w)$, represents the predictability of w; i.e., when $\bar{r}(w)$ is greater than unity, w is predictable while w is unpredictable when $\bar{r}(w)$ is less than unity.

It is seen from Fig. 9 that $\bar{r}(w)$ is greater than unity in most parts of the parameter space. It is also found that $\bar{r}(w)$ reaches a maximum between L = 10 and 20 km, and $\bar{r}(w)$ decreases with L for L $\geq$ 10 km. For instance, $\bar{r}(w)$ is usually less than unity when L is less than 5 km; however, $\bar{r}(w)$ is larger than 10 when L is between 10 and 20 km. In other words, in contrast to the homogeneous case (i.e., L = 0), boundary forcing (i.e., mesoscale circulations due to horizontal differential heating over inhomogeneous surface) significantly increases the predictability of w. For L = 150 km with small wind, $\bar{r}(w)$ is less than unity. The reason is that, when L is large, except near the coastal region, the domain is dominated by boundary-layer turbulence, just like the homogeneous case.

The actual vertical velocity fields in the control and perturbed simulations are shown in Fig. 10 for the case with U = 7 m s^{-1} and L = 70 km. It is seen that the w field in the control simulation is very close to that in the perturbed simulation due to strong surface modulation. Because the updraft in the convergence zone provides a favourable condition for the development of convection, and the location of the convergence zone is unchanged from the control simulation to the perturbed simulation, the possible initiation location of convection due to the convergence zone is actually predictable in Fig. 10, as subjectively concluded in Pielke (1974). In contrast, the initiation location of convection over a homogeneous surface is unpredictable.

An important conclusion from this Section is that atmospheric features that are strongly forced by well-resolved spatially structured forcing can be adequately

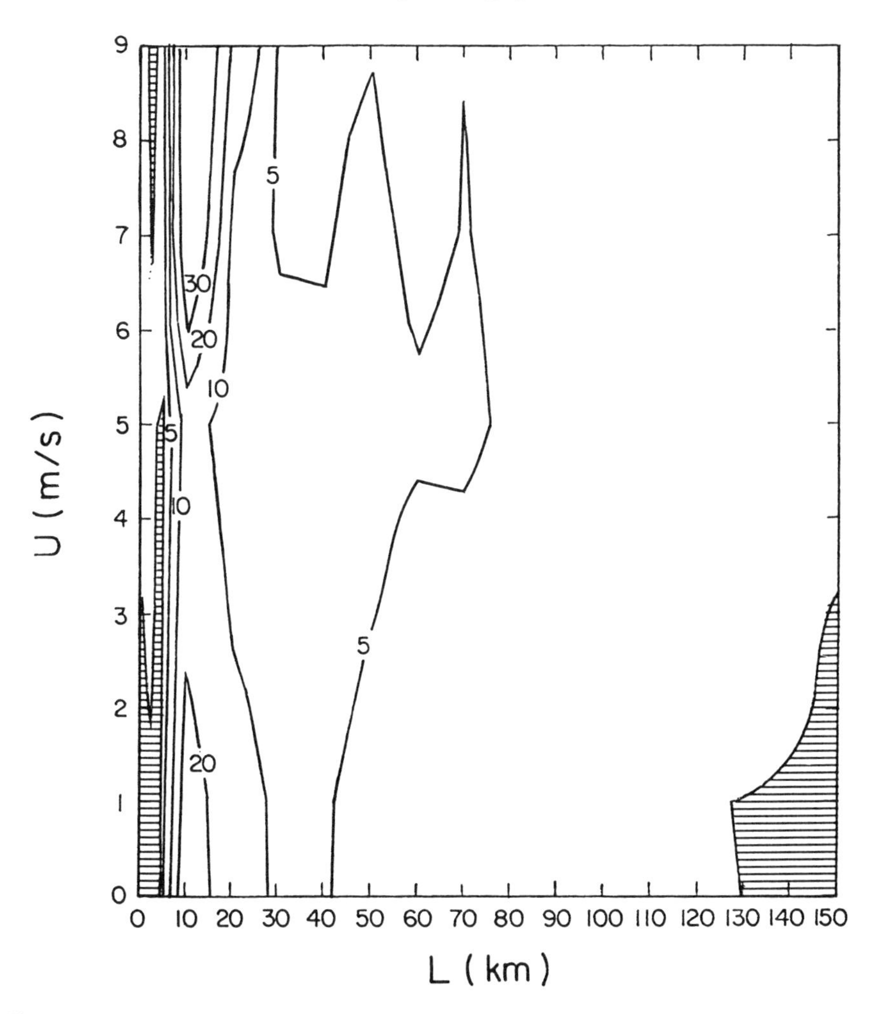

Fig. 9 The time-averaged RMSC(w)/RMSP(w) as a function of the large-scale wind U and the patch size L. Values that are less than unity are shaded. The isolines are drawn with values of 1, 5, 10, 20 and 30, and the maximum (minimum) value is 49.2 (0.45).

represented using an ensemble based subgrid-scale parametrization. A stochastic subgrid-scale representation, such as the "backscatter" for LES modelling need not be used for these strongly forced atmospheric systems.

5 The influence of compressibility in the presence of diabatic heating

In this Section, we address another issue in numerical modelling of atmospheric flow, i.e., the compressibility of the atmosphere. This geophysical process (i.e.,

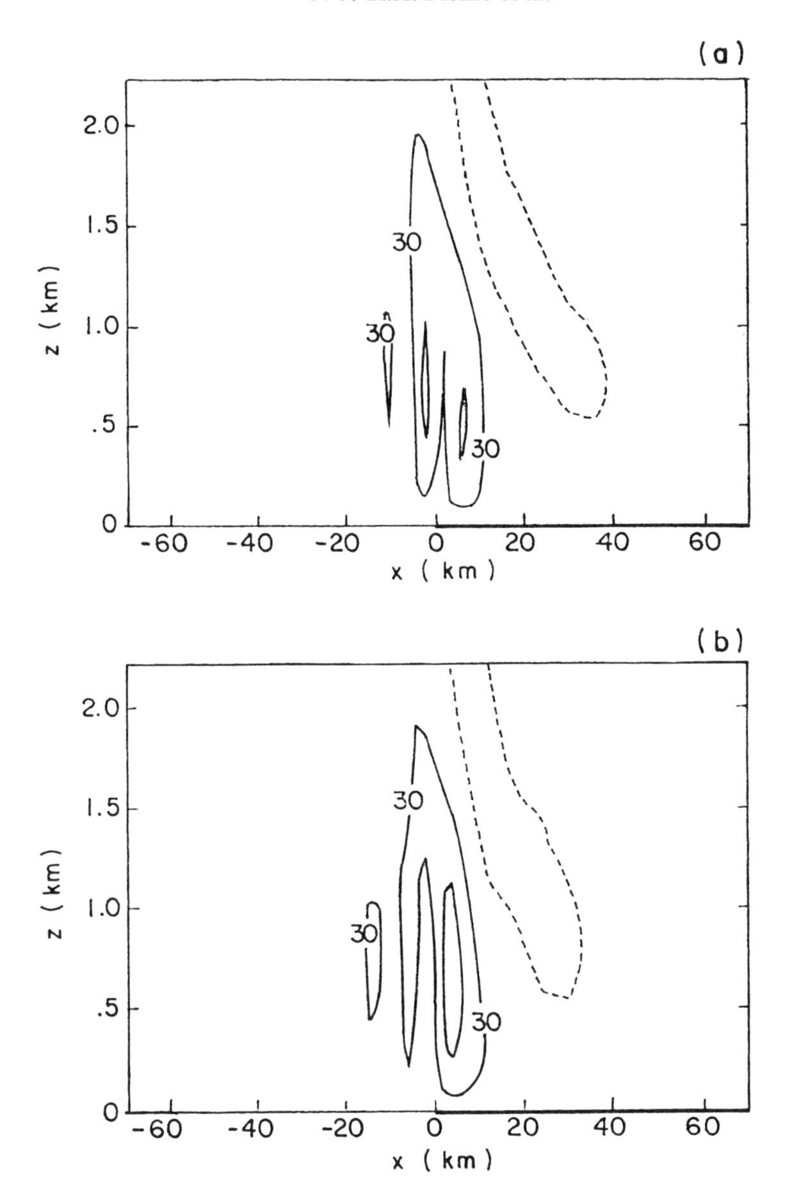

Fig. 10 The vertical velocity field at time $t = 8$ hours after sunrise for the case with L = 70 km and U = 7 m s^{-1}. The location of water strips is indicated by heavy lines on the x-axis, and negative velocities are represented by dotted contours. The unit of the contour values is 10^{-3} m s^{-1}, and the contour interval is 0.06 m s^{-1}. (a) Control simulation, and (b) Perturbed simulation.

compressibility) has generally been ignored in atmospheric modelling. When an air parcel is heated the pressure increases and the parcel expands, while the adjacent region is compressed. This results in a compression wave which propagates away from the heated region at the speed of sound. The amplitudes of the pressure perturbation and air motions associated with these waves are very small for typical diabatic heating rates that occur naturally in the atmosphere. Moreover, these thermally-induced acoustic modes have very low frequencies which makes it difficult to detect them. Numerical modelling studies of thunderstorms typically use soundproof equations or neglect terms which allow the existence of thermally-induced acoustic modes. Interest in these waves lies in the physical interpretation of mass adjustment and total energy transfer produced by diabatic effects and, more practically, in the possibility of observing them, which could result in a new passive detection technique for determining the location and intensity of thunderstorms (Pielke et al., 1993).

Recent numerical modelling studies have investigated some of the properties of thermal compression waves using a fully compressible model (Nicholls and Pielke, 1994a, b). Nicholls and Pielke (1994b) discuss a two-dimensional simulation of a heated land surface that is surrounded by ocean. For this simulation there are 150 horizontal grid points and 25 vertical grid points. The horizontal grid increment is 50 km and the vertical grid increment is 1 km. A heat source of magnitude 0.001 K s^{-1} is prescribed at the central 20 grid points and at the lowest level above the surface. It is applied for 1 h after which it is turned off. The simulation is then run for a further 1 h. Figures 11a, b, c, d and e, show the horizontal mass flux, vertical mass flux, perturbation temperature, perturbation pressure and perturbation density, respectively, at $t = 900$ s. Sea-breeze circulations have begun to develop at the coasts (in this and subsequent figures the land is indicated by the broad solid line). However, broader deeper regions of off-shore flow are evident which are propagating at the speed of sound. These regions are propagating inland as well as offshore. The vertical mass flux is strongest within the sea-breeze circulations. However, there is a deep region of weak upward motion above the land surface associated with thermal compression waves. The upward mass flux due to thermal compression waves is much weaker than the horizontal flux (note that different contour intervals are used). The upward propagating acoustic mode produced by the low-level heating results in an adjustment towards hydrostatic balance (Nicholls and Pielke, 1994b; Bannon, 1995). The surface temperature over land has increased uniformly. The perturbation pressure is positive everywhere; however, relative surface lows occur just onshore. The positive pressure perturbations extend off shore and delineate the position of the compression-wave front. The perturbation density has increased exterior to the heat source. The contour interval used is too small to show the variation within the heat source region, but it is similar, though of opposite sign, to the perturbation temperature. Although the density has decreased over the centre of the land at low levels owing to an upward mass flux, the surface-pressure perturbations are still positive at this time.

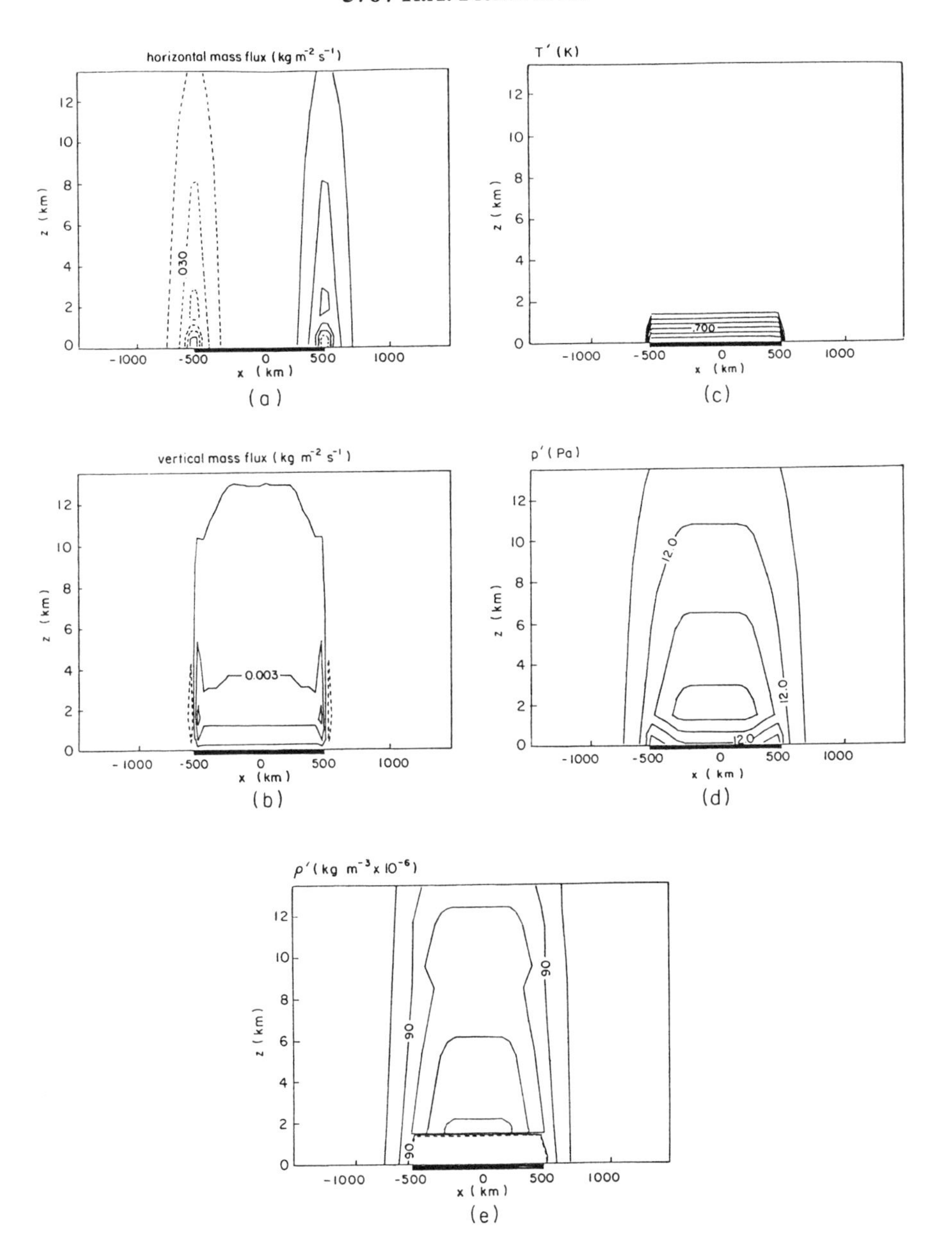

Fig. 11 Results for the heated land surface at t = 900 s. (a) Horizontal mass flux; the contour interval is 0.02 kg m^{-2} s^{-1}. (b) Vertical mass flux; the contour interval is 0.002 kg m^{-2} s^{-1}. (c) Perturbation temperature; the contour interval is 0.2 K. (d) Perturbation pressure; the contour interval is 8 Pascals. (e) Perturbation density; the contour interval is 60 × 10^{-6} kg m^{-3}. The label scale is 10^{6}.

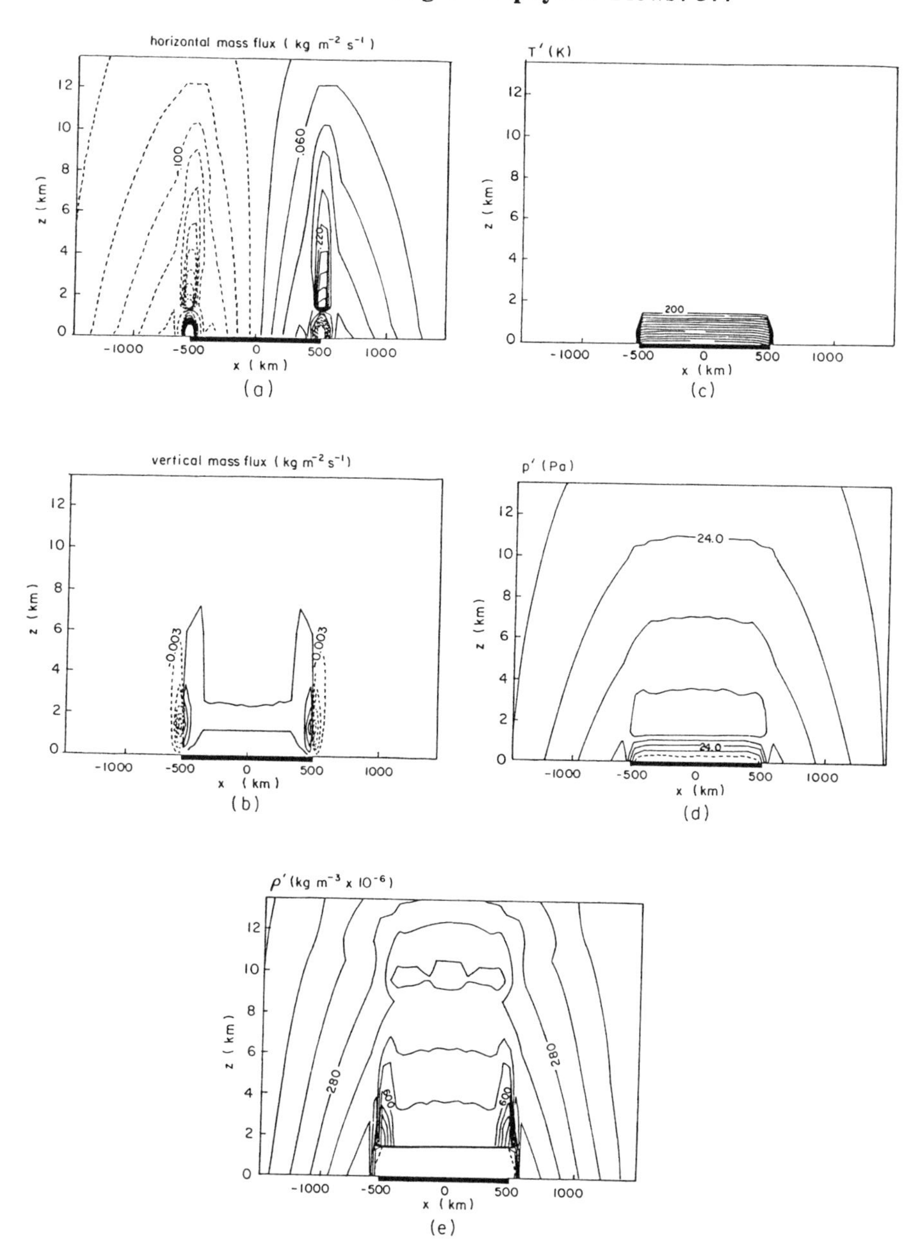

Fig. 12 Results for the heated land surface at $t = 3600$ s. (a) Horizontal mass flux. The contour interval is 0.04 kg m^{-2} s^{-1}. (b) Vertical mass flux. The contour interval is 0.006 kg m^{-2} s^{-1}. The label scale is 10^4. (c) Perturbation temperature. The contour interval is 0.4 K. (d) Perturbation pressure. The contour interval is 16 Pascals. (e) Perturbation density. The contour interval is 80×10^{-6} kg m^{-3}. The label scale is 10^6.

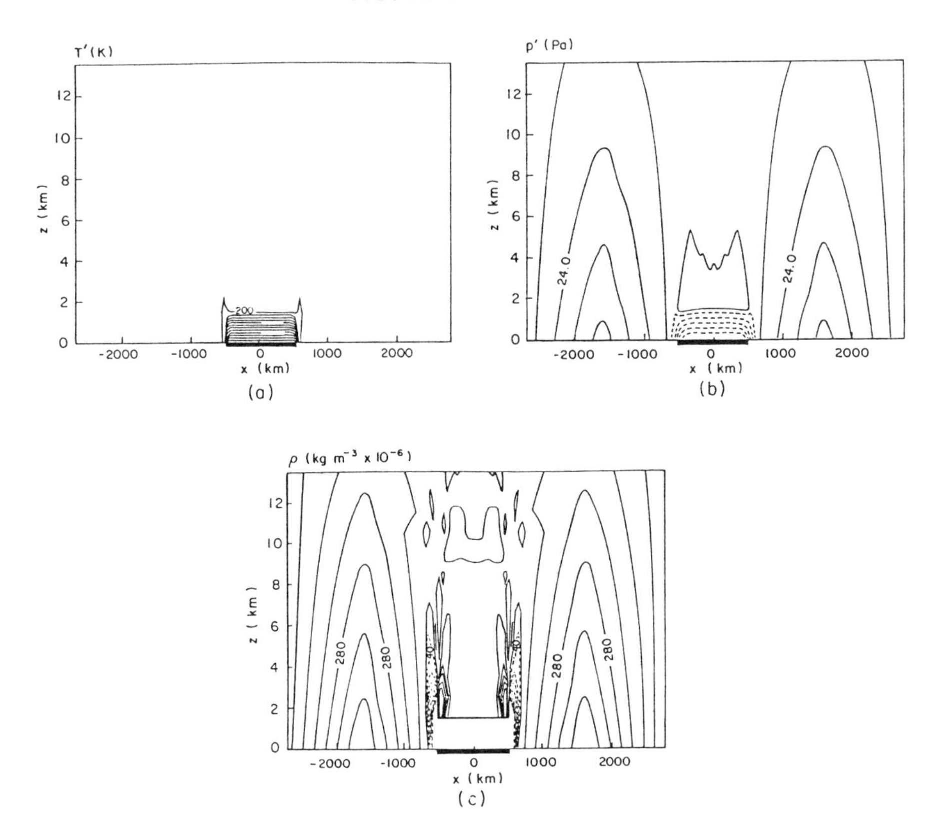

Fig. 13 Results for the heated land surface at t = 7200 s. (a) Perturbation temperature. (b) Perturbation pressure. The contour interval is 16 Pascals. (c) Perturbation density. The contour interval is 60×10^{-6} kg m^{-3}. The label scale is 10^6.

Figures 12a, b, c, d and e show the horizontal mass flux, vertical mass flux, perturbation temperature, perturbation pressure and perturbation density, respectively, at t = 3600 s. Note that the contour intervals have been increased from those used at t = 900 s. The sea-breeze circulations have strengthened considerably, the surface temperature has continued to increase, and the surface pressure has decreased over the whole of the land surface. The perturbation density shows relatively strong positive perturbations at the top of the sea-breeze circulation updrafts.

After an hour the heating is terminated. Figures 13a, b and c show the perturbation temperature, perturbation pressure and perturbation density, respectively, at t = 7200 s. The perturbation temperature field is very similar to that an hour earlier. The perturbation pressure field shows that two oppositely moving compression waves have formed, leaving an enhanced region of low pressure at the centre of the domain compared with an hour earlier. The on-shore directed pressure gradient continues to drive sea-breeze circulations (not shown). The perturbation density within the compression waves is largest near the surface. The strongest perturbations occur within the heated region and in the sea-breeze circulations. The horizontally propagating compression waves are Lamb waves with mass and pressure fields in hydrostatic balance.

Latent heat release within thunderstorms can be expected to be a source of thermal compression waves. The amplitude of the perturbations associated with these waves should be directly related to the latent heat release. It may be possible to detect these waves hundreds of kilometres from the thunderstorm using arrays of infrasound detectors. Preliminary simulations suggest that a rapidly developing convective cell would result in a pressure jump of $\approx$0.02 mb at $\approx$200 km from a thunderstorm. The time scale for the pressure jump to occur should be approximately the time taken for an intense convective cell to develop ($\approx$20 minutes). Due to the spatial and temporal variability of latent heat release in thunderstorms one would expect there also to be considerable higher frequency components of the signal superimposed on this low frequency component. It is possible that these higher frequency components have already been observed in the studies by Beasley et al. (1976) and Bedard et al. (1986). Numerical modelling studies could provide valuable information on the power spectra of thermally-induced infrasound waves emanating from thunderstorms and aid in the design of infrasound arrays to detect them.

6 Conclusions

This paper discusses several issues in the numerical modelling of atmospheric flow which have been left unresolved. These include the influence of partial model resolution of features that are represented in a model by only a few grid points and the implication of this inadequacy in terms of the parametrization of subgrid-scale processes. As a result of nonlinear interactions, this lack of resolution requires the introduction of new parametrization techniques such as "backscatter", as well as an assessment of the predictability and parametrizability of atmospheric flows as a result of the limited spatial resolution. Examples of these issues are presented in the paper. Among the conclusions is that for large eddy simulations a stochastic subgrid-scale parametrization is needed, while for well-defined surface-forced mesoscale systems such as sea breezes, an ensemble-based subgrid-scale representation is adequate. Also introduced in this paper is an additional physical realism to atmospheric models (i.e., the compressibility effect) which provides an additional modelling degree of freedom to internal and externally imposed perturbations.

7 Acknowledgments

R. Pielke and R. Walko contributions were supported by National Science Foundation (NSF) Contract #ATM-9306754. M. Nicholls was supported by NSF Cooperative Agreement #BCS-8821542 with Texas Tech. University and a Grant from the Department of Defense through the U.S. Army Research Office for A Center for

Geosciences Research – Phase II from the Cooperative Institute for Research in the Atmosphere. T. Nygaard was partially supported by the Institut for Energiteknikk in Norway and by the Norwegian Research Council. X. Zeng was partially supported by the Department of Energy Grant DE-FG02-91ER61216 to R.E. Dickinson. The authors would like to thank Bryan Critchfield and Dallas McDonald who prepared the manuscript.

References

BANNON, P.R. 1995. Hydrostatic adjustment: Lamb's problem. *J. Atmos. Sci.* **52**: 1743–1752.

BEASLEY, W.H., T.M. GEORGES and M.W. EVANS. 1976. Infrasound from convective storms: An experimental test of electrical source mechanisms. *J. Geophys. Res.* **81(18)**: 3133–3140.

BEDARD, A.J., JR.; J. INTRIERI and G.E. GREENE. 1986. Infrasound originating from regions of severe weather. Paper J2-5 presented at the 12th International Congress on Acoustics, Toronto, Canada.

CHASNOV, J.R. 1991. Simulation of the Kolmogorov inertial subrange using an improved subgrid model. *Phys. Fluids*, **A3**: 188–200.

COSTIGAN, K.R. 1992. Large eddy simulations of the atmospheric boundary layer east of the Colorado Rockies. Ph.D. Thesis, Department of Atmospheric Science, Colorado State University, 142 pp.

DERICKSON, R. 1992. Finite difference methods in geophysical flow simulations. Ph.D. Thesis, Department of Civil Engineering, Colorado State University, 320 pp.

GAGE, K.S. 1979. Evidence of a $k^{-5/3}$ low inertial range in mesoscale two-dimensional turbulence. *J. Atmos. Sci.* **36**: 1950–1954.

LAPRISE, R. 1992. The resolution of global spectral models. *Bull. Am. Meteorol. Soc.* **73**: 1453.

LEONARD, A. 1974. Energy cascade in large-eddy simulations of turbulent fluid flows. *In*: Proc. Symposium on Turbulent Diffusion in Environmental Pollution, 8–14 April 1973, F.N. Frenkiel and R.E. Munn (Eds), *Adv. Geophysics*, **18A**: 237–248.

LESLIE, D.C. and G.L. QUARINI. 1979. The application of turbulence theory to the formulation of subgrid modelling procedures. *J. Fluid. Mech.* **91**: 65–91.

LILLY, D.K. 1967. The representation of small-scale turbulence in numerical simulations experiments. *In*: Proc. 10th IBM Scientific Computing Symp. on Environmental Sciences, Thomas J. Watson Research Centre, Yorktown Heights, pp. 195–210.

———. 1983. Stratified turbulence and the mesoscale variability of the atmosphere. *J. Atmos. Sci.* **40**: 749–761.

LUMLEY, J.L. and H.A. PANOFSKY. 1964. *The structure of atmospheric turbulence*. Wiley and Sons, New York, 239 pp.

MASON, P.J. and D.J. THOMSON. 1987. Large-eddy simulations of the neutral-static-stability planetary boundary layer. *Q. J. R. Meteorol. Soc.* **113**: 413–443.

——— and ———. 1992. Stochastic backscatter in large-eddy simulations of boundary layers. *J. Fluid Mech.* **242**: 51–78.

MORAN, M.D. 1992. Numerical modelling of mesoscale atmospheric dispersion. Ph.D. Thesis, Colorado State University, 758 pp.

NICHOLLS, M.; R. PIELKE and R. MERONEY. 1993. Large eddy simulation of microburst winds flowing around a building. *J. Wind Eng. Indus. Aerodyn.* **46 & 47**: 229–237.

———; ———, J.L. EASTMAN, C.A. FINLEY, W.A. LYONS, C.J. TREMBACK, R.L. WALKO and W.R. COTTON. 1993. Applications of the RAMS numerical model to dispersion over urban areas. In: *Wind Climate in Cities*, J.E. Cermak et al. (Eds), Kluwer Academic Publishers, The Netherlands, pp. 703–732.

——— and ———. 1994a. Thermal compression waves. I: Total energy transfer. *Q. J. R. Meteorol. Soc.* **120**: 305–332.

——— and ———. 1994b. Thermal compression waves. II: Mass adjustment and vertical transfer of total energy. *Q. J. R. Meteorol. Soc.* **120**: 333–359.

PIELKE, R.A. 1974. A three-dimensional numerical model of the sea breezes over south Florida. *Mon. Weather Rev.* **102**: 115–139.

———. 1984. *Mesoscale Metereological Mod-*

elling. Academic Press, New York, New York, 612 pp.

———. 1991. A recommended specific definition of ' resolution". *Bull. Am. Meteorol. Soc.* **72**: 1914.

———; W.R. COTTON, R.L. WALKO, C.J. TREMBACK, W.A. LYONS, L.D. GRASSO, M.E. NICHOLLS, M.D. MORAN, D.A. WESLEY, T.J. LEE and J.H. COPELAND. 1992. A comprehensive meteorological modeling system – RAMS. *Meteorol. Atmos. Phys.* **49**: 69–91.

———; M.E. NICHOLLS and A.J. BEDARD. 1993. Using thermal compression waves to assess latent heating from clouds. *EOS*, **74**: 493.

ROBERT, A. 1982. A semi-Lagrangian and semi-implicit numerical integration scheme for the primitive meteorological equations. *J. Meteorol. Soc. Jpn.* **60**: 319–325.

SMAGORINSKY, J. 1963. General circulation experiments with the primitive equations: 1. The basic experiment. *Mon. Weather Rev.* **91**: 99–164.

STANIFORTH, A. and J. CÔTÉ. 1991. Semi-Lagrangian integration schemes of atmospheric models – A review. *Mon. Weather Rev.* **119**: 2206–2223.

TANGUAY, M.; E. YAKIMIW, H. RITCHIE and A. ROBERT. 1992. Advantages of spatial averaging in semi-implicit semi-Lagrangian schemes. *Mon. Weather Rev.* **120**: 113–123.

TENNEKES, H. 1978. Turbulent flow in two and three dimensions. *Bull. Am. Meteorol. Soc.* **59**: 22–28.

ZENG, X. and R.A. PIELKE. 1993. Error-growth dynamics and predictability of surface thermally-induced atmospheric flow. *J. Atmos. Sci.* **50**: 2817–2844.

——— and ———. 1995a. Further study on predictability of surface thermally-induced circulations. *J. Atmos. Sci.* **52**: 1680–1698.

——— and ———. 1995b. Landscape-induced atmospheric flow and its parameterization in large-scale numerical models. *J. Clim.* **8**: 1156–1177.

Supplement

This section contains four previously unpublished manuscripts written by Dr. Robert. They are all pedagogical in nature. The editors feel that these papers represent a fitting way to conclude this memorial volume.

The first paper, "An anomaly in the behaviour of a time filter used with the leapfrog scheme in atmospheric models", was written jointly with Mario Lépine in 1986. The remaining three papers were authored solely by Dr. Robert. They were retrieved by René Laprise from Dr. Robert's documents at the Université du Québec à Montréal. The paper "Incompressible homogeneous fluids" gives a brief overview of the numerical integration of the governing equations of incompressible homogeneous fluids.

Two short papers are reproduced in their original handwritten French version, as prepared by Dr. Robert in 1988. English translations of these papers follow the French versions. They deal with different aspects of the semi-Lagrangian scheme, with titles "Erreurs induites par le schème Lagrangien dans les ondes stationnaires" and "Comportement du schème Lagrangien dans une équation différentielle relativement simple". These papers exemplify Dr. Robert's ability to distill a problem to its essence using simple equations.

Supplément

Cette section contient quatre manuscrits inédits du Dr Robert. Ils sont tous de nature pédagogique. Les éditeurs pensent que ces articles représentent une façon appropriée de conclure ce volume commémoratif.

Le premier article, "An anomaly in the behaviour of a time filter used with the leapfrog scheme in atmospheric models", a été écrit conjointement avec Mario Lépine en 1986. Le Dr Robert est l'auteur unique des trois autres articles. Ils ont été récupérés par René Laprise des documents du Dr Robert à l'Université du Québec à Montréal. L'article intitulé "Incompressible homogeneous fluids" fait un bref survol de l'intégration numérique des équations régissant les fluides homogènes incompressibles.

Nous reproduisons deux petits manuscrits dans leur version originale française, tels que préparés par le Dr Robert en mai 1988. La traduction anglaise de ces manuscrits suit les versions françaises. Ils traitent de différents aspects du schème semi-Lagrangien, avec les titres "Erreurs induites par le schème Lagrangien dans les ondes stationnaires" et "Comportement du schème Lagrangien dans une équation différentielle relativement simple". Ces articles démontrent le talent du Dr Robert pour distiller l'essence d'un problème à l'aide d'équations simples.

An Anomaly in the Behaviour of the Time Filter Used with the Leapfrog Scheme in Atmospheric Models

André Robert and Mario Lépine
Recherche en prévision numérique
Atmospheric Environment Service
2121 Trans-Canada Highway
Dorval, QC, Canada H9P 1J3

[manuscript written 1986, with revisions by Andrew Staniforth and Harold Ritchie, 24 April 1995]

ABSTRACT *Integration techniques such as the explicit leapfrog scheme or the semi-implicit method carry out their calculations over three time levels. Because of this characteristic, these algorithms can produce computational modes that appear as superimposed onto the physical solution of the meteorological equations. These computational modes can cause problems and in order to avoid these problems or remove them, a common practice is to use a time filter. The time filter that is frequently used for this purpose efficiently damps the computational mode while it has little effect on the physical solution.*

This filter does not always behave exactly as expected and in some cases, strange results are observed. A series of simple numerical experiments are carried out in order to show that the filter can display an anomalous behaviour under certain conditions. An attempt is made to explain this behaviour by numerical analysis. This analysis gives some information as to how to avoid this anomalous behaviour in atmospheric models.

RÉSUMÉ *Les techniques d'intégration telles que le schéma explicite saute-mouton ou la méthode semi-implicite effectuent leurs calculs sur trois niveaux temporels. À cause de ce fait, ces algorithmes peuvent engendrer des modes numériques qui se superposent à la solution physique des équations météorologiques. Ces modes numériques peuvent causer des problèmes et, afin de les éliminer, une pratique courante consiste en l'utilisation d'un filtre temporel. Le filtre temporel qui est utilisé fréquemment dans ce but amortit efficacement le mode numérique tout en ayant peu d'effet sur la solution physique.*

Ce filtre ne se comporte pas toujours exactement comme prévu et, dans certains cas, on observe des résultats étranges. Une série d'expériences numériques simples ont été réalisées afin de montrer que le filtre peut présenter un comportement anormal sous certaines conditions. On utilise l'analyse numérique pour tenter d'expliquer ce comportement. L'analyse nous renseigne sur la façon d'éviter ce comportement anormal dans les modèles atmosphériques.

S4 / André Robert and Mario Lépine

1 Introduction

In the early years of numerical weather prediction, it was common practice to use time integration schemes that were as simple as possible. A wide variety of simple schemes were used and most of these turned out to be dissipative. This was a useful property at a time when it was hard to achieve stable integrations.

As time evolved, better spatial finite difference formulations were developed and eventually it was found that stability was no longer a major problem. In stable models, it was found that the dissipative characteristics of time integration schemes were generally too strong. During this period some models started using the so-called leapfrog scheme and gradually this algorithm became popular for a number of reasons. Among these reasons, there is the fact that the amplitude of a pure oscillation is preserved perfectly well when the leapfrog scheme is used. This absence of computational damping is a property that is highly desirable when long integrations are considered.

Another characteristic of this algorithm is that the right-hand sides of the meteorological tendency equations need to be computed only once per time step. This is in contrast with most other stable schemes where the right hand side must be computed twice per time step. For this reason, the leapfrog scheme is considered to be very efficient. On the other hand, the calculations associated with the leapfrog scheme involve three time levels, unlike other methods that generally involve only two time levels.

When a three time level scheme is applied to a simple set of linear differential equations, analysis shows that the integration scheme generates two modes of motion for every mode generated by the differential equations. One of these modes is *physical*, meaning that it corresponds closely to one of the solutions permitted by the differential equations while the other mode is *computational*, implying that it is artificial and bears no relation to the differential equations that are being integrated. The presence of computational modes may not even be noticed in some integrations but on occasion, integrations will be carried out where these are found to be a real plague. In atmospheric models, the computational modes are generally hard to detect in short integrations, but as soon as we decide to carry out integrations beyond a few days, the computational modes become a problem.

In a long integration, Robert (1966) encountered computational modes and proposed a time filter designed to eliminate this noise without significantly affecting the meteorological waves. This filter was investigated by Asselin (1972) and used later by Bleck (1974), Haltiner and McCullock (1975), Daley et al. (1976) and Gordon and Stern (1982). We can also find two additional references to this filter in a study carried out by Lange and Hellsten (1986) indicating that it is used in the global model of the European Centre for Medium-Range Weather Forecasts and also in the spectral model of the French Meteorological Service. Two additional studies of this filter were published recently by Schlesinger et al. (1983) and Dèque and Cariolle (1986).

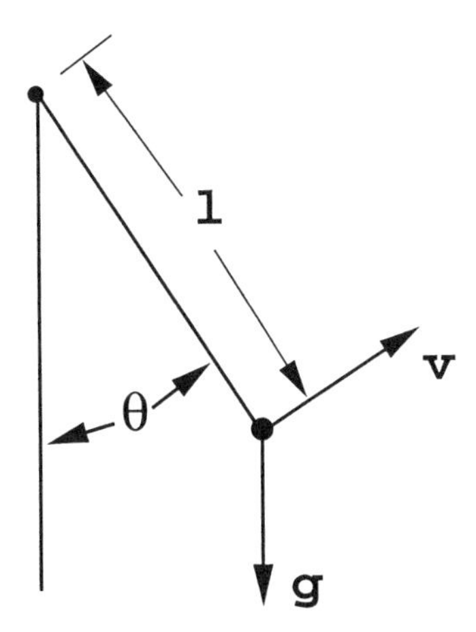

Fig. 1 Variables and constants used to represent the motion of a pendulum. v is the velocity, θ is the angle formed with the local vertical, l is the length of the rod and g is the acceleration of gravity.

In the following sections, we will apply this filter to a simple problem and we will show that it can produce unexpected results. We will attempt to analyze the observed behaviour and relate this to the characteristics of the filter. In conclusion, we will find that it is necessary to be careful when using this filter if we want the integrations to be meaningful.

2 Integrations with a time filter

In order to train young people, it is generally advisable initially to avoid the use of sophisticated and complicated atmospheric models because the study of problems in the context of these models is always quite confusing. Furthermore, the problems that occur in big models can nearly always be reproduced in simple differential equations where they can be investigated with great ease. This is the method that will be used here and it will be applied to the differential equations that describe the motion of a simple pendulum. The corresponding equations will be given as follows:

$$\frac{dv}{dt} = -g \sin \theta \tag{1}$$

$$\frac{d\theta}{dt} = \frac{v}{l}. \tag{2}$$

In these equations, v is the velocity of the pendulum and θ is the angle that it forms with the local vertical as shown in Fig. 1. The thin rod has a length l and g is the acceleration of gravity. If we use centred finite differences to evaluate the derivatives, we get the following approximation.

$$\frac{v(t+\Delta t) - v(t-\Delta t)}{2\Delta t} = -g \sin \theta(t) \tag{3}$$

$$\frac{\theta(t+\Delta t) - \theta(t-\Delta t)}{2\Delta t} = \frac{v(t)}{l}. \tag{4}$$

For simplicity, we will use as argument the counter n that is used to count the time steps and we will rewrite these equations as follows:

$$v_{n+1} = v_{n-1} - 2g\,\Delta t \sin \theta_n \tag{5}$$

$$\theta_{n+1} = \theta_{n-1} + \frac{2\Delta t}{l} v_n. \tag{6}$$

We now have the familiar form of the leapfrog scheme applied to both equations. These are recursive equations that can be used to compute new values of v and θ from values available at two time levels. In order to start the integration, we note that we normally have v_0 and θ_0 at initial time. A frequent practice is to use a forward scheme to generate another pair of values:

$$v_1 = v_0 - g\,\Delta t \sin \theta_0 \tag{7}$$

$$\theta_1 = \theta_0 + \frac{\Delta t}{l} v_0. \tag{8}$$

The recursive equations (5) and (6) can now be used to generate as many values as we wish. This procedure will immediately be applied to examine a case of gradual decoupling that is quite common in atmospheric models. For this case, we will use $g = 10$ m s^{-2}, $l = 0.9$ m, $\Delta t = 0.05$ s, $v_0 = 5$ m s^{-1} and $\theta_0 = 0$.

The result is given in Fig. 2. Two separate curves are starting to show up after 10 s. The values of v at even time steps form one curve while the values at odd time steps form the other curve. These curves are completely out of phase after 85 s and they come back in phase after 175 s. This phenomenon is a characteristic of the leapfrog integration scheme. This is the visual aspect taken by the computational mode that was mentioned earlier.

The application of a time filter represents one out of a variety of techniques that can be used to control computational modes. The particular filter proposed by Robert (1966) gives the following set of equations:

$$v^*_{n+1} = v_{n-1} - 2g\,\Delta t \sin \theta^*_n \tag{9}$$

$$\theta^*_{n+1} = \theta_{n-1} + \frac{2\Delta t}{l} v^*_n \tag{10}$$

$$v_n = v^*_n + \alpha(v^*_{n+1} + v_{n-1} - 2v^*_n) \tag{11}$$

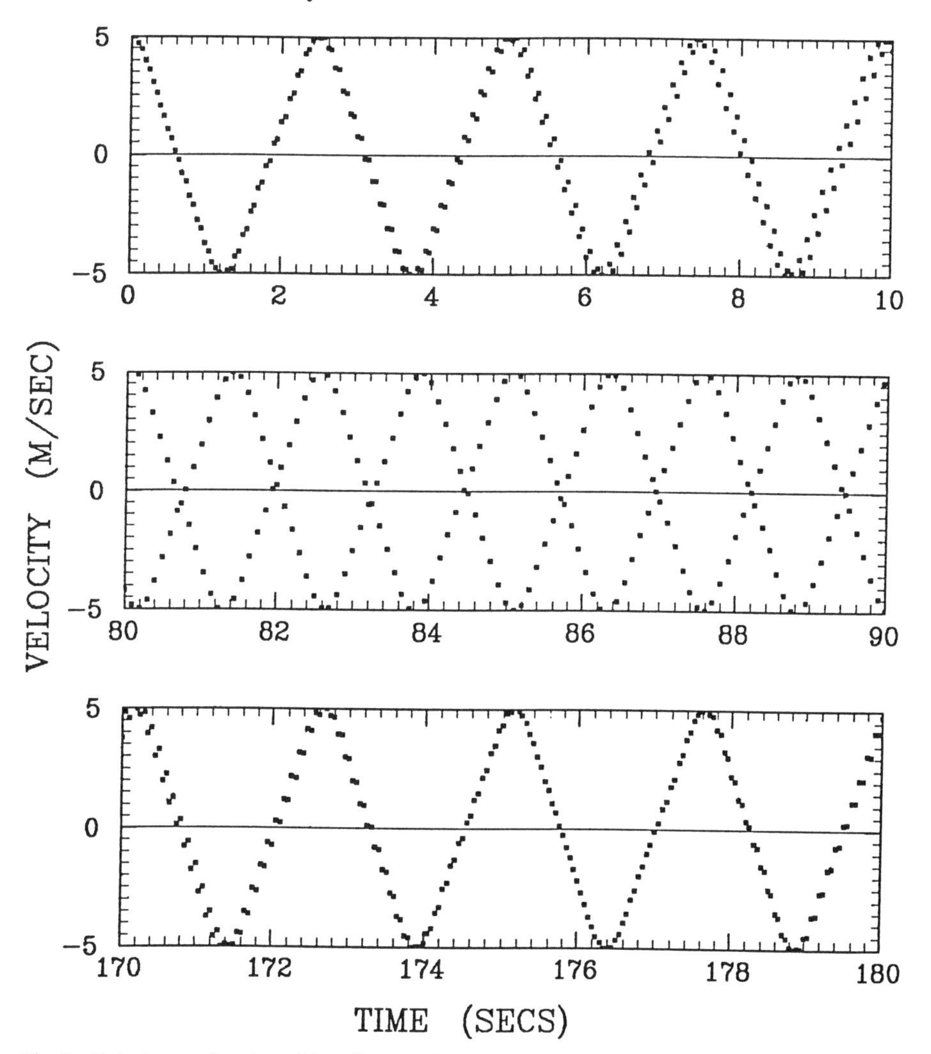

Fig. 2 Velocity as a function of time for an unfiltered leapfrog integration with a time step of 0.05 s. Initial velocity is 5 m s^{-1}.

$$\theta_n = \theta_n^* + \alpha(\theta_{n+1}^* + \theta_{n-1} - 2\theta_n^*). \tag{12}$$

In these equations, the variables with an asterisk represent intermediate values that are retained for only two time steps. Here, the usual leapfrog scheme represented by (9) and (10) is used to generate intermediate or temporary values at time $n+1$ while the filter represented by (11) and (12) is used to finalize the values at time n.

We will apply this technique to our last integration using $\alpha = 0.01$. The cor-

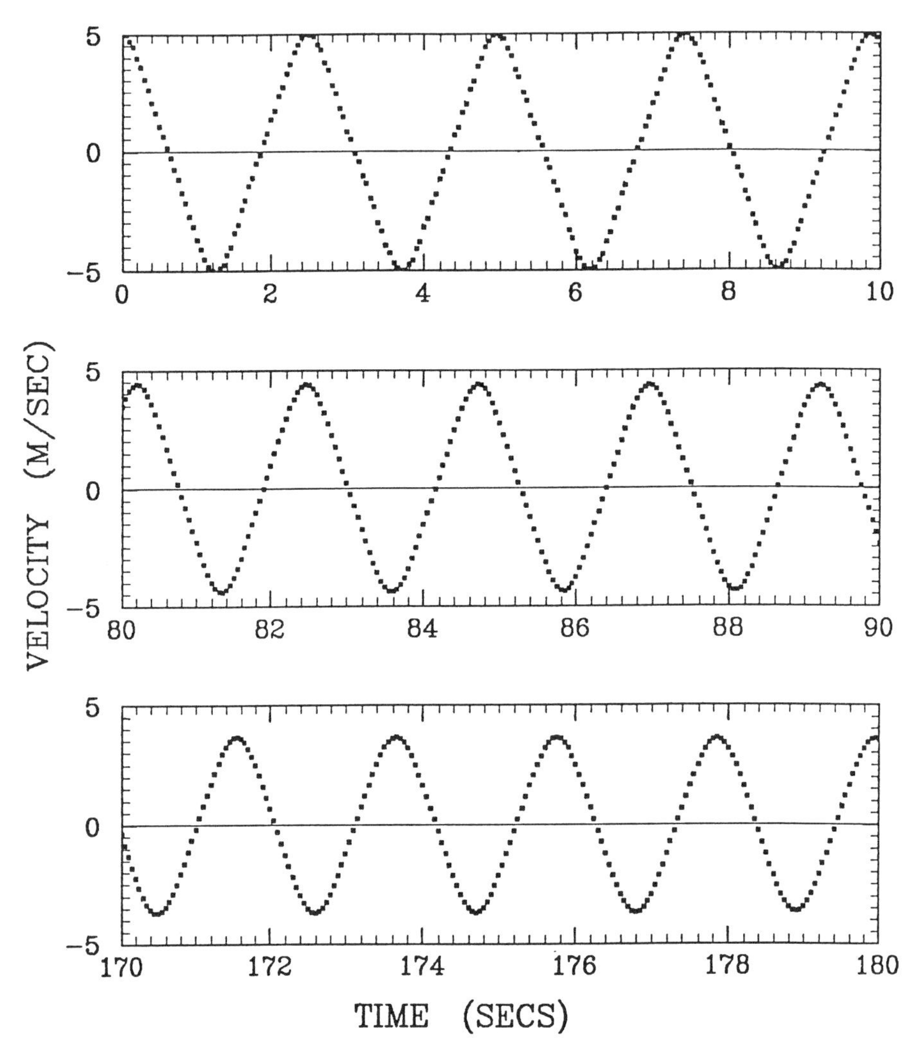

Fig. 3 Same as Fig. 2 with a time filter using a coefficient of $\alpha = 0.01$.

responding result is presented in Fig. 3. We can easily see that the computational mode has been completely eliminated. On the other hand, we can see also, that the filter has affected the physical mode. The amplitude of the oscillation of the velocity has been reduced by 30%. This reduction in amplitude may or may not be acceptable depending on the standard that we want to achieve.

Since the time filter has an arbitrary coefficient α, we can either increase or reduce this coefficient to control the amount of damping that will take place during the integration. A run will be performed with a smaller value $\alpha = 0.001$ in order

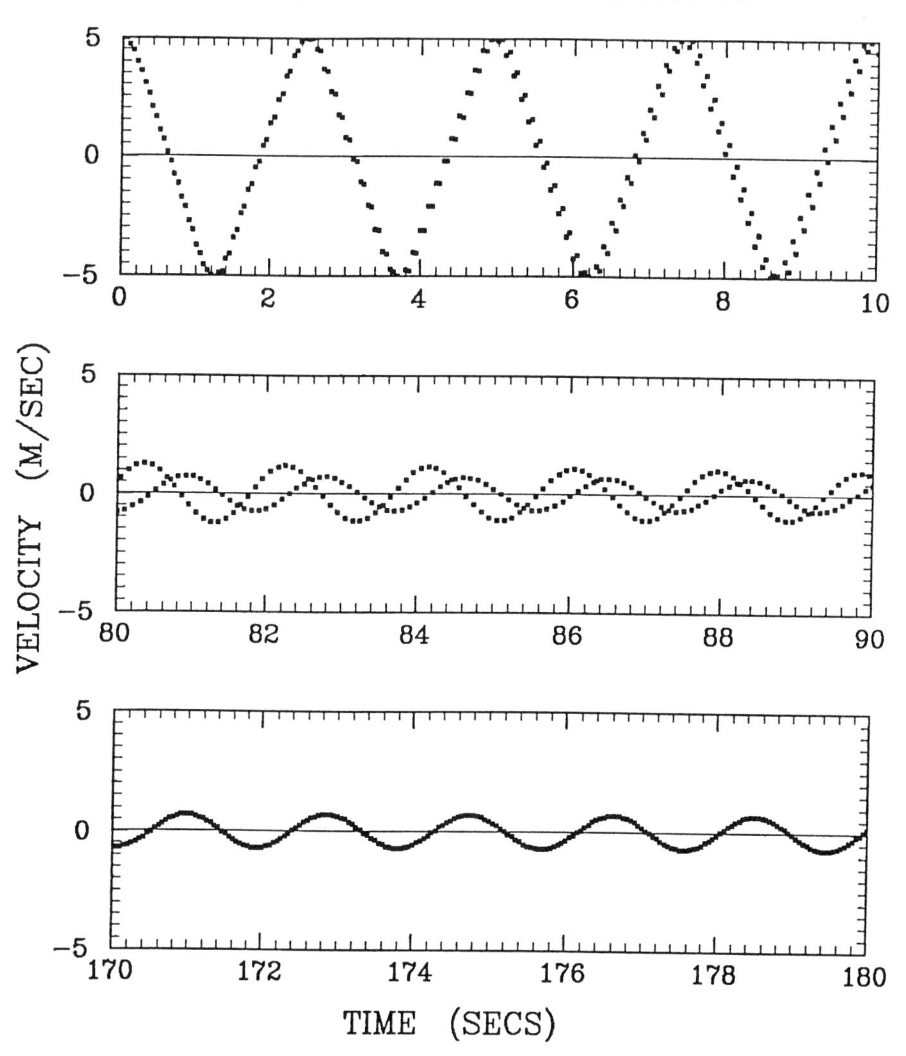

Fig. 4 Same as Fig. 2 with a time filter using a coefficient of $\alpha = 0.001$.

to reduce the damping of the physical mode. The resulting curve is presented in Fig. 4. Surprisingly, we observe a much stronger damping of the physical mode. Near the end of the integration, the amplitude of the oscillation is less than 1 m s^{-1}. This behaviour is unexpected and indicates that we should be careful when we use this time filter.

3 Analysis of the filter

At this point, we will return to the recursive equations given in (5) and (6). It should be noted that these equations contain two completely separate solutions. One of these solutions is formed from the values of v at even time steps and the values of θ at odd time steps. Conversely, the other solution is formed from the values of v at odd time steps and the values of θ at even time steps. These two solutions are completely independent of each other. With an appropriate staggering of the variables in time we could easily eliminate one of the solutions without affecting the other one.

As shown in Fig. 2 both solutions oscillate with an amplitude of 5 m s^{-1} but their periods are different. One of these goes through 70 cycles in 175 seconds while the other one completes 71 full cycles. This is the important feature that must be accounted for in the analysis. For simplicity, we will assume that we have pure oscillations with frequencies ω_1 and ω_2.

$$\frac{\partial F}{\partial t} = i\omega_1 F \tag{13}$$

$$\frac{\partial G}{\partial t} = i\omega_2 G. \tag{14}$$

In these equations, the variables F and G are independent of each other. In the numerical integration of these equations, a coupling will be established if we proceed as we did for the pendulum. That is, we consider a time-series consisting of two pure oscillations with slightly different frequencies, which get coupled together through a time filtering of the time-series.

$$F^*_{n+1} = F_{n-1} + 2iW_1 F^*_n \tag{15}$$

$$G^*_{n+1} = G_{n-1} + 2iW_2 G^*_n \tag{16}$$

$$F_n = F^*_n + \alpha(G^*_{n+1} + G_{n-1} - 2F^*_n) \tag{17}$$

$$G_n = G^*_n + \alpha(F^*_{n+1} + F_{n-1} - 2G^*_n) \tag{18}$$

where

$$W_1 = \omega_1 \Delta t \tag{19}$$

$$W_2 = \omega_2 \Delta t. \tag{20}$$

The coupling takes place in the time filter as evidenced in (17) and (18). The recursive equations given can be expressed in terms of an operator X defined in such a way that

$$P_{n+1} = XP_n \tag{21}$$

for each of the four variables F, G, F^* and G^*. Dropping the subscript n, we may then rewrite (15) to (18) as follows:

$$XF^* = X^{-1}F + 2iW_1F^* \tag{22}$$

$$XG^* = X^{-1}G + 2iW_2G^* \tag{23}$$

$$F = F^* + \alpha(XG^* + X^{-1}G - 2F^*) \tag{24}$$

$$G = G^* + \alpha(XF^* + X^{-1}F - 2G^*). \tag{25}$$

This is an homogeneous system of algebraic equations. It will produce a non-trivial solution only if the determinant of these equations vanishes. This property generates the following frequency equation:

$$(X^2 - 2iWX - 1 + 2\alpha)^2 = 4q^2 \tag{26}$$

where

$$W = \tfrac{1}{2}(W_1 + W_2) \tag{27}$$

$$q^2 = \alpha^2[(X - iW)^2 + \delta^2] - X^2\delta^2 \tag{28}$$

$$\delta = \tfrac{1}{2}(W_1 - W_2). \tag{29}$$

If we consider only small values of α and δ, then q will also be a small quantity and we can produce the following approximate solution

$$X = e^{i\theta} - \frac{\alpha \pm q}{\cos\theta} \tag{30}$$

where

$$\sin\theta = W \tag{31}$$

$$q^2 = \alpha^2(\delta^2 + \cos^2\theta) - \delta^2 e^{2i\theta}. \tag{32}$$

For small values of W, the damping rate D associated with this wave can be approximated by:

$$D = \frac{\alpha \pm [\alpha^2(1 + \delta^2 - \frac{1}{2}W^2) - \delta^2(1 - 2W^2)]^{1/2}}{\Delta t}. \tag{33}$$

This expression is valid only if α and δ are significantly different from each other. At this point, we will try to represent D as a function of α. For this purpose,

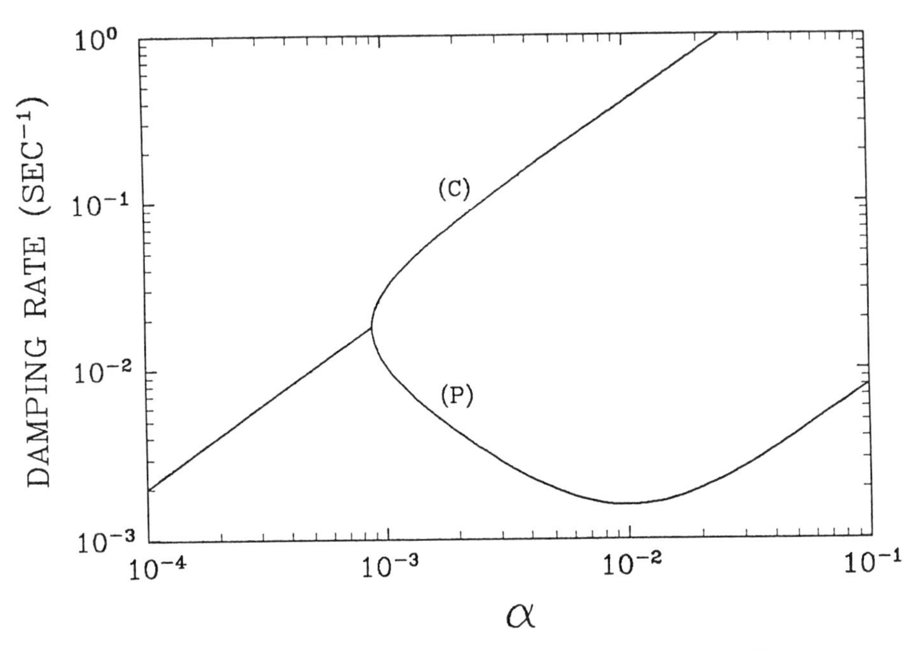

Fig. 5 Damping rate as a function of the filtering coefficient α. The upper branch (C) represents the computational mode and the lower branch (P) represents the physical mode.

we will use

$$W = \frac{2\pi}{50} \tag{34}$$

$$\delta = \frac{1}{140} W. \tag{35}$$

These values are taken from Fig. 2. With these values and the corresponding Δt, we produce the curve presented in Fig. 5. For the larger values of α, we have two values of D, the larger value represented by (C) in Fig. 5 applies to the computational mode while the smaller value (P) applies to the physical mode. For $\alpha < 10^{-3}$, there is only one value of D that applies to both waves.

When we increase the coefficient of the filter (α), the damping rate of the computational mode increases. This is not always true for the physical mode. When we increase this coefficient from $\alpha = 10^{-3}$, to $\alpha = 10^{-2}$, the damping rate drops by an order of magnitude. This is the anomalous behaviour that was depicted in Fig. 3.

We can see from Fig. 5 that there is an optimum value ($\alpha = 10^{-2}$) for which the impact on the physical mode is a minimum. In general, this minimum occurs near the value

$$\alpha_0 = \frac{\delta\sqrt{2}}{W} = \frac{0.35}{N_c} \tag{36}$$

where N_C represents the number of elapsed cycles at the time when the computational mode reaches its maximum amplitude. In the integration presented in Fig. 2, it took 35 cycles to reach this point. In order to avoid problems, we should always use $\alpha > \alpha_0$.

In atmospheric models, the computational mode rarely reaches its maximum amplitude in less than 20 cycles so that with $\alpha > 0.02$ there should be no problem. In the formulation given by Asselin (1972), this corresponds to a value of 0.04. The smallest value reported in the literature seems to be the value of 0.06 used by Gordon and Stern (1982).

4 Experimental verification

Our analysis seems to confirm the results presented in Figs 2 and 3. In order to get a better picture of the relative validity of this analysis, we will compare it to the results of several integrations with different values of α. For each value of α, we will estimate the residual amplitude after 175 seconds and we will plot this value on a graph. This gives us the series of points appearing in Fig. 6. The corresponding computed amplitude of the physical mode is also presented as a solid line. This amplitude is computed from Eqs (30)–(32) without making any assumption on W as was done to generate (33).

The solid line has the same overall shape as the curve produced by the experimental values but it differs in the details. In order to attempt to explain these differences, it should be noted that the period of the pendulum is a function of the amplitude of the oscillation. For an amplitude of 5 m s^{-1}, the period of 2.5 s used in (34) is representative, but for low amplitudes this period reduces to 1.9 s.

The constant δ associated with the computational mode is also a problem. In Fig. 2, we found that the computational mode reaches its maximum amplitude after 35 cycles. If we run with $\alpha = 0.01$ for 50 seconds and then continue the integration without a time filter in order to re-evaluate δ, we find that its value has dropped by a factor of 3.

Since both W and δ vary in the numerical integrations, we must be careful in interpreting the results of an analysis that assumes that these quantities are constants. On the other hand, the approximate nature of our analysis does not alter our conclusion that the time filter does not work well for small values of α.

5 Concluding remarks

The time filter described and discussed in the preceding sections effectively damps the computational modes associated with the leapfrog integration scheme and has virtually no effect on the physical modes. This is what we find from practical experience in using this filter.

If we have a close look at the integrations, especially in cases where there are

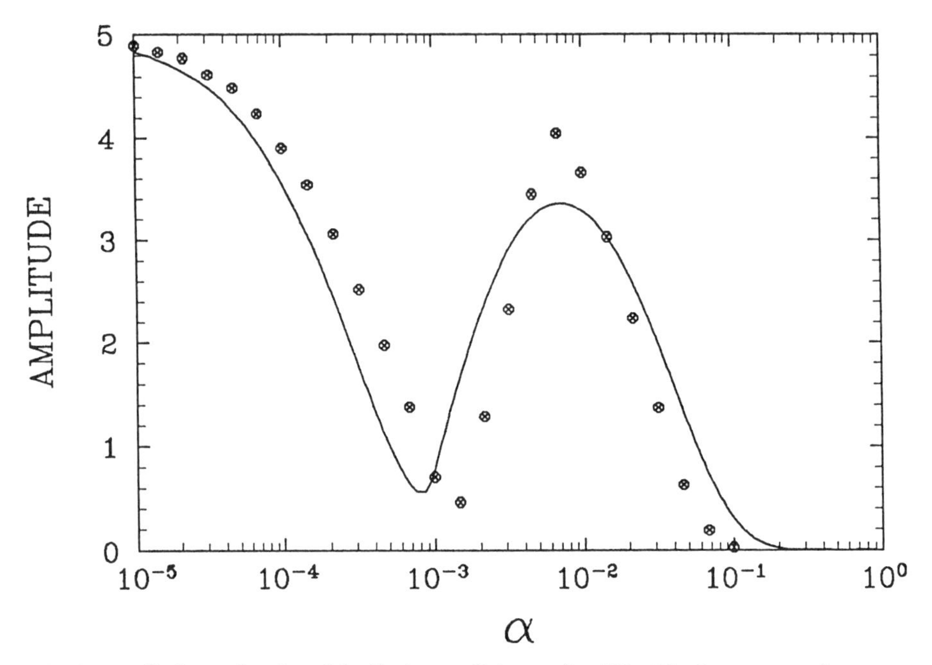

Fig. 6 Amplitude as a function of the filtering coefficient α after 175 s. The dots represent the experimental values and the full line represents the analytical results.

no other forms of damping in the model, then we find that the physical modes are damped slightly. In these situations, one might be tempted to reduce the coefficient of the filter. This is an action that must be given careful consideration. It must be realized that at some point, as we keep on reducing α, the damping of the physical mode will start being enhanced, which is exactly the opposite of what is being sought.

Normally, the damping associated with the time filter is fairly small compared to the damping associated with some of the physical processes that are included in the model. In other words, the time filter introduces only small inaccuracies in numerical integrations. The level of these errors cannot be controlled easily but this is of little importance as long as these errors remain at an acceptable low level. In any experiment where these errors cannot be reduced to the desired level, the time filter should be either abandoned or applied in conjunction with other techniques that can be used to control computational modes.

References

ASSELIN, R. 1972. Frequency filter for time integration. *Mon. Weather Rev.* **100**: 487–490.

BLECK, R. 1974. Short-range prediction in isentropic coordinates with filtered and unfiltered numerical models. *Mon. Weather Rev.* **102**: 813–829.

DALEY, R.; GIRARD, C., HENDERSON, J. and I. SIMMONDS. 1976. Short term forecasting with a multi-level spectral primitive equation model. *Atmosphere*, **14**: 98–134.

DÈQUE, M. and D. CARIOLLE. 1986. Some destabilizing properties of the Asselin time filter. *Mon. Weather Rev.* **114**: 880–884.

GORDON, C.T. and W.F. STERN. 1982. A description of the GFDL global spectral model. *Mon. Weather Rev.* **110**: 625–644.

HALTINER, G.J. and J.M. MCCULLOCH. 1975. Experiments in the initialization of a global primitive equation model. *J. Appl. Meteorol.* **14**: 281–288.

LANGE, A. and E. HELLSTEN. 1986. Results of the WMO/CAS NWP data study and intercomparison project for forecasts for the Northern Hemisphere in 1994. PSMP Report Series No. 16, WMO Technical Document No. 60, World Meteorological Organization, Case postale No. 1, CH-1211 Geneva 20, Switzerland, 82 pp.

ROBERT, A. 1966. The integration of a low order spectral form of the primitive meteorological equations. *J. Meteorol. Soc. Jpn.* **45**: 237–245.

SCHLESINGER, R.E.; L.W. UCCELLINI and D. JOHNSON. 1983. The effects of the Asselin time filter on numerical solutions to the linearized shallow-water wave equations. *Mon. Weather Rev.* **III**: 455–467.

Incompressible Homogeneous Fluids

André Robert

ABSTRACT *The four equations required for the study of motion in incompressible homogeneous fluids are given. Problems associated with the integration of these equations are examined. The balance equation and the vorticity equation are derived. Three integration procedures are described. The spatial finite difference discretization of these equations is examined. Integrations with the Crank-Nicolson scheme are presented.*

RÉSUMÉ *Les quatre équations nécessaires à l'étude du mouvement des fluides homogènes incompressibles sont données. On examine les problèmes associés à l'intégration de ces équations. On dérive l'équation de balance et l'équation du tourbillon. On décrit trois procédures d'intégration. On examine la discrétisation spatiale en différences finies de ces équations. On présente des intégrations utilisant le schéma de Crank-Nicholson.*

1 Introduction

It has been common practice over the past 30 years to condense the dynamic properties of the atmosphere into simple models. These are easier to understand and manipulate than the complete set of meteorological equations. Simple models of the atmosphere provide us with a practical approach to the problem of predicting motion in a fluid and these models can easily be generalized into more complex and more complete time extrapolation procedures.

The analogue most widely used as an introduction to the field of dynamic meteorology is still the non-divergent vorticity equation in two dimensions frequently called the barotropic model. The earth has a radius of the order of 6000 km and close to 99% of the atmosphere is contained in the first 30 km above sea level. This means that the atmosphere is a very thin spherical shell comparable roughly to the peel on an apple. It seems obvious from this comparison that the large-scale motion in the atmosphere will be primarily horizontal. This is the condition that leads to the barotropic model and this analogue will be examined as a first approximation to the atmosphere.

2 The modelling equations

If a fluid is both incompressible and homogeneous then $\rho = \rho^*$ (a constant) and the equations that describe motion in this fluid reduce to

$$\frac{du}{dt} = -\frac{1}{\rho^*}\frac{\partial p}{\partial x} \tag{1}$$

$$\frac{dv}{dt} = -\frac{1}{\rho^*}\frac{\partial p}{\partial y} \tag{2}$$

$$\frac{dw}{dt} = -g - \frac{1}{\rho^*}\frac{\partial p}{\partial z} \tag{3}$$

$$\frac{\partial u}{\partial x} + \frac{\partial v}{\partial y} + \frac{\partial w}{\partial z} = 0. \tag{4}$$

The four equations given here contain the four dependent variables u, v, w and p. It forms a complete system that we should be able to solve by analytical or by numerical methods.

Let us assume that these four variables are given at $t = 0$ for all x, y and z. Obviously, these given quantities must be such that

$$\frac{\partial u_0}{\partial x} + \frac{\partial v_0}{\partial y} + \frac{\partial w_0}{\partial z} = 0.$$

The next step is to attempt to produce a forecast at time $t = \Delta t$ where Δt is a very small quantity. If we can attain this goal successfully, then a simple repetition of the process will generate forecasts valid at any desired time. For simplicity, we will try to use a forward integration scheme. For the first three equations, we get

$$u_1 = u_0 - \Delta t\left[A(u_0) + \frac{1}{\rho^*}\frac{\partial p_0}{\partial x}\right]$$

$$v_1 = v_0 - \Delta t\left[A(v_0) + \frac{1}{\rho^*}\frac{\partial p_0}{\partial y}\right]$$

$$w_1 = w_0 - \Delta t\left[A(w_0) + g + \frac{1}{\rho^*}\frac{\partial p_0}{\partial z}\right].$$

In these equations, A is the advection operator defined as follows:

$$A(F) = u\frac{\partial F}{\partial x} + v\frac{\partial F}{\partial y} + w\frac{\partial F}{\partial z}. \tag{5}$$

These equations give us a prediction for u, v and w, but we still have to produce a forecast for p if we want to be able to continue the integration. This does not seem possible since we do not have a predictive equation for this variable. This is the problem that we will start investigating immediately.

3 The balance equation

When we look at the procedure described in the preceding section, we find another problem. We have no guarantee that u, v and w will satisfy the equation of continuity. In order to clarify this point, all we have to do is to substitute these variables into the equation of continuity. This gives:

$$\frac{\partial u_1}{\partial x}+\frac{\partial v_1}{\partial y}+\frac{\partial w_1}{\partial z}=\frac{\partial u_0}{\partial x}+\frac{\partial v_0}{\partial y}+\frac{\partial w_0}{\partial x}-\Delta t\,\frac{\partial}{\partial x}\left[A(u_0)+\frac{1}{\rho^*}\,\frac{\partial p_0}{\partial x}\right]$$

$$-\Delta t\,\frac{\partial}{\partial y}\left[A(v_0)+\frac{1}{\rho^*}\,\frac{\partial p_0}{\partial y}\right]-\Delta t\,\frac{\partial}{\partial z}\left[A(w_0)+g+\frac{1}{\rho^*}\,\frac{\partial p_0}{\partial z}\right].$$

Here we can easily see that if there is no divergence at $t=0$ and that this property must be maintained at $t=\Delta t$, then the following condition must be satisfied

$$\frac{\partial}{\partial x}\left[A(u_0)+\frac{1}{\rho^*}\,\frac{\partial p_0}{\partial x}\right]+\frac{\partial}{\partial y}\left[A(v_0)+\frac{1}{\rho^*}\,\frac{\partial p_0}{\partial y}\right]+\frac{\partial}{\partial z}\left[A(w_0)+g+\frac{1}{\rho^*}\,\frac{\partial p_0}{\partial z}\right]=0.$$

It is obvious that this condition must be satisfied at all times so that we will omit the reference to $t=0$. Also, we will keep the terms containing the pressure on the left hand side and we will move all other terms to the right hand side giving:

$$\frac{1}{\rho^*}\left(\frac{\partial^2 p}{\partial x^2}+\frac{\partial^2 p}{\partial y^2}+\frac{\partial^2 p}{\partial z^2}\right)=-\frac{\partial}{\partial x}A(u)-\frac{\partial}{\partial y}A(v)-\frac{\partial}{\partial z}A(w). \tag{6}$$

This equation is called the "balance equation". It is a diagnostic relation between the pressure field and the wind field. If the three wind components are given at time t, then we can use this equation to compute the pressure distribution at time t. We do not know yet how to do this. This is a problem that will be examined later.

The balance equation is not a new (independent) equation. It has been deduced from equations (1) to (4). It is obtained by taking

$$\frac{\partial}{\partial x}(1)+\frac{\partial}{\partial y}(2)+\frac{\partial}{\partial z}(3)$$

and then using (4) to simplify the result.

With the balance equation, we now have all the tools we need in order to carry out an integration. If the wind components are given at $t=0$, we first use (6) to determine the pressure distribution at that time. Then we use (1), (2) and (3) to compute the wind components at time $t=\Delta t$. We simply have to keep on repeating this procedure until we get the desired forecast. This can be done with the leapfrog scheme or any other integration scheme. If the wind field is non-divergent at $t=0$, it should maintain this property at all times.

4 The vorticity equation

The balance equation was generated by applying the divergence operator ($\nabla\cdot$) to the equation of motion written in its vectorial form. This suggests that it might be useful to examine other possible manipulations for the equations given in (1) to (4). Another such transformation that is popular in the field of fluid dynamics consists of eliminating pressure from two scalar equations. As an example, we can compute

$$\frac{\partial}{\partial x}(2) - \frac{\partial}{\partial y}(1).$$

This gives

$$\frac{\partial}{\partial x}\left(\frac{dv}{dt}\right) - \frac{\partial}{\partial y}\left(\frac{du}{dt}\right) = 0$$

since the terms containing pressure vanish. This equation will now be transformed as follows:

$$\frac{\partial}{\partial x}\left[\frac{\partial v}{\partial t} + A(v)\right] - \frac{\partial}{\partial y}\left[\frac{\partial u}{\partial t} + A(u)\right] = 0$$

$$\frac{\partial}{\partial t}\left(\frac{\partial v}{\partial x} - \frac{\partial u}{\partial y}\right) + A\left(\frac{\partial v}{\partial x} - \frac{\partial u}{\partial y}\right) + \frac{\partial A(v)}{\partial x} - \frac{\partial A(u)}{\partial y} = 0.$$

The last two partial derivatives apply only to the variables that appear in the operator A.

At this point, we will define the vorticity vector as follows:

$$\mathbf{Q} = Q_x\hat{\mathbf{i}} + Q_y\hat{\mathbf{j}} + Q_z\hat{\mathbf{k}}$$

with

$$Q_x = \frac{\partial w}{\partial y} - \frac{\partial v}{\partial z}$$

$$Q_y = \frac{\partial u}{\partial z} - \frac{\partial w}{\partial x}$$

$$Q_z = \frac{\partial v}{\partial x} - \frac{\partial u}{\partial y}$$

so that the preceding equation becomes

$$\frac{d}{dt}Q_z + \frac{\partial u}{\partial x}\frac{\partial v}{\partial x} + \frac{\partial v}{\partial x}\frac{\partial v}{\partial y} + \frac{\partial w}{\partial x}\frac{\partial v}{\partial z} - \frac{\partial u}{\partial y}\frac{\partial u}{\partial x} - \frac{\partial v}{\partial y}\frac{\partial u}{\partial y} - \frac{\partial w}{\partial y}\frac{\partial u}{\partial z} = 0$$

or finally

$$\frac{d}{dt}Q_z + Q_z\left(\frac{\partial u}{\partial x} + \frac{\partial v}{\partial y}\right) + \frac{\partial w}{\partial x}\frac{\partial v}{\partial z} - \frac{\partial w}{\partial y}\frac{\partial u}{\partial z} = 0. \tag{7}$$

Similar equations can be produced for the two other components of the vorticity vector.

$$\frac{d}{dt}Q_x + Q_x\left(\frac{\partial v}{\partial y} + \frac{\partial w}{\partial z}\right) + \frac{\partial u}{\partial y}\frac{\partial w}{\partial x} - \frac{\partial u}{\partial z}\frac{\partial v}{\partial x} = 0 \tag{8}$$

$$\frac{d}{dt}Q_y + Q_y\left(\frac{\partial u}{\partial x} + \frac{\partial w}{\partial z}\right) + \frac{\partial v}{\partial z}\frac{\partial u}{\partial y} - \frac{\partial v}{\partial x}\frac{\partial w}{\partial y} = 0. \tag{9}$$

These are the predictive equations that apply to each of the three components of the vorticity vector. A description of the physical meaning of vorticity may be found in most textbooks covering the subject of fluid dynamics. It is a measure of the angular velocity of rotation of an infinitesimal parcel of fluid on itself (spin). These equations are of interest to us because they may be used to generate a prediction. If the wind components are given at initial time but pressure is unknown, we can use (7) and (8) to compute $Q_z(\Delta t)$ and $Q_x(\Delta t)$ at $t = \Delta t$ and then we must solve the following equations

$$\frac{\partial v}{\partial x} - \frac{\partial u}{\partial y} = Q_z(\Delta t) \tag{10}$$

$$\frac{\partial w}{\partial y} - \frac{\partial v}{\partial z} = Q_x(\Delta t) \tag{11}$$

$$\frac{\partial u}{\partial x} - \frac{\partial v}{\partial y} + \frac{\partial w}{\partial z} = 0$$

for u, v and w. In order to find a solution, we differentiate the last equation with respect to y

$$\frac{\partial}{\partial x}\left(\frac{\partial u}{\partial y}\right) + \frac{\partial^2 v}{\partial y^2} + \frac{\partial}{\partial z}\left(\frac{\partial w}{\partial y}\right) = 0$$

and we substitute from the two other equations

$$\frac{\partial}{\partial x}\left(\frac{\partial v}{\partial x} - Q_z(\Delta t)\right) + \frac{\partial^2 v}{\partial y^2} + \frac{\partial}{\partial z}\left(\frac{\partial v}{\partial z} + Q_x(\Delta t)\right) = 0$$

giving

$$\frac{\partial^2 v}{\partial x^2} + \frac{\partial^2 v}{\partial y^2} + \frac{\partial^2 v}{\partial z^2} = \frac{\partial Q_z(\Delta t)}{\partial x} - \frac{\partial Q_x(\Delta t)}{\partial z}$$

and we can solve this equation for v. After this, we can solve (10) for u and then (11) for w. At first glance, this procedure appears to be relatively simple, but we must be careful before we draw any conclusions, as the boundary conditions might generate some difficulties. In any case, we have here another method of integrating the equations. It does not require pressure and we do not need to use the balance equation.

5 The barotropic model

We will now examine motion in two dimensions (horizontal). We will use (7) and (4) for this purpose. Since there is no vertical motion in this case, these equations reduce to

$$\frac{d}{dt}\left(\frac{\partial v}{\partial x} - \frac{\partial u}{\partial y}\right) + \left(\frac{\partial v}{\partial x} - \frac{\partial u}{\partial y}\right)\left(\frac{\partial u}{\partial x} + \frac{\partial v}{\partial y}\right) = 0 \tag{12}$$

$$\frac{\partial u}{\partial x} + \frac{\partial v}{\partial y} = 0. \tag{13}$$

Because of (13), equation (12) reduces to

$$\frac{d}{dt}\left(\frac{\partial v}{\partial x} - \frac{\partial u}{\partial y}\right) = 0. \tag{14}$$

Since $Q_x = Q_y = 0$, vorticity reduces to a variable in one dimension, oriented along the vertical axis, so that we may write the above equation as

$$\frac{dQ}{dt} = 0 \tag{15}$$

where

$$Q = \frac{\partial v}{\partial x} - \frac{\partial u}{\partial y}. \tag{16}$$

Now, we can use (13) to replace u and v by a single variable ψ defined as follows

$$u = -\frac{\partial \psi}{\partial y} \tag{17}$$

$$v = \frac{\partial \psi}{\partial x} \tag{18}$$

where ψ is called the stream function. In this case we get

$$Q = \frac{\partial^2 \psi}{\partial x^2} + \frac{\partial^2 \psi}{\partial y^2}. \tag{19}$$

We have only one predictive equation given in (15). In order to integrate this equation, we can use an Eulerian leapfrog discretization

$$Q(t+\Delta t) = Q(t-\Delta t) - 2\Delta t A(Q). \tag{20}$$

Given the stream function ψ at time t, we compute first u, v and Q from (17), (18) and (19) at that time. Then we compute

$$A(Q) = u\frac{\partial Q}{\partial x} + v\frac{\partial Q}{\partial y}$$

and we use this result to compute $Q(t+\Delta t)$ from (20). Finally, we solve

$$\frac{\partial^2\psi}{\partial x^2} + \frac{\partial^2\psi}{\partial y^2} = Q(t+\Delta t)$$

for ψ at $(t+\Delta t)$. We can repeat this procedure until we get the desired forecast.

6 Solving the elliptic equation

In association with the barotropic vorticity equation, we have seen that we have to solve the following equation for ψ with Q given

$$\frac{\partial^2\psi}{\partial x^2} + \frac{\partial^2\psi}{\partial y^2} = Q(x, y).$$

As a first step, we will produce a discretized version of this equation. A simple, second-order finite difference formula for a second derivate may be given as follows

$$\frac{\partial^2\psi}{\partial x^2} = \frac{\psi_{i+1,j} + \psi_{i-1,j} - 2\psi_{i,j}}{\Delta x^2}$$

$$\frac{\partial^2\psi}{\partial y^2} = \frac{\psi_{i,j+1} + \psi_{i,j-1} - 2\psi_{i,j}}{\Delta y^2}$$

and we will assume that $\Delta y = \Delta x$ for clarity. A small grid network will be used; in this particular case, it will be the 5×5 point network presented in Fig. 1.

Since the variable Q is generally available only for the 9 interior points of this grid, we will write

$$\psi_6 + \psi_8 + \psi_2 + \psi_{12} - 4\psi_7 = \Delta x^2 Q_7$$

$$\psi_7 + \psi_9 + \psi_3 + \psi_{13} - 4\psi_8 = \Delta x^2 Q_8$$

$$\psi_8 + \psi_{10} + \psi_4 + \psi_{14} - 4\psi_9 = \Delta x^2 Q_9$$

$$\psi_{11} + \psi_{13} + \psi_7 + \psi_{17} - 4\psi_{12} = \Delta x^2 Q_{12}$$

$$\psi_{12} + \psi_{14} + \psi_8 + \psi_{18} - 4\psi_{13} = \Delta x^2 Q_{13}$$

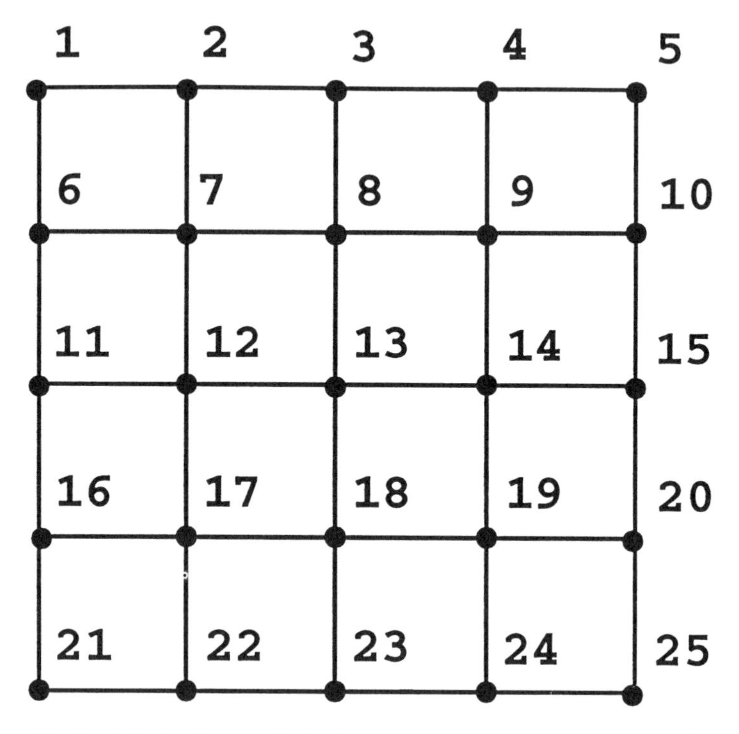

Fig. 1 A 5 × 5 grid network. The label refers to the ordering of the grid points.

$$\psi_{13} + \psi_{15} + \psi_{9} + \psi_{19} - 4\psi_{14} = \Delta x^2 Q_{14}$$

$$\psi_{16} + \psi_{18} + \psi_{12} + \psi_{22} - 4\psi_{17} = \Delta x^2 Q_{17}$$

$$\psi_{17} + \psi_{19} + \psi_{13} + \psi_{23} - 4\psi_{18} = \Delta x^2 Q_{18}$$

$$\psi_{18} + \psi_{20} + \psi_{14} + \psi_{24} - 4\psi_{19} = \Delta x^2 Q_{19}.$$

We have here 21 unknown values of ψ and only 9 equations. On the other hand, if we consider that the boundaries of our grid are solid walls, then the stream function must have the same value at all boundary points, otherwise there would be a flow through the boundary. We will set this value to zero at all boundary points, and we get:

$$\psi_{8} + \psi_{12} - 4\psi_{7} = \Delta x^2 Q_{7}$$

$$\psi_{7} + \psi_{9} + \psi_{13} - 4\psi_{8} = \Delta x^2 Q_{8}$$

$$\psi_{8} + \psi_{14} - 4\psi_{9} = \Delta x^2 Q_{9}$$

$$\psi_{13} + \psi_{7} + \psi_{17} - 4\psi_{12} = \Delta x^2 Q_{12}$$

$$\psi_{12} + \psi_{14} + \psi_8 + \psi_{18} - 4\psi_{13} = \Delta x^2 Q_{13}$$

$$\psi_{13} + \psi_9 + \psi_{19} - 4\psi_{14} = \Delta x^2 Q_{14}$$

$$\psi_{18} + \psi_{12} - 4\psi_{17} = \Delta x^2 Q_{17}$$

$$\psi_{17} + \psi_{19} + \psi_{13} - 4\psi_{18} = \Delta x^2 Q_{18}$$

$$\psi_{18} + \psi_{14} - 4\psi_{19} = \Delta x^2 Q_{19}.$$

In this case the system contains only 9 unknowns and can be solved. If we used a 42×42 grid network we would have 1600 interior points and this would yield a system of 1600 equations with the same number of unknowns. To solve such a system by the method of elimination would be a difficult task even on a fast computer. The best approach to this problem is to use a method that gives a good approximation to the answer rapidly. Such methods have been devised and they will converge to within 1% of the exact solution in less than 1/1000 of the time required by the direct method. The relaxation method is of particular interest and will be discussed now.

This method is approximate but it has the advantage that it always converges to the exact solution. An approximate solution Ψ^k is first used. Substitution of this function into the differential equation yields the residuals S_k

$$S_k = \nabla^2 \psi^k - Q.$$

The residuals are computed for all the interior points and then the trial solution is replaced by another one where:

$$\psi^{k+1} = \psi^k + \alpha S_k.$$

This process is repeated over and over until convergence is achieved. If we wish, we may start with $\psi^0 = 0$ at all points. This process converges if

$$0 < \alpha < 0.25$$

and a proof of convergence will be given in another set of notes.

The procedure described here converges rapidly in the short waves but the rate of convergence is rather slow in the long waves. For very large grids, this could become a serious problem. Much better methods exist and are effectively used commonly in fluid dynamics. One such method is the ADI relaxation scheme. Because of its widespread use, we will describe it now.

7 The ADI relaxation scheme

The ADI (alternating-directions implicit) method of Peaceman and Rachford will be described here. It is well suited for the solution of the Poisson equation. We will use it to compute the stream function from the vorticity.

$$\frac{\partial^2 \psi}{\partial x^2} + \frac{\partial^2 \psi}{\partial y^2} = Q(x, y).$$

For the ADI scheme, we will present this equation in two different forms

$$\frac{\partial^2 \psi}{\partial x^2} - \gamma\psi = Q - \gamma\psi - \frac{\partial^2 \psi}{\partial y^2}$$

$$\frac{\partial^2 \psi}{\partial y^2} - \gamma\psi = Q - \gamma\psi - \frac{\partial^2 \psi}{\partial x^2}$$

and we will use these to generate an iterative procedure:

$$\frac{\partial^2 \psi^*}{\partial x^2} - \gamma_n \psi^* = Q - \gamma_n \psi^n - \frac{\partial^2 \psi^n}{\partial y^2} \tag{21}$$

$$\frac{\partial^2 \psi^{n+1}}{\partial y^2} - \gamma_n \psi^{n+1} = Q - \gamma_n \psi^* - \frac{\partial^2 \psi^*}{\partial x^2}. \tag{22}$$

Here, we assume that ψ^n is given. We use it to compute the right hand side of (21) and then we solve for ψ^*. The next step consists of computing the right hand side of (22) and then we solve for ψ^{n+1}. We repeat this procedure until convergence is achieved.

In order to establish the convergence of this procedure, we will represent the difference between ψ^n and the true solution ψ by E^n where:

$$\psi^n = \psi + E^n.$$

Substitution in (21) and (22) gives

$$\frac{\partial^2 E^*}{\partial x^2} - \gamma_n E^* = -\gamma_n E^n - \frac{\partial^2 E^n}{\partial y^2}$$

$$\frac{\partial^2 E^{n+1}}{\partial y^2} - \gamma_n E^{n+1} = -\gamma_n E^* - \frac{\partial^2 E^*}{\partial x^2}.$$

It is possible to find an exact particular solution of these equations. We propose the following

$$E^n = A^n e^{i(kx+ly)}$$

and we can show that this is a solution by substitution

$$(k^2 + \gamma_n)A^* = (\gamma_n - l^2)A^n$$

$$(l^2 + \gamma_n)A^{n+1} = (\gamma_n - k^2)A^*.$$

These algebraic equations give

$$A^{n+1} = rA^n$$

where

$$r = \frac{(\gamma_n - k^2)(\gamma_n - l^2)}{(\gamma_n + k^2)(\gamma_n + l^2)}.$$

We can see from this result that if $\gamma_n > 0$ then $|r| < 1$. In other words, this procedure always converges for our particular solution. If we choose γ in such a way that $\gamma_n = k^2$ or $\gamma_n = l^2$, then we get the exact solution in one iteration.

If our particular solution is made up of two waves with wavenumbers (k_1, l_1) and (k_2, l_2), then we can find two values γ_1 and γ_2 that will generate an exact solution in two iterations. If our particular solution is made up of a large number of waves, then we will need many values of γ (many iterations) in order to get an exact solution. On the other hand, we might be able to produce a reasonably accurate solution with a limited number of values of γ. We will examine this question now. For this purpose, we will look at the following quantity:

$$G = \frac{(\gamma_1 - k^2)(\gamma_2 - k^2)}{(\gamma_1 + k^2)(\gamma_2 + k^2)}.$$

Since $G = 0$ for $k^2 = \gamma_1$ and $G = 0$ for $k^2 = \gamma_2$, there must be a maximum somewhere between these two values. In order to determine this maximum (G_M), we will proceed as follows:

$$G = \left[\frac{2\gamma_1}{\gamma_1 + k^2} - 1\right]\left[\frac{2\gamma_2}{\gamma_2 + k^2} - 1\right]$$

$$\frac{\partial G}{\partial k^2} = -\frac{2\gamma_1}{(\gamma_1 + k^2)^2}\left[\frac{2\gamma_2}{\gamma_2 + k^2} - 1\right] - \frac{2\gamma_2}{(\gamma_2 + k^2)^2}\left[\frac{2\gamma_1}{\gamma_1 + k^2} - 1\right] = 0$$

$$2\gamma_1(\gamma_2^2 - k^4) + 2\gamma_2(\gamma_1^2 - k^4) = 0$$

$$\gamma_1\gamma_2(\gamma_1 + \gamma_2) - k^4(\gamma_1 + \gamma_2) = 0$$

$$k^2 = \sqrt{\gamma_1\gamma_2}$$

and this value gives

$$G_M = -\left(\frac{\sqrt{\gamma_1} - \sqrt{\gamma_2}}{\sqrt{\gamma_1} + \sqrt{\gamma_2}}\right)^2.$$

We will consider the following geometric progression

$$\gamma_n = \gamma_0 a^{2n}$$

and substitution gives

$$G_M = -\left(\frac{1-a}{1+a}\right)^2.$$

For $a = 2$, we get

$$G_M = -\frac{1}{9}$$

In two dimensions, we have

$$r_1 r_2 \leqslant \frac{1}{81}$$

for all values of k^2 and l^2 that lie between γ_1 and γ_2.

From this, we can see that if we cover the entire spectrum of admissible wavenumbers with values one octave apart from each other, we will obtain an accuracy of the order of 1%. Under these conditions, for a grid with 512 by 512 points, we need 10 iterations[1] to attain this level of accuracy.

We can see from this, that we can construct an iterative procedure that remains quite efficient even for very large grids. Such procedures are used extensively in the field of computational fluid dynamics.

8 Numerical tests with the SDI scheme

Since the ADI scheme is one of the most popular of the techniques used to solve a Poisson equation, we will carry out an experiment in order to evaluate its performance. For this purpose, we will define the following stream function

$$\psi_0 = -Uy\left[B - \exp\left(1 - \frac{r^2}{s^2}\right)\right]$$

with $U = 10$, $B = 1$ and $r^2 = x^2 + y^2$. We will represent this stream function on a 51 by 51 grid, $s = 8\Delta x$ and $\Delta x = \Delta y = 0.1$. The resulting stream function is presented in Fig. 2.

Now we will use this stream function to compute the corresponding vorticity

$$Q = \nabla^2 \psi_0$$

and the corresponding result is presented in Fig. 3.

[1]Here we choose the variation of the wavelength λ from $2\Delta x$ to $2L$ (where $L = (NI - 1)\,\Delta x$ is the domain length). We get $(\pi/L)^2 \leqslant k^2 \leqslant (\pi/\Delta x)^2$. We can use for $a = 2$ the equation $\gamma_n = \gamma_0 a^{2n} = (\pi/\Delta x)^2 = (\pi/L)^2 a^{2n}$ to compute the number of iterations n. Furthermore, we obtain $n = \ln\,(NI-1)^2/\ln 4$.

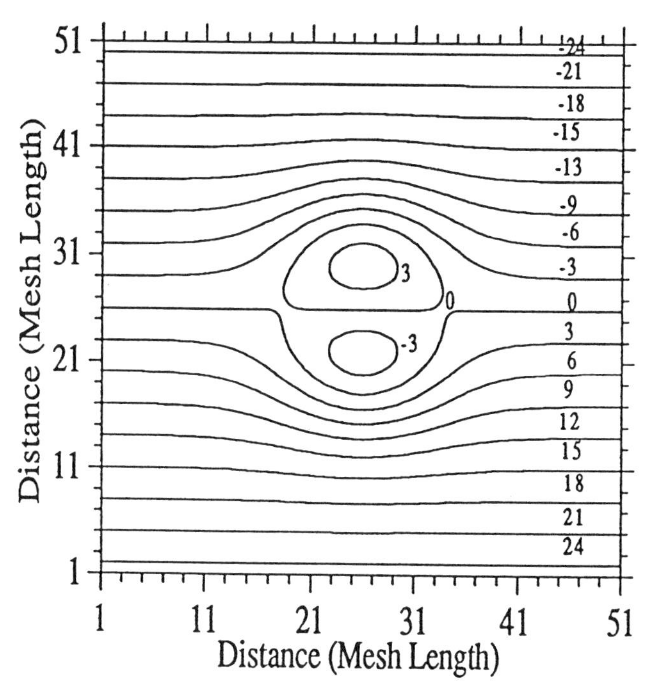

Fig. 2 The stream function ψ_0 on a 51×51 grid. The contour interval is 3 m^2/s with a maximum and minimum value of 25 and -25 m^2/s respectively.

At this point, we will solve the following equation for ψ_1

$$\nabla^2 \psi_1 = Q.$$

This is where we use the ADI scheme. Six iterations will be required for the grid that we selected. The corresponding solution is presented in Fig. 4. This stream function ψ_1 may be compared with ψ_0. We cannot see any differences. Finally, we compute the difference field D

$$D = 1000(\psi_1 - \psi_0)$$

and this result is presented in Fig. 5. As we can see, the differences are in the fourth decimal place. We conclude that the ADI scheme is both accurate and efficient.

9 An implicit model formulation

At this point, we will consider the possibility of using the Crank-Nicolson integration scheme. For this purpose, our equations will be written as follows:

$$\frac{Du}{Dt} = -\frac{1}{\rho^*} \overline{\frac{\partial p^t}{\partial x}}$$

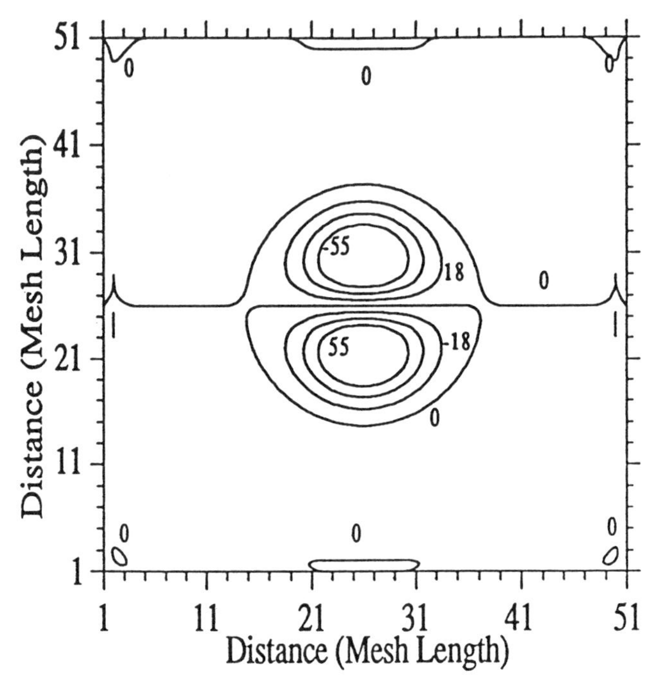

Fig. 3 The vorticity Q computed from the stream function ψ_0 on a 51×51 grid. The contour interval is 18.34 s^{-1}, with a maximum and minimum value of 91.71 and -91.71 s^{-1} respectively.

$$\frac{Dv}{Dt} = -\frac{1}{\rho^*}\overline{\frac{\partial p'}{\partial y}}$$

$$\frac{Dw}{Dt} = -g - \frac{1}{\rho^*}\overline{\frac{\partial p'}{\partial z}}$$

$$\frac{\partial u}{\partial x} + \frac{\partial v}{\partial y} + \frac{\partial w}{\partial z} = 0$$

and we will apply the following definitions:

$$\frac{DF}{Dt} = \frac{F^+ - F^-}{2\Delta t}$$

$$\bar{F}^t = \frac{F^+ - F^-}{2}$$

In these expressions, we have

$$F^+ = F(X, Y, Z, t + \Delta t)$$

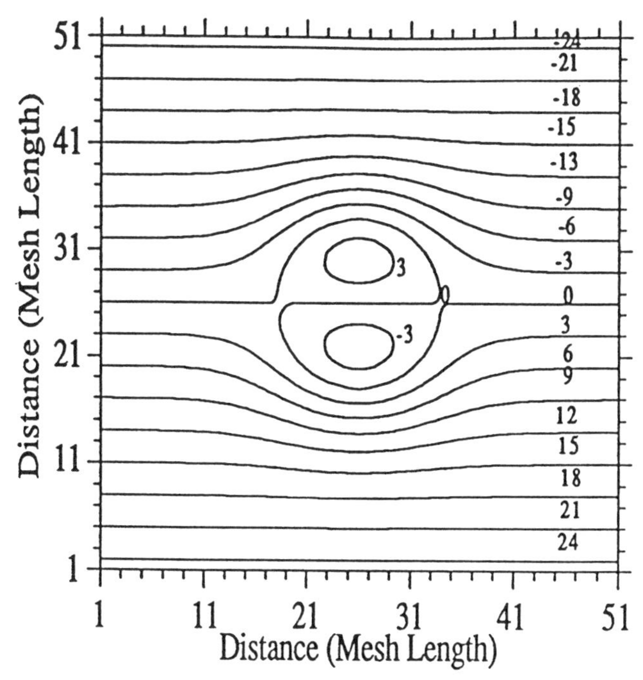

Fig. 4 The stream function ψ_1 computed from the vorticity Q using the ADI scheme on a 51×51 grid. The contour interval is 3 m^2/s, with a maximum and minimum value of 25 and -25 m^2/s respectively.

where X, Y, Z represents a grid point, and

$$F^{-} = F(X - 2\alpha, Y - 2\beta, Z - 2\gamma, t - \Delta t).$$

Also

$$\alpha = \Delta t\, u^0$$

$$\beta = \Delta t\, v^0$$

$$\gamma = \Delta t\, w^0$$

and

$$F^0 = F(X - \alpha, Y - \beta, Z - \gamma, t).$$

From this definition, we can see that we are considering three points on the trajectory followed by a parcel of fluid. It is an approximate trajectory since the

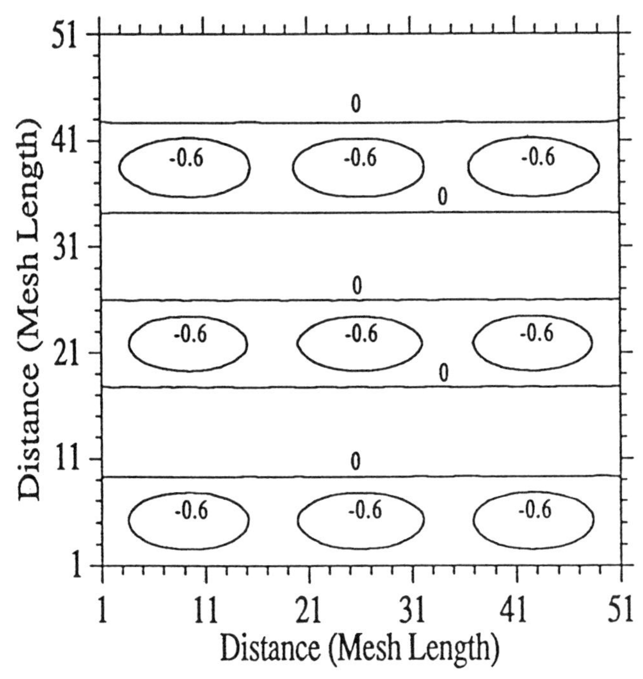

Fig. 5 The difference field $D = 1000\ (\psi_1 - \psi_0)$ of stream function with units of m^2/s.

three points lie on a straight line. The wind at the mid-point is used to determine the orientation and the length of this segment of a straight line.

Since the mid-point and the upstream point are not grid points, we will use interpolation to evaluate variables at these points. For large time steps, the three points might be far apart. Interpolation is always carried out from the grid points that are in the immediate neighbourhood of the mid-point or the upstream point respectively. Cubic interpolation (fourth-order Taylor series) will be used in this model in all three dimensions.

The displacements that are needed above must be computed by an iterative procedure. The following procedure is used

$$\alpha^{\kappa+1} = \Delta t\, u(X - \alpha^{\kappa}, Y - \beta^{\kappa}, Z - \gamma^{\kappa}, t)$$

$$\beta^{\kappa+1} = \Delta t\, v\, (X - \alpha^{\kappa}, Y - \beta^{\kappa}, Z - \gamma^{\kappa}, t)$$

$$\gamma^{\kappa+1} = \Delta t\, w(X - \alpha^{\kappa}, Y - \beta^{\kappa}, Z - \gamma^{\kappa}, t)$$

This procedure is usually initiated with $\alpha^0 = \beta^0 = \gamma^0 = 0$ and in general, two iterations provide a sufficently accurate result.

In order to integrate the equations, we will manipulate them so that they appear as follows:

$$\left(u + \frac{\Delta t}{\rho^*}\frac{\partial p}{\partial x}\right)^+ = Q_1$$

$$\left(v + \frac{\Delta t}{\rho^*}\frac{\partial p}{\partial y}\right)^+ = Q_2$$

$$\left(w + \frac{\Delta t}{\rho^*}\frac{\partial p}{\partial z}\right)^+ = Q_3$$

where

$$Q_1 = \left(u - \frac{\Delta t}{\rho^*}\frac{\partial p}{\partial x}\right)^-$$

$$Q_2 = \left(v - \frac{\Delta t}{\rho^*}\frac{\partial p}{\partial y}\right)^-$$

$$Q_3 = \left(w - \frac{\Delta t}{\rho^*}\frac{\partial p}{\partial z} - g\right)^- .$$

The quantities in parentheses are computed at grid points and the result is interpolated to the appropriate upstream point. We see that we can easily compute Q_1, Q_2 and Q_3. For the next step, we will write the corresponding equations as follows:

$$u = Q_1 - \frac{\Delta t}{\rho^*}\frac{\partial p}{\partial x}$$

$$v = Q_2 - \frac{\Delta t}{\rho^*}\frac{\partial p}{\partial y}$$

$$w = Q_3 - \frac{\Delta t}{\rho^*}\frac{\partial p}{\partial z}$$

and we substitute in the equation of continuity to get:

$$\frac{\Delta t}{\rho^*}\nabla^2 p = \frac{\partial Q_1}{\partial x} + \frac{\partial Q_2}{\partial y} + \frac{\partial Q_3}{\partial z}.$$

From this equation, we can easily compute the pressure p and then calculate the three wind components at time $t + \Delta t$.

In order to test this model formulation, we will compare its results to those of a laboratory experiment. Here, we will observe flow past a circular cylinder.

Fig. 6 Results from laboratory experiment of the von Karman vortex street of flow past a circular cylinder. The cylinder is located at the extreme left of the figure. Water is used with an injected colloidal substance to make visible the motion. The Reynolds number is $R_e = 140$. (See Fig. 94 in Van Dyke, Milton. 1982. An Album of Fluid Motion. Parabolic Press, Stanford, California. pp. 56.)

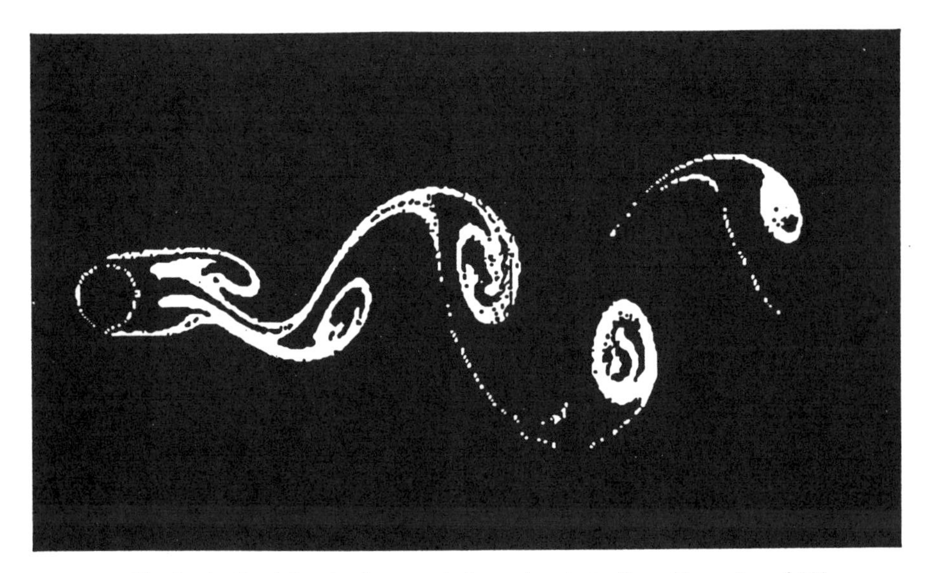

Fig. 7 As Fig. 6, but for the numerical experiment at a Reynolds number of 140.

Water is used and a colloidal substance is injected in it in order to make the motion visible. This experiment is carried out with a Reynolds number (R_e) of 140, where $R_e = UD/\mu$. Here, U is the speed far upstream from the cylinder, D is the diameter of the cylinder and μ is the molecular viscosity of water. As we can see from Fig. 6 vortices are shed from the cylinder and they move downstream with the current. This is the well known von Karman vortex street experiment. Observing this phenomenon in the laboratory is quite spectacular.

The corresponding numerical experiment is presented in Fig. 7. The resemblance with the laboratory experiment is quite striking.

8 Conclusion

The equations that describe motion in an incompressible fluid cannot be integrated directly. In order to carry out the integrations, it was common practice to derive the vorticity equation in the early days of computational fluid dynamics. Integrating the vorticity equation in two dimensions is not a difficult problem. This equation was used for more than two decades and it is still being used by some scientists.

Nowadays, a common technique consists of applying implicit integration schemes to the primitive Euler equations. This technique really pays off when it is used to study compressible fluid motion. This is something that we will examine in detail at a later time.

We have seen that numerical models can produce spectacular results. In some cases, they are more accurate than laboratory experiments. There is no doubt that as they continue to improve they will become widely used in most fields of experimentation.

Erreurs induites par le schème Lagrangien dans les ondes stationnaires.

André Robert

19 mai 1988

On commencera par l'étude de l'équation suivante:

$$\frac{\partial F}{\partial t} + U \frac{\partial F}{\partial x} = G \qquad (1)$$

Et on supposera qu'il existe une solution particulière telle que

$$F = A e^{ikx} \qquad (2)$$

et

$$G = B e^{ikx} \qquad (3)$$

On peut voir que (2) et (3) représentent une solution exacte de (1) par simple substitution

$$ikUAe^{ikx} = Be^{ikx} \qquad (4)$$

et on voit qu'on a une solution exacte si:

$$A = \frac{B}{ikU} \qquad (5)$$

En général, l'utilisation d'un schème numérique pour intégrer (1) produira une solution légèrement différente. Dans ce cas, on représentera la solution par:

$$F^* = A^* e^{ikx} \qquad (6)$$

Si on considère un schème Eulérien avec des différences centrées d'ordre deux, notre équation sera remplacée par

$$\overline{F}^t_t + U\,\overline{F}^x_x = G \qquad (7)$$

avec

$$\overline{F}^t_t = \frac{F(x, t+\Delta t) - F(x, t-\Delta t)}{2\Delta t} \qquad (8)$$

et

$$\overline{F}^x_x = \frac{F(x+\Delta x, t) - F(x-\Delta x, t)}{2\Delta x} \qquad (9)$$

Pour la fonction donné dans (6), on a

$$\overline{(F^*)}^x_x = \frac{A^* e^{ik(x+\Delta x)} - A^* e^{ik(x-\Delta x)}}{2\Delta x}$$

$$= A^* e^{ikx}\left[\frac{e^{ik\Delta x} - e^{-ik\Delta x}}{2\Delta x}\right]$$

$$= \frac{iA^* e^{ikx}}{\Delta x}\left[\frac{e^{ik\Delta x} - e^{-ik\Delta x}}{2i}\right]$$

$$\overline{(F^*)}_x^x = \frac{i A^* e^{ikx}}{\Delta x} \sin k\Delta x$$

$$= ikF^* \frac{\sin k\Delta x}{k\Delta x}$$

finalement, on obtient

$$\overline{(F^*)}_x^x = ikSF^* \qquad (10)$$

$$S = \frac{\sin\theta}{\theta} \qquad (11)$$

$$\theta = k\Delta x \qquad (12)$$

Si on substitue maintenant dans (7), on obtient

$$ikSUF^* = G$$

$$ikSUA^* = B$$

$$A^* = \frac{B}{ikUS}$$

$$A^* = \frac{A}{S} \qquad (13)$$

Si l'angle θ est très petit, on aura $S=1$ et la solution numérique A^* sera la même que la solution analytique A. Par contre, à l'autre bout du spectre où θ s'approche de π, on voit que S s'approche de zéro et dans ce cas la solution numérique est beaucoup plus intense que la solution analytique.

L'erreur dans cette solution est donnée par:

$$\epsilon = (A^* - A)\, e^{ikx}$$

et si on effectue l'intégrale sur tout le domaine, on obtient la variance de l'erreur

$$\overline{\epsilon^2} = \frac{1}{2}(A^* - A)^2$$

$$\overline{\epsilon^2} = \frac{1}{2} A^2 \left(\frac{1-S}{S}\right)^2 \qquad (14)$$

Pour $S < \frac{1}{2}$, on a

$$\overline{\epsilon^2} > \frac{1}{2} A^2$$

Dans ce cas, on serait mieux d'utiliser $A^* = 0$ qui donne $\overline{\epsilon^2} = \frac{1}{2} A^2$. On voit ici qu'on ferait mieux de se débarrasser de toutes les ondes pour lesquelles $S < \frac{1}{2}$ et qu'en agissant ainsi, on obtiendra une prévision plus précise.

Le schème Lagrangien

Dans ce cas, (1) prend la forme suivante:

$$\frac{F(x+\alpha,\ t+\Delta t) - F(x-\alpha,\ t-\Delta t)}{2\Delta t} = G(x,t) \qquad (15)$$

où

$$\alpha = U\Delta t \qquad (16)$$

Si on effectue les substitutions comme on l'a fait pour le schème Eulérien, on obtient encore

$$i k S U F^* = G$$

ce qui produit de nouveau la solution donnée dans (13). La définition de S est toujours celle qui est donnée dans (13) mais dans ce cas ici on a

$$\theta = k\alpha$$

$$\theta = k U \Delta t$$

$$\theta = \frac{U\Delta t}{\Delta x} k \Delta x$$

$$\theta = C k \Delta x \qquad (17)$$

On voit que pour le schème Lagrangien, la définition de θ donnée dans (17) diffère légèrement de celle produite par le schème Eulérien dans (12). On voit qu'il faudra maintenant tenir compte du nombre de Courant

$$C = \frac{U \Delta t}{\Delta x} \qquad (18)$$

Si on revient maintenant à la valeur critique de $S = 1/2$ calculée auparavant, cette valeur correspond à $\theta = 108.5°$ ou à la longueur d'onde

$$L_c = 3.3 \, \Delta x \qquad (19)$$

ce qui signifie qu'on devrait se débarrasser de toutes les ondes plus courtes que L_c quand on utilise un schème Eulérien. Si on utilise un schème Lagrangien avec $C = 3$, dans ce cas on obtient

$$L_c = 10 \, \Delta x \qquad (20)$$

Errors Produced by the Lagrangian Scheme in Stationary Waves (Translation)

André Robert
19 May 1988

Let us start by studying the following equation:

$$\frac{\partial F}{\partial t} + U\frac{\partial F}{\partial x} = G. \tag{1}$$

And let us suppose that there exists a particular solution such that

$$F = Ae^{ikx} \tag{2}$$

and

$$G = Be^{ikx}. \tag{3}$$

We can see that (2) and (3) represent an exact solution of (1) by simple substitution

$$ikUAe^{ikx} = Be^{ikx} \tag{4}$$

and we see that we have an exact solution if

$$A = \frac{B}{ikU}. \tag{5}$$

In general, the use of a numerical scheme to integrate (1) will produce a slightly different solution. In this case, let us represent the solution by

$$F^* = A^*e^{ikx}. \tag{6}$$

If we consider an Eulerian scheme with second-order centred differences, our equation will be replaced by

$$\bar{F}_t^t + U\bar{F}_x^x = G \tag{7}$$

with

$$\bar{F}_t^t = \frac{F(x, t+\Delta t) - F(x, t-\Delta t)}{2\Delta t} \tag{8}$$

and

$$\bar{F}_x^x = \frac{F(x+\Delta x, t) - F(x-\Delta x, t)}{2\Delta x}. \tag{9}$$

For the function given in (6), we have

$$\begin{aligned}
(\overline{F^*})_x^x &= \frac{A^* e^{ik(x+\Delta x)} - A^* e^{ik(x-\Delta x)}}{2\Delta x} \\
&= A^* e^{ikx}\left[\frac{e^{ik\Delta x} - e^{-ik\Delta x}}{2\Delta x}\right] \\
&= \frac{iA^* e^{ikx}}{\Delta x}\left[\frac{e^{ik\Delta x} - e^{-ik\Delta x}}{2i}\right] \\
&= \frac{iA^* e^{ikx}}{\Delta x}\sin k\Delta x \\
&= ikF^* \frac{\sin k\Delta x}{k\Delta x}
\end{aligned}$$

finally, we get

$$(\overline{F^*})_x^x = ikSF^* \tag{10}$$

$$S = \frac{\sin\theta}{\theta} \tag{11}$$

$$\theta = k\Delta x. \tag{12}$$

If we now substitute into (7), we obtain

$$ikSUF^* = G$$

$$ikSUA^* = B$$

$$A^* = \frac{B}{ikUS}$$

$$A^* = \frac{A}{S}. \tag{13}$$

If the angle θ is very small, we have $S = 1$ and the numerical solution A^* is the same as the analytical solution A. However, at the other end of the spectrum where θ approaches π, we see that S approaches zero and in this case the numerical solution is much larger than the analytical solution.

The error in this solution is given by

$$\varepsilon = (A^* - A)e^{ikx}$$

and if we integrate over the whole domain, we obtain the variance of the error

$$\overline{\varepsilon^2} = \tfrac{1}{2}(A^* - A)^2$$

$$\overline{\varepsilon^2} = \frac{1}{2}A^2\left(\frac{1-S}{S}\right)^2. \tag{14}$$

For $S < 1/2$, we have

$$\overline{\varepsilon^2} > \tfrac{1}{2}A^2.$$

In this case, it would be better to use $A^* = 0$ which gives $\overline{\varepsilon^2} = (1/2)A^2$. We see from this that it would be better to discard all waves for which $S < 1/2$ in order to obtain a more accurate forecast.

The Lagrangian Scheme

In this case, (1) takes the following form:

$$\frac{F(x+\alpha, t+\Delta t) - F(x-\alpha, t-\Delta t)}{2\Delta t} = G(x, t) \tag{15}$$

where

$$\alpha = U\Delta t. \tag{16}$$

If we repeat the substitutions as we did for the Eulerian scheme, we get again

$$ikSUF^* = G$$

which gives us once more solution (13). The definition for S is still the same as in (13) but in this case we have

$$\begin{aligned}
\theta &= k\alpha \\
\theta &= kU\Delta t \\
\theta &= \frac{U\Delta t}{\Delta x}k\Delta x \\
\theta &= Ck\Delta x.
\end{aligned} \tag{17}$$

We see that for the Lagrangian scheme, the definition of θ given in (17) is slightly different from the one given in (12) for the Eulerian scheme. We now see that we need to take into consideration the Courant number:

$$C = \frac{U\,\Delta t}{\Delta x}. \tag{18}$$

If we now return to the previously calculated critical value of $S = 1/2$, this value corresponds to $\theta = 108.5°$ or to wavelength

$$L_c = 3.3\,\Delta x \tag{19}$$

which means that we should discard all waves shorter than L_c when we use the Eulerian scheme. If we use a Lagrangian scheme with $C = 3$, we obtain in this case

$$L_c = 10\,\Delta x. \tag{20}$$

Comportement du schème Lagrangien dans une équation différentielle relativement simple.

André Robert

19 mai 1988

Nous allons maintenant commencer l'étude de l'équation suivante:

$$\frac{dF}{dt} = G - qF \qquad (1)$$

Celle-ci ne diffère de l'équation précédente que par le dernier terme que nous venons d'ajouter. Dans ce terme, q est une constante

Encore une fois, on supposera qu'il existe une solution particulière telle que

$$F = A\,e^{ikx} \qquad (2)$$

et

$$G = B\,e^{ikx} \qquad (3)$$

Dans ce cas, on aura une solution exacte si

$$A = \frac{B}{q+ikU} \qquad (4)$$

Dans les équations météorologiques, on verra qu'en général, l'équivalent de la constante q prendra une valeur très grande de telle façon qu'on aura presque toujours

$$q >> kU$$

C'est pour cette raison que nous voulons étudier cette equation. Cela signifie aussi que nous devrons appliquer un traitement semi-implicite au dernier terme

$$\overline{F}^{t} = \frac{F(x+\alpha, t+\Delta t) + F(x-\alpha, t-\Delta t)}{2} \qquad (5)$$

Dans ce cas, nous aurons

$$\overline{F}^{t} = R\,F \qquad (6)$$

avec $\qquad R = \cos\theta \qquad (7)$

$$\theta = Ck\Delta x \qquad (8)$$

$$C = \frac{U\Delta t}{\Delta x} \qquad (9)$$

Dans ce cas-ci, notre solution devient

$$A^{*} = \frac{B}{qR + ikUS} \qquad (10)$$

Ici, on a toujours

$$S = \frac{\sin \theta}{\theta} \qquad (11)$$

Puisque nous sommes intéressés à avoir une solution numérique A^* qui sera très proche de la solution analytique A, nous écrirons :

$$A^* = P A \qquad (12)$$

et nous avons

$$P = \frac{q + ikU}{qR + ikUS} \qquad (13)$$

Idéalement, nous aimerions avoir $P = 1$, mais cette valeur n'est réalisée que lorsque θ est très petit. En pratique, on pourrait se considérer satisfait si on obtient $P < 2$ comme on l'a fait dans les notes précédentes.

Si on considère le cas ou kU est beaucoup plus petit que q, on peut alors écrire $P = R^{-1}$ et nous serons satisfaits si $\theta < 60°$. Dans ce cas, nous obtenons la formule

$$L_c = 6 C \Delta x \qquad (14)$$

Si on fixe le nombre de courant à $C = 3$, on obtient

$$L_c = 18 \Delta x \qquad (15)$$

et on voit qu'on devra se débarrasser de toutes les ondes plus courtes que cette valeur critique.

On doit noter que ce résultat est encore pire que celui que nous avons obtenu précédemment.

Behaviour of the Lagrangian Scheme for a Relatively Simple Differential Equation (Translation)

André Robert
19 May 1988

Let us start now by studying the following equation:

$$\frac{dF}{dt} = G - qF. \tag{1}$$

This equation is different from the previous one only by the last term that we have just added on. In this term, q is a constant.

Once more, let us suppose that there exists a particular solution such that

$$F = Ae^{ikx} \tag{2}$$

and

$$G = Be^{ikx}. \tag{3}$$

In this case, we see that we have an exact solution if

$$A = \frac{B}{q + ikU}. \tag{4}$$

Generally in meteorological equations, the equivalent of the constant q will take on a very large value such that we will almost always have

$$q \gg kU.$$

It is for this reason that we want to study this equation. This also means that we will need to treat the last term semi-implicitly:

$$\bar{F}^t = \frac{F(x + \alpha, t + \Delta t) + F(x - \alpha, t - \Delta t)}{2}. \tag{5}$$

In this case, we have

$$\bar{F}^t = RF \tag{6}$$

with

$$R = \cos\theta \tag{7}$$

$$\theta = Ck\Delta x \tag{8}$$

$$C = \frac{U\Delta t}{\Delta x}. \tag{9}$$

In this case, our solution becomes

$$A^* = \frac{B}{qR + ikUS}. \tag{10}$$

Here we still have

$$S = \frac{\sin\theta}{\theta}. \tag{11}$$

Since we are interested in obtaining a numerical solution A^* which is very close to the analytical solution A, we will write

$$A^* = PA \tag{12}$$

and we get

$$P = \frac{q + ikU}{qR + ikUS}. \tag{13}$$

Ideally, we would like to have $P = 1$, but this value is only obtained when θ is very small. In practice, we could be satisfied with $P < 2$ as in the previous notes.

If we consider the case in which kU is much smaller than q, we can thus write $P = R^{-1}$ and be satisfied if $\theta < 60°$. In this case, we obtain the formula

$$L_c = 6C\Delta x. \tag{14}$$

If we fix the Courant number to $C = 3$, we obtain

$$L_c = 18\Delta x \tag{15}$$

and we see that we will need to discard all waves shorter than this critical value.

We must note that this result is worse than the one obtained previously.